CIRCUITS AT THE NANOSCALE

Communications, Imaging, and Sensing

CIRCUITS AT THE NANOSCALE
Communications, Imaging, and Sensing

Edited by
Krzysztof Iniewski

CRC Press
Taylor & Francis Group
Boca Raton London New York

CRC Press is an imprint of the
Taylor & Francis Group, an **informa** business

CRC Press
Taylor & Francis Group
6000 Broken Sound Parkway NW, Suite 300
Boca Raton, FL 33487-2742

© 2009 by Taylor & Francis Group, LLC
CRC Press is an imprint of Taylor & Francis Group, an Informa business

No claim to original U.S. Government works
Printed in the United States of America on acid-free paper
10 9 8 7 6 5 4 3 2

International Standard Book Number-13: 978-1-4200-7062-0 (Hardcover)

Library of Congress Cataloging-in-Publication Data

Circuits at the nanoscale : communications, imaging, and sensing / editor, Krzysztof Iniewski.
    p. cm.
  "A CRC title."
  Includes bibliographical references and index.
  ISBN 978-1-4200-7062-0 (alk. paper)
  1. Nanoelectronics--Materials. 2. Integrated circuits--Design and construction. 3. Radio circuits. 4. Image converters. 5. Detectors. 6. Image processing--Equipment and supplies. I. Iniewski, Krzysztof. II. Title.

TK7874.84.C57 2008
621.3815--dc22   2008022715

Visit the Taylor & Francis Web site at
http://www.taylorandfrancis.com

and the CRC Press Web site at
http://www.crcpress.com

Contents

PART I   CMOS Technology at the Nanoscale

PART II   Emerging Design Techniques

PART III   Mixed-Signal CMOS Circuits

PART IV   Circuits for Communications

PART V   Circuits for Imaging and Sensing

Editor

Krzysztof (Kris) Iniewski is a director at CMOS Emerging Technologies Inc., a consulting company in Vancouver, British Columbia. His research interests are in VLSI circuits for medical applications.

From 2004 to 2006, he was an associate professor at the electrical engineering and computer engineering department of the University of Alberta, Edmonton where he conducted research on low-power wireless circuits and systems. During his tenure at the University of Alberta, he edited *Emerging Wireless Technologies: Circuits, Systems and Devices* (CRC Press, 2007).

From 1995 to 2003, he was with PMC-Sierra and held various technical and management positions in R&D and strategic marketing areas. He led several developments of high-speed networking chips from early feasibility to volume production. Before joining PMC-Sierra, he was an assistant professor at the University of Toronto's electrical engineering and computer engineering department from 1990 to 1994, researching high-speed semiconductor devices and circuits.

Dr. Iniewski has published over 100 research papers in international journals and conferences. He holds 18 international patents granted in the United States, Canada, France, Germany, and Japan. He is a frequent invited speaker and has consulted for multiple organizations internationally. He received his PhD in electronics (honors) from the Warsaw University of Technology, Poland in 1988. Together with Carl McCrosky and Dan Minoli, he authored *Network Infrastructure and Architecture*: *Designing High-Availability Networks* to be published by John Wiley & Sons in 2008.

Contributors

Jim Adkisson
IBM Microelectronics Division
Essex Junction, Vermont

Amit Agarwal
Intel Corporation
Hillsboro, Oregon

David J. Allstot
Department of Electrical Engineering
University of Washington
Seattle, Washington

Kazutami Arimoto
Renesas Technology Corporation
Tokyo, Japan

Farrokh Ayazi
School of Electrical and Computer
    Engineering
Georgia Institute of Technology
Atlanta, Georgia

Artur Balasinski
Cypress Semiconductor
San Diego, California

Tony Bonaccio
IBM Systems & Technology Group
Essex Junction, Vermont

Anthony Chan Carusone
Department of Electrical and Computer
    Engineering
University of Toronto
Toronto, Ontario, Canada

Shuo-Wei Mike Chen
Atheros Communications
Santa Clara, California

Hsiu-Yu Cheng
University of Oxford
Oxford, United Kingdom

Giovanni Cherubini
IBM Zurich Research Laboratory
Rüschlikon, Switzerland

Bhaskar Choubey
University of Oxford
Oxford, United Kingdom

Steve Collins
Department of Engineering Science
University of Oxford
Oxford, United Kingdom

Vijay Dhanasekaran
Analog and Mixed-Signal Center
Texas A&M University
College Station, Texas

Evangelos Eleftheriou
IBM Zurich Research Laboratory
Rüschlikon, Switzerland

John Ellis-Monaghan
IBM Microelectronics Division
Essex Junction, Vermont

Alexander Fish
University of Calgary
Calgary, Alberta, Canada

Manisha Gambhir
Analog and Mixed-Signal Center
Texas A&M University
College Station, Texas

Roman Genov
Department of Electrical and Computer
    Engineering
University of Toronto
Toronto, Ontario, Canada

Pawel Grybos
Department of Measurement and
    Instrumentation
AGH University of Science and Technology
Krakow, Poland

Ronald J. Gutmann
Center for Integrated Electronics
Rensselaer Polytechnic Institute
Troy, New York

Christoph Hagleitner
IBM Zurich Research Laboratory
Rüschlikon, Switzerland

Donhee Ham
School of Engineering and Applied Sciences
Harvard University
Cambridge, Massachusetts

James W. Haslett
Department of Electrical and Computer
  Engineering
The University of Calgary
Alberta, Canada

Sitaraman V. Iyer
Intel Corporation
Santa Clara, California

Mark Jaffe
IBM Microelectronics Division
Essex Junction, Vermont

Albert Jerng
Ralink Technology
Cupertino, California

Hyejung Kim
Department of Electrical Engineering
  and Computer Sciences
Korea Advanced Institute of Science
  and Technology (KAIST)
Daejeon, South Korea

Jan G. Korvink
IMTEK-University of Freiburg
Freiburg, Germany

Ram Krishnamurthy
Intel Corporation
Hillsboro, Oregon

Hasnain Lakdawala
Intel Corporation
Hillsboro, Oregon

Jan Lienemann
IMTEK-University of Freiburg
Freiburg, Germany

Xiaofeng Li
School of Engineering and Applied Sciences
Harvard University
Cambridge, Massachusetts

Jian-Qiang Lu
Rensselaer Polytechnic Institute
Troy, New York

Douglas A. Mercer
Analog Devices
Wilmington, Massachusetts

Fukashi Morishita
Renesas Technology Corporation
Hyogo, Japan

Byeong-Gyu Nam
Department of Electrical Engineering
  and Computer Sciences
Korea Advanced Institute of Science
  and Technology (KAIST)
Daejeon, South Korea

Borivoje Nikolić
Department of Electrical Engineering
  and Computer Sciences
University of California at Berkeley
Berkeley, California

Ashkan Olyaei
Marvell Semiconductors
Santa Clara, California

Stephen Otim
University of Oxford
Oxford, United Kingdom

Liang-Teck Pang
Department of Electrical Engineering
  and Computer Sciences
University of California at Berkeley
Berkeley, California

Yannis Papananos
National Technical University of Athens
Athens, Greece

Marco Racanelli
Jazz Semiconductor
Newport Beach, California

David S. Ricketts
Department of Electrical and Computer
  Engineering
Carnegie Mellon University
Pittsburgh, Pennsylvania

Hugo Rothuizen
IBM Zurich Research Laboratory
Rüschlikon, Switzerland

Paul Rousseau
Taiwan Semiconductor Manufacturing
  Company
San Jose, California

Edgar Sánchez-Sinencio
Analog and Mixed-Signal Center
Texas A&M University
College Station, Texas

Jafar Savoj
Qualcomm Incorporated
Campbell, California

Hanspeter Schmid
Institute of Microelectronics
University of Applied Sciences
Northwestern Switzerland
Windisch, Switzerland

Ajit Sharma
Texas Instruments Inc.
Dallas, Texas

Sudip Shekhar
Department of Electrical Engineering
University of Washington
Seattle, Washington

Jose Silva-Martinez
Analog and Mixed-Signal Center
Texas A&M University
College Station, Texas

Robert Sobot
University of Western Ontario
London, Ontario, Canada

Charles G. Sodini
Massachusetts Institute of Technology
Cambridge, Massachusetts

Robert Bogdan Staszewski
Texas Instruments Inc.
Dallas, Texas

Gerasimos Theodoratos
National Technical University of Athens
Athens, Greece

Georgios Vitzilaios
National Technical University of Athens
Athens, Greece

Khurram Waheed
Texas Instruments Inc.
Dallas, Texas

Jeffrey S. Walling
Department of Electrical Engineering
University of Washington,
Seattle, Washington

Dorothea Wiesmann
IBM Zurich Research Laboratory
Rüschlikon, Switzerland

Orly Yadid-Pecht
The VLSI Systems Center
Ben-Gurion University
Beer-Sheva, Israel

Kazuya Yamamoto
Mitsubishi Electric Corporation
Hyogo, Japan

Hoi-Jun Yoo
Department of Electrical Engineering
  and Computer Science
Korea Advanced Institute of Science
  and Technology (KAIST)
Daejeon, South Korea

Abdel-Fattah S. Yousif
Department of Electrical and Computer
  Engineering
The University of Calgary
Alberta, Canada

Fei Yuan
Department of Electrical
  and Computer Engineering
Ryerson University
Toronto, Ontario, Canada

Mohammad F. Zaman
Qualtré Inc.
Atlanta, Georgia

Part I

CMOS Technology at the Nanoscale

1 CMOS: An Emerging Technology System Driver

Paul Rousseau

CONTENTS

1.1 CMOS RISE TO DOMINANCE: A VERY BRIEF HISTORY

This is not a real history of complementary metal oxide semiconductor (CMOS), but rather a rapid overview of the steps in getting to CMOS. Whereas key inventions are important, their applications to manufacturing are really the milestones in this march. For example, although the field-effect transistor (FET) may have been the first proposed in the 1930s, it could not be built. The first transistor honors and the Nobel Prize went to the bipolar device. Whereas the 1960s was the decade of the bipolar, the early 1970s saw the introduction of P-channel metal oxide semiconductor (PMOS) followed by the higher performance N-channel mixed oxide semiconductor (NMOS) in the late 1970s. By the 1980s, CMOS was establishing itself as the technology of the future and by the 1990s, unquestionably CMOS was the king of logic, a position it has only strengthened today.

Several factors contributed to this rise. First and foremost, the low standby power that CMOS offered as originally proposed by Wanlass of Fairchild in 1963 [1]. Second, the ability to scale MOS efficiently as articulated by Dennard et al. of IBM [2]. Third, the simple fact that an affordable and manufacturable process could be developed for CMOS.

Today, as CMOS gate oxide scales to dimensions between 10 and 15 Å and gate leakage approaches that of a bipolar base, there is still no turning back. As a voltage-controlled current device, CMOS is ideally suited as a low standby power circuit. It is much harder to imagine this property with bipolar, as it is a current-controlled device. Some may argue that this distinction is arbitrary as one can consider base–emitter voltage (V_{be}) as a voltage control. But can I have V_{be} high (0.6–0.9 V range) without significant current? No, yet in CMOS, I can have high gate-source voltage (V_{gs}) and still have no current flow.

Still, there are clouds on the horizon. As gate leakage and source drain leakage became too important, the scaling laws were abandoned. The power supply voltage has stopped scaling, and the power reductions afforded by such scaling were lost. The power problem grew, implying that designers needed to add power closure on top of the timing closure.

Nevertheless, human ingenuity has kept going and as CMOS marches down the scaling curve, the focus has been to keep tweaking it, even when this involves fairly fundamental structural changes. Strained silicon was introduced to improve performance, and scaling continued in its modified way, still helping reduce the power problem. Perhaps most dramatic has been the recent introduction of high-k metal gate (HKMG) as reported by Intel [3], a complete reengineering of the very heart of the transistor. This solves the gate leakage problem, as least for now.

Process and design have to work hand in hand. Multithresholds methodologies have become very common. Multivoltage domains, dynamic voltage scaling, and sleep mode back biasing methodologies were also introduced to manage the power, although always at the cost of additional complexity.

Whereas these problems left the door open for a contender to CMOS, the task is formidable. For example, electron spin device make interesting scientific papers, but how do we connect a million of them, let alone a billion together to produce a useful system? The beauty of transistors is they react to charge and so to connect one transistor to the next, I just need a wire, thus allowing charge transfer. And, as we shall see, wires are very cheap. How can I do this for spin devices?

Still, there are potential viable candidates waiting for their chance. Fundamentally, the MOS device is flawed by the fact that once in inversion, the current drive increases linearly with supply voltage (this holds for deep submicron device in velocity saturation mode). The device of the future will need an exponential turn on, similar to emitter-coupled logic (ECL) and this will allow scaling of the power supply voltage, hence the power. But that is the potential for future, for now, CMOS reigns supreme.

1.2   CMOS IS EVERYWHERE

Worldwide monthly semiconductor sales from 1991 through 2006 are shown in Figure 1.1. Although memory almost caught up with logic in 1996, the data shows the clear dominance of logic wafers, more than two times bigger than the memory sales. Within logic, Figure 1.2 shows wafer starts from 1997 to 2007. The data illustrate the dominance of CMOS logic over bipolar that only keeps growing, from ~90% of wafer starts in 1997 to ~95% of wafer starts in 2007. The total numbers are also very significant. With over $200 billion in logic wafer sales, it is clear that this supports a significant R&D effort and capital expenditure that nothing else can compete with. Therefore the semiconductor processing tools are focused on CMOS logic and the foundries, the integrated device manufacturers (IDMs) are focused on CMOS logic as well as the design tool providers and the intellectual property (IP) industry. Basically the sheer size of CMOS logic means that an entire and well-oiled ecosystem has been built on supporting CMOS logic first and making it the technology driver.

One reason for CMOS success has been cost. Scaling, as described by Moore's observation, has doubled the density with each generation, with the associated cost reduction. Applications such as graphical processing unit (GPU), which are "Moore law–friendly," can take full advantage of these benefits. Nvidia is a good case in point; they have maintained close to $2\times$ doubling of transistor count every year [4].

In 65 nm technology, $1 can purchase well over 10 million transistors. What can you accomplish with 10 million transistors? How about with 1 billion; this is within reach now. And this cost includes eight levels of copper wiring to interconnect these transistors together, meaning each wire costs a fraction of a microcent.

Now that we have seen that logic is the 800 lb gorilla in the room, and memory is the 300 lb chimpanzee, it is instructive to further explore how these differ in what they enable. There is no

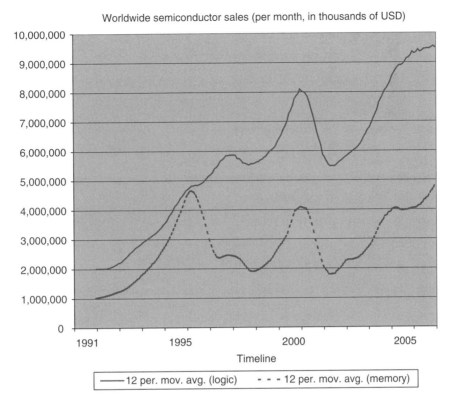

FIGURE 1.1 Worldwide semiconductor sales by month from 1991 to 2006 using 12-month moving average to remove seasonal variations.

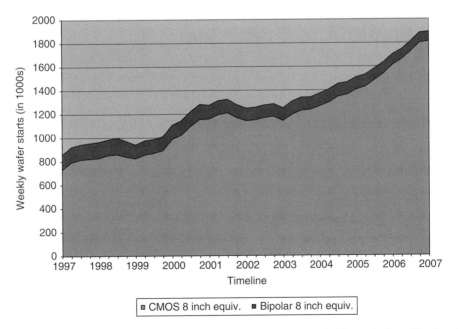

FIGURE 1.2 Weekly logic wafer production from 1997 to 2007 by quarters in 8 in. equivalents (Semiconductor Industry Association data).

question that memory has been a separate technology driver and its products can enable whole new systems; one only has to think of the explosion of digital photography, or portable digital music in recent years. However, memory remains a somewhat standardized and therefore commoditized market, dominated by a few large manufacturers in flash and dynamic random access memory (DRAM). The reasons behind this are fairly straightforward. Although I can enable interesting new systems with memory, it is hard to build memory chips with new functions. Higher density, speed, lower power consumption are the standard fare of memory scaling. In fact, memory technologies are so poor at implementating logic functions that there is a proliferation of logics chip that, built into multichip module (MCM) with memory, aid in optimizing this memory.

Logic on the other hand is the Lego set of designers. Given a logic process with appropriate design tools and technology files, designers have endless possibilities as to how to connect these transistors together to perform new and useful functions. Eventhough for advanced logic, the standard metrics of higher density, faster, lower power still hold, we have not finished exploiting the potential of older technologies and really the only limiting factor is our imagination. What other killer applications remain to be invented? Grant it, this has become harder with time as the low hanging fruit gets picked, but I am always struck by the numerous possibilities that still remain.

CMOS logic by its ubiquity and versatility becomes the obvious foundation for a whole range of emerging technologies. First, there is plain vanilla CMOS as a platform for a whole set of logic applications. Second, there is the flavored CMOS I refer to as CMOS+, which enables a whole new set of possibilities. The industry has used this concept to graft on many varieties onto the CMOS root stock. One only has to look at the Global Semiconductor Alliance (GSA) membership to see the dominance of CMOS and CMOS+. Their membership includes the entire ecosystems from foundries, assembly and test, Electronic Design Automation (EDA), IP vendors, and fabless semiconductors. In the last category, the overwhelming majority are using CMOS followed by CMOS+. Here is a brief list of CMOS varieties: logic base–embedded DRAM, logic base–embedded flash, high voltage, CMOS image sensor (CIS), and RF CMOS and SiGe BiCMOS. Let us now review each CMOS+ option.

1.3 LOGIC-BASED EMBEDDED DRAM CMOS

The distinction between DRAM-based logic and logic-based DRAM is important. DRAM-based logic had its day, but as technology continued to scale, this approach lost steam. Fundamentally this came down to a divergence in transistor capabilities between the two processes.

For a classical 1T-1C DRAM cell, there are four types of leakage, all important to consider to keep refresh rates under control: the capacitor itself, the drain of the transistor (junction leakage), the transistor leakage itself (source–drain leakage), and the gate leakage of the transistor (gate–drain leakage). The latter has limited gate oxide scaling in a DRAM process to somewhere between a 0.25 and a 0.18 μm process (gate oxide thickness of 30–40 Å). As gate oxide is the heart of the transistor, it is easy to see that one will suffer large penalties on power, performance, and density when limited by an older generation transistor. Such an approach can only be used for a chip that is almost exclusively memory with little logic.

In logic-based DRAM, the logic suffers no penalty beyond additional cost. This embedded DRAM, although approximately four times less dense than a pure DRAM, is two to three times denser than a static random access memory (SRAM). The advantages of embedded DRAM over external DRAM are the following:

- *Performance*: Huge memory bandwidth is possible since it is not I/O limited. Remember, wires are cheap.
- *Power*: Lower active and standby power versus SRAM at worse case corners. Power savings are achieved by avoiding large I/O transistors going to off-chip memory.

- *Cost*: Higher density of embedded memory can sometimes save chip memory, especially important as the minimum sizes of DRAM memory readily available in the market keep increasing.

The higher bandwidth enables applications such as high-performance games and graphics. The lower power enables storage applications while lower cost and power help portable applications. Interestingly, in the industry, we find logic-based DRAM starting with 0.25 μm and down to 90 nm and beyond. The timing of this ramp up is in part due to past technical integration difficulties of a logic-based DRAM process, and partly due to the crash of DRAM prices around 1998, which removed any cost incentives for an embedded solution at the time (this crash is visible in Figure 1.1).

1.4 LOGIC-BASED EMBEDDED FLASH CMOS

Similar to DRAM-based logic, the problems with flash-based logic come down to gate oxide. Flash gate oxides are around 80–100 Å thick and have not been able to scale. This gives a transistor equivalent to roughly a 0.5 μm technology, with all the inherent penalties of power, performance, and some penalties in density. Actually the transistor problems of pure flash become so significant in advanced nodes that it hampers new product introduction and other logic solutions have to be found. Again, we find logic-based embedded flash really taking off in 0.35 μm and smaller geometries.

This technology enables microcontrollers, smart cards, and early product introductions when read-only memory (ROM) code is not frozen. Additionally, another category of memory has emerged here as logic identical OTP/MTP (one-time programming [OTP], multitime programming [MTP]). By logic identical, I mean a pure CMOS logic process with no additional steps for an embedded flash structure. I use this term to distinguish it from the confusing "logic compatible," which sometimes can just mean logic-based embedded flash and refers to the fact that the logic does not change, even though there are additional process steps.

The OTP solutions include either polyfuse, oxide breakdown, or floating gate mechanisms, whereas the MTP requires a floating gate using an unfolded rather than the typical flash-stacked structure. In keeping with discussion above, the floating gate must use a minimum 70 Å gate oxide, thus making it only possible for 3.3 V or higher I/O transistors. These solutions enable low-cost memory that may be used for chip ID, security, analog trimming, redundancy, and even small one-time programmable ROM.

1.5 HIGH-VOLTAGE CMOS

High-voltage CMOS has breathed new life into more mature CMOS manufacturing lines. The plethora of high-voltage applications requiring tighter geometries or integration has led high-voltage technology to merge with CMOS logic all the way down to 0.18 μm.

There are two distinct offerings using high-voltage CMOS:

- Double-diffused drain (DDD) MOS or lateral double-diffused MOS (LDMOS) structure used for low-current applications such as LCD or microelectromechanical system (MEMS) drivers
- Bipolar CMOS DMOS (BCD) technology for high-current applications such as power management or the upcoming LED driver market

Again, by leveraging logic-compatible processes, one benefits from a robust existing technology as well as a complete suite of silicon-proven IP. An innovator in this field could focus on their core high-voltage competency and use off-the-shelf CMOS IP components to complete their total solution offering.

1.6 CMOS IMAGE SENSOR

The rise of CIS is an area where completely new markets have been opened for CMOS technology. Charge-coupled device (CCD) has done a reasonable job so far of keeping market share in the pure camera/camcorder space, but the ubiquity of cameras on cell phones has been enabled by the low power of CIS and easier system integration. With a voltage less than half of that used in CCDs, CIS offer a huge power benefit. Also, CIS offer a significant speed advantage, especially as the number of pixels keeps growing. It seems only a matter of time before the market dominance of CIS is complete.

CIS technologies were initially built on a proven CMOS platform with all those previously mentioned advantages. Recently there has been a shift as the back-end metal oxide stack has been thinned to allow for better light efficiency. At that point, the performance required by CIS had increased, and the market had grown sufficiently to justify this. Nevertheless, the basic premise still holds that this technology leverages the existing CMOS platforms and makes as little changes as possible. This affords another advantage to CIS over CCD: it does not require a dedicated fab and can thereby leverage existing manufacturing facilities to enable the rapid ramp up witnessed in the last 5 years.

1.7 RF CMOS AND SiGe BiCMOS

With a flourish of wireless applications, CMOS radios are in hot demand. Though SiGe BiCMOS offers typically better RF performance, the cost advantage of CMOS and integration capability are the key factors. WLAN 802.11 and Bluetooth 802.13 are ubiquitous, and WIMAX 802.16 and ultrawide band (UWB) are upcoming applications based on standard protocols. Beyond that, a myriad of applications exist from low-power sensors communicating wirelessly, computer and consumer peripherals streaming both video and audio [5]. Finally, high-speed wired communication is also enabled by RF solutions.

Leveraging the CMOS baseline, one needs no process change but just to characterize the process with RF-calibrated models and add a few passives such as inductors and capacitors. This creates a technology that can enable all these high-volume applications. Adding a SiGe BiCMOS on top of the existing CMOS further enables some demanding RF applications such as power amplifiers and silicon TV tuners. Yes, CMOS can provide fast transistors with cutoff frequencies over 200 GHz but with the biasing of about 1 V. SiGe can provide the same speed at a higher power supply, enabling a larger dynamic range. From this comparison, one can immediately understand why power amplifiers are one of the SiGe's niche areas. These solutions enable tremendous performance in a silicon solution, and even CMOS radios operating at 60 GHz are becoming a reality.

1.8 OTHER TECHNOLOGY CONSIDERATIONS

1.8.1 AUTOMOTIVE

The automotive market appetite for silicon wafers keeps growing. Applications range from the various creature comforts, safety systems associated with vehicle handling, navigation through global positioning system (GPS), and simple things such as air tire pressure monitors (which require a radio) or backup cameras that use CIS. Future applications look at collision avoidance systems. Will we someday have completely automatic pilots for those long freeway drives? In sum, automotive market utilizes all the previously mentioned technologies. With the appropriate and stringent quality controls, an automotive grade CMOS is a perfect solution for this wide variety of applications.

1.8.2 MEMS

Finally, we come to something that is not CMOS: MEMS. MEMS enables several key markets, such as front projection systems, inkjet printers, accelerometers and gyroscopes, and silicon resonators.

Each MEMS product is usually a custom process, although the concept of "process libraries" enabling some integration is possible. In these cases, MEMS is in many ways married to CMOS. For example, in a micromirror array, the control of the millions of mirrors is accomplished by a CMOS driver chip, which must be integrated with the array. Similarly, in a resonator, CMOS is used to correct for temperature and process variations. So whereas MEMS is truly non-CMOS, CMOS as an enabler for MEMS system remains key.

1.8.3 SoC versus SiP

One key decision in building new systems is whether to use system-on-a-chip (SoC) or system-in-package (SiP). SoC using CMOS+ is thought to enable low power, low cost, and higher reliability, but not always. Add-on technologies to CMOS cost more because of added process complexity, and schedule readiness may not be a match.

Typically plain vanilla CMOS is available first on a new node. Still, CMOS+ enables elegant solutions. Each implementation requires a careful trade-off of all aspects such as system needs, readiness, risk, cost, and power. The order here is somewhat arbitrary; the priorities are dictated by each system itself.

1.9 CONCLUSIONS

CMOS cannot do everything and there will always be room for specialized technologies. Nevertheless, I have hopefully convinced you that CMOS is an incredible platform for enabling emerging technologies. A good rule of thumb when inspiration strikes is to first ask the question, "Can it be done in CMOS, or CMOS+?" If not, which pieces will be done in CMOS? By doing so, you would be leveraging the semiconductor industry's most significant investments and basically have access to high performance, power efficiency, and low-cost technology. Your competition will also have access to CMOS; therefore another good question to ask is "what happens if my competition succeeds in doing this in CMOS." In sum, one can imagine CMOS is a bit like the interstate highway system while CMOS+ is the local highway network. They both allow you to explore new territory in a very fast and efficient way. Of course, you will still need to take the dirt tracks on the side and even walk a bit to create something truly new.

REFERENCES

1. Wanlass, F.M., Low stand-by power complementary field effect circuitry. U.S. Patent 3,356,858 (filed June 18, 1963; issued December 5, 1967).
2. Dennard, R.H., Gaensslen, F.H., Yu, H.-N., Rideout, V.L., Bassous, E., and LeBlanc, A.R., Design of ion-implanted MOSFETs with very small physical dimensions, *Proc. IEEE*, SC-9(5), 256–268, 1974.
3. Bohr, M.T., Chau, R.S., Ghani, T., and Mistry, K., The high-k solution, *IEEE Spectrum*, 44(10), 29–35, 2007.
4. Design Automation Conference (DAC), 2002, http://vlsicad.ucsd.edu/~abk/TALKS/dac2002_session26_malachowsky.pdf
5. Iniewski, K. (Ed.), *Wireless Technologies: Circuits, Systems and Devices*, CRC Press, Boca Raton, FL, 2007.

2 CMOS Manufacturability

Artur Balasinski

CONTENTS

2.1   INTRODUCTION: TWO DIRECTIONS FOR TECHNOLOGY DEVELOPMENT

These are challenging and exciting times for semiconductor manufacturers. While this has generally been true for the recent 40 "Mooreish" years [1], recent developments have added a new spin to it. Looking down from an altitude of the total available market of $230 billion per year [2], the silicon industry is gradually abandoning the successful shrink path. In order to reduce cost, it is moving the manufacturing research and development to offshore foundries. Why then be concerned about complementary metal oxide semiconductor (CMOS) manufacturing issues if they are no longer considered being worthy of prime-time interest? In addition, the brick wall of scalability [3] is casting an increasingly deep shadow: it may not be economically viable to build and utilize atomic-scale metal oxide semiconductor field-effect transistors (MOSFETs). How can we make sure that CMOS still has future worth exploring before it becomes a textbook staple? We will try to answer this question.

Semiconductor technology is an art of building transistors. Throughout the last 50 years, it was driven by field-effect transistor (FET) development. It was FET gate CD, FET gate oxide thickness, and FET junction depths and profiles that had to be scaled down according to the Dennard rules [4] to make faster and cheaper devices. Accordingly, leading device manufacturers, product lines, and pure-play foundries have recently started implementing the 65 nm node and are looking forward to the 45 nm process. The underlying paradigm used to be that the shrink path would eventually provide enough transistors to capture any information in the digital form, be it originally analog or digital. This paradigm, despite all the years of following Moore's law, proved to be correct only to some degree. There are still new applications and economic opportunities ahead for analog and radiofrequency (RF) devices within the existing process capabilities that have only been marginally comprehended.

Operating MOSFETs for digital applications should be considered relatively simple. The devices switch from off-state to saturation and the key effort is directed to maximize their switching speed. In the absence of significant power supply limitations, leakage control used to be required to reduce heat dissipation. However, the market for countless, small, and relatively leaky FETs for home applications is saturating. Now everybody wants to take FETs on a field trip, power them from a tiny battery, and instantaneously transfer the captured information to a remote base unit. The trade-off between high speed and low leakage is swinging towards the low leakage but not such that the speed can be quite forgotten. At the same time, digitization of the incoming information has to start at the source, i.e., in an arbitrary environment. Therefore, an increasing importance is given to interface devices: sensors of light, temperature, pressure, chemical structure, digital-to-analog converter (DAC), and analog-to-digital converter (ADC) circuits. In order for them to work reliably, novel design and computer-aided design (CAD) procedures are needed.

Such system architecture, as in powerful central processing units (CPUs) and diversified peripherals, has been visible in integrated circuit (IC) development for many years. However, technology was focused on strengthening the CPU by adding more of the same types of MOSFETs, only with ever-smaller dimensions. Now the approach is shifting: two main trends decide about technology development. One trend is to keep following Moore's law and continuously increase the density of core logic and memories, while the other one is to develop new devices for diversified consumer applications, mostly exploiting the existing technology knowledge base. As more companies exit the Moore's shrink path as too expensive, their research interest is changing rapidly from the CPU-centric to the periphery- or auxiliary-device-centric. This is because everyday applications can no longer benefit much from the growing computing power of personal computers (PCs). In addition, market-driven information transfer in its analog, rather than digital, form necessitates development of systems on chip with significant analog/RF content.

Given the old and the new challenges, it is not possible to properly cover all aspects of CMOS manufacturability issues in one chapter. The choice of material was based on the subjective vision of the current problems. For a more detailed view, the reader may want to refer to many excellent works such as Wong et al. [5] and Wolf [6].

2.2 CONTINUOUS SHRINKING: MEMORIES AND LOGIC

Scaling rules meet more obstacles every year, as technology is approaching its fundamental physical limits. This has implications for both new product development and manufacturing. At design time, designers need to accept device models with increasingly wider parameter distributions and assume correspondingly larger margins for circuit simulation, as the deep-sub-100 nm manufacturing process is becoming increasingly less deterministic. While device models should reflect the across-wafer, across-lot, and lot-to-lot process variations, the impact of mask CD control, deficiencies of the imaging process, or etch-related macroloading are hard to characterize and often neglected. Electronic design automation (EDA) and mask industry are ramping up complex design-to-wafer simulation flows with emphasis on horizontal and vertical CD control [7]. For example, in order to minimize the impact of mask-error-enhancement factor (MEEF) [8] while reducing the line CD without changing the patterning wavelength, on pattern resolution and the resulting device parametric variation, optical proximity correction (OPC) techniques evolved from manual sizing into model-based OPC. The key manufacturing issues related to the classical shrink path discussed below include photo imaging, device leakage control, and novel back-end materials.

2.2.1 LITHOGRAPHY AND OPC: CLOSING THE SUBWAVELENGTH GAP

Lithography is the key to the shrink path. For several recent technology generations, the imaging wavelength used to print features on silicon has been larger than the linewidth of these features. This difference, known as the subwavelength gap [9], requires aggressive development of patterning techniques. Next-generation lithography solutions competing for manufacturing viability are the dual-patterning technology (DPT), the extreme ultraviolet (EUV), and the nano-imprint lithography (NIL) (Table 2.1). Their success is still far from predictable. Inverse lithography, a technique that practically negates OPC, seems to be also making progress [10]. The key manufacturing task is to

TABLE 2.1

Progress in Patterning Techniques

Name	Issues	Resolution Limits	Timeframe	Target k_1
Immersion	Max n value ∼1.7	45 nm	Current (2007)	0.25
DPT	Doubles the cost of the photo process (time, masks) Requires layout modifications	32 nm	2008–2010	0.125
NIL	Poor overlay	22 nm or less	2009–2012 (?)	N/A
EUV	Mirrors, photoresist, light sources	10 nm or less	2009-?	0.8

optimize the resolution by controlling the (dimensionless) lithography resolution factor k_1 defined by the guidelines for a diffraction-limited projector according to the Raleigh scaling equation [11]:

$$R = k_1 \lambda / NA \tag{2.1}$$

where R is the minimum-resolved linewidth, λ is the imaging wavelength, and NA the numerical aperture of the projection optics system. The k_1 factor depends on the process conditions and aggressiveness of OPC. In the past, when the subwavelength gap was small, the IC manufacturing line was expected to print minimum features with $k_1 = 0.8$. Gradually, k_1 has been driven down to the values beyond 0.5, and innovative techniques such as phase-shift mask, off-axis illumination, and OPC potentially allow $k_1 < 0.4$. The ultimate k_1 limit for the standard lithography was expected to be ∼0.25 but it can be reduced further down to ∼0.13 or less, for example, using DPT. Therefore, a lot of progress is likely to be accomplished without reducing the wavelength.

While the role of OPC is to ensure a stable printability of the critical shapes (especially minimum CD features) to improve lithography process window for corners and line ends, and to prevent from line breaking or pattern bridging, OPC itself is highly sensitive to process variations and design style. Arbitrary layout geometries, even adhering to design rules, do not always lend themselves to processing at high yield and OPC can only help to a certain point. Recently, OPC was extended into design-for-manufacturability (DfM) methodology, which intends to optimize layout style for yield, reliability, or lithography (litho-friendly-design, LfD). Along with these efforts, EDA tools have been developed to assist with dummy pattern generation via doubling, wire spreading, proximity pattern placement, and extraction of all geometries printable on wafers. The DfM process is ongoing and starts resembling the classical, industrial concept originally developed for the automotive industry [12], as the CMOS technology reaches new levels of maturity.

2.2.2 MOSFETs

For isolated, straight channel MOSFETs, defined without lithography limitations, critical issues include all aspects of leakage, for example, isolation leakage, gate leakage, and gate-induced drain leakage. Improvement of the MOSFET ON/OFF current ratio, a performance boost is accomplished by the use of strained silicon [13,14].

For the 45 nm technology and below, standard MOSFET isolation scheme consists of two independent steps: because creating shallow trench isolation (STI) first, followed by gate definition, does not ensure sufficient overlay tolerance for the gate end caps. As a result, either the circuit (i.e., static random access memory [SRAM] cell) area or the end cap misalignment leakage is compromised.

An isolation scheme with oxide trenches self-aligned to poly became therefore a necessity to further reduce the cell size [15]. The process starts from defining active areas of buried diffusion and the gate pattern with a single mask. A protective oxide is then formed between the polysilicon lines. Next, a second mask with noncritical alignment selects polysilicon lines and defines self-aligned etch regions. The trenches are etched using a high-selectivity recipe, which cuts through polysilicon and the silicon substrate in the selected lines faster than through the protective oxide. Thus, a single mask defines the diffusion regions, the first layer of polysilicon, and the isolation trenches. The second mask defining isolation structures does not need to be critically aligned, which increases the margin of the isolation structures in the memory array.

Even with such advanced STI process, the intrinsic problem is the compressive stress, which causes energy band gap narrowing. As a consequence, the effective barrier heights for band-to-band tunneling and trap-assisted tunneling become lower and the higher intrinsic carrier concentration increases the leakage current. An asymmetric cell layout has been proposed to reduce the gate-induced drain leakage current [16].

Last but not the least, MOSFET leakage is due to the low gate oxide thickness. The desire to stay with SiO_2 as the gate dielectric for as long as possible has recently been abandoned [17], as its thickness beyond the 45 nm node would have to become on the order of a monoatomic layer making the leakage prohibitively large, even if it is nitrided to increase permittivity, reduce boron penetration, and improve hot carrier resistance. Low gate leakage process uses high-k (dielectric constant) hafnium oxides [17] and new metal gate materials to eliminate threshold voltage pinning and phonon scattering, which lower the performance of MOSFETs with high-k dielectric under polysilicon gates. A metal gate (different for NMOS and PMOSFETs to control the work function), combined with the high-k dielectric, ensures their very low leakage and high drive currents. Uniaxial strained silicon with fully silicided (FUSI) NiSi metal gate (Figure 2.1) enables drive currents in saturation IDSAT = 1.75 mA/μm (NMOS) and IDSAT = 1.06 mA/μm (PMOS), at drain voltage VDD = 1.2 V and off-state current IOFF = 100 nA/μm [18]. The FUSI process requires precise optimization of poly-Si, Ni film thickness, and silicide thermal budget, to minimize process variations due to incomplete or excessive NiSi formation. The first effect can be misinterpreted as a change in the gate work function and result in threshold voltage VT variations, the second one can lead to Ni penetration through the thin gate oxide.

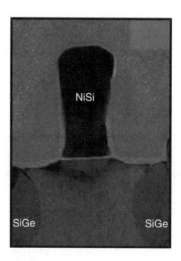

FIGURE 2.1 Cross section of a MOSFET with NiSi gate. (From Ranade, P., Ghani, T., Kuhn, K., Mistry, K., Pae, S., Shifren, L., Stettler, M., Tone, K., Tyagi, S., and Bohr, M., IEDM, 2005. With permission.)

2.2.3 New Materials for Interconnects

In order to take advantage of high-drive current MOSFETs using narrow metal buses, one needs new materials to reduce contact and interconnect resistance as well as electromigration risk. One also needs to control dielectric permittivity to avoid capacitive coupling in the circuits with increasingly higher switching speeds. Low contact resistance to meet the International Technology Roadmap for Semiconductors (ITRS) expectations [19] can be accomplished by reducing Schottky barrier height of NiSi/Si contact by interface engineering. Electromigration in copper (Cu) damascene interconnect lines is preventable by selective surface doping with calcium (Ca) ions. The process restricts Cu diffusion pathways along the interconnect surface by selectively forming a $Cu-Ca$ film on the Cu interconnect line and argon sputtering to remove the contaminants [20].

To reduce capacitive coupling, a variety of low-k materials with dielectric constant about 2.5 can be deposited using traditional spin-on and chemical vapor deposition (CVD). Ultralow-k materials with a dielectric constant around 2.0 can demand a trade-off with the mechanical strength or thermal stability of the device, as most such materials would be porous.

Microprocessor companies integrate fluorinated silica glass (FSG) first because its properties are similar to those of the oxide, and it is deposited using existing fab tools, to an obvious cost advantage. However, this does not mean that FSG is a simple "drop in" low-k solution because fluorine containment and adhesion are still concerns. There are several viable low-k materials with dielectric constants between 2.5 and 3.0 deposited using spin-on and CVD, such as organic or thermosetting polymers and organosilicate glass.

The dielectric constant is minimized by reducing the number or the polarizability of chemical bonds (density) and minimizing moisture content in the material [21,22]. One way to decrease the k value is fluorine incorporation. However, fluorine is very corrosive to metal if not tightly bound and fluorinated films often show adhesion problems. Chemical bonds with low polarizability tend to be weaker, limiting thermal stability, and less stiff, limiting the thermomechanical properties [21]. Organic polymers with linear, flexible molecular backbones can have very low dielectric constants but they are often much softer than the rigid, network bonded oxide-like materials. They also decompose at lower temperatures.

Porous oxides may be formed in several ways, for example, using solgel chemistry to form aerogels and xerogels [23] with the volume fraction of solids as low as 1%. Porosity in a material is induced by a labile component (e.g., an organic compound), later removed by thermal decomposition, leaving behind a void [24].

2.3 TECHNOLOGIES FOR NEW APPLICATIONS: SYSTEM-ON-CHIP AND ANALOG/RF

2.3.1 Process Challenges

Memory technology focuses on small area cells to achieve extreme packing density. Logic process optimizes devices with more random geometries but with line CDs also close to the minimum printable values to build MOSFETs, with low leakage and high speed. To make Systems-on-Chip (SoC's) with digital (memory, logic) and analog (RF) content, one needs to add passive devices. Large differences in geometries among the digital and analog/RF layouts, such as solid, parallel plate, or vertical multifin capacitors, large area inductors, precision resistors, and MOSFETs of different sizes, translate into electrical and optical macro- and microloading at wafer level, impacting lithography, etch, and planarization (Table 2.2). This enhances the need for precise models but may also significantly challenge fab process capability due to the new macroscopic layout properties. For a fab running a diversified product family, there may be only very few new process modules required to make a successful SoC product (e.g., nonstandard metal layers may need to be added to create vertical capacitors for RF applications), and these modules may need to be integrated in a new sequence.

TABLE 2.2
SoC as Technology Driver

Device	Features	Issues	Mitigations
MOSFETs	Fast	CD control	Strained silicon
	Low leakage	Resolution	OPC and dummy features
	Analog-matching	Variability reduction	Tight process control
	RF		Reduction of parasitics
	High current		Low contact resistance
	High voltage		Drain extended architecture
Resistors	Standard	CD control	Large CD
	Precision	Electromigration	
		Contact resistance	
Capacitors	MIM (Metal-Insulator-Metal)—vertical	Large capacitance	Thin ILD
	Horizontal	Variability reduction	Circuit solutions
Inductors	Q-factor	Planar architecture	Thick top metal

According to the ITRS roadmap [25], it is the variability of device parameters such as the threshold voltage, local power supply, and current distribution, which actually is becoming a key challenge to the IC functionality and yield, especially for the SoC products with high content of analog circuitry. The analog/RF devices require minimal device-to-device variation to match input resistances of differential amplifiers by ensuring their symmetry, reduce leakage to enable handheld applications, eliminate capacitive coupling to the nonextractable dummy features, etc. Device variability control in the sub-100 nm technology range becomes necessary at several stages of IC development [26]:

- At design stage, by using correct by construction (CBC) primitives with parameterized and standardized layout with known characteristics
- At layout stage, by observing DfM rules and methodology
- At postlayout stage, by applying CAD processing to optimize layout distribution at mask level
- At manufacturing stage, by tightening the manufacturing process control, for example, by manufacturability-for-design (MfD) equipment upgrades or tuneups

2.3.2 Variability Reduction for MOSFETs and Passive Elements

The most predictable approach to minimize device variability is to start at the electrical design phase by allowing only standard cells representing devices with preapproved, highly manufacturable CBC layouts. For device primitives such as individual MOSFETs or inverters, one can propose two approaches to reduce the within-die variability:

- Increase minimum device geometries to improve printability
- Change layout aspect ratio to reduce proximity effects

For analog/RF applications, device sensitivity to the parasitic resistance, capacitance (and recently, also inductance) is becoming a higher priority for layout techniques as compared to footprint reduction or pattern fidelity. As an example, Figure 2.2 shows three inverter layouts designed to be identical from the DC parametric standpoint [27]. However, one can expect that their frequency responses would be significantly different. While by changing the layout from 2a to 2b (area increase by ~130%), one trades the die footprint for lithography yield on the one hand, and for point defect yield on the other, the layout 2c (75% bigger than 2a), which trades line CD uniformity for the immunity

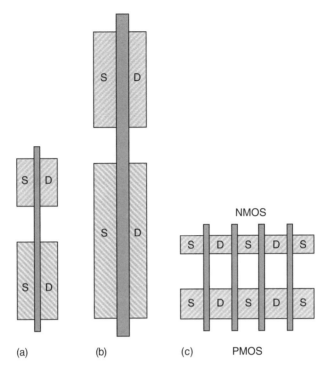

(a)   (b)   (c)   PMOS

FIGURE 2.2 Three options of inverter layout. (From Balasinski, A., Cetin, J., and Karklin, L., *J. Micro/Nanolithogr., MEMS, MOEMS*, 6(3), 003008-1, 2007. With permission.)

to optical proximity, should be preferred for RF applications due to the lower parasitics of folded gates. On the other hand, CBC layout optimization requires significant precharacterization effort, for example, by test chips, to model parasitic effects.

Variability reduction at layout time may require adjustments to the die footprint with respect to the one dictated by minimum design rules. It may be difficult to explore how such layout modifications affect the yield [28]. Yield gain on the order of 5% would need to be verified on at least one split lot and at a 1% level, multiple lots are needed, often making die area increases on the order of few percent unacceptable, as it would be very costly to prove their benefit. For example, by increasing the gate length for 50% MOSFETs in the die, to reduce their leakage by a factor of 2, one would only reduce the total die leakage by 25%, which may not suffice to consider the product suitable for a handheld application.

Standard design rules guarantee process capabilities within the 10% variability margin. Reducing this margin, for example, down to the 1%–2% range as may be required by analog/RF product applications, calls for low variability rules, based on photo and etch bias and proximity effects. Accordingly, reducing MOSFET CD variation from 10% to 2% could mean size increase by 5×.

The lowest variability can be achieved for the layout of on-grid, one-dimension (1D) long, straight lines. By contrast, line discontinuity rules require a lot of area to compensate for the minimum extension of poly beyond active (end cap), spacing of the gate to the inside or outside active corner, and spacing of the gate to the line of extended width (Figure 2.3).

Unfortunately, these rules typically would not scale with the linewidth but with the wavelength of the imaging light. For the 65 nm process, they can require as much room to comply with, as the actual device area. They also ensure that the channel would not only be free of the pullback or rounding effects, but, in the case of the sensitive analog devices, also free of any "ringing" due to overcompensated OPC features, such as aggressive hammerheads or corner serifs.

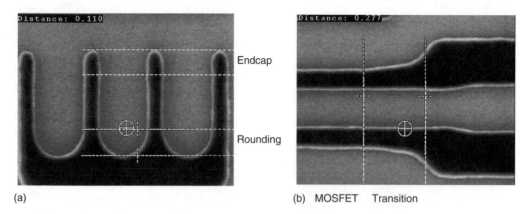

FIGURE 2.3 SEM images showing line end effects ((a) rounding and (b) pullback) requiring design rules to mitigate rounding design rules. (From Balasinski, A., Cetin, J., and Karklin, L., *J. Micro/Nanolithogr., MEMS, MOEMS*, 6(3), 003008-1, 2007. With permission.)

To mitigate the 2D variability, dummy structures are added around the layout matrix, with a minimum width rule of a dummy cell or feature ring (DRW), checked by simulation (Figure 2.4):

$$DRW = 4\lambda \qquad (2.2)$$

As discussed in Section 2.2.1, OPC which can help reduce the k_1 factor down to the theoretically lowest value of ~0.25 would result in the lithography process window, potentially too small to ensure high yield of sensitive layouts. Variability reduction can only be accomplished by increasing k_1 to ~0.6–1.0 range. This would correspond to increasing minimum CDs by a factor of ~4 vs. their minimum process capability values. One therefore can propose a rule of thumb for low variability layout by which any geometry should be 4× larger compared to the minimum process capability CD.

One should keep in mind that even carefully drawn and fully symmetrical analog layout can potentially be affected by lens aberrations such as astigmatism or coma and fail to deliver the expected performance. Modern lithographic printing devices (scanners) are equipped with integrated lens interferometers. They allow capturing full-field interferometric data to in situ measure scanner's

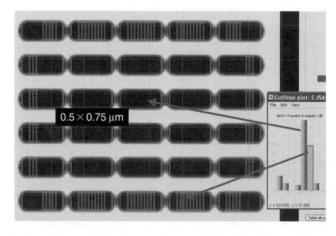

FIGURE 2.4 Simulation of impact of the dummy ring around MOSFET inverter input array on the gate CD variation: layout with cutlines (left), CD distribution (right), with CD variation below 2 nm for 500 nm gates. (From Balasinski, A., Cetin, J., and Karklin, L., *J. Micro/Nanolithogr., MEMS, MOEMS*, 6(3), 003008-1, 2007. With permission.)

lens aberrations. Information about scanner's field uniformity could be passed to the designer to place the most distortion sensitive geometries (such as analog/digital converters) into "clean" imaging field areas, i.e., areas with minimum root mean square (RMS) wave front errors (aberrations).

CAD processing helps control horizontal and vertical CD variation by OPC and fill pattern. Lowering the aggressiveness of OPC improves the linearity of mask pattern transfer to the silicon. Figure 2.5 shows an example of model-based OPC for a 65 nm process with photoresist trimming, applied to a comb structure. The uncorrected lines have CD uniformity of ±4 nm vs. the corrected ones of ±5 nm. While the corrections do improve significantly the line end pullback—from ±30 to ±8 nm—in the event that even a very small CD variation across MOSFET width cannot be accepted, OPC should be reduced to minimum, sometimes to the disadvantage of the device footprint.

Simple CAD solutions could also be rule based such as gate sizing to selectively increase channel length of leaky MOSFETs. However, as opposed to the drawn device changes, the automated sizing takes device performance control away from design and requires extensive silicon verification.

Pattern density of conducting and isolating layers of the die, which create its physical, vertical structure, i.e., active, poly, interlayer dielectric (ILD), and metals, impacts its electrical parameters. Technology shrinks are more pronounced in the x-direction than in the y-direction, increasing the

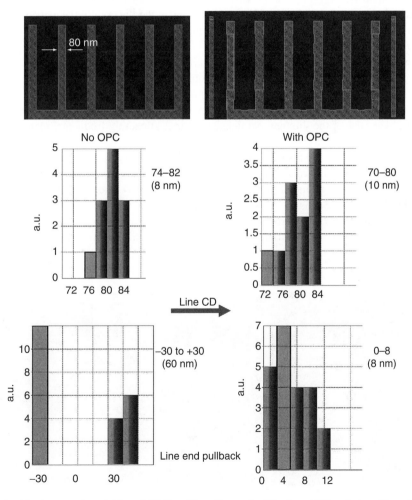

FIGURE 2.5 The impact of model-based OPC on 65 nm line end pullback and CD variation. (From Balasinski, A., Cetin, J., and Karklin, L., *J. Micro/Nanolithogr., MEMS, MOEMS*, 6(3), 003008-1, 2007. With permission.)

cross-sectional aspect ratio of devices and interconnects and necessitating vertical CD control. At active level, variations in pattern density translate into wide distributions of punch-through or break-down voltages. At poly and metal, nonuniform pattern density resulting in poor planarity gives rise to high contact or via resistances, as all the openings etched through the ILD are expected to clear at the same time. Another effect is variations of the interlayer capacitive coupling. Therefore, pattern density distribution of interconnects is restricted to preserve CD control of the ILD deposited over them.

The optical and etch macroloading and surface dishing during the chemical–mechanical polishing (CMP) process depend on both pattern density and definition of the patterned layer. For example, the metal damascene process would call for a different type of dummy pattern as compared to the subtractive metal etch, because in the first case, the material being removed is the metal (Cu), and in the second case, it is the oxide (ILD).

In order for the waffling to remain in design control, one can add waffles by manual drawing at the IP block level, which is time consuming (requires multiple features to be hand-drawn) and iterative, as designers working at block level do not have the visibility on the target level die PD, until die integration is complete. The second, most common option for waffling is to add it by an automated procedure at die level in the open locations. Its key drawback is that uniform density often comes at the expense of a significant increase of the total PD beyond what was intended by design or process, which may not be acceptable for many regions in the die. While the sensitive signal paths can be protected by the keep outs, their neighboring areas would be overwaffled to compensate it and to keep the average PD on target.

Figure 2.6 shows an example of a die region waffled using two algorithms, producing sparse and dense fill pattern. The sparse waffling (Type 1) would have been sufficient to adjust pattern density at the local level. However, due to the large number of keep out regions (black areas in Figure 2.7), to compensate pattern density at the global scale, more waffles are added in the adjacent areas (Type 2). As a result, over 50% of the die may get dense waffling, impacting frequency response of an SoC signal path.

New types of waffling are also being developed (Table 2.3). Figure 2.8 shows die-level PD data using a simple geometric approach of fixed density and intelligent waffling, which creates dummy structures only as needed to match the pre-existing die-level PD as closely as possible. The latter one lowering both the PD range and its absolute value has double advantage for the CMP process as well as the extraction procedure as it has the lowest waffle to space ratio (0.32:1) and largest waffle-to-waffle space (1.2 μm) and helps reduce the ILD thickness difference D from the prewaffle

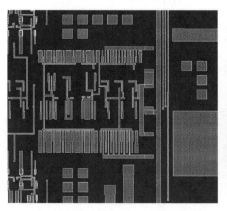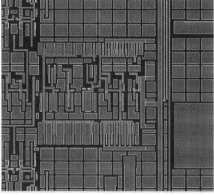

FIGURE 2.6   A sample of metal layout with two different fill patterns: sparse and dense. (From Balasinski, A. and Cetin, J., *6th IEEE Int. Workshop on SoC-RT*, Cairo, December 2006, pp. 156–159. With permission.)

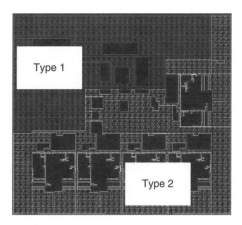

FIGURE 2.7 Division of die area into the regions of (1) sparse and (2) dense waffles. (From Balasinski, A. and Cetin, J., *6th IEEE Int. Workshop on SoC-RT*, Cairo, December 2006, pp. 156–159. With permission.)

value of >400 nm down to <200 nm. In comparison, the geometric approach is far less efficient in driving down both pattern density and ILD variation.

Updating process technology to improve product manufacturability is often preferred as the single-point and downstream action, advantageous for all designs for the given technology node. As an example, tightening postetch CD distribution by improving gate etch profile, involving the use of a new plasma etch source, can minimize pitch dependency of the etch bias and gate profile and reduce the 3σ leakage across wafer by a factor of 4 [29].

In Table 2.4, we show examples of SoC product issues related to parametric variability and the proposed mitigation options. The CBC, DfM, CAD, and MfD solutions, respectively, call for test chip development, part redesign (e.g., using devices with longer channels), sizing up the gates of MOSFETs exhibiting high leakage, or implementation of new process equipment. In the last column, we show examples of process issues along with the proposed grades for their preferred solutions. As discussed above, for the issues #1, #2, and #4, change in equipment or its settings would be preferred over CAD or design solutions requiring edits to the database and potentially affecting product performance. Assuming that the timely ramp-up of the product volume would compensate the cost of the new tool, the advantage of MfD for issue #1 can be as high as 45:25:20:10, depending on the confidence level and the risk involved. For the issues #2 and #4, the preference for MfD over DfM or CAD solutions could be even higher as there is no extra equipment cost. In contrast, to resolve issue #3, a manufacturing solution may not work without mask pattern using dummy features, which help achieve uniform pattern density but create the capacitive coupling problem at the same time. The edits to the database or modified CAD algorithms to account for this

TABLE 2.3
Summary of Yield Improvement Methodologies at Different Phases of Product Development

Stage	Approach	Implementation
Design	CBC layout	Precharacterized standard cells
Layout	DfM rules	Layout with reduced randomness
Mask	CAD postprocessing	Random layout with OPC and fill pattern
Fab	MfD	Upgraded equipment or dedicated processing

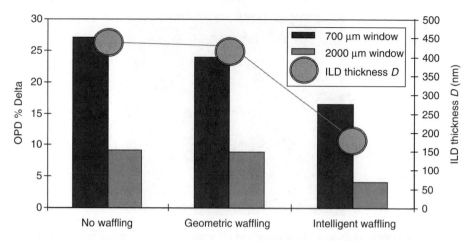

FIGURE 2.8 Impact of waffling on oxide pattern density calculated for different PD window sizes using two algorithms: geometric and intelligent. (From Balasinski, A. and Cetin, J., *6th IEEE Int. Workshop on SoC-RT*, Cairo, December 2006, pp. 156–159. With permission.)

parasitic effect are the preferred options. For legacy technologies, the low cost and risk of MfD due to the increasingly high maturity of the equipment would outcompete the upstream approaches. This trend is changing for the 65 nm technology node and below as the increasing cost of MfD solutions makes it necessary to proactively define the low variability layout options based on CBC principles and new generation of EDA tools.

TABLE 2.4
Examples of Process Issues and Graded Product Development Responses

	Issue		Response	Cost	Preference
1	High leakage incompatible with handheld applications	CBC	Sleep circuits	Design complexity	10
		DfM	Use MOSFETs with longer channels	Slower parts	20
		CAD	New OPC and fill to reduce CD variation for multiple CD targets	IP invalidation CAD execution	25
		MfD	New etch for better line profile	New equipment	45
2	Device mismatch across die	CBC	Cells with predefined pattern density	Design iterations	5
		DfM	Design rules for pitch and orientation	Model calibration	15
		CAD	Addition of dummy devices	Model calibration	15
		MfD	Exposure correction to tighten CD budget	Process development	65
3	Capacitive coupling for RF as dangerous as resistive path for DC	CBC	Dummy rings	Dummy models	10
		DfM	Relaxed layout	Larger die	10
		CAD	Low density of fill pattern	Waffling tool	50
		MfD	New type of CMP slurry or reverse mask	Material use or more masking steps	30
4	Poor frequency response due to high R (poly, via)	CBC	New routing methodology	Area penalty	15
		DfM	Increase device dimensions	Area penalty	15
		CAD	Via doubling and poly widening	Tool and die area	15
		MfD	New contact materials New CMP slurry	Cost of slurry Time to etch	55

2.4 CONCLUSIONS

Historically, circuits with highest device density were used to drive technology development. For many years, dynamic random access memories (DRAMs) were the flagships of semiconductor industry. Sideshows were the SRAMs used by the logic companies to debug the process. This trend still continues at the largest semiconductor makers, which can afford the increasing risk of ripping benefits of scaling transistors into single atomic dimensions. The risk is related to the growing pressure of both technical and economical issues—can they all find a viable solution? We mentioned several top-level challenges, such as lithography, device leakage, and back-end processing. These create such a large spectrum of problems that completely novel device architectures, such as the nanotubes, are being explored instead. At the same time, memory products, even built with up-to-date technology, tend to generate low and unstable profit margins. Therefore, the increasing cost and uncertainty of technology made it necessary to broaden the product portfolio. This new trend was aligned with customer expectations driving the increasing importance of SoC. The requirements of SoC product family make it necessary to add circuit elements with diversified geometries, such as the inductors, capacitors, and precision resistors. To ensure that these devices can be reliably manufactured, according to their electrical models, new layout rules and approaches were necessary. While the standard memory technology took advantage mainly of the linewidth reduction, the SoC products have to deal with a number of other technologies, CAD, and design issues.

Layout techniques to reduce the impact of process variations on IC parameters are of critical importance for new product development, especially for analog/RF and low-power applications. Design rules and product flow methodologies for low variability need to be developed, such as CBC, DfM, and MfD. Best protection against die-level lithography-related variation would require using feature CDs $4\times$ larger than the minimum CD corresponding to process capabilities. A separate set of rules is required for lines, corners, and line ends, to control the k_1 factor, as the rounding does not scale with technology node but with the wavelength. Proximity effects can be reduced by adding dummy lines up to the screening distance of 4λ. At the same time, changing layout geometries would not only impact its footprint but also device parasitics. Therefore, the optimal layout geometry depends on the product application, making it less practical to propose a set of deterministic rules to ensure high parametric yield for every situation. It is intuitive that to reduce the variation, parameterized layout of a large footprint is an attractive option at all stages of the design process. For designers, it is easy to control and yield predictable models. For DfM, it should enable simpler OPC and pattern density control. For manufacturing, it helps product characterization, reproducibility, and quality. An alternative, model-based DfM is still under development and its cost and practicality cannot be properly judged at this time.

REFERENCES

1. G. Moore, Cramming more components onto integrated circuits, *Electronics*, 38(8), 1965, 114–117.
2. P. Rieppo, TAM for semiconductors to respond to changes in the semiconductor value chain, *Gartner Res.*, 13 May 2005.
3. Semiconductor Industry Association Issue Backgrounders: ITRS Roadmap, http://public.itrs.net, 2003.
4. R.H. Dennard, F.H. Gaensslen, V.L. Rideout, E.Bassous, and A.R. LeBlanc, Design of ion-implanted MOSFETs with very small physical dimensions, *IEEE J. Solid-State Circuits*, 9, 1974, 256–265.
5. B.P. Wong, A. Mittal, Y. Cao, and G. Starr, *Nano-CMOS Circuit and Physical Design*, Wiley-Interscience, New York, 2005.
6. S. Wolf, *Silicon Processing for the VLSI Era*, Vols. 1–4, 2002. Lattice Press, Sunset Beach, CA.
7. D. Perry, M. Nakamoto, N. Verghese, P. Hurat, and R. Rouse, Model-based approach for design verification and co-optimization of catastrophic and parametric-related defects due to systematic manufacturing variations, *Proc. SPIE*, 6521, 2007, 65210E1–E10.

8. P.J.M. van Adrichem, F.A.J.M. Driessen, K. van Hasselt, and H.-J. Brück, Mask error enhancement-factor (MEEF) metrology using automated scripts in CATS (*Proc. of Int. Conf. on Photomask Technol.*) BACUS, 2002, pp. 551–557.

9. L. Karklin, Phase shifting and OPC address subwavelength challenges, *EETimes*, May 1999, 1.

10. P.M. Martin, C.J. Progler, G. Xiao, and R. Gray, Manufacturability study of masks created by inverse lithography technology (ILT), *Proc. SPIE*, 5992, 2005, 921–929.

11. B. Lin, New λ/NA scaling equations for resolution and depth-of-focus, *Proc. SPIE*, 4000, 2000, 759–764.

12. D.M. Anderson, *Design for Manufacturability and Concurrent Engineering*, CIM Press, Cambria, CA, 2004.

13. T. Ema, Process technologies for SOCs, *Fujitsu Sci. Tech. J.*, 36(1), 2000, 91–98.

14. V. Chan, K. Rim, M. Ieong, S. Yang, R. Malik, Y.W. Teh, M. Yang, and Q. (Ch.) Ouyang, Strain for CMOS performance improvement, *Proc. IEEE Custom Integrated Circuits Conf., 2005*, pp. 667–674.

15. Y. Chang, Self-aligned isolation and planarization process for memory array, US Patent 5763309, 1998.

16. W. Yang, G. Qin, X. Shao, Z. Yu, and L. Tian, Analysis of GIDL dependence on STI-induced mechanical stress, 2005 *IEEE Conf. on Electron Devices and Solid-State Circuits*, Hong Kong, Dec. 2005, pp. 769–772.

17. R. Chau, J. Brask, S. Datta, G. Dewey, M. Doczy, B. Doyle, J. Kavalieros, B. Jin, M. Metz, A. Majumdar, and M. Radosavljevic, Application of high-*k* gate dielectrics and metal gate electrodes to enable silicon and non-silicon logic nanotechnology, *J. Microelectronic Eng.*, 80, 2005, 1–6.

18. P. Ranade, T. Ghani, K. Kuhn, K. Mistry, S. Pae, L. Shifren, M. Stettler, K. Tone, S. Tyagi, and M. Bohr, High performance 35 nm LGATE CMOS transistors featuring NiSi Metal Gate (FUSI), uniaxial strained silicon channels and 1.2 nm gate oxide, IEDM, Washington, D.C., 2005.

19. P. Majhi, P. Kalra, H.R. Harris, J. Oh, M. Mustafa Hussain, H.-H. Tseng, and R. Jammy, CMOS scaling beyond high-*k* and metal gates (1/9/2007), *Future Fab Intl.*, 22.

20. S. Lopatin, Method of reducing electromigration in copper lines by calcium-doping copper surfaces in a chemical solution, U.S. Patent 6509262, 2003.

21. E.T. Ryan, A.J. McKerrow, J. Leu, and P. S. Ho, Synthesis and characterization of porous polymeric low dielectric constant films, *MRS Bull.*, 22, 1997, 49.

22. M. Morgen, E.T. Ryan, J.-H. Zhao, T. Cho, and P.S. Ho, Low dielectric constant materials for ULSI interconnects, *Annu. Rev. Mater. Sci.*, 30, 2000, 645–680.

23. C.J. Brinker and G.W. Scherer, *Sol-Gel Science: The Physics and Chemistry of Sol-Gel Processing*, Academic Press, San Diego, CA, 1990.

24. J. Wen, B. Dhandapani, S.T. Oyama, and G.L. Wilkes, Low-*k* dielectric materials for advanced interconnect applications, *Chem. Mater.*, 9, 1997, 1968.

25. ITRS roadmap, www.itrs.net.

26. A. Balasinski, Question: DRC or DfM? Answer: FMEA and RoI, *Proc. Int. Symp. on Quality of Electronic Design*, ISQED-2006, pp. 789–794.

27. A. Balasinski, J. Cetin, and L. Karklin, Layout techniques and rules to reduce process related variability, *J. Micro/Nanolithogr., MEMS, MOEMS*, 6(3), 2007, 003008-1–003009-8.

28. A. Balasinski and J. Cetin, Intelligent fill pattern and extraction methodology for SoC, *6th IEEE Int. Workshop on SoC-RT*, Cairo, Dec. 2006, pp. 156–159.

29. N.D. Arora, L. Song, S.M. Shah, K. Joshi, K. Thumaty, A. Fujimura, L.C. Yeh, and P. Yang, Interconnect characterization of X architecture diagonal lines for VLSI design, *IEEE Trans. Semi. Mfg.*, 18(2), 2005, 262–271.

3 Variability in Deeply Scaled CMOS

Liang-Teck Pang and Borivoje Nikolić

CONTENTS

3.1 INTRODUCTION

CMOS device scaling has increased the impact of process variability to the point where it is now regarded as a major roadblock to further scaling [1]. The control of process fluctuations has not kept pace with rapidly shrinking device dimensions. Furthermore, the drive to improve performance has enticed device and circuit designers to operate at conditions that are more sensitive to variability. Oxide thickness, threshold voltage, and gate length variations have become more significant as the manufacturing process has entered sub-100 nm technology nodes. Currently, the technology is pushing the limits of photolithography with complex resolution enhancement techniques to continue printing ever smaller subwavelength features.

All this has compounded the impact of variability and increased the amount of design margin to cope with worst-case scenarios. Since the tolerances on the process parameters do not track the scaling of their nominal values, designers are experiencing a proliferation of various design corners. Characterizing variability in a more detailed way would allow designers to use the right amount of margins to obtain an optimal design that maximizes performance, power, and yield.

In this chapter, various test structures are used to measure variability and analyze its systematic, random, within-die (WID), and die-to-die (D2D) components. The impact of different layout topologies on variation is investigated; the spatial correlation of ring oscillator frequencies and leakage current is characterized [2].

3.1.1 CHARACTERIZATION OF VARIATIONS

Process variations can be systematic or random, and are generally characterized as WID, D2D, and wafer-to-wafer (W2W) [3] (Table 3.1). WID and D2D classifications reflect some of the spatial characteristics of the variations. Those which vary rapidly over distances smaller than the dimension

TABLE 3.1

Random and Systematic Sources of Process Variations

Parameter	Random	Systematic
Channel dopant concentration (N_{ch})	Affects $\sigma_{V_{th}}$ [6]	Nonuniformity in the process of dopant implantation, dosage, diffusion
Gate oxide thickness (T_{ox})	Si/SiO$_2$ and SiO$_2$/poly-Si interface roughness [5]	Nonuniformity in the process of oxide growth
Gate length (L)	LER [4]	Lithography: Proximity effects, RET, OPC, PSM Resist development, etching, etc.

45 nm lines/90 nm pitch

of a die result in WID variations whereas variations that change gradually over the wafer will cause D2D variations. W2W variations reflect both the spatial as well as temporal characteristics of the process and cause different wafers to have different properties. Systematic variations are deterministic shifts in process parameters, whereas random variations change the performance of any individual instance in the design. In practice, although many of the systematic variations have a deterministic source, they are not known at the design time, or are too complex to model and are treated as random. The resulting random variation component will have a varying degree of spatial correlation. The primary sources of variability are the transistors themselves, interconnects, and the operating environment (supply and temperature) [4].

Many sources of systematic variability can be attributed to the particular characteristics of the manufacturing process. Deviations in nominal widths and lengths are due to photolithography and etching. Variation in film thicknesses (e.g., oxide thickness, gate stacks, wire, and dielectric layer height) is due to the deposition and growth process, as well as the chemical–mechanical planarization (CMP) step. Additional electrical properties of CMOS devices are affected by variations in the dosage of implants, as well as the temperature of annealing steps. Finally, random device parameter fluctuations stem from line-edge roughness (LER) [5], Si/SiO$_2$ and polysilicon (poly-Si) interface roughness [6], and doping fluctuations [7].

Presently, variability is captured in the design process through the use of simulation corners whereby certain transistor parameters are varied by three standard deviations from their nominal values. Even though the corner cases represent the total D2D and W2W variations, designers typically use them as worst-case D2D variations, with all devices on a chip having correlated process parameters. As the corner spread increases with scaling, simultaneously satisfying performance, power and corner requirements become challenging. In order to better account for the variability, it is necessary to develop a better understanding of the contributions of random and systematic variations and distinguish between WID and D2D components. It is also important to determine the spatial correlation distance of process parameters; uncorrelated variations can be averaged out, which benefits the chip yield [8]. Characterizing variations in the manner proposed here would allow circuit designers to incorporate design methodologies that mitigate the effects of systematic variations and correctly account for random variations.

3.1.2 Lithography-Induced Variations

Present lithography systems employ a step-and-scan method, where the stepping is used to move the wafer between the exposure fields. Within an exposure field, a narrow slit of light illuminates the mask and the mask pattern is optically projected onto a wafer [9]. The mask and the wafer are moved simultaneously in such a way that the slit of light scans the entire mask and projects the image onto the wafer. This is illustrated in Figure 3.1.

In subwavelength lithography, the effective linewidth depends on the surrounding features [10–12]. The process step of fabricating poly-Si gates is shown in Figure 3.2. When exposed beyond a certain light intensity, positive resist will dissolve in the developer fluid. The exposed poly-Si gate stack layer will then be etched away leaving behind the transistor gates. Narrow poly-Si lines with varying pitch will have different channel lengths when exposed with the 193 nm wavelength light, as illustrated in Figure 3.3. Dense lines also have higher depth of focus, and are more immune to defocusing of the optical system [13]. Optical proximity correction techniques in the mask processing add sublithographic assist features to control the printed critical dimensions (CDs). However, their effect is limited due to the shallow depth of focus.

After the resist is spun onto the wafer and exposed, it undergoes postexposure bake (PEB). This step is essential to activate the photoactive compound and set the resist exposure threshold [14]. If the temperature over the entire wafer is not even, this will result in different resist exposure thresholds over the wafer and cause D2D gate length variation. This effect is mostly systematic and wafers have exhibited a radial temperature profile during baking.

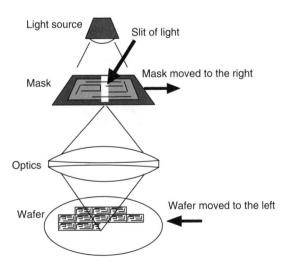

FIGURE 3.1 Step-and-scan photolithography using a slit of light. (From B. Nikolic and L.-T. Pang, *8th IEEE International Conference on Solid-State and Integrated Circuit Technology*, 505, 2006. With permission.)

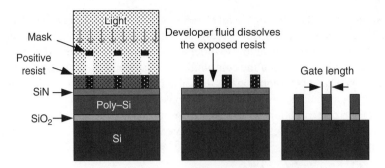

FIGURE 3.2 Formation of poly-Si gates. (From B. Nikolic and L.-T. Pang, *8th IEEE International Conference on Solid-State and Integrated Circuit Technology*, 505, 2006. With permission.)

Lens imperfections are often characterized through aberrations (Figure 3.4). Aberrations create optical path differences for each pair of rays through the imaging system. Effects such as astigmatism and spherical aberrations cause differences in exposed patterns at the level of a reticle. Coma effect [15] is an aberration due to lens imperfection, which causes a gate surrounded by nonsymmetrical structure to print differently from its mirror image [16]. Flare results from the scattering and reflection of light through the projection system and causes variations in the effective CDs (Figure 3.5). In general, the amount of flare is dependent on the local pattern density in the mask [16]. Proximity effects, aberrations, and flare are usually not captured in the design process, and they induce layout-dependent systematic variations in the design.

Another source of variability comes from e-beam mask stitching [17] discontinuity. E-beam lithography is employed in writing the optical mask. In order to cover the large reticle field of the mask with a small e-beam field, it is necessary to construct the mask by drawing smaller e-beam fields and stitching the whole image together. Even though optics will reduce imperfections on the mask by a few times, aggressive scaling has made stitching discontinuities more significant.

3.2 TEST CHIP

A test chip has been implemented in a general-purpose 90 nm CMOS technology to evaluate the effects of lithography-induced variations, and to measure WID and D2D variations, as well as WID spatial correlation. This is done by measuring ring-oscillator (RO) frequencies and transistor source–drain leakage currents (I_{LEAK}) of an array of test structures [18].

The chip contains 10×16 tiles, with a total area of 1×1 mm. Each tile has twelve 13-stage ROs and twelve transistors in the off-state, each with a different layout (Figure 3.6). Several transistors in

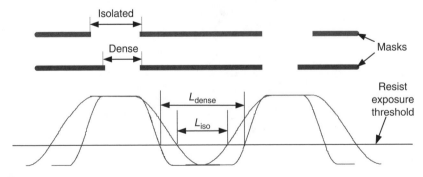

FIGURE 3.3 Isolated and dense lines. (From B. Nikolic and L.-T. Pang, *8th IEEE International Conference on Solid-State and Integrated Circuit Technology*, 505, 2006. With permission.)

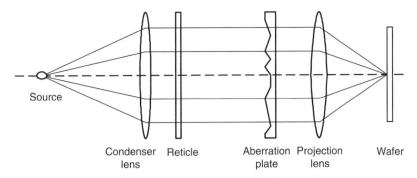

FIGURE 3.4 Lens imperfections. (From B. Nikolic and L.-T. Pang, *8th IEEE International Conference on Solid-State and Integrated Circuit Technology*, 505, 2006. With permission.)

both the ROs and those in the off-state consist of a single poly finger, while the rest are constructed with a stack of three fingers. The poly-Si pitch of neighboring dummy poly-Si lines is varied in the test structures to capture proximity effects. Poly orientations, together with the properties of the two-dimensional tile array, are used to characterize spatial correlation. Non symmetrical structures and their mirror image target measurements of the coma effect. The first-layer metal coverage over certain gates is also varied in the layout to investigate the effects of anneal [19].

The test chip shown in Figure 3.7 includes an on-chip single-slope analog-to-digital converter (ADC) to measure transistor off-currents from 1 nA to 1 μA with 1 nA resolution. The ADC (Figure 3.8) consists of a high-gain folded-cascode amplifier implemented with 2.5 V devices, a large on-chip metal fringe capacitor, and comparators. The switches, controlled by signals P1, P2, and P1b, provide the correct bias to the input and output of the amplifier and determine whether to measure the reference current (I_{ref}) only or the sum of I_{ref} and I_s. The latter is the sum of all parasitic gate and substrate leakage currents and the I_{LEAK} of the selected device. By not selecting any of the devices, it is possible to measure the parasitic leakage currents and subtract them from the measurement. I_{ref} is obtained from an external source through current mirrors that divide the current down by a factor of 10.

During current integration, the output of the op-amp will ramp up, and as it crosses the voltages V_1 and V_2, the start and stop signals will be generated. Measuring the time interval between start and stop signals will give the current (Figure 3.9). The selection of the device to be measured is done with row and column bits obtained from a chain of registers in order to reduce pad count [20–22].

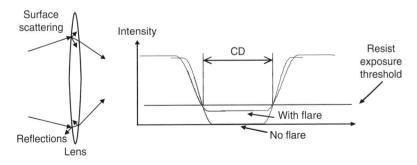

FIGURE 3.5 Effect of flare. (From B. Nikolic and L.-T. Pang, *8th IEEE International Conference on Solid-State and Integrated Circuit Technology*, 505, 2006. With permission.)

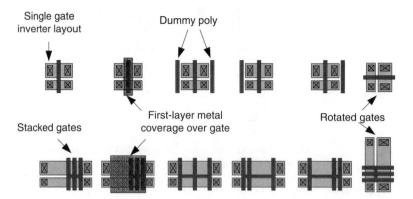

FIGURE 3.6 Layout variations in the test chip, pictured in the RO inverter stages. (From L.-T. Pang and B. Nikolic, *Symposium on VLSI Circuits*, 2006. With permission.)

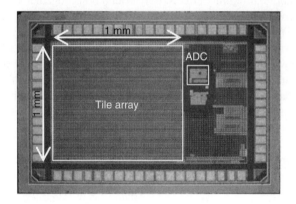

FIGURE 3.7 Test chip die photo. (From L.-T. Pang and B. Nikolic, *Symposium on VLSI Circuits*, 2006. With permission.)

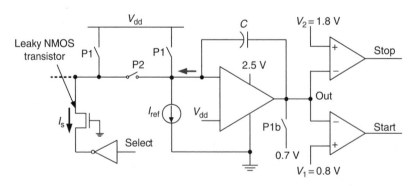

FIGURE 3.8 Single-slope ADC for current measurement. (From L.-T. Pang and B. Nikolic, *Symposium on VLSI Circuits*, 2006. With permission.)

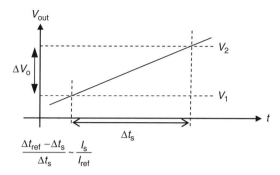

$$\frac{\Delta t_{ref} - \Delta t_s}{\Delta t_s} \sim \frac{I_s}{I_{ref}}$$

Δt_{ref} = Integration time when measuring I_{ref}

Δt_s = Integration time when measuring $I_{ref} + I_s$

FIGURE 3.9 Current measurement procedure. (From L.-T. Pang and B. Nikolic, *Symposium on VLSI Circuits*, 2006. With permission.)

3.3 MEASUREMENT RESULTS

In the analysis of the measurement results, tiles at the perimeter of the array are ignored to eliminate edge effects. The ROs with single isolated polyfingers often exceed the fast corner and are excluded in the measurements.

To distinguish systematic from the random effects, the results for different layouts are compared for each die, from die to die. Measurements for identical structures within a die are compared to obtain WID correlations. Averages for various dies are used to indicate the D2D spread. Averages across several dies are used to analyze the mask writing effects. Measured data show several important trends, which are analyzed in detail in this section. The largest impact of layout on performance comes from gate poly-Si density, which causes a systematic shift in frequency of up to 10%. D2D variation is significant resulting in a 3× standard deviation/median ($3\sigma/\mu$) of 15% over half a wafer. Finally, WID variation for identical structures is relatively small ($3\sigma/\mu \sim 3.5\%$). WID spatial correlation of RO frequency shows a dependency on the direction of spacing and the orientation of the gates. Each of these observations can be attributed to a particular step in the manufacturing process.

3.3.1 EFFECTS OF LAYOUT ON FREQUENCY AND LEAKAGE

In order to investigate the effects of layout, we compare the RO frequency and I_{LEAK} distribution of different layout structures. The mean value for each layout configuration for each die is compared to observe the systematic effects. The variations are divided into WID and D2D categories. Proximity effects due to different poly-Si gate pitch cause a shift in frequency of over 10%, and a 20× shift in I_{LEAK} (Figures 3.10 and 3.11). More isolated gates have shorter gate lengths and hence higher RO frequencies and higher leakage currents. This is much larger than the 1.1% frequency shift predicted by Spice simulations of the extracted layout, which capture only changes in parasitic capacitances. D2D and WID leakage current variations are reduced with increased poly-Si density. There is also a similar but smaller effect of poly-Si density on stacked gates. Figure 3.12 plots the RO frequency distribution for stacked gates. Extracted simulation results show that the RO frequencies for different gate densities vary (position of the vertical red lines). Measurement shows that for high-density gates, the RO frequency falls approximately between the TT and FF corners (top plot). However, for low-density gates, RO frequency is increased by around 5% with respect to the TT and FF corners (bottom two plots). By using the simulation results as a reference in order to account for different parasitics, the frequency of the densest stacked gate configuration is 5% slower than the least dense stacked gates configuration.

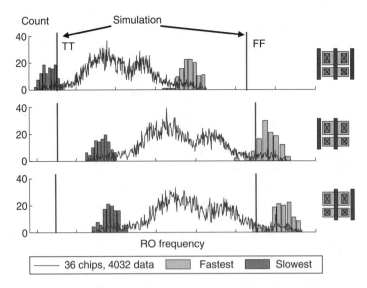

FIGURE 3.10 Frequency distribution for single-finger configurations. Vertical lines correspond to typical and fast corner simulation results. Bar plots correspond to the WID distribution of the fastest and slowest chips. (From B. Nikolic and L.-T. Pang, *8th IEEE International Conference on Solid-State and Integrated Circuit Technology*, 505, 2006. With permission.)

The coma effect is present but small, causing a roughly 1%–2% shift in the mean RO frequency with single gate stages as shown in the second and third plots of Figure 3.10. The effect of M1 coverage on gates is small and due mainly to differences in parasitic capacitances. Large threshold voltage (V_{th}) shifts due to metal coverage over gate were not observed here. It was reported in Ref. [19] that metal-covered gates have a larger number of interface states because the metal coverage lowers the temperature on the gate during annealing. In this layout, metal 1 is in close proximity to nonmetal-covered gate since the gate length is small and the source and drain are fully covered with metal 1

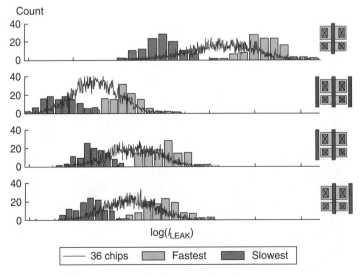

FIGURE 3.11 Log(I_{LEAK}) distribution for single-finger configuration. (From B. Nikolic and L.-T. Pang, *8th IEEE International Conference on Solid-State and Integrated Circuit Technology*, 505, 2006. With permission.)

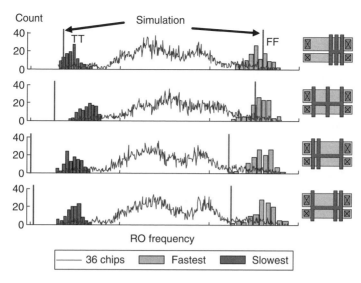

FIGURE 3.12   Frequency distribution for stacked gates. Less dense gates show a 5% increase in frequency with respect to extracted simulation results, whereas dense gates stay within the simulation corners. (From B. Nikolic and L.-T. Pang, *8th IEEE International Conference on Solid-State and Integrated Circuit Technology*, 505, 2006. With permission.)

contacts. Hence, it is likely that the temperature difference during annealing between metal-covered gates and non metal-covered gates is too small to create a significant shift in V_{th}. This result is likely of general value for small, digital gates.

3.3.2   D2D VARIATIONS

The wafer maps of mean frequency and leakage of each die for the layout configuration shown in Figure 3.13 show a radial pattern that can be attributed to nonuniform resist development [23,24]. Faster and leakier chips are located at the center of the wafer. D2D variation is significant resulting in a $3\sigma/\mu$ of 15% over half a wafer for the densest single gate structure. For the other single gate structures, it increases slightly to around 17%.

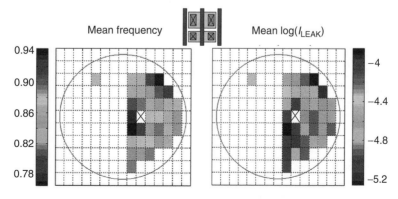

FIGURE 3.13   Wafer maps of mean RO frequency and mean $\log(I_{LEAK})$. X marks a defective chip. Location of dies on only half the wafer is known. (From B. Nikolic and L.-T. Pang, *8th IEEE International Conference on Solid-State and Integrated Circuit Technology*, 505, 2006. With permission.)

3.3.3 WID VARIATIONS

WID variation of identical layout structures is small ($3\sigma/\mu \sim 3.5\%$) and weakly dependent on the layout. For each layout structure, the data are normalized to zero mean and unit variance for each chip before being used to compute its spatial correlation. Confidence intervals for the correlation are computed using Fisher's z-transformation to convert Pearson's correlation to a normally distributed random variable. In the frequency measurements, the spatial correlation depends only on the direction of spacing and orientation of the gates.

For horizontally spaced ROs (Figure 3.14a), correlation is higher for vertically oriented gates whereas for vertically spaced ROs (Figure 3.14b), correlation is higher for horizontally oriented gates. Hence, we find that gates placed parallel to each other have higher correlation. Horizontally spaced ROs with vertically oriented gates have a spatial correlation of 0.4 between immediate neighbors whereas horizontally spaced ROs with horizontally oriented gates have a spatial correlation of 0.25. ROs that are aligned in the horizontal direction have stronger spatial correlation than those that are aligned vertically (Figure 3.14).

This dependence on direction could be explained by the step-and-scan photolithography. The horizontal direction is along the slit of light in the stepper and is subject to lens aberrations and curvature, resulting in more correlated features. The vertical direction is along the scan direction, which is subject to variation in scan speeds, stage vibration, and light dosage, resulting in less correlated features. Leakage current, which is less sensitive to gate length (L) but more sensitive to V_{th} than RO frequency, has much smaller spatial correlation.

3.3.4 INFERRING PROCESS PARAMETERS

The scatter plot of $\log(I_{LEAK})$ versus frequency shows a strong positive correlation between I_{LEAK} and frequency for D2D and layout-to-layout (L2L) variations (Figure 3.15). This means that these variations are dominated by process parameters, which cause both $\log(I_{LEAK})$ and frequency to vary

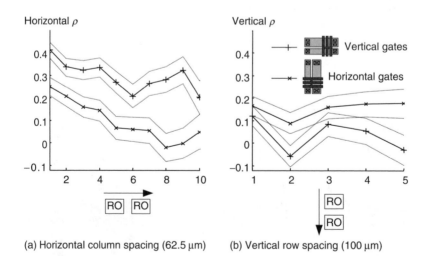

(a) Horizontal column spacing (62.5 μm)   (b) Vertical row spacing (100 μm)

FIGURE 3.14 Spatial correlation coefficient for vertical and horizontal gates versus (a) horizontal column spacing and (b) vertical row spacing. Dotted lines represent 99% confidence bounds. (From B. Nikolic and L.-T. Pang, *8th IEEE International Conference on Solid-State and Integrated Circuit Technology*, 505, 2006. With permission.)

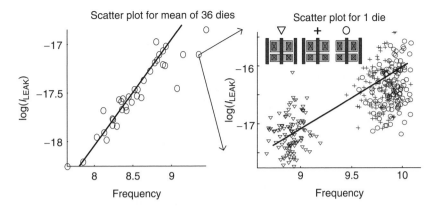

FIGURE 3.15 Scatter plot of $\log(I_{LEAK})$ versus frequency.

in the same manner. WID scatter plot shows no significant correlation, suggesting that it is caused by a combination of process parameters.

In order to relate these variations to process parameters, we use the least squares method and the BSIM3 model to infer the variation in gate length (L), gate oxide thickness (T_{ox}), and channel dopant concentration (N_{ch}) from the frequency (F) and leakage current ($I = \log I_{LEAK}$) measurements. Since the sensitivities of the measured data to process parameters vary with different conditions of supply voltage (V_{dd}), substrate voltage (V_{bs}), and temperature (T), the accuracy of the inference can be improved by taking the measurements under different conditions.

Figure 3.16 shows the distribution of normalized L, T_{ox}, and N_{ch} obtained from the inference process. The plot of normalized L shows a strong correlation to the plots in Figures 3.10 and 3.11, suggesting that faster and leakier transistors have shorter L.

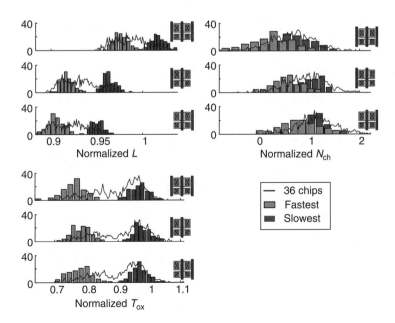

FIGURE 3.16 De-embedded distribution of L, T_{ox}, and N_{ch}.

3.4 CONCLUSION

Layout has a significant impact on lithography-induced variability in 90 nm technology. Analysis of the inferred process parameters indicates that poly-Si gate pitch has the strongest effect on the effective transistor gate length, resulting in up to 10% variation in RO frequency for inverters laid out with isolated poly lines. This systematic effect can be compensated by using layout extraction tools that account for proximity effects [16]. A simpler method would be to use regular layouts [25] or more restrictive layout rules that only allow transistors with a few possible gate pitches, together with an extractor that maps each gate pitch to its respective gate length.

The use of step-and-scan lithography induces stronger correlation between gates placed parallel to the direction of slit of light than those parallel to the direction of scan. This effect can be exploited in the layout of regular data paths and memory [1,26]. By placing gates of the same path in the low correlation direction and by placing parallel paths in the high correlation direction, we can obtain a tighter performance spread.

Analysis of the spatial correlation of ROs can be used to model the spatial correlation coefficient of circuit paths and gates, allowing for more accurate statistical timing analysis. For the same layout, systematic D2D variations dominate the total variations for small chips, indicating that the use of die-level adjustable supply voltages and substrate biasing can improve parametric yield significantly [3].

ACKNOWLEDGMENTS

The authors wish to acknowledge the contributions of the students, faculty, and sponsors of the Berkeley Wireless Research Center, the National Science Foundation Infrastructure Grant No. 0403427, wafer fabrication donation of STMicroelectronics, and the support of the Center for Circuit and System Solutions (C2S2) Focus Center, one of the five research centers funded under the Focus Center Research Program, a Semiconductor Research Corporation program. The authors would also like to thank Professor Andrew Neureuther and his students for helpful discussions, and Christopher Siow for his help with the measurements.

REFERENCES

1. K.A. Bowman, S.G. Duvall, and J.D. Meindl, Impact of die-to-die and within-die parameter fluctuations on the maximum clock frequency distribution for gigascale integration, *IEEE Journal of Solid-State Circuits*, 37, 183, 2002.
2. B. Nikolic and L.-T. Pang, Measurements and analysis of process variability in 90 nm CMOS, *8th International Conference on Solid-State and Integrated Circuit Technology*, Shanghai, China, 505–508, 2006.
3. J.W. Tschanz, J.T. Kao, S.G. Narendra, R. Nair, D.A. Antoniadis, A.P. Chandrakasan, and V. De, Adaptive body bias for reducing impacts of die-to-die and within-die parameter variations on microprocessor frequency and leakage, *IEEE Journal of Solid-State Circuits*, 1396–1402, 2002.
4. K. Bernstein, D.J. Frank, A.E. Gattiker, W. Haensch, B.L. Ji, S.R. Nassif, E.J. Nowak, D.J. Pearson, and M.J. Rohrer, High-performance CMOS variability in the 65-nm regime and beyond, *IBM Journal of Research and Development*, 50, 2006.
5. P. Oldiges, Qimghuamg Lin, K. Petrillo, M. Sanchez, M. Ieong, and M. Hargrove, Modeling line edge roughness effects in sub 100 nm gate length devices, *SISPAD 2000*, 131, 2000.
6. A. Asenov, S. Kaya, and J.H. Davies, Intrinsic threshold voltage fluctuations in decanano MOSFETs due to local oxide thickness variations, *IEEE Transactions on Electronic Devices*, 49, 112, 2002.
7. D.J. Frank, Y. Taur, M. Ieong, and H.-S.P. Wong, Monte Carlo modeling of threshold variation due to dopant fluctuations, Digest of Technical papers, *1999 Symposium on VLSI Circuits*, Kyoto, Japan, 171, 1999.
8. H. Masuda, S. Ohkawa, A. Kurokawa, and M. Aoki, Challenge: Variability characterization and modeling for 65 to 90 nm processes, *Proceedings of CICC*, Vol. 593, 2005.
9. J.D. Plummer, M.D. Deal, and P.B. Griffin, *Silicon VLSI Technology*, Prentice-Hall, Englewood Cliffs, NJ, 2000.
10. A.K.-K. Wong, *Resolution Enhancement Techniques in Optical Lithography*, Bellingham: SPIE Press, 2001.

11. M. Orshansky, L. Milor, P. Chen, K. Keutzer, and C. Hu, Impact of spatial intrachip gate length variability on the performance of high-speed digital circuits, *IEEE Transactions on CAD*, 21, 544, 2002.
12. M. Orshansky, L. Milor, and C. Hu, Characterization of spatial intrafield gate CD variability, its impact on circuit performance, and spatial mask-level correction, *IEEE Transactions on Semiconductor Manufacturing*, 17, 2, 2004.
13. T. Brunner, D. Corliss, S. Butt, T. Wiltshire, and C.P. Ausschnitt, Laser bandwidth and other sources of focus blur in lithography, *Proceedings of SPIE*, Vol. 6154, 2006.
14. D. Steele, A. Coniglio, C. Tang, and B. Singh, Characterizing post exposure bake processing for transient and steady state conditions, in the context of critical dimension control, In: D.J. Herr (Ed.), *Metrology, Inspection and Process Control for Microlithography XVI, Proceedings of SPIE*, Vol. 4689, 2002.
15. T.A. Brunner, Impact of lens aberrations on optical lithography, *IBM Journal of Research and Development*, 41, 57, 1997.
16. M. Choi, and L. Milor, Impact on circuit performance of deterministic within-die variation in nanoscale semiconductor manufacturing, *IEEE Transactions on CAD*, 25, 1350, 2006.
17. T. Kjellberg and R. Schatz, Effect of stitching errors on the performance of DFB lasers fabricated using e-beam lithography, *Journal of Lightwave Technology*, 10(9), 1256–1266, 1992.
18. L.-T. Pang and B.Nikolic, Impact of layout on 90 nm process parameter fluctuations, Digest of Technical papers, *Symposium on VLSI Circuits*, Hawaii, USA, 2006.
19. H. Tuinhout, M. Pelgrom, R. Penning de Vries, and M. Vertregt, Effects of metal coverage on MOSFET matching, *Technical digest, International Electron Devices Meeting*, San Francisco, USA, 735–738, 1996.
20. K. Gonzalez-Valentin, Extraction of variation sources due to layout practices, MS Thesis, Massachusetts Institute of Technology, Cambridge, MA, 2002.
21. J. S. Panganiban, A ring oscillator based variation test chip, MS Thesis, Massachusetts Institute of Technology, Cambridge, MA, 2002.
22. D. Boning, J. Panganiban, K. Gonzalez-Valentin, C. McDowell, A. Gattiker, and F. Liu, Test structures for delay variability, presented at 2002, *ACM/IEEE Tau Workshop*, Montercy, USA, 2002.
23. J.P. Cain and C.J. Spanos, Electrical linewidth metrology for systematic CD variation characterization and causal analysis, *Proceedings of SPIE*, Vol. 5038, 2003.
24. P. Friedberg, Y. Cao, J. Cain, R. Wang, J. Rabaey, and C. Spanos, Modeling within-field gate length spatial variation for process-design co-optimization, *Proceedings of SPIE*, Vol. 5756, 2005.
25. V. Kheterpal, V. Rovner, T.G. Hersan, D. Motiani, Y. Takegawa, A.J. Strojwas, and L. Pileggi, Design methodology for IC manufacturability based on regular logic-bricks, *Proceedings of DAC*, Vol. 353, 2005.
26. H. Onodera, Variability: Modeling and its impact on design, *IEICE Transactions on Electronics*, E89-C, 342, 2006.

4 Wafer-Level Three-Dimensional Integration for Advanced CMOS Systems

Ronald J. Gutmann and Jian-Qiang Lu

CONTENTS

4.1   INTRODUCTION

Silicon complementary metal oxide semiconductor (CMOS) integrated circuits (ICs) continue to achieve increasing levels of integration for digital applications such as microprocessors and high-performance application-specific integrated circuits (ASICs). However, manufacturing cost and copper-low-k interconnect delay may constrain future scaling past the 22 nm technology node. In addition, integration of high-performance, highly integrated digital ICs with other technologies such as analog/mixed-signal ICs, imagers, sensors, and wireless transceivers can be limited by packaging technologies, both conventional planar (or 2D) technologies and more recent three-dimensional (3D) stacking (either packaged die, bare die with full wafer thickness, or thinned die). While 3D packaging solutions offer increased functional density, electrical performance and interstrata interconnectivity are limited by wire bonding at the edge of the die. Micron-sized through-silicon vias (TSVs) offer significantly increased interstrata interconnectivity, along with significantly decreased interconnect parasitics (particularly inductance). By incorporating monolithic processes of wafer-to-wafer alignment, wafer bonding, wafer thinning, and interwafer interconnection, novel high-performance designs and unique system architectures become possible, combined with low manufacturing costs inherent with full-wafer IC processing.

This chapter is organized into five major sections, each including a summary perspective followed by a specific design and/or system architectural example. First, alternative wafer-level 3D technology platforms are presented, with emphasis on dielectric adhesive bonding using benzocyclobutene (BCB). Second, digital applications such as microprocessors and ASICs are discussed, with emphasis on performance prediction of static random access memory (SRAM) stacks for high-density memory. Third, analog/mixed-signal applications are discussed, with emphasis on wireless transceivers for software radios. Fourth, unique system architectures enabled by wafer-level 3D integration are discussed, with emphasis on both (1) point-of-load (PoL) DC–DC converters for power delivery with wide control bandwidth and (2) a novel system architecture using optical clocking for synchronous logic without latency. Fifth, technology and system drivers, which need to be addressed before wafer-level 3D technology platforms are implemented for high-volume manufacturing, are projected.

The authors believe that many unique system architectures will be realized with wafer-level 3D integration and hope that this chapter motivates IC system architects and designers to extend their horizons to include 3D integration with low-parasitic interstrata interconnects. Two edited books on 3D integration are scheduled for publication in 2008, with contributions from many research groups in academia and industry [1,2]. They will provide comprehensive coverage of all high-performance 3D integration technologies and applications.

4.2 WAFER-LEVEL 3D TECHNOLOGY PLATFORMS

The purpose of 3D integration is to vertically stack and interconnect devices and circuits to form multifunctional, high-performance systems. The various wafer-level 3D technology platforms that have been investigated can be classified as (1) front-end-of-the-line (FEOL) platforms, (2) back-end-of-the-line (BEOL) platforms, and (3) wafer-level packaging (WLP) platforms. In FEOL-based platforms, strata of Si device layers are formed early in the IC process by techniques such as epitaxial overgrowth and recrystallization of polysilicon. In BEOL-based platforms, active device layers are formed by wafer-to-wafer alignment and wafer bonding of two separately fabricated functional wafers, followed by wafer thinning and interwafer interconnect processing. In WLP-based platforms, fully processed wafers are aligned and interconnected using packaging-based processes such as flip-chip soldering.

FEOL platforms offer the highest density of interstrata interconnects and minimize increased complexity of IC interconnect processing and/or packaging. The potential for unique cell structures for digital designs is attractive, interstrata vias of 100 nm are feasible at technology nodes below 22 nm, and no new interconnect or packaging processes are required. However, high-quality devices in upper layers of the strata have not been achieved to date, primarily because of FEOL thermal budgets. FEOL platforms have been demonstrated for SRAMs and used in commercial products for nonvolatile memories (NVMs). However, the potential for heterogeneous integration, especially with different semiconductors requiring heteroepitaxy growth, remains a tremendous challenge. The future of FEOL 3D platforms is limited by the lack of a materials technology for achieving multiple strata with high-quality single-crystal silicon.

BEOL platforms offer a very high density of interstrata interconnects and allow fully interconnected device wafers to be processed similar to conventional planar 2D ICs. Interstrata vias of 1 μm and below are achievable today with a variety of platforms outlined later in this section. Both heterogeneous integration of wafers with diverse technologies and high-density digital 3D stacks are feasible, both present and near-term technologies such as imagers with pixel processing and emerging technologies are described in the last section of this book. Most research and development into 3D integration during the past decade has focused on BEOL platforms.

WLP-based platforms offer a more modest density of interstrata interconnects, while offering similar advantages for heterogeneous integration as BEOL platforms. Interstrata vias of 20–40 μm are achievable using micro flip-chip solder bumps. The via diameter limits the performance for advanced digital designs that require a large density of interstrata interconnects. While useful with 65 nm technology node CMOS digital ICs, the technology may not be scalable at the 22 nm node and beyond.

BEOL platforms, the focus of the remainder of this section, can be classified in many ways. Wafer bonding is either oxide-to-oxide, metal-to-metal (mostly copper-to-copper or Cu-to-Cu), or with dielectric adhesives (such as polyimides or BCB). The wafers to be bonded can be either fully interconnected (with ~ four to ten interconnect layers) or include only local interconnect (silicide, salicide, or tungsten) with extensive interconnect levels after wafer bonding. The bonding process can form the interwafer via directly such as with Cu-to-Cu bonding (via-first process flow) or wafer bonding can occur with blanket films on each wafer surface, such as bonding oxide-to-oxide or with dielectric adhesives, followed by a damascene process for interstrata interconnect (via-last process flow). A handling wafer can be used to enable a back-to-front bond with wafer thinning occurring on the handling wafer or the wafers can be directly bonded to simplify the process flow and form a face-to-face bond. These alternative process flows and the enabling unit processes are described elsewhere.

Platforms based upon oxide-to-oxide bonding have been proposed by Guarini and colleagues at IBM using a glass handling wafer to simplify wafer-to-wafer alignment, a via-last process flow and face-to-back bonding with wafers processed through local interconnects [3]. Thus the requirement of surface roughness to within 1 nm and surface planarity required for BEOL-compatible oxide bonding can be obtained. A modified version of this approach has been recently adapted by Keast and colleagues at the Lincoln Laboratory in establishing a research-based 3D foundry capability [4]. These approaches with bonding occurring after local interconnects are formed in each wafer are suitable for integrated device manufacture (IDM). However, Enquist and colleagues at Ziptronix are applying oxide-to-oxide bonding after each wafer has a multilevel interconnect structure, both for wafer-to-wafer and die-to-wafer platforms [5]. Issues such as oxide surface roughness, planarity after multilevel interconnect processing, and subsequent bond strength need further evaluation.

The fundamentals of copper-to-copper bonding have been explored by Reif and colleagues at MIT using a handling wafer, a via-first process flow and face-to-back bonding [6,7]. This group has demonstrated copper grain growth across the bonding interface at BEOL-compatible temperatures. Morrow and colleagues at Intel have demonstrated the compatibility of Cu-to-Cu bonding with TSV density suitable for microprocessors using face-to-face bonding without a handling wafer [8]. Patti and colleagues at Tezzaron are in pilot scale manufacturing of both stacked memories and memories stacked with a compatible processor wafer [9]. With such Cu-to-Cu bonding, the field dielectric is recessed prior to bonding in order to insure the intimate contact of the Cu surfaces (Cu via and bonding pad) to be bonded as well as the high down force required for Cu bonding to be applied only to the Cu vias. A seal ring is probably needed around each die to avoid moisture interacting with the copper TSVs.

Dielectric adhesive bonding using polyimide and a via-last process flow have been investigated by Keast and colleagues at Lincoln Laboratories [10], Ramm and colleagues at Fraunhofer Institute [11], and Koyanagi and colleagues at Tohoku University [12] using polyimide adhesives, a via-last process flow, and face-to-face bonding without handling wafers. The authors and colleagues at Rensselaer [13–17] have established a viable process flow using BCB, which is widely used as a passivation layer, interlevel dielectrics (ILDs), and microelectromechanical system (MEMS) bonding material. Besides the traditional application in a via-last process flow, a via-first process flow has recently been established using Cu damascene patterned vias with partially cured BCB to simplify the process flow [18]. Compared to Cu-to-Cu bonding, this via-first platform has higher bond strength (due to the BCB bonding) and will not require seal rings around each die in most applications (as BCB-to-BCB contact is maintained across the bonding interface) at the expense of increased wafer planarity requirements.

To illustrate the status of these technology platforms, the via-last and via-first platforms with BCB adhesive bonding is described in more detail. A three-wafer stack of the via-last 3D platform is depicted in Figure 4.1, with summary comments describing the four key unit processes listed in Table 4.1. Partially cured BCB has been used in the bonding process to improve wafer-to-wafer

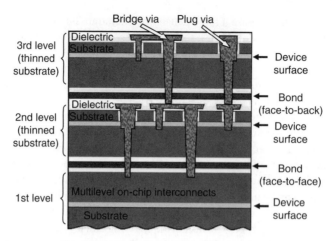

FIGURE 4.1   Schematic cross-section of three-strata stack of via-last 3D platform with BCB wafer bonding, showing bonding interface and vertical interstrata vias.

TABLE 4.1

Summary of Four Key Unit Processes

- Precise alignment of 200 mm wafers (≤ 1 μm accuracy)
- Thin adhesive-layer bonding at low temperature ($\leq 400°C$)
- Precision thinning and leveling of top wafer (~ 1 μm thick)
- Interwafer connection with high aspect ratio ($\sim 5:1$) vias

---

alignment reproducibility with high bonding strength. Bonding of silicon to glass wafers, approximately thermal coefficient of expansion (TCE) matched to silicon, allows optical inspection of bonding defects, both immediately after bonding and after silicon thinning by grinding, chemical–mechanical polishing (CMP) and, if desired, selective wet etching to an oxide layer using trimethyl ammonium hydroxide (TMAH). Collaborations with the University at Albany have demonstrated the capability for low-resistance interstrata interconnects with such a process flow [14].

In collaborations with both SEMATECH (wafers having two-level copper interconnect test structures with either oxide or low-k ILDs) and Freescale (wafers having four-level copper interconnects and fully functional 130 nm technology node silicon-on-insulator (SOI) CMOS devices), the robustness of the BCB bonding and three-step thinning process on advanced interconnect structures and CMOS devices and circuits have been demonstrated [15,16]. A cross-section of a CMOS test die with four levels of Cu interconnect, which has undergone two bonding and thinning processes and ashing of the initial BCB, is shown in Figure 4.2; only the four-level interconnect, the ~ 150 nm SOI layer, and part of the buried oxide (BOX) layer remain, bonded to a prime Si wafer. Moreover, bonding integrity is maintained after liquid-to-liquid thermal shock (LLTS), high-temperature high-humidity (120°C and 100% relative humidity) conditions, and standard dicing and packaging operations [15,17].

Recently we have established a via-first process that provides direct electrical interconnects during the bonding operation, while achieving BCB–BCB bonding in the field region. A three-wafer stack of the via-first 3D platform is depicted in Figure 4.3. Vias and redistribution layers are formed by copper damascene patterning with partially cured BCB, with a cross-section of the bonded interface shown in Figure 4.4. Excellent BCB-to-BCB interfaces are apparent and good Cu-specific contact resistance has been demonstrated [18]. Planarity requirements are stringent, but appear to

FIGURE 4.2   Cross-section of CMOS SOI test structure with four-level copper interconnect after two bonding and thinning processes and BCB ashing.

be reasonable [19]. Further research and development is necessary to fully evaluate this promising platform.

In addition to these platforms using Cu or W interstrata interconnects with various process flows, several groups have explored capacitive interstrata coupling [20] with face-to-face bonding or inductive interstrata coupling [21] with either face-to-face or face-to-back bonding for noncontacting interstrata signal interconnects. While processing complexity is reduced, these alternatives may be limited to RF/microwave systems with limited signal bandwidth. The application to high-speed digital ICs is more questionable since the interconnect impedance is frequency-dependent. Moreover, power distribution to, and grounding between, strata require interwafer electrical connections. These

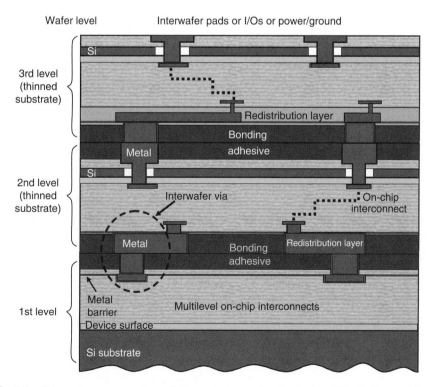

FIGURE 4.3   Schematic cross-section of three-strata stack of via-first 3D platform with patterned metal/adhesive (e.g., Cu/BCB) bonding.

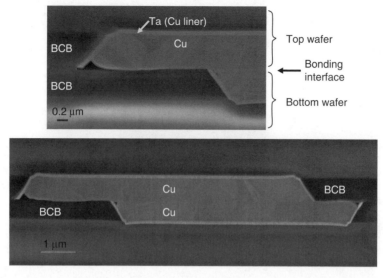

FIGURE 4.4 Cross-sectional FIB/SEM image of bonded damascene-patterned Cu/BCB wafers for via-first 3D platform, showing well bonded Cu-to-Cu, BCB-to-BCB interfaces.

noncontacting techniques appear limited with BEOL-based platforms, but could be attractive with WLP-based 3D platforms.

While appreciable development is needed to fully evaluate these BEOL-based technology platform alternatives, no technology showstoppers have been demonstrated to date. The technology platforms that become selected for high-volume IC manufacturing will depend upon the following: (1) the two key application drivers (specifically, high-density digital ICs and heterogeneous integration of diverse process flows), (2) process compatibility (such as thermal and mechanical stress) with FEOL processing, die stack packaging, and diverse application environments, and (3) manufacturing metrics such as die yield, reliability, and cost.

4.3 DIGITAL IC APPLICATIONS

Digital ICs have been the key driver for CMOS scaling, and digital CMOS performance optimization drives the International Technology Roadmap for Semiconductors (ITRS). Analog/mixed signal, RF/microwave, sensors (including imagers), and power ICs leverage the process development and digital CMOS devices as much as possible. As mentioned previously, reduction of interconnect delay may be a key driver for 3D IC integration past the 22 nm technology node. Certainly, any 3D integration platform must be compatible with scaled digital CMOS memory, microprocessors, ASICs, and digital signal processors (DSPs) in order to achieve high-volume manufacturing.

The low-hanging fruit for 3D integration is memory stacks, such as dynamic random access memory (DRAM), SRAM, and flash. Both the regular structure of memories and compatibility with redundancy to achieve high within-wafer yield are desirable attributes for wafer-level 3D integration. In addition to a higher memory size proportional to the number of strata, the electrical performance (both access time and power dissipation) can be enhanced as discussed later in this section.

More aggressive 3D integration includes processor and memory stack(s) for logic ICs such as microprocessors, ASICs, DSPs, and graphics chips. These planar ICs currently consist of large die often with large embedded memory, which can be partitioned into smaller die and realized with simpler process flows in multiple strata. In each cases, die yield of the individual wafers will be increased, compensating yield loss in the additional processes for 3D integration described in the previous section.

TABLE 4.2

Performance Comparison of 2D and 3D Implementations for 130 nm Technology Node, 16 Mb L2 Cache ($\lambda = 0.5$)

	Original Design	Original Configuration 3D	Optimized 3D
Strata	1	3	3
NumBank	8	8	16
Ndwl/Ndbl/Nspd	8/8/8	8/8/8	64/16/8
Address-in routing delay (ns)	1.21	0.09	0.13
Decoder delay (ns)	1.01	1.01	0.80
Wordline delay (ns)	1.00	1.00	0.40
Bitline delay (ns)	1.12	1.12	0.98
Output driver delay (ns)	0.56	0.56	0.52
Data output delay (ns)	0.90	0.10	0.12
Access time (ns)	5.80	3.88	2.96

An example of the performance advantages of 3D integration for SRAMs used for a 16 Mb L2 cache for real-time image processing has been analyzed at Rensselaer. An analytical model of the access time, cycle time, and dynamic power dissipation with circuit models was confirmed using Cadence Spectre simulations. The analytic model, named PRACTICS (which stands for predictor of access time and cycle time for cache stack), calculates performance parameters for all cache configurations with design parameters either fixed or variable within a certain range. Design parameters include number of strata, number of banks, associativity, word line parameters, and bit line parameters. Cycle time and/or dynamic power dissipation can be minimized by varying an evaluation parameter, λ (equal to 0.0 to minimize cycle time, 1.0 to minimize dynamic power, and 0.5 for equal weighting) [22].

An optimized (with $\lambda = 0.5$) performance comparison using 130 nm technology node CMOS technology (the most aggressive technology that could be fully evaluated with device and circuit parameters implemented in PRACTICS) is shown in Table 4.2 for three implementations: planar 2D, two-wafer stack, and four-wafer stack [22]. The four-wafer stack reduces the number of repeaters in the critical I/O channels by more than 80% (1344 to 224), decreases access time by 58% (5.09 ns to 2.04 ns), and achieves a cycle time of 0.5 ns, which provides real-time decoding of fast-acting videos without enhanced decoding algorithms. In addition, 3D implementations provide more uniform delay distribution with increasing number of strata, which improves pipelining efficiency.

While such memory stacks are very attractive in both embedded and stand-alone implementations, digital products such as microprocessors, ASICs, DSPs, and system-on-chips (SoCs) require integration of logic functions. While embedded SRAM and DRAM are becoming more of the die area for all CMOS ICs at newer technology nodes, high-performance processor cores are a particular concern in 3D integration. The interconnectivity is less regular, thermal concerns (already a major issue in 2D microprocessors) are enhanced, and prime power distribution can become more complicated. In a flip-chip packaged 3D die stack, heat sinking is through the bottom wafer in the original (i.e., before flipping) stack; the thermal path is lowest for this die. In a single-core processor, the wafer with the processor core is in contact with heat sink, but for multiple core processors, alternative 3D implementations need to be considered (i.e., all cores in the wafer to be contacted to the heat sink or processor cores in all strata).

The electrical performance advantage for processors with 3D integration has also been analyzed as well. To first order for SoCs, the clock frequency increases by $N^{3/2}$ and the interconnect power decreases by $1/N^{1/2}$ [23,24]; therefore, in a four-strata implementation, the clock frequency increases by a factor of 8 and the interconnect power is reduced by a factor of 2. This first-order model result indicates the significant enhancement that can be potentially achieved in SoCs, but is probably somewhat optimistic. Further discussion is beyond the scope of this chapter.

4.4   ANALOG/MIXED-SIGNAL APPLICATIONS

Analog/mixed-signal ICs do not have the same scaling advantages as digital ICs. In fact, analog design issues such as dynamic range and power handling have become more difficult to achieve using a CMOS digital IC process at technology nodes below 130 nm. SoC realizations become more complex as metal oxide semiconductor field-effect transistor (MOSFET) sizing and bias requirements for analog and RF/microwave applications increasingly deviate from digital CMOS devices at decreasing technology nodes. Wafer-level 3D provides an opportunity to partition analog and digital functions into different strata, with process flows and parameters selected to optimize each functional requirement. For example, Si-based microwave transceivers above 30 GHz usually incorporate silicon–germanium (SiGe) heterojunction bipolar transistors (HBTs) because of their higher cutoff frequency compared to Si MOSFETs. Incorporating baseband DSPs in high-performance digital CMOS in a BiCMOS process for a single-chip solution is not feasible (at best, the approach is not economical). However, a SiGe transceiver can be bonded to, and interconnected with, a Si CMOS DSP, forming a 3D-stacked SoC. In less cost-sensitive applications demanding highest electrical performance, higher cost GaAs-based or InP-based HBTs can also be used in the transceiver stratum.

A long-range objective is to achieve a software radio, which requires an analog-to-digital converter (ADC) well beyond the current state-of-the-art. A conceptual three-strata future implementation is depicted in Figure 4.5a. Three strata are indicated, a top-surface passive-component only stratum with an on-die antenna and high-Q passives, a SiGe BiCMOS stratum including a transceiver and ADC, and a digital CMOS processor and memory stratum. The top stratum could contain high-permeability, low-loss magnetic thin films for compact, high-Q inductors and high-k dielectric films for compact, high-Q capacitors, as well as a compact on-die antenna [25]. A key need in a true software radio in the long term or an effective software radio covering multiple frequency bands in the short term is the ADC.

We designed and evaluated a conventional pipelined ADC using gain-of-2 sample/hold (S/H) amplifiers with an operational transconductance amplifier (OTA) in a negative feedback loop using precise-value capacitors. An IBM 6HP process providing 47 GHz SiGe HBTs and 250 nm CMOS was used to design a wide-bandwidth, high-gain, fast-settling OTA using SiGe NPN HBTs in place of the usual NFETs in a cascade configuration. The 34 MS/s sampling rate with 12 bit resolution was limited by capacitive mismatch and lack of self-calibration techniques. An improved OTA with a triple cascade architecture should achieve a 115 MS/s sampling rate with 12 bit resolution; with digital self-calibration using a 7 bit pipeline seed, 205 MS/s is predicted, corresponding to an ADC figure-of-merit (FoM) of 4000 GHz/W for the 6HP process introduced in 2000 [25]. A microphotograph of the prototype ADC is shown in Figure 4.5b. In comparison, the ITRS predicts that a CMOS ADC will achieve such a FoM in 2012. While SiGe BiCMOS ADCs are more expensive than CMOS ADCs in a stand-alone product, such designs are attractive when integrated with a SiGe BiCMOS transceiver.

Other analog/mixed-signal architectures will be more readily realized in 3D with TSV interstrata interconnects, whether the technologies are Si MOS/Si CMOS, Si BiCMOS/Si CMOS, GaAs/Si CMOS, InP/Si CMOS, or any innovative 2D planar IC (such as a nanotechnology memory) with Si CMOS. Wafer-level 3D extends the reach of digital CMOS into monolithic-stacked ICs with embedded DSP and processor capabilities. One principal constraint on SoC implementation is eliminated with such 3D platforms.

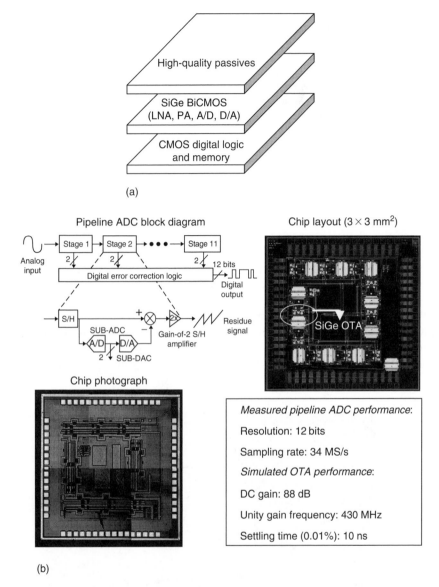

FIGURE 4.5 Software radio 3D architecture showing (a) three-die stack of SiGe BiCMOS transceiver and ADC with CMOS processor/memory and passive components and (b) prototype BiCMOS pipelined ADC.

4.5 UNIQUE SYSTEM ARCHITECTURES

While SoC stacks of conventional architectures discussed in the previous section are currently being pursued as demonstration vehicles with near-term product potential, future drivers of wafer-level 3D technology with TSVs for interstrata interconnect may be novel system architectures not envisioned with planar 2D ICs. We have explored two such innovations: (1) unique prime power delivery systems using a cell-based PoL DC–DC converter [26] and (2) optical clock delivery across die to maintain synchronous logic with low latency [27]. Both examples are briefly described here.

Prime power delivery to ICs such as microprocessors, ASICs, and DSPs are an increasing system-level concern. A 3D stack with low-parasitic interwafer interconnects enables DC–DC converters to be distributed across one stratum. Power is delivered to the packaged 3D die stack at relatively high

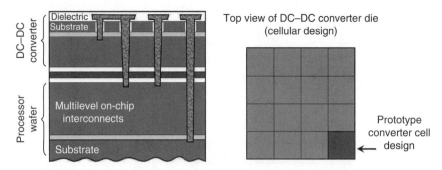

FIGURE 4.6 Cell-based 3D power delivery with DC–DC converter wafer.

voltage and to the signal electronics in a cell-based manner as depicted in Figure 4.6. From a power systems perspective, the PoL converter is within a few millimeters of the load with low-parasitic interconnection. A large control bandwidth ($\sim$200 MHz) can be obtained, enabling dynamic power delivery to signal electronics with low transient time. With power input to a die-stack package at relatively high voltage, the large number of power and ground pins required for microprocessor and SoC packages can be significantly reduced.

A fully monolithic interleaved buck converter with linear feedback control was designed in a 180 nm SiGe BiCMOS process to operate at 200 MHz switching frequency and deliver 500 mA output current at 1.0 V. A microphotograph of the test chip, which contains a MOSFET dummy load to simplify testing, is shown in Figure 4.7 [26]. Even though the positive channel mixed oxide semiconductor (PMOS) control switch has a total gate width of 16.6 mm and the N-channel mixed oxide

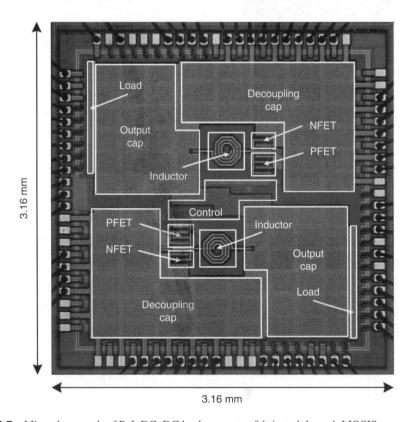

FIGURE 4.7 Microphotograph of PoL DC–DC buck converter fabricated through MOSIS.

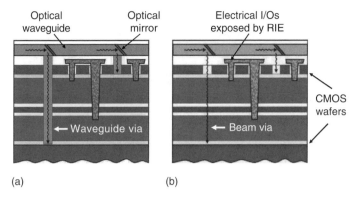

FIGURE 4.8 Clock distribution using 2D optical waveguide H-tree and 45° reflectors with (a) optical waveguide vias and (b) optical beam vias.

semiconductor (NMOS) synchronous switch has a total gate width of 11.0 mm, the die area is mostly used by input decoupling capacitors (31%), output capacitors (27%), and bond pads and electrostatic discharge (ESD) protection (31%). The low conversion efficiency of 64% obtained is limited by the available monolithic inductor and MOSFET dissipation in this first-generation prototype.

Scaling calculations indicate that this basic converter design with interleaved cells requires a die area of 250 mm^2 to power the Intel Core Duo processor [26]. While additional design effort is needed to achieve the 80%–85% desired, the wide control bandwidth possible (limited to 10 MHz with the linear feedback control in this initial prototype, but with ~100 MHz anticipated) will enable effective dynamic power delivery with advanced logic ICs such as multicore processors.

Optical interconnects is one technique to distribute clocks across a synchronous-logic die in a manner to minimize latency. In conventional approaches using planar 2D ICs, optical waveguide fabrication must be integrated with digital CMOS fabrication. With a 3D implementation, an optical clock can be distributed with an H-tree network to perhaps 16 sections of a processor die, with only photodetectors required in the digital CMOS strata. With 45° mirrors in the H-tree network to orient the optical clock vertically where desired, the optical vias can either be an optical waveguide or an optical beam as depicted in Figure 4.8. While the optical waveguide approach confines the optical beams, the fabrication requirements are difficult to achieve. Fortunately, the optical beam approach has acceptable performance with a variety of materials and dimensions appropriate for BEOL-compatible processing and 3D platform requirements [27].

We envision that implementation of such unique architectures with a high-volume density of electronic devices may be limited in the future by signal integrity issues rather than the cost of such integration. With further scaling of digital CMOS almost assured past the 22 nm technology node, increasing capability of analog/mixed-signal technologies, opportunities for unique system architectures and ability to integrate optical and nanotechnology planar ICs, the integration capability of wafer-level 3D technology platforms will be almost mind-boggling. The limitation imposed by signal integrity constraints could become a fundamental constraint, and the use of optical interconnects could alleviate electronic signal integrity constraints.

4.6 FUTURE DRIVERS FOR WAFER-LEVEL 3D IN IC MANUFACTURING

Present emphasis on FEOL device technology such as strained layers, high-k gate dielectrics, metallic gates, and wrap-around gate structures is resulting in enhanced digital CMOS devices at the 45 and 32 nm technology nodes. By the 22 nm node, these FEOL enhancements will result in a second-generation interconnect bottleneck (the first being in the 1990s when aluminum (Al) lines, tungsten (W) plugs as interlevel vias, oxide ILDs, and reactive ion etching (RIE) of Al as a patterning process were replaced with Cu lines and interlevel vias, low-k ILDs, and in-laid Cu as a patterning process (Damascene patterning)). Wafer-level 3D is the only near-term alternative to planar (2D) ICs with Cu/lowest-k interconnects, as well as enabling heterogeneous integration of different planar technologies for innovative SoCs (at least further levels of system integration). The technology and infrastructure needs, which must be overcome before, or soon after, a decision to move to large-volume manufacturing, are discussed in this section, and split into technology drivers, design drivers, equipment infrastructure drivers, and industry infrastructure drivers.

4.6.1 TECHNOLOGY DRIVERS

Selection of the product base to drive wafer-level 3D is paramount as the technology requires establishment of additional infrastructure compared to 2D ICs. Digital CMOS with cell-to-cell interconnectivity offers the most design freedom, but requires the highest interstrata interconnectivity and probably the most involved design tools. Heterogeneous integration drivers such as mixed-signal SoCs and smart sensors (e.g., imagers with pixel processing) would impact selection of the technology platform. Certainly any 3D technology platform (BEOL or WLP based) offers different processing challenges and different opportunities. Perhaps a nanotechnology innovation to be integrated with digital CMOS will be such a driver, rather than a product that can be envisioned now. Down-selection of technology platform options will be needed prior to large-scale manufacturing.

4.6.2 DESIGN DRIVERS

A key design driver is the constraint imposed by wafer-level 3D integration, a common die size (as well as a common wafer size) for full utilization of semiconductor area in each stratum. This constraint is somewhat limiting for an IDM or a 3D ASIC where die size is not constrained initially. However, for SoC implementation with some standard die being reused, a set of standardized die sizes are needed. In addition, design rules must be established that accommodate processing, and electrical and mechanical constraints. These will require simulation software as well to insure signal integrity with semiconductor devices in multiple strata.

4.6.3 EQUIPMENT INFRASTRUCTURE DRIVERS

Wafer-level 3D integration requires new tools for IC processes, particularly wafer-to-wafer alignment and wafer bonding. While these tools have been available for many years from both EV Group and Suss Microtech, and have been used in product manufacturing, they do not meet the requirements of the IC industry for high-volume 24/7 manufacturing. TSV technology is being developed for 3D packaging and can be readily modified for wafer-level 3D integration. In addition, the wafer bonding processes developed to date can take an hour or more; shorter process times are certainly desirable. Other process requirements are technology platform–dependent as described earlier.

4.6.4 INDUSTRY INFRASTRUCTURE DRIVERS

We are not certain whether wafer-level 3D technology platforms that align and bond wafers with complete interconnect structures will be integrated with the IC back-end or be incorporated into the expanding WLP infrastructure. Many major IDMs are actively involved in wafer-level 3D research and development; these IDMs would probably extend their BEOL technology base accordingly. We expect that major silicon foundries with multiple process flows will extend their BEOL technology base as well. However, second-tier silicon foundries with limited IC process flows will probably not extend their manufacturing base; major packaging foundries will probably extend their WLP and system-in-package (SiP) capabilities. Such industry infrastructure considerations could impact the acceptance of wafer-level 3D for high-volume mainstream products, at least in the near term.

4.6.5 PREDICTIONS

We believe that wafer-level 3D platforms (possibly die-to-wafer and probably wafer-to-wafer) will be driven by both high-speed digital CMOS at the 22 nm node and beyond and by heterogeneous integration for SoCs (or to more closely achieve SoCs). We believe that this will occur within 3 to 5 years as (1) current FEOL technologies such as high-k gate dielectrics, metal gates, and wrap-around gate structures become fully incorporated, (2) on-chip interconnect delay becomes a serious limitation to CMOS synchronous-logic performance, (3) novel IC features and products become feasible in wafer-level 3D demonstrations, and (4) novel nanotechnology planar ICs will be established that need to be integrated with digital CMOS processors.

ACKNOWLEDGMENTS

The authors gratefully acknowledge faculty colleagues, postdoctoral associates, and graduate students for the 3D interconnect research results: Faculty contributors include Professors T.S. Cale (technology platform), M. Hella (RF/microwave transceiver), J.F. McDonald (test mask design), P.D. Persans (optical interconnects), K. Rose (SiGe ADC and 3D memory performance prediction), and J. Sun (power delivery and DC/DC converter); visiting scientists and postdoctoral associates include A. Jindal [now at Micron], R.J. Kumar [now at Intel], K. Lee [now at Samsung], and F. Niklaus [on leave from KTH] (all in the technology platform); graduate students include S. Devarajan (SiGe ADC) [now at Analog Devices], D. Giuliano (DC/DC converter) [now at MIT], Y. Kwon (technology platform) [now at Samsung], J.J. McMahon (technology platform) [at RPI], J. Yu (technology platform) [now at IBM], and A. Zeng (3D memory performance prediction) [now at Freescale]. This research was sponsored principally through the Interconnect Focus Center, funded by MARCO, DARPA, and NYSTAR.

REFERENCES

1. P. Garrou, C. Bower, and P. Ramm (Eds.), *Handbook of 3D Integration: Technology and Applications*, Wiley, New York, 2008.
2. C.N. Tan, R.J. Gutmann, and L.R. Reif (Eds.), *Wafer-Level Three-Dimensional (3D) IC Process Technology*, Springer, Berlin, 2008.
3. K.W. Guarini, A.W. Topol, M. Ieong, R. Yu, L. Shi, M.R. Newport, D.J. Frank, D.V. Singh, G.M. Cohen, S.V. Nitta, D.C. Boyd, P.A. O'Neil, S.L. Tempest, H.B. Pogge, S. Purushothaman, and W.E. Haensch, Electrical integrity of state-of-the-art 0.13 mm SOI CMOS devices and circuits transferred for three-dimensional (3D) integrated circuit (IC) fabrication, *Digest of International Electron Device Meeting*, pp. 943–945, 2002.
4. J.A. Burns, B.F. Aull, C.K. Chen, C.-L. Chen, C.L. Keast, J.M. Knecht, V. Suntharalingam, K. Warner, P.W. Wyatt, and D.-R.W. Yost, A wafer-scale 3-D circuit integration technology, *IEEE Transactions on Electron Devices*, 53(10), 2507–2516, 2006.
5. P. Enquist, Room temperature direct wafer bonding for three dimensional integrated sensors, *Sensors and Materials*, 17(6), 307, 2005.
6. A. Fan, K.N. Chen, and R. Reif, Three-dimensional integration with copper wafer bonding, *Proceedings Electrochemical Society: ULSI Process Integration Symposium*, ECS PV 2001–02, pp. 124–128, 2001.
7. K.N. Chen, A. Fan, C.S. Tan, and R. Reif, Microstructure evolution and abnormal grain growth during copper wafer bonding, *Applied Physics Letters*, 81(20), 3774–3776, 2002.
8. P. Morrow, C.-M. Park, S. Ramanathan, M.J. Kobrinsky, and M. Harmes, Three-dimensional wafer stacking via Cu–Cu bonding integrated with 65 nm strained-Si/low-k CMOS technology, *IEEE Electron Device Letters*, 27(5), 335–337, 2006.
9. R. Patti, Three-dimensional integrated circuits and the future of system-on-chip designs, *Proceedings of the IEEE*, 94(6), 1214–1222, 2006.
10. J. Burns, L. McIlrath, C. Keast, A. Loomis, K. Warner, and P. Wyatt, Three-dimensional integrated circuits for low-power, high-bandwidth systems on a chip, *Proceedings of IEEE International Solid-State Circuits Conference Technical Digest*, pp. 268–269, 2001.

11. P. Ramm, D. Bonfert, H. Gieser, J. Haufe, F. Iberl, A. Klumpp, A. Kux, and R. Wieland, InterChip via technology for vertical system integration, *Proceedings IEEE International Interconnect Technology Conference 2001 (IITC 2001)*, pp. 160–162, 2001.

12. K.W. Lee, T. Nakamura, T. One, Y. Yamada, T. Mizukusa, H. Hasimoto, K.T. Park, H. Kurino, and M. Koyanagi, Three dimensional shared memory fabricated using wafer stacking technology, *Digest of International Electron Device Meeting*, pp. 165–168, 2000.

13. J.-Q. Lu, Y. Kwon, R.P. Kraft, R.J. Gutmann, J.F. McDonald, and T.S. Cale, Stacked chip-to-chip interconnections using wafer bonding technology with dielectric bonding glues, *Proceedings of the 2001 IEEE International Interconnect Technology Conference (IITC)*, pp. 219–221, IEEE, June 4–6, 2001.

14. J.-Q. Lu, K.W. Lee, Y. Kwon, G. Rajagopalan, J. McMahon, B. Altemus, M. Gupta, E. Eisenbraun, B. Xu, A. Jindal, R.P. Kraft, J.F. McDonald, J. Castracane, T.S. Cale, A. Kaloyeros, and R.J. Gutmann, Processing of inter-wafer vertical interconnects in 3D ICs, in B.M. Melnick, T.S. Cale, S. Zaima, and T. Ohta (Eds.), *Advanced Metallization Conference in 2002 (AMC 2002)*, MRS Vol. V18, pp. 45–51, 2003.

15. J.-Q. Lu, A. Jindal, Y. Kwon, J.J. McMahon, M. Rasco, R. Augur, T.S. Cale, and R.J. Gutmann, Evaluation procedures for wafer bonding and thinning of interconnect test structures for 3D ICs, *2003 IEEE International Interconnect Technology Conference (IITC)*, pp. 74–76, June 2003.

16. R.J. Gutmann, J.-Q. Lu, S. Pozder, Y. Kwon, D. Menke, A. Jindal, M. Celik, M. Rasco, J.J. McMahon, K. Yu, and T.S. Cale, A wafer-level 3D IC technology platform, *Proceedings of Advanced Metallization Conference*, pp. 19–26, 2003.

17. S. Pozder, J.-Q. Lu, Y. Kwon, S. Zollner, J. Yu, J.J. McMahon, T.S. Cale, K. Yu, and R.J. Gutmann, Back-end compatibility of bonding and thinning processes for a wafer-level 3D interconnect technology platform, *2004 IEEE International Interconnect Technology Conference (IITC04)*, pp. 102–104, June 2004.

18. J.J. McMahon, J.-Q. Lu, and R.J. Gutmann, Wafer bonding of damascene-patterned metal/adhesive redistribution layers for via-first 3D interconnect, *Proceedings of the IEEE Electronic Components and Technology Conference*, pp. 331–336, 2005.

19. J.J. McMahon, R.J. Gutmann, and J.-Q. Lu, Three dimensional (3D) integration, Chapter 14 in Y. Li (Ed.), *Microelectronic Applications of Chemical Mechanical Planarization*, Wiley Interscience, New York, 2008.

20. Q. Gu, Z. Xu, J. Kim, J. Ko, and M.F. Chang, Three-dimensional circuit integration based on self-synchronized RF-interconnect using capacitive coupling, *Digest of 2004 Symposium on VLSI Technology*, pp. 96–97, June 2004.

21. J. Xu, J. Wilson, S. Mick, L. Luo, and P. Franzon, 2.8 Gb/s inductively coupled interconnect for 3D ICs, *Digest of 2005 Symposium on VLSI Circuits*, pp. 352–355, June 2005.

22. A.Y. Zeng, J.-Q. Lu, K. Rose, and R.J. Gutmann, First-order performance prediction of cache memory with wafer-level 3D integration, *IEEE Design & Test of Computers*, 22(6), 548–555, 2005.

23. J.D. Meindl, R. Venkatesan, J.A. Davis, J.W. Joyner, A. Naeemi, P. Zarkesh-Ha, M. Bakir, T. Mulé, P.A. Kohl, and K.P. Martin, Interconnecting device opportunities for gigascale integration (GSI), *International Electronic Devices Meeting*, pp. 525–528, 2001.

24. J.W. Joyner and J.D. Meindl, Opportunities for reduced power dissipation using three-dimensional integration, *2002 IEEE International Interconnect Technology Conference (IITC02)*, pp. 148–150, June 2002.

25. R.J. Gutmann, A.Y. Zeng, S. Devarajan, J.-Q. Lu, and K. Rose, Wafer-level three-dimensional monolithic integration for intelligent wireless terminals, *Journal of Semiconductor Technology and Science*, 4(3), 196–203, 2004.

26. J. Sun, J.-Q. Lu, D. Giuliano, P. Chow, and R.J. Gutmann, 3D power delivery for microprocessors and high-performance ASICs, *Proceedings of 22nd Annual IEEE Applied Power Electronics Conference and Exposition (APEC 2007)*, pp. 127–133, 2007.

27. P.D. Persans, M. Ojha, R. Gutmann, J.-Q. Lu, A. Filin, and J. Plawsky, Optical interconnect components for wafer level heterogeneous hyper-integration, in R.J. Carter, C.S. Hau-Riege, G.M. Kloster, T.-M. Lu, and S.E. Schulz (Eds.), *Materials, Technology, and Reliability for Advanced Interconnects and Low-k Dielectrics*, 2004 MRS Proc., Vol. 812, pp. F6.11.1–F6.11.5, 2004.

5 CMOS SOI Memory Design Technology

Kazutami Arimoto and Fukashi Morishita

CONTENTS

5.1 BACKGROUND AND FUTURE TRENDS

Advanced complementary metal oxide semiconductor (CMOS) devices have been facing several kinds of physical limitations. The extremely slow voltage scaling down and the process parameter variation (threshold voltage, gate oxide thickness, etc.) induce performance saturation of the bulk CMOS system-on-a-chip (SoC). The silicon-on-insulator (SOI) device has the advantage of suppressing this problem. The smaller junction capacitance provides high-speed switching and the soft error immunity gives high reliability. Additionally, the SOI device can be applied to bulk CMOS circuit design methodologies, suppressing the transistor leakage current and variation of the process parameter, and can enhance performance, utilizing the SOI device structure.

Figure 5.1 shows the SOI device advantage and milestone. The current SOI transistor is a partially depleted (PD) type because of the precise threshold voltage (V_{th}) controllability for the "analog" and "I/O" etc., even if the fully depleted (FD) transistor can provide higher speed switching for the "logic portion." The device technologies will improve from the PD type to the hybrid one (PD and FD) with strain Si to get high-performance logic. The higher speed logic portion uses the FD and the analog and the I/O portion use the PD in future SoC.

The current SOI application is limited, for example, a microprocessing unit (MPU) and full custom design game machines, which can get high-speed, low-power, and system merits from the SOI device. To enlarge the SOI application, open design environments should be prepared. By solving this problem, the SOI application will enlarge from high- to low-end SoC.

The key technology of SOI circuit design is the floating body controlling one. The SOI transistor's body region is isolated perfectly from another transistor. Therefore, the dynamic body-control circuit

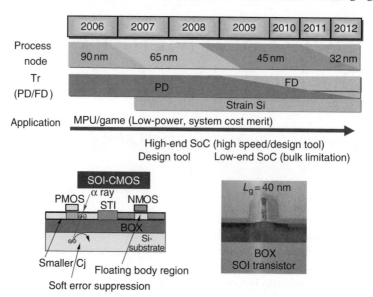

FIGURE 5.1   Advantages of SOI device and milestones.

technologies in the transistor level, for example, the power gating and dynamic voltage scaling (DVS)/dynamic voltage and frequency scaling (DVFS), can be applied.

In this chapter, several kinds of low-voltage SOI circuit designs will be introduced for the SOI CMOS logic gate, static random access memory (SRAM), and higher density memory with a CMOS-compatible process. (Figure 5.2 shows the memory technology trend and Table 5.1 shows the comparison of high-density memory.) These techniques enhance system performance and provide the breakthrough of the voltage scaling down.

5.2   LOW-POWER AND HIGH-SPEED SOI CIRCUIT DESIGN

5.2.1   DEMANDS OF BOTH LOW-POWER AND HIGH-SPEED OPERATION

SOI devices have several advantages compared to bulk Si devices as follows: (a) small junction capacitance, (b) V_{th} controllability with forward bias, and (c) a better Q factor. These features provide ultralow voltage operation with high speed and small noise, and are suitable for future

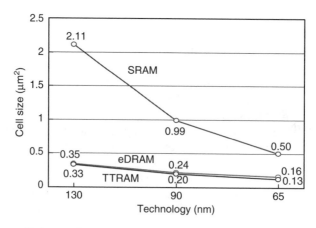

FIGURE 5.2   Memory cell size trend.

TABLE 5.1

Comparison of High-Density Memory

	eDRAM	1T(FBC)	TTRAM
Cell size	1	0.77	0.94
CMOS compatibility	Plus eDRAM cell process step	Plus protection structure against noise	No additional structure
Library compatibility with CMOS	Tuning for hard IP	Tuning for hard IP	Fully compatible
Performance (power)	Low-voltage operation by large Cs	Higher boosting voltage	1/2 VDD BL swing/ wide operating margin

multimedia applications used in, for example, mobile and wireless networks. Many techniques have been proposed to speed up large-scale integration (LSI) performance and to lower power supply voltage and power consumption [1–5]. Some dual-threshold CMOS schemes for high-speed and low-power circuits have been reported [4,5]. In the SOI CMOS, the low threshold voltage and high-performance transistor is realized by reducing and optimizing the channel concentration in the body-tied scheme [6,7]. However, the off-state current of such a low threshold voltage transistor cannot be sufficiently suppressed. Therefore, to reduce the off-state current, a variable threshold (VT) CMOS has previously been reported [8]. This variable threshold scheme requires recovery time from the standby mode, and the scheme is realized using body-tied transistors with an additional area for back-gate contacts. In this section, several-body potential-controlling techniques for the fully-depleted body-floating SOI CMOS are introduced [9,10]. Using these techniques, a transitional dual-threshold SOI CMOS is realized, which maintains both low-power and high-speed operation. CMOS logic operates at a high speed during the active period and maintains a low leakage current during the standby period.

As for active current, DVFS is an effective technique, and is particularly useful for controlling microprocessor energy and performance. In recent years, in spite of demands for power reduction, embedded systems require higher performance and more functions to support enormous data contents for computing and networking. In order to reduce power consumption, it is essential to decrease the supply voltage depending on the operating status. However, due to various reasons such as the requirements of the control software or technique, it is difficult to use DVFS for general logic circuits. Therefore, the PVT (process, voltage, temperature)–free power management system for general purposes is also introduced in this section [10].

5.2.2 DYNAMIC BODY-CONTROL TECHNIQUE

A method for improving the data retention characteristics of SOI dynamic random access memory (DRAM) has been reported [11,12]. In this technique, the threshold voltage becomes high transitionally by pulling out accumulated carriers from the floating body region by the forward biased current of the p–n junction. This threshold voltage increase is applied to the logic gates as well [9]. In the case of the logic gates, this operation is performed as shown in Figure 5.3. Figure 5.3a shows the initial state in active mode. Under this condition, the threshold voltage is low. The source line voltage is then pulled down to $-V_1$, the accumulated holes are discharged from the floating body region to the source line, and the body potential decreases immediately, as shown in Figure 5.3b. Though the source line is pulled up to 0 V, the body potential is kept low, as shown in Figure 5.3c. Therefore, the threshold voltage is maintained at a high level. In this way, dual-threshold voltages are achieved through these operations (active to sweep-out and sweep-out to standby). However, according to research on the floating body potential [11], the body potential increases inevitably as time passes due to various leakage mechanisms. In the CMOS logic, the leakage current at the

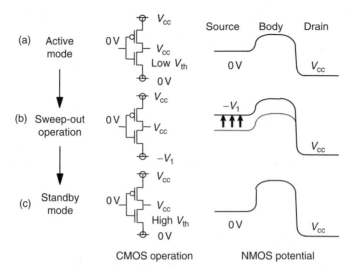

FIGURE 5.3 Sweep-out operation.

drain–body junction, which has a voltage difference of full V_{cc}, causes the body potential and standby current to increase. This indicates that the body-control operation should be performed at fixed intervals determined by the device characteristics. These leakage characteristics of the body region are similar to those of a DRAM cell [11,12]. As for the DRAM cell, the leakage characteristics are equal to the refresh characteristics and they are dominated by a worst-case memory cell. However, in this body-control scheme, even if some bad transistors exist, a fatal error will not occur unless the standby current slightly increases.

The technique mentioned above was verified by using a 100-stage inverter chain, which is a switched-source voltage CMOS, as shown in Figure 5.4. In the measurement setup, the V_{bcc}/V_{bss} power lines are driven through the switching transistors. The signature (SIG) pad is activated during the sweep-out operation. This circuit is operated at $V_{cc} = 1.0\,V$, and the power line for the body control (V_{bcc}/V_{bss}) is normally set at the same level as the main line. However, during the sweep-out

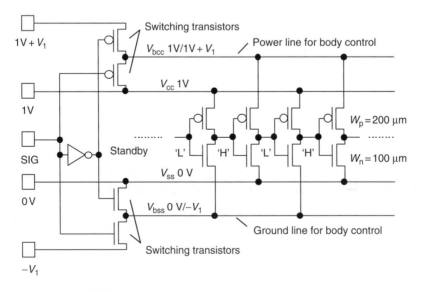

FIGURE 5.4 Schematic of 100-stage inverter chain.

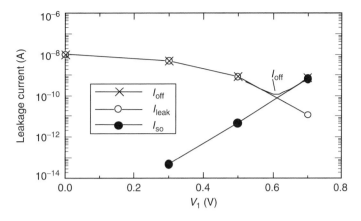

FIGURE 5.5   Measured leakage current.

operation, V_{bcc} is changed to $1.0\,V + V_1$ (V_1: the pull-down/pull-up level of the source node [SN]) and V_{bss} is changed to $-V_1$. This operation is performed every 1 s for a period of only 10 ns, as an excess sweep-out period increases power consumption due to the subthreshold leakage current. Figure 5.5 shows the measured leakage current I_{off} as a function of V_1. This current includes the current for the sweep-out operation. The static leakage current of the transistors cannot be separated from the sweep-out current because it is decreased by the dynamic sweep-out operation. The leakage current I_{so} during the sweep-out period (10 ns) is also shown in this figure as the average current over the sweep-out interval (1 s). In addition, the leakage current I_{leak} without the sweep-out current is also shown, since it is easily estimated from I_{off} and I_{so}. Because I_{so} is mainly determined by the forward biased current at $V_{gs} = V_1$ during the short sweep-out period, the average value of I_{so} over the long sweep-out interval becomes small. It can be seen that the total current I_{off} ($= I_{leak} + I_{so}$) at $V_1 = 0.6\,V$ is suppressed to less than 1/50th of the value of the non-sweep-out current at $V_1 = 0\,V$.

5.2.3   AUTOMATIC SOURCE/BODY-CONTROL TECHNIQUE

Another body-control technique for the 0.5 V-level ultralow-voltage operation is introduced. This technique uses gate–body connecting transistors [13] called actively body-bias controlled (ABC) transistors, shown in Figure 5.6. By using this structure, the body potential is increased with the gate

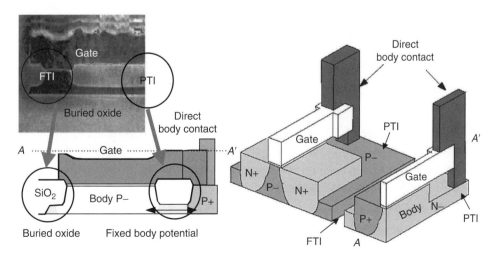

FIGURE 5.6   Gate–body connecting transistor.

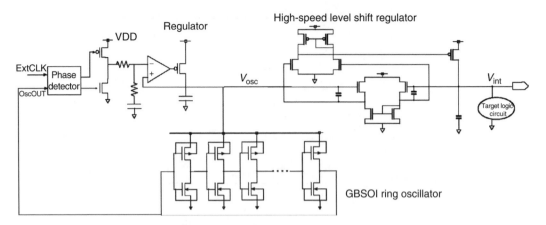

FIGURE 5.7   Self-compensating PVT-free system.

activation, and thereby the threshold voltage of the transistor is decreased transitionally (high-speed state). This threshold-voltage decreasing returns to the original state immediately with gate voltage lowering and the off-leakage current never increases (low-leakage state). However, in consideration of the PVT variations, the operation characteristics vary greatly in this structure. In order to improve the variation, a self-compensating PVT-free system has been proposed [10], as shown in Figure 5.7. This system automatically realizes dynamic voltage and frequency management and stabilizes the logic circuitry composed of the gate–body connecting CMOS under ultralow voltage, such as 0.5 V. This system supplies V_{osc} ($=V_{int}$) to the target CMOS logic and the replica oscillator. V_{osc} changes in accordance with the frequency of the output signal OscOUT and tunes the frequency to a stable speed by itself. For tuning the frequency, OscOUT is compared to the external clock at the phase detector. The phase detector pumps up V_{osc} if the OscOUT is slower than the external clock, and vice versa. In this way, V_{osc} rises by degrees and then levels off at the voltage, which makes the oscillation frequency the same with ExtCLK. Finally, V_{osc} is transferred to V_{int} by the regulator.

By simulation with a 90 nm technology parameter, 1.0 GHz operation can be achieved by using the 17-stage replica oscillator. Figure 5.8 shows the simulation result. The operation voltage V_{osc} can

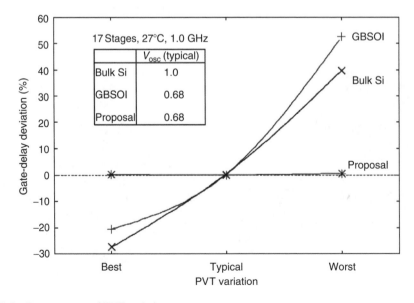

FIGURE 5.8   Improvement of PVT variation.

be decreased to 0.68 V compared with the bulk-Si of 1.0 V for realizing the same 1.0 GHz operation, and the gate-delay deviation due to PVT variation also can be reduced to 1% compared with the bulk-Si of 40% and the simple gate–body connecting SOI of 58%. By these results, this system presents an attractive solution for GHz-speed and ultralow-voltage operation.

5.3 SOI MEMORY TECHNOLOGIES

5.3.1 CAPACITORLESS MEMORY DESIGN

In an advanced SoC, an embedded DRAM is used as high-density memory intellectual proper-ties (IPs). However, further miniaturization of the memory-cell capacitor having a stack-type or trench-type structure is difficult, and therefore miniaturization of the DRAM is also going to show a limitation. Given this circumstance, a one-transistor gain cell [14–20] is one of the candidates for compact and scalable DRAM cells in the future generation. Those papers disclose the one-transistor gain-cell structure in which electric charges are accumulated in a floating body region of an SOI transistor. In the one-transistor gain cell, by applying a high voltage between a source and a drain, impact ionization is caused in the vicinity of the drain, and the holes that are thereby generated are accumulated in the floating body region. The one-transistor gain cell is smaller, less complex to make and more scalable than the conventional one-transistor one-capacitor DRAM.

However, there is a limit in generating holes by impact ionization, and therefore the difference of a threshold voltage cannot be increased between the state of data "1" and the state of data "0." In order to increase the difference of the "1" and "0" state, higher voltage than the memory array operation voltage is desirable, but this means that a lower voltage operation cannot then be performed. Furthermore, various kinds of power supply voltages are required for controlling reading and writing, and a driver for supplying voltages of three values is required to control a word line and a bit line. Furthermore, when the memory cell is constituted by only one memory transistor, in some cases, word-line noise and bit-line (BL) noise become a fatal issue for data retention characteristics.

Another issue for the advanced SoC is that leakage current problems by the physical limitations can occur. To overcome these problems, system power management techniques such as DVFS and compact memory IPs are expected to achieve the high-performance SoC. However, it is difficult to apply the DVFS to the memory IPs because voltage scaling induces the degradation of the static noise margin in the SRAM and signal margin in the DRAM.

In order to solve the problems described above, twin-transistor random access memory (TTRAM) has been proposed [21,22]. Although the TTRAM cell requires two transistors, simple array control is realized compared with the 1T gain cell, and therefore TTRAM can achieve a high-speed array operation like that of the conventional DRAM. In addition, the active power consumption can also be reduced because (1) neither boosted voltage nor negative voltage is required, (2) the BL voltage is suppressed to 1/2VDD swing, and (3) the standard SOI structure can be applied. To expect further enhancement of the operating characteristic, this paper describes the DVFS controllable and higher area efficiency of configurable macro for system-level power management unified memory, which is based on TTRAM using SOI technology.

5.3.2 TTRAM CELL OPERATION

The proposed TTRAM cell is shown in Figure 5.9. The TTRAM cell is composed of two serial transistors. Both transistors have floating body structures and are isolated by full trench isola-tion. A storage transistor is able to change the storage data state in accordance with its floating body potential. An access transistor limits the excessive read current of the storage transistor in the read operation, controls the purge node (PN) state in the write operation, and blocks the BL noise dynamically.

The principle of data storage is shown in Figure 5.10. "0"-data storage is achieved by sweeping out the accumulated holes from the body region of the storage transistor. The storage transistor is set

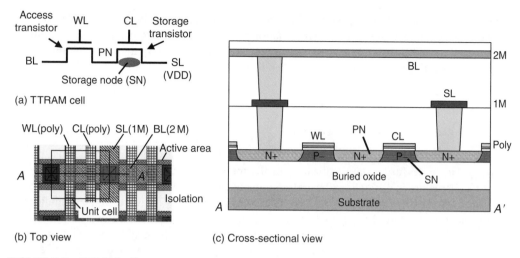

FIGURE 5.9   TTRAM cell.

in high-V_{th} state by lowering the voltage of the BL side from the stable state. In addition, in about "1"-data storage, hole injection is the dominant factor, and is controlled by changing the charge-line (CL) voltage of the storage transistor. CL voltage lowering causes a gate-induced drain leakage (GIDL) current [23] and can accumulate holes in the body region; hole injection is accelerated by the CL body coupling after that. The leakage current by GIDL raises the SN potential level when the "1" data is transferred to the BL at the same timing of word-line activation until the CL rises. By this operation, the storage transistor is set in a low-V_{th} state. It is desirable to compare the current difference of the storage transistor to detect a difference of these two states.

The memory cell operation of TTRAM is described in detail next. Four operation states exist: (1) write operation, (2) read operation, (3) hold operation, and (4) refresh operation. First, the write operation is shown in Figure 5.11. In the write operation, a BL and a word line (WL) is set at 0 V and 1/2VDD, respectively. In this state, the SN data is fixed to the balanced potential level between the forward biased current of the pn junction (SN to PN) and the GIDL current (SL [source line] to SN). Write data is continuously stored by the gate coupling of a CL. In the case of "0"-write, BL is kept at 0 V. Although CL is raised to VDD, the channel of the access transistor is formed and the body potential rising is thus suppressed [22]. In the case of "1"-write, BL is raised to 1/2VDD previous to CL rising. In this state, because the access transistor becomes an off state, PN becomes floating. While CL is kept at 0 V, holes are injected to the body region by the GIDL current from SL [23]. In

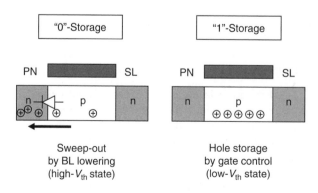

FIGURE 5.10   Principle of data storage in access transistor.

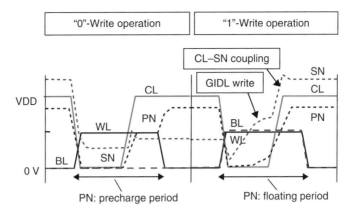

FIGURE 5.11 Write operation of TTRAM cell.

addition, when CL is changed to VDD, both of the floating nodes (SN and PN) are raised to near VDD by the capacitive coupling. In this way, two storage states are realized.

Figure 5.12 shows the effect of the GIDL current in the write operation. Generally, the GIDL current has wide variations due to its complex mechanisms. Therefore, the quantity of injection charge for "1" data cannot be controlled with high accuracy. As shown in Figure 5.12, V_f is defined as the balanced point of the p–n diode current (SN to PN) and GIDL current (CL to SN), and the variations of the GIDL current generate different balanced points (V_{f1} and V_{f2}). In the write operation, although the difference of V_{f1} and V_{f2} occurs during the off-state of CL, p–n forward bias current due to the capacitive coupling at CL rising decreases the difference by the write completion time.

Figure 5.13 shows the other memory cell operations. The read and hold operations are shown in Figure 5.13a, and the refresh operation is shown in Figure 5.13b. In the read operation as shown in Figure 5.13a, CL is fixed to VDD and the TTRAM cell flows the read current toward 0 V-level BL. The "0" data and "1" data are identified as a difference of the read current. WL-level limitation to 1/2VDD prevents the destruction of the "1" data from the excessive current. In the read operation,

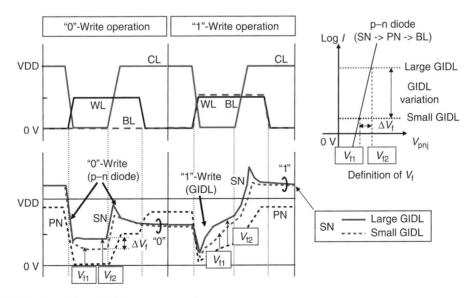

FIGURE 5.12 Effect of GIDL current in write operation.

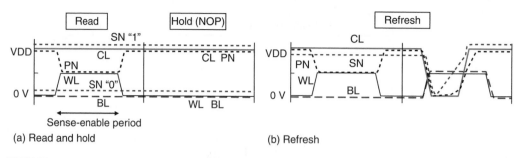

FIGURE 5.13 Memory cell operation (read/hold/refresh).

the drastic change of the floating body potential never occurs because the charge node CL is fixed. In the hold (no operation [NOP]) operation as shown in Figure 5.13a, both BL and WL are kept at 0 V, and therefore the access transistor is not turned on. Accordingly, the current does not flow from SL to BL and the previous data are held.

The refresh operation as shown in Figure 5.13b is explained. The refresh operation can be achieved by performing the write operation immediately after the read operation, while it requires two cycle times. These cell operations can be realized by using only VDD and 1/2VDD, and the complex control voltage is not necessary.

5.3.3 TTRAM ARRAY ARCHITECTURE

TTRAM adopts two types of array architecture depending on the application. The first type is the current-sense type shown in Figure 5.14. The 64 kb subarray has 64 rows. One BL is connected to 64 memory cells and a mirror cell (MCmrr). The mirror cell can carry away the cell current at the same level as that of the memory cells, and in this way, the BL voltage V_{BL} is determined. One subarray has two reference BLs (RBL0, RBL1), which always writes "0" and "1," respectively. Four BLs are selected by a BL selector and connected to a sense amplifier (SA) and a write driver (WD). At the time of sensing, the read-out current of BL and reference BLs is compared, and the memory cell data are fixed. In this configuration, although the negative voltage of VDD is required for the mirror cells, it is used only for the bias voltage of the sense operation, and the memory cell operation itself is not necessarily complicated.

Figure 5.15 shows the current-sense amplifier and word driver. Memory cells are connected to a negative voltage–VDD, whose voltage is supplied by the charge pump circuit via memory cell mirrors. The memory cell mirrors are arranged with the purpose of setting each read voltage of the bit line BL and the reference bit line RBL1, RBL0 at the value in the vicinity of 0 V. The write driver is driven with the supply voltage of 1/2VDD, and outputs 1/2VDD at "1"-write and 0 V at "0"-write. In addition, the write drivers leading to RBL1 and RBL0 lead to output "1" and "0" data, respectively. The sense amplifier includes a differential amplifier with two pairs of parallel inputs. The RBL1 voltage of VBL1 and the RBL0 voltage of VBL0 are input into one of the parallel inputs, and the BL voltage VBL is input as another parallel input. Because "1" and "0" data are always retrieved from RBL1 and RBL0, respectively, the relation of VBL1 > VBL0 is satisfied. In the case of "1"-read, VBL is equal to VBL1. Conversely, in the case of "0"-read, VB0 is equal to VBL0. VBL1, VBL0, and VBL are compared in the differential amplifier. The sense amplifier outputs the amplified storage data detecting this voltage difference with 4.3 ns after the WL activation.

Another type of array architecture similar to the DRAM and cross-coupled sense amplifiers was adopted. In this architecture, the ABC sense amplifier [22] with a higher sensing margin, compact size, and lower voltage operation was used. The sense amplifier of TTRAM is the read/write separated type, because the write mechanism and read mechanism are completely different mechanisms. The gate–body connecting transistor shown in the previous section (Figure 5.6) solves this problem.

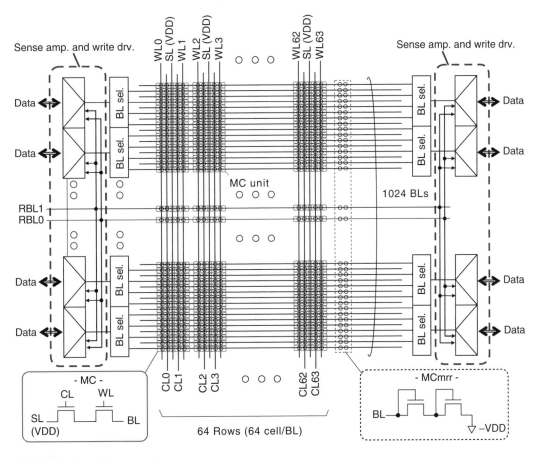

FIGURE 5.14 64 kb array structure.

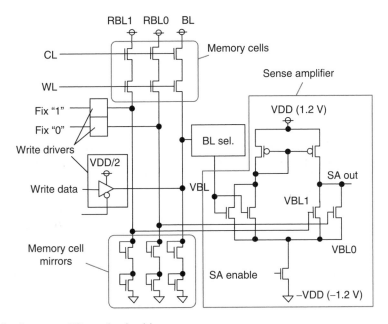

FIGURE 5.15 Sense amplifier and write driver.

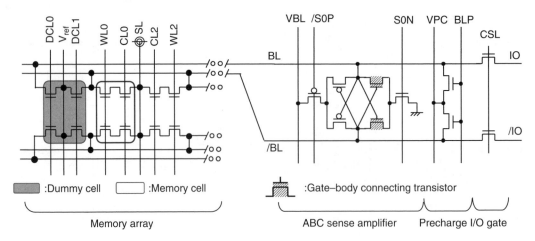

FIGURE 5.16   Memory array and ABC sense amplifier.

To realize the direct gate–body connection, we used hybrid trench isolation (HTI) [13] composed of full trench isolation (FTI) and partial trench isolation (PTI). The point of this HTI structure process is to add FTI etching to the PTI process, which is the same as the standard shallow trench isolation (STI) process in the bulk CMOS process. Therefore, the cost of this process is not very different from the standard CMOS process. The FTI and PTI support the lower cross-talk noise, and latch-up free and body-tied transistors with no floating body effects. In the FTI area, the whole of the SOI layer is replaced by the field oxide. The FTI is applied to memory cell blocks and sense amplifier blocks. In the PTI area, a part of the SOI layer remains under the field oxide. The PTI is used to separate the standard transistors. The bodies of the logic transistors are fixed using this isolation. The PTI structure also enables the compact size and smaller parasitic capacitance of the gate–body connecting transistor.

Figure 5.16 shows the memory array and ABC sense amplifier. The ABC sense amplifier consists of n-ch cross-coupled gate–body connecting transistors and the BL node is directly connected to the body of the cross-coupled transistors. The middle level signal between the "1" data and "0" data is read out from dummy cells by programming the SL level using reference voltage bias. The body bias of the S0P and S0N transistors, which are synchronized with own control signals, also enhanced the sense operation. As the read-out signal on the BL enhanced by CP-SN GIDL writing changes the V_{th} of the n-ch cross-coupled transistor, it can be sensed with high speed and a lower voltage operation. The ABC sense amplifier has the following merits compared to the current-sense amplifier: (1) the size of the compact sense amplifier is the same as the conventional read/write common amplifier, (2) simple dummy cell configuration, and (3) a nonnegative voltage requirement.

Figure 5.17 shows the configurable cell/bit mode architecture. Even if applying the CL-SN GIDL writing method and the ABC sense amplifier, it is difficult to handle the high reliability memory operation for mass production. The advanced SoC requires handling several kinds of power supply voltages for system-level power management. It means both full VDD operation and low-voltage operation modules are operated at the same time. A 1-cell/bit and 2-cell/bit configurable macro is proposed to solve the problems described above. The 2-cell/bit DVFS mode with a half size of memory capacity can operate at 0.5 V. The configurable 1-cell/bit and 2-cell/bit macro can be easily achieved. As shown in Figure 5.17a, in the 1-cell/bit mode, bit lines of even memory cells are connected to BL, and bit lines of odd memory cells are connected to /BL. Both BL and /BL connected to the memory cells (lower bit lines) are laid out with a second metal layer, and upper bit lines for dummy cell connection are laid out with a fourth metal layer. The upper and lower bit lines are twisted at the middle point. In contrast, in the 2-cell/bit mode, the odd memory cells of the

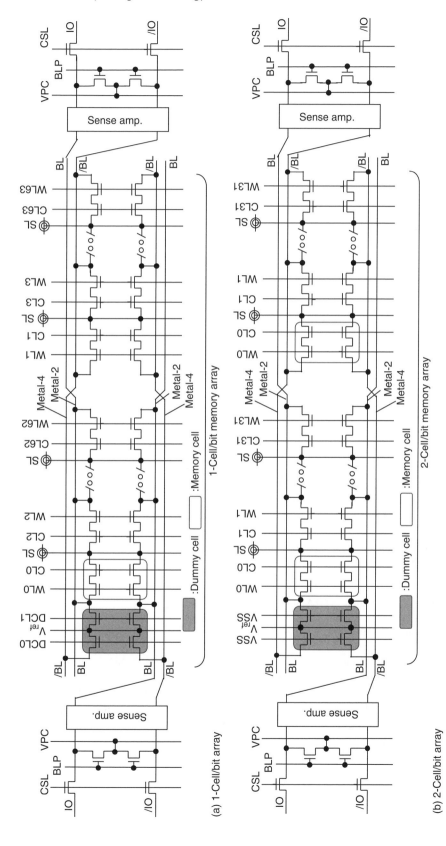

(a) 1-Cell/bit array

(b) 2-Cell/bit array

FIGURE 5.17 Configurable 1-cell/bit and 2-cell/bit mode memory.

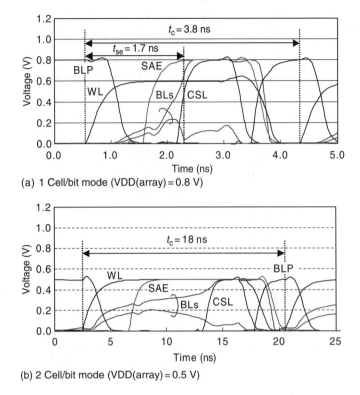

(a) 1 Cell/bit mode (VDD(array) = 0.8 V)

(b) 2 Cell/bit mode (VDD(array) = 0.5 V)

FIGURE 5.18   Simulation waveform.

1-cell/bit mode replace the complementary cells for the even ones. The dummy cell is not needed in the 2-cell/bit mode.

The simulation waveforms of the configurable ABC-sense architecture are shown in Figure 5.18. The performance is obtained as 263 MHz (random-cycle time = 3.8 ns) at a 0.8 V memory array operation in the 1-cell/bit mode. The sense time from WL activation (t_{se}) is 1.7 ns, which is 2.6 ns faster than the sense time of the current-sense amplifier, as shown in Figure 5.15. The active power is 108 mW. In the 2-cell/bit mode, the performance is obtained as 56 MHz (random-cycle time = 18 ns) at the 0.5 V memory array operation. In this mode, the active power is reduced to 10.2 mW.

5.3.4   IMPLEMENTATION OF **TTRAM**

Figure 5.19 shows the experimental 4 Mb ET2RAM macro with 90 nm SOI CMOS process technology. This macro consists of an array of eight-bank blocks, row decoders, a data path, and power supply circuits. Each bank has 512 kb memory cells at the 1-cell/bit mode. In one bank, there are four slim subblocks, and each subblock has 128 WLs and 1k BLs. The column select is 8-way, and therefore 128 data come to and from this macro. Table 5.2 shows the macro features. The 90 nm standard SOI-CMOS process is used and the macro size is 4 Mb. It has eight banks, which can be operated by different power supply voltage. The cell size of 0.20 μm² is smaller than the 90 nm embedded DRAM [24] of 0.24 μm².

5.3.5   SCALABLE FUNCTIONS OF **TTRAM**

The TTRAM can be applied to many kinds of applications by WV/WOV mode. The CL-SN GIDL writing method and ABC sense amplifier permit the read operation with restore named WV mode

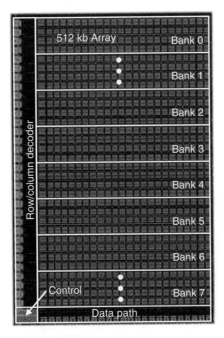

FIGURE 5.19 Experimental 4 Mb ET2RAM macro.

same as the conventional DRAM and the read operation without restore named WOV mode as shown in Figure 5.20. The WOV mode is suitable for the replacement of the short cycle time/high-speed SRAM, and the WV mode is suitable for mobile applications with a long data retention time. The body potential of the storage transistor is refreshed with an ABC sense amplifier in every WV mode, and the V_{th} of the storage transistor is verified with CL–SN coupling write and CL-SL GIDL write at the read-verify-write cycle, which has a longer cycle time than the read cycle. The WV mode provides about 10 times long data retention time (Figure 5.21). The reason for the difference of the data retention time between the WV mode and WOV mode in the capacitorless SOI memory is shown in Figure 5.22. The memory cell of TTRAM consisting of the series transistors is protected from the bit line noise induced by the dynamic operation of the bit line node. However, there are two hole-injection mechanisms degrading the data retention time. One is the hole injection from the SL

TABLE 5.2
Macro Features

Technology	90 nm SOI-CMOS Process	
Macro size	4 Mb (=512 kb × 8 banks)	
I/O	128 b	
Cell size	0.20 μm² (cf. eDRAM 0.24 μm²)	
	1-Cell/bit mode	2-Cell/bit mode
Logic voltage	1.2 V	0.8 V
Array voltage	0.8 V	0.5 V
Operating frequency	263 MHz	56 MHz
(Random cycle)	(3.8 ns)	(18 ns)
Power consumption	108 mW	10.2 mW
Data retention	130 ms@80°C	

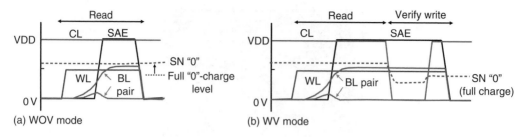

FIGURE 5.20   WOV and WV modes in TTRAM.

node biased at VDD. The other is the read current from the SL node (VDD bias). It induces the hole injection by impact ionization under the CL gate of the storage transistor and slightly raises the body potential of the storage transistor. The verify operation is very effective to improve this phenomena. The read-verify-write operation in the WV mode can provide a stable body potential and V_{th} of the storage transistor, because the body potential is refreshed in every read cycle.

The scalable TTRAM can support the high-speed data transfer mode of the page/burst mode operation. The read and write data access is done for the ABC sense amplifier in the page/burst mode. The page/burst mode also has the WOV mode and the WV mode. The page/burst WOV mode is the same as the conventional DRAM mode. In contrast, the page/burst WV mode can support both a high-speed data transfer rate and a long data retention time by adding only one write verify operation at the end of the cycle, as shown in Figure 5.23. This mode seems like a write-back operation in the cache memory system. This page/burst WV mode is suitable for the L2/L3 cache memory for MPU, graphic buffer memory for games and display applications with higher data transfer and wider data valid windows by long data retention times. The scalable TTRAM has three main active modes: WV, WOV, and page/burst WV.

Table 5.3 summarizes the application targets of each scalable TTRAM access characteristic programmed by the mode register. The WVO mode for high-speed SRAM replacement, WV mode for low operating power SRAM, and page/burst WV mode for L2/L3 cache and graphics buffer are useful for digital consumer, mobile, and MPU/game applications, respectively.

The scalable TTRAM prepares the additional function of a long data retention (LDR) mode for a standby mode suitable for the battery-used mobile applications. The concept of the LDR mode is shown in Figure 5.24. This figure shows a cross-sectional view of the scalable TTRAM memory cell

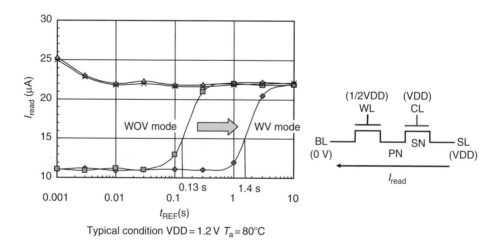

FIGURE 5.21   Data retention characteristics.

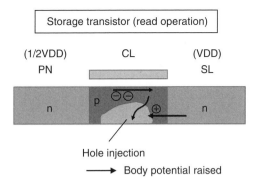

FIGURE 5.22   Data degradation mechanism.

in the normal and long retention mode. The bias level of SL controlling the read current depends on the access time. However, it does not need to keep the bias level of SL at VDD in the long data retention mode. The bias level of SL strongly depends on the data retention characteristics. The electric field between the SL and body region of the storage transistor decides the leakage current, which raises the body potential.

The optimum voltage level to suppress the leakage current is half VDD. This operation suppresses the hole injection to the body region of the storage transistor. The SL bias level is set at the initial state of the data retention mode, and is reset after receiving the sleep-out commands. The recovery time from the sleep-out command is less than 1 µs. By controlling the SL bias level, the data retention time becomes four times longer. Figure 5.25 shows the timing diagram of the LDR mode.

5.3.6   ACTIVELY BODY-BIAS CONTROLLED SRAM

This chapter introduces SRAM technology utilizing the SOI device structure. The conventional 6T SRAM in the advanced process technology (beyond 65 nm CMOS) suffers from the scaling down of the operating voltage because of the degradation of the static noise margin and the fluctuation of V_{th}. The ABC SRAM can relax the V_{cc} minimum operation margin compared to the conventional one. Figure 5.26 shows the schematic of the memory cell. The memory cell is composed of the ABC transistors shown in Figure 5.6. The access and driver transistor bodies are connected to the word line. When the memory cell is activated, the word line turns on and the access and driver transistor bodies

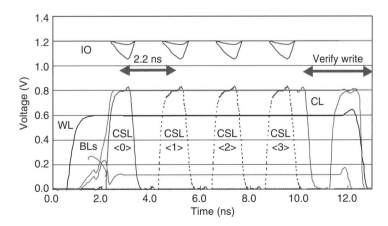

FIGURE 5.23   Simulated waveform of page/burst WV mode.

TABLE 5.3
Application Targets of TTRAM

Mode	Access	Application	Target
WOV mode	Fast random cycle	Digital consumer	High-speed SRAM replacement
WV mode	Random cycle	Mobile	Low operating power SRAM
Burst/page WV mode	Page/burst cycle	MPU, games	L2/L3 cache, graphics

are forward biased. The V_{th} of these transistors becomes low. In the inactivated state, the word line and access and driver transistor bodies become 0V and the V_{th} becomes high. This structure provides high-speed switching and lower voltage operation in the active state, and suppresses the subthreshold leakage current in the inactivated state. The automatic body bias control functions are achieved in the SOI structure. Figure 5.27 shows the SRAM operation voltage in each process technology node. There is no operating voltage difference between the bulk and SOI conventional SRAM. The ABC SRAM can keep the operating voltage scaling down following the process technology improvement and support 35% speed enhancement and 40% power savings compared to the same technology of the conventional one. Additionally, no area overhead of memory cell size is available by the SOI structure.

5.4 FEATURE SOI TECHNOLOGY TREND

The SOI technology has provided high-speed, low-power dissipation, and low-voltage operation compared to bulk Si technology. Initially, the parasitic capacitance reduction in the SOI device was the main merit. However, Si technology was strongly required from the system applications to improve the device characteristics and circuit design, as follows:

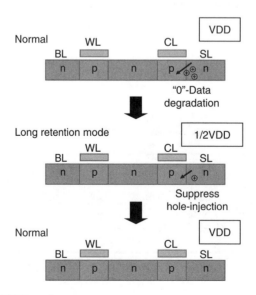

FIGURE 5.24 Concept of LDR mode.

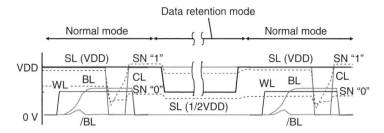

FIGURE 5.25 Timing diagram of LDR mode.

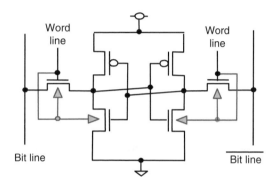

FIGURE 5.26 SOI ABC SRAM.

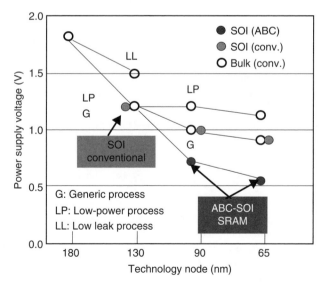

FIGURE 5.27 SRAM operating voltage trend.

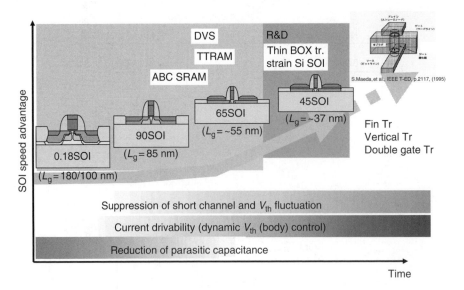

FIGURE 5.28 SOI device road map.

- *Device issues*: Suppression of the leakage current and fluctuation of V_{th}, and higher current drivability
- *Design issues*: Power management technique (power gating, DVS, DVFS)

The SOI technology has the advantage of the above issues; in particular, the dynamic body-control technique combining power gating and DVS/DVFS for logic, ABC-SRAM, and capacitorless high-density TTRAM, as described. These circuit technologies should have the device scaling down scalability to keep the SOI advantage for system application requirements.

The bulk Si technologies announce the new device structure of strain Si and the three-dimensional transistor; for example, a fin-transistor to overcome the physical limitations. These new structures can adapt to SOI. Additionally, the SOI has another parameter of buried oxide (BOX) layer thickness. Figure 5.28 shows the SOI device road map. In the 65 nm era, the strain SOI device is used, and beyond the 45 nm era, the double gate transistor utilizes the thin BOX film [25], fin transistor for high current drivability [26], precise V_{th} control, and low leakage current. New circuit technologies such as the dynamic body-control technique will be supported utilizing these new device technologies. The SOI device will provide the highest performance CMOS and expand the system application fields using SOI SoC.

REFERENCES

1. R. Jejurikar, C. Pereira, and R. Gupta, Leakage aware dynamic voltage scaling for real-time embedded systems, *Proc. Des. Automation Conf.*, 2004, pp. 275–280.
2. T. Ishihara and H. Yasuura, Voltage scheduling problem for dynamically variable voltage processors, *Proc. of Int. Symp. on Low Power Electron. and Des.*, 1998, pp.197–202.
3. M. Y. Lim and V. W. Freeh, Determining the minimum energy consumption using dynamic voltage and frequency scaling, *IEEE Int. Parallel and Distributed Processing Symp.*, 2007, pp. 1–8.
4. S. Mutoh, T. Douseki, Y. Matsuya, T. Aoki, S. Shigematsu, and J. Yamada, 1-V power supply high-speed digital circuit technology with multithreshold-voltage CMOS, *IEEE J. Solid-State Circuits*, 30(8), 1995, 847–854.

5. F. Assaderaghi, A dynamic threshold voltage MOSFET (DTMOS) for ultra-low voltage operation. *IEDM Tech. Dig.*, 1994, 809–812.
6. Y. Omura and K. Izumi, Simplified analysis of body-contact effect for MOSFET/SOI, *IEEE Trans. Electron. Devices*, 35(8), 1988, 1391–1399.
7. M. Patel, P. Ratnam, and C.A.T. Salama, A novel body contact for SIMOX based SOI MOSFETs, *Solid-State Electron.*, 34(10), 1991, 1071–1078.
8. T. Kachi, T. Kaga, S. Wakahara, and D. Hisamoto, Variable threshold-voltage SOI CMOSFETs with implanted back-gate electrodes for power-managed low-power and high-speed sub-1V ULSIs, *Symp. VLSI Tech. Dig.*, 1996, pp. 124–125.
9. F. Morishita, M. Tsukude, and K. Arimoto, Dynamic floating body control SOI CMOS for power managed multimedia ULSIs, *IEEE Custom Integr. Circuits Conf.*, 1997, pp. 263–266.
10. L. Okamura, F. Morishita, K. Dosaka, K. Arimoto, and T. Yoshihara, An automatic source/body level controllable 0.5 V level SOI circuit technique for mobile and wireless network applications, *Proc. ISCIT*, 2006, pp. 771–774.
11. F. Morishita, K. Suma, M. Hirose, T. Tsuruda, Y. Yamaguchi, T. Eimori, T. Oashi, K. Arimoto, Y. Inoue, and T. Nishimura, Leakage mechanism due to floating body and countermeasure on dynamic retention mode of SOI-DRAM, *Symp. VLSI Technol. Dig.*, 1995, pp. 141–142.
12. S. Tomishima, F. Morishita, M. Tsukude, T. Yamagata, and K. Arimoto, A long data retention SOI-DRAM with the body refresh function, *Symp. VLSI Circuit Dig.*, 1996, pp. 198–199.
13. Y. Hirano, T. Ipposhi, H. Dang, T. Matsumoto, T. Iwamatsu, K. Nii, Y. Tsukamoto, T. Yoshizawa, H. Kato, S. Maegawa, K. Arimoto, Y. Inoue, M. Inuishi, and Y. Ohji, Impact of actively body-bias controlled (ABC) SOI SRAM by using direct body contact technology for low-voltage application, *IEDM Tech. Dig.*, December, 2003, 35–38.
14. H. Wann and C. Hu, A capacitorless DRAM cell on SOI substrate, *IEDM Tech. Dig.*, 1993, 635–638.
15. S. Okhonin, M. Nagoga, J.M. Sallese, and P. Fazan, A SOI capacitor-less 1T-DRAM concept, *Int. SOI Conf.*, Oct. 2001, pp. 153–154.
16. P. Fazan, S. Okhonin, M. Nagoga, J.M. Sallese, L. Portmann, R. Ferrant, M. Kayal, M. Pastre, M. Blagojevic, A. Borschberg, and M. Declercq, Capacitor-less 1-transistor DRAM, *Proc. Int. SOI Conf.*, 2002, pp. 10–13.
17. T. Ohsawa, K. Fujita, T. Higashi, Y. Iwata, T. Kajiyama, Y. Asao, and K. Sunouchi, Memory design using a one-transistor gain cell on SOI, *IEEE J. Solid-State Circuits*, 37(11), 2002, 1510–1522.
18. K. Inoh, T. Shino, H. Yamada, H. Nakajima, Y. Minami, T. Yamada, T. Ohsawa, T. Higashi, K. Fujita, T. Ikehashi, T. Kajiyama, Y. Fukuzumi, T. Hamamoto, and H. Ishiuchi, FBC (floating body cell) for embedded DRAM on SOI, *Symp. VLSI Tech. Dig. Tech. Papers*, 2003, pp. 63–64.
19. P. Fazan, S. Okhonin, and M. Nagoga, SOI floating body memories for embedded memory applications, *Proc. Ext. Abs. Solid State Devices Mater. (SSDM)*, 2004, pp. 228–229.
20. E. Yoshida and T. Tanaka, A capacitorless 1T-DRAM technology using gate-induced drain-leakage (GIDL) current for low-power and high-speed embedded memory, *IEICE Trans. Electron.*, 53(4), 2006, pp. 692–697.
21. F. Morishita, H. Noda, I. Hayashi, T. Gyohten, M. Okamoto, T. Ipposhi, S. Maegawa, K. Dosaka, and K. Arimoto, A capacitorless twin-transistor random access memory (TTRAM) on SOI, *IEICE Trans. Electron.*, E90-C(4), 2007, 765–771.
22. F. Morishita, I. Hayashi, T. Gyohten, H. Noda, T. Ipposhi, H. Shimano, K. Dosaka, and K. Arimoto, A configurable enhanced TTRAM macro for system-level power management unified memory, *IEEE J. Solid-State Circuits*, 42(4), 2007, 853–861.
23. J. Chen, F. Assaderaghi, P.-K. Ko, and C. Hu, The enhancement of gate-induced drain leakage (GIDL) current in short-channel SOI MOSFET and its application in measuring lateral bipolar current gain, *IEEE Electron Device Lett.*, 13(11), 1992, pp. 572–574.
24. M. Iida, N. Kuroda, H. Otsuka, M. Hirose, Y. Yamasaki, K. Ohta, K. Shimakawa, T. Nakabayashi, H. Yamauchi, T. Sano, T. Gyohten, M. Maruta, A. Yamazaki, F. Morishita, K. Dosaka, M. Takeuchi, and K. Arimoto, A 322 MHz random-cycle embedded DRAM with high-accuracy sensing and tuning, *J. Solid-State Circuits*, 40(11), 2005, 2296–2304.
25. R. Tsuchiya, M. Horiuchi, S. Kimura, M. Yamaoka, T. Kawahara, S. Maegawa, T. Ipposhi, Y. Ohji, and H. Matsuoka, Silicon on thin BOX: A new paradigm of the CMOSFET for low-power and high-performance application featuring wide-range back-bias control, *IEDM Tech. Dig.*, 2004, 631–634.

26. C.Y. Kang, R.Choi, S.C. Song, K. Choi, B.C. Ju, M.M. Hussain, B.H. Lee, G. Bersuker, C. Yang, D. Heh, P. Kirsch, J. Barnet, J-W. Yang, W. Xiong, H-H Tseng, and R. Jammy, A novel electrode-induced strain engineering for high performance SOI finFET utilizing Si (110) channel for both N and PMOSFETs, *IEDM Tech. Dig.*, 2006, 885–888.

6 SiGe Technology

Marco Racanelli

CONTENTS

6.1 INTRODUCTION

SiGe bipolar complementary metal oxide semiconductor (BiCMOS) has become a dominant technology for the implementation of radio-frequency (RF) circuits by providing performance, power consumption, and noise advantages over standard CMOS while leveraging the same manufacturing infrastructure. Today, most cell phones, wireless LAN devices, global positioning system (GPS) receivers, and digital TV tuners employ SiGe BiCMOS for RF receive and transmit functions because of these advantages. In recent times, nanometer CMOS has achieved performance levels that enable some of these applications to be realized in CMOS for trailing edge products but SiGe continues to provide advantages for the most leading edge products. These markets, as well as emerging applications in the use of SiGe for power amplifiers and millimeter-wave products, continue to drive SiGe technology development.

In this chapter, we will review SiGe BiCMOS technology and its most significant applications. First we will provide a basic understanding of how SiGe devices achieve a performance advantage over traditional bipolar and CMOS devices. Next, we review historical application drivers for SiGe technology and project a roadmap of SiGe performance well into the future. Then, we discuss many of the components built around SiGe devices that are part of modern SiGe BiCMOS technologies and make them useable for advanced RF product design. Finally, we review how SiGe provides an advantage over CMOS and GaAs in key circuit blocks such as low-noise amplifiers (LNAs) and power amplifiers and provide a glimpse of future, higher speed applications with a survey of some recent results in 60 GHz receivers, 40 Gb/s circuits, and high-frequency radar.

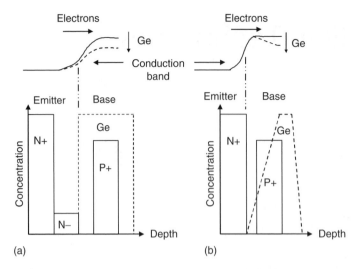

FIGURE 6.1 Common SiGe doping and germanium profiles shown along with resulting band diagrams for (a) a box Ge profile and (b) a graded Ge profile.

6.2 SiGe DEVICES: THE BASICS

SiGe devices are silicon bipolar devices created by using a thin epitaxial base with 8%–30% atomic germanium content fabricated alongside CMOS devices with the addition of four to seven mask layers to a core CMOS process. SiGe devices derive part of their performance benefits from heterojunction effects and part from their epitaxial base architecture. Heterojunction effects were first described in the 1950s by Kroemer and more recently summarized by the same author in Ref. [1] and arise from the combination of different materials (in this case SiGe alloy and Si) to create bandgap changes in the device that can be manipulated to improve performance. Two common techniques for using the heterojunction effect to improve performance are depicted in Figure 6.1 in which typical doping and germanium profiles are shown along with their resulting band diagrams. The first technique (Figure 6.1a) uses a box Ge profile to create an offset in bandgap at the emitter–base junction, lowering the barrier for electron current flow into the base, increasing the collector current, and resulting in higher current gain. The high current gain can then be traded off for increased base doping or lower emitter doping to improve base resistance and emitter–base capacitance, resulting in higher speed. The second technique (Figure 6.1b) uses a graded Ge profile to create an electric field in the base that accelerates electrons, reducing base transit time and improving high-frequency performance. Today's SiGe bipolar devices make use of these two techniques to varying degrees to create a performance advantage over conventional bipolar devices.

Use of an epitaxially grown base rather than the one formed by ion implantation is another reason SiGe devices have better performance than conventional bipolar devices. The base of a conventional bipolar device is formed by implanting base dopant into silicon, which results in a relatively broad base. Epitaxy allows one to grow the base through deposition of doped and undoped Si and SiGe layers controlled to nearly atomic dimensions. This allows the device designer to create an arbitrary base profile. Most typically, this technique is used to distribute the same base dose in a narrower base width, improving transit time through the base, and resulting in better high-frequency performance.

In more recent times, another material is being added in the epitaxial base of SiGe devices: carbon. A small amount of carbon is placed in the SiGe base during epitaxial deposition such that electrical behavior is not changed but the material properties are altered to reduce boron diffusion (typically <1% atomic concentration of carbon is used). Boron is typically used as the base dopant species and without carbon can diffuse during high-temperature processing steps required to complete wafer processing and widen the final width of the base. The addition of carbon reduces boron diffusion, maintaining the as-deposited profile intact through the completion of wafer processing. The electrical effect is a faster transit time due to a narrower base width improving high-frequency performance.

6.3 APPLICATIONS DRIVING SiGe DEVELOPMENT

Several applications have driven advances in SiGe technology since the first high-speed SiGe bipolar devices were demonstrated in the late 1980s [2]. Initially, SiGe devices were conceived as a replacement to the Si bipolar device for emitter-coupled logic (ECL), high-speed digital integrated circuits (ICs) where SiGe transistors promised higher f_T, improving gate delay relative to their Si bipolar or CMOS counterparts. CMOS advancements in density, performance, and power consumption, however, quickly made it the logical choice for all but a very few of these applications and so in the mid-1990s, SiGe technology appeared to have a limited application base in only specialized very high-speed digital functions.

With the boom in wireless communications that began in the mid-1990s however, a new application emerged as the primary driver for SiGe technology: the transceiver of a cellular phone. This application is tailor-made for SiGe BiCMOS as it requires good high-frequency performance to support carrier frequencies in the 900 MHz to 2.4 GHz range, together with very low-noise operation as very small signals must be received and amplified, along with large dynamic range as large output signals are required to drive the power amplifier and the antenna communicating with a far away base station. In addition to wireless transceivers, high-speed fiber optic transceivers also provided a good application for SiGe transistors as these pushed to even higher speeds moving from 3 to 10 Gb/s and targeting 40 Gb/s data rates. The transition from 3 to 10 Gb/s has, in fact, provided a strong market for SiGe devices as many of the same characteristics required for wireless transceivers are important in these transceivers (high speed, low noise, large dynamic range) but the transition from 10 to 40 Gb/s was delayed with the busting of the dot-com bubble and is just now beginning to be discussed again. Despite 40 Gb/s not becoming a reality, it was this expected transition in the late 1990s and early 2000s that pushed researchers to invest in creating very high-speed SiGe transistors (with f_T and f_{MAX} of 200 GHz and above) that are now poised to take advantage of perhaps other emerging applications.

Today, deep-submicron CMOS is challenging SiGe for some of these applications for two reasons: the speed of CMOS is adequate for many applications (although SiGe maintains an advantage in noise and an even wider advantage in dynamic range) and the density of CMOS is now high enough to enable new architectures that rely more heavily on digital signal processing rather than high-fidelity analog manipulation. In many cases, however, SiGe technology still offers a cost, performance, and power advantage and will continue to play a strong role in both the wireless and wire-line transceiver market. At the same time, today's SiGe transistor performance is adequate to serve these applications and therefore these markets are becoming less important as drivers for future technology advancements.

Looking forward, two applications are primarily driving SiGe performance advancements today: higher frequency mmWave and higher power but lower frequency products. mmWave applications include, for example, a proposed 60 GHz wireless local area network (WLAN) standard, 77 GHz automotive collision avoidance systems, and 40 to 100 Gb/s optical networking communications and will drive the speed of the SiGe transistor to higher and higher levels. At the same time, high-power applications include, for example, the power amplifier for wireless devices and laser driver for wire-line transceivers and will drive improved trade-off between speed and breakdown voltage in SiGe transistors. In the next section, we will review in more detail the design of SiGe transistors and see how improved speed and improved high-power performance are being realized.

6.4 SiGe PERFORMANCE ROADMAP

Two figures of merit are typically used to benchmark high-frequency device performance: the cutoff frequency (f_T), which for a bipolar device, is defined as the frequency at which the current gain is unity, and the maximum frequency of oscillation (f_{MAX}), which for a bipolar device, is defined as the frequency at which the power gain is unity (specifically, the unilateral power gain).

For a silicon bipolar device, f_T and f_{MAX} are related to basic device parameters by the commonly used equations:

$$f_T = \frac{1}{2\pi \tau_F} \tag{6.1}$$

$$\tau_F = (C_{be} + C_{bc})\left(R_e + \frac{kT}{qI_c}\right) + \frac{W_b^2}{2D_b} + \frac{W_c}{2v_s} + R_c C_{bc} \tag{6.2}$$

$$f_{MAX} = \sqrt{\frac{f_T}{8\pi R_b C_{bc}}} \tag{6.3}$$

where

τ_F is forward transit time
C_{be} is emitter–base capacitance
C_{bc} is base–collector capacitance
R_e is emitter series resistance
I_c is the collector current
W_b is vertical base width
D_b is electron diffusion length in the base
v_s is electron saturation velocity
W_c is collector–base vertical depletion width
R_c is collector resistance

At high current, W_b becomes a function if I_c as the charge from electrons transiting through the base–collector depletion region adds to the background collector charge and "pushes" the depletion region away from the base, effectively widening the base. This "base-push-out" effect is responsible for f_T to decrease at high current rather than continue to increase as would otherwise be predicted by Equations 6.1 and 6.2 and so both f_T and f_{MAX} peak at a specific current density.

Figure 6.2 shows peak f_T of bipolar devices compared against CMOS devices at each major technology node. From Figure 6.2, the Si bipolar device was expected to run out of steam versus

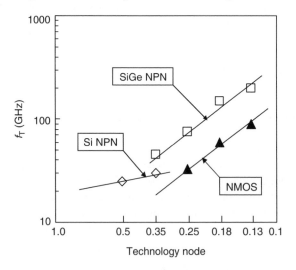

FIGURE 6.2 SiGe NPN, Si NPN, and NMOS f_T plotted as a function of technology node. All data are plotted for Jazz Semiconductor foundry technology but is representative of industry trends. (From Racanelli, M. and Kempf, P., *IEEE Trans. Electr. Dev.*, 52, 1259, 2005. With permission.)

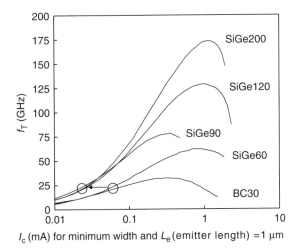

FIGURE 6.3 Peak f_T of Jazz BiCMOS technology plotted as a function of I_c for a minimum width and unit length NPN. BC30 is a 0.35 μm Si BiCMOS process, SiGe60 is a 0.35 μm SiGe BiCMOS process, SiGe90 and SiGe120 are 0.18 μm SiGe BiCMOS processes, and SiGe200 is a SiGe BiCMOS process implemented both in 0.18 and 0.13 μm nodes. In addition to higher peak f_T, subsequent technology node lowers power consumption even when biasing at low f_T as indicated by the arrow. (From Racanelli, M. and Kempf, P., *IEEE Trans. Electr. Dev.*, 52, 1259, 2005. With permission.)

N-channel metal oxide semiconductor (NMOS) performance near the 0.35 μm generation, prompting many to begin investigating the migration of RF functions to RF CMOS. The introduction of SiGe, however, placed the bipolar device on a performance learning curve higher than that of NMOS devices, enabling a continued roadmap of RF circuits implemented in BiCMOS technology.

Higher peak f_T and f_{MAX} are important because, while today's volume RF applications target modest operating frequencies relative to the peak f_T, shown in Figure 6.2, high peak f_T (and f_{MAX}) can be traded off for other benefits including reduced power consumption, higher breakdown voltage, and reduced noise. Figure 6.3 shows an example of the power savings that can be achieved with higher f_T SiGe technology even when operating at relatively low frequencies. In this example, current is reduced by about a factor of 4 when biasing at f_T of 25 GHz in a higher f_T 0.18 μm SiGe device relative to a lower f_T 0.35 μm SiGe device. Similarly, minimum noise figure has been expressed by [3]

$$\text{NF}_{\min} = 1 + \frac{n}{\beta} + \frac{f}{F_t} \cdot \sqrt{\frac{2qI_c}{kT} \cdot (R_e + R_b) \cdot \left(1 + \frac{F_t^2}{\beta f^2}\right) + \frac{n^2 F_t^2}{\beta f^2}} \qquad (6.4)$$

where
  n is the collector current quality factor
  β is the current gain

From Equation 6.4, it is seen that with a high β as typically seen in SiGe devices, the term $f/F_t\sqrt{(2qI_c/kT)\,(R_e + R_b)}$ is important and a higher f_T and lower R_b (or higher f_{MAX} from Equation 6.3) can result in lower noise figure. Finally, f_T can be traded off for higher breakdown voltage by modulating the collector-doping concentration through a collector implant mask such that multiple devices spanning a range of f_T and breakdown are made available on the same wafer. Figure 6.4 shows the family of devices realized by this technique across four generations of Jazz technology. Each subsequent generation supports devices with higher f_T but also improves the trade-off between f_T and breakdown voltage, improving large signal performance for applications such as integrated drivers and power amplifiers. This is in contrast with CMOS in which each new generation makes the integration of power devices more difficult due to the more brittle gate oxide, forcing lower voltage ratings.

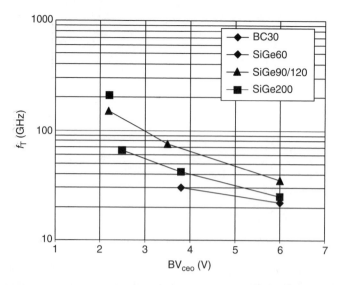

FIGURE 6.4 f_T versus BV_{ceo} plotted for all devices available in the four generations of BiCMOS technology shown in Figure 6.3 (Figure 6.3 plots performance for only the highest speed device in each technology generation). Multiple devices that trade off f_T for higher breakdown voltages are created in most generations by optimizing the collector doping, and the trade-off between speed and breakdown is improved with each subsequent generation. (From Racanelli, M. and Kempf, P., *IEEE Trans. Electr. Dev.*, 52, 1259, 2005. With permission.)

Having established the importance of f_T and f_{MAX} as figures of merit for RF and analog circuits, how then can we continue to improve f_T and f_{MAX} in bipolar devices to continue the trajectory shown in Figure 6.3? The answer, much as it is with CMOS, is that performance can be improved by improving lithography to shrink lateral dimensions of the device. To first order with CMOS devices, a lateral shrink in gate length and subsequent vertical shrink in gate oxide and junction depth result in improved performance. Similarly, to first order with bipolar devices, a lateral shrink in emitter width and subsequent vertical shrink in base width result in improved performance. Without shrinking the emitter laterally, a shrink in base width results in higher base resistance and so f_T can be improved but typically at the expense of f_{MAX}. Smaller emitter width reduces base resistance and thus enables a higher f_T and a higher f_{MAX}. Similarly, increased collector-doping concentration could be used to increase f_T (i.e., trade-off for lower breakdown voltage as shown in Figure 6.4) but higher collector capacitance would degrade f_{MAX}. By reducing lateral dimensions through improved lithography, collector capacitance can also be reduced improving both f_T and f_{MAX}.

Figure 6.5 shows the 2007 ITRS Roadmap for SiGe bipolar devices [4]. The roadmap predicts that the same lithography advancements responsible for the CMOS roadmap will enable improved SiGe performance for the foreseeable future.

To realize useful RF and analog circuits, however, more than just high-speed SiGe devices are necessary. In the next section, we will discuss modules integrated with SiGe transistors that help create a more complete modern platform for RF and analog IC design.

6.5 MODERN SiGe BiCMOS RF PLATFORM

Technology features integrated with SiGe transistors that make them useful for product design include active elements such as high-density CMOS, high-voltage CMOS, high-performance PNPs as well as passive elements such as high-density metal-insulator-metal (MIM) capacitors and high-quality inductors.

Today, most SiGe development is done in the context of a BiCMOS process in a CMOS node that typically trails the most advanced digital node by one to two generations. The critical hurdle to

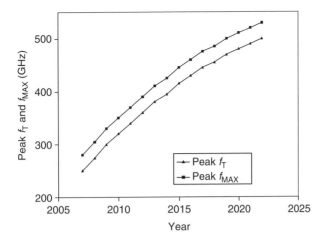

FIGURE 6.5 2007 SIA ITRS Roadmap for peak f_T and f_{MAX} of SiGe bipolar devices plotted through 2022. (From *International Technology Roadmap for Semiconductors*, International SEMATECH, Austin, TX, 2007. With permission.)

integrating advanced CMOS and SiGe devices is to marry their respective thermal budget without degrading either of the devices. The addition of carbon to SiGe layers has been used as a partial solution to this problem as it helps reduce boron diffusion [5], allowing for higher thermal budget after SiGe deposition. This, along with careful optimization of the integration scheme, has resulted in demonstrations of SiGe integration down to the 90 nm node [6].

Power management of the PA (Power Amplifier) or even of the cell phone can be enabled with higher voltage CMOS devices (typically requiring tolerance of 5 to 8 V). In smaller geometries that support only lower core voltage levels, these are enabled by introducing drain extensions to the CMOS devices that can enable higher drain bias than supported in the native transistor. An example of such devices is shown in Figure 6.6 and these are becoming common modules in SiGe technology offerings, often not costing additional masking layers to create.

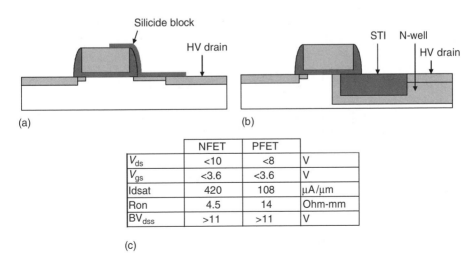

	NFET	PFET	
V_{ds}	<10	<8	V
V_{gs}	<3.6	<3.6	V
Idsat	420	108	µA/µm
Ron	4.5	14	Ohm-mm
BV_{dss}	>11	>11	V

(c)

FIGURE 6.6 Sketch of two types of commonly used extended drain devices: (a) silicide block extension, (b) STI extension, and (c) a table showing characteristics of high-voltage devices available in a 0.18 µm SiGe BiCMOS technology using approach (b). Idsat is quoted for 3.3 V V_{gs} and 5 V V_{ds}. (From Racanelli, M. and Kempf, P., *IEEE Trans. Electr. Dev.*, 52, 1259, 2005. With permission.)

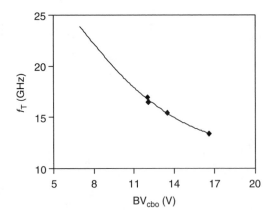

FIGURE 6.7 Performance (f_T) and breakdown (BV_{cbo}) trade-off of vertical silicon PNP integrated with SiGe NPNs to form a complementary pair.

A high-speed vertical PNP (VPNP) can form a complementary pair with the SiGe NPN and is important for certain high-speed analog applications such as fast data converters and push-pull amplifiers and drivers used, for example, in hard disk drive pre-amps. A VPNP can be made very fast by the use of a separate SiGe deposition step and f_T of 100 GHz has been reported [7], but the cost associated with such a VPNP is prohibitive for most applications today. A more popular approach re-uses many of the steps needed to create the NPN and CMOS devices while adding specialized implants to optimize the performance of the VPNP and achieve speeds of 15 to 25 GHz depending on the desired breakdown voltage as shown in Figure 6.7.

In addition to active components, high-quality passive components are necessary to enable advanced RF circuits. The most critical passive elements for RF design are capacitors and inductors as these can consume significant die area and at times limit performance of RF and analog circuits. MIM capacitors are available in most commercial SiGe BiCMOS and RF CMOS processes as they achieve excellent linearity and matching. The density of MIM capacitors has been steadily increasing, helping to shrink RF and analog die. Figure 6.8 shows a timeline of capacitance density for Jazz-integrated MIM capacitors. An initial improvement in density from <1 to 2fF/μm^2 was enabled by a move from oxide to nitride dielectrics [8]. More recently, a move from 2 to 4fF/μm^2 has been enabled by the stacking of a 2fF/μm^2 capacitor on two metal layers and finally a further

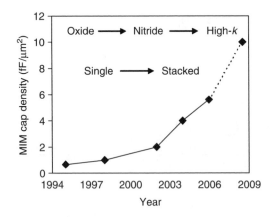

FIGURE 6.8 MIM capacitor density plotted as a function of year of first production (actual or planned) showing progression in dielectric technology (from oxide to nitride to high-k) and in integration (single to stacked capacitors).

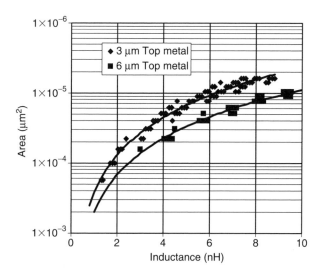

FIGURE 6.9 Inductor area as a function of inductance for a four-turn inductor with peak Q of 10 built in 3 and 6 μm top metal.

optimization of the dielectric has resulted in density of $5.6fF/\mu m^2$. Today, high-k dielectrics are being investigated to enable even higher densities and it is conceivable that in the next few years, densities of 10 to $20\,fF/\mu m^2$ will be achieved.

Integrated inductor performance, measured as the quality factor (Q), is improved by the reduction of metal resistance made possible by thicker metal layers. Inductor Q can be traded off for reduced footprint such that a thicker metal layer can also help reduce chip area. This concept is demonstrated in Figure 6.9 where the area required to realize an inductor with Q of 10 is compared between use of a 6 and 3 μm top metal in a four-layer metal process. A 6 μm metal inductor consumes half the die area of a 3 μm metal inductor while achieving the same Q in this example.

Die scaling enabled by the advanced passive elements described in this section can often more than pay for the additional processing cost such that an optimized process can in many cases not only provide better performance than a digital CMOS process but also lower die cost. Similarly, the integration of advanced active modules described in this section can help integrate more analog functionality on fewer die reducing overall system level costs. In the next section, we will review application examples that take advantage of SiGe performance as well as many of the integrated features described so far.

6.6 KEY APPLICATION TRENDS AND EXAMPLES

In this section, we will review some key circuit blocks realized with SiGe technology and discuss the advantages SiGe brings relative to CMOS or GaAs implementations. First we begin with a review of key wireless transceiver circuits: the LNA and the power amplifier. Next we scan recent results that extend these concepts to higher speed applications such as 60 GHz WLAN, 40 Gb/s networks, and high-frequency radar.

6.6.1 KEY WIRELESS RF CIRCUIT BLOCKS: LOW-NOISE AMPLIFIER

LNA design involves the trade-off of noise figure, linearity, power consumption, and gain. High gain with low-noise figure enables high sensitivity while high linearity allows wider fluctuation in input signal strength without introducing distortion and low power consumption helps extend battery life. SiGe transistors offer three significant advantages over CMOS for LNA design: lower inherent noise

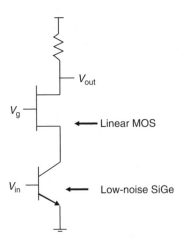

FIGURE 6.10 LNA topology utilizing a low-noise SiGe input transistor cascoded with a linear MOS transistor used to achieve 16 dB of gain at 2 GHz with a noise figure of 1.67 dB, an input IP3 of -6.5 dB m while consuming only 3 mW of power. (From Ma, P., Racanelli, M., Zheng, J., and Knight, M., *RFIC Symp. Dig.*, pp. 237–240, June 2003.)

figure at bias conditions and input match that also maximize gain (thus either higher gain at a given noise figure or lower noise figure at a given gain relative to CMOS), larger dynamic range (which can be used to achieve better linearity), and an input impedance that can be more easily matched (this often translates into a die size advantage as CMOS LNAs often require on-chip inductors that consume more die area). Given these intrinsic advantages, SiGe LNAs today offer generally better performance over LNAs realized in CMOS. Looking to future generations of CMOS, the lower supply voltages will limit their ability to handle large unwanted signals such that scaling beyond a 0.25 or 0.18 μm generation is not likely to improve the CMOS LNA design trade-off significantly without further circuit design innovation.

With SiGe BiCMOS, it is also possible to combine the benefits of CMOS and SiGe in the same circuit [9]. Figure 6.10 shows one such LNA circuit where a bipolar input stage is used to achieve low-noise figure and a MOS output stage is used to achieve higher linearity in a cascode configuration, which minimizes power consumption. The results of this LNA realized in a 0.18 μm SiGe BiCMOS process rival those of commercial, discrete, III–V LNAs and are far superior to the best reported performance for RF CMOS implementations [10].

6.6.2 KEY WIRELESS RF CIRCUIT BLOCKS: POWER AMPLIFIER

Power amplifiers are used to drive the antenna of wireless devices and can be subdivided into three categories: low power (<500 mW) for applications such as WLAN, personal handphone systems (PHS), and cordless telephones; high power (>1 W) for cellular handset applications; and very high power for base stations. Base-station PAs are typically discrete devices normally built in large geometry RF lateral double-diffused MOSFET (LDMOS) or III–V technology. SiGe devices can be competitive for these applications but do not offer a significant advantage relative to the existing technology and will therefore not be discussed further in this chapter.

One of the major constraints in power amplifier design is the fact that the antenna is exposed to an unpredictable environment and can present large swings in output impedance potentially reflecting large amounts of RF power back into the device. The reflected RF power can raise the voltage of the output node significantly above the supply voltage. For this reason, a high breakdown voltage (typically three to four times higher than the supply voltage) is required to pass what is typically referred to as "ruggedness" tests for the power amplifier where it is subjected to large impedance mismatch at full RF power. SiGe transistors offer a significant advantage relative to CMOS transistors in performance at a given breakdown voltage. Figure 6.11 shows this comparison at the 0.18 μm node where drain extensions are used to extend the breakdown voltage of 0.18 μm CMOS and collector doping is used to vary the breakdown voltage of a SiGe transistor as described in prior Sections 6.4 and 6.5. SiGe transistors have a higher f_T (and f_{MAX}) for the same breakdown voltage or higher breakdown voltage for the same f_T (or f_{MAX}). Figure 6.4 shows how scaling to future generations is only expected to widen this gap as SiGe speed and breakdown trade-off will improve while CMOS breakdown will continue to worsen. Recent reports have in fact concluded that the optimum node for power amplifier design in CMOS is the 0.25 μm node [11] and future generations of CMOS will be less competitive without significant circuit innovation.

Table 6.1 compares performance of power amplifier devices in current state-of-the-art production GaAs heterojunction bipolar transistor (HBT), SiGe, and CMOS technologies. As indicated above, CMOS suffers from low breakdown voltages relative to the alternative technologies. SiGe also has a lower breakdown voltage at same performance levels relative to GaAs, but with a breakdown voltage of 18 V is adequate for most power amplifier applications and therefore it is not at a significant disadvantage. From a material standpoint, GaAs has an advantage in being a semi-insulating substrate, which minimizes substrate losses (important, for example, for inductor Q and isolation) while the silicon substrate has better thermal conductivity important in reducing self-heating, which can induce performance or reliability constraints on the design. The primary advantage of SiGe relative to GaAs is cost and ability to achieve high levels of integration. In the next few paragraphs, we discuss how these trade-offs are being played out in both low-power and high-power PA designs.

Increasingly, low-power PAs such as those used in Bluetooth, WLAN, cordless telephones, and PHS are being built in SiGe or CMOS because breakdown constraints are not as severe as those of high-power PAs and because these can more easily be integrated with transceiver functions, which are always built in either SiGe or CMOS, offering a significant cost advantage to stand-alone GaAs solutions. Because of the advantage in speed and breakdown of SiGe over CMOS, SiGe PAs typically offer higher power levels at lower overall power consumption than CMOS PAs and so CMOS tends

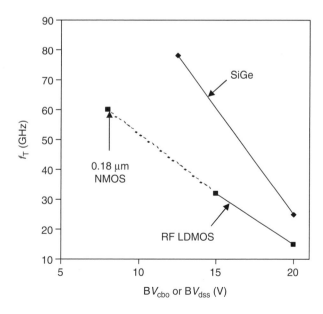

FIGURE 6.11   Performance and breakdown voltage trade-off of 0.18 μm SiGe devices compared with that of 0.18 μm CMOS devices. Voltage in CMOS devices is increased above the 1.8 V core voltage by extended drain techniques shown in Figure 6.6.

TABLE 6.1

State-of-the-Art Performance of Production III–V (GaAs-based) HBT, SiGe Bipolar, and CMOS Transistors Designed for Power Amplifier Applications from the 2007 SIA ITRS Roadmap

	III–V HBT	SiGe	CMOS
f_{MAX} (at V_{cc}) (GHz)	45	60	45
BV_{CBO} (V)	25	18	12
Linear efficiency (%)[a]	52	50	45
Area (mm²)[b]	2.5	2.5	6

[a] Linear efficiency, power-added efficiency of the final PA stage under personal communication service (PCS) CDMA (IS-95) modulation.

[b] Area, total semiconductor area necessary for the implementation of the quad-band GSM/general packet radio service (GPRS)/enhanced data rates for GSM evolution (EDGE) PA function, including matching/filtering.

to be used for the shorter reach standards (i.e., Bluetooth) while SiGe for longer reach standards (i.e., WLAN, cordless telephones, and PHS).

Figure 6.12 and Tables 6.2 and 6.3 show examples of low-power PAs built in SiGe technologies. Figure 6.12 shows a schematic of a PHS PA realized with two gain stages and a power amplifier output stage built in a 0.35-μm SiGe technology. The process offers three NPN transistors with increasing breakdown voltages and multiple emitter widths (from 0.3 to 0.9 μm). The design makes use of narrow emitter devices in the initial gain block to improve base resistance and small signal performance and wide emitter devices in the final stage to optimize large signal behavior and power handling capability. In addition, devices with increasing breakdown voltage are used in the initial and final stages to optimize the gain/breakdown trade-off of each stage. Measured results are show in Table 6.2. PAE (Power Added Efficiency) of 48% is achieved with 25 dBm of maximum output power and PAE of ∼35% is achieved with 23 dBm of linear power, which is competitive with performance achieved in typical III–V implementations. Such a PA could be readily reconfigured to operate at 2.4 GHz and meet requirements for 802.11b,g and Bluetooth. Table 6.3 compares performance of SiGe WLAN power cells operating at 802.11b (11 Mbps) and 802.11g (54 Mbps) realized in 0.35 and 0.18 μm SiGe technology from the same foundry source. The 0.35 μm power amplifiers are today integrated with the 802.11b/g transceiver in many common products [12] and meet all requirements of these standards. The 0.18 μm results are experimental at this time but demonstrate the ability to

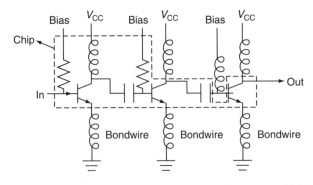

FIGURE 6.12   Schematic of a three-stage PHS power amplifier. (From Racanelli, M. and Kempf, P., *IEEE Trans. Electr. Dev.*, 52, 1259, 2005.)

TABLE 6.2

Room Temperature Measured Results for the 1.9 GHz PHS PA at V_{cc} = 3 V Shown in Figure 6.12 and Built in 0.35 μm SiGe BiCMOS

Parameter	Symbol	Typical	Units
Small signal power gain	$|S21|$	34.5	dB
Linear power gain	P_g	34.0	dB
Output power	P_{out}	23.0	dB m
Linear efficiency	PAE	34.5	%
Alternative ACPR	ACPR	−50	dB c
Input return loss	$|S11|$	−11.9	dB
Output return loss	$|S22|$	−16.4	dB
Idle current	I(idle)	210	mA

Source: From Racanelli, M. and Kempf, P., *IEEE Trans. Electr. Dev.*, 52, 1259, 2005.

increase power levels and improve performance, further broadening the appeal of SiGe for these low power standards in the next generation products.

High-power PAs targeting cellular applications are dominated today by GaAs and discrete RF LDMOS implementations. SiGe PAs are emerging and are likely to take some share over coming years due to their cost advantage over GaAs and discrete RF LDMOS. SiGe can be cost-effective, because it enables greater levels of integration relative to discrete RF LDMOS and GaAs PAs, which typically cannot integrate sophisticated power control functions important in the intelligent biasing of the PA and thus require a companion CMOS chip. Table 6.4 summarizes results from 0.35 μm SiGe power amplifier cells modulated in global system for mobile communication (GSM) and code division multiple access/personal communication system (CDMA/PCS) bands. The CDMA/PCS standard requires higher linearity from the power amplifier in addition to operating at higher frequency and thus is a more challenging standard relative to the GSM band. When compared to GaAs performance, in general the 0.35 μm SiGe performance shown in Table 6.4 is comparable to that of GaAs in the GSM band but worse than GaAs in the CDMA/PCS band. As demonstrated in Table 6.3, it is expected that more advanced SiGe technology (e.g., 0.18 μm SiGe) will close this gap and compete

TABLE 6.3

Power Amplifier Cell Performance Compared between 0.35 and 0.18 μm Jazz SiGe Technology When Modulated to Comply with 802.11b (11 Mb/s) and 802.11 g (54 Mb/s) WLAN Standards

		0.35 μm SiGe	0.18 μm SiGe
802.11b	Output power (dB)	20	23
	EVM (%)	5.2	2
	Current consumption (mA)	190	160
802.11g	Output power (dB)	16	19
	EVM (%)	5.3	3.7
	Current consumption (mA)	178	210

TABLE 6.4

Performance for 0.35 μm SiGe Power Amplifier Cells Measured at GSM and CDMA/PCS Bands

Modulation and Frequency Band	V_{cc} (V)	P_{out} (dBm)	PAE (%)	ACPR (dBc)
GSM (897.5 MHz)	3	35.2	62.8	NA
CDMA/PCS (1.88 GHz)	2.5	28.2	29.1	−46, −47

Source: From Racanelli, M. and Kempf, P., *IEEE Trans. Electr. Dev.*, 52, 1259, 2005.

more effectively with GaAs across all bands and cellular standards. This does not mean an immediate replacement of GaAs PAs by SiGe PAs since die cost differential is only one component of the overall cost of a PA module and GaAs PAs are the established, proven solution, but SiGe PAs are sure to play a significant future role in this market in the future.

6.6.3 EMERGING APPLICATIONS

The same attributes that make SiGe attractive for lower frequency applications (namely low noise, high speed, large voltage handling capability) become even more important for high-frequency applications in which CMOS either does not have enough speed or support high enough voltage to be attractive. By the same token, InP and other III–V compounds maintain a performance advantage at high frequency relative to SiGe. The newest SiGe technologies with >200 GHz f_T, however, close this gap and, given the economic and integration benefits of working with silicon, are serious contenders for these emerging applications.

Potential applications include emerging WLAN, automotive radar, and collision avoidance products in the 10 to 100 GHz range as well as wire-line communications at 40 Gb/s and beyond. In this section, we provide three examples of millimeter-wave circuits to give a flavor of the performance levels achievable with SiGe. All of these were realized in a production, 150 GHz f_T, 200 GHz f_{MAX} 0.18 μm SiGe foundry process [13].

The first example circuit is a 60 GHz transceiver with on-chip antenna intended for a newly proposed high–data rate WLAN standard. Figure 6.13 shows a die photograph and Ref. [14] provides

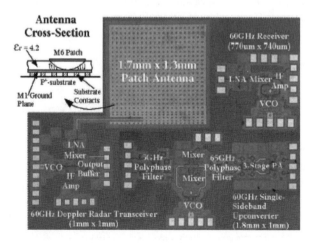

FIGURE 6.13 Die photograph shown for 60 GHz transceiver building blocks with patch antenna built in 0.18 μm SiGe technology as reported. (From Voinigescu, S.P., Chalvatzis, T., Yau, K.H.K., Hazneci, A., Garg, A., Sharahmian, S., Yao, T., Gordon, M., Dickson, T.O., Laskin, E., Nicolson, S.T., Carusone, A.C., Tchoketch, L., Yuryevich, O., Ng, G., Lai, B., and Liu, P., *BCTM Proc.*, Sept. 2006. With permission.)

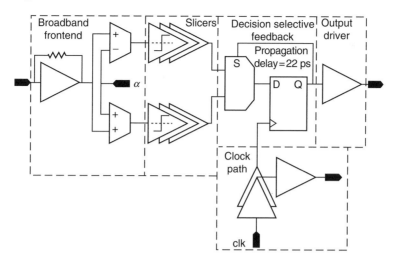

FIGURE 6.14 Schematic diagram of 40 Gb/s equalizer architecture as reported. (From Voinigescu, S.P., Chalvatzis, T., Yau, K.H.K., Hazneci, A., Garg, A., Sharahmian, S., Yao, T., Gordon, M., Dickson, T.O., Laskin, E., Nicolson, S.T., Carusone, A.C., Tchoketch, L., Yuryevich, O., Ng, G., Lai, B., and Liu, P., *BCTM Proc.*, Sept. 2006. With permission.)

details of the implementation. The transceiver achieves very respectable conversion gain of 20 dB with noise figure of 12 dB and more than 45 dB of image rejection in the transmitter at a supply voltage of 3.3 V. WLAN at 60 GHz promises to bring gigabit data rates to wireless communications, but is hampered by high system costs. SiGe can help reduce at least part of this cost by enabling a cost-effective high-frequency radio solution.

The second example circuit is a 40 Gb/s equalizer intended for next-generation fiber optic communications. Figure 6.14 shows a schematic diagram of the equalizer and Figure 6.15 shows the results, while Ref. [14] provides details of the implementation. The circuit is able to equalize correctly a pseudorandom bit sequence (PRBS) signal passing through a 9-ft Sub-Miniature version A (SMA) cable from 5 to 40 Gb/s. While today's fiber optic networks are designed around a 10 Gb/s standard, the 40 Gb/s standard promises to quadruple the capacity without requiring additional fiber deployment, making it a cost-effective option to significantly increase bandwidth and meet future requirements.

The third and final example is a phased array, beam-steering receiver operating up to 18 GHz. Figure 6.16 is a die photo and Table 6.5 shows performance of the chip, while Refs. [15,16] give details of the implementation. Beam steering is an effective way of improving the efficiency of the antenna and SiGe is making beam steering at these high frequencies cost-effective as the beam can be shaped purely electronically rather than mechanically as previously done. Applications for these techniques are broad and include both military as well as commercial wireless communications and have recently been extended to higher frequency using the same 0.18 μm SiGe technology with demonstrations in the 40 to 50 GHz Q-band [15,16].

With processes today being announced in the 200 to 300 GHz range versus the 150 to 200 GHz process used in the examples above, it is clear that even better performance than presented here is likely to be reported in the near future for SiGe millimeter-wave circuits. While deep-submicron CMOS (65 nm and beyond) is also a contender for some of these applications, many of the millimeter-wave circuits cannot scale aggressively due to transmission lines and other analog elements occupying significant die space as can be seen by some of the die photos reviewed here. This will make it likely that SiGe will continue to be a cost-effective technology for the implementation of many circuits in emerging millimeter-wave markets.

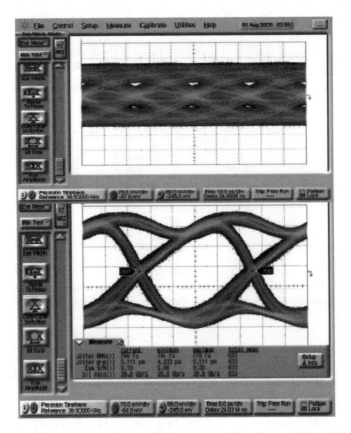

FIGURE 6.15 40 Gb/s input and output eye diagrams of the equalizer shown in Figure 6.14 when realized in a 0.18 μm SiGe technology after passing through a 9 ft long SMA cable, demonstrating good signal recovery. (From Voinigescu, S.P., Chalvatzis, T., Yau, K.H.K., Hazneci, A., Garg, A., Sharahmian, S., Yao, T., Gordon, M., Dickson, T.O., Laskin, E., Nicolson, S.T., Carusone, A.C., Tchoketch, L., Yuryevich, O., Ng, G., Lai, B., and Liu, P., *BCTM Proc.*, Sept. 2006. With permission.)

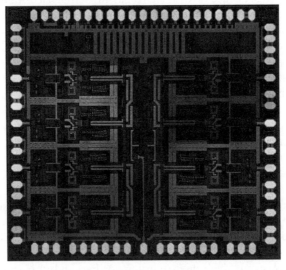

FIGURE 6.16 Die photograph of an eight-element 6 to 18 GHz phased array receiver as described. (From Koh, K. and Rebeiz, G., *Microwave J.*, 270, May 2007 and Koh, K. and Rebeiz, G.M., *IEEE Custom Integr. Circuits Conf.*, San Jose, CA, pp. 761–764, Sept. 2007. With permission.)

TABLE 6.5

Performance of Phased Array Receiver Depicted in Figure 6.16 and Realized in a 0.18 μm SiGe Technology

	0.18 SiGe	
Frequency (GHz)	6 to 18	
Supply voltage (V)	3.3	
Current consumption (mA)	100 to 200	Function of power gain
Power gain at 12 GHz (dB)	2 to 24	At 12 GHz, variable gain
Phase resolution (bit)	4	6 to 18 GHz
Phase error (RMS deg)	<6	6 to 18 GHz
Noise figure (dB)	3.9	At 12 GHz with 19 dB of gain
Size (mm)	2.2 × 2.4	

Source: From Koh, K. and Rebeiz, G., *Microwave J.*, 270, May 2007. With permission.

6.7 CONCLUSIONS

In this chapter, we have reviewed SiGe BiCMOS technology and discussed how it has become important for many RF applications by providing a performance advantage over CMOS while sharing its manufacturing infrastructure to provide integration and cost advantages over III–V technology. In addition to higher speed, we have seen that an intrinsic advantage of SiGe over CMOS is its ability to maintain higher breakdown voltages and therefore support applications that require higher dynamic range. This gap will widen with more advanced generations of both CMOS and SiGe as each new generation of CMOS results in lower breakdown voltages while each new generation of SiGe results in a better trade-off between speed and breakdown. In addition, we have seen that performance of SiGe devices can be improved with advanced lithography much in the same way as with CMOS devices such that a raw performance gap will continue to exist between SiGe and CMOS as more advanced nanometer nodes are created in the future. This will continue to enable a market for SiGe at the bleeding edge of performance, which today is translating into interest for SiGe in several millimeter-wave applications and very high-speed networks as reviewed in the prior chapter.

The biggest threat to SiGe advancements is the failure to identify high-speed, high-volume applications that take advantage of these benefits in the future but, much like Moore's law for CMOS has held true for decades and applications have taken full advantage, the imagination of the industry has never let us down before and is not likely to do so in this case.

ACKNOWLEDGMENTS

The author would like to acknowledge the help of current and former coworkers at Jazz Semiconductor including Volker Blaschke, Dieter Dornisch, David Howard, Chun Hu, Paul Hurwitz, Amol Kalburge, Arjun Karroy, Paul Kempf, Lynn Lao, Zachary Lee, Pingxi Ma, Edward Preisler, Greg U' Ren, Jie Zheng, and Bob Zwingman, as well as contributions from many of Jazz Semiconductor's customers and partners.

REFERENCES

1. H. Kroemer, Heterostructure bipolar transistors and integrated circuits, *Proc. IEEE*, 70, 13–25, 1982.
2. G. L. Patton, D. L. Harame, J. M. C. Stork, B. S. Meyerson, G. J. Scilla, and E. Ganin, Sige-base, poly-emitter heterojunction bipolar transistors, *Proc. Symp. VLSI Technol.*, pp. 35–36, 1989.

3. S. P. Voinigescu, M. C. Maliepaard, J. L. Showell, B. E. Babcock, D. Marchesan, M. Schroter, P. Schvan, and D. Harame, A scalable high-frequency noise model for bipolar transistors with application to optimal transistor sizing for low-noise amplifier design, *J. Solid-State Circuits*, 32(9), 1430–1439, 1997.

4. *Int. Technol. Roadmap for Semiconductors*, International SEMATECH, Austin, TX, 2007.

5. W. Winkler, J. Borngraber, He. Erzgraber, Ha. Erzgraber, B. Heinemann, D. Knoll, H. J. Osten, M. Pierschel, K. Pressel, and P. Schley, Wireless communication integrated circuits with CMOS-compatible SiGe HBT technology modules, *CICC Proc.*, pp. 351–358, 1998.

6. K. Kuhn, M. Agostinelli, S. Ahmed, S. Chanbers, S. Cea, S. Christensen, P. Fischer, J. Gong, C. Kardas, T. Letson, L. Henning, A. Murthy, H. Muthali, B. Obradovic, P. Packan, S.W. Pae, I. Post, S. Putna, K. Raol, A. Roskowski, R. Soman, T. Thomas, P. Vandervoorn, M. Weiss, and I. Young, A 90 nm communication technology featuring SiGe HBT transistors, RF CMOS, precision R-L-C RF elements and 1 μm^2 6-T SRAM cell, *IEDM Tech. Dig.*, pp. 73–76, Dec. 2002.

7. B. Heinemann, R. Barth, D. Bolze, J. Drews, P. Formanek, O. Fursenko, M. Glante, K. Lowatzki, A. Gregor, U. Haak, W. Hoppner, D. Knoll, R. Kurps, S. Marschmeyer, S. Orlowski, H. Rucker, P. Schley, D. Schmidt, R. Scholz, W. Winkler, and Y. Yamamoto, A complementary BiCMOS technology with high speed npn and pnp SiGe:C HBTs, *IEDM Tech. Dig*, pp.117–120, Dec. 2003.

8. A. Kar-Roy, C. Hu, M. Racanelli, C. A. Compton, P. Kempf, G. Jolly, P. N. Sherman, J. Zheng, Z. Zhang, A. Yin, High density metal insulator metal capacitors using PECVD nitride for mixed signal and RF circuits, *Interconnect Technol. IEEE Int. Conf.*, pp. 245–247, 1999.

9. P. Ma, M. Racanelli, J. Zheng, and M. Knight, A 1.4 mA and 3 mW, SiGe90, BiFET low noise amplifier for wireless portable applications, *RFIC Symp. Dig.*, pp. 237–240, June 2003.

10. V. Aparin and L. Larson, Modified derivative superposition method for linearizing fet low noise amplifiers, *RFIC Symp. Dig.*, pp. 105–108, 2004.

11. J. Scholvin, D. R. Greenburg, and J. A. del Alamo, Performance and limitations of 65 nm CMOS for integrated RF power applications, *IEDM Tech. Dig*, pp. 381–384, Dec. 2005.

12. M. Racanelli and P. Kempf, SiGe BiCMOS technology for RF circuit applications, *IEEE Trans. Electr. Dev.*, 52(7), 1259–1270, 2005.

13. M. Racanelli and P. Kempf, SiGe BiCMOS technology for communication products, *CICC Proc.*, pp. 331–334, Sept. 2003.

14. S.P. Voinigescu, T. Chalvatzis, K. H. K. Yau, A. Hazneci, A. Garg, S. Sharahmian, T. Yao, M. Gordon, T. O. Dickson, E. Laskin, S. T. Nicolson, A. C. Carusone, L. Tchoketch, O. Yuryevich, C. Ng, B. Lai, and P. Liu, SiGe BiCMOS for analog, high speed digital and millimetre-wave applications beyond 50 GHz, *BCTM Proc.*, pp. 223–230, Sept. 2006.

15. K. Koh and G. Rebeiz, An eight-element 6 to 18 GHz SiGe BiCMOS RFIC phased-array receiver, *Microwave J.*, 270–274, May 2007.

16. K. Koh and G. M. Rebeiz, An X- and Ku-band 8-element linear phased array receiver, *IEEE Custom Integr. Circuits Conf.*, San Jose, CA, pp. 761–764, Sept. 2007.

Part II

Emerging Design Techniques

Emerging Design Techniques

7 Offset, Flicker Noise, and Ways to Deal with Them

Hanspeter Schmid

CONTENTS

7.1 INTRODUCTION

After almost a century of research into flicker noise, we still do not know as much about it as we would like to: we do not know enough about its origin, nor do we know everything about its behavior, nor has the last word about good methods to fight it been spoken: effectively, we still are like Alice standing in front of the rabbit hole, before she enters the Wonderland ...

So the intent of this chapter is to give the reader an idea of what flicker noise is, how it is connected to other low-frequency noise effects, and what today's designers do to fight it. This chapter will just give a broad overview, focusing on concepts and design philosophy, providing just as much mathematics as is strictly necessary. Interested readers will have to follow the cited references to find out details about mathematics and design.

In this chapter, Section 7.2.1 on the nature of flicker noise is followed by Section 7.3 on switched-capacitor techniques and noise sampling. Three more sections (Sections 7.4 through 7.6) deal with the three main techniques used against flicker noise, which are large-scale excitation, chopping, and correlated double sampling (CDS). An appendix contains information on how to simulate flicker noise in MATLAB®, and finally, a short annotated literature list is given, inviting the readers to find out for themselves how deep the rabbit hole really is.

7.2   WHAT IS FLICKER NOISE?

Flicker noise, or $1/f$-noise, seems to be so easy to define: it is noise whose power spectral density (PSD) has the form

$$S(f) = S(1)\frac{1}{f^x}$$

where x typically is around 1. In most circuits, this means that white noise dominates above a certain frequency, and we will see a behavior as in Figure 7.1. While this definition looks so simple, it immediately begs the question: does flicker noise really go down all the way to $f = 0$? And what would such behavior actually mean?

One thing this would mean is that flicker noise would then have infinite power over a finite frequency band, because

$$\int_0^1 S(1)\frac{1}{f^x}\mathrm{d}f \to \infty$$

The problem we are facing with flicker noise is actually rather simple: we are looking at it now in the frequency domain only, without thinking about what integrating from $f = 0$ upwards actually means: it means that we are looking at a process that takes an infinite time to happen, and this is not realistic at all. Looking at spectra is normally very helpful for understanding amplifiers, filters, regulators and the like, but we should never forget that the time and the frequency domains are only equivalent mathematically, but in reality, signals are varying in time, and frequency is only an abstract, if helpful, tool we use for our convenience [1].

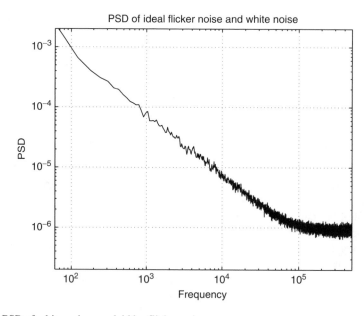

FIGURE 7.1   PSD of white noise overlaid by flicker noise.

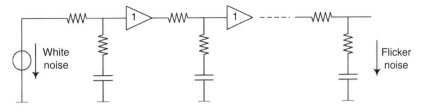

FIGURE 7.2   Flicker noise generated from white noise.

7.2.1   THE NATURE OF FLICKER NOISE

Looking at processes generating flicker noise in the time domain instead of the frequency domain gives us much more insight into the nature of flicker noise. We have no problems in finding the flickering systems in nature and science; it seems that flicker noise is the rule rather than the exception. It can be observed in systems like vacuum tubes, diodes, transistors, thin films, quartz oscillators, the average seasonal temperature, the annual amount of rainfall, the rate of traffic flow, the loudness and pitch of music, the pressure in lakes, search engine hits on the Internet, and so on [2,3].

Keshner showed in 1982 [2] that a system flickers when it has memories whose time constants are distributed evenly over logarithmic time. Therefore, an easy way to produce flicker noise in simulation is to concatenate many stages of first-order filters with one pole and one zero each, and let it filter white noise, as shown in Figure 7.2, where four first-order filters are used per decade. The number of filters per decade decides how far the simulated $1/f$ curve deviates from the ideal curve. The poles and zeros must be spaced evenly on a logarithmic scale. For the simulations shown in this chapter, we have used the spacing shown in Figure 7.3, as described in Appendix 7.A1.

This system gives the very nice $1/f$ behavior in Figure 7.3, and it is amazing to see that the number of memory blocks needed to make flicker noise is relatively small. According to Bloom [4], mixed oxide semiconductor field-effect transistors (MOSFETs) show flicker noise behavior from, for example, 10^{-8} up to 10^5 Hz, which would require only 25 memory cells with time constants distributed evenly on a logarithmic scale.

Making simulations with this model of flicker noise, we soon find funny effects. Figure 7.4 shows, for example, the variance of the output signal of the circuit in Figure 7.2 as a function of time.

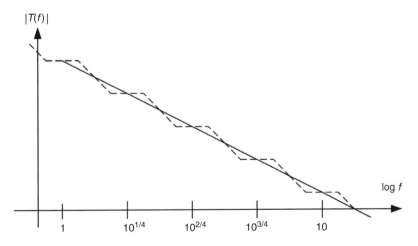

FIGURE 7.3   Transfer function of the flicker noise generator.

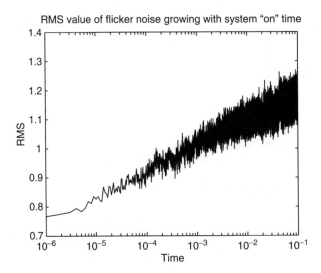

FIGURE 7.4   Variance of flicker noise as a function of system "on" time.

It is immediately apparent that this variance rises with log t. In other words, the random signal we are looking at is not stationary. The theoretical consequences of this have been discussed in Ref. [2], and measurements of practical problems coming from this nonstationarity have been shown in Ref. [5], so the nonstationarity is not a problem of our model, but an inherent feature of flicker noise: it means that if you have a system with long time constants in its memories, then that system takes a long time to reach its steady state. To return to Bloom [4], the 10^{-8} Hz he mentioned corresponds to a time of three years, so normally we will never really see the steady state in MOSFET circuits. However, as long as we do not do CDS, this does not concern us.

7.2.2 MEMORY IN SYSTEMS

Each of the systems mentioned above has memory of some sort. For example, it is described in Ref. [3] that the number of vowels in words like "aargh" and "loooove" and the number of hits (the frequency) when these words are entered as search terms in Internet search engines are related by an $1/f^x$ law. The absolute numbers are different for each word, but the exponent x only depends on the nature of the memory, which is as follows: when a person sees a word like "looooove" on a web page, that person may feel inclined to write "love" with even more o's in an attempt to express stronger feelings. So in this case the memory are Internet pages interacting with users' memories, and x is the same for all words.

Lakes also show flicker noise; in this case the behavior is close to $1/f^{5/3}$ for every lake in the world, but only the magnitude is different. What happens there is that the Coriolis force (from earth rotation) causes whirls of big dimensions; these whirls transfer their energy to smaller whirls, and so on, until their energy is dissipated at molecular level. This cascade of whirls is not very much unlike the filter cascade shown in Figure 7.2, and Kolmogorow showed long ago that simply having such a cascade of whirls already determines the exponent 5/3, but again not the magnitude of the flicker noise.

7.2.3 MEMORIES IN MOSFETs AND OTHER ELECTRONIC DEVICES

Almost every electronic device shows some flicker noise: vacuum tubes, resistors, diodes, BJTs, and MOSFETs; but in MOSFETs, the magnitude is by far the largest. The reason for this is that there are several different effects causing flicker noise in electronic devices, in every case with $1/f^x$ and $x \approx 1$, but these effects can be divided into volume effects and surface effects [6].

The two main volume effects are Bremsstrahlung and carrier scattering. Bremsstrahlung is a German word used in quantum mechanics that roughly means "deceleration radiation." Whenever an electron is accelerated, it will emit low-frequency Bremsstrahlung, and will be slowed down by its own Bremsstrahlung, as will other electrons in its vicinity. Thus we again have low-frequency energy and a cascade that remembers it, giving $1/f$ noise. This is the main source of $1/f$ noise in vacuum tubes.

The second volume effect is scattering, when electrons are scattered at the silicon lattice, or at impurities in the material, or by acoustical or optical phonons, and so on. In all cases, the scattering will interact with the lattice, generating phonons, which will later cause more scattering, and again we will have $1/f$ noise. This is the dominant source in most solid-state devices.

The effect that dominates in MOSFETs, though, is something quite different: in MOSFETs, electrons tunnel from traps in the oxide to the gate and the conducting channel, and vice versa. If there is only one single trap (which may indeed happen in minimum-size deep-submicron transistors), then this causes a PSD of the drain current

$$S(f) \approx \frac{S(o)}{1 + (2\pi f)^2 \tau^2}$$

with a certain trap time constant τ. This is $1/f^2$ behavior, as white noise fed through a one-pole low-pass filter would give, but due to the quantum nature of the electron trapping, this noise signal will only have two current levels. Such noise is called "random telegraph noise" [5]. Now what happens if we have several traps? It can be shown that the time constant for a trap at a distance z from the interface is

$$\tau = \tau_0 \exp\left(\frac{10^{10}}{m}z\right) \tag{7.1}$$

for some process-dependent time constant τ_0, so if traps are uniformly distributed over $z = 0 \cdots z_g$, we will have memories with time constants that are uniformly distributed over a logarithmic scale, as in Figures 7.2 and 7.3. The difference is that we drive the filter in these figures with Gaussian white noise instead of a two-level signal with white frequency characteristic. We also see from (7.1) that even for a gate with thickness $z_g = 1$ nm, the time constants of the flicker noise are spread over more than three orders of magnitude.

Experiments with large-scale excitation of MOSFETs—where part of the memory is deleted and therefore flicker noise is reduced intrinsically—show that flickering occurs even when the transistor is switched off completely. It can then just not be measured directly, because what we can measure are just the effects caused by electron trapping: electrons tunneling in and out of traps will cause both carrier number fluctuations and also fluctuations of the carrier mobility μ [5], which in turn make the drain current of the MOSFET flicker. This is also reflected in one of the widely used simple flicker noise models of the MOSFET,

$$\overline{V_g^2} = \frac{K}{WLC_{ox}f}$$

where K and C_{ox} are technology parameters, and W and L the transistor dimensions: this formula does not depend on the bias conditions of the device, meaning it does not depend on any current flowing through the MOSFET.

7.2.4   MEMORY AND CORRELATION

Turning back to the mathematics of flicker noise, the Fourier transform of a PSD is the autocorrelation function, which, for $1/f^x$ noise, is [2]

$$R(\tau) \sim |\tau|^{x-1}$$

So for $x = 1$, $R(\tau)$ is constant, meaning the present value of the flicker noise signal correlates very well with *all other values* of the same signal, and so flicker noise can be removed effectively with techniques that operate on correlated samples of the flickering signals (e.g., CDS).

7.2.5   FLICKER NOISE IS OFFSET EXTENDED IN FREQUENCY

If we extend our view of flicker noise down to $f = 0$, we look at an error signal that is constant in time: offset. While this is not mathematically inspiring, it still means something in practice: most techniques removing flicker noise will also cancel offset, and vice versa.

7.2.6   TECHNIQUES TO REDUCE FLICKER NOISE

Considering all that has been said until here, we end up with three techniques to fight flicker noise:

- Knowing that flicker noise comes from memory, we attempt to reset this memory. This is known as large-signal excitation (LSE).
- Knowing that flicker noise has a flat autocorrelation function, we attempt to remove it by subtracting two correlated samples. This is known as CDS.
- Knowing that flicker noise is a low-frequency effect, we attempt to modulate it into a frequency band outside the signal band. This is known as chopping.

Except for chopping, these techniques only work on sampled signals, so we must first have a look at switched-capacitor techniques and noise sampling.

7.3   SWITCHED-CAPACITOR TECHNIQUES

Figure 7.5 shows a very simple switched-capacitor circuit. The two switches are closed during the clock phases ϕ_1 and ϕ_2, and the two clock signals do not overlap, such that the two switches are never closed simultaneously.

When ϕ_1 is closed, the capacitor is charged to V_1, storing the charge $Q = CV_1$. When ϕ_2 is closed, $Q = CV_2$. Therefore, in every clock cycle, the charge $\Delta Q = C(V_1 - V_2)$ is transferred. The mean current through this circuit is then $I_{12} = \Delta Q/T_{\text{clk}} = f_{\text{clk}} C(V_1 - V_2)$, so we have a resistor with equivalent resistance $R_{\text{eq}} = 1/(f_{\text{clk}} C)$.

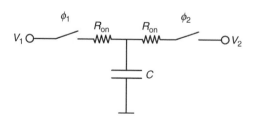

FIGURE 7.5   Switched-capacitor resistor.

The interesting thing about SC filters is that they become *much* faster with technology scaling. This can be shown as follows [7]: For good settling, we require $T_{clk}/2 > 5R_{on}C$, where R_{on} is the on-resistance of the switches. So we want

$$f_{clk} < \frac{1}{10R_{on}C} \tag{7.2}$$

The on-resistance of a MOSFET switch is

$$R_{on} = \frac{1}{\mu C_{ox}(W/L)V_{eff}} \tag{7.3}$$

where
μ is the carrier mobility
C_{ox} is the gate oxide capacitance density
W/L is the width over the length
V_{eff} is the gate overdrive voltage

In addition, we know that when a switch is opened, approximately half of the channel charge $Q_{ch} = -WLC_{ox}V_{eff}$ will go into the capacitor and cause a voltage error

$$|\Delta V| = \frac{|Q_{ch}|}{2C} = \frac{WLC_{ox}|V_{eff}|}{2C}$$

So the C we have to use for a certain switch and some given $|\Delta V|_{max}$ is

$$C = \frac{WLC_{ox}|V_{eff}|}{2|\Delta V|_{max}} \tag{7.4}$$

Replacing R_{on} in (7.2) according to (7.3) and C according to (7.4) gives a very simple result:

$$f_{clk} < \frac{\mu|\Delta V|_{max}}{5L^2} \tag{7.5}$$

$|\Delta V|_{max}$ depends on the maximum signal and therefore on V_{dd}. The product $\mu|\Delta V|_{max}$ does not change a lot as technology scales, so, to the first order, (7.5) means that the maximum speed of SC circuits scales as does the number of transistor per area, which means that Moore's law is also valid for the speed of SC circuits.

The main advantage of SC techniques can be shown in Figure 7.6. This is an integrator with time constant

$$\tau = R_{eq}C_2 = \frac{C_1}{C_2}\frac{1}{f_{clk}}$$

So we have a time constant derived from a ratio of capacitors, which can be made precise to within less than 1%, and a clock frequency, which is even more precise.

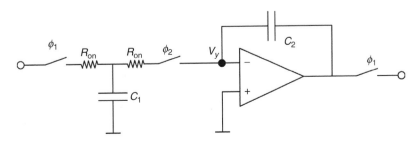

FIGURE 7.6   Switched-capacitor integrator.

7.3.1 SAMPLED NOISE IN SC CIRCUITS

This great advantage is paid with more aliasing though. The precise calculation is quite difficult even for the simple circuit in Figure 7.6—see Ref. [8] for details—because, at the output, one simultaneously sees direct noise from the op-amp as well as sampled noise from the earlier stages. Fortunately, aliased broadband noise often dominates, and a simplified analysis can be made.

What noise sampling means can be shown using the very simple circuit in Figure 7.5. When ϕ_1 closes, and we wait for the system to reach the thermal equilibrium, then the energy stored in the capacitor is $\frac{1}{2}CV_c^2$. Similarly, the noise energy coming from a noise voltage $\overline{V}_{c,\mathrm{rms}}$ is $\frac{1}{2}C\overline{V}_{c,\mathrm{rms}}^2$. We also know from thermodynamics that the energy in a system with one degree of freedom is $\frac{1}{2}kT$, so it directly follows that the variance of the thermal noise is

$$\frac{1}{2}C\overline{V}_{c,\mathrm{rms}}^2 = \frac{1}{2}kT \Rightarrow \overline{V}_{c,\mathrm{rms}}^2 = \frac{kT}{C} \tag{7.6}$$

This can also be shown in a different way: the noise caused by R_{on} is $\overline{V}_r^2 = 4kTR_{\mathrm{on}}$, and the bandwidth of the filter consisting of R_{on} and C is $1/R_{\mathrm{on}}C$. Integrating the filter's noise over the bandwidth will also give the result in (7.6).

So, essentially, as long as R_{on} is low enough such that the circuit in Figure 7.5 reaches equilibrium at the end of the clock phase, the integrated noise power depends on C only. To the first order, this noise is white noise. So what goes into the node V_y of Figure 7.6 is essentially sampled white noise with a PSD of

$$S_n(f) = \frac{kT}{Cf_{\mathrm{clk}}} \quad \text{for} \quad -\frac{1}{2}f_{\mathrm{clk}} \leq f \leq \frac{1}{2}f_{\mathrm{clk}}$$

We also have to look at sampled white noise. Assume that the inputs of the circuit in Figures 7.5 and 7.6 are driven by a preamplifier producing white noise up to a noise bandwidth f_{nbw} that is related to the amplifier bandwidth, so that its single-sided PSD is approximately

$$S_a(f) = \frac{\overline{V}_{\mathrm{amp,rms}}^2}{f_{\mathrm{nbw}}} \quad \text{for} \quad 0 \leq f \leq f_{\mathrm{nbw}}$$

The square root of the level of this noise PSD would be in the unit $\mathrm{nV}/\sqrt{\mathrm{Hz}}$ often found in op-amp data sheets. Since the amplifier must be fast enough to settle well within one clock period, we normally have $f_{\mathrm{nbw}} \gg f_{\mathrm{clk}}$ and therefore the noise is aliased. Through aliasing, the noise is compressed from a range $0 \cdots f_{\mathrm{nbw}}$ to a range $-\frac{1}{2}f_{\mathrm{clk}} \cdots \frac{1}{2}f_{\mathrm{clk}}$, so the aliased noise is scaled up

$$S_{a,\mathrm{aliased}}(f) = \frac{f_{\mathrm{nbw}}}{f_{\mathrm{clk}}} \frac{\overline{V}_{\mathrm{amp,rms}}^2}{f_{\mathrm{nbw}}} \quad \text{for} \quad -\frac{1}{2}f_{\mathrm{clk}} \leq f \leq \frac{1}{2}f_{\mathrm{clk}}$$

or, if we use single-sided spectra for the sampled signals,

$$S_{a,\mathrm{aliased}}(f) = 2 \cdot \frac{f_{\mathrm{nbw}}}{f_{\mathrm{clk}}} \frac{\overline{V}_{\mathrm{amp,rms}}^2}{f_{\mathrm{nbw}}} \quad \text{for} \quad 0 \leq f \leq \frac{1}{2}f_{\mathrm{clk}}$$

This means that sampling 10 MHz wide white noise at 1 MHz gives 20 times higher noise power. In Figure 7.6, this noise is then integrated by the SC integrator.

With this way of thinking, we can identify all noise sources, calculate their noise transfer functions to the output of the circuit, and add all contributions. Gobet and Knob [8] show this using Figure 7.5 as an example. A general method using matrix equations and including white noise, flicker noise, and amplifier noise was presented in Refs. [9,10] describes the simplified noise analysis of choppers

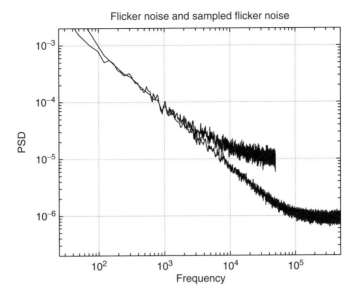

FIGURE 7.7   PSD of white noise overlaid by flicker noise, sampled with 1 MHz and 100 kHz.

and correlated double samplers; this will be discussed again briefly in the following sections of this chapter.

Fortunately, in SC applications that do not attempt to cancel flicker noise, sampled white noise normally dominates, which makes an analysis simpler. To illustrate this, the lower curve in Figure 7.7 is a (sampled) signal with a white-noise and a flicker-noise component. The flicker noise corner frequency is at approximately one-fifth of the signal bandwidth. If this signal is undersampled 10 times, the upper curve results, with the same flicker noise, but 10 times more white noise, so the flicker noise corner frequency still is at approximately one-fifth of the signal bandwidth. So sampling generally reduces the flicker noise corner frequency.

7.4   BIAS SWITCHING AND LARGE-SCALE EXCITATION

Figure 7.8 shows a switched current source. If this circuit is operated with a variable-duty-cycle clock ϕ and its inverse $\bar{\phi}$, then the current can be tuned by a factor of 2. It has been observed that for duty cycles between 0% and 100%, this circuit is much less noisy than the circuit simulator predicts [5]. The reason for this is that switching a transistor off deletes some of its flickering memory by kicking some of the trapped electrons out of their traps.

Figure 7.9 shows another MATLAB simulation in which the memory of the flicker noise is deleted almost completely once every 10 µs. The flicker noise disappears almost completely in this example; normally, some flicker noise remains at low frequencies because it is not possible to delete

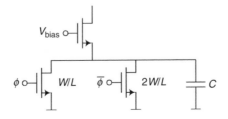

FIGURE 7.8   Switched current source.

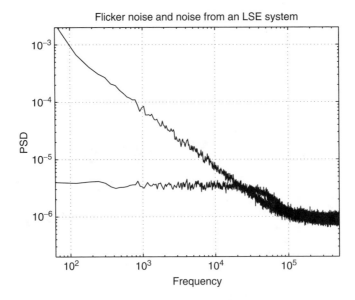

FIGURE 7.9   Spectrum of flicker noise before and after large-scale excitation (LSE).

all of the memory. This effect can be calculated [5], but not simulated; there is as yet no circuit simulator that takes flicker noise memory effects into account. However, there are already many applications other than Figure 7.9 in which LSE is used.

For example, Koh et al. [11] present an op-amp with a switched input differential pair as in Figure 7.10. The two transistors are used alternatively; the clock switches the unused one off, deleting its flicker noise memory. This will of course introduce spikes in the output voltage at multiples of f_{clk}, but it also reduces the flicker noise of the op-amp. In Ref. [11] the measured noise at low frequencies was reduced by 5 dB.

Another place where such memory effects are observed are oscillators. In oscillators, transistor flicker noise will cause low-frequency phase noise, which is narrow-band noise around the oscillator center frequency that is not less paradox in nature than flicker noise itself [12]. Periodically switching off MOSFETs in oscillators should reduce such low-f phase noise because it reduces flicker noise. This has been shown experimentally both for complementary metal oxide semiconductor (CMOS) ring oscillators, where the measured phase noise often is lower than simulated [13], and for RF LC oscillators, where flicker noise can be reduced by using two alternatively switched tail transistors, similar to what has been done in Figure 7.10 [14].

Figure 7.11 shows a pixel of an image sensor [5]. In this circuit, the photo diode accumulates charge while it is exposed to light. To read out, M1 is switched on, charging the floating diffusion to

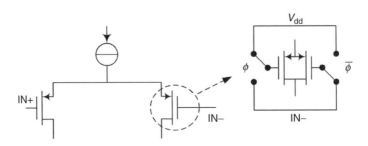

FIGURE 7.10   Switched differential pair.

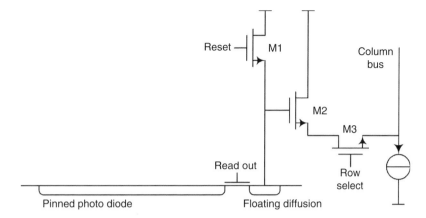

FIGURE 7.11 Image sensor pixel.

a high potential. This voltage is read out by activating M3, "row select." In a second step, the read out transistor between the wells is activated, transferring the photo charge to the floating diffusion. Then a second read out is made. The difference of the two measurements is formed, removing offset and also flicker noise. Flicker noise in this circuit comes mainly from M2, and it is possible to reduce the intrinsic flicker noise of M2 by resetting it after each read out through pulling the column bus. This, however, can be a bad idea, as will be discussed in the section on CDS (Section 7.6).

7.5 CHOPPING

Chopping is one of the two fundamentally different ways to remove flicker noise from the signal. Chopping can be done whenever it is possible to feed a signal through the flickering amplifier with different signs in every other clock period. This chopping operation can then be reversed at the output, after the amplifier, as shown in Figure 7.12.

Essentially this system modulates the input signal up to the frequency f_{chop}, and also $3f_{\text{chop}}$, $5f_{\text{chop}}$, and so on. Then the signal goes through the amplifier, where it picks up flicker noise and also offset. After the amplifier, the signal is modulated back to the base band, but at the same time, the flicker noise and the offset are modulated up to the multiples of f_{chop}. So, as long as f_{chop} is far enough above the signal band, the signal is not disturbed by flicker noise [10].

The formulas for the chopped noise spectrum can be found in Ref. [10], but Figure 7.13 shows that the relations between the amplifier output noise and the spectrum after the second multiplier in Figure 7.12 are really simple: below f_{chop}, the noise is white and on the level of the amplifier output

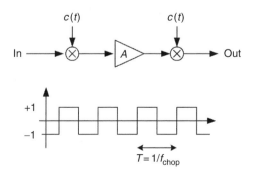

FIGURE 7.12 Principle of chopping.

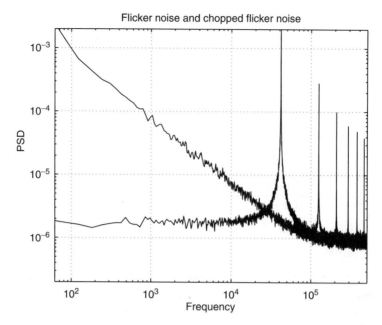

FIGURE 7.13 Flicker noise and chopped flicker noise.

noise at frequency f_{chop}. This makes it advisable to choose the chopper frequency f_{chop} at the $1/f$-noise corner frequency, or higher.

Note that chopping is just a modulation, it does not involve sampling. So while it is possible to use chopping in a sampled-data system, it is just as well possible to use it in a continuous-time system, where it will not do any noise aliasing.

It is equally important to note that chopping does not remove offset and flicker noise. For example, if the amplifier has an input offset of 1 mV, a gain of 100, no input signal, and $f_{chop} = 10\,kHz$, then its output will be a rectangular signal with frequency f_{chop} and a magnitude of $200\,mV_{pp}$. This means that when a signal is present, the signal will be added to this huge rectangular wave, and may well saturate the following stages, which is why most chopping systems have low-pass filters after the second chopper.

7.5.1 CONVENTIONAL CHOPPER AMPLIFIER

Figure 7.14 shows a conventional amplifier. Although we draw a multiplier in Figure 7.12, the chopper section is very simple to realize, all that is needed are four switches that cross the lines of the balanced amplifier during ϕ_2, or do not cross them during ϕ_1 [10]. The design constraints on such a system are the following:

- f_{chop} should be higher than the $1/f$-noise corner frequency and must be at least twice the signal band's upper frequency, f_{sig}.
- The amplifier will process the signal in the frequency band $f_{chop} \pm f_{sig}$, so it must work well and with sufficient slew rate in this frequency range.
- It is advisable to remove the energy of the chopped signal after the second chopper using a low-pass filter with passband up to f_{sig} and stop band below f_{chop}.
- The switches must be designed such that they result in as little charge injection as possible (see Section 7.3); such charge injection will cause residual offset.

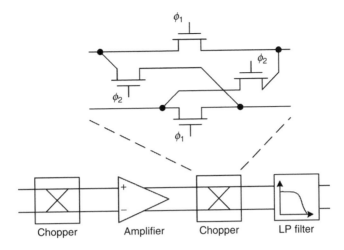

FIGURE 7.14   Chopper amplifier.

One way to reduce residual offset due to charge injection is shown in Figure 7.15. In this amplifier, the inner chopper is designed at a frequency above the $1/f$ corner frequency, thus moving $1/f$ noise out of the signal band. A second outer chopper can then operate on a frequency below the $1/f$ corner frequency; it will remove the residual offset of the inner chopper and will cause a low residual offset itself, because it operates at a low frequency. A 100 nV-offset nested chopper amplifier was reported by Bakker et al. [15]. Note that in such amplifiers, f_{sig} must be lower than half of the lower chopper frequency.

Very good results can also be obtained with tackling the residual offset at its source, for example, by staggering the clock edges of the second chopper in Figure 7.14 slightly behind the edges of the first chopper, leaving a small time gap in which the error pulses of the first chopper can die away [16].

7.5.2   MULTIPATH CHOPPER AMPLIFIERS

Nevertheless, in all these examples, the chopper frequency must be above twice the maximum signal frequency. This limitation can be overcome by building a multipath amplifier, as in Figure 7.16.

If g_{m4} is chosen such that both the DC gain of the lower path and its unity-gain frequency are much lower than those of the upper path, a situation as in Figure 7.17 occurs: the transfer functions of the two paths will cross at the frequency f_{cross}; below this frequency, the lower path dominates the op-amp's behavior and above f_{cross}, the upper path dominates.

So it becomes possible to replace the lower path by a chopper amplifier as in Figure 7.14, and operate it on a very low chopper frequency. Witte et al. [17] present a chopper amplifier that has 1 μV offset, $f_{chop} = 4$ kHz, and a unity-gain frequency of 1.3 MHz with 50 pF load. This amplifier

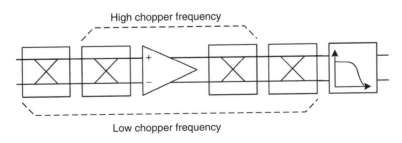

FIGURE 7.15   Nested chopper amplifier.

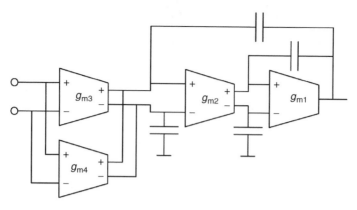

FIGURE 7.16 Multipath amplifier.

has more residual offset than the ones in Refs. [15,16], but the upper signal frequency of 1.3 MHz is large compared to the 5.6 kHz of Huang and Menolfi [16] and huge compared to the 8 Hz [sic!] of Bakker et al. [15]. This shows that the main frequency limitation of chopper amplifiers can be overcome, although with considerable circuit design effort.

7.5.3 CHOPPING IN SAMPLED-DATA SYSTEMS

Finally, chopping can also be used in sampled-data systems. For example, Figure 7.18 shows the cross-section of a microelectromechanical system acceleration sensor and a block diagram of the read out electronics.

The sensor is capacitive, with two rigid plates at the top and the bottom, and one plate that hangs in free space, attached by a spring, in the centre. When accelerated, the centre plate will move up or down, resulting in a different distribution of the capacitances toward the top and bottom plates. Since this is a linear electrical system, the position can be read out by measuring V_{center} while either setting $V_{top} = V_{DD}$, $V_{bottom} = V_{SS}$; or setting $V_{top} = V_{SS}$, $V_{bottom} = V_{DD}$. This will give the same value with opposite sign, which can be read out by a switched-capacitor low-noise amplifier (LNA). So performing the two possibilities alternatively amounts to chopping at the input of the amplifier (LNA).

If the offset and flicker noise are not very high in such a system, the output of the LNA can be digitized and the second chopper can be a simple digital sign change on the sampled value. However, if the offset or flicker noise are so high that the analog stages after the LNA are saturated, then it is necessary to add an analog second chopper and a filter after the LNA as in Figure 7.14.

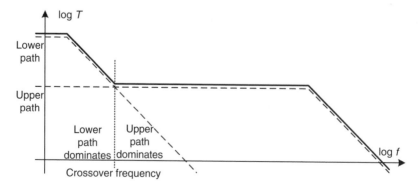

FIGURE 7.17 Open loop transfer function of the multipath amplifier.

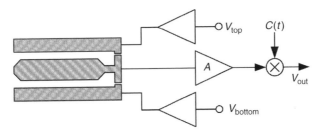

FIGURE 7.18   Acceleration sensor with SC LNA of gain A.

7.6   CORRELATED DOUBLE SAMPLING AND AUTO-ZERO TECHNIQUES

The third idea to deal with flicker noise is to remove it after it has occurred. Techniques doing this are called auto-zeroing or CDS. Both are fundamentally the same; the technique is to first sample without a signal (i.e., only the offset) and then sample again with a signal, and subtract the two values.

The effect on offset, ideally, is that it is removed, because the offset of the sampling amplifier will be the same for both samples. Flicker noise will mostly be removed, because two samples of a flicker noise process correlate well (see Section 7.2 and the discussion of the autocorrelation function of $1/f$ noise). White noise, however, does not correlate with earlier samples of itself, so the power of the white noise of the amplifier will simply be doubled.

This can be seen well in Figure 7.19, which shows the spectrum of a process with flicker noise and white noise (bottom); the same process under sampled has 10 times as much white noise and double sampled, 20 times as much white noise, but no flicker noise.

Figure 7.19 shows the CDS performed on a signal that had already been sampled. Sampling a continuous-time signal gives different results. We will now look at the white-noise and the flicker-noise contributions independently. For white noise whose bandwidth $B = \frac{\pi}{2}f_c$ is much larger than the input sampling frequency $2f_s$, the spectrum after CDS is [10]

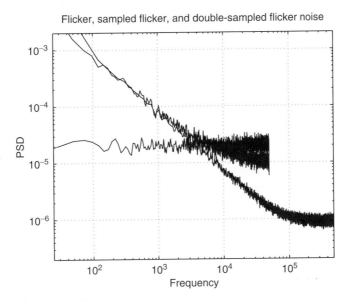

FIGURE 7.19   Flicker noise subjected to sampling and CDS.

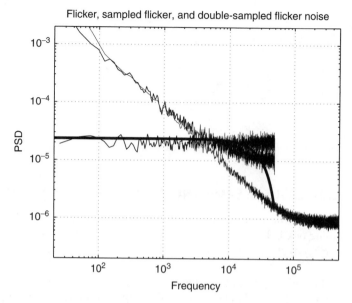

FIGURE 7.20 Continuous-time flicker noise subjected to CDS.

$$S_{\text{CDS,white}} \approx \left(\pi \frac{f_c}{2f_s} - 1 \right) S_0 \, \text{sinc}^2 \left(\frac{\pi f}{2f_s} \right)$$

where
  f_c is the corner frequency of the white noise
  S_0 is the DC noise level.

Note that here we choose f_s to be the sampling frequency at the output of the CDS block, after two samples have been subtracted.

Similarly, the flicker noise at low frequencies will not disappear completely; a fold-over component will dominate at low frequencies:

$$S_{\text{fold,}1/f} \approx \frac{S_0 f_{1/f}}{f_s} \left[1 + \ln \left(\frac{1}{3} \frac{f_c}{f_s} \right) \right] \text{sinc}^2 \left(\frac{\pi f}{2f_s} \right)$$

where $f_{1/f}$ is the corner frequency of the $1/f$ noise. The shape of the two spectra is exactly the same, and the different factors in front of the sinc function mean that as long as the flicker noise corner frequency $f_{1/f}$ is sufficiently far below the sampling frequency, aliased white noise will dominate the behavior at low frequencies.

Figure 7.20 shows the sum of these aliased components superimposed on Figure 7.19. The effect of using CDS on a continuous-time signal is that while a simple calculation as in Section 7.3 or a simulation with sampled signals estimate the total noise correctly, both underestimate the low-f noise by a factor of $\frac{\pi}{2} = 1.57 = 4\,\text{dB}$. On the other hand, they overestimate noise at the upper end of the frequency band, where almost only white noise of power S_0 will be seen in reality.

7.6.1 SWITCHED-CAPACITOR COMPARATOR WITH CDS

CDS is not very difficult to implement, and it is used in many applications. Our first example, Figure 7.21, is a comparator that can be used in Flash A/D converters [18]. The operation of this comparator is simple: in phase ϕ_1 (Figure 7.22 left), the input voltage V_{in} is sampled onto the

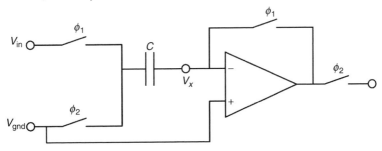

FIGURE 7.21   Comparator with CDS.

capacitor C. Because of the closed negative feedback loop, the negative input of the amplifier settles to the offset voltage V_{os}, so C is charged to the voltage $V_{in} - V_{os}$. In the phase ϕ_2 (Figure 7.22 right), the voltage $V_{in} - V_{os}$ between the negative input and V_{gnd} is compared to the new value of V_{os}, so the comparator actually tests whether

$$V_{os}|_{\phi_2} - \left(V_{os}|_{\phi_2} - V_{in}|_{\phi_1}\right) > 0$$

The difference $V_{os}|_{\phi_2} - V_{os}|_{\phi_1}$ is formed; this is CDS that removes offset and a lot of $1/f$ noise as explained above.

As with all switched-capacitor circuits, the main difficulties of this circuit are parasitic charges injected when switches open. Apart from that, such a system can remove so much offset that the comparator can even be a simple CMOS inverter, as shown in Figure 7.23 [18].

For a high-resolution comparator, an inverter will not have sufficient gain in the transition region, so an op-amp must be used, for example a Miller op-amp. The problem there is that during ϕ_1, the amplifier must be stable in the feedback loop, while during ϕ_2 it just has to be as fast as possible. A switchable compensation as shown in Figure 7.24 will take care of this, and with proper scaling of the switch transistor, this switch, while on, will introduce a compensating zero in the Miller amplifier (e.g., Ref. [7]).

7.6.2   SWITCHED-CAPACITOR AMPLIFIER WITH CDS

The comparator in Figure 7.21 can readily be modified to give an SC amplifier [10] by adding a capacitor that is switched into the signal path in phase ϕ_2, as shown in Figure 7.25.

Then, in ϕ_1, C_1 will be charged to $V_{in} - V_{os}$, and C_2 to $-V_{os}$; in ϕ_2 the difference is formed and the resulting output voltage $V_{out} = V_{in} (C_1/C_2)$ with the offset and flicker noise removed.

While this circuit works well in practice, it has two problems: first, ϕ_1 and ϕ_2 must not overlap. This means that while neither is active, the amplifier is in an open-loop configuration and care must be taken that the output does not jump to a supply rail during that time and pushes the op-amp into a state from which it takes long to recover. Second, during ϕ_1, the output is always V_{os}, so the output jumps forth and back between the signal voltage and the small voltage V_{os}, and thus the amplifier needs to have a high slew rate. Koh et al. [10] present a good overview on SC amplifiers in which the amplifier needs to have only a modest slew rate.

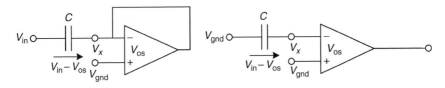

FIGURE 7.22   CDS comparator in phases ϕ_1 (left) and ϕ_2 (right).

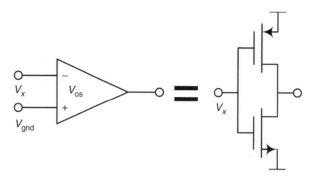

FIGURE 7.23   Inverter-based comparator with CDS.

7.6.3 CORRELATED DOUBLE SAMPLING IN SAMPLED SYSTEMS

In Section 7.5, we introduced an example of an acceleration sensor—in Figure 7.18—and discussed chopping. This system can easily be transformed into a CDS system: the sensor is operated with the same sequence as in Section 7.5, but instead of only changing the sign of every second sample, we then also form the difference of two consecutive samples.

The advantage of this system is that offset and flicker noise are removed—and not only modulated out of the signal band—so amplifiers and A/D converters after the CDS stage are not in danger of being saturated. The clear disadvantages are that white noise is doubled, and also that now two input samples are needed to provide one output sample. The latter means that either the time available for sampling has to be cut in two pieces, requiring faster amplification than in the chopper system, or that two circuits forming differences are operated in parallel, one making $V[2n + 1] - V[2n]$, the other making $V[2n + 2] - V[2n + 1]$.

7.6.4 CORRELATED DOUBLE SAMPLING COMBINED WITH LARGE-SCALE EXCITATION

In Figure 7.11, we showed a simple photo diode read out circuit, in which LSE was used to reduce the intrinsic noise of the read out transistor. Simultaneously, CDS is also used.

The problem is that doing both at the same time can give more flicker noise instead of less flicker noise [5]. LSE resets the memory of the transistors, so after they return into their operating point, the memory starts to fill up again, and the variance of the flicker noise will start to increase as shown in Figure 7.4. So if the two samples used for CDS are taken at two times when the variance of the flicker noise is very different, the two samples do not really correlate and CDS can increase the flicker noise

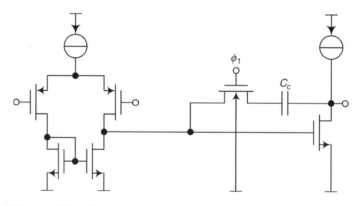

FIGURE 7.24   Miller amplifier with switchable compensation.

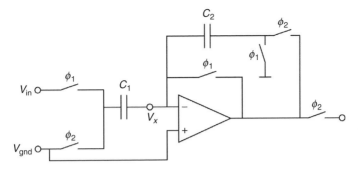

FIGURE 7.25 SC amplifier with CDS.

instead of cancelling it. This is extremely difficult to simulate, but has been shown by measurement in Ref. [5].

This means in general it is not enough that the transistor biasing is the same at both sample times, the history of the biasing also needs to be as similar as possible at both sampling instants.

7.7 CONCLUSION

Not all three methods to fight flicker noise can be used in every system. LSE is mostly used— or happens by itself—in sensor circuits with low transistor count, and in oscillators: it cannot be simulated and calculating it is also difficult. CDS is mostly used in systems that process sampled data, or are designed to sample data. Chopping is mostly used in continuous-time systems.

This chapter has given an introduction into all three techniques, together with a description of the nature of flicker noise and of noise sampling. The cited literature was chosen carefully to give the interested reader starting points for going deeper into different aspects of flicker noise; the four main papers to read would be Slepian [1] for the mathematics of flicker noise, Van der Wel et al. [5] and Van der Ziel [6] for its physics, and Van der Wel et al. [5] and Enz and Temes [10] for cicruit solutions.

7.A1 APPENDIX

Figure 7.A1 shows the MATLAB/Simulink® model use for the simulations in this chapter. The figure was made with the following three scripts:

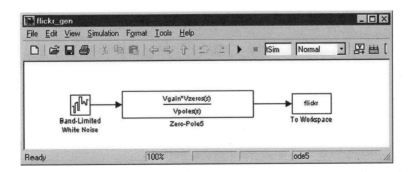

FIGURE 7.A1 Simulink model generating flicker noise.

flickr_fig01.m

```
%
% Hanspeter Schmid, June 2007
%
% Draw the flicker noise / white noise spectrum
%
clear

poles_and_zeros

sim('flickr_gen')

save data_fig01
```

poles_and_zeros.m

```
%% Poles and Zeros for the flicker noise generator
divideDecade=8;
fMax=4e5;
iMax=floor(log10(fMax)*divideDecade)

Vpoles = [];
Vzeros = [];

tSim=1;
rSeed=26649;

for k=1:4:iMax
    Vpoles=[Vpoles -10^((k+1)/divideDecade)];
    Vzeros=[Vzeros -10^((k+3)/divideDecade)];
end

format compact
Vpoles
Vzeros
Vgain=1/10^(1/divideDecade);
```

plot_fig01.m

```
load data_fig01

[Pxx,w]=pwelch(flickr.signals.values(1:1e6+1),hann(2^14),2^13,2^14,1e6);
loglog(w,Pxx,'r');
axis([min(w) max(w) 2e-7 2e-3])
grid
grid minor

xlabel ('Frequency')
ylabel ('PSD')
title ('Power spectral density of ideal flicker noise and white noise')

print -deps2 matlab_fig01.eps
```

REFERENCES

1. D. Slepian, On bandwidth, *Proceedings of the IEEE*, 64(3), 292–300, etc. 1976.
2. M. Keshner, 1/f noise, *Proceedings of the IEEE*, 70(3), 212–218, 1982.
3. H. Schmid, Aaargh! I just loooove flicker noise, *IEEE Circuits and Systems Magazine*, 1, 32–35, 2007.
4. I. Bloom and Y. Nemirovsky, 1/f noise reduction of metal-oxide-semiconductor transistors by cycling from inversion to accumulation, *Applied Physics Letters*, 58(15), 1664–1666, 1991.
5. A. P. van der Wel, E. A. M. Klumperink, J. S. Kolhatkar, E. Hoekstra, M. F. Snoeij, M. Cora Salm, H. Wallinga, and B. Nauta, Low-frequency noise phenomena in switched MOSFETs, *IEEE Journal of Solid-State Circuits*, 42(3), 540–550, 2007.

6. A. van der Ziel, Unified presentation of $1/f$ noise in electronic devices: Fundamental $1/f$ noise sources, *Proceedings of the IEEE*, 76(3), 233–258, 1988.
7. D. A. Johns and K. Martin, *Analog Integrated Circuit Design*. New York: John Wiley & Sons, 1997.
8. C.-A. Gobet and A. Knob, Noise analysis of switched capacitor networks, *IEEE Transactions on Circuits and Systems*, 30(1), 37–43, 1983.
9. L. Tóth, I. Yusim, and K. Suyama, Noise analysis of ideal switched-capacitor networks, *IEEE Transactions on Circuits and Systems-I*, 46(3), 349–363, 1999.
10. C. C. Enz and G. C. Temes, Circuit techniques for reducing the effects of op-amp imperfections: Autozeroing, correlated double sampling, and chopper stabilization, *Proceedings of the IEEE*, 84(11), 1584–1615, 1996.
11. J. Koh, D. Schmitt-Landsiedel, R. Thewes, and R. Brederlow, A complementary switched MOSFET architecture for the $1/f$ noise reduction in linear analog CMOS ICs, *IEEE Journal of Solid-State Circuits*, 42(6), 1352–1361, 2007.
12. F. M. Gardner, Can analog PLLs hold lock? A paradox explored, *IEEE Circuits and Systems Magazine*, 3, 46–52, 2007.
13. S. L. J. Gierkink, E. A. M. Klumperink, A. P. van der Wel, G. Hoogzaad, E. van Tuijl, and B. Nauta, Intrinsic $1/f$ device noise reduction and its effect on phase noise in CMOS ring oscillators, *IEEE Journal of Solid-State Circuits*, 34(7), 1022–1025, 1999.
14. C. C. Boon, M. A. Do, K. S. Yeo, J. G. Ma, and X. L. Zhang, RF CMOS low-phase-noise LC oscillator through memory reduction tail transistor, *IEEE Transactions on Circuits and Systems–II*, 51(2), 85–90, 2004.
15. A. Bakker, K. Thiele, and J. H. Huijsing, A CMOS nested-chopper instrumentation amplifier with 100-nV offset, *IEEE Journal of Solid-State Circuits*, 35(12), 1877–1883, 2000.
16. Q. Huang and C. Menolfi, A 200 nV offset 6.5 nV/$\sqrt{\text{Hz}}$ noise psd 5.6 kHz chopper instrumentation amplifier in 1μm digital CMOS, *Proceedings of the IEEE International Solid-State Circuits Conference* (San Francisco), p. 23.3, February 2001.
17. J. F. Witte, K. A. A. Makinwa, and J. H. Huijsing, A CMOS chopper offset-stabilized opamp, in *Proceedings of the European Solid-State Circuits Conference* (Montreux, Switzerland), pp. 360–363, September 2006.
18. R. Gregorian, *Introduction to CMOS Op-Amps and Comparators*. New York: John Wiley, 1999.

8 Design of Wideband Amplifiers in CMOS

David J. Allstot, Sudip Shekhar,
and Jeffrey S. Walling

CONTENTS

8.1 INTRODUCTION

Techniques for extending the bandwidth of amplifiers were proposed as early as the 1930s for audio and television applications encompassing passive filtering techniques such as inductive peaking [1], and later, distributed amplification [2]. Interest in these techniques was fueled again for transistor-based oscilloscope design at Tektronix, Inc. in the 1970s [3]. The emergence of deep-submicron complementary metal oxide semiconductor (CMOS) technology has created the need for wide bandwidth amplifiers that can be used in mixed-signal/radiofrequency (RF) wireless and wire-line applications for high-speed data communication with low error rates, low cost, and low power. For example, the merging ultrawideband (UWB) standard affords many high-bandwidth wireless local area network (WLAN) applications in the 3.1–10.6 GHz range that require techniques that facilitate the design of broadband amplifiers such as a low-noise amplifier (LNA) (Figure 8.1a) [4]. Electrical and optical communication links operating at 10–40 GB/s require broadband techniques in the design of preamplifiers, drivers, transimpedance amplifiers, multiplexers/de-multiplexers, etc. (Figure 8.1b) [5].

Some of the older design techniques are not optimally suited for integrated implementations in CMOS. For example, die area concerns limit the number of on-chip inductors that can be used, thus precluding the use of inductive techniques that use (generally) more than two inductors. Large parasitics in CMOS limit the bandwidth improvement that can be achieved through the use of simple techniques like shunt and series peaking and fail to enable the desired data rate. Distributed amplifiers

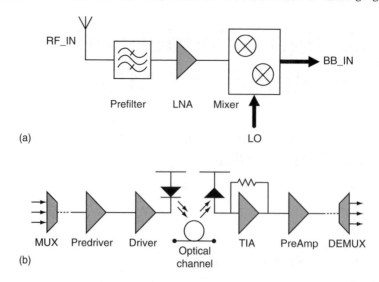

(a)

(b)

FIGURE 8.1 (a) UWB receiver front-end. (From Roovers, R., Leenaerts, D. M. W., Bergervoet, J., Harish, K. S., van de Beek, R. C. H., van der Weide, G., Waite, H., Zhang, Y., Aggarwal, S., and Razzell, C., *IEEE J. Solid-State Circuits*, 40, 2563, 2005.) (b) Typical optical communication transceiver. (From Galal, S. and Razavi, B., *IEEE J. Solid-State Circuits*, 39, 2389, 2004.)

consume large chip area and high power in CMOS (especially in pure digital processes), and require extensive modeling and simulation for accurate characterization.

The purpose of this chapter is to describe techniques that build on conventional inductive peaking techniques to achieve greater enhancements in bandwidth [6]. These methods cleverly exploit the parasitic capacitances in CMOS toward that goal. Passive filtering techniques are attractive for CMOS implementations as they do not require high-Q inductors, and bandwidth extension is therefore achieved without added power consumption or complexity.

Most of the broadband amplifiers in communication circuits, be it a differential amplifier in a wireline transceiver or a common-source/common-gate LNA in a wireless transceiver, can be viewed as a simple common-source amplifier (Figure 8.2) to describe the basic principles of operation. Here, R is the load resistance, and C_1 and C_2 represent the drain parasitic and load capacitances, respectively. Differential amplifiers are usually cascaded in several stages in wireline applications to achieve high gain. Thus, C_2 includes the gate capacitance of the next stage, whereas in a wireless receiver, C_2 includes the gate capacitance of the buffer or mixer that typically follows the LNA.

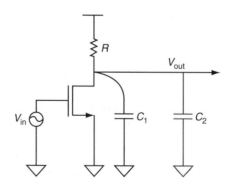

FIGURE 8.2 Simple common-source amplifier.

The ratio C_1/C_2 depends on the transistor sizes in the two stages, which in turn, are dictated by performance specifications including gain, voltage swing, noise figure, bias current, etc. Hence, it is advantageous to define a design parameter $k_C = C_1/C = C_1/(C_1 + C_2)$; its value is typically 0.2–0.5. Exceptions where $k_C > 0.5$ include the case of a large photodiode junction capacitance (C_1) followed by a smaller capacitive load looking into a transimpedance amplifier (C_2) (Figure 8.1b). Using the techniques described herein, it is shown that a certain k_C value necessitates a certain peaking technique to achieve optimum performance. Consequently, different stages in a cascaded multistage amplifier may require different peaking techniques for superior overall performance.

8.2 SHUNT PEAKING AND BRIDGED-SHUNT PEAKING

The simplest peaking technique, shunt peaking, utilizes an inductor, L, connected in series with the load resistor, R, shunting the output capacitor, $C = (C_1 + C_2)$ (Figure 8.3a, $C_B = 0$) [3,7]. Assuming the small-signal transconductance, g_m, is constant, and the gain is the product of g_m and the transimpedance, $Z(s)$, then the frequency response is clearly governed by $Z_{sh}(s)$:

$$Z_{sh}(s) = \frac{V_{out}}{I_{in}} = \left(\frac{1}{sC}\right) \| (R + sL) = \frac{R + sL}{1 + sRC + s^2LC} \tag{8.1}$$

Compared to the transimpedance function, $Z_0(s) = R/(1 + sRC)$, of the amplifier in Figure 8.2, the series inductor in Figure 8.3a adds a left-half-plane zero and pole to its transimpedance function. By

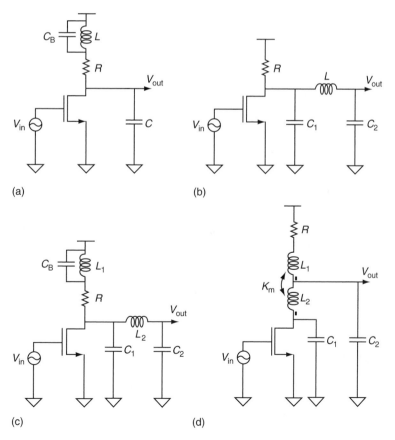

FIGURE 8.3 Common-source amplifier with (a) bridged-shunt peaking, (b) series peaking, (c) bridged-shunt-series peaking, and (d) asymmetric T-coil peaking.

properly sizing the inductance, the zero can be used to increase the impedance at higher frequencies, which mitigates the decreasing impedance of C, which extends the overall -3 dB bandwidth. Increased bandwidth in the frequency domain is equivalent to reduced rise time in the step response in the time domain. For a step input to the amplifier in Figure 8.2, the drain current initially (after $t = 0^+$) flows through the resistive branch rather than the capacitive branch. In contrast, in the case of shunt peaking (Figure 8.3a, $C_B = 0$), current flow to the resistor is delayed due to the series inductor, and thus the drain current initially charges C, which reduces rise time [7].

For the common-source amplifier of Figure 8.2, the -3 dB bandwidth is $\omega_0 = 1/RC$. To determine the value of L for a given R and C, another design variable is defined as the ratio of time constants, $m = (RC)/(L/R) = R^2C/L$. Substituting ω_0 and m into Equation 8.1 and normalizing to the impedance at DC (R):

$$Z_{N,\text{sh}}(s) = \frac{1 + s/m\omega_0}{1 + s/\omega_0 + s^2/m\omega_0^2} \tag{8.2}$$

Defining the bandwidth extension ratio (BWER) as the ratio of the -3 dB bandwidths of the inductive peaked amplifier to the reference amplifier of Figure 8.2, a maximum BWER of 1.84 is obtained for shunt peaking for $m = \sqrt{2}$ with 1.5 dB of peaking in the gain response [3,7,8]. A maximally flat gain is obtained for $m = 1 + \sqrt{2}$ but BWER is reduced to 1.72.

By noting that the peaking in the response is due to the increased impedance of the inductor over a specific range of frequencies, the inductor can be tuned with a shunt capacitor (C_B) over certain frequencies to eliminate the peaking and retain the BWER of 1.84. This leads to the bridged-shunt network shown in Figure 8.3a [1,9] whose transimpedance is

$$Z_{B-\text{sh}}(s) = \frac{V_{\text{out}}}{I_{\text{in}}} = \left(\frac{1}{sC}\right) \parallel \left(R + sL \parallel \frac{1}{sC_B}\right) = \frac{R + sL + s^2RC_BL}{1 + sRC + s^2L(C + C_B) + s^3RCC_BL} \tag{8.3}$$

The normalized transimpedance with a design parameter defined as $k_B = C_B/C$ is

$$Z_{N,B-\text{sh}}(s) = \frac{1 + \left(\dfrac{1}{m}\right)\dfrac{s}{\omega_0} + \left(\dfrac{k_B}{m}\right)\dfrac{s^2}{\omega_0^2}}{1 + \dfrac{s}{\omega_0} + \left(\dfrac{k_B + 1}{m}\right)\dfrac{s^2}{\omega_0^2} + \left(\dfrac{k_B}{m}\right)\dfrac{s^3}{\omega_0^3}} \tag{8.4}$$

Magnitude and phase plots for several values of k_B are shown in Figure 8.4. Compared to simple shunt peaking ($k_B = 0$), bridged-shunt peaking attains a BWER of 1.83 for a smaller inductance (larger m), and achieves smaller die area and higher self-resonant frequency. As any inductor in silicon has significant parasitics, its location in the L-R branch is also important. If connected to the drain node, its parasitics add to C and degrade bandwidth. However, by connecting it to the power supply (Figure 8.3a), its parasitic forms part of C_B. In a differential implementation, this configuration also permits the use of symmetrical inductors to save area.

8.3 SERIES PEAKING

Series peaking is suited to amplifiers whose drain parasitic, C_1 (Figure 8.2), is significant compared to the total capacitance (C) [10,11]. By inserting an inductor in series with the load capacitor (C_2) (Figure 8.3b), the total load capacitance is separated into two constituent components, thereby enabling *capacitor splitting*. Consider the step response of the series-peaked amplifier. In the absence of inductor L, the transistor charges the total capacitance C ($C_1 + C_2$); with the inductor in place, the current is delayed into the L-C_2 branch, and thus the transistor charges only C_1 for a while [7].

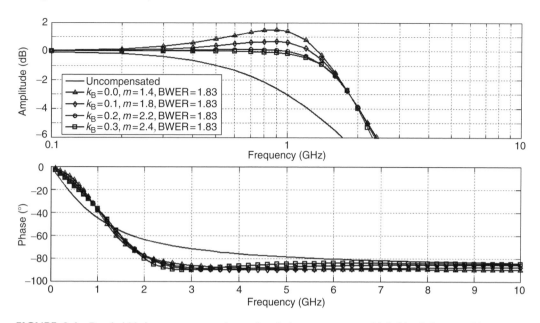

FIGURE 8.4 Bandwidth improvement and associated phase responses with bridged-shunt peaking versus $k_B = C_B/C$.

The smaller charging capacitance in turn implies a shorter rise time and a higher bandwidth. The transimpedance function is

$$Z_{se}(s) = \frac{V_{out}}{I_{in}} = \left[R \,\middle\|\, \frac{1}{sC_1} \,\middle\|\, \left(sL + \frac{1}{sC_2} \right) \right] \left[\frac{1/sC_2}{sL + 1/sC_2} \right] = \frac{R}{1 + sRC + s^2\, LC_2 + s^3\, LC_2RC_1}$$

(8.5)

The normalized transimpedance with $k_C = C_1/C$ is

$$Z_{N,se}(s) = \frac{1}{1 + \dfrac{s}{\omega_0} + \left(\dfrac{1 - k_C}{m} \right) \dfrac{s^2}{\omega_0^2} + \left(\dfrac{k_C(1 - k_C)}{m} \right) \dfrac{s^3}{\omega_0^3}}$$

(8.6)

The addition of the series inductor adds a pair of poles in the transimpedance function. Depending upon the value of k_C, an optimum inductance can be determined for maximum BWER as tabulated in Table 8.1. If a greater tolerance on passband peaking is specified, a larger BWER is achieved. Figure 8.5 shows the gain and phase responses for several typical values of k_C.

TABLE 8.1
Series Peaking Summary

$k_C = C_1/C$	Peak (dB)	$m = R^2C/L$	BWER
0.0	0.0	2.0	1.41
0.1	0.0	1.8	1.58
0.2	0.0	1.8	1.87
0.3	0.0	2.4	2.52
0.4	1.0	1.9	2.75
	2.0	2.5	3.17
0.5	3.3	1.5	2.65

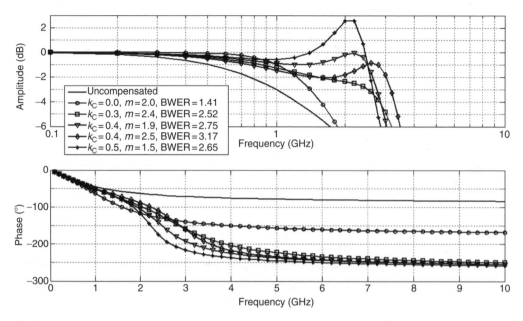

FIGURE 8.5 Bandwidth improvement and associated phase responses with series peaking versus k_C and m.

8.4 BRIDGED-SHUNT-SERIES PEAKING

Consider once again the step response of the series-peaked network of Figure 8.3b. The series inductor delays the flow of current into the L-C_2 branch; however, the initial drain current flows only through the R branch as described earlier. Hence, a method to further improve the rise time is to incorporate bridged-shunt peaking in the R branch as illustrated in Figure 8.3c. Thus, the drain current initially flows into C_1 only for a longer time, considerably improving rise time and bandwidth. This is the bridged-shunt-series-peaked network of Figure 8.3c. It uses two inductors but provides larger BWER values than its shunt-series-peaked counterpart [6]. The transimpedance function of the bridged-shunt-series (BSS) network is

$$
\begin{aligned}
Z_{B\text{-sh-se}}(s) &= \frac{V_{\text{out}}}{I_{\text{in}}} = \left[\left(R + sL_1 \left\|\frac{1}{sC_B}\right.\right) \left\|\frac{1}{sC_1}\right\| \left(sL_2 + \frac{1}{sC_2}\right)\right]\left[\frac{1/sC_2}{sL_2 + 1/sC_2}\right] \\[6pt]
&= \frac{R + sL_1 + s^2 L_1 R C_B}{\begin{aligned}&1 + sRC + s^2(L_1 C + L_1 C_B + L_2 C_2) + s^3(L_2 C_2 R C_1 + L_1 C_B R C) \\ &+ s^4 L_1 (C_1 + C_B)L_2 C_2 + s^5 R C_B L_1 C_1 L_2 C_2\end{aligned}}
\end{aligned}
\tag{8.7}
$$

With $m_1 = R^2 C/L_1$ and $m_2 = R^2 C/L_2$, the corresponding normalized transimpedance function is

$$
Z_{N,B\text{-sh-se}}(s) = \frac{1 + \left(\dfrac{1}{m_1}\right)\dfrac{s}{\omega_0} + \left(\dfrac{k_B}{m_1}\right)\dfrac{s^2}{\omega_0^2}}{\begin{aligned}&1 + \dfrac{s}{\omega_0} + \left(\dfrac{1+k_B}{m_1} + \dfrac{1-k_C}{m_2}\right)\dfrac{s^2}{\omega_0^2} + \left(\dfrac{k_B}{m_1} + \dfrac{k_C(1-k_C)}{m_2}\right)\dfrac{s^3}{\omega_0^3} \\ &+ \left(\dfrac{(k_C + k_B)(1-k_C)}{m_1 m_2}\right)\dfrac{s^4}{\omega_0^4} + \left(\dfrac{k_B k_C(1-k_C)}{m_1 m_2}\right)\dfrac{s^5}{\omega_0^5}\end{aligned}}
\tag{8.8}
$$

Table 8.2 shows results for several values of k_C and passband peaking values; for $k_C = 0.4$, a BWER of 4 is realized. Figure 8.6 shows gain and phase responses for a range of k_C. Note that the amplitude response with no gain peaking is achieved for $m_1 = 8$ and $m_2 = 2.4$, which affords

TABLE 8.2

BSS Peaking Summary

$k_C = C_1/C$	Peak (dB)	$m_1 = R^2 C/L_1$	$m_2 = R^2 C/L_2$	$k_B = C_B/C$	BWER
0.4	0	8	2.4	0.3	3.92
	2	6	2.4	0.2	4.00
0.5	2	6	2.0	0.2	3.53

pole-zero cancellation. However, such a cancellation requires precise component values that are difficult to realize due to distributed parasitic effects and process, voltage, and temperature (PVT) variations.

8.5 ASYMMETRIC T-COIL PEAKING

For amplifiers with a large load capacitance (C_2) compared to the drain capacitance (C_1), bridged-shunt-series peaking is not the optimum architecture for maximum BWER. For these cases ($k_C \leq 0.3$), series-shunt-series peaking is cleverly obtained via the magnetic coupling action of a transformer (i.e., a single spiral). In an asymmetric ($L_1 \neq L_2$) T-coil-peaked amplifier (Figure 8.3d) [9], the coils are wound to achieve a negative mutual inductance. Consider the response of the asymmetric T-coil (ATC) amplifier to a step input rise (fall) as shown in Figure 8.7. The initial current discharges (charges) only C_1 due to the presence of the secondary inductor L_2 (capacitor splitting). After some delay, the drain current begins to flow from (to) L_2, which causes a proportional amount of current to flow from (to) C_2. The negative magnetic coupling allows for an initial boost in the current flow from (to) the load capacitance C_2, because the capacitor is effectively connected in series with the negative mutual inductance ($-M$) element of the T-coil. This allows for an improvement in fall-time

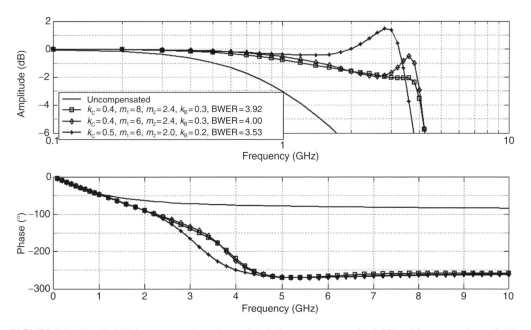

FIGURE 8.6 Bandwidth improvements and associated phase responses with BSS peaking versus $k_C = C_1/C$.

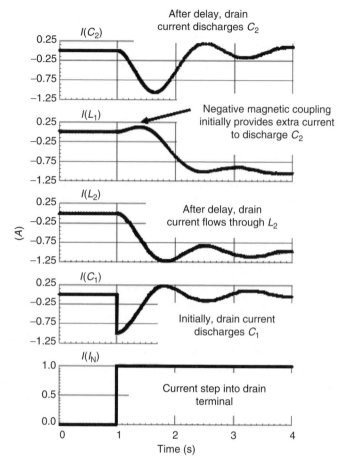

FIGURE 8.7 Detailed transient response waveforms to a step input current in an ATC peaked amplifier.

(rise-time) and bandwidth. The coupling constant, k_m, is related to the mutual inductance, M, as $k_m = k_m = M / \sqrt{L_1 L_2}$. Hence, the transimpedance is

$$Z_{\text{atc}}(s) = \frac{R + s(L_1 + M)}{1 + sRC + s^2(L_1 C_2 + \{L_1 + L_2 + 2M\} C_1) + s^3 RC_1 C_2 L_2 + s^4 C_1 C_2 (L_1 L_2 - M^2)} \quad (8.9)$$

Substituting ω_0, m_1, m_2, k_m, and k_C, the normalized transimpedance is

$$Z_{N,\text{atc}}(s) = \frac{1 + \left(\dfrac{1}{m_1} + \dfrac{k_m}{\sqrt{m_1 m_2}} \right) \dfrac{s}{\omega_0}}{1 + \dfrac{s}{\omega_0} + \left(\dfrac{1}{m_1} + \dfrac{k_C}{m_2} + \dfrac{2k_C k_m}{\sqrt{m_1 m_2}} \right) \dfrac{s^2}{\omega_0^2} + \left(\dfrac{k_C(1 - k_C)}{m_2} \right) \dfrac{s^3}{\omega_0^3} + \left(\dfrac{k_C(1 - k_C)(1 - k_m^2)}{m_1 m_2} \right) \dfrac{s^4}{\omega_0^4}}$$

$$(8.10)$$

Table 8.3 shows results for several values of k_C and passband ripple. Although the nonpeaked cases show large BWER, they are difficult to implement for the reasons mentioned earlier. For 2 dB peaking, a BWER of 5.59 is obtained for $k_C = 0.1$. Figure 8.8 plots the magnitude and phase improvements for various values of k_C.

TABLE 8.3
ATC Peaking Summary

$k_C = C_1/C$	Peak (dB)	$m_1 = R^2C/L_1$	$m_2 = R^2C/L_2$	$k_m = \frac{M}{\sqrt{L_1 L_2}}$	BWER
0.1	0	4.0	1.6	−0.7	4.63
	1	3.5	1.2	−0.6	4.92
	2	3.5	1.6	−0.6	5.59
0.2	0	5.5	2.4	−0.6	4.14
	1	3.0	2.0	−0.6	4.51
	2	4.0	2.4	−0.5	4.86
0.3	0	4.0	2.8	−0.5	3.93
	1	3.5	2.0	−0.4	3.98
	2	4.0	2.8	−0.4	4.54

Employing an ATC and properly utilizing the drain capacitance (C_1) lead to pole-zero locations that are optimized for a large BWER. The classical bridged T-coil network (with $L_1 = L_2$) does not include the effect of drain capacitance and attains a BWER of only 2.83.

8.6 DESIGN OF HIGH-SPEED WIDEBAND DIFFERENTIAL AMPLIFIERS

When designing wideband amplifiers, a trade-off between gain and bandwidth is usually required due to the constant and finite gain-bandwidth (GBW) product. In deep-submicron CMOS, the trade-off is exacerbated because the GBW product is not fixed, but actually decreases with increased gain owing to parasitic effects [12]. This is mostly due to increased small-signal drain conductance (g_{ds}), with increased transistor W/L ratio. In order to simultaneously achieve both high gain and wide bandwidth, a compromise is made by cascading multiple stages, each designed for low gain and wide bandwidth, to provide the desired overall gain. This trade-off is unpleasant because of the large

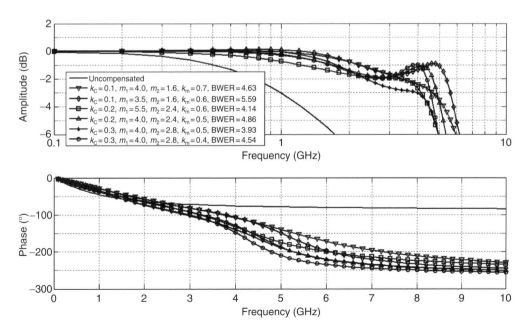

FIGURE 8.8 Bandwidth improvements and associated phase responses with ATC peaking versus $k_C = C_1/C$.

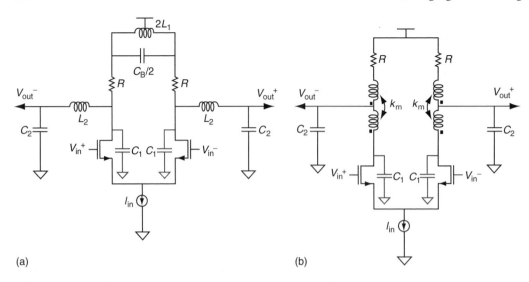

(a)   (b)

FIGURE 8.9   (a) BSS-peaked amplifier and (b) ATC-peaked amplifier.

power consumption and the need to design each stage for even larger bandwidth than required due to the bandwidth shrinkage effects associated with cascading.

Using the techniques presented earlier, an exceptionally large BWER is obtained for a variety of load values. Because of this characteristic, it is possible to favor gain in the GBW trade-off, thus increasing the gain per stage, while still meeting the desired gain and bandwidth goals with the minimum number of stages. To this end, single-stage amplifiers with different k_C values are designed to achieve gains greater than 10 dB with corresponding bandwidths of approximately 10 GHz. A BSS amplifier with $k_C = 0.4$ (Figure 8.9a) and an ATC-peaked amplifier with $k_C = 0.3$ (Figure 8.9b) are designed along with unpeaked reference amplifiers (Figure 8.2) for comparison.

For simplicity, the BSS amplifiers utilize standard spiral inductors available from a standard design kit, whereas the ATC amplifier requires the design of a custom T-coil. Due to parasitic effects, after choosing design values from the appropriate tables, the amplifier designs are optimized to ensure maximum bandwidth, gain, and gain-flatness; it is noted that the pole-zero placement of the amplifiers is substantially affected not only by transistor parasitics, but also by parasitics associated with the spiral inductor structure. Parasitic-aware optimization is performed to determine the final design parameters [13]. As gain is increased, the effect of g_{ds} increases and BWER decreases from its theoretical value. BWER approaches its ideal value as gain is decreased, but never achieves it due to the effect of parasitics. It is shown in the final design results that the BSS amplifier designed for 14 dB gain exhibits a larger deviation from the calculated (BWER = 4) to simulated (BWER = 3.3) than the ATC amplifier, which is designed for a smaller gain of 12 dB. It has calculated and simulated BWER values of 4.6 and 4.2, respectively. The single-stage amplifiers of Figure 8.9 achieve the largest combined gain and BWER values reported to date, and the total current consumption in each differential amplifier is only 15 mA.

8.7   EFFECTS OF OTHER DEVICE PARASITICS

In order to fully exploit scaling in CMOS, it is desirable to realize the desired voltage gain in as few stages as possible. This serves to reduce die area and power consumption, and mitigates the bandwidth shrinkage effects when several stages are cascaded [6]. To meet these objectives, the size of M_1 must be increased because the transconductance, g_m, of the amplifier scales with device width [12]. In earlier derivations, the parasitic gate-to-drain capacitance, C_{gd}, and the small-signal

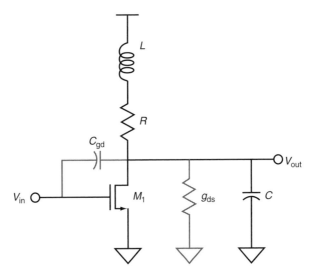

FIGURE 8.10   Shunt-peaked common-source amplifier with small-signal driver device parasitics C_{gd} and g_{ds}.

channel conductance, g_{ds}, were ignored and the gain expressions of Equations 8.1 and 8.2 were obtained; however, as the device width is increased to increase g_m, deleterious increases in the C_{gd} and g_{ds} parasitics follow. To examine the effects of these increasing parasitics, a simple shunt-peaked common-source amplifier is considered. Specifically, the shunt-peaked common-source amplifier is modified to include the parasitic components C_{gd} and g_{ds} of the driver device, M_1, as shown in Figure 8.10.

8.7.1   EFFECTS OF C_{gd}

An expression for the voltage gain of the shunt-peaked amplifier including the gate-to-drain capacitance C_{gd} (Figure 8.10) is

$$A_V(s) = \frac{\left(sC_{gd} - g_m\right)(R + sL)}{s^2 L \left(C + C_{gd}\right) + sR \left(C + C_{gd}\right) + 1} \tag{8.11}$$

From this expression it is seen that C_{gd} introduces a RHP (Right Half Plane) zero, and as it is increased, it degrades the settling response of the amplifier. As in previous analyses, a design parameter relating C_{gd} to the load capacitance is introduced, $k_p = C_{gd}/(C_{gd} + C)$, and with $\omega_z = g_m/C_{gd}$ substituted into Equation 8.10:

$$A_V(s) = -g_m R \frac{(1 - s/\omega_z)(1 + s/m\omega_0)}{1 + s/\omega_0 \left(1 - k_p\right) + s^2/m\omega_0 \left(1 - k_p\right)} \tag{8.12}$$

From Equation 8.12, if C_{gd} is small (i.e., $k_p = 0$) the RHP zero is at a very high frequency. However, as C_{gd} is increased the zero moves to a lower frequency where it adversely impacts the frequency response and ultimately the step response of the amplifier. The normalized frequency–domain responses for several values of k_p are shown in Figure 8.11.

It should be noted that as the amplifier is scaled to increase gain, C_{gd} is also increased thus bringing the RHP zero to a lower frequency. Because of this, there is always some trade-off between latency and gain in wideband amplifier design.

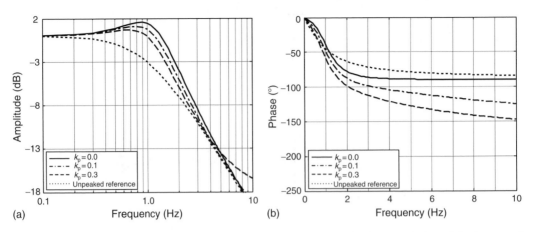

FIGURE 8.11   Normalized (a) gain responses and (b) phase responses of a shunt-peaked amplifier for several values of k_{p}.

8.7.2   EFFECTS OF g_{ds}

An expression for the voltage gain of a shunt-peaked amplifier including the drain-to-source conductance, g_{ds}, (Figure 8.10) is

$$A_{\mathrm{V}}(s) = -g_{\mathrm{m}} \frac{R + sL}{(1 + Rg_{\mathrm{ds}}) + s(RC + g_{\mathrm{ds}}L) + s^2 LC} \tag{8.13}$$

Substituting $k_{\mathrm{r}} = 1/(1 + Rg_{\mathrm{ds}})$ and $\omega_{\mathrm{g}} = g_{\mathrm{ds}}L$ into Equation 8.13 gives

$$A_{\mathrm{V}}(s) = -g_{\mathrm{m}} R k_{\mathrm{r}} \frac{(1 + s/m\omega_0)}{1 + s\left(1/\omega_0 + \omega_{\mathrm{g}}\right) k_{\mathrm{r}} + s^2 k_{\mathrm{r}}/m\omega_0^2} \tag{8.14}$$

It can be seen from Equation 8.14 that g_{ds} has two important effects on the performance of the circuit. If g_{ds} is small (i.e., $k_{\mathrm{r}} = 1$), the response is that of the simple shunt-peaked amplifier. Increasing g_{ds} results in a reduction of the DC gain and a repositioning of the poles. The normalized frequency–domain responses of this system for several values of k_{r} are shown in Figure 8.12. It is clear that

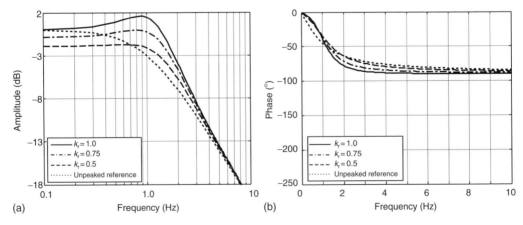

FIGURE 8.12   Normalized (a) gain responses and (b) phase responses of a shunt-peaked amplifier for several values of k_{r}.

finite g_{ds} has adverse effects on the DC gain of the amplifier, as expected. Furthermore, there is a reduction of the bandwidth. In a simple shunt-peaked design, the only degree of freedom is the size of the peaking inductance. Because of this lack of degrees of freedom, it is difficult to mitigate the effects of the C_{gd} and g_{ds} parasitics without using cascade configurations; however, other peaking techniques provide more degrees of freedom that can be used to exert control over the frequency- and time-domain responses.

The analysis of these parasitic effects is straightforward for the shunt-peaked amplifier; the adverse effects of the parasitics are also evident for other peaking techniques, but the increased complexity in the transfer functions obscures intuition. Thus, for more complex peaking techniques such as ATC, BSS, etc., parasitic-aware optimization is essential to optimize the design [13].

8.8 MEASUREMENT RESULTS

8.8.1 Measurement Methodology

Due to the possibility of corruption of measurements, the circuits are designed as stand-alone amplifiers with neither buffers nor matching circuitry. This is done for several reasons. First, the design of a high-speed buffer is nearly as difficult as the design of a peaked amplifier because its frequency performance must be similar to that of the peaked amplifier. By not including buffer amplifiers, there is no possibility for the corruption of the measurement data by a buffer that is only included for measurement convenience. Second, these amplifiers are rarely designed with matching networks at either the input or output. These circuits add unnecessary loss due to the poor quality factors of the on-chip passives and also add unnecessary size to the design of the wideband amplifier. As with the buffer amplifier, adding a reactive passive matching network could corrupt the measurement of the peaked amplifiers. Third, a resistive match is avoided because of its negative impact on the gain and noise performance (not described here) of the amplifier [6].

By not including this circuitry, some difficulty is introduced in the measurements because most measurement equipment is specifically designed for 50 Ω interface impedances. One possibility for measurement characterization of unmatched devices is to use a network analyzer to find the S-parameters of the data. From there, it is straightforward to manipulate the data into other suitable forms using mixed mode S-parameters and general two-port network theory [14]:

$$\begin{bmatrix} S_{DD} & S_{DC} \\ S_{CD} & S_{CC} \end{bmatrix} = \begin{bmatrix} S_{DD11} & S_{DD12} & S_{DC11} & S_{DC12} \\ S_{DD21} & S_{DD22} & S_{DC21} & S_{DC22} \\ S_{CD11} & S_{CD12} & S_{CC11} & S_{CC12} \\ S_{CD21} & S_{CD22} & S_{CC21} & S_{CC22} \end{bmatrix} \tag{8.15}$$

To find the differential-mode gain of the amplifier, we use the upper left quadrant of the mixed-mode S-parameter matrix of Equation 8.15. From this differential-mode S-parameter data, a differential-mode Z-parameter matrix is obtained using a reference impedance of 50 Ω as constrained by the reference impedance of the measurement equipment.

$$\begin{bmatrix} S_{DD11} & S_{DD12} \\ S_{DD21} & S_{DD22} \end{bmatrix} \overset{Z_0=50}{\Leftrightarrow} \begin{bmatrix} Z_{DD11} & Z_{DD12} \\ Z_{DD21} & Z_{DD22} \end{bmatrix} \tag{8.16}$$

The terms Z_{DD11} and Z_{DD21} represent the differential input impedance and differential forward transimpedance, respectively, while Z_{DD22} and Z_{DD12} represent the differential output impedance and differential reverse transimpedance, respectively. These terms are defined as follows:

$$Z_{DD11} = \frac{V_{DD1}}{I_{DD1}}\bigg|_{I_{DD2} = 0} \tag{8.17}$$

$$Z_{DD12} = \frac{V_{DD1}}{I_{DD2}}\bigg|_{I_{DD1} = 0} \tag{8.18}$$

$$Z_{DD21} = \frac{V_{DD2}}{I_{DD1}}\bigg|_{I_{DD2} = 0} \tag{8.19}$$

$$Z_{DD22} = \frac{V_{DD2}}{I_{DD2}}\bigg|_{I_{DD1} = 0} \tag{8.20}$$

From this it is easy to see that the voltage gain is

$$A_{V_{DD}} = \frac{Z_{DD21}}{Z_{DD11}} \tag{8.21}$$

8.8.2 Measured Data

To verify the bandwidth enhancements that are possible with the aforementioned peaking techniques, the BSS-peaked and ATC-peaked amplifiers are fabricated in a commercially available 0.18 μm CMOS process, featuring six AlCu metal layers, with a top metal thickness of 2 μm. The chip microphotographs are shown in Figure 8.13. The S-parameter data are obtained using a Cascade probe station and a four-port Agilent network analyzer.

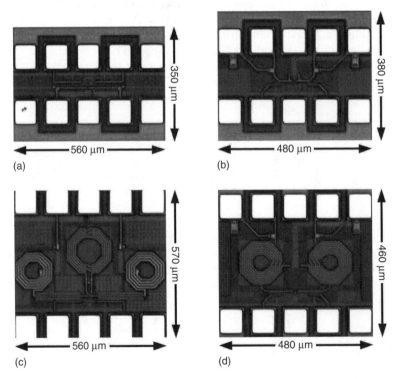

FIGURE 8.13 Chip microphotographs of (a) reference amplifier with $k_C = 0.4$, (b) reference amplifier with $k_C = 0.3$, (c) BSS-peaked amplifier with $k_C = 0.4$, and (d) ATC-peaked amplifier with $k_C = 0.3$.

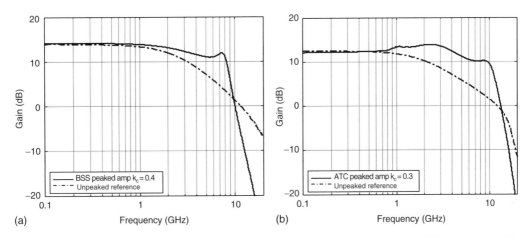

FIGURE 8.14 Gain responses of (a) BSS-peaked amplifier with $k_C = 0.4$ and (b) ATC-peaked amplifier with $k_C = 0.3$.

Many wideband amplifiers are designed to be cascaded, and thus are comprised of many individually peaked stages [5,10,15]. This approach consumes greater area and typically greater power. In order to minimize power and area consumption, single-stage amplifiers with large gain are demonstrated. The circuits are each designed to draw 15 mA from a 2 V supply. The BSS amplifier (Figure 8.13c) with $k_C = 0.4$ achieves a voltage gain of 14.1 dB with a 3 dB bandwidth of 8 GHz. The ATC-peaked amplifier achieves a voltage gain of 12 dB with a 3 dB bandwidth of 10.4 GHz. When compared to their respective unpeaked reference amplifiers, the BSS-peaked amplifier shows a BWER of 3.0, while the ATC-peaked amplifier achieves a BWER of 4.1, the largest measured BWER for a low-pass peaking technique. The measured gain and phase responses are shown in Figures 8.14 and 8.15, respectively.

It is desirable to achieve a linear phase response to avoid pulse dispersion. One way to judge the linearity of the phase response is via the group delay. Group delay, τ_g, is defined as

$$\tau_g = -\frac{\delta\phi(\omega)}{\delta\omega} \tag{8.22}$$

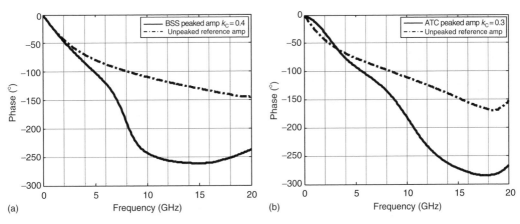

FIGURE 8.15 Phase responses of (a) BSS-peaked amplifier with $k_C = 0.4$ and (b) ATC-peaked amplifier with $k_C = 0.3$.

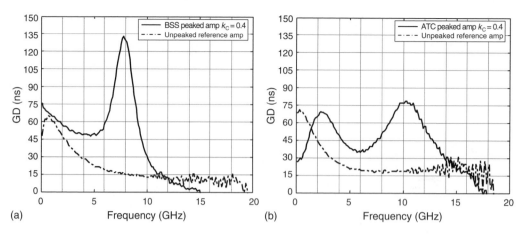

FIGURE 8.16 Group delay responses of (a) BSS-peaked amplifier with $k_C = 0.4$ and (b) ATC-peaked amplifier with $k_C = 0.3$.

Thus, if the group delay response is constant, the phase response is linear. The measured group delays of the peaked amplifiers and their respective reference amplifiers are plotted in Figure 8.16.

8.9 CONCLUSIONS

The design of high-speed cascaded amplifiers desires to achieve wide bandwidth simultaneously with high gain, which reduces the total number of cascaded stages. This saves power and area. Passive peaking techniques like BSS and ATC are attractive for large BWER values. A judicious design selection is optimal as different techniques achieve maximum BWER for different k_C values: specifically, BSS is best for $0.3 > k_C > 0.5$ and the ATC is best for $k_C < 0.3$.

REFERENCES

1. H. Wheeler, Wide-band amplifiers for television, *Proc. of the I.R.E.*, 429–438, July 1939.
2. D. G. Sarma, On distributed amplification, *Proc. of the I.R.E.*, 102B, 689–697, September 1955.
3. B. Hofer, *Amplifier Freq. and Transient Response (AFTR) Notes*, Tektronix, Inc., Beaverton, OR, 1982.
4. R. Roovers, D. M. W. Leenaerts, J. Bergervoet, K. S. Harish, R. C. H. van de Beek, G. van der Weide, H. Waite, Y. Zhang, S. Aggarwal, and C. Razzell, An interference-robust receiver for ultra-wideband radio in SiGe BiCMOS technology, *IEEE J. Solid-State Circuits*, 40, 2563–2572, 2005.
5. S. Galal and B. Razavi, 40 Gb/s amplifier and ESD protection circuit in 0.18 μm CMOS technology, *IEEE J. Solid-State Circuits*, 39, 2389–2396, 2004.
6. S. Shekhar, J. S. Walling, and D. J. Allstot, Bandwidth extension techniques for CMOS amplifiers, *IEEE J. Solid-State Circuits*, 41, 2424–2439, 2006.
7. T. H. Lee, *Planar Microw. Eng.*, Cambridge University Press, Cambridge, UK, 2004.
8. S. S. Mohan, Md. M. Hershenson, S. P. Boyd, and T. H. Lee, Bandwidth extension in CMOS with optimized on-chip inductors, *IEEE J. Solid-State Circuits*, 35, 346–355, 2000.
9. F. A. Muller, High-frequency compensation of RC amplifiers, *Proc. of the I.R.E.*, 1271–1276, 1954.
10. B. Analui and A. Hajimiri, Bandwidth enhancement for transimpedance amplifiers, *IEEE J. Solid-State Circuits*, 39, 1263–1270, 2004.
11. C.-H. Wu, C.-H. Lee, W.-S. Chen, and S.-I. Liu, CMOS wideband amplifiers using multiple inductive-series peaking technique, *IEEE J. Solid-State Circuits*, 40, 548–552, 2005.
12. E. Crain and M. Perrott, A numerical design approach for high speed, differential, resistor-loaded, CMOS amplifiers, *IEEE Int. Symp. on Circuits and Syst.*, pp. 508–511, 2004.

13. D. J. Allstot, K. Choi, and J. Park, *Parasitic-Aware Optimization of CMOS RF Circuits*, Kluwer Academic, Dordrecht, 2003.
14. D. E. Bockelman and W. R. Eisenstadt, Combined differential and common-mode scattering parameters: theory and simulation, *IEEE Trans. Microw. Theory Techniques*, 43, 1530–1539, 1995.
15. K. Kanda, D. Yamakazi, T. Yamamoto, M. Horinaka, J. Ogawa, H. Tamura, and H. Onodera, 40 Gb/s 4:1 MUX/1:4 DEMUX in 90 nm standard CMOS, *IEEE Int. Solid-State Circuits Conf.*, pp. 152–153, 2005.

9 CMOS Active Transformers and Their Applications

Fei Yuan

CONTENTS

9.1 INTRODUCTION

The use of complementary metal oxide semiconductors (CMOS) on-chip spiral transformers in radiofrequency (RF) and high-speed data communications has emerged recently [1–6]. As compared with spiral inductors, spiral transformers offer the advantages of a reduced silicon area with an increased inductance and quality factor. The drawbacks of spiral inductors such as a low quality factor, a low self-resonant frequency, a nontunable inductance, and the need for a large silicon area, however, remain. Although micromachined inductors and transformers, also known as microelectromechanical system (MEMS) inductors and transformers, offer a very large quality factor and tunable inductance, the need for monolithic CMOS–MEMS processes, however, significantly increases the cost to fabricate these devices. CMOS active inductors that are synthesized using active devices offer many unique advantages of CMOS active inductors over their passive spiral counterparts including virtually no chip area requirement, a large and variable inductance, a tunable quality factor, and full compatibility with digitally oriented CMOS technologies. These inductors have been used successfully in many high-speed applications such as optical receivers [7] and multi-Gbps clock and data

135

recovery [8]. This chapter extends the design methodology of CMOS active inductors to CMOS active transformers. We show that for each CMOS active inductor, there exists a set of corresponding CMOS active transformers. The characteristics of active inductors are inherited by active transformers. The chapter starts with a brief review of the fundamentals of active inductors. It is followed by the introduction of active transformers with a detailed mathematical treatment. The characterization of active transformers is presented. The nonidealities of transconductors synthesizing active transformers and their effect on the performance of the active transformers are investigated. CMOS implementation of four active transformers derived from the same CMOS active inductor is presented. The application of active transformers is exemplified using an active transformer quadrature oscillator.

9.2 ACTIVE INDUCTORS

9.2.1 CONFIGURATION AND CHARACTERISTICS

A gyrator consists of a pair of back-to-back connected transconductors. When one port of a gyrator is terminated with a capacitive load, as shown in Figure 9.1, the other port of the gyrator-C network exhibits an inductive characteristic. The synthesized inductor is known as gyrator-C active inductors. The input admittance of the active inductor is given by

$$Y = sC_2 + g_{o2} + \frac{1}{s\left(\frac{C_1}{g_{m1}g_{m2}}\right) + \frac{g_{o1}}{g_{m1}g_{m2}}}, \qquad (9.1)$$

where

g_{mj} is the transconductance of the transconductor j of the gyrator

g_{oj} and C_j are the conductance and capacitance encountered at node j of the active inductor, respectively

g_{oj} and C_j include both the input and output capacitances and resistances of the transconductors connected to the node. Equation 9.1 can be represented by the resistance, inductance, capacitance (RLC) network shown in Figure 9.1 with its parameters given by

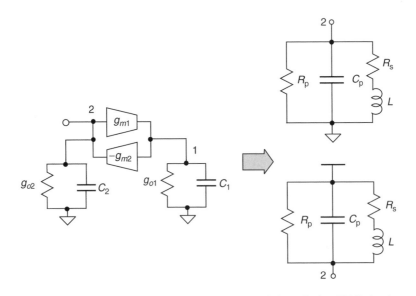

FIGURE 9.1   Configuration of single-ended active inductors and their equivalent RLC circuits.

$$R_p = \frac{1}{g_{o2}},$$

$$C_p = C_2,$$

$$R_s = \frac{g_{o1}}{g_{m1}g_{m2}}, \tag{9.2}$$

$$L = \frac{C_1}{g_{m1}g_{m2}}.$$

Note that the inductance of the active inductor is determined by the capacitance at node 1 and the transconductance of the transconductors of the gyrator. It is independent of both the conductance encountered at nodes 1 and 2, and the capacitance at node 2.

9.2.2 FREQUENCY RANGE

To find out the frequency range over which the gyrator-C network exhibits an inductive characteristic at port 2, we examine the impedance of the equivalent RLC network of the active inductor

$$z_{in} = \left(\frac{R_s}{C_p L}\right) \frac{s\frac{L}{R_s} + 1}{s^2 + s\left(\frac{1}{R_p C_p} + \frac{R_s}{L}\right) + \frac{R_p + R_s}{R_p C_p L}}. \tag{9.3}$$

The input impedance has a zero at the frequency

$$\omega_z = \frac{R_s}{L} = \frac{g_{o1}}{C_1}. \tag{9.4}$$

The frequency of the zero is solely determined by the conductance and capacitance at node 1 and is independent of the conductance and capacitance at node 2. In case complex conjugate poles are encountered, which give the largest bandwidth, the impedance has its pole resonant frequency at

$$\omega_p = \sqrt{\frac{R_p + R_s}{LC_p R_p}} \approx \sqrt{\frac{1}{LC_p}} = \sqrt{\omega_{t1}\omega_{t2}}, \tag{9.5}$$

where

$$\omega_{tj} = \frac{g_{mj}}{C_j}. \tag{9.6}$$

is the cutoff frequency of the capacitively loaded transconductor j. This frequency is set by both the load capacitance and the transconductance of the transconductor. It is seen that the pole resonant frequency is approximately the same as the self-resonant frequency of the RLC equivalent circuit of the active inductor. This frequency is the self-resonant frequency of the active inductor. The preceding results demonstrate that the self-resonant frequency of the active inductor is set by the cutoff frequency of the transconductors constituting the active inductor. The Bode plots of the input impedance of the active inductor are sketched in Figure 9.2. It is evident that the gyrator-C network is inductive only when $\omega_z < \omega < \omega_p$. It should be emphasized that ω_t of the transconductors can be made close to the intrinsic cutoff frequency of transistors when the transconductors are primitively configured and C_j is the intrinsic capacitance of the devices, the self-resonant frequency of active inductors can thus be made close to the cutoff frequency of transistors, much higher than the self-resonant frequency of spiral inductors, which has a typical value of a few GHz.

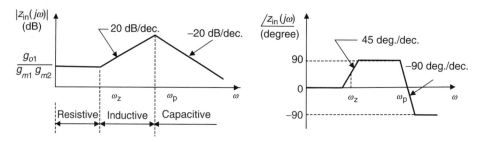

FIGURE 9.2   Bode plots of the input impedance of active inductors.

9.2.3 QUALITY FACTOR

The parasitic resistances of an active inductor give rise to the ohmic loss of the active inductor. There are two parasitic resistances associated with an active inductor, namely the parasitic parallel resistance R_p and the parasitic series resistance R_s. Both affect the quality factor of the active inductor. If the quality factor of an active inductor is dominated by the series resistance, the quality factor of the active inductor at frequencies below the cutoff frequency of the transconductors synthesizing the active transformer is given by

$$Q_s \approx \frac{\omega L}{R_s}. \tag{9.7}$$

If the quality factor of the active inductor is determined mainly by the parasitic parallel resistance of the active inductor, it is evaluated from

$$Q_p \approx \frac{R_p}{\omega L}. \tag{9.8}$$

Whether Equation 9.7 or 9.8 should be used to evaluate the quality factor of an active inductor depends upon the configuration of the active inductor.

9.2.4 CMOS ACTIVE INDUCTORS

A large number of CMOS active inductors have been developed recently. In this section, we limit our attention to a specific set of CMOS active inductors proposed by Wu et al. [9,10]. The schematics of Wu current reuse active inductors are shown in Figure 9.3. Because the biasing current source behaves as an open circuit in the small-signal analysis, both active inductors are single-ended.

9.3   ACTIVE TRANSFORMERS

9.3.1 CONFIGURATION

A spiral transformer is constructed by coupling two spiral inductors via a magnetic link. In the same way, an active transformer can be constructed by coupling two active inductors via an electrical link, as shown in Figure 9.4. Each active inductor forms a winding of the active transformer. The coupling network is transconductor-based to take advantage of the fact that transconductors can be constructed using simple circuit configurations, such as a single mixed oxide semiconductor field-effect transistor (MOSFET), and that there is no direct current path between the two active inductors constituting the active transformer such that both active inductors of the active transformer can be biased and tuned independently. The coupling of the two active inductors can be either from the interface node of one active inductor to the internal node of the other or vice versa.

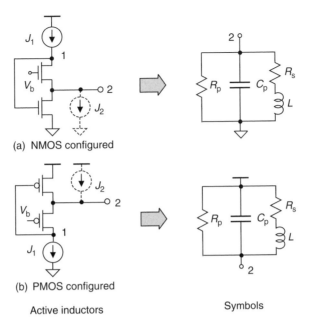

(a) NMOS configured

(b) PMOS configured

Active inductors

Symbols

FIGURE 9.3 Wu current reuse active inductors.

9.3.2 CHARACTERIZATION

There are a number of figures-of-merit that quantify the performance of transformers. In what follows, we examine some of the most important figures-of-merit that quantify the performance of active transformers.

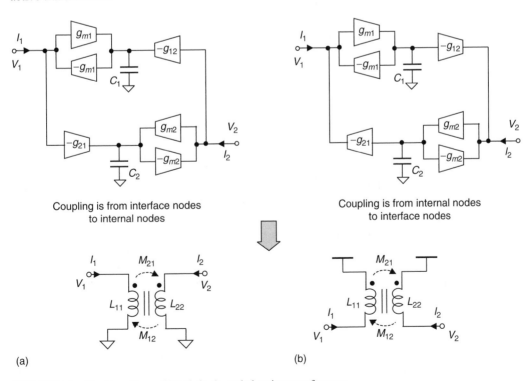

Coupling is from interface nodes
to internal nodes

Coupling is from internal nodes
to interface nodes

(a)   (b)

FIGURE 9.4 Configurations of ideal single-ended active transformers.

9.3.2.1 Self and Mutual Inductances

Because an active transformer is constructed by coupling two active inductors using a transconductor network, both windings synthesized by active inductors exhibit self inductances. In addition, mutual inductances between the two winding also exist. In this section, we examine these quantities in detail. Assume that all blocks of the active transformer are ideal and linear. It can be shown that the voltage of the secondary winding of the active transformer is given by

$$V_2(s) = \frac{sC_2}{g_{m2}^2 \Delta} I_2(s) + \left(\frac{g_{21}}{g_{m1}} \right) \frac{sC_1}{g_{m1}^2 \Delta} I_1(s), \tag{9.9}$$

where

$$\Delta = 1 - \frac{g_{12}g_{21}}{g_{m1}g_{m2}}. \tag{9.10}$$

Because for an ideal linear transformer, we have

$$V_2(s) = sM_{21}I_1(s) + sL_{22}I_2(s), \tag{9.11}$$

where L_{22} and M_{21} denote the self-inductance of the secondary winding and the mutual inductance from the primary winding to the secondary winding of the active transformer, respectively. A comparison of the preceding two equations yields the self-inductance of the secondary winding and the mutual inductance from the primary winding to the secondary winding of the active transformer

$$L_{22} = \frac{sC_2}{g_{m2}^2 \Delta}, \tag{9.12}$$

$$M_{21} = \left(\frac{g_{21}}{g_{m2}} \right) \frac{sC_1}{g_{m1}^2 \Delta}. \tag{9.13}$$

In a very similar way one can show that

$$L_{11} = \frac{sC_1}{g_{m1}^2 \Delta}, \tag{9.14}$$

$$M_{12} = \left(\frac{g_{12}}{g_{m1}} \right) \frac{sC_2}{g_{m2}^2 \Delta}, \tag{9.15}$$

where L_{11} and M_{12} denote the self-inductance of the primary winding and the mutual inductance from the secondary winding to the primary winding. Note that in order to have positive inductances, $g_{12}g_{21} < g_{m1}g_{m2}$ is required. If $g_{12}g_{21} \ll g_{m1}g_{m2}$, L_{11} and L_{22} are independent of g_{12} and g_{21} whereas M_{12} and M_{21} are linearly proportional to g_{12} and g_{21}. The mutual inductance can thus be tuned by varying the transconductance of the coupling transconductor network without affecting the self-inductances.

It is observed that the mutual inductance from the primary winding to the secondary winding can be made different from that from the secondary winding to the primary winding. This differs from ideal transformers where there is only one mutual inductance. Also, if both the signs of g_{12} and g_{21} are changed, L_{11} and L_{22} will remain unchanged. M_{12} and M_{21}, on the other hand, change their signs. Active transformers with negative mutual inductances can be constructed in this way.

9.3.2.2 Turn Ratio

The turn ratio of an ideal transformer quantifies the ratio of the number of the turn of one winding of the transformer to that of the other winding. It is closely related to the self-inductances of the transformer as per the following equation:

$$n_{21} = \sqrt{\frac{L_{22}}{L_{11}}},$$

$$n_{12} = \sqrt{\frac{L_{11}}{L_{22}}}. \tag{9.16}$$

Substituting the expression of the self and mutual inductances of active transformers into Equation 9.16 yields the turn ratio of the active transformers:

$$n_{21} = \frac{g_{m1}}{g_{m2}}\sqrt{\frac{C_2}{C_1}},$$

$$n_{12} = \frac{g_{m2}}{g_{m1}}\sqrt{\frac{C_1}{C_2}}. \tag{9.17}$$

It is seen that both the conductance encountered at nodes 1 and 2, and the coupling transconductances g_{12} and g_{21} have no effect on the turn ratio. They are solely determined by the load capacitance and the transconductance of the transconductors of the gyrator.

9.3.2.3 Coupling Factors

The coupling factors between the primary winding and the secondary winding of an ideal linear transformer are defined as

$$k_{21} = \frac{M_{21}}{\sqrt{L_{11}L_{22}}},$$

$$k_{12} = \frac{M_{12}}{\sqrt{L_{11}L_{22}}}. \tag{9.18}$$

Substituting the expression of the self and mutual inductances of active transformers into Equation 9.18 yields the coupling factors of active transformers:

$$k_{21} = \frac{g_{21}}{g_{m1}}\sqrt{\frac{C_1}{C_2}},$$

$$k_{12} = \frac{g_{12}}{g_{m2}}\sqrt{\frac{C_2}{C_1}}. \tag{9.19}$$

k_{21}, k_{12} can be tuned by varying g_{21}, g_{12} without affecting L_{11}, L_{22}.

9.3.2.4 Voltage Ratio

The ratio of the voltage of the primary winding to that of the secondary winding of an active transformer can be derived by applying an ideal voltage source to the primary winding of the active transformer while leaving the secondary winding of the active transformer open-circuited ($I_2(s) = 0$). Because $V_2(s) = sM_{21}I_1(s)$ and $V_1(s) = sL_{11}I_1(s)$, we have

$$V_2(s) = \frac{M_{21}}{L_{11}}V_1(s) = \frac{g_{21}}{g_{m2}}V_1(s). \tag{9.20}$$

The voltage ratio of the active transformer can be changed by varying the coupling transconductance between the active inductors without affecting the self-inductances of the transformer. The voltage ratio of the active transformers can also be represented using the coupling factors

$$V_2(s) = \frac{g_{21}}{g_{m2}} V_1(s) = k_{21} \frac{g_{m1}}{g_{m2}} \sqrt{\frac{C_2}{C_1}} V_1(s) = k_{21}[n_{21} V_1(s)]. \quad (9.21)$$

9.3.2.5 Current Ratio

To obtain the ratio of the current of the primary winding to that of the secondary winding of an active transformer, a test voltage source can be applied to the primary winding of the transformer while the secondary winding of the transformer is short-circuited ($V_2(s) = 0$):

$$I_2(s) = -\frac{M_{21}}{L_{22}} I_1(s) = -\frac{g_{21}}{g_{m2}} \frac{L_{11}}{L_{22}} I_1(s). \quad (9.22)$$

Making use of the expression of the turn ratio and coupling factor yields

$$I_2(s) = -k_{21} \left[\frac{I_1(s)}{n_{21}} \right]. \quad (9.23)$$

The preceding results show that unlike an ideal linear transformer whose voltage and current ratios are given by

$$V_2(s) = n_{21} V_1(s),$$

$$I_2(s) = -\frac{I_1(s)}{n_{21}}, \quad (9.24)$$

the voltage and current transfer characteristics of an active transformer are scaled by both the turn ratio and the coupling factors.

9.3.3 Nonideal Active Transformers

In this section, we examine the characteristics of nonideal active transformers where both the input and output capacitances and resistances of the transconductors synthesizing the active transformers are considered. The block diagram of a nonideal transformer is shown in Figure 9.5. The capacitance and conductance of each node are the total capacitance and resistance encountered at the node. It can be shown that the voltage of the secondary winding is given by

$$V_2(s) = \left(\frac{g_{21}}{g_{m1}^2 g_{m2}} \right) \frac{A(s)I_1(s) + B(s)I_2(s)}{C(s)}, \quad (9.25)$$

where

$$A(s) = sC_{11} + g_{o11},$$

$$B(s) \approx \frac{g_{m1}^2}{g_{21}g_{m2}}(sC_{22} + g_{o22}), \quad (9.26)$$

$$C(s) \approx \Delta.$$

Note that we have utilized (i) $g_o \ll g_m$ for devices biased in the saturation and (ii) active transformers are operated at frequencies below the cutoff frequency of the transconductors synthesizing the transformers in simplification of the results. The self-impedance of the secondary winding is obtained from

$$Z_{22} = \left. \frac{V_2(s)}{I_2(s)} \right|_{I_1=0} = sL_{22} + R_{22}, \quad (9.27)$$

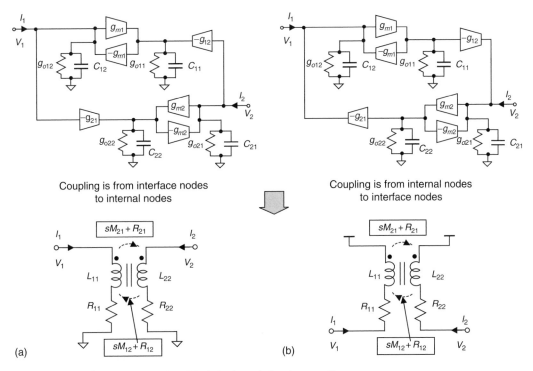

FIGURE 9.5   Configurations of nonideal single-ended active transformers.

where

$$L_{22} = \frac{sC_{22}}{g_{m2}^2 \Delta},$$ (9.28)

$$R_{22} = \frac{g_{o22}}{g_{m2}^2 \Delta}.$$ (9.29)

The above results indicate that the secondary winding can be represented by an inductor of inductance L_{22} in series with a resistor of resistance R_{22}. Similarly, one can show that the mutual impedance from the primary winding to the secondary winding is given by

$$Z_{21} = \left. \frac{V_2(s)}{I_1(s)} \right|_{I_2=0} = sM_{21} + R_{21},$$ (9.30)

where

$$M_{21} = \left(\frac{g_{21}}{g_{m2}} \right) \frac{C_{11}}{g_{m1}^2 \Delta},$$ (9.31)

$$R_{22} = \left(\frac{g_{21}}{g_{m2}} \right) \frac{g_{o11}}{g_{m1}^2 \Delta}.$$ (9.32)

The self-inductance and resistance of the primary winding, the mutual inductance and resistance from the secondary winding to the primary winding of the active transformer can be obtained in a similar manner and the results are given by

$$L_{11} = \frac{sC_{11}}{g_{m1}^2 \Delta},$$ (9.33)

$$R_{11} = \frac{g_{o11}}{g_{m1}^2 \Delta}, \tag{9.34}$$

$$M_{12} = \left(\frac{g_{12}}{g_{m1}}\right)\frac{C_{22}}{g_{m2}^2 \Delta}, \tag{9.35}$$

$$R_{12} = \left(\frac{g_{12}}{g_{m1}}\right)\frac{g_{o22}}{g_{m2}^2 \Delta}. \tag{9.36}$$

The preceding results reveal that when the conductance of each transconductor of an active transformer is accounted for, the transformer exhibits both self and mutual inductances, and parasitic self and mutual series resistances. These parasitic series resistances give rise to finite quality factors of the active transformers.

9.3.4 QUALITY FACTOR

It was shown earlier that if the quality factor of an active inductor is dominated by its parasitic series resistance, the quality factor of the active inductor at frequencies below the cutoff frequency of the transconductors synthesizing the active inductor is given by

$$Q \approx \frac{\omega L}{R_s}. \tag{9.37}$$

This expression can be used to evaluate the self quality factor of the primary and secondary windings of an active transformer and the mutual quality factor from one winding of an active transformer to the other.

9.3.5 CMOS ACTIVE TRANSFORMERS

As pointed out earlier, a CMOS active transformer is constructed by coupling two CMOS active inductors using a coupling transconductor network. For each CMOS active inductor, there exist at least two corresponding active transformers, depending upon whether the coupling is from the interface node of one active inductor to the internal node of the other, or vice versa. In this section, we use the CMOS current reuse active inductors proposed by Wu et al. in Refs. [9,10] and reviewed earlier to demonstrate the circuit implementation of active transformers.

The active transformers shown in Figures 9.6 and 9.7 are derived from Wu current reuse active inductors. The coupling is from the interface node of one active inductor to the internal node of

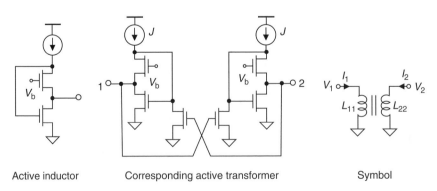

Active inductor   Corresponding active transformer   Symbol

FIGURE 9.6 Wu current reuse active inductor (NMOS) and corresponding active transformer. The coupling is from the interface node of one active inductor to the internal node of the other.

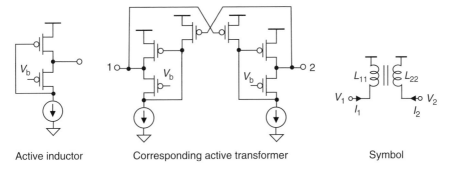

Active inductor   Corresponding active transformer   Symbol

FIGURE 9.7 Wu current reuse active inductor (PMOS) and corresponding active transformer. The coupling is from the interface node of one active inductor to the internal node of the other.

the other. The active transformers shown in Figures 9.8 and 9.9 are also derived from Wu current reuse active. The coupling in these cases is from the internal node of one active inductor to the interface node of the other. The preceding four active transformers have the following common characteristics:

- Coupling networks consist of only two transistors. The simple configuration of the coupling networks ensures a large frequency range of the active transformers, a low level of power consumption, and a small layout area.
- Configuration of the primary winding and that of the secondary winding are identical.
- Both the primary and secondary windings offer a tunable self-inductance.
- Mutual inductances are nontunable as the transconductance of the coupling transconductor network cannot be tuned in this case.

9.4 APPLICATIONS OF CMOS ACTIVE TRANSFORMERS

Active transformers have been used for applications where spiral transformers are usually employed to significantly reduce the silicon area requirement. These applications include current-mode

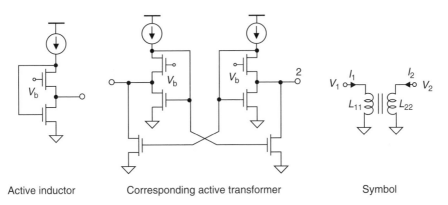

Active inductor   Corresponding active transformer   Symbol

FIGURE 9.8 Wu current reuse active inductor (NMOS) and corresponding active transformer. The coupling is from the internal node of one active inductor to the interface node of the other.

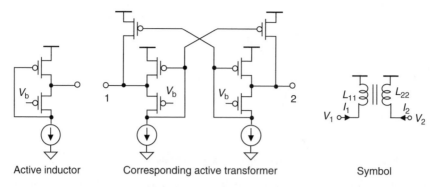

FIGURE 9.9 Wu current reuse active inductor (PMOS) and corresponding active transformer. The coupling is from the internal node of one active inductor to the interface node of the other.

phase-locked loops [11], quadrature phase-shift keying (QPSK) modulators for Bluetooth applications [13,14], and voltage-controlled oscillators [15]. In this section, a 1.6 GHz active transformer quadrature oscillator implemented in Taiwan Semiconductor Manufacturing Company (TSMC)- 0.18 μm 1.8 V CMOS technology is used to demonstrate the usefulness of active transformers. The schematic of the quadrature oscillator employing two active transformers is shown in Figure 9.10. The oscillator consists of two active transformer LC-tank oscillators. Each tank oscillator employs two negative resistors to improve the quality factor of the transformer, subsequently the phase noise of the oscillator. The coupling between the two tank oscillators is via four N-channel metal oxide semiconductor (NMOS)-based transconductors. The oscillation frequency of the oscillator is tuned by varying the dc biasing condition of the active inductors of the transformer. Figure 9.11 shows the phase noise of the quadrature oscillator at the nominal supply voltage of 1.8 V and with ±10% supply voltage variation. They are obtained from SpectreRF with BSIM3v3 RF device models. It is observed that the phase noise of the oscillator is sensitive to supply voltage fluctuation. Replica-biasing technique can be employed to minimize this effect, as demonstrated in Ref. [9].

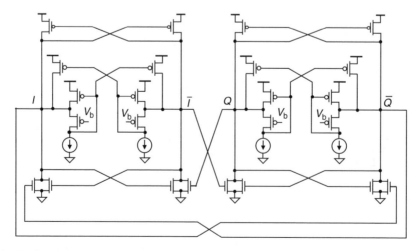

FIGURE 9.10 Quadrature voltage-controlled oscillator.

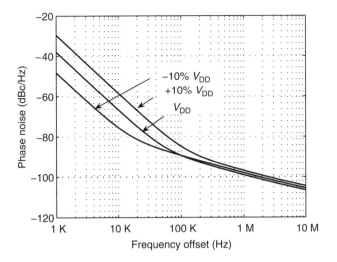

FIGURE 9.11 Phase noise of the quadrature voltage-controlled oscillator.

9.5 CONCLUSIONS

A systematic approach to synthesize active transformers using MOSFETs only has been presented. We have shown that an active transformer can be constructed by coupling two active inductors using a coupling transconductor network. The self and mutual inductances of active transformers have been obtained. The intrinsic relationship between the self and mutual inductances of active transformers has also been established. The voltage and current ratios have also been derived. Nonideal active inductor and their characteristics have been investigated. We have shown that the finite conductance of each node of active transformers gives rise to parasitic series resistances that limit the quality factors of active transformers. Four CMOS active transformers derived from Wu current reuse active inductors have been developed. The applications of CMOS active transformers have been exemplified using a quadrature voltage-controlled oscillator.

REFERENCES

1. G. Kelmens, M. Ghagat, D. Jessie, and N. Frederick, Analysis and circuit modeling of on-chip transformers, *Proc. Topical Meeting on Silicon Monolithic Integr. Circuits in RF Syst.*, pp. 167–170, 2004.
2. K. Kwok and H. Luong, Ultra-low-voltage high-performance CMOS VCOs using transfer feedback, *IEEE J. Solid-State Circuits*, 40(3), 652–660, 2005.
3. J. Zhou and D. Allstot, Monolithic transformers and their application in a differential CMOS RF low-noise amplifier, *IEEE J. Solid-State Circuits*, 33(12), 2020–2027, 1998.
4. A. Zolfaghari, A. Chan, and B. Razavi, Stacked inductors and transformers in CMOS technology, *IEEE J. Solid-State Circuits*, 36(4), 620–628, 2001.
5. Y. Mayevskiy, A. Watson, P. Francis, K. Hwang, and A. Weisshaar, A new compact model for monolithic transformers in silicon-based RFICs, *IEEE Microw. Wireless Components Lett.*, 15(6), 419–421, 2005.
6. K. Chong and Y. Xie, High-performance on-chip transformers, *IEEE Electron Device Lett.*, 26(8), 557–559, 2005.
7. E. Sackinger and W. Fischer, A 3-GHz 32-dB CMOS limiting amplifier for SONET OC-48 receivers, *IEEE J. Solid-State Circuits*, 35(12), 1884–1888, 2000.
8. S. Song, S. Park, and H. Yoo, A 4 Gb/s CMOS clock and data recovery circuit using 1/8-rate clock technique, *IEEE J. Solid-State Circuits*, 38(7), 1213–1219, 2003.
9. Y. Wu, X. Ding, M. Ismail, and H. Olsson, RF bandpass filter design based on CMOS active inductors, *IEEE Trans. Circuits Syst. II*, 50(12), 942–949, 2003.

10. Y. Wu, M. Ismail, and H. Olsson, CMOS VHF/RF CCO based on active inductors, *IEEE Electron. Lett.*, 37(8), 472–473, 2001.

11. D. DiClemente and F. Yuan, Current-mode phase-locked loops: a new architecture, *IEEE Trans. Circuits Syst. II*, 54(4), 303–307, 2007.

12. D. DiClemente, F. Yuan, and A. Tang, Current-mode phase-locked loops with CMOS active transformers, *IEEE Trans. Circuits Syst. II*, Accepted for publication in June 2007.

13. A. Tang, F. Yuan, and E. Law, A new CMOS active transformer QPSK modulator with optimal bandwidth control, *IEEE Trans. on Circuits Syst. II*, 55(1), 11–15, 2008.

14. A. Tang, F. Yuan, and E. Law, A new CMOS BPSK modulator with optimal transaction bandwidth control, *Proc. IEEE Int. Symp. Circuits Syst.*, New Orleans, 2007, in press.

15. A. Tang, F. Yuan, and E. Law, Low-noise CMOS active transformer voltage-controlled oscillators, *Proc. IEEE Mid-West Symp. Circuits Syst.*, Montreal, in press.

10 High-Performance Leakage/Variation-Tolerant Circuit Technologies for 45 nm and Below

Amit Agarwal and Ram Krishnamurthy

CONTENTS

10.1 INTRODUCTION

As complementary metal oxide semiconductor (CMOS) technology continues to scale, power dissipation and robustness to leakage, noise, and process variations are becoming major obstacles for circuit design in these nanoscale regimes. Due to increased density of transistors in a die and higher frequencies of operation, the power consumption is reaching cooling capacity limits. On the other hand, due to increased leakage, noise, and process variations, the predictability and therefore the design yield is threatened. This chapter discusses challenges and design solutions for high-performance energy-efficient register file circuit design.

In modern general purpose microprocessors, register files are one of the most performance-critical components. They are required to meet two constrains: (1) single clock cycle read/write latency to support back-to-back read/write operation and (2) multiple read/write port capability to enable the simultaneous access of several execution units in a superscalar processor. To meet these requirements, high fan-in OR wide domino circuits are typically used in local bitline (LBL) and global bitline (GBL) of multiported register files. However, as CMOS devices are being scaled down aggressively in each technology generations to achieve higher integration density and performance, the gate and subthreshold leakage current has increased drastically [1], resulting in poor noise tolerance trend of wide domino gates (Figure 10.1) [2]. This increase in leakage with technology scaling also limits the number of bitcells merged on a single bitline. Circuit robustness can be recovered by upsizing the keeper or using high-V_t devices. However, resulting performance loss due to keeper upsizing or high-V_t devices is too prohibitive. Therefore, alternate bitline circuit techniques that curtail this technology scaling trend are required to achieve high noise immunity while sustaining high performance.

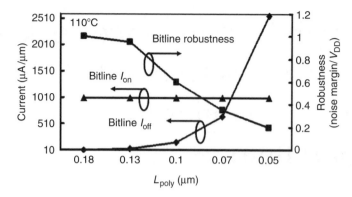

FIGURE 10.1 Bitline robustness versus technology scaling. (From R. Krishnamurthy, A. Alvandpour, G. Balamurugan, N.R. Shanbhag, K. Soumyanath, and S.Y. Borkar, *IEEE J. Solid State Circuits*, 37, 624, 2002. With permission.)

Moreover, controlling the variation in device parameters during fabrication is becoming a great challenge for scaled technologies. The leakage in a device depends on the transistor geometry (gate length, oxide thickness, width, the doping profile, and "halo" doping concentration, etc.), the flat-band voltage, and the supply voltage. Any statistical variation in each of these parameters results in significant spread in the transistor leakage current. Figure 10.2 shows $20\times$ variation in die-to-die N-channel mixed oxide semiconductor (NMOS) leakage in 150 nm technology [3]. Designing for the worst-case leakage (e.g., reducing the skew of static stages or increasing the keeper size for worst-case leakage condition) may cause large number of low-leakage dies to suffer from performance loss due to unnecessary guard-banding. However, underestimating leakage variation will cause unnecessarily low yield, as dies that are free from manufacturing defects are discarded for violating the noise robustness criteria. Hence, there is a need of process variation compensating bitline circuit technique, which improves robustness and delay variation by restoring robustness of worst-case leakage dies and improving performance of low-leakage dies. This chapter reviews novel high-speed and leakage/process-tolerant register file circuits.

The remainder of this chapter is organized as follows. Section 10.2 describes the organization of the 4-read, 4-write ported $128 \times 64b$ register file, the fully time-borrowable two-phase timing plan, and clock generation circuits. Section 10.3 reviews the bitline active leakage scaling issues of conventional low-V_t and dual-V_t dynamic bitline schemes. In Section 10.4, different leakage/process-tolerant circuit techniques are presented, followed by conclusions in Section 10.5.

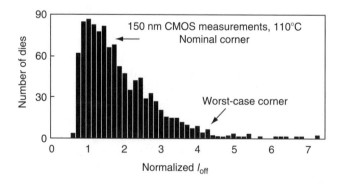

FIGURE 10.2 Normalized NMOS I_{off} variation distribution. (From Kim, C.H., Roy, K., Hsu, S.K., Alvandpour, A., Krishnamurthy, R., and Borkar, S., *IEEE Symposium on VLSI Circuits Digest of Technical Papers*, 205, 2003. With permission.)

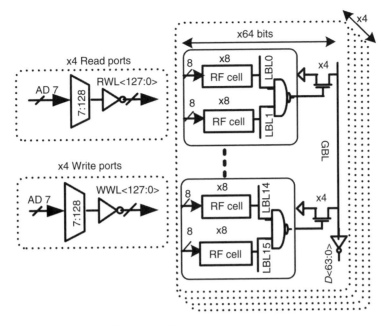

FIGURE 10.3   128 × 64b register file organization.

10.2   REGISTER FILE ORGANIZATION

Figure 10.3 shows the organization of the 4-read, 4-write ported 128 × 64b register file. The complete read/write operation is performed in two cycles. A 7-bit read/write address per port is decoded in the first cycle to deliver the 128 read/write (per port) select signals into the register file array. The decoder cycle is noncritical and implemented in static CMOS, first with a predecoder to minimize interconnect and then with a final decode stage for each read port wordlines. In the next cycle, which is performance-critical, wordline buffers drive across the 64-bit array and bitline evaluation is performed. This cycle includes the wordline delay, LBL delay, merging multiplexer (mux) logic, and GBL delay. Figure 10.4 shows the 2Φ domino timing plan for this cycle. Clock generation and driver circuits are shown in Figure 10.5. Φ1d is a delayed version of incoming clock Φ1. Φ2 clock is locally generated and is an inverted version of Φ1 clock, offering an automatic stretching capability. Thus, when Φ1 core clock period is reduced for slow frequency test/debug, Φ2 clock stretches out

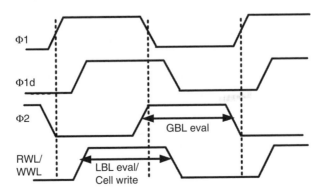

FIGURE 10.4   2Φ domino timing plan. (From Krishnamurthy, R., Alvandpour, A., Balamurugan, G., Shanbhag, N.R., Soumyanath, K., and Borkar, S.Y., *IEEE J. Solid State Circuits*, 37, 624, 2002. With permission.)

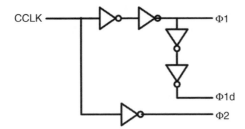

FIGURE 10.5 Clock generation circuits. (From Krishnamurthy, R., Alvandpour, A., Balamurugan, G., Shanbhag, N.R., Soumyanath, K., and Borkar, S.Y., *IEEE J. Solid State Circuits*, 37, 624, 2002. With permission.)

in proportion. $\Phi 2$'s rising edge is positioned slightly ahead of $\Phi 1$'s falling edge to achieve good tolerance to $\Phi 1$ clock skew, jitter, and duty-cycle variations.

At the beginning of the read cycle, 128 D1 footed-domino buffers per port are triggered by $\Phi 1$ and drive the decoded read/write wordlines across the 64-bit array width. The LBL evaluation or bitcell write operation occurs during the $\Phi 1$ portion of the cycle. Figure 10.6 shows the register file bitcell, with symmetric loading of two read ports on each side of the storage cell for optimal cell write stability. Matched pass transistors are used on each side of the storage cell to enable single-ended full-swing write operation. The input data D_{in} is locally inverted to get its complement for the write operation. Figure 10.7a and 10.7b shows the full-swing LBL and GBL scheme. Each LBL (1 per read port) supports 8 bitcells and a two-way merge via a static NAND gate. Data from the storage cell is read by two access transistors per word (M1 and M2) on each LBL and thus forms a dynamic 8-way NAND–NOR. The LBL is triggered by a delayed version of $\Phi 1$, shown in Figure 10.4 as $\Phi 1d$ and is fully time-borrowable, enabling the read/write select signals to arrive into the LBL evaluate phase. The GBL is a dynamic 4-way OR, which merges the LBL 2-input static NAND outputs to deliver a 64-bit word per read port. GBL operation is conducted during $\Phi 2$ of the cycle and to avoid precharge races and crowbar currents at the phase boundary, the GBL domino is footed by the clock transistor. GBL phase is also fully time-borrowable, enabling the LBL NAND-merge outputs to arrive into $\Phi 2$. This register file organization enables the complete LBL and GBL operation to be performed in four gate stages. An equivalent static CMOS LBL and GBL implementation will require seven stages of

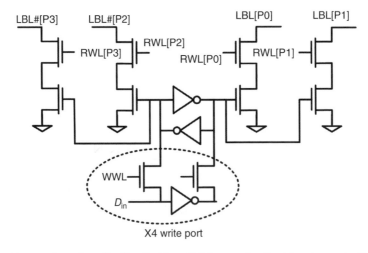

FIGURE 10.6 Symmetric register file bitcell. (From Krishnamurthy, R., Alvandpour, A., Balamurugan, G., Shanbhag, N.R., Soumyanath, K., and Borkar, S.Y., *IEEE J. Solid State Circuits*, 37, 624, 2002. With permission.)

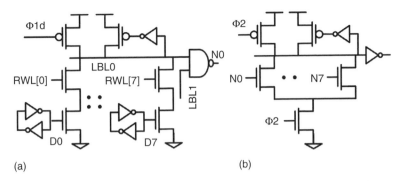

(a)   (b)

FIGURE 10.7   (a) LBL scheme and (b) GBL scheme. (From Krishnamurthy, R., Alvandpour, A., Balamurugan, G., Shanbhag, N.R., Soumyanath, K., and Borkar, S.Y., *IEEE J. Solid State Circuits*, 37, 624, 2002. With permission.)

2- and 3-input NAND–NOR gates. Single-ended read/write selects and bitline signaling are used throughout to reduce wiring congestion and enable a dense layout.

10.3   BITLINE LEAKAGE ROBUSTNESS

In a dynamic LBL or GBL, at the time of evaluation, clock goes high, which turns off the precharge transistor. During this period, LBL and GBL are susceptible to noise due to high active leakage if the dynamic node is supposed to stay high. LBL is more sensitive to noise than GBL due to smaller domino node stored charge and a wider dynamic OR structure (8-way for LBL vs. 4-way for GBL). The worst-case leakage for the conventional LBL scheme that maximizes bitline leakage is when RWL[7:0] = V_{SS} and D0–D7 = V_{DD} after precharge is done and Φ1d clock transitions high (Figure 10.7a). With DC noise on RWL[7:0] signals and the worst-case process leakage corner, the all low-V_t implementation does not meet the minimum noise floor set by realistic supply/ground bounce and coupling noise specification. In this scenario, the bitline keeper is sized to be 10% of the effective NMOS transistor pulldown width in order to replenish the dynamic node during the evaluation phase. A straightforward solution to improve bitline robustness is to strengthen the positive-channel metal oxide semiconductor (PMOS) keeper transistor for a given effective NMOS pulldown strength. However, this increases bitline contention during evaluate operation, resulting in increased bitline short-circuit power consumption and degraded read delay [4].

A more effective alternative is to replace the low-V_t transistors with less leaky high-V_t transistors in a dual-V_t technology [5]. The conventional dynamic dual-V_t register file optimized for high performance uses high-V_t on the LBL access transistors. Low-V_t is used on all other transistors for best performance. For the LBL input setup for worst-case noise described earlier, the high-V_t access transistors now limit the bitline leakage. This implies that the worst-case domino node active leakage is reduced by 10 times, offering significant improvement in robustness. However, this robustness benefit is achieved at the cost of increase in critical path delay compared to the all low-V_t best speed design due to the reduced drive currents of high-V_t transistors.

Reducing the P/N skew of static stage (ratio of PMOS vs. NMOS size of NAND merge) is another way to improve robustness. The P/N skew determines the switching threshold voltage of static NAND merge. This in turn determines the amount of DC droop (in dynamic node of LBL due to leakage), which gets propagated to the output of static NAND merge as noise and hence DC noise margin of LBL. The P/N skew of static stage is design such that the worst-case leakage dies meet the robustness criteria. However, reducing the P/N skew of static stages for worst-case leakage corner causes (i) a large number of low-leakage dies to suffer from performance loss due to unnecessary reduction in skew and (ii) the excess-leakage dies still cannot meet the robustness requirement.

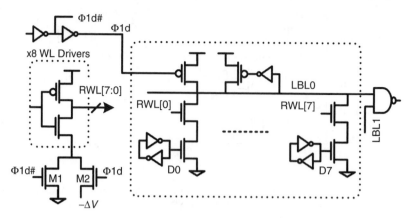

FIGURE 10.8 Wordline underdrive (WLUND) LBL scheme. (From Agarwal, A., Roy, K., Hsu, S., Krishnamurthy, R., and Borkar, S., *IEEE Symposium on VLSI Circuit Digest of Technical Papers*, 386, 2004. With permission.)

10.4 LEAKAGE/PROCESS-TOLERANT REGISTER FILE CIRCUITS

10.4.1 WORDLINE UNDERDRIVE LOCAL BITLINE

Noise immunity of wide dynamic circuits used for driving LBL degrades as V_t is lowered because of excessive leakage in the bitline pull-down devices. In conventional designs, leakage-sensitive devices are made high V_t to achieve required DC robustness. Hence, they cannot take advantage of the large drive current available from low-V_t devices in dual-V_t technology. In the WLUND scheme (Figure 10.8), two extra high-V_t NMOS transistors (M1, M2) are added, which are shared among eight-wordline drivers, driving a common LBL [6]. The source of these transistors is connected to ground and a small negative voltage ($-\Delta V$), respectively. $-\Delta V$ is generated using an efficient off-chip power supply or an internal negative voltage generator [7]. The gates of both the transistors are driven by the LBL clock. The maximum voltage across read-select transistors ($\Delta V + V_{DD}$) is limited to within the gate-oxide reliability limits of the process. During precharge, when clock is low, all the wordlines are strapped to GND (Figure 10.9), M2 is partially turned on due to $V_{GS} = \Delta V$, which results in a short-circuit power between GND and $-\Delta V$. However, since M1 and M2 are shared

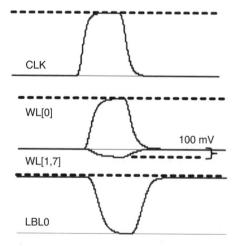

FIGURE 10.9 Operation of WLUND LBL technique. (From Agarwal, A., Roy, K., Hsu, S., Krishnamurthy, R., and Borkar, S., *IEEE Symposium on VLSI Circuit Digest of Technical Papers*, 386, 2004. With permission.)

among eight wordline drivers, overall power overhead is 0.01%. On evaluation, all the wordlines except the accessed entry (which goes to V_{DD}) are pulled to a negative supply through M2. This eliminates the active bitline leakage through the LBL keeper because of the negative V_{GS} applied to the domino pulldown devices regardless of stored value, read-select voltage, and V_t. Therefore, performance-critical devices can be made low-V_t without impacting robustness. This technique does not add any evaluate read/write delay overhead since added transistors are in a noncritical path. The ΔV is chosen such that WLUND LBL achieves the robustness comparable to conventional dual-V_t design, while achieving high performance due to use of low-V_t transistors.

10.4.2 NAND MERGE SKEW LOCAL BITLINE

In conventional design, the P/N skew is designed such that the fast corner dies meet the required noise margin floor constraints. However, for the slow dies, the surplus robustness can be traded off for higher performance via up-skewing the NAND merge. Similarly, the robustness failing dies with excess leakage can be salvaged by the opposite robustness–delay trade-off. The idea of proposed NAND merge skew (NMS) scheme is to make P/N skew programmable based on leakage [6]. Figure 10.10 shows the proposed NMS scheme with a digitally programmable NMS of an LBL. This scheme adds extra PMOS transistors blocks, which can be attached to NAND merge by using programmable bits. The sizes of PMOS blocks are chosen to achieve the NAND skew of 0.5s, s, 1.5s, and 2.0s by asserting appropriate globally routed signals b[1:0], where s is the P/N skew of conventional design. In low-leakage dies, NAND merge is up-skewed, trading off robustness for performance. High-leakage dies NAND merge is down-skewed salvaging the robustness failing dies to improve yield.

Excess dynamic node capacitance due to added extra PMOS transistor causes LBL delay and energy penalty. However, this penalty is more than offset by the opportunistic speedup achieved by up-skewing the NAND merge on low-leakage dies, resulting in an overall performance improvement and reduced die-to-die delay variation. Optimal skew is programmed once via fuses based on die-to-die leakage measurement. The NMS technique can be further improved to compensate within-die variations as well by locally generating control bits b[1:0] using a self-contained on-die leakage sensor [8].

10.4.3 PROCESS VARIATION COMPENSATING DYNAMIC LOCAL BITLINE

Increasing I_{off} with process scaling has forced designers to upsize the LBL keeper in order to obtain an acceptable robustness for worst-case leakage corner dies. However, a large number of low-leakage

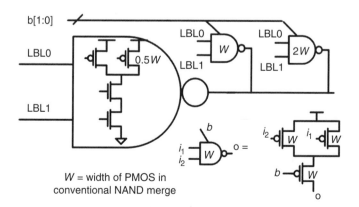

FIGURE 10.10 NMS LBL technique. (From Agarwal, A., Roy, K., Hsu, S., Krishnamurthy, R., and Borkar, S., *IEEE Symposium on VLSI Circuit Digest of Technical Papers*, 386, 2004. With permission.)

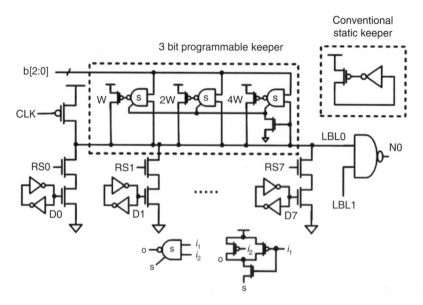

FIGURE 10.11 Process variation compensating dynamic (PCD) LBL technique. (From Kim, C.H., Roy, K., Hsu, S.K., Alvandpour, A., Krishnamurthy, R., and Borkar, S., *IEEE Symposium on VLSI Circuits Digest of Technical Papers*, 205, 2003. With permission.)

dies suffer from the performance loss due to the contention current with the unnecessarily strong keeper, while the excess leakage dies still cannot meet the robustness requirements even with a keeper sized for the fast corner leakage. Rather than using a fixed-strength keeper that will sacrifice the performance of low-leakage dies, just like NMS technique, a postsilicon keeper technique that optimally programs the keeper strength based on the measured die leakage improves robustness and delay variation spread [3].

Figure 10.11 shows the PCD circuit scheme with a digitally programmable 3-bit keeper applied on an 8-way dynamic circuit. Each of the three binary-weighted keepers with respective widths W, 2W, and 4W can be activated or deactivated by asserting appropriate globally routed signals b[2:0]. A desired effective keeper width can be chosen among {0, W, 2W, . . . , 7W}. Excess dynamic node capacitance of PCD scheme due to the keeper circuitry causes a delay penalty. However, this penalty is offset by the opportunistic speedup achieved by keeper downsizing on low-leakage dies, resulting in an overall performance improvement. The PCD technique can be used to compensate both die-to-die variation or within die variation by one-time programming via fuses based on die leakage measurements or by locally generating the control bits b[2:0] using a self-contained on-die leakage sensor [8].

10.5 SUMMARY

Continuous scaling of CMOS devices causes the leakage current to increase every generation, which degrades the noise tolerance of wide domino gates, which are extensively used in high-performance register files. Moreover, the increasing statistical variation in the process parameters has led to significant variation in the transistor leakage current across and within different dies. Designing for the worst-case leakage may cause excessive guard-banding, resulting in a lower performance. This chapter describes bitline leakage impact on high-performance register file circuits with technology scaling and presents different leakage/process-tolerant LBL techniques.

REFERENCES

1. K. Agawa, H. Hara, T. Takayanagi, and T. Kuroda, A bitline leakage compensation scheme for low-voltage SRAMs, *IEEE Journal of Solid State Circuits*, 36, 726–734, 2001.
2. R. Krishnamurthy, A. Alvandpour, G. Balamurugan, N.R. Shanbhag, K. Soumyanath, and S. Y. Borkar, A 130-nm 6 GHz 256 × 32 bit leakage-tolerant register file, *IEEE Journal of Solid State Circuits*, 37(5), 624–632, 2002.
3. C. H. Kim, K. Roy, S. K. Hsu, A. Alvandpour, R. Krishnamurthy, and S. Borkar, A process variation compensation technique for sub-90 nm dynamic circuits, *IEEE Symposium on VLSI Circuits Digest of Technical Papers*, pp. 205–206, June 2003.
4. A. Alvandpour, R. Krishnamurthy, K. Soumyanath, and S. Borkar, A conditional keeper technique for sub-130 nm wide dynamic gates, *IEEE Symposium on VLSI Circuit Digest of Technical Papers*, pp. 29–30, June 2001.
5. M. Anders, R. Krishnamurthy, K. Soumyanath, and R. Spotten, Robustness of sub-70 nm dynamic circuits: Analytical techniques and scaling trends, *IEEE Symposium on VLSI Circuits Digest of Technical Papers*, pp. 23–24, June 2001.
6. A. Agarwal, K. Roy, S. Hsu, R. Krishnamurthy, and S. Borkar, A 90 nm 6.5 GHz 128 × 64b 4-read 4-write ported parameter variation tolerant register file, *IEEE Symposium on VLSI Circuit Digest of Technical Papers*, pp. 386–387, June 2004.
7. H. Tanaka, M. Aoki, S. Kimura, N. Sakashita, H. Hidaka, T. Tachibana, and K. Kimura, A precise on-chip voltage generator for a giga-scale DRAM with a negative word-line scheme, *IEEE Symposium on VLSI Circuits Digest of Technical Papers*, pp. 94–95, June 1998.
8. C. H. Kim, K. Roy, S. K. Hsu, R. Krishnamurthy, and S. Borkar, On-die CMOS leakage current sensor for measuring process variation in sub-90 nm generations, *IEEE Symposium on VLSI Circuits Digest of Technical Papers*, pp. 250–251, June 2004.

11 Soliton and Nonlinear Wave Electronics

David S. Ricketts, Xiaofeng Li, and Donhee Ham

CONTENTS

11.1 INTRODUCTION

Solitons are a unique class of pulse-shaped waves that propagate in nonlinear dispersive media. They maintain spatial confinement of wave energy in a pulse shape in the course of propagation (no dispersion) and exhibit singular nonlinear dynamics [1–4]. Balancing between nonlinearity and

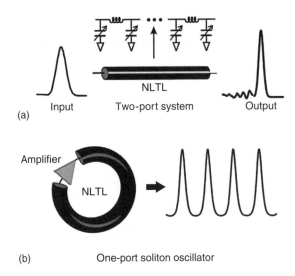

(a)   Input   Two-port system   Output

(b)   One-port soliton oscillator

FIGURE 11.1   (a) Two-port nonlinear transmission line (NLTL) and (b) one-port electrical soliton oscillator. (From Ricketts, D.S., Li, X., Sun, N., Woo, K., and Ham, D., *IEEE J. Solid-State Circuits*, 42, 1657, 2007.)

dispersion causes the soliton phenomenon. Solitons are encountered throughout nature. Shallow water and optical fibers are only a couple of examples of nonlinear dispersive media where solitons (hydrodynamic and optical) are found [4,5].

In electronics, the NLTL, a one-dimensional lattice of inductors and varactors (Figure 11.1a), serves as a nonlinear dispersive medium where electrical solitons can propagate in the form of voltage waves [6,7]. Electrical solitons are especially interesting from an engineering standpoint. Their narrow pulse width directly translates to high temporal resolution and, as such, they have played important roles in ultrafast time-domain metrology: they are used to sample rapidly varying signals or as probe signals in ranging radars and time-domain reflectometry [8–10]. Short pulses have also found active usage in communication, notably, in impulse radio [11].

Due to this engineering promise, the generation of electrical solitons on the NLTL has been extensively studied over the past 40 years [6,7]. The significant advance made in the field is seen in today's state-of-the-art NLTL that can achieve a rise time down to 480 fs [12]. In these previous NLTL studies, however, the NLTL has been used almost exclusively as a "two-port" (input + output) system (Figure 11.1a) where an external high-frequency input is required to produce a soliton pulse output.

Reporting in Ref. [1], the authors introduced the first one-port (output-only) electrical circuit that robustly self-generates a periodic, stable train of electrical solitons with no high-frequency input. This electrical soliton oscillator marks a distinctive departure from the two-port NLTL, as the former is autonomous and self-contained (no input needed), and provides an improved pulse quality control due to self-regulatory mechanisms inherent in any stable oscillator. Construction of such a one-port system has been historically difficult [13,14] due to the oscillation instabilities caused by the nonlinear dynamics of the solitons. Our soliton oscillator was made possible by combining the NLTL with a unique amplifier in a circular topology (Figure 11.1b). The amplifier provides not only gain but also mechanisms to "tame" the instability-prone soliton dynamics on the NLTL, achieving a stable soliton oscillation. At the same time, we have shown in Ref. [15] that the "unruly" behavior of solitons may also be exploited in the soliton oscillator for chaotic signal generation in encrypted communication [17,18]. The chaotic soliton oscillator is potentially advantageous over existing chaos generators due to its ultrawideband enabled by the NLTL [12].

This chapter begins with a review of the NLTL and electrical solitons in Section 11.2 and then the soliton oscillator concept in Section 11.3. Section 11.4 outlines the design procedure for the soliton oscillation waveform in the oscillator. Sections 11.5 through 11.7 demonstrate the concepts of the soliton oscillator with three prototypes: two discrete prototypes (Sections 11.5 and 11.6) that elucidate the operating and stabilizing principles of the soliton oscillator and one chip-scale prototype (Section 11.7) that demonstrates the viability for integrated circuit (IC) applications. These experimental sections are then followed by a summary to date of our soliton oscillator work and future directions. Finally, Section 11.9 concludes this chapter by exploring the polar opposite of a stable soliton oscillator, discussing the promotion of unruly soliton dynamics to create extreme oscillation instability, i.e., chaos.

11.2 NONLINEAR TRANSMISSION LINE AND ELECTRICAL SOLITONS

In this section, we review the properties of electrical solitons and the medium on which we propagate them, the NLTL.

11.2.1 SOLITONS

In a dispersive medium, a pulse will be generally broken into multiple sinusoidal waves in the course of propagation as different Fourier components of the pulse travel at different speeds. In other words, the energy initially locally confined in the pulse cannot maintain the spatial localization as the pulse travels.

In a medium where nonlinearity coexists with dispersion, however, the nonlinearity can balance out the dispersion, and a unique pulse that exhibits no dispersion can be propagated. This pulse is known as a soliton [2–4]. In the absence of loss, the soliton preserves its exact shape while traveling. In the presence of loss, the soliton has to change its shape in the course of propagation as it has to lose energy, but it still maintains spatial localization of energy in the changing pulse shape, not breaking into multiple sinusoids [19]. Solitons also possess other singular properties, as will be described shortly in the context of electrical solitons.

Solitons are encountered throughout nature [20]. Water, plasma, mechanical lattices, optical fibers, and magnetic films are only a few examples that can act as nonlinear dispersive media where solitons can be created and propagated in the form of hydrodynamic waves, charge-density waves, lattice waves, light waves, and spin waves.

11.2.2 ELECTRICAL SOLITONS

In electronics, the NLTL serves as a nonlinear dispersive medium where electrical solitons can propagate in the form of voltage waves [7]. The NLTL is constructed by periodically loading a linear transmission line with varactors such as reverse-biased pn junction diodes or mixed oxide semiconductor (MOS) capacitors (Figure 11.2a). Alternatively, the NLTL can be obtained by arranging inductors and varactors in a 1D ladder network (Figure 11.2b). The nonlinearity of the NLTL originates from the varactors whose capacitance changes with the applied voltage, while the dispersion of the NLTL arises from its structural periodicity.

Figure 11.2c shows the general soliton wave formed on an NLTL in the absence of loss, which is a periodic train of voltage solitons. This waveform is generally known as a cnoidal wave, and is a solution to the Korteweg-de Vries (KdV) wave equation [3]. There are an infinite number of possible cnoidal waves that can form on a given NLTL by varying the amplitude, A, pulse spacing, Λ, and the pulse width, W_z (Here, A, Λ, and W_z are with reference to Figure 11.2c). Initial or boundary conditions will determine the specific cnoidal waves that can propagate on the nonlinear line.

Many other physical manifestations of solitons, e.g., water wave solitons, plasma solitons, and mechanical lattice solitons are solutions to the KdV equation as well, and therefore, the essential

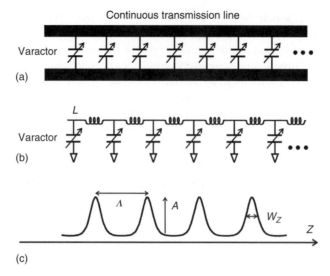

FIGURE 11.2 (a) An NLTL consisting of a linear transmission line periodically loaded with varactors. (From Rodwell, M.J., Allen, S.T., Yu, R.Y., Case, M.G., Bhattacharya, U., Reddy, M., Carman, E., Kamegawa, M., Konishi, Y., Pusl, J., Pullela, R., *Proc. IEEE*, 82, 1037, 1994). (b) An artificial NLTL. (c) A general soliton waveform on an NLTL, which is a periodic soliton train, a.k.a., a cnoidal wave. (From Ricketts, D.S., Li, X., Sun, N., Woo, K., and Ham, D., *IEEE J. Solid-State Circuits*, 42, 1657, 2007.)

properties of electrical solitons described in Sections 11.2.3 and 11.2.4 are common among these KdV-type solitons.*

Understanding of our electrical soliton oscillator described in Section 11.3 does not warrant a detailed mathematical description of the KdV equation and the cnoidal wave (soliton train) on the NLTL. In the following two subsections, we will rather focus on the essential physical properties of electrical solitons, which are pertinent to this paper.

11.2.3 PROPAGATION AND COLLISION OF ELECTRICAL SOLITONS

In addition to their ability to maintain spatial localizations of energy in pulse shapes, electrical solitons propagating and colliding on the NLTL exhibit other unique dynamics [2,3]. In this section, we review three relevant dynamical properties of electrical solitons.

First, a taller soliton travels faster than a shorter one on the NLTL. Due to this amplitude-dependent speed, if a taller soliton is originally placed behind a shorter one, as shown in Figure 11.3a, the taller one will catch up and collide with the shorter one, and move ahead of it after the collision (Figure 11.3a and 11.3b). Two other important properties are seen in this collision process. During the collision (Figure 11.3c), the two solitons do not linearly superpose and experience a significant amplitude modulation (nonlinear collision). After the nonlinear collision (Figure 11.3a and 11.3b), the two solitons recover their original shapes, however, they have acquired a permanent time (phase) shift, ϕ_κ. The three soliton properties mentioned above, i.e., (1) amplitude-dependent speed, (2) amplitude modulation during the collision, and (3) phase modulation after the collision, are what makes solitons unruly and cause difficulties in constructing a stable soliton oscillator [Section 11.3].

11.2.4 FORMING AND DAMPING OF ELECTRICAL SOLITONS

In the previous subsection, we have assumed that electrical solitons are already formed on the NLTL. But how do we form solitons on the NLTL in the first place? If a nonsoliton wave is launched onto the

---

* There are other classes of solitons. For instance, light wave solitons in optical fibers are described not by the KdV equation but by the nonlinear Schrodinger equation [20].

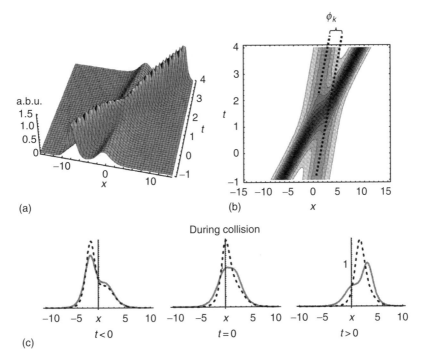

FIGURE 11.3 (a) Simulation of amplitude-dependant soliton speed; taller soliton is overtaking shorter soliton. (b) Contour plot of collision. Phase shift, ϕ_κ, is shown. (c) Before and after pictures of a soliton collision. Gray is simulated soliton collision; dashed is linear pulses traveling at the same velocity as the solitons (taller pulse faster than shorter pulse). The difference between linear superposition (dashed) and soliton dynamics (gray) is evident.

NLTL, it will change its waveform in the course of propagation to form into a soliton or solitons [3]. More specifically, a nonsoliton input close to soliton shape will be sharpened into a soliton while shedding extra energy into a tail ringing (Figure 11.4a). A nonsoliton input significantly different from soliton shape will break up into multiple solitons of different amplitudes: Figure 11.4b illustrates an example where a square pulse input on the NLTL breaks up into multiple solitons. It is these transient soliton-forming processes (Figure 11.4a and 11.4b) that have been widely exploited for the past 40 years in the traditional two-port NLTL work to generate short-duration electrical pulses [6,7,12]. In our one-port soliton oscillator design (Section 11.3), the process shown in Figure 11.4a is beneficial (Section 11.5) while the process shown Figure 11.4b can be detrimental (Section 11.3.2).

Once a soliton is formed through the transient process described above, it does not undergo any further sharpening or breakup. In the absence of loss, the formed soliton will maintain its exact shape in the course of further propagation. When loss is present, the formed soliton will change its shape as it has to lose energy. The relationships among the amplitude A, width W, and velocity v of a nondamping soliton* are preserved even in the damping soliton, and since the damping lowers A in the course of propagation, the damping soliton exhibits increasing W and decreasing v while traveling as shown in Figure 11.4c. This unique soliton damping process is well known as reported by Ott and Sudan [19]. An important notion is that even in the damping case, the spatial localization of energy is maintained in the changing pulse shape: the pulse does not break into multiple sinusoids, i.e., no dispersion occurs. In addition, the damping solitons also retain all the propagation and collision properties described in Section 11.2.3.

---

* A larger A corresponds to a larger v and a smaller W_z with specific mathematical relations, where the A–v relation is what we previously referred to as amplitude-dependent speed.

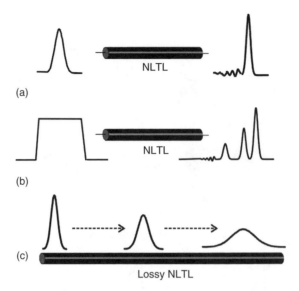

FIGURE 11.4 Transient soliton-forming processes on an NLTL. (a) Pulses close to a soliton shape sharpen into a soliton and a dispersive tail, (b) pulses significantly different from a soliton, such as a wide square pulse, break up into multiple solitons, and (c) soliton damping on a lossy NLTL. (From Ricketts, D.S., Li, X., Sun, N., Woo, K., and Ham, D., *IEEE J. Solid-State Circuits*, 42, 1657, 2007.)

It should be finally mentioned that the distinctive dynamics between the soliton's damping (increasing pulse width) and the nonsoliton's soliton-forming transient process (sharpening of pulse width) provide an important criterion to determine when a soliton has actually formed on the lossy NLTL, as will be seen in Section 11.5.

11.3 THE ELECTRICAL SOLITON OSCILLATOR

This section reviews the core operating principles of the electrical soliton oscillator [1].

11.3.1 Soliton Oscillator Topology

The soliton oscillator consists of a ring NLTL and a noninverting amplifier inserted in the ring (Figure 11.5a) [1]. The ring NLTL supports certain soliton circulation modes determined by the periodic boundary condition, $l = m\Lambda$ ($m = 1, 2, 3, \ldots$ and l is circumference of the ring NLTL, Λ is the spacing between two adjacent solitons) (Figure 11.5b). The amplifier provides gain to enable initial oscillation startup from noise and to compensate for system loss in steady state (as is commonly done in sinusoidal oscillators [21,22]). The ultimate goal of this topology is to self-generate and self-sustain one of the soliton circulation modes of Figure 11.5b.

11.3.2 Oscillation Instability Mechanisms

The topology described above does indeed lead to oscillations, self-starting from noise. However, when standard amplifiers are used in the topology the oscillations tend to be plagued with instability problems, exhibiting significant variations in pulse amplitude and repetition rate [1].

Generally speaking, the oscillation instabilities arise because the circular loop topology of Figure 11.5a not only generates the desired soliton circulation mode, but can also excite other parasitic solitons. The desired and parasitic solitons continually collide while circulating in the loop due to their generally different amplitudes and resultant speed difference (due to solitons' amplitude-dependent speed, Section 11.2.3). It is these soliton collision events that cause the significant modulations in the

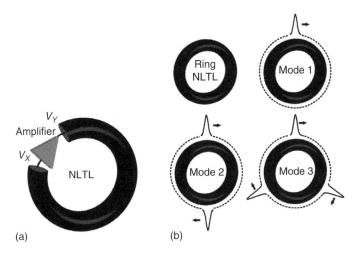

FIGURE 11.5 (a) Soliton oscillator topology and (b) ring NLTL. Mode 1 $(l = \Lambda)$, mode 2 $(l = 2\Lambda)$, mode 3 $(l = 3\Lambda)$. (From Ricketts, D.S., Li, X., Sun, N., Woo, K., and Ham, D., *IEEE J. Solid-State Circuits*, 42, 1657, 2007.)

pulse amplitude and repetition rate (these undesirable effects of the soliton collisions were explained in Section 11.2.3).

Let us examine two concrete cases to better understand the above general statement on the oscillation instabilities, that is, to see what causes the generation of multiple colliding solitons.

- *Case I—Saturating amplifier*: First consider the case where a standard saturating, noninverting amplifier of Figure 11.6a is used in the soliton oscillator (Figure 11.6c). The amplifier is biased at a fixed operating point. Assume that a soliton pulse appears at the input of the amplifier at a certain time (V_X, Figure 11.6c). This soliton pulse, after passing through the amplifier, will turn into a square pulse due to the clipping of the amplifier (V_Y, Figure 11.6c). As explained in Section 11.2.4, the square pulse will break up into several solitons with different amplitudes and differing speeds while propagating down on the NLTL,

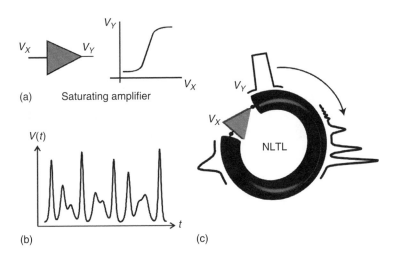

FIGURE 11.6 (a) Transfer function of a standard saturating, noninverting amplifier, (b) simulated unstable oscillation, and (c) impact of signal saturation. (From Ricketts, D.S., Li, X., Sun, N., Woo, K., and Ham, D., *IEEE J. Solid-State Circuits*, 42, 1657, 2007.)

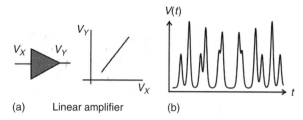

(a)   Linear amplifier   (b)

FIGURE 11.7   (a) Transfer function of a linear, noninverting amplifier and (b) depiction of oscillations with pulse amplitude and pulse repetition rate variations [13,14]. (From Ricketts, D.S., Li, X., Sun, N., Woo, K., and Ham, D., *IEEE J. Solid-State Circuits*, 42, 1657, 2007.)

eventually reappearing at the input of the amplifier at different times. This process repeats itself, creating many solitons with various amplitudes in the loop. These solitons circulate with different speeds owing to their amplitude-dependant speeds, and, therefore, continually collide with one another, causing time shifts and amplitude variations. The result is an unstable oscillation, as confirmed by the simulation shown in Figure 11.6b. The amplifier saturation is a clear cause of the soliton collision events and the resultant oscillation instability.

- *Case II—Linear amplifier*: The discussion above suggests that one might be able to attain a stable soliton oscillation if signal saturation is mitigated by using a linear amplifier of Figure 11.7a in the soliton oscillator. Ballantyne and coworkers [13,14] indeed implemented such a system, where a periodic train of solitons was seen. With minor changes to loop parameters such as gain or termination ($\sim$2%), however, multiple pulses appeared in the oscillator and collided with one another, causing once again pulse amplitude and repetition rate variations (Figure 11.7b). This shows the lack of robustness, reproducibility, and controllability of the soliton oscillator topology when a linear amplifier is used.

A consideration of this second case suggests that the saturation reduction is a necessary but not a sufficient condition to completely stabilize the oscillation and ensure robustness. There must be other mechanisms that cause the soliton collision events. Two other mechanisms we identified in the development of the stable soliton oscillator [1] are perturbations and multimode oscillations. First, any small ambient perturbation (e.g., noise, tail ringings) can grow into parasitic solitons in the soliton oscillator due to the gain provided by the amplifier. The desired soliton circulation mode and parasitic solitons will propagate at different speeds due to their generally different amplitudes, colliding and building up oscillation instabilities. Second, various soliton circulation modes of Figure 11.5b with generally different amplitudes can co-circulate in the loop at different speeds, leading to soliton collision events again and hence unstable oscillation.

Summarizing, the soliton oscillator of Figure 11.5a readily lends itself to the production of multiple colliding solitons that cause oscillation instabilities. The three main causes of the multiple soliton collision events are

- Signal saturation
- Ambient perturbations
- Multimode oscillation

The key to success in our stable soliton oscillator design in Ref. [1] was to identify these three instability mechanisms and to develop a special amplifier for the soliton oscillator topology of Figure 11.5a, which not only provides gain but also incorporates functionalities to suppress the three instability mechanisms. The following subsection details the operation of this stabilizing amplifier.

11.3.3 TAMING ELECTRICAL SOLITONS WITH AN AMPLIFIER

In Ref. [1], we attained the stabilizing amplifier that simultaneously suppresses the three instability mechanisms by incorporating an adaptive bias control in a standard saturating amplifier. With Figure 11.8 we will explain how this is achieved.

Figure 11.8a shows the transfer curve of a saturating amplifier, which can be divided into the attenuation, gain, and saturation regions based on the curve's tangential slopes. At the initial startup, the amplifier is biased at A in the gain region so that ambient noise can be amplified to initiate the oscillation startup. As the oscillation grows and forms into a soliton train, the dc component of the oscillation signal increases (Figure 11.8b). This increase in the dc component is used to adaptively lower the amplifier bias (dashed arrow, Figure 11.8c). The reduced bias corresponds to an overall gain reduction, since a portion of the pulse enters the attenuation region. The bias point continues to move down on the curve until the overall gain becomes equal to the system loss, settling at the steady-state bias B.

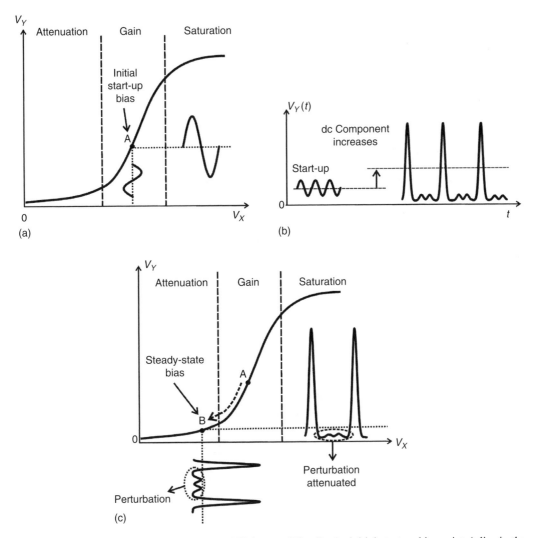

FIGURE 11.8 (a) Transfer curve of our stabilizing amplifier. In the initial startup, bias point A lies in the gain region, (b) dc component of V_Y increases as the oscillation grows into a soliton train, and (c) increased dc component is used to lower the bias point of the amplifier, leading to a steady-state bias point B. (From Ricketts, D.S., Li, X., Sun, N., Woo, K., and Ham, D., *IEEE J. Solid-State Circuits*, 42, 1657, 2007.)

In steady state (Figure 11.8c) with the bias at B situated in the attenuation region, the three instability mechanisms are all prevented simultaneously. First, the reduced bias ensures that the peak portions of the input pulses do not enter the saturation region, mitigating signal saturation (the first instability mechanism is mitigated). Second, with the reduced bias in steady state (Figure 11.8c), the input soliton train is placed across the attenuation and gain regions, causing small perturbations such as noise and tail ringings around the bias to be attenuated (the second instability mechanism is hence suppressed). Note that the perturbation rejection is accomplished while maintaining gain for the main portions of the soliton pulses to compensate loss. This threshold-dependent gain-attenuation mechanism is widely employed in modelocked lasers in optics, where it is known as saturable absorption [23], but was originally introduced into the electronics domain by Cutler for his linear pulse oscillator [1,24]. Note that the nonlinearity in the lower portion of the transfer curve is exploited to suppress the second instability mechanism (perturbations) while the nonlinearity in the upper portion of the curve is avoided to mitigate the first instability mechanism (signal saturation).

The third instability mechanism (multimode oscillation) is also suppressed via the adaptive bias control. In steady state, a higher mode (among various soliton modes shown in Figure 11.5b) has a larger dc component and therefore its corresponding steady-state bias sits farther down on the transfer curve, making the mode receive less overall gain. One can take advantage of this mode-dependent gain to select a particular mode. Only those modes with sufficient gain to overcome the loss of the system can be sustained in steady-state oscillations. When more than one mode has sufficient gain, only the highest mode is stable since any small perturbation to a lower mode will grow into a soliton, resulting in a higher mode oscillation. Consequently, the mode-dependent gain allows only one soliton pulse train mode.

11.4 DESIGN PROCEDURE

This section describes the design procedure for determining the oscillating soliton parameters shown in Figure 11.9. (Note: this figure shows the waveform with respect to time, t.) This methodology was used to develop three prototypes, which are detailed in Sections 11.5 through 11.7.

In the absence of loss, the temporal soliton width W_t (in seconds), the soliton propagation velocity v (measured as number of sections per second), and the soliton repetition rate f (in Hertz), of the steady-state soliton train on a ring NLTL of $N\,LC$ sections are given by [20]

$$W_t = \frac{3 - 2bA}{\sqrt{6bA}}\sqrt{LC_0} \tag{11.1}$$

$$v = \frac{1}{\sqrt{LC_0}}\sqrt{\frac{3}{3 - 2bA}} \tag{11.2}$$

$$f = v\frac{n}{N} \tag{11.3}$$

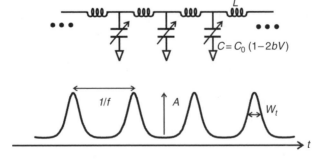

FIGURE 11.9 Lumped NLTL: component and design variables. *Note:* Waveform is with respect to time, t, rather than space, z, as in Figure 11.2.

where A is the soliton amplitude; n represents the nth soliton circulation mode on the ring NLTL (Figure 11.5b); L is the inductance of a single inductor in the NLTL (Figure 11.9); and b and C_0 are derived from the varactor model $C(V) = C_0(1 - 2bV)$ (Figure 11.9) where V is the applied voltage [4]. Note that the varactor model requires $2bV < 1$.

Equations 11.1 and 11.2 quantitatively describe the already qualitatively mentioned relations between W_t (or spatial width W_z), v, and A (Section 11.2.4), i.e., as the soliton amplitude increases, the pulse width decreases while the propagation speed increases. Equation 11.3 is derived from a simple length–velocity–time relation.

The equations above serve as useful, first-order guidelines to set the initial values of W_t and f close to the desired values in the soliton oscillator design. They, however, should be used in conjunction with circuit simulations to obtain a proper oscillator design producing the desired W_t and f for the following reasons:

- W_t and f are functions of not only the circuit component values, L, C_0, and b, but also the soliton amplitude, A. The soliton amplitude is determined by the complicated nonlinear dynamics of the entire soliton oscillator, and is difficult to express analytically.
- The delay in the amplifier will result in the pulse repetition rate f smaller than what is predicted by Equation 11.3.
- The equations above assume that the soliton waveform is the same everywhere on the NLTL in steady-state oscillations. This is untrue, however, as we have discussed in Section 11.2.4 and will see experimentally in Figure 11.16 in Section 11.5.2. The steady-state signals on the NLTL in the soliton oscillator always exhibit spatial variations due to the soliton forming and damping processes, as we expound upon in Section 11.5.3.

11.5 FIRST PROTOTYPE—DISCRETE PROOF OF CONCEPT

To demonstrate the concept of the stable electrical soliton oscillator, we constructed a low megahertz prototype first at the discrete board level (Figure 11.10) [1]. Lower frequency contents (the soliton pulse repetition rate of around 1 MHz and the soliton pulse width of about 100 ns) are chosen to facilitate the explicit oscilloscope measurement of various circuit nodes for rigorous proof of concept.

11.5.1 PROTOTYPE IMPLEMENTATION

The prototype is composed of an NLTL and amplifier in a ring topology, i.e., Figure 11.5a. An artificial NLTL consisting of discrete inductors and varactors (pn junction diodes) is used. The entire circuit schematic is shown in Figure 11.11, where the amplifier is within the dashed box. This amplifier incorporates the adaptive bias control to perform the functionality described with Figure 11.8. The amplifier consists of two functionally equivalent, complementary inverting stages, one built around an N-channel metal oxide semiconductor (NMOS) transistor, M_1, and the other built around a positive channel metal oxide semiconductor (PMOS) transistor, M_2, which, when taken together, form a noninverting amplifier. The PMOS stage functions as follows. The waveform at the amplifier output, V_Y, is sensed by the voltage divider consisting of the two resistors, R_A and R_B, and then is filtered by the $R_1 - C_1$ low pass filter. The filtered voltage, V_1, represents a scaled dc component of $V_Y(t)$. This dc component is fed back to the gate of M_2 through the feedback resistor, R_F, to set its bias. As the dc level of V_Y increases, V_1 will rise with respect to ground. The increase in V_1 corresponds to a reduction in the gate-source voltage of M_2, effectively lowering its bias. A similar argument applies to the NMOS stage. Combining the two stages, the effective bias of the amplifier at the input is reduced as the dc component of V_Y increases, performing the adaptive bias control to achieve the amplifier functionality of Figure 11.8.

Note in Figure 11.11 that the soliton oscillator includes a termination at the input of the amplifier to absorb energy delivered by solitons arriving from the NLTL. The termination is required since

FIGURE 11.10   Low MHz soliton oscillator prototype. (From Ricketts, D.S., Li, X., and Ham, D., *IEEE Trans. Microw. Tech.*, 54, 373, 2006.)

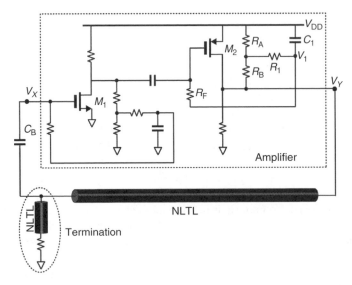

FIGURE 11.11   Circuit schematic of the first soliton oscillator prototype. (From Ricketts, D.S., Li, X., Sun, N., Woo, K., and Ham, D., *IEEE J. Solid-State Circuits*, 42, 1657, 2007.)

it is not the energy but the voltage that goes through the amplifier for signal amplification. The design of a perfect termination for an NLTL is a challenging task since the characteristic impedance of the NLTL changes with voltage. In our soliton oscillator, the termination consists of a high-loss section of NLTL (loss was added intentionally) and a resistor whose value is the average of the characteristic impedance seen by a desired signal. Due to the inevitable imperfect termination, reflections will occur, acting as perturbations to the oscillator, but fortunately they can be attenuated by the perturbation rejection mechanism (saturable absorption) incorporated in the amplifier as described in Section 11.3.3.

11.5.2 EXPERIMENTAL RESULTS

This subsection demonstrates the operating principles of the oscillator with experimental results from the first, low-MHz prototype.

11.5.2.1 Adaptive Bias Control

Figure 11.12a shows the voltage signal measured at the output of the amplifier during the oscillation startup transient while Figure 11.12b is the corresponding, measured bias adjustment in the amplifier. As the oscillation grows and forms into soliton pulses, the amplifier self-adjusts to lower its bias according to the adaptive bias control scheme explained in Section 11.3.3. The bias point eventually settles to B, at which time the net overall gain of the amplifier becomes equal to the system loss. Figure 11.8c is repeated in Figure 11.12 for convenience, and hypothetically illustrates the bias adjustment in the transient process. Note that the bias adjustment exhibits a familiar underdamped response.

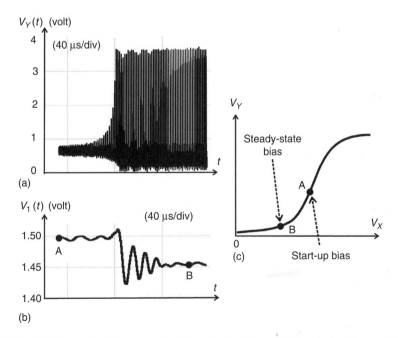

FIGURE 11.12 (a) Measured voltage signal at the output of the amplifier during the oscillation start-up transient, (b) measured bias response, $V_1(t)$, of the amplifier ($V_1(t)$ is with reference to Figure 11.11) filtered by the oscilloscope with a 500 kHz BW. $V_1(t)$ and $V_X(t)$ in the amplifier of Figure 11.11 have the same dc component, and (c) redrawing of Figure 11.8c. (From Ricketts, D.S., Li, X., and Ham, D., *IEEE Trans. Microw. Tech.*, 54, 373, 2006.)

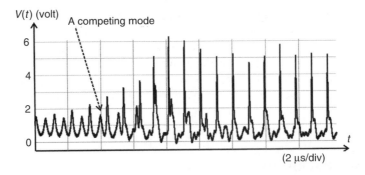

FIGURE 11.13 Measured start-up transient. (From Ricketts, D.S., Li, X., and Ham, D., *IEEE Trans. Microw. Tech.*, 54, 373, 2006.)

11.5.2.2 Startup Soliton Dynamics

Figure 11.13 shows a detailed view of the oscillation startup measured after the eighth section on the NLTL (a total of 22 *LC* sections were used in this specific experiment to best illustrate the startup dynamics). The oscillator starts by amplifying ambient noise creating a small oscillation, eventually growing into a steady-state soliton pulse train. During this process, another competing mode is clearly seen: it first grows with time, but is eventually suppressed by the stabilizing mechanism of the amplifier. In the figure, one can also observe that the shorter pulse (competing mode) propagates at a different speed than the taller pulse (main mode that survives): in this time-domain measurement at the fixed point on the NLTL, the shorter pulse originally behind the taller pulse catches up with the taller pulse and eventually moves ahead of it after collision. In the space domain, this corresponds to the taller pulse propagating faster than the shorter pulse, which is a key signature of solitons on the NLTL as explained in Section 11.2.3. The variation in pulse amplitude seen in Figure 11.13 is a result of the transient underdamped bias adjustment, similar to what is shown in Figure 11.12b. The waveform eventually reaches a stable oscillation with constant pulse amplitude and spacing as will be seen shortly.

11.5.2.3 Perturbation Rejection

Figure 11.14 shows input–output measurements of the stand-alone amplifier when the amplifier is biased at its steady-state bias point B. The test input signal consists of two main pulses and perturbations between them. The perturbations are significantly attenuated at the output of the amplifier while

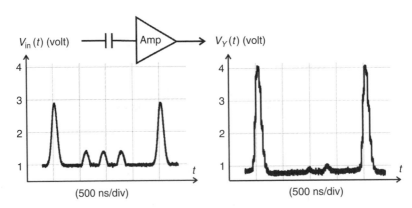

FIGURE 11.14 Input–output measurements of the stand-alone amplifier biased at B. (From Ricketts, D.S., Li, X., and Ham, D., *IEEE Trans. Microw. Tech.*, 54, 373, 2006.)

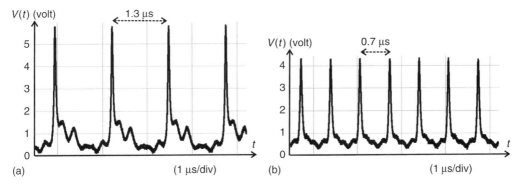

FIGURE 11.15   Measured stable soliton oscillation: (a) $l = \Lambda$ oscillation and (b) $l = 2\Lambda$ oscillation. (From Ricketts, D.S., Li, X., and Ham, D., *IEEE Trans. Microw. Tech.*, 54, 373, 2006.)

the main pulses are amplified, due to the threshold-dependent gain-attenuation mechanism (saturable absorption, Section 11.3.3). Note that in addition to the perturbation attenuation, the amplifier has sharpened the main pulses. This is because the saturable absorption mechanism attenuates the lower portion of the input main pulses while amplifying their higher portion.

11.5.2.4 Steady-State Soliton Oscillation

Figure 11.15a shows a steady-state soliton pulse train measured after the eighth section on the NLTL (a total of 30 LC sections). This waveform corresponds to the $l = \Lambda$ mode (mode 1) shown in Figure 11.5b: in this mode, there exists only one pulse propagating around the NLTL. The measured period, T_{mode1}, for this particular mode was 1.3 µs.

By tuning the gain (bias), $l = 2\Lambda$ mode (mode 2) oscillation shown in Figure 11.15b was controllably obtained. This mode corresponds to the $l = 2\Lambda$ mode (mode 2) shown in Figure 11.5c, and two pulses co-propagate in the NLTL. The measured period, T_{mode2}, for this mode was 0.7 µs. It is noteworthy that $2T_{\text{mode2}} \neq T_{\text{mode1}}$ while $2\Lambda_{\text{mode2}} = \Lambda_{\text{mode1}}$. This is because the two modes have different amplitudes, and hence propagate with different speeds, which is another signature of the soliton propagation.

In both oscillation modes, the amplitude and pulse repetition rate remained stable, showing no discernable variation. Even when deliberately perturbed with large external signals, the oscillation always returned to the same steady-state soliton pulse train. Additionally, for a given set of circuit parameters, every start-up led to the same steady-state oscillation. These experiments demonstrate the level of robustness and controllability as found in traditional LC oscillators.

11.5.2.5 Spatial Soliton Propagation in Steady State

One of the most fascinating dynamics of the soliton oscillator can be observed by following the pulse around the oscillator loop in the steady state. Figure 11.16 shows such spatial dynamics measured at three different positions on the NLTL. At the output of the amplifier, the pulse (width = 100 ns) is not an exact soliton, and, hence, sharpens into a soliton propagating down the NLTL. Once the soliton is formed at the eighth section (width = 43 ns), it does not further sharpen since it is now a soliton. Instead, the loss on the NLTL becomes the dominant process, and the soliton exhibits damping (Section 11.2.4) as it further travels down the NLTL, lowering in amplitude and velocity while increasing in width. At the end of the NLTL, the pulse width has increased to 110 ns. It is this clear existence of the transition point (eight section) between the pulse sharpening and the pulse widening (damping) that suggests the formation of the soliton at that transition point.

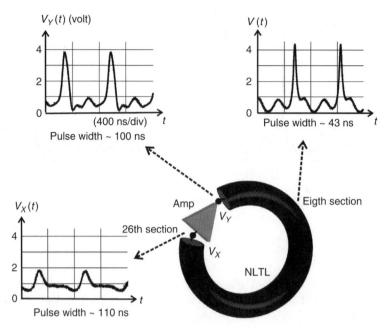

FIGURE 11.16 Measured stable soliton oscillation at various points with Mode 2. Pulse width measured is the full width of the pulse at half of the maximum amplitude (FWHM). (From Ricketts, D.S., Li, X., Sun, N., Woo, K., and Ham, D., *IEEE J. Solid-State Circuits*, 42, 1657, 2007.)

The measurement in Figure 11.16 also shows that the amplifier sharpens the pulse from 110 ns FWHM (full width at half maximum) to 100 ns FWHM. This narrowing by the amplifier is due to the saturable absorption as mentioned earlier. However, it is important to note that in the oscillator, the pulse sharpening by the NLTL (from 100 to 43 ns) is much more significant than the sharpening by the amplifier.

11.5.3 In-depth Experimental Studies on the Soliton Oscillator Dynamics

In this subsection, we take a closer look at two important dynamics of the soliton oscillator: existence of a limit cycle and soliton spatial dynamics. These in-depth analyses further demonstrate the stability of the oscillator and the generation of solitons on the NLTL.

11.5.3.1 Existence of a Limit Cycle

The essential contribution of Ref. [1] was the development of the stabilizing amplifier to tame the instability-prone dynamics of solitons in the circular soliton oscillator topology of Figure 11.5a. The stability of the soliton oscillator was demonstrated by steady-state oscillations that were robust to ambient noise and external perturbations (Section 11.5.2).

The key character of any stable oscillator is the existence of a limit cycle in the phase space. For instance, for an LC oscillator in steady state, the voltage, V, across the LC tank and its time-derivative, dV/dt, which is proportional to the current I in the LC tank, circulate along on an ellipse-like closed trajectory in the two-dimensional plane whose x-axis is V and y-axis is dV/dt [25]. This two-dimensional space is called phase space, and the closed trajectory is called a limit cycle. Regardless of the initial condition, the oscillation point (V, dV/dt) always ends up being on the limit cycle in steady state. If the oscillation is significantly perturbed and the oscillation point (V, dV/dt) is displaced far off of the limit cycle as a result, it will always trace back to the limit cycle. This makes

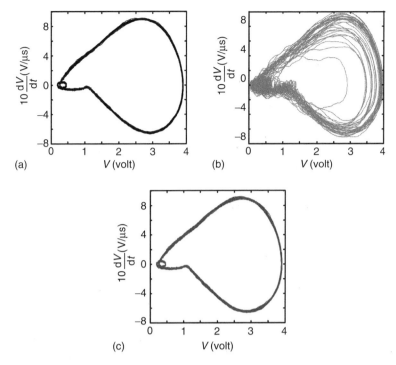

FIGURE 11.17 (a) Steady-state trajectory of $(V, dV/dt)$, (b) transient response to a significant disturbance, and (c) steady-state trajectory after the effect of perturbation has settled. (From Ricketts, D.S., Li, X., Sun, N., Woo, K., and Ham, D., *IEEE J. Solid-State Circuits*, 42, 1657, 2007.)

an oscillator an excellent engineering system: the character of the steady-state oscillation signal is independent of the initial condition or of perturbation.

Here in this section, we experimentally demonstrate the existence of a limit cycle in the soliton oscillator using the first discrete prototype (Section 11.5.2). The low-frequency discrete prototype is used in order to facilitate a deliberate, large-signal perturbation of the system.

Figure 11.17a shows the trajectory of a steady-state oscillation point $(V, dV/dt)$ in the two-dimensional phase space* where V is the voltage taken from the seventh node on the NLTL of the discrete prototype. After a deliberate, significant perturbation of the soliton oscillator, using an external pulsed source, the oscillation trajectory in the phase space deviates from the steady-state trajectory of Figure 11.17a for many cycles as depicted in Figure 11.17b, but eventually settles back to the same original steady-state trajectory as shown in Figure 11.17c. This experiment clearly demonstrates that the steady-state trajectory of Figure 11.17a is indeed the limit cycle of the soliton oscillator, and re-confirms the robustness and stability of the soliton oscillator. Figure 11.18 shows a time-domain picture of recovery from a similar disturbance as in Figure 11.17, once again demonstrating the robustness of the soliton oscillator.

11.5.3.2 Spatial Dynamics Revisited

The spatial dynamics of the first prototype shown with Figure 11.16 in Section 11.5.2 exhibited a clear transition from a pulse-sharpening process to a pulse-widening process. We had argued that pulses in these two processes are in two entirely different states, that is, during pulse sharpening, the

---

* Generally speaking, the phase space has a dimension of $2N$ where N is the number of LC section in the NLTL. The oscillation dynamics in the two-dimensional "pseudo"-phase-space, however, provides a good representation of the dynamics of the entire system.

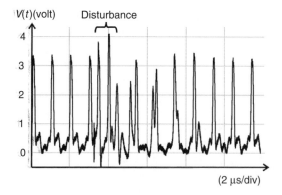

FIGURE 11.18 Recovery of steady-state oscillation to a large signal disturbance. Two additional pulses were artificially added by capacitively coupling a signal onto the NLTL. The oscillation recovers in just a few cycles.

pulse is nonsoliton, which is forming into a soliton, and during pulse widening, the pulse is a soliton, which is undergoing a damping process.

To confirm this reasoning more firmly, we measured the temporal pulse width, W_t, and pulse amplitude, A, at each section of the NLTL in the first soliton oscillator prototype from Ref. [1] in its steady state. Here again, we use the discrete low-frequency prototype since it facilitates the direct measurement of the pulse width and amplitude at every section. Figure 11.19 shows the measured results,* where the x-axis is A and the y-axis is $W_t\sqrt{A}$. The reason for this specific choice of the y-axis has a mathematical reason as will be seen shortly, but at this point in time, it is sufficient to just know that Figure 11.19 represents how the width and amplitude change as the pulse circles around the oscillator loop.

In Figure 11.19, we observe two distinct trajectories, V_Y to T (dashed line) and T to V_X (solid line), which correspond to the initial pulse sharpening and the following pulse widening, respectively. The two entirely different relations between A and W_t in these two processes shown by the two entirely different trajectories indicate that the pulses in the two processes are in different states.

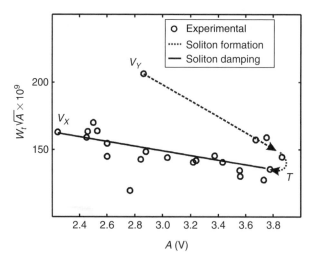

FIGURE 11.19 $W_t\sqrt{A}$ versus A. (From Ricketts, D.S., Li, X., Sun, N., Woo, K., and Ham, D., *IEEE J. Solid-State Circuits*, 42, 1657, 2007.)

---

* The set of circuit parameters used in this specific experiment is different from the set of circuit parameters used for the experiment to obtain Figure 11.16, even though the same physical prototype was employed in both experiments.

To show that the pulse in the widening/damping process (solid line) is indeed a soliton, let us first note that for a soliton on the NLTL, A and W_t are related by the following formulas [20]:

$$W_t\sqrt{A} = -\frac{2\sqrt{b}\sqrt{LC_0}}{\sqrt{6}}A + \frac{3}{\sqrt{6b}}\sqrt{LC_0} \qquad (11.4)$$

As mentioned earlier in Section 11.2.4, even in the damping process, the A–W_t relation should be preserved [19]. This relation states that $W_t\sqrt{A}$ and A will draw a straight line, which is why we chose the y-axis as such in Figure 11.19. As can be seen from the figure, the data points in the damping process can be placed roughly on a straight line. Extraction of b, C_0, and L from this straight line deviates from the real values by 20%. This deviation arises from the inaccurate first-order modeling of the varactor, but the general trend clearly shows that the pulses in the widening/damping process are indeed solitons.

11.6 SECOND PROTOTYPE—DISCRETE MICROWAVE SOLITON OSCILLATOR

The soliton oscillator concept is general and the oscillator can be scaled in frequency. We built the second prototype at a higher frequency range (the lower part of the microwave frequency range), once again at the discrete board level [1].

The overall circuit schematic is shown in Figure 11.20, where the amplifier, using bipolar junction transistors this time, is shown within the dashed box. The amplifier consists of a common emitter

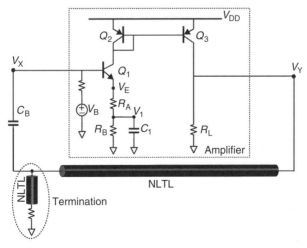

FIGURE 11.20 Microwave soliton oscillator prototype and circuit schematic. (From Ricketts, D.S., Li, X., Sun, N., Woo, K., and Ham, D., *IEEE J. Solid-State Circuits*, 42, 1657, 2007; Ricketts, D.S., Li, X., and Ham, D., *IEEE Trans. Microw. Tech.*, 54, 373, 2006.)

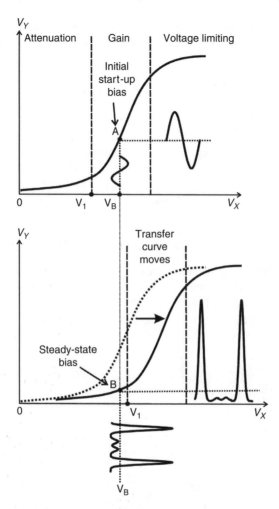

FIGURE 11.21 Effective adaptive bias control in the amplifier of the second soliton oscillator prototype in Figure 11.20. (From Ricketts, D.S., Li, X., Sun, N., Woo, K., and Ham, D., *IEEE J. Solid-State Circuits*, 42, 1657, 2007.)

amplifier with emitter degeneration and a signal inversion stage. Differently from the MOS amplifier of Figure 11.11, this amplifier senses its input signal V_X to produce the input signal's scaled dc component at V_1 via the resistive divider R_A–R_B and the low-pass filter of R_B–C_1. Also differently from the MOS amplifier, this amplifier's bias at the base of the bipolar heterojunction transistor (BJT) Q_1 is fixed. Now as the oscillation grows and forms into a pulse train, V_1 increases, which moves the amplifier transfer curve toward right (Figure 11.21). In the initial start-up transient, the fixed bias of the amplifier is in the gain region. As the oscillation grows into a pulse train, the translation of the transfer curve toward the right direction places the fixed bias of the amplifier in the attenuation region (Figure 11.21). As a result, in steady state, the input pulse is placed across the gain-attenuation regions. Summarizing, while this amplifier configuration moves the transfer curve instead of the amplifier bias point, in steady state, it places the pulse train just like in Figure 11.8c, hence achieving all of the three stability mechanisms.

The measured steady-state soliton oscillation from the second prototype is shown in Figure 11.22. A periodic soliton pulse train is clearly seen. The soliton pulse width of 827 ps is well into the microwave region.

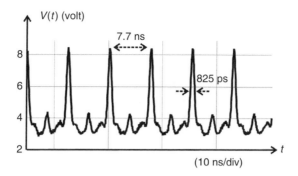

FIGURE 11.22 Steady-state soliton oscillation from the second soliton oscillator prototype. (From Ricketts, D.S., Li, X., Ham, D., *IEEE Trans. Microw. Theory. Tech.*, 54, 373, 2006; Ricketts, D.S., Li, X., Sun, N., Woo, K., and Ham, D., *IEEE J. Solid-State Circuits*, 42, 1657, 2007.)

11.7 THIRD PROTOTYPE—CMOS ELECTRICAL SOLITON OSCILLATOR

To demonstrate a chip-scale operation of the soliton oscillator, we implemented a CMOS soliton oscillator prototype using a 0.18 μm CMOS technology. The schematic of this oscillator is identical to that of the first discrete prototype, i.e., Figure 11.11, following the same design procedure, albeit at much higher frequencies ($\approx$1 GHz).

11.7.1 CHIP-SCALE IMPLEMENTATION

Figure 11.23 shows a micrograph of the implemented chip-scale soliton oscillator, which consists of an IC mounted on an in-house fabricated glass substrate [26]. The IC implemented in a 0.18 μm CMOS technology contains the stabilizing amplifier and the varactors (pn-junction diodes) of the NLTL. The NLTL inductors were created by bonding gold wires back and forth between the pads on the IC and metallic pads on the glass substrate. Bonding was done by using an automated bonding

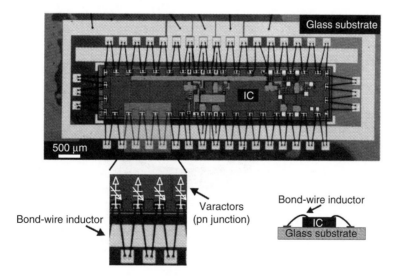

FIGURE 11.23 Micrograph of the chip-scale soliton oscillator, which consists of a CMOS IC mounted on a glass substrate. NLTL inductors are created by bonding gold wires back and forth between the IC and substrate. The IC contains the rest of the soliton oscillator. (From Ricketts, D.S., Li, X., Sun, N., Woo, K., and Ham, D., *IEEE J. Solid-State Circuits*, 42, 1657, 2007.)

machine to ensure consistency in wire length. It was estimated that the variation from inductor to inductor was less than 5%. Magnetic couplings between adjacent bonding wires can be estimated via electromagnetic simulations to choose proper lengths of the bonding wires to generate a desired inductance.

The utilization of the bonding wires as NLTL inductors is to test various different NLTL characteristics (inductance value and NLTL length) in the soliton oscillator via postfabrication adjustment in this very initial stage of the chip-scale prototype development. However, the pads on the IC necessary for bonding introduce parasitic linear capacitance and hence compromise the nonlinearity of the NLTL (resulting in a reduced pulse sharpening by the NLTL as will be seen in Section 11.7.2). As the chip-scale operation has been firmly demonstrated (Section 11.7.2), in the future, on-chip planar transmission lines may be used as the NLTL inductors as has been done in the ultrafast two-port NLTL work [7,12] in order to improve nonlinearity and speed.

11.7.2 MEASUREMENT RESULTS

The chip-scale system of Figure 11.23 self-starts from noise, leading to a self-sustained stable soliton oscillation. Figure 11.24a shows the soliton train produced, measured at one point on the NLTL using

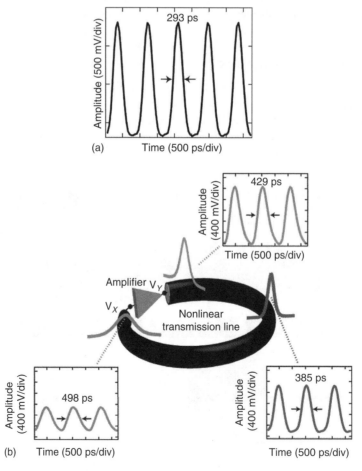

FIGURE 11.24 (a) Measured, steady-state soliton oscillation at a node on the NLTL of the chip-scale soliton oscillator of Figure 11.23 and (b) measured, spatial dynamics of the chip-scale soliton oscillator in steady state. (From Ricketts, D.S., Li, X., Sun, N., Woo, K., and Ham, D., *IEEE J. Solid-State Circuits*, 42, 1657, 2007.)

an Agilent 54855A real-time oscilloscope. The soliton train has a pulse repetition rate of 1.14 GHz and a pulse width of 293 ps at FWHM. Several different samples of the chip-scale prototype were obtained by varying the NLTL characteristics (different lengths and inductance values) via bond wire adjustment. These samples had a pulse repetition rate ranging from 0.9 to 1.9 GHz and pulse width ranging from 293 to 400 ps.

Figure 11.24b shows the steady-state voltage signals measured at three different points on the NLTL using the real-time oscilloscope. This is to show how the pulse shape changes as it circulates in the chip-scale oscillator loop in steady-state oscillation, as we showed in the first discrete prototype in Ref. [1] (Figure 11.16). The spatial dynamics seen in Figure 11.16 in the discrete prototype, i.e., soliton forming first, followed by soliton damping, is again observed here, suggesting the universality of these dynamics. The pulse at the immediate output of the amplifier, V_Y, which is 429 ps wide, is not an exact soliton, and hence, sharpens into a soliton as it propagates down the NLTL. The soliton is formed at the eighth section, achieving the minimum pulse width of 385 ps. Once the soliton is formed, as it propagates farther down the NLTL, it no longer sharpens. Rather, the loss on the NLTL becomes the dominant process: the soliton undergoes Ott's damping [19] where its amplitude and speed decrease and its width increases; the spatial confinement of wave energy is still maintained in the soliton pulse. At the end of the NLTL, the pulse width has been reduced to 498 ps.

Note that, in this specific chip-scale implementation, the degree of the pulse compression in the soliton forming process (429 ps → 385 ps: width reduction by a factor of 89%) is not as significant as in the case of the first discrete prototype (Figure 11.16: 100 ns → 43 ns, that is, width reduction by a factor of 43%). This is because the nonlinearity of the NLTL is compromised by the parasitic linear capacitance introduced by the bonding pads, which were necessary in our specific implementations using bonding wires to construct the NLTL (Section 11.7.1). In future implementations, the inductive elements can be obtained using on-chip coplanar waveguides [22] as is done in ultrafast two-port NLTL work [7,12], which can significantly reduce the undesired stray elements, enhancing the soliton compression ability.

11.8   SOLITON OSCILLATOR SUMMARY AND FUTURE DIRECTIONS

Table 11.1 summarizes the temporal pulse width, W_t, and pulse repetition rate, f, of each of the three prototypes discussed in the preceding sections.

The minimum pulse width of 293 ps achieved in our latest chip-scale CMOS soliton oscillator prototype is not a record number as compared to the state-of-the-art pulse generation circuits, including the ultrafast two-port NLTL [12]. The value of our work so far, rather, lies in the clear demonstration of the stable electrical soliton oscillator concept and its feasibility for chip-scale operation. Now with both the concept and chip-scale operation firmly demonstrated, the soliton oscillator, especially its NLTL, can be quickly scaled to a significantly smaller size to enhance the speed. For instance, an ultrafast NLTL, such as the two-port GaAs NLTLs in [12] (480 fs pulse rise time), can be incorporated in our integrated soliton oscillator to significantly reduce the soliton pulse width.

TABLE 11.1
Summary of Soliton Oscillator Prototypes

Reference	Implementation	f	W_t
[1]	Discrete MOS	1.4 MHz	43 ns
[1]	Discrete BJT	130 MHz	827 ps
[26]	CMOS IC	1.14 GHz	293 ps

Placing such an ultrafast NLTL in the electrical soliton oscillator raises an important question on the impact of the amplifier bandwidth on the minimum soliton pulse width. While the propagation of a 1 ps wide pulse on the stand-alone NLTL is feasible as demonstrated in Ref. [12], amplifiers, even in the state-of-the-art solid-state technologies, cannot provide bandwidth for such a sharp pulse. The experimental results shown with Figures 11.16 and 11.24b clearly suggest, however, that the soliton-forming process (pulse-sharpening process) on the NLTL after the immediate output of the amplifier may be able to overcome the bandwidth limitation of the amplifier, and, hence, it may be feasible to achieve a 1 ps pulse width using the NLTL of Ref. [12] despite the relatively slower amplifier. The explicit demonstration of this interesting possibility remains an open question, and would be a natural future extension of this work.

Finally, we would like to remark that the original soliton oscillator topology of Figure 11.5a, which we have used so far to demonstrate the stable soliton oscillation, uses a lumped gain element in the sense that the amplification takes place only at one point of the NLTL. Therefore, the design of a distributed soliton oscillator where the gain is provided all along the NLTL and the lumped parasitic capacitance (that slows the speed) of the lumped amplifier is distributed all along the NLTL could be an important future direction to further enhance the speed.

11.9 CHAOTIC SOLITON OSCILLATOR

The focal point of this chapter has been on the description of how we attained stable soliton oscillations by controlling the inherently unruly behaviors of electrical solitons on the NLTL using our uniquely engineered amplifier. As explained in Section 11.3.2, lack of such soliton control functionalities easily leads to continual soliton collision events, building up oscillation instabilities with significant variations in the pulse amplitude and pulse repetition rate, as shown with the waveforms in Figures 11.6 and 11.7. This tendency toward instability when our controlling mechanisms are not present suggests that one may attain extreme oscillation instability, i.e., chaos, by promoting the soliton's unruly dynamics. The discussion in Section 11.3.2 with Figure 11.6 suggests that one way of promoting the soliton's unruly behaviors is to increase the signal saturation in the amplifier in the soliton oscillator topology of Figure 11.5a and at the same time to increase the nonlinearity of the NLTL by using highly nonlinear varactors. This combination of higher signal saturation and increased nonlinearity will increase the number of soliton collision events, increasing the possibility for a chaotic oscillation.

The preliminary simulation results presented in Ref. [15] suggest that this is indeed possible, offering another direction for soliton-based oscillators as chaotic signal generators. Important applications for chaotic signal generation exist, most notably, encrypted communication (e.g., [17,18]). The theory is to utilize chaotic synchronization [17] to "privately" synchronize a transmitter and a receiver. Just as two coupled sinusoidal oscillators can be frequency-synced via injection locking, two coupled chaotic oscillators can synchronize their chaotic outputs, converging from distant initial conditions. This is a remarkable phenomenon, as a key property of isolated chaotic systems is their exponentially fast divergence in evolution, even with very close initial conditions. While circuits capable of chaotic signal generations already abound, the chaotic signal produced by the chaotic soliton oscillator can potentially have a bandwidth as large as 1 THz due to the extremely short soliton pulse in an ultrafast NLTL [12], which gives an edge to the chaotic soliton oscillator.

11.10 CONCLUSION

While the hegemony of sinusoidal signals for high-frequency signal processing in electronic circuits will undoubtedly last into the foreseeable future, certain nonsinusoidal signals may further enrich the scope and capacity of modern electronic circuits. Two notable nonsinusoid signals are short-duration pulses and chaotic signals. The short-duration pulses are beneficial for ultrafast time-domain metrology as well as pulse-based wireless system applications. The complex nature of chaotic signals offers a new means of encrypted communication. Over the last two years, we have developed a new class of circuits, electrical soliton oscillator, which can self-generate (1) stable short-duration electrical soliton trains by taming the inherently instability-prone dynamics of solitons and (2) chaotic signals by promoting the unruly behaviors of solitons. The soliton's superb capability of pulse width compression and resultant large bandwidth give the edge to the electrical soliton oscillator over other pulse and chaotic signal generation circuits. This prospect is brightened by the fact that nature's most intricate and brilliant circuit, the human brain, utilizes soliton-like neuron impulses, and often, their chaotic behaviors.

ACKNOWLEDGMENTS

The authors would like to thank Matt DePetro of Harvard University for his significant efforts in fabricating the first discrete prototype, and Lawrence DeVito, Susan Feindt, and Rick Sullivan of Analog Devices Inc. for their continued support and for their help with the bonding-wire NLTL construction. Professor T. H. Lee of Stanford University shared his invaluable insights, and inspired us to initiate the research on the chaotic soliton oscillator. We would like to thank Professor Charles G. Sodini, Kyungbum Kevin Ryu, and Dr. Anh Pham of Massachusetts of Institute of Technology for their continued support, discussions, help with the varactor measurement, and help with the IC layout. William F. Andress, Dr. Ali Belabbas, and Professor Roger W. Brockett, all of Harvard University, gave us constructive criticisms. Sonnet and AnSoft Corporation donated their electromagnetic field solvers and Agilent EEsof donated ADS.

REFERENCES

1. D. S. Ricketts, X. Li, and D. Ham, Electrical soliton oscillator, *IEEE Trans. Microw. Theory Tech.*, 54(1), 373–382, 2006.
2. A. C. Scott, F. Y. F. Chu, and D. W. McLaughlin, The soliton: A new concept in applied science, *Proc. IEEE*, 61(10), 1443–1483, 1973.
3. P. G. Drazin and R. S. Johnson, *Solitons: an Introduction*, Cambridge, UK: Cambridge University Press, 1989.
4. M. Remoissenet, *Waves Called Solitons: Concepts and Experiments*, New York: Springer, 1999.
5. A. Hasegawa and F. Tappert, Transmission of stationary nonlinear optical pulses in dispersive dielectric fibres. I. Anomalous dispersion, *Appl. Phys. Lett.*, 23(3), 142–144, 1973.
6. J. R. Alday, Narrow pulse generation by nonlinear transmission lines, *Proc. IEEE*, 22(6), 739, 1964.
7. M. J. Rodwell, S. T. Allen, R. Y. Yu, M. G. Case, U. Bhattacharya, M. Reddy, E. Carman, M. Kamegawa, Y. Konishi, J. Pusl, and R. Pullela, Active and nonlinear wave propagation devices in ultrafast electronics and optoelectronics, *Proc. IEEE*, 82(7), 1037–1059, 1994.
8. M. Kahrs, 50 years of RF and microwave sampling, *IEEE Trans. Microw. Theory Tech.*, 51(6), 1787–1805, 2003.
9. R. Y. Yu, M. Reddy, J. Pusl, S. T. Allen, M. Case, and M. J. W. Rodwell, Millimeter-wave on-wafer waveform and network measurements using active probes, *IEEE Trans. Microw. Theory Tech.*, 43(4), 721–729, 1995.
10. R. Y. Yu, J. Puls, K. Yoshiyuki, M. Case, K. Masayuki, and M. Rodwell, A time-domain millimeter-wave vector network analyzer, *IEEE Microw. Guided Wave Lett.*, 2(8), 319–321, 1992.
11. G. F. Ross, U.S. patent 3 728 632, April 17, 1973.
12. D. W. van der Weide, Delta-doped schottky diode nonlinear transmission lines for 480-fs, 3.5-V transients, *Appl. Phy. Lett.*, 65(7), 881–883, 1994.
13. G. J. Ballantyne, P. T. Gough, and D. P. Taylor, Periodic solutions of Toda lattice in loop nonlinear transmission line, *Electron. Lett.*, 29(7), 607–609, 1993.
14. G. J. Ballantyne, Periodically amplified soliton systems, PhD Dissertation, University of Canterbury, New Zealand, 1994.
15. D. Ham, X. Li, S. A. Denenberg, T. H. Lee, and D. S. Ricketts, Ordered and chaotic electrical solitons: Communication perspectives, *IEEE Commun. Magazine*, 44(12), 126–135, December 2006.

16. D. S. Ricketts, X. Li, N. Sun, K. Woo, and D. Ham, On the self-generation of electrical soliton pulses, *IEEE J. Solid-State Circuits*, 42(8), 1657–1665, 2007.
17. L. M. Pecora and T. L. Carroll, Synchronization in chaotic systems, *Physical Rev. Lett.*, 64(8), 821–825, Feb. 1990.
18. K. M. Cuomo and A. V. Oppenheim, Synchronization of Lorenz-based chaotic circuits with applications to communications, *IEEE Trans. Circ. Sys.-II*, 40(10), 626–633, October 1993.
19. E. Ott and R. N. Sudan, Damping of solitary waves, *The Physics of Fluids*, 13(6), 1432–1434, June 1970.
20. D. Ricketts, The electrical soliton oscillator, PhD dissertation, Harvard University, 2006.
21. D. Ham and A. Hajimiri, Concepts and methods in optimization of integrated *LC* VCOs, *IEEE J. Solid-State Circuits*, 36(6), 896–909, 2001.
22. W. Andress and D. Ham, Standing wave oscillators utilizing wave-adaptive tapered transmission lines, *IEEE J. Solid-State Circuits*, 40(3), 638–651, 2005.
23. H. A. Haus, Mode-locking of lasers, *IEEE J. Sel. Top. Quant. Elect.*, 6(6), 1173–1185, 2000.
24. C. C. Cutler, The regenerative pulse generator, *Proc. IRE*, 43(2), 140–148, 1955.
25. D. Ham and A. Hajimiri, Virtual damping and Einstein relation in oscillators, *IEEE J. Solid-State Circ.*, 38(3), 407–418, 2003.
26. D. S. Ricketts and D. Ham, A chip-scale electrical soliton modelocked oscillator, *IEEE Int. Solid-State Circuits Conf.*, pp. 432–433 February 2006.
27. E. Ott, *Chaos in Dynamical Systems*, Cambridge, UK: Cambridge University Press, 1993.

Part III

Mixed-Signal CMOS Circuits

12 Current Steering Digital-to-Analog Converters

Douglas A. Mercer

CONTENTS

12.1 DIGITAL-TO-ANALOG CONVERTER BASICS

Real-world analog signals such as temperature, pressure, sound, or images are routinely converted to a digital representation that can be easily processed in modern digital systems. In many systems, this digital information must be converted back to an analog form to perform some real-world function. The circuits that perform this step are digital-to-analog converters (DACs), and their outputs are used to drive a variety of devices. Loudspeakers, video displays, motors, mechanical servos, radiofrequency (RF) transmitters, and temperature controls are just a few diverse examples. DACs are often incorporated into digital systems in which real-world signals are digitized by analog-to-digital converters (ADCs), processed, and then converted back to analog by DACs. In these systems, the performance required of the DACs will be influenced by the capabilities and requirements of the other components in the system.

A DAC produces a quantized (discrete step) analog output in response to a binary digital input code. The transfer function for an example 3 bit DAC is shown in Figure 12.1. The digital input may

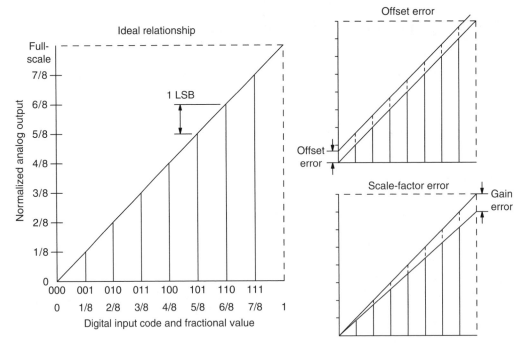

FIGURE 12.1 3 Bit DAC transfer function.

be TTL, ECL, complementary mixed oxide semiconductor (CMOS), or Low Voltage Differential Signalling (LVDS), while the analog output may be either a voltage or a current. To generate the output, a reference quantity (either a voltage or a current) is divided into binary and/or linear fractions. Then the digital input drives switches that combine an appropriate number of these fractions to produce the output. The number and size of the fractions reflect the number of possible digital input codes, which is a function of converter resolution or the number of bits (N) in the input code. For N bits, there are 2^N possible codes. The analog output of the DAC output is the digital fraction represented as the ratio of the digital input code divided by 2^N times the analog reference value.

$$A_o = \frac{D_i}{2^N} R_{ef}$$

where
  A_o is the analog output
  D_i is the digital input code
  N is the number of digital input bits (resolution)
  R_{ef} is the reference value (full-scale)

Analog signals are continuous time-domain signals with infinite resolution and possibly infinite bandwidth. However, the DAC's output is a signal constructed from discrete values (quantization) generated at uniform, but finite, time intervals (sampling). In other words, the DAC output attempts to represent an analog signal with one that features finite resolution and bandwidth. Quantization and sampling impose fundamental, yet predictable, limits on DAC performance. Quantization determines the maximum dynamic range of the converter and results in quantization error or noise in the output. Sampling determines the maximum bandwidth of the DAC output signal according to Nyquist criteria. The Nyquist theory states that the signal frequency (that is, the DAC output) must be less than or equal to one-half the sampling frequency to prevent sampling images from occurring in the frequency band of the DAC output. In an ideal DAC, the analog outputs are exactly one least significant bit (LSB) apart, where one LSB is the full-scale analog output amplitude divided by 2^N, and N is the DAC resolution expressed in number of bits. In addition, DAC operation is also affected by nonideal effects beyond those dictated by quantization and sampling. These errors are characterized by a number of AC and DC performance specifications that determine the converter's static and dynamic performance.

A number of factors affect static or DC performance. Gain error is the deviation of the slope of the converter's transfer function from that of the ideal transfer function (see Figure 12.1). Offset error is the deviation of the DAC output from that of the ideal transfer function when gain error is zero. Offset error is thus constant for all input codes. Differential nonlinearity (DNL) is the deviation of the actual step size at each input code from the ideal 1-LSB step. DNL errors can result in additive noise and spurs beyond quantization effects. Integral nonlinearity (INL) is the deviation of the actual output voltage from the ideal output voltage on a straight line drawn between the end points of the transfer function. INL is calculated after offset and gain errors are removed. INL error can also result in additive harmonics and spurs in the output. A DAC is monotonic if its output increases or remains the same for an increment in the digital input code. Conversely, a DAC is nonmonotonic if the output decreases for an increment in the digital code.

There are a number of time-domain specifications often provided for DACs. Settling time is defined as the time for the analog output to settle to a value within its specified error limits in response to a step change in the digital input. Glitch is the amount of charge injected into the converter output from the inputs when they change state. Output noise from digital feed-through can be caused by high-frequency logic signals leaking through to the converter's output.

Frequency-domain or AC performance can be characterized by a number of parameters, such as spurious-free dynamic range (SFDR), total harmonic distortion (THD), and signal-to-noise ratio (SNR). Another parameter, THD + N, is the ratio of the rms sum of the harmonics plus noise to the amplitude of the fundamental.

12.1.1 COMMON D/A ARCHITECTURES

Conceptually, the simplest DACs use a binary-weighted architecture, where N-binary-weighted elements (current sources, resistors, or capacitors) are combined to provide an analog output (N = DAC resolution). Digital decoding circuits are minimized, but the difference or scale factor between the most significant bit (MSB) and the LSB-weighted elements increases with increasing resolution, making accurate matching difficult. High-resolution D/As using this architecture are difficult to manufacture and are sensitive to mismatch errors.

12.1.2 VOLTAGE DIVIDER

The voltage divider architecture, shown in Figure 12.2, consists of 2^N equal value resistors, simplifying matching compared with the binary-weighted approach. All the resistors are of equal value, so the input must be decoded. The output is determined by decoding 1 of 2^N switches to tap into

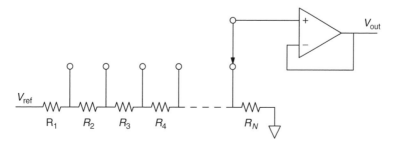

FIGURE 12.2 The voltage divider architecture, equal value resistors.

a particular location on the resistor string. This architecture has the advantage of being completely monotonic, voltage output, and low glitch (as only two switches operate during each code transition). It is also linear if all the resistors are of equal value. A related current-output architecture uses 2^N equal current sources connected in parallel between a supply voltage and an output node where the currents are summed. The major disadvantage of this architecture is the large number or resistors or current sources required for higher resolutions. This becomes prohibitive in terms of size (and matching) for resolutions above 8 bits. Nevertheless, while not practical for higher resolutions, these architectures (known as "fully decoded") are often used as building blocks for high-resolution "segmented" DACs.

12.1.3 SEGMENTED DACS

Segmented architectures, where the full resolution of the converter is spread across two or more sub-DACs, can be used for both current- and voltage-output DACs. The voltage across the decoded resistor in a resistor string divider circuit can be further subdivided to build a voltage-segmented DAC. This subdivision can be achieved through a second voltage divider circuit or with even a different architecture, as shown in Figure 12.3. The output of the overall DAC remains monotonic as long as the individual segments are monotonic and the offsets of the two buffer amplifiers in Figure 12.3 are less than one LSB. Monotonicity is easy to achieve because the individual segments have lower resolution. Segmentation has the added benefit of reducing the number of resistors (or current sources) required to achieve a given resolution, allowing smaller die sizes. Thus, it is common for high-resolution DACs to be segmented. Overall linearity is still determined by the matching of individual elements.

12.1.4 R–2R LADDER DACS

The R–2R, or ladder, architecture simplifies resistor-matching requirements since only two resistor values are required in a 2:1 ratio. The R–2R architecture can be used as a voltage- or current-mode DAC. Most R–2R current-mode architectures are based on the circuit shown in Figure 12.4a. An external reference is applied to the V_{ref} pin. The R–2R ladder divides the input current into binary-weighted currents. These currents are steered to node 1 or node 2 depending on the digital input. The

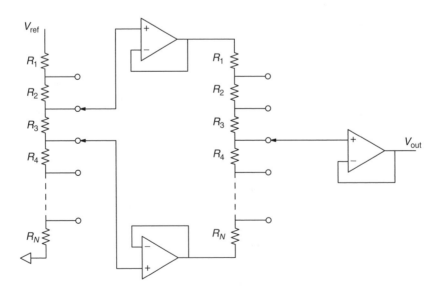

FIGURE 12.3 Monotonicity is easily achieved because the individual segments have lower resolution.

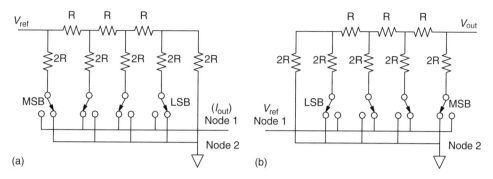

FIGURE 12.4   The R-R2 architecture: (a) current-mode and (b) voltage-mode.

current-output node is often connected to an op amp configured as a current-to-voltage converter. For matching reasons, the op-amp feedback resistor is usually included on the DAC chip. The switches are always at ground potential, and their voltage rating does not affect the reference voltage rating. If the switches are designed to carry current in either direction, a variable or ac signal may be used as the reference, resulting in a multiplying DAC. The input impedance of V_{ref} is constant and equal to R. The disadvantages of this architecture include the inversion introduced by the op amp requiring both positive and negative power supplies, and the complicated stabilization of the op amp, as the DAC output impedance, seen at node 1, varies with digital input. Current-mode operation also results in higher glitch, since the switches connect directly to the output.

Voltage-mode R–2R DACs switch resistors between V_{ref} and ground. The reference voltage is applied to node 1. Each rung on the ladder provides a binary-scaled value with the output taken as the cumulative voltage at the end of the ladder as shown in Figure 12.4b. The output voltage has constant impedance, simplifying amplifier stabilization. A positive reference voltage will provide a positive output, so single supply operation is possible. Glitch generated by switch capacitance is minimized. The drawback is that the reference input impedance varies widely, so a low-impedance reference must be used. Also, the switches operate from ground to V_{ref}, restricting the allowed range of the reference.

For high-resolution DACs, it is common to combine an R–2R ladder architecture with a fully decoded DAC in a segmented architecture. For example, the 16 bit AD7546 was one of the first DACs to use a fully decoded 4 bit resistor string combined with a 12 bit R–2R. The 65,536 output levels were divided down into 16 groups of 4096 steps. The 4 bit section is monotonic by design, so the 12 bit R–2R D/A determines the overall monotonicity. Matching and trimming are much easier than for a full 16 bit DAC. Segmentation reduces the overall number of resistors and simplifies trimming for higher resolution DACs.

12.1.5   DELTA–SIGMA ARCHITECTURE

A delta–sigma architecture (also called oversampling) can be used for DACs where linearity is preferred over bandwidth (e.g., in audio DACs). The architecture consists of a digital interpolation filter, delta–sigma modulator, and a 1 bit DAC, shown in Figure 12.5. The interpolation filter accepts an input data stream at a low rate and inserts zeros to increase the overall number of words in a particular time period, thus increasing the sampling rate of the D/A. The filter interpolates to assign values to the inserted words so that the noise in the output spectrum is concentrated at high frequency. This has the effect of pushing noise out of band, thus reducing the in-band noise and increasing the resolution. The modulator acts as a low-pass filter to the signal, converting it to a high-speed bit stream that is fed into a 1 bit DAC. Depending on the average number of ones or zeros in the bit stream, the DAC output will lie between the positive and negative reference voltages. Very high linearity can be achieved from the 1 bit DAC, which is theoretically perfectly linear. A major part of the converter uses digital circuits, so the chip area and power consumption can be kept small.

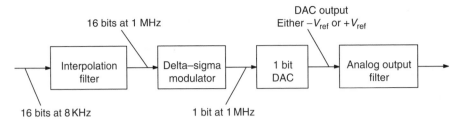

FIGURE 12.5   Delta–sigma architecture, output bandwidth ≪ sample rate.

12.1.6 MANUFACTURING PROCESSES

Architecture is not the sole contributor to DAC performance. DACs are made up from a combination of switches, resistors, amplifiers, and logic. Building a monolithic DAC in a bipolar process can provide good device matching, which yields good DC performance. However, device scaling is difficult, so an R–2R architecture is required for higher resolution. Also, this approach typically consumes higher power and it cannot be integrated easily with digital signal processing. CMOS processes are ideal for making high-density low-power logic and switches, but are less suitable for amplifiers. CMOS processes are often preferred for DACs requiring low power and small packages. For a DAC implemented in a CMOS process, scaling issues are simplified, so there is no need for an R–2R network and its drawbacks. Moreover, CMOS allows integration of digital signal processing and still offers good device matching for 12 bit linearity. But calibration is often required for higher resolution.

12.2 CURRENT-MODE DACs IN CMOS

Submicron CMOS technologies have become the process of choice for high sample rate switched current DAC design [1–9]. Switching speeds of submicron gate length MOS transistors have allowed sample rates of many hundreds of MHz and at the extreme beyond a giga-sample-per-second. Unlike switched capacitor circuits used in many ADCs, which require mixed-signal process variants with high-quality poly–poly or metal–metal capacitors, switched current DACs can make use of the standard CMOS processes. The designs have marched down the process generations from 0.8 to 0.18 μm and beyond. There are certain common features of these designs that have become givens. Because of this, it is important to note that many of the best circuit techniques are protected intellectual property and special care should be taken when developing a commercial product. The data converter area is a mine field of patents.

Almost universally, DACs with resolutions from 8 to 16 bits are split into two or more segments. The MSB segment is nearly always made from unit-weighted elements and is thermometer-coded. The number of bits in the MSB segment can vary from as few as 4 bits to as many as 8 bits, with 5 and 6 bits being the slightly more common choice. The rest of the bits may be binary coded but are often further segmented into a thermometer-coded intermediate significant bit (ISB) section and an LSB binary-coded section. A notable exception to the use of thermometer coding is proposed in Ref. [10]. Here competitive performance is achieved using unit elements but combined and switched in binary fashion. It also seems that P-channel metal oxide semiconductor (PMOS) currents and switches are used more often than N-channel metal oxide semiconductor (NMOS) currents, especially when the next circuit block in the signal chain does not reside on the same die as the DAC. Using PMOS devices on a standard twin-well process on P-type wafers provides the opportunity to isolate the back gates of the devices and bias them at some potential other than a power supply or ground. Within a system with only positive power supplies, PMOS provides the convenience of having the output load referenced to ground as well. Newer triple well processes have become available in deep submicron, which provide the ability to isolate the NMOS devices but poorer $1/f$ noise performance, as well as needing positive supply referred output loads, has limited the appeal of going with NMOS

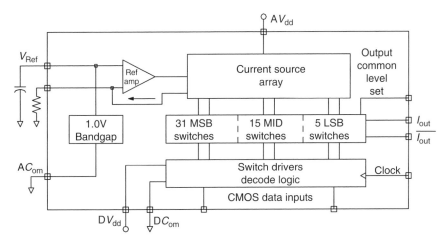

FIGURE 12.6 Basic structure of typical CMOS current-mode DAC.

currents. Depicted in Figure 12.6 is the basic structure of a typical CMOS DAC [1,8]. This example provides 14 bits of overall resolution. The five MSBs are composed of 31 unit-weighted elements and are thermometer-coded. Each unit element consists of a cascoded PMOS current source and a PMOS differential current switch pair. The remaining bits of the DAC are further segmented into thermometer-decoded intermediate bits with the five LSBs being binary coded. Because just five of the MSBs are thermometer-coded, leaving 9 bits remaining, the inclusion of the thermometer coding for the 4 intermediate bits helps insure these 9 bits have sufficient INL and DNL accuracy for the overall resolution of the DAC.

12.3 POWER DISSIPATION

Power or supply current in a CMOS switch current DAC can be divided into three categories. The first comes from the digital logic and clock section and often directly scales with the sample frequency and the data pattern. CMOS has the advantage that the design will benefit from advances in process and supply voltage scaling. By way of illustration, the digital logic portion of DAC in Ref. [1] consumes 60 μA/MSPS (Mega-Samples-Per-Second) at 5 V, up to a reported 125 MSPS maximum. A slightly more optimized logic block of a low-power design in Ref. [8] consumes 56 μA/MSPS at 3.3 V, up to a 200 MSPS maximum. At the same time, the CMOS logic can be operated from 1.8 V and the current drops to less than half at 22 μA/MSPS while the maximum frequency drops by one-half to 100 MSPS. This process and supply voltage scaling results in an 86% reduction in the digital power consumed. The DAC in Ref. [10] takes this one step further by eliminating the binary to thermometer decoding logic in favor of using the binary code directly. This lower power binary coding approach can be used when 10 bit performance levels are sufficient for the target application.

The second and third supply current categories are analog in nature. The full-scale output current is the major single contributor to the current in the analog supply. A popular output current for many designs is 20 mA because it provides 1 V signal swings in 50 ohm systems. An obvious place to trade off power for signal amplitude is to lower the full-scale current. The third part of the analog supply current is overhead and comes from the bandgap reference and various bias circuits. Inclusion of these bias circuits can have a direct effect on the spurious performance of the DAC (SFDR). For example, with an analog supply overhead current of 5 mA in addition to the 20 mA full-scale output, the DAC in Ref. [1] does not include a cascode in the PMOS current sources and achieves 61 dBc at $F_{out} = 10$ MHz. Whereas a similar DAC in Ref. [31] includes a cascode and has SFDR = 73 dBc, an improvement of 12 dB over the DAC in Ref. [1]. The addition of the cascode bias circuit increases the analog supply overhead current by 7 mA to a total of 12 mA. The challenge in power-efficient designs is to implement these performance-enhancing parts of the circuit while using a minimum of current. In the example from Ref. [8], two levels of cascode are included in a total analog supply current of 2.5 mA in addition to the reduced 2 mA full-scale output while maintaining an SFDR = 78 dBc at 10 MHz. In some cases, this overhead current can be made to at least partially scale with the full-scale output. Designs such as these often make use of mixed voltage process options to allow the analog sections to be powered from higher supply voltages than the digital decoding logic and provide larger voltage swings on the output.

12.4　STATIC ERRORS AND MATCHING

Device matching in CMOS processes has been studied and is well documented starting with the often cited work by Pelgrom in Ref. [11]. By taking advantage of statistical averaging, layout techniques and random switching order, accuracy of up to 14 bits has been reported in Ref. [6]. The PMOS devices, which made up the main current sources in previous designs in 0.6 μm CMOS such as Ref. [1], which operate from a 5 V power supply, are sized to provide sufficient yield to 12 bit linearity without calibration. An important aspect of design for good current source matching is the level of $V_{gs} - V_t$ at which the devices operate. The larger this gate overdrives the less effect the random variations in V_t have on the current sources. With the available headroom a 5 V supply affords, it is possible to size the transistors with a generous $V_{gs} - V_t$ of around 600 mV in the 0.6 μm process. As the supply voltage shrinks such as in a 0.35 μm 3.3 V design [32] the $V_{gs} - V_t$ is reduced to approximately 450 mV. For the DAC in Ref. [8], which can operate from 1.8 V, the $V_{gs} - V_t$ is further reduced to 250 mV. It is also important to point out that the value of V_t has scaled by 260 mV in these examples going from the 0.6 μm process (V_t = 935 mV) to the 0.18 μm process (V_t = 675 mV). Table 12.1 lists the untrimmed DNL and INL normalized to 14 bit level LSBs for various reported designs. It can be seen that the number of unit currents used for the MSB segment has a strong effect on the resulting linearity.

The use of statistical averaging across a large collection of smaller devices will result in improved matching performance. There are a number of approaches to how to arrange the current sources and the individual devices that compose them. Figure 12.7 shows one possible floor plan where each unit cell includes the output switch pair, current source, possibly with a cascode, along with the final re-timing latch and final decoding logic gate [2]. These unit cells are arranged in a two-dimensional array or matrix. The area required by the extra devices in each unit cell increases the distance between current source devices. This results in an accuracy disadvantage as we can see from Table 12.1. The two examples in Refs. [2,12] each use 256 unit elements for the 8 MSBs; however, there is nearly a factor of 10 difference in the reported linearity. Another issue with a matrix configuration such as this is that it forces distributing the final latch clock and the row, column data lines through cells, and may result in undesired coupling into the analog output and current-source bias nodes.

Another possible floor plan more or less used by the rest of the examples in Table 12.1 is shown in Figure 12.8. Here the circuit blocks are arranged by functional block. All the data latches and binary

TABLE 12.1
Segmentation Comparison

Reference	Segmentation	Process Node (μm)	14 Bit DNL	14 Bit INL
[1]	5–4–3	0.6	+4.0 LSB	−3.6 LSB
[8]	5–4–5	0.18	−2.6 LSB	+3.0 LSB
[5]	6–8	0.18	−0.7 LSB	−1.2 LSB
[2]	8–2	0.35	−1.6 LSB	−3.6 LSB
[12]	8–6	0.5	+0.15 LSB	+0.3 LSB

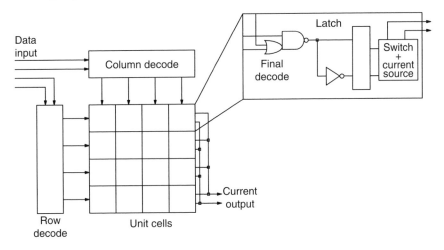

FIGURE 12.7 Row–column floor plan.

to thermometer decode logic are placed together in one block. The output switches are arranged in a single row with the analog currents entering on one side and the switch gate drive signals entering on the other. By placing all the current source devices close together, the best matching can be achieved. Clock and data routing can be kept away from the analog output and current-source bias nodes.

The individual devices, which make up the unit current sources in the matrix, can be broken up and distributed around the matrix to cancel out process-induced gradients across the array. Figure 12.9a and 12.9b shows two possible layout techniques to minimize matching errors in the current sources. The individual transistors, which make up each cell in the matrix, consists of two gate stripes sharing common source and drain diffusions, thus minimizing the overall area. Often included, but not shown here, are rows of dummy devices around the periphery. This insures that the local environment is uniform when the polygates are patterned. In the examples shown, the 64 elements of the 8×8 matrix are combined into eight current sources. In Figure 12.9a, the units are combined along diagonals of the matrix as proposed (by Reynolds) in Ref. [23]. This is a simple interconnect method and requires the fewest number of metal layers because each combination of elements has at least two members on an edge and all interior cells are adjacent to another member of their group.

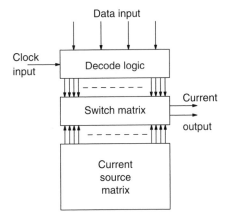

FIGURE 12.8 Floor plan arranged by functional block.

1	2	3	4	5	6	7	8
8	1	2	3	4	5	6	7
7	8	1	2	3	4	5	6
6	7	8	1	2	3	4	5
5	6	7	8	1	2	3	4
4	5	6	7	8	1	2	3
3	4	5	6	7	8	1	2
2	3	4	5	6	7	8	1

(a)

1	2	3	4	4	3	2	1
2	8	5	6	7	5	8	2
3	5	1	7	6	1	5	3
4	7	6	8	8	6	7	4
4	6	7	8	8	7	6	4
3	5	1	7	6	1	5	3
2	8	5	6	7	5	8	2
1	2	3	4	4	3	2	1

(b)

FIGURE 12.9 (a) Unit devices arranged along diagonals and (b) unit devices arranged around common centroid.

In Figure 12.9b, the units are combined around a common centroid where the average distance from the center of the matrix is the same for all eight combinations. A number of minor variations on this basic concept have been proposed in Ref. [12]. This is a more complex interconnect method and requires a larger number of metal layers. Many of the combinations are land locked so to speak with no members on an edge of the matrix.

In order to insure optimal differential linearity the carry from an MSB to the sum of the remaining LSBs should be addressed. For low-resolution LSB segments (4 to 5 bits) individual transistors in the matrix could be combined in a binary fashion to generate the desired current values. For higher resolution LSB segments, a popular way to accomplish this is by subdividing an extra MSB unit current source to provide the remaining lower bits of the DAC. This insures that the current from the total of all the LSBs closely matches the MSBs. A current-splitting array of transistors (sub-DAC) can be used in place of a single cascode device as used in an MSB cell. For example, a 9 bit sub-DAC splitter could be further segmented into a 4 bit thermometer-coded upper segment and with binary-weighted elements for the five LSBs. The current splitter gate rail could be driven by a control loop separate from that of the MSB cascodes as shown in Figure 12.10 [28], closing the loop on the drain of the current source. In this way, all the main MSB unit current source operating

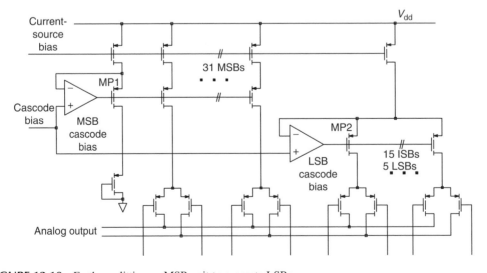

FIGURE 12.10 Further splitting an MSB unit to generate LSBs.

points now match even though the effective cascodes, MP1 for the MSBs and MP2 for the LSBs, have potentially different V_{gs} operating points.

12.5   SELF-CALIBRATION

To achieve even higher accuracy or to increase yields with less layout area, trimming or calibration techniques are often used as in the literature [4,8,13,20,22]. There are two basic approaches to implementing self-calibration, foreground, and background. A converter, which is foreground-calibrated, must be taken off line and not used while being calibrated because as each current source is measured, it is removed from the output. In background calibration, an additional current source is used to replace each current source as it is calibrated. This allows the DAC to be in use while being continuously calibrated. However, the operation of removing and replacing current sources from the output could cause extra disturbances. The calibration clock may operate synchronous or asynchronous to the main DAC clock.

There are also two basic approaches to storing the correction factors for the individual current sources. One technique, as proposed in Ref. [13], is shown in Figure 12.11. In this figure, a correction voltage is stored on the gate capacitance of a MOS transistor MN1. In calibration mode, MN1 is diode connected through S2 and the gate will settle to a value such that the sum of I_m and the current in MN1 will equal I_{ref}. In normal operation S1 switches the current to the output and switch S2 is opened holding the voltage on C_{gs}, the gate of MN1. In the process of opening switch S2, a small error charge may be dumped onto C_{gs} so care must be taken in how S2 is implemented. A dynamic technique such as this needs to be constantly refreshed and lends itself to background calibration. There is also a minimum clock rate requirement and the calibration will be lost if the clock is turned off during power saving modes.

There are certain systematic sources of error in this technique. The operating conditions of trim device MN1 and the devices in the main current source I_m potentially vary between the calibration and normal operation modes. The V_{ds} of the devices is set equal to the V_{gs} of MN1 while in calibration (Figure 12.11a). However, while in normal operation (Figure 12.11b), the V_{ds} is set by whatever circuitry is connected to the terminal OUT. This could be a cascode device or the output switches of the DAC. Because of the finite output impedance of the current source devices, the current that results in the operating mode will be different than that flowing in the calibration configuration. Each cell is slightly different and the amount of adjustment needed, i.e., the gate voltage of MN1, will be different. The change in current between calibration and operation modes will depend on the level of the adjustment. This will limit the accuracy of the calibrated result. In order to maximize the

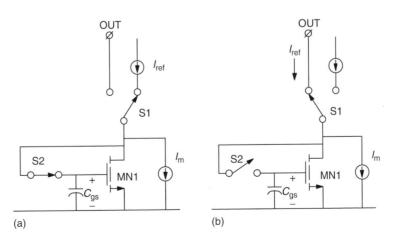

FIGURE 12.11   Dynamic storage correction.

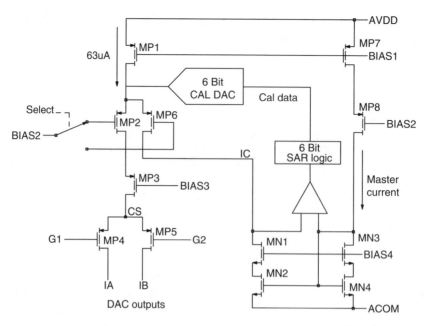

FIGURE 12.12   Static digital storage correction.

headroom in designing current sources, the devices are often biased such that the V_{ds} is marginally larger than the V_{dsat} of the device. In this configuration, the V_{ds} of MN1 and I_m must be equal to the V_{gs} of MN1 when in calibration.

Static storage of a digital correction value as used in Refs. [4,8] does not need to be refreshed and would lend itself to foreground calibration. In this approach, shown in Figure 12.12, the unit element to be calibrated, inside the dashed box, is measured against a master reference current and the difference adjusted as close to zero as possible through the successive approximation register (SAR) logic. A CAL DAC injects a small correction current in parallel with the main current source at the drain of MP1. The switches, which redirect the current either to the output node or the calibration hardware, act as the cascode devices and thus fix the drain voltage of the main current source device, MP1, to be the same, within the matching of the V_{gs} of the two cascode switches, in both cases [27]. This can result in a much more accurate calibration. The additional circuitry used for the calibration is not clocked during normal operation and does not consume power or inject noise into the main signal path.

The calibration algorithm in its first cycle calibrates the master current from MP7 to be the same as MSB current segment number 1 with its calibration DAC set to mid-scale. This in effect trims out any systematic offsets from either the NMOS mirror (MN1–4) or the voltage comparator. The trim range of the CAL DAC used to adjust the master I_{ref} (not shown in Figure 12.12) is twice that of the other CAL DACs to allow for these offsets. In the second cycle, MSB current number 1 is trimmed against the now adjusted master current. This readjustment must be done because, as just indicated, the master current CAL DAC step size is twice that of all other CAL DACs and may result in a systematic difference between the value of MSB number 1 and the other MSBs. In the following cycles, the remaining MSB currents are adjusted in sequence to equal the master current.

The configuration of the 6 bit calibration DAC is shown in Figure 12.13. The weight of MSB current MP1 is equal to 512 14 bit LSBs. A current equal to 16 LSBs is generated by the combination of MP2 and the splitting devices. From the typical raw DNL values from Table 12.1, we see a typical DNL of around than 3 LSB. Providing a trim range of ±8 LSB or ±1.5% is sufficient to cover the worst-case matching errors. MP2 operates in the linear region and serves as a degeneration resistance for the devices in the splitting array. These devices are weighted as shown totaling 63 units. Each unit

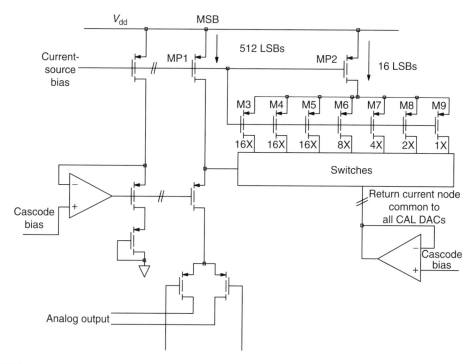

FIGURE 12.13 6 Bit resolution calibration DAC.

is equal to approximately one-fourth of the 14 bit main DAC LSB. To keep the area of the splitting array small and insure monotonicity at the 6 bit level, the top two bits are thermometer-coded and generated from three equal 16× fractions. Switches are configured to direct the currents to either the drain of MP1 increasing the output current of the cell or discarded to a return current node common to all calibration DACs. The voltage to which the discarded currents are returned is forced to be approximately equal to the drain of MP1 by a buffer amplifier, which is driven from the same cascode bias used in the cascode for MP1. This insures that the splitting action is unaffected by how the switches are set.

12.6 FINITE OUTPUT IMPEDANCE

An INL mechanism that results from the use of a switched multiple current source architecture is code-dependent output impedance (Figure 12.14a). As the number of current source elements is switched to the output, the resistance R_{sw} of that element's current source appears in parallel with the load resistor R_L. As the number of elements turned on increases, the effective output impedance of the DAC in total decreases. The varying impedance in parallel with the load resistor results in a nonlinear output voltage across the load. For cases where R_{sw} is much larger than R_L, the maximum single-ended INL error with respect to the full-scale voltage ($I_{FS} \times R_L$) can be approximated using the following formula:

$$\text{INL} = \frac{I_{unit}R_L^2 N_u^2}{4R_{sw}}$$

where

I_{unit} is the magnitude of the unit current source

R_L is the load impedance

N_u is the number of unit current elements

R_{sw} is the impedance of a unit current source

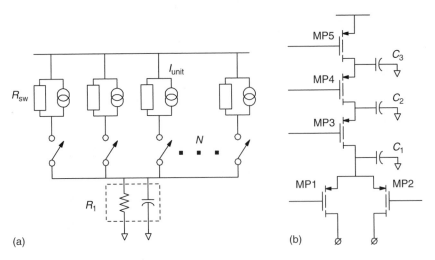

(a)   (b)

FIGURE 12.14 Code-dependent output impedance.

What we actually need to know is R_{sw} to design the DAC unit element. This formula can be rearranged to give us the required R_{sw} for a given overall DAC resolution and $1/2$ LSB INL error:

$$R_{sw} = R_L N_u 2^{N_R - 1}$$

where
  R_L is the load impedance
  N_u is the number of unit current elements
  N_R is the number of bits for the overall DAC

While it is true that switch output resistance requirements are greatly reduced for fully differential output configurations, as pointed out in Ref. [14], it is important to design the output switches and their gate voltages so as to keep the output switches in saturation. This maximizes the attenuation of the output swing seen at the common source nodes of the differential switches. The small signal attenuation of the switches is given by the ratio of the device g_m to g_{ds}. Typical values of this ratio can be in the range of 20 to 50.

The parasitic capacitances shown in Figure 12.14b reduce the current source impedance as the output frequency increases [21]. As indicated in the figure, one or more cascode stages can be included to improve low-frequency output impedance, and extend the frequency range over which the current source output impedance is acceptable. The simulated output impedance versus frequency for an example unit cell in a standard 0.18 μm CMOS process is shown in Figure 12.15. The triangle curve is for the total of the drain to gate and drain to bulk junction capacitance of switches MP1 and MP2, which always appear on the output nodes independent of whether the switch is off or on. The other three curves are the impedance seen when the switch is on (excluding the fixed drain capacitance). The circle is for the case where the main current source devices connect directly to the switch pair. The square curve includes one cascode and the x curve includes two levels of cascode. For the two cascoded cases, the drain capacitance dominates the impedance until the DC resistance is reached. We can use the same INL formula to gauge at what frequency the distortion will cross the required specification level. Again for differential output configurations, the even order distortion terms are greatly attenuated. At some point, the unavoidable nonlinearity of the drain to bulk junction capacitance will dominate.

Early work on a 16 bit DAC in BiCMOS in Ref. [18] pointed out that high-frequency operation requires not only that the switch common source node capacitance be small but also linear. A switch

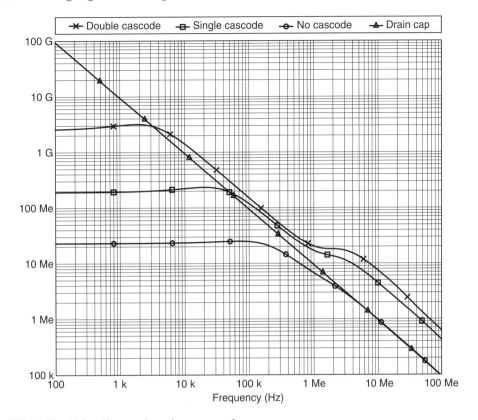

FIGURE 12.15 Unit cell output impedance versus frequency.

unit element will see an attenuated output signal on the switch common source node, and any nonlinear back gate capacitance, depicted as C1 in Figure 12.14b, on this node will produce odd order output distortion. Tying the switch and cascode transistors' back gates to the supply reduces nonlinear capacitances, but for a large array, the total nonlinear capacitance can be significant. The input of a unity-gain level shifting amplifier can be connected to the switch common source node and used to drive the back gate of the switches and cascode [4]. The nonlinear back gate capacitances now see the signal on both plates, thereby bootstrapping the well capacitances and leaving small linear parasitic capacitances. The amplifier's dc level shift should set high to minimize the switch's nonlinear capacitance.

12.7 SIMILARITIES BETWEEN DAC AND FLASH ADC

The thermometer-coded segments of a switched current DAC are very much analogous to the full parallel flash ADC. The complexity and hardware of both double for each bit of resolution. In the ADC, the distribution of the analog input signal to the comparators with matched delays is much the same as the collection and combining of the individual unit current outputs of the DAC. Also, as in the Flash ADC where the delays in the clock distribution to the individual comparators must be tightly matched, the clock distribution network driving the final stage of re-timing latches in the DAC is equally important.

One possible approach to this is propagation delay matching [4] illustrated in Figure 12.16a. Here if we assume that each cell has the same delay Δ and the delay along the clock distribution line from bottom cell 1 to top cell n is δ_1 and the delay along the output signal line is δ_2 then the sample timing is preserved if $\delta_1 = \delta_2$.

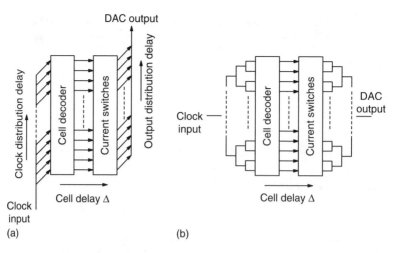

FIGURE 12.16   Signal distribution by (a) propagation matching and (b) constant wavefront matching.

Binary tree distribution structures are often used to match these delays as well as done in Ref. [5] for a DAC with a 1.4 Gsample/s clock rate. This results in a constant wavefront as illustrated in Figure 12.16b. The clock distribution tree is arranged to have equal lengths from the driver to each cell. Likewise the output collection tree is arranged with equal length from each cell to the output pad(s). The clock tree delay does not need to match the output tree delay. The physical placement of the unit cells in the layout is an important consideration and geometric shuffling of the placement is often used to break up any linear gradients in the cell delay Δ (Figure 12.16) that might be present [33].

12.8   DIGITAL DATA PATTERN-DEPENDENT NOISE

The observation was made in Ref. [15] that noise generated by the data passing through the digital logic portions, specifically the thermometer decode section, of a DAC can cause spurious tones and distortion in the analog output. The U.S. patent [24] teaches us that it is possible to concentrate this noise at the clock frequency F_s or $F_s/2$. This is accomplished by including a shadow or mirror data path with a one-to-one correspondence to the main data path. This shadow data path is driven by a data pattern in such a way that for each node in the main data path that does not change value at a given clock transition the corresponding node in the shadow path does. Likewise, when a node in the main path does change the corresponding shadow node does not change. This makes the sum total of all nodes changing at each clock transition constant and independent of the data pattern.

An example of this technique in an oversampled switched current audio DAC is proposed in Ref. [19]. In this approach, a dummy data shift register creates constant local digital edge activity on the supply, ground, and substrate. NMOS switch devices, driven by full rail swings, are used to switch the cascoded PMOS current sources. The use of the dummy data to drive dummy switch devices balances the switching activity injected into the output stage thus minimizing the demodulation of out-of-band noise into the base band.

A similar notion, referred to as modified mismatch shaping (MMS), is proposed in Ref. [16]. The idea is to set the number of elements or cells switching per clock period to a constant. This turns the errors caused by nonideal element dynamics into a dc offset and energy at $F_s/2$. An oversampling converter is assumed, where the maximum output bandwidth is reduced. The choice of what fraction of the total number elements to set the constant to is problematic and the optimum is a function of the nature of the signals being converted, however. In any case, the constant can never be set to more than one-half the number of elements. This limits either the maximum amplitude or the maximum output frequency to only one-half of what it would have been otherwise. Therefore, we conclude that, for a Nyquist rate converter, to make use of this constant element switching concept, we will need twice as many elements.

12.9 DATA-DEPENDENT CLOCK LOADING

As pointed out in the previous section, due to the mixed-signal nature of a DAC, digital data activity on the die will cause interference in the analog and clock sections of the device. This becomes an important performance issue as the output signal power is reduced or the frequency of the reconstructed output increases. A special case of data pattern-dependent interference comes from the varying load seen by the final clock buffer, which drives the final rank of re-timing latches in the DAC [7]. The now popular six-transistor latch topology first used in Ref. [1] is shown in Figure 12.17 [24]. True and complements of the data are provided to inputs at D and DB and are allowed to change only when clock is low, i.e., NMOS transistors MN1,2 are off. When a rising edge transition occurs on the clock input, the value of D is passed to Q and DB is passed to QB. When the clock signal transitions back to a low state, falling edge, and MN1,2 turn off, the state of Q and QB is held by the positive feedback around the weak inverters.

An example of this effect is shown in Figure 12.18. The effect is made more pronounced in this simulation for clarity by using a relatively weak clock buffer. The simulated time when the rising edge of the clock signal crosses mid supply, 1.8 V in this case, is plotted for the case of a single latch when the input data are not changing and when the input data are changing. This simulation shows a 4 ps difference between the x curve when the data do not change and the square curve when the data do change. Given the finite strength of the final clock buffer, the effect is magnified when a large number of latches are driven by the same common clock buffer and is proportional to the number of latches, which change their state. In the case of thermometer-coded data, the number of unit MSB cells switching is proportional to the absolute value of the rate of change of the reconstructed output waveform. The time-shift of the output samples is thus proportional to this rate of change and so results in odd-order distortions, mainly third order. A strong clock buffer can be used, which minimizes the time differences thus the effect of the data-dependent clock loading is most prominent at high-output frequencies.

We can get a workable solution by taking the shadow or mirror data paths concept of Ref. [25] and combining it with what we concluded from Ref. [16] and realize that by doubling the number of latches by simply adding the second as a mirror path for each original latch and driving the mirror latch in such a way as to cause it to change state only when the main data latch does not. One way this mirror data can be generated is shown in Figure 12.19 [29]. By combining the main data signal with a clock signal at $F_{clock}/2$, or one-half the main clock rate, with an exclusive OR gate the mirror data signal is created such that it changes only when the main data do not. By doubling the number of latches, we have doubled the load on the clock driver, but it is now independent of the incoming data pattern.

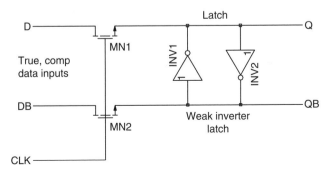

FIGURE 12.17 Final latch circuit.

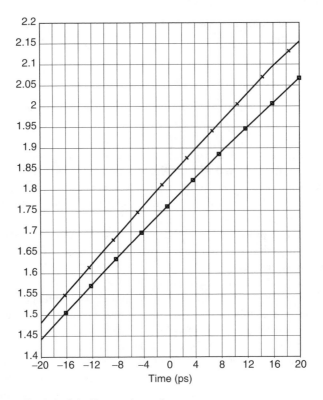

FIGURE 12.18 Normalized clock buffer crossing point.

Additional power is consumed to generate and distribute the $F_{clock}/2$ signal. Some of this $F_{clock}/2$ energy may leak onto the main F_{clock} signal and cause spurious outputs centered around $F_{clock}/2$.

A more area- and power-efficient solution, which addresses this problem, is shown in Figure 12.20 [8,30]. The top portion, which includes transistors MN1,2 and INV1,2, is the standard latch from Figure 12.17. The bottom portion is the compensating load, which provides, through NMOS transistors MN3,4, a load that varies in a way opposite to the load provided by MN1,2.

The gate current that the buffer driving CLK needs to supply to transistor MN1 is a function of the relative voltage levels present at input D and output Q. If the voltage on D is the same as on Q a slightly smaller amount of charge is needed to turn on MN1 than if D is not equal to Q. MN3 shares

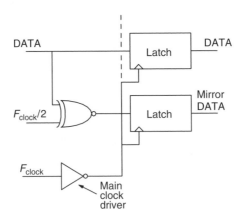

FIGURE 12.19 Constant clock load data path.

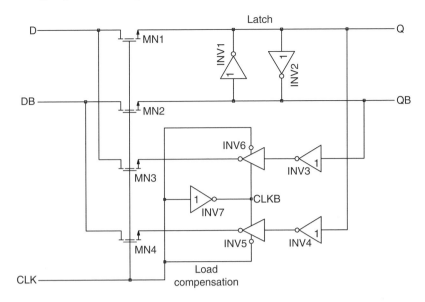

FIGURE 12.20 Constant clock load latch.

its drain connection with MN1 at input D, but the source is connected to the output of INV6. The voltage on the output of INV6 will be opposite INV1 because the input of INV6 is connected to the output of INV3, an inverted version of output QB. INV6 is a gated inverter and the output will be in a high impedance state when CLK is high and will be driven high or low when CLK is low.

It can be seen that, for all possible combinations of the inputs and state of the latch, of the four switches MN1,2,3,4, the first will have high to low across source to drain (S-D), the second will have low to high across S-D, the third will have high to high across S-D, and the fourth will have low to low across S-D. Therefore, as far as the charge that is needed to be supplied by the clock driver to turn on these four switches, it should be invariant with the data pattern. There is no energy at $F_{clock}/2$ and actually the supply current in INV5,6 has energy at $2F_{clock}$.

Shortly after the CLK line goes high, the latch formed by cross-connected INV1 and INV2 will have regenerated, making the signal levels across S-D of MN1,2 the same and INV5,6 will now be in tri-state, equalizing the voltages across S-D of MN3,4. When CLK returns low MN1–4 turn off and INV5,6 come out of tri-state and the cycle is ready to repeat. It is necessary to balance the relative strengths of the weak inverters INV1,2, used in the latch, with the gated inverters INV5,6 to insure data-independent loading on the clock driver. Gated inverters INV5,6 are sized such that their delay entering into tri-state is about the same as the regeneration time of the INV1,2 latch.

12.10 SWITCH GATE DRIVE

The differential output switch pair (MP1,MP2; Figure 12.21) could be driven directly with full supply rail swing outputs of the CMOS logic, the Q, QB nodes in Figure 12.17 or Figure 12.20. This would be the lowest power solution. However, it is well known that for the best SFDR performance, the crossing point for the gate drive signals of the output current switch pair needs to be optimized [1,2]. The circuit that drives the differential switch should ensure that both switches are never completely off at the same time so that the current from the current source is always flowing at a constant value. This minimizes the excursion of the voltage on the switch common source node, C_s, during a transition. Any current lost to parasitic capacitor C1 causes output distortions. The disturbance on C_s should be symmetric around the nominal DC value as indicated in Figure 12.21. To the extent that the disturbance cannot be completely eliminated, as pointed out earlier, it is important that C1

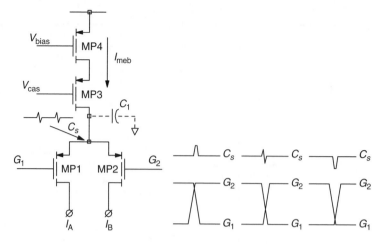

FIGURE 12.21 Differential current switch.

be minimized [21]. It is also important to point out that it is not necessary to bring the gates of the switch devices any higher than the voltage on the common source node C_s, when turning off the device ($V_{gs} = 0$). This reduces any feed-through of the gate drive signals to the outputs or the common source node.

Another source of dynamic error relates back to the fact that a small attenuated amount of the output signal leaks through the g_{ds} of the differential switch onto node C_s. The amplitude of the signal seen at node C_s is typically about 1/20 that seen at outputs I_A and I_B or 50 mV for a 1 V swing at the output. Each switch element turns on at a different point in the transfer function and as a result will have a different wave shape on node C_s. In Figure 12.22a, the complementary outputs IA and IB are shown. Referring to Figure 12.21, node C_s will have the attenuated version of IA when MP1

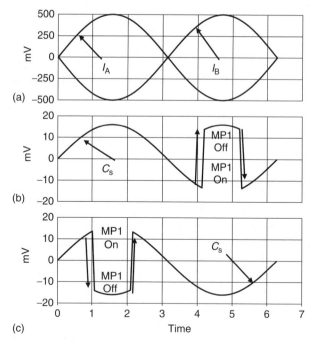

FIGURE 12.22 Common source node waveforms.

is on and the attenuated version of IB when MP2 is on. In Figure 12.22b, we see what the signal on C_s will look like when a switch element is near the lower end of the transfer function. Similarly, for Figure 12.22c, we see the signal on C_s when a switch element is near the top of the transfer function.

In Figure 12.20, note the point at which MP1 and MP2 switch is determined by the crossing point of gate drive signals G1, G2 with respect to the value of node C_s. If the relative value of C_s is modulated by the output swing and where in the transfer function the switch element is, the actual time point when the switches change will also be a function of the output swing and their position in the transfer function. This will result in a signal-dependent timing error seen in the output. As indicated in Figure 12.22b, MP1 switches from on to off when C_s is at its low point and MP1 switches from off to on when C_s is near the high point. For the case shown in Figure 12.22c, just the opposite happens. The amount of timing error depends on the magnitude of the signal on C_s and the rise/fall time of the gate drive signals.

Simulation results of an example case is shown in Figure 12.23, where the normalized zero crossing point of the differential output voltage at I_A, I_B is shown for three cases. The horizontal axis is 5 ps per division and the vertical axis spans 1 mV. The three curves are for cases where the difference between I_A and I_B when the switch flips is -333 mV, 0 V, and $+333$ mV. For these three cases, the node C_s has shifted its nominal value by a total of 32 mV or approximately 1/20 of the output. We see a shift in time of 4 ps, which results from a differential slew rate on the gate drive signals G1,G2 of 125 ps/V in this simulation. This could be a significant source of error when generating high-frequency outputs.

The circuit, which produces the appropriate signals at the gates, is shown in Figure 12.24. The full supply rail swing outputs from the final latch, Q,QB (Figure 12.20), are used to turn on and off NMOS devices MN1–4, which connect the two outputs G1 and G2 to either the node, which sets the output common mode level or the VSB bias node. The output common mode level is most often ground but in this example circuit can be adjusted, external to the die, to accommodate interfacing to other circuits, which may require that the common mode voltage be as much as 1.2 V or more such as a mixer or modulator. The amount of output shift can be traded off with increasing or decreasing the analog supply voltage. The VSB node is driven to a voltage approximately the V_{gs} of the output switches above the output common mode level.

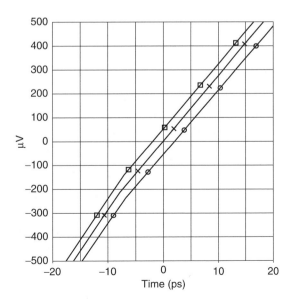

FIGURE 12.23   Switch timing delay versus output swing.

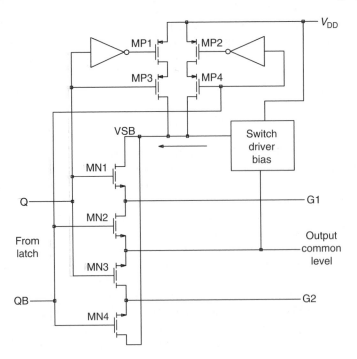

FIGURE 12.24 Output switch gate driver.

For each transition of the data, a large narrow spike of current is drawn from the VSB node by device MN1 or MN4. Normally the switch driver bias block would need to be designed to supply this current and have sufficiently low impedance to settle back to its nominal value within one clock cycle. This often requires considerable static DC current, increasing the power consumed in the circuit as in Refs. [1,7]. PMOS devices MP1-4 are added to supply a similar narrow spike of current from the V_{DD} power supply when a data transition occurs. This allows the switch driver bias block to be designed with much smaller static current. The power consumed by the circuit is now much more a function of the clock frequency and the data pattern. To save power, two of the inverters, INV3,4, in Figure 12.20 can also serve as the two inverters which drive the gates of MP1,2 in Figure 12.24.

The switch driver bias generator is shown in Figure 12.25. PMOS device MP1 is scaled to mimic one of the output switches. The voltage on node V_{SB} will be equal to the voltage on node OT_CM

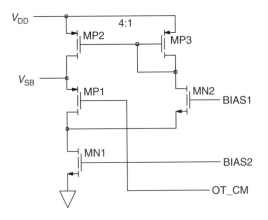

FIGURE 12.25 Switch driver bias circuit.

(output common mode level from Figure 12.24), plus the V_{gs} of MP1. NMOS device MN1, driven from BIAS2, determines the overall current level in the block. A portion of MN1's current is diverted by MN2 and through the mirror gain of MP3 to MP2 (approximately 3 in this case) supplies the current to the source of MP1. This feedback provides some degree of regulation and lowers the dynamic impedance at node VSB.

12.11 RETURN-TO-ZERO SWITCHING

Another way to re-time the data samples at the output of the DAC is to use a return-to-zero output stage as proposed in Ref. [6]. In this case, a set of additional switches have been added in the output path between the data-driven current switches and the external output load as shown in Figure 12.26, for one of the differential outputs. Switch MN1 is switched off for one-half of the clock period while the DAC current switches (IDAC) change and then back on after the currents have settled. Resistor R1 provides a load while MN1 is off. As well, there is switch MN2 to short the output to ground, through resistor R2, when the other switch is off thus the return-to-zero operation. This effectively reduces the timing skew between the various DAC switches. There is, however, a loss of one-half of the signal amplitude, 6 dB, due to the return-to-zero output.

Return-to-zero switching can reduce the distortion from digital data noise-induced timing errors, but for very high sample rates if the output does not completely settle in each half of the cycle then the history effect or intersample interference is not eliminated and can result in signal-dependent distortions. It is important to note that this scheme is not totally free of signal-dependent timing errors. The turn-on and turn-off points for MN1 and MN2 will depend on the signal levels seen at the node OUT. It is difficult to tell if the reported SFDR results for this method are really any better than those reported in Ref. [4] because both seem, for output frequencies above 25 MHz, to be limited to about −75 dBc, which seems to be the measurement limit of many spectrum analyzers.

12.12 QUAD SWITCHING/CONSTANT DATA ACTIVITY

Dynamic element matching or distortion spreading techniques are popular methods of improving SFDR by smearing the distortion into a noise-like component in the output of the DAC. Random spreading produces a more white noise-like result. Other approaches can shape the noise characteristic

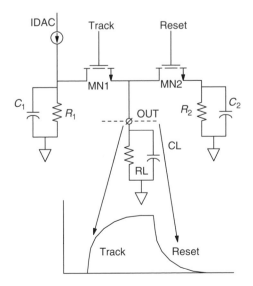

FIGURE 12.26   Return-to-zero switching.

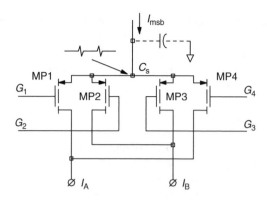

FIGURE 12.27 Quad switch.

to place it out of the band of interest, if there is some amount of oversampling in the system. While DEM (Dynamic Element Matching) will increase the amount of data activity, it is not constant for each clock cycle. When constant switching techniques are used, the distortion or noise is concentrated as a tone at the sampling frequency.

Ordinary differential current switching results in some data-dependent distortions arising from the jump or glitch on the common source node of the switch pair. This ordinary switch does not toggle every clock transition, and as a result, the switching event is dependent on the data pattern, introducing distortion in the band of interest. Another approach to the data pattern-dependent dynamic errors pointed out in Ref. [12] is a quad differential current switch proposed in Ref. [17] for an oversampling DAC and again in Ref. [5] for a multibit Nyquist DAC.

By using four switches instead of the normal two, we are in effect interleaving two return-to-zero switches. The configuration of the quad switch is shown in Figure 12.27. There are four switch devices MP1, MP2, MP3, and MP4, which share a single common source connection C_s, which nearly doubles the parasitic capacitance compared to the conventional two switch scheme. The unit element current I_{msb} is supplied to node C_s as in the ordinary differential switch. Only one of the four switches is on at any given time as indicated by the switching waveforms of Figure 12.28. The gate of each switching transistor is driven by a signal shown in Figure 12.28, three of the four gates

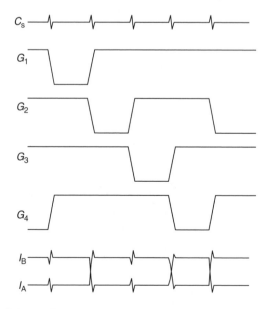

FIGURE 12.28 Quad switch waveforms.

will be high and one low for a given clock cycle. For each clock cycle the gate that is low will transition high and another gate will transition low. At the bottom of the figure, the outputs I_A and I_B indicate where the current I_{msb} is being directed. The output at I_A and I_B is logically the same as with ordinary differential switching. There will be switching glitches as indicated even if the current does not change outputs. Switching in this manner eliminates the nonlinearity due to uneven pulse duration, as in RZ switching, because every pulse has the same width. There are at least two and only two signal transitions, one rising and one falling, per clock transition. Switching noise is now moved to the sample clock frequency by the constant toggling of both sides of the switch. It is also important to note that the switching disturbance seen on the common source node C_s is constant and independent of the input data pattern. However, the voltage on node C_s will still be effected by the swing seen at IA and IB just as in the ordinary differential switch.

Quad switching like this incorporates some of the good points of both RZ switching and ordinary differential switching is suitable for high sample rates, and reduces transition-dependent noise. A drawback of quad switching is an increase in complexity, four gate signals need to be generated and the increased dynamic power consumption due to the fact that one pair of the four switches each cycle.

12.13 CONCLUSIONS

A number of major contributors to errors and distortion in modern switched current DACs have been discussed. Static device matching can be addressed either through statistical averaging or calibration. One or more cascodes can be included, along with insuring that the output switches remain in saturation, to reduce the effect of output impedance variation. The importance of gate drive signals was explored. Much like the flash ADC, clock distribution is a key factor. Digital data pattern generated noise needs to be addressed and the effect on clock noise can be a major source of distortion. Return-to-zero switching can be employed to retime the output sampling time. The use of a quad switch and constant data activity switching techniques can shift spurious outputs to the sampling frequency.

REFERENCES

1. D. Mercer et al., 12-b 125 MSPS CMOS D/A designed for spectral performance, *ISLPED 1996 Dig. of Tech. Pap.*, pp. 243–246, 1996.
2. C.H. Lin et al., A 10b 500-Msample/s CMOS DAC in 0.6 mm², *IEEE J. Solid-State Circuits*, 33(12), 1948–1958, 1998.
3. B. Tesch et al., A 14-b, 125 MSPS digital to analog converter and bandgap voltage reference in 0.5 μm CMOS, ISCAS 1999 *Dig. of Tech. Pap.*, pp. II-452–II-455.
4. W. Schofield et al., A 16b 400MS/s DAC with <-80 dBc IMD to 300 MHz and <-160 dBm/Hz noise power spectral density, *ISSCC Dig. Tech. Pap.*, pp. 126–127, February 9, 2003.
5. B. Schafferer et al., A 14b 1.4 GS/s 3 V CMOS DAC for multi-carrier applications, *ISSCC Dig. of Tech. Pap.*, February 2004.
6. A.R. Bugeja et al., A 14-b, 100-MS/s CMOS DAC designed for spectral performance, *IEEE J. Solid-State Circuits*, 34(12), 1719–1732, 1999.
7. D. Mercer, A study of error sources in current steering digital-to-analog converters, *CICC 2004 Conf. Proc.*, pp. 185–190.
8. D. Mercer, A low power current steering digital to analog converter in 0.18 μm CMOS, *ISLPED 2005 Dig. of Tech. Pap.*, pp. 72–77.
9. D. Mercer, Low power approaches to high speed CMOS current steering DACs, *CICC 2006 Conf. Proc.*, pp. 153–160.
10. J. Deveugele et al., A 10b 250 MS/s binary-weighted current-steering DAC, *IEEE J. Solid-State Circuits*, 41(2), 320–329, 2006.
11. M.J.M. Pelgrom et al., Matching properties of MOS transistors, *IEEE J. Solid-State Circuits*, 24(5), 1433–1440, 1989.

12. G.A.M. Van der Plas et al., A 14-bit intrinsic accuracy Q^2 random walk CMOS DAC, *IEEE J. Solid-State Circuits*, 34(12), 1708–1718, 1999.
13. D.W.J. Groeneveld et al., A self-calibration technique for monolithic high-resolution D/A converters, *IEEE J. Solid-State Circuits*, 24(6), 1517–1522, 1989.
14. S. Luschas et al., Output impedance requirements for DACs, *Proc. of the 2003 ISCAS*, Vol. 1, pp. I-861–I-864, May 25–28, 2003.
15. J.L. Gonzalez et al., Clock-jitter induced distortion in high speed CMOS switched-current segmented digital-to-analog converters, *ISCAS 2001 Dig. of Tech. Pap.*, pp. I-512–I-515, May 2001.
16. T. Shui et al., Mismatch shaping for a current-mode multibit delta–sigma DAC, *IEEE J. Solid-State Circuits*, 34(3), 331–333, 1999.
17. S. Park et al., A digital-to-analog converter based on differential-quad switching, *IEEE J. Solid-State Circuits*, 37(10), 1335–1338, 2002.
18. D. Mercer, A 16b D/A converter with increased spurious free dynamic range, *IEEE J. Solid-State Circuits*, 29(10), 1180–1185, 1994.
19. T. Rueger, A 110 dB ternary PWM current-mode audio DAC with monolithic 2 Vrms driver, *ISSCC Dig. of Tech. Pap.*, February 2004.
20. M. Tiilikainen, A 14-bit 1.8 V 20 mW 1 mm^2 CMOS DAC, *IEEE J. Solid-State Circuits*, 36(7), 1144–1147, 2001.
21. A. Van den Bosch et al., SFDR-bandwidth limitations for high-speed high-resolution current-steering CMOS D/A converters, *Proc. IEEE Int. Conf. Electron., Circuits, and Syst.*, pp. 1193–1196, 1999.
22. Y. Cong et al., A 1.5 V 14-bit 100 MSPS self-calibrated DAC, *IEEE J. Solid-State Circuits*, 38(12), 2051–2060, 2003.
23. D. Reynolds, MOS current source layout technique to minimize deviation, U.S. Patent 5,568,145, October 22, 1996.
24. D. Mercer et al., Skewless differential switch and DAC employing the same, U.S. Patent 5,689,257, November 18, 1997.
25. X.M. Gong, Digital signal processor with reduced pattern dependent noise, U.S. Patent 5,719,572, February 17, 1999.
26. D. Mercer, Differential current switch, U.S. Patent 6,031,477, February 29, 2000.
27. D. Mercer and W. Schofield, Calibrated current source, U.S. Patent 6,583,740, June 24, 2003.
28. D. Mercer and W. Schofield, Digital/analog converter including gain control for a sub-digital/analog converter, U.S. Patent 6,738,006, May 18, 2004.
29. D. Mercer and W. Schofield, Current DAC code independent switching, U.S. Patent 6,768,438, July 27, 2004.
30. D. Mercer, Latch with data jitter free clock load, U.S. Patent 7,023,255, April 4, 2006.
31. AD9754 data sheet.
32. AD9744 data sheet.
33. T. Chen et al., The analysis and improvement of a current-steering DACs dynamic SFDR—I: The cell-dependent delay differences, *IEEE Trans. Circuits Syst.-I*, 53(1), 3–15, 2006.

13 High-Speed, Low-Power CMOS A/D Converter for Software Radio

James W. Haslett and Abdel-Fattah S. Yousif

CONTENTS

13.1 INTRODUCTION

High-speed analog-to-digital converter (ADC) architectures have been developing for the past two decades to support the increasing requirements for data processing. Today, high sampling rate converters are in great demand due to the advances in software-defined radio (SDR) architectures and the drive toward direct radiofrequency/intermediate frequency (RF/IF) sampling. Performance specifications for these communication systems dictate that ADC designs, in a low-cost process technology such as CMOS, must achieve high spurious-free dynamic range (SFDR) and low-power consumption [1]. On the other hand, the sampling rate must meet or exceed the Nyquist rate to ensure the correct reconstruction of the signal information. ADC designs have been gaining performance due to process technology scaling. Current demand for ADC performance has exceeded the potential performance limits of existing high-speed ADC architectures, such as the flash ADC [2]. This chapter reviews the current high-speed architectures and predicts their performance as the CMOS process technology scales down further. The focus in this chapter will be on ADC architectures that can be beneficial in meeting the industry's application needs.

It is important to highlight some of the applications that require high-speed ADC components in their architecture. Table 13.1 lists some of these applications with emphasis on sampling rate, bit resolution, and power consumption. There are a number of other important specifications that are unique to some applications, such as the dynamic performance of an ADC over the signal's bandwidth in SDR transceivers. We will be discussing some of these specifications as different ADC designs are analyzed throughout the chapter. As Table 13.1 shows, requirements for high-speed ADCs range from low-bit resolution (up to 6 bit resolution) and high sampling rate (up to 1–1.5 GS/s) to very high resolution (more than 10 bit) and multi-GHz sampling rate. Some of these requirements, such as the low resolution and high-speed ADCs in read channel applications, are achievable today with better resolution ADC designs expected in the next few years. Other ADC future requirements such

TABLE 13.1

ADC Applications and Performance Requirements

Application	Input Signal Bandwidth	Resolution/Sampling Rate
Read channels	250–500 MHz	6 bit/1 GS/s, 10s of mW
Optical links	>1 GHz	2–3 bit/5 GS/s, 100s of mW
SDR communications	10 MHz–1 GHz	10 bit/2+ GS/s, low power
Industrial applications	1–10 MHz	16 bit/100 MHz, low power

as the 10+ bit resolution, 2+ GS/s sampling rate, and less than 100 mW power consumption need extensive research work before they can be attained.

In order to improve future ADC static and dynamic performance, designers are looking closely at new design techniques and at the process technology improvements. On the other hand, it is not possible to look at these expected improvements in ADC performance in isolation of the changes at the system level in the semiconductor industry. After all, one of the main goals of this industry is to reduce system cost by integrating as many functions as possible on a single silicon chip. The move toward the 90 nm process technology and beyond has helped the industry in many ways but caused new issues that were in conflict with the industry's early goals. Static power consumption due to leakage has soared to levels that practically halted the progress of many high-end industry projects, such as microprocessors and digital signal processing (DSP) chips. Instead of focusing on the growth in the frequency of operation of these chips, a migration to the multi-CPU on a chip path has been chosen by the industry [3,4]. This seems to be the best solution in the short term, but in reality any future scaling that is based on accommodating more modules on the same die area is not practical as long as the underlying issues such as leakage power and interconnects scaling are not solved. On the IF/RF circuit design side, even though process technology scaling helped increase the transistor's performance, power consumption in RF circuits has not improved as much as in digital designs. Cost reduction through integration and favorable power consumption trends were the main premises for moving RF circuits (in the 1 to 5 GHz range of operation) to the CMOS technology. Therefore, much research work, especially in the SDR area, focuses on converting analog signals to digital signals as close as possible to the transceiver's antenna. This has led to resurgence in the research efforts in the design of a new generation of high-speed ADCs that meet these system requirements.

13.2 ADC PERFORMANCE SCALING

ADC performance can be characterized in terms of speed, resolution, power consumption, and the dynamic performance over a specified range of input frequencies. In order to pinpoint the required ADC performance in high-speed applications, future trends in communication systems must be reviewed. Figure 13.1 shows the trends in communication systems as a function of the system's bandwidth and the expected mobility of its application. The system signal-to-noise ratio (SNR) requirement directly impacts the ADC dynamic performance while the system mobility signifies the need for the lowest possible ADC power consumption. By the year 2010, the data rate for high-mobility systems will reach over 100 Mbit/s and about 1 Gbit/s for low-mobility systems. High-volume products will be centered around high-mobility devices. With a minimum bandwidth of 100 Mbit/s at the low end of the application range, the ADC sampling rate should be at least 200 MS/s. There are a number of commercial ADC designs that achieve 10 or more bits of resolution at this sampling rate. These designs, however, consume close to 0.5 W. It is evident that these designs cannot fill the need for future low-power ADC designs in mobile devices. Process technology improvement gains are offset by second-order design issues that degrade performance as higher sampling rate demand increases.

As process technology scales down, supply voltage reduction and the increase in CMOS channel noise tend to decrease the dynamic range achievable by any mixed-signal circuit design, including

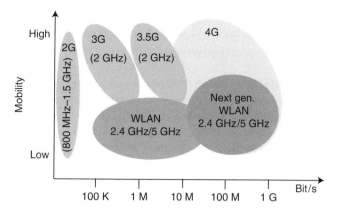

FIGURE 13.1 Trends in wireless communication systems.

ADCs. To maintain the ADC dynamic range, noise must be reduced by lowering *KT/C*, and hence increasing the input capacitance in the sample and hold circuit and any switched-capacitor circuit in an ADC design, leading to higher power consumption [5]. This trend has been evident in ADC designs over the past two decades. Figure 13.2 shows published ADC performance as a function of the sampling rate. It is clear that as the ADC sampling rate approaches the point of interest (200 MS/s and beyond), the achievable bit resolution is diminished significantly (8 or less bits). As the ADC sampling rate approaches 1 GS/s, the resolution achievable is 6 bit or less. At 1 GS/s or higher sampling rate, only the flash ADC architecture can achieve this performance at the expense of high-power consumption.

The figure-of-merit (FOM) of an ADC is used to describe its performance in relation to power consumption, resolution, and sampling rate. Equations 13.1 and 13.2 express the two commonly used types of FOMs. Equation 13.1 expresses the FOM as a function of the sampling frequency (f_s), while Equation 13.2 expresses the FOM as a function of effective resolution bandwidth (ERBW) (the 3 dB point below the low frequency SFDR point):

$$\text{FOM1} = \frac{P}{2^{\text{ENOB}} \times f_s} \text{ pJ/conversion step} \tag{13.1}$$

$$\text{FOM2} = \frac{P}{2^{\text{ENOB}} \times 2\text{ERBW}} \text{ pJ/conversion step} \tag{13.2}$$

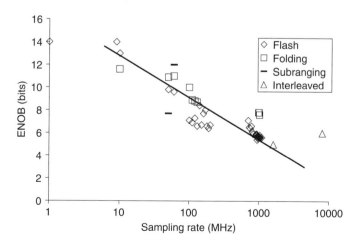

FIGURE 13.2 ENOB versus sampling frequency.

where
  ENOB is the effective number of resolution bits of an ADC
  P is the ADC's total power consumption

FOM2 is particularly used to describe the ADC performance in communication transceivers as the ADC bandwidth plays an important role in meeting system level specifications [6]. Generally, designs based on the same architecture (flash and averaging ADCs, for instance) produce correlated FOM values, which emphasize the possible design trade-offs in these ADC architectures. Equations 13.1 and 13.2 also indicate that for a given FOM value, an increase of 1 bit in an ADC resolution would result in doubling the ADC's power consumption (with same sampling frequency or ADC bandwidth). This fact explains why all of today's high-speed and low-power ADC designs (beyond 1 GS/s) have a limited number of resolution bits.

13.3   HIGH-SPEED SUBSAMPLING AND OVERSAMPLING ADCs

The two well-known design approaches to ADC design are subsampling and oversampling architectures. A typical application where these architectures are used is the narrow band, direct conversion transceivers where the input signal to the ADC is centered around the IF. Subsampling applies the Nyquist theory literally; by sampling the input signal at a rate twice that of the bandwidth of the signal of interest. The obvious advantage of subsampling is that the sampling rate is usually much smaller than the IF of the input signal. Therefore, large bit resolution can be achieved using this approach. The use of a relatively low sampling rate in the subsampling architecture makes the design of the antialiasing filter much more difficult because noise from the full spectrum is aliased into the first Nyquist zone where the ADC baseband data exist. There is a direct correlation between the filter's stopband attenuation and the achievable dynamic range of the ADC design. Increasing the sampling frequency relaxes the design requirements for the antialiasing filter. In other words, for the same filter design, higher sampling frequency leads to higher dynamic range for the ADC design.

At the low end of the high-speed conversion rate, direct IF (below 100 MHz of signal bandwidth) conversion to baseband using subsampling ADCs is slowly moving into some transceiver designs. At a sampling rate around 100 MS/s and 10 or more bits of resolution, ADC designs of commercially available products are showing good dynamic performance but suffer from high-power consumption and high prices. ADC architectures used for subsampling designs are mostly a combination of pipelined and flash ADCs.

Subsampling ADC architectures represent the only choice today for system manufacturers of the wireless communications industry. The adoption of subsampling ADCs has been very slow though because of the excessive power consumption of these converters. A typical 200 MS/s, 12 bit, subsampling, and industry-ready ADC can easily draw 500 mW of power. An indicator of the slow adoption of subsampling ADCs is the price premium (typically over $100 per ADC), which translates into low-volume production. Perhaps the most important question that highlights the migration path of any ADC architecture is how easy it will be to integrate the ADC function on-chip with other mixed-signal blocks. The trend toward system-on-chip (SoC) integration and mobile-ready systems that support wireless communication technologies, such as WiMax, will dictate the overall system performance and the power envelope that these systems can handle. It becomes apparent that the subsampling ADC architecture will continue to serve niche markets where power consumption, cost, and integration are not a concern. Military applications, such as radar signal detection, and base station receivers are some of the niche areas where subsampling ADCs can be used reliably.

Oversampling ADC architectures are defined as converters with a sampling rate greater than twice the bandwidth of the converted signal or more than twice the IF of a communication signal. As indicated earlier, future communications system trends will make it possible for a signal bandwidth to approach the IF range. The implication of this trend is that not only will oversampling ADCs converge into a single mainstream definition, but the blur line between subsampling and oversampling will converge as well. As such, future high-speed ADC designs will converge on a single and dominant sampling rate specification that meets or exceeds the Nyquist rate relative to the IF. Commercial and recently published ADC designs confirm this trend. Future ADC architectures will focus more on power consumption issues as bit resolution growth stagnates around the 8 bit point. Only when industry-ready ADC architectures with manageable power consumption levels are designed, will the quest toward higher bit resolution ADC designs continue.

13.4 HIGH-SPEED AND LOW-RESOLUTION ADC ARCHITECTURES

The issue of ADC performance and power consumption in the nanotechnology era arises from the fact that integrating multiple communication system standards on a single chip would require many highly performing ADCs on the chip. Since the semiconductor industry is moving strongly toward digital processes, mainly CMOS, high-speed ADC designers must adapt to the scaling trends of feature size. With smaller transistor features, device mismatch issues exacerbate the complexity of high-speed ADC designs. The performance of high-speed flash ADC architecture is mainly dominated by device matching issues. Therefore, as sampling speeds increase into the GHz range and the scaling trend continues to dominate the industry, the dynamic performance of current ADC architectures will not fully benefit from new process technology nodes.

Low-resolution ADC architectures with 1 GHz or more of sampling rate have many applications in the market. Figure 13.3 shows a typical read channel for a storage system where clock recovery and timing adjustment block are used alongside a high-speed ADC to recover data from a hard disk drive or any higher speed devices such as data flash cards. The bandwidth of the output signal from the read channel filter is in excess of 250 MHz. Such an application requires a high SFDR, 6 bit ADC at conversion rates of 1 GHz and beyond. The flash ADC architecture is commonly used in these applications to provide a single cycle latency operation.

13.4.1 FLASH ADC ARCHITECTURE

The flash ADC architecture, in its simplest form, is a single cycle, high-speed converter where the input signal is compared to all the quantization levels of the ADC. An n-bit flash ADC requires $2^n - 1$ comparators to complete the comparisons and generate the thermometer code, which is later decoded into the output ADC bits. Figure 13.4 shows a diagram of a flash ADC converter, where the different reference voltages corresponding to the quantization levels are generated using a resistor ladder. The performance of the flash ADC architecture at GHz sampling rates is limited by two sources: device mismatches, which result in offset voltages at the comparator inputs, and the large silicon area needed to build the ADC, which also results in high-power consumption. At low resolution (6 bit or less) and sampling rate below 1.5 GHz, these issues can be resolved using a number of design techniques that are now commonly used.

Figure 13.5 shows one of the techniques used to reduce the offset voltage at the preamplifier or comparator inputs of a flash ADC. The offset voltage is stored on a capacitor every conversion cycle to zero out the differential offset voltage at the input terminals when the input signal is applied. This approach was recently used to design a 4 bit, 1.25 GS/s flash ADC with a FOM of 0.16 pJ

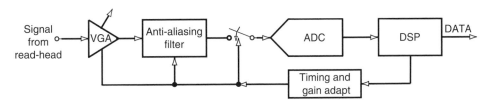

FIGURE 13.3 A read channel model.

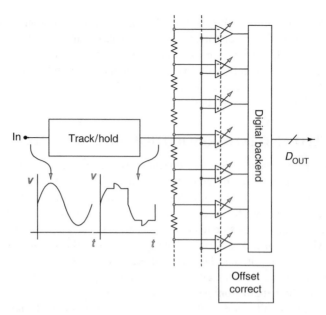

FIGURE 13.4 Flash ADC architecture.

per conversion step [7]. The low-power consumption was achieved by combining functions such as preamplification and sample/hold into the design of a digitally switched comparator. The resistor ladder was also replaced by a set of built-in threshold levels using CMOS transistor–based capacitors. Other approaches such as averaging and interpolation of the preamplifier's outputs are also used to minimize the effect of the offset voltage on the ADC performance. These techniques are covered extensively in the literature [8–10] and will not be discussed in this chapter.

The issue with offset voltage elimination and ADC calibration is that they require mixed-signal processing steps that complicate the design and add significant power consumption to the design. For instance, DAC (Digital-to-Analog Converter) trimming for ADC calibration has been shown to improve the nonlinearity of ADC designs at the expense of more than doubling the power consumption. Figure 13.6 shows the DAC trimming and calibration technique applied to a 4 bit, 4 GS/s flash ADC design [11]. When the inputs to the comparators are shorted during calibration, a differential current is applied at the outputs of the comparators. A self-timed circuit detects the correct bias point that guarantees that the ADC output is calibrated. The power consumption for the analog circuitry is less than 90 mW, while the digital circuit power consumption is around 500 mW at 4 GS/s. Despite the excessive power consumption, background digital calibration approaches will

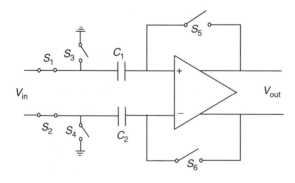

FIGURE 13.5 Input offset cancellation.

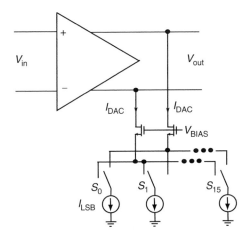

FIGURE 13.6 Weighted DAC trimming and calibration.

benefit future ADC designs as processes scale down and the digital circuit power consumption scales down as well. Background digital calibration has been proven to significantly improve the dynamic behavior of ADC performance. Figure 13.7 shows the performance data of a 12 bit and 75 MS/s pipelined ADC design, with and without calibration of the residue amplifiers between the pipeline stages [12]. An 18 dB improvement in the SFDR was achieved using digital calibration techniques for the ADC. This trend will continue in future ADC designs with more emphasis on controlling the power consumption growth in these digital calibration circuits.

Another contributor to ADC performance degradation is related to the inaccuracies in the timing of the sampling clock. Clock phase jitter in the sampling clock causes voltage error at the output of the sample and hold circuit. This results in degradation in the SNR of the output signal of the ADC. Equation 13.3 quantifies this degradation as a function of the input signal frequency f and the time jitter amount T_j [13]:

$$\text{SNR} = 20\log_{10}\left(\frac{1}{2\pi f T_j}\right) \tag{13.3}$$

Clock jitter poses a substantial problem when large bandwidth signals (100 Mb/s and beyond) are converted by an ADC. As sampling rate increases, clock jitter becomes a larger percentage of the

Process, area	0.35 µm CMOS, 7.9 mm²	
VDD	3 V	
Resolution	12 b	
Conversion rate	75 MS/s	
	Without postproc.	With postproc.
SNR	48 dB	68.2 dB (f_{in} = 1 MHz) 67 dB (f_{in} = 40 MHz)
THD	−50 dB	−74 dB (f_{in} = 1 MHz) −76 dB (f_{in} = 40 MHz)
SFDR	58 dB	80 dB (f_{in} = 1 MHz) 76 dB (f_{in} = 40 MHz)
DNL	−1, 0.6 LSB	−0.5, +0.5 LSB
INL	−19, +16 LSB	−0.9, +0.6 LSB
Power dissipation ADC core output drivers	290 mW 24 mW	

FIGURE 13.7 ADC performance with and without digital calibration.

clock cycle and contributes to bigger voltage errors and lowers SNR. Fortunately, device switching performance improves with process scaling, which will lead to keeping clock jitter to clock cycle ratio steady for high sampling rate converters.

13.4.2 TIME-BASED ADC ARCHITECTURE

New and innovative ADC architectures are being investigated as alternatives to conventional approaches that existed in the literature for the past 40 years. Even with today's advancements in process technology, there is hardly any new approach to solving analog to digital conversion issues at very high sampling rate without excessive power consumption. New CMOS process technology nodes at 90 nm and below have made it possible to achieve very high time-domain resolution of a digital signal edge transition. This time resolution is a result of superior device-switching characteristics. This fact led the authors to investigate a novel, time-based ADC architecture where high-speed analog signals are converted into high-resolution time-domain signals. Specifically, signal amplitude can be converted into a pulse width–modulated signal or a pulse position–modulated signal. This signal is fed to a digital block, which translates the timing information coded in the signal into binary digits. Figure 13.8 shows this basic architecture, with the voltage-to-time converter (VTC) and the time-to-digital converter (TDC) comprising the two blocks for an ADC [14,15]. The VTC converts an analog signal into a PWM (Pulse Position Modulation) signal, while the TDC generates the ADC digital data bits by measuring the time difference between the rising edges of the reference clock and the PWM signal at the output of the VTC. The TDC resolution is defined by the lowest possible delay value it can measure. This delay value is also referred to as the LSB (Least Significant Bit) for the TDC. The LSB value scales with process technology, making it possible for time-based ADC architectures to provide higher resolution ADCs in the future.

There are a number of advantages to the time-based ADC approach. Most of the ADC design will be digital and therefore benefits directly from process advancements and scaling. This includes low power consumption, low cost, and relative insensitivity to lower supply voltages since the ADC's dynamic range will be defined in the time resolution domain rather than the voltage domain. Sample-and-hold circuits are not needed in the time-based ADC architecture, because the input signal is not sampled by the VTC. Figure 13.9 shows a design example of a VTC that is based on current-starving inverters and the voltage-controlled delay line approach [14]. The main current-starving device, M5, is biased in saturation when the inverter consisting of M1 and M2 begins to make the transition from a logic high output to logic low. M5 is linearized by using source degeneration implemented with M7. Additional current-starving devices M11 and M13 are used in parallel with M5. These additional current-starving devices are biased in the subthreshold region and enter the heavy inversion region when the input signal is sufficiently large. This mitigates the compression of the pulse delay time versus input voltage characteristic at high input voltages. The additional parallel current-starving devices also increase the voltage sensitivity of the VTC. The transfer characteristic of this VTC was shown to be susceptible to process variations. Further research is being done to calibrate the VTC against process variations as well as improving the linearity of the VTC to expand its dynamic range.

In its simplest form, a TDC design is a counter that gets reset with the clock's high phase and stops counting when the PWM signal rising edge arrives at the counter. This approach can be used for

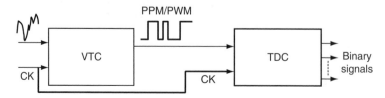

FIGURE 13.8   Time-based ADC architecture.

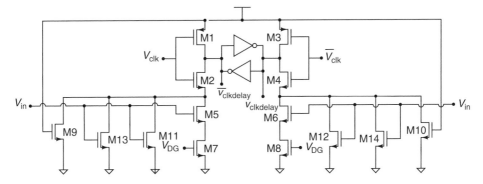

FIGURE 13.9   VTC design example.

high resolution and very low sampling rates. For high-speed sampling rate, the TDC must calculate the time difference in a single cycle and must run at high clock rates. Figure 13.10 shows the design principle of a new generation of TDC architectures that can be used in time-based ADC designs [15]. The hierarchical nature of this TDC design lends itself well to process technology scaling.

13.5 HIGH-RESOLUTION AND MULTI-GHz ADC ARCHITECTURES

So far, this chapter covered low-resolution (6 bit or less) and moderately high-speed (less than 1.5 GS/s) ADC architectures. For multi-GHz sampling, large bandwidth (500 MHz and more) and bit resolution greater than 6 bit, no single ADC architecture can provide an optimal solution without sacrificing a lot of power consumption. In order to break the quadratic relationship between bit resolution, frequency, and power consumption, time-interleaving ADC architecture is often cited as the most promising solution for multi-GHz sampling speed and bit resolution above the 6 bit limit of

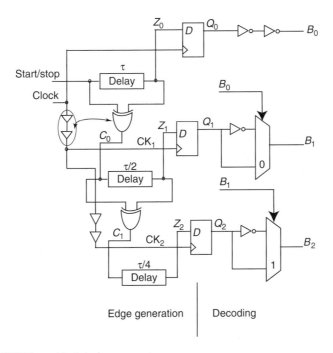

FIGURE 13.10   TDC hierarchical design concept.

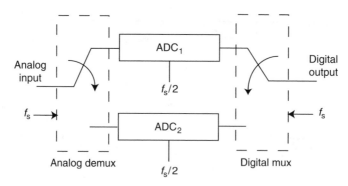

FIGURE 13.11   Time-interleaved ADC architecture.

today's designs. Figure 13.11 shows a block diagram of a time-interleaved ADC architecture. The underlying ADC architecture in each interleaving channel is selected based on the overall system resolution and speed [16].

For very high-speed applications, 8 bit and 4 GS/s for instance, an eight-channel time-interleaved ADC can be built around eight flash ADCs, each generating 8 bits and running at 500 MHz. An 8 bit flash ADC can be built using the subranging technique, as shown in Figure 13.12, where the signal conversion occurs in two cycles instead of a single cycle. Subranging also reduces the ADC hardware by allowing the use of only 15 comparators in each stage, instead of the 255 comparators that would have been used in a single cycle conversion. With each flash ADC running at 500 MHz, power consumption can be limited per channel to less than 10 mW, using optimization techniques described in previous sections. The overall power consumption is now linear with the number of channels in the time-interleaved ADC. Such design can achieve 8 bit of resolution at 4 GS/s rate, with less than 100 mW power consumption.

Unfortunately, the performance of time-interleaved ADCs is sensitive to offset and gain mismatches as well as aperture errors between the interleaved channels. Different offsets in the ADC channels contribute to a dc value as well as a periodic additive pattern in the output of the ADC array. In the frequency domain, the periodic pattern appears as a tone at the channel sampling rate. Gain mismatches between the parallel channels cause amplitude modulation of the input samples by the sequence of channel gains. In the frequency domain, this error causes a copy of the input signal spectrum to appear centered around the channel sampling rate. Figure 13.13 shows the correlation between channel mismatch errors and the output SFDR of time-interleaved ADCs.

Deviations from the ideal sampling instants can be represented as a sequence of sample-time errors that introduce errors in the input samples. For a sinusoidal input, the input samples are phase modulated by the sequence of sample-time errors in the ADC channels. In the frequency domain, this error produces copies of the input signal spectrum at the same frequencies as the spurious components

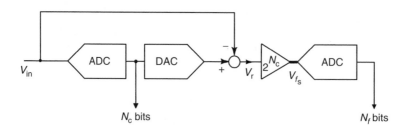

FIGURE 13.12   Subranging ADC implementation.

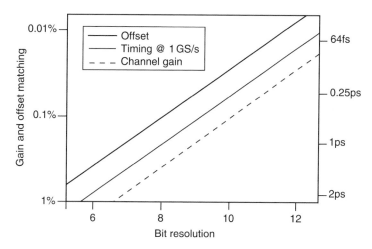

FIGURE 13.13 Channel matching requirements for time-interleaved ADCs.

stemming from gain mismatch. The sample-time error can be corrected by digitally processing the ADC outputs to interpolate the sample values that would have occurred at the ideal sample times.

Much work has been done on calibration to correct for offset and gain mismatches [17,18]. These techniques, especially digital calibration, will continue to benefit the scaling of time-interleaved ADC architectures in the future. Calibration of sample-time errors requires both detection and correction of timing errors. Sample-time errors are commonly detected using both foreground and background detection techniques. A sinusoidal or a ramp signal is applied to the ADC for calibration and the ADC output is compared to the ideal output. Once the sample-time errors have been measured, there are two main options for correcting sample-time error. The sampling clock for each ADC channel could be adjusted to eliminate the sample-time error or by digitally processing the ADC outputs to interpolate the sample values that would have occurred at the ideal sample times.

13.6 CONCLUSIONS

In summary, high-speed ADC architectures in the next decade will continue to evolve using a subset of today's architectures and technologies. Emphasis on power consumption reduction will determine which approaches will make it to the high-volume marketplace and which will be relegated to niche applications. There is no doubt that the static and dynamic performance requirements of future ADC applications are very demanding and will put more burden on designers to meet these specifications. Fortunately, some of the techniques used today, such as background digital calibration and jitter control through Phase Locked Loop (PLL) and Delay Locked Loop (DLL) circuits, will help designers reach their goals. However, new innovations are needed to break the power consumption envelope of current ADC architectures and to leverage the advances in new process technology nodes. Time-based and time-interleaved ADC architectures will dominate the ADC designs in the future since they provide the best opportunity to break the power envelope and the chance to scale in performance with new processes.

REFERENCES

1. K. Uehara, Trends in broadband wireless communication systems and software defined radios, *Interdisciplinary Information Sciences Journal*, 12(2), 163–172, 2006.
2. I. Lu, N. Weste, and S. Parameswaran, ADC precision requirements for digital ultra-wideband receivers with sub-linear front ends, a power and performance perspective, *Proceedings of the 19th International Conference on VLSI Design*, pp. 575–580, Hyderabad, India, 2006.

3. J. Schutz and C. Webb, A scalable X86 CPU design for 90 nm process, *ISSCC*, Session 3, San Francisco, 2004.
4. P. Gepner and M. Kowalik, Multi-core processors: New way to achieve high system performance, *International Symposium on Parallel Computing*, pp. 9–13, 2006.
5. R. Walden, Analog-to-digital converter survey and analysis, *IEEE Journal on Selected Areas in Communications*, 17(4), 539–550, 1999.
6. A. Varzaghani and C. Yang, A 600 MS/s 5-bit pipeline A/D converter using digital filter calibration, *IEEE Journal of Solid State Circuits*, 41(2), 310–319, 2006.
7. G. Van der Plas, S. Decoutere, and S. Donnay, A 0.16 pJ/conversion-step 2.5 mW 1.25 GS/s 4b ADC in a 90 nm digital CMOS process, *ISSCC*, Paper 31.1, San Francisco, 2006.
8. M. Choe, B. Song, and K. Bacrania, A 13b 40 MS/s CMOS pipelined folding ADC with background offset trimming, *IEEE Journal of Solid-State Circuits*, 35(12), 1781–1790, 2000.
9. K. Makigawa, K. Ono, T. Ohkawa, K. Matsuura, and M. Segami, A 7b 800 Msps 120 mW folding and interpolation ADC using a mixed-averaging scheme, *Digest of Technical Papers*, Symposium on VLSI Circuits, pp. 138–139, 2006.
10. M.P. Flynn and B. Sheahan, A 400 MS/s, 6b CMOS folding and interpolating ADC, *IEEE Journal of Solid-State Circuits*, 33(12), 1932–1938, 1998.
11. S. Park, Y. Palaskas, and M. Flynn, A 4 GS/s 4-bit flash ADC in 0.18 μm technology, *IEEE Journal of Solid-State Circuits*, 42(9), 1865–1872, 2007.
12. B. Murmann and B. Boser, A 12-bit 75-MS/s pipelined ADC using open-loop residue amplification, *IEEE Journal of Solid-State Circuits*, 38(12), 2040–2050, 2003.
13. B. Brannon, Aperture uncertainty and ADC system performance, *Analog Devices Application Note* AN-501, 1998.
14. H. Pekau, A. Yousif, and J. Haslett, A CMOS integrated, linear voltage to pulse delay time converter to time-based ADC, *ISCAS*, Greece, 2373–2376, 2006.
15. A. Yousif and J.W. Haslett, A fine resolution TDC for next generation PET imaging, *IEEE Transactions on Nuclear Sciences*, 54(5), 1574–1582, 2007.
16. Y. Jang, J. Bae, S. Park, J. Sim, and J. Park, An 8.8 GS/s 6-bit time-interleaved flash analog-to-digital converter with multi-phase clock generator, *IEICE Transactions on Electronics*, E90-C(6), 1156–1164, 2007.
17. Y. Yang, A. Chokhawala, M. Alexander, J. Melanson, and D. Hester, A 114 dB 68 mW chopper-stabilized stereo multi-bit audio A/D converter, *IEEE International Solid-State Circuits Conference*, vol. XLVI, pp. 56–57, February 2003.
18. E. Siragusa and I. Galton, A digitally enhanced 1.8-V 15-bit 40-MSample/s CMOS pipelined ADC, *IEEE Journal of Solid-State Circuits*, 39, 2126–2138, 2004.

14 Energy-Efficient ADC Topology Enabled with Asynchronous Techniques

Shuo-Wei Mike Chen

CONTENTS

14.1 INTRODUCTION

Enabled by the continuous technology scaling and growing system requirements, the state-of-the-art integrated circuits (IC) have been pushed toward more system integrations, programmable configurations, higher speed, and bandwidth. In most cases, the system flexibility naturally comes from digital signal processing techniques. Some examples of such highly flexible systems include software-defined radio [1] and cognitive radio [2], which requires operations in multiple frequency bands and communication standards. Some high-bandwidth systems such as ultra-wideband (UWB) radio and high-speed serial links also require a wide analog bandwidth and thus a faster sampling speed. As a result, an analog-to-digital converter (ADC) has inevitably become one of the most critical blocks of the system, where higher input bandwidth, sampling rate, and sometimes higher dynamic range are desired. From a system architecture perspective, much efforts [3,4] have been made to minimize analog processing while allowing digital processing to do most of the work, as they are also favored by technology scaling. However, the feasibility and cost (i.e., technology cost, silicon area, and power consumption) of a high-speed and high-bandwidth ADC often dominate overall system trade-off. Before the booming of complementary metal oxide semiconductor (CMOS) technology, SiGe and GaAs were a popular choice for making high-speed ADCs [5,6], which could be often seen in many radar systems and instruments. In the past decade or so, CMOS ADC has benefited from faster CMOS devices, and the higher sampling speed was achieved with a much lower power compared to III-V technologies. In the future, further lowering the cost of high-speed ADCs will potentially enable a novel system architecture and result in much more efficient system solutions for many cost- and power-sensitive applications, such as most consumer electronics, which accounts for a significant portion in the IC markets nowadays.

In this chapter, we will approach from an ADC architecture level to explore the cost limit of CMOS ADC for going toward higher speed and bandwidth.

14.2 POWER EFFICIENCY OF ADC ARCHITECTURE

Figure 14.1 shows the three commonly used Nyquist ADC topologies: flash, pipeline, and successive approximation converter (SAR). A first-order estimation of power and conversion speed of these conventional topologies is performed to identify the best entry point for further power efficiency improvement. Traditionally, flash ADCs are favored for high-speed N-bit converter since $2^N - 1$ comparators are utilized to make a fully parallel comparison with the entire quantization levels within one clock cycle. The decoding circuits solving sparkle and metastability issues and thermometer-to-binary code conversion also dissipate extra power. The total power consumption of a flash ADC therefore roughly scales as 2^N. In Figure 14.1, the conversion speed is normalized to one for comparison with other architectures corresponding to the fact that the full conversion is complete within one sample clock cycle. An approach to breaking the exponential dependence of the number of comparators on the number of bits is the use of a pipeline ADC. Instead of fully parallel comparison, it divides the process into several comparison stages, the number of which is proportional to the number of bits. Therefore, the total number of required comparators is greatly reduced, with only N comparators required for a 1-bit per stage, N-bit pipeline ADC. However, due to the pipeline structure of both analog and digital signal path, interstage residue amplification is needed, which consumes considerable power and limits high-speed operation. While it is possible to make use of open-loop residue amplification [7], an extra calibration loop is needed, increasing the overall complexity and power consumption. Therefore, the total power consumption of a pipeline ADC increases as N with a speed <1.

For low conversion speeds, an SAR approach is often used since it also divides a full conversion into several comparison stages in a way similar to the pipeline ADC, except the algorithm is executed

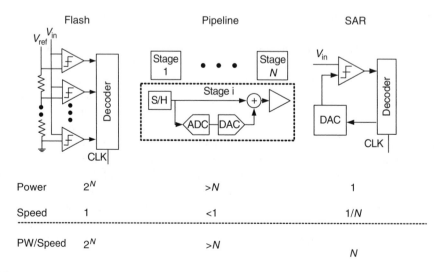

	Flash	Pipeline	SAR
Power	2^N	$>N$	1
Speed	1	<1	$1/N$
PW/Speed	2^N	$>N$	N

FIGURE 14.1 Conventional architectures for Nyquist ADCs.

sequentially rather than in parallel as in the pipeline case. An N-bit SAR converter utilizes only one comparator with N clock cycles to complete a full conversion. Thus, the total power consumption is normalized to approximately 1, while speed is now $1/N$. Since the ratio of power and speed represents the energy consumption per conversion sample, SAR converters clearly have a power efficiency advantage over the other approaches. Due to the fact that the power efficiency difference between SAR and flash topologies increases exponentially with the number of bits, N, an SAR converter provides a promising starting point for achieving the most power-efficient solution. However, the sequential operation of the SA algorithm has traditionally been a limitation in achieving high-speed operation, so in the following section, an architecture based on asynchronous processing will be used to yield high-speed operation with a normalized power/speed ratio $\ll N$.

14.3 ASYNCHRONOUS PROCESSING

The conventional implementation of the SA algorithm, such as an SAR converter, relies on a synchronous clock to divide time into signal-tracking and conversion phases, which progress from the most significant bits (MSB) to the least significant bits (LSB) as shown in Figure 14.2. For an N-bit converter with conversion rate of F_s, a synchronous approach would require a clock running at least $(N + 1)F_s$. Since an SAR converter is traditionally used in lower conversion rate regime, therefore, clock generation is less of an issue. However, for a high-speed converter, the clock generation of this high-speed internal clock is a significant overhead. For example, a 300 MS/s and 6-bit SAR would require a 2.1 GHz clock. Synthesizing such a high-frequency clock plus clock distribution network would likely consume more power than the ADC itself. From the speed perspective, every clock cycle has to tolerate the worst-case comparison time, which is composed of maximum digital-to-analog converter (DAC) settling and comparator-resolving time depending on the minimum resolvable input level. This is due to the fact that the conventional clock has 50% duty cycle to partition the time into comparator-resolving phase and DAC-settling phase. A self-timed comparator technique was proposed in Ref. [8] to effectively change the duty cycle of the clock and achieves a better time partition between the two phases. It potentially improves the clocking rate, but does not entirely change its nature of synchronous conversion. Additionally, every clock cycle requires margin for the clock jitter, which will either slow down the conversion speed or impose a stringent jitter requirement on the clock generator.

Therefore, the power and speed limitations of a synchronous SA design come largely from the high-speed internal clock. By using asynchronous processing of the internal comparisons, it removes the need for such a clock and substantially improves the power efficiency compared to a synchronous design. On the top level, a global clock running at the sample rate is still used for a uniform sampling, since most of the digital baseband to date remains in a synchronous world. The concept of asynchronous processing is to trigger the internal comparison from MSB to LSB like dominoes. Whenever the current comparison is complete, a ready signal is generated to trigger the following comparison is shown in Figure 14.3.

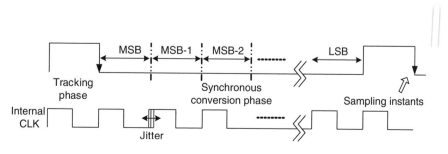

FIGURE 14.2 Synchronous conversion for SAR ADCs.

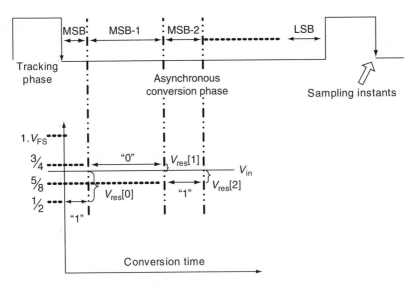

FIGURE 14.3   Asynchronous processing concept.

The voltage difference (V_{res}) between input signal and reference level determines the comparator-resolving time. For example, a typical regenerative latch has the following trade-off between input voltage (V_{res}) and resolving time (T_{cmp}) [9].

$$T_{cmp} = \frac{\tau}{A_o - 1} \ln \frac{V_{FS}}{V_{res}} = K \ln \frac{V_{FS}}{V_{res}} \tag{14.1}$$

where
  A_o is the small-signal gain of the internal inverting amplifier
  τ is the time constant at the latch outputs
  V_{FS} is the full logic swing level

Depending on the comparator topology, the resolving time and input voltage trade-off will change. Nevertheless, this simple regenerative latch model provides intuition into how asynchronous processing helps to improve the conversion speed. For an N-bit converter, the total resolving time of both synchronous and asynchronous design can be expressed as

$$T_{async} = \sum_{i=0}^{N-1} K \ln \frac{V_{FS}}{V_{res}[i]} \tag{14.2}$$

$$T_{sync} = NK \ln \frac{V_{FS}}{V_{min}}$$

where
  $V_{res}[i]$ denotes the input voltage of the comparator at ith stage (Figure 14.3)
  V_{min} is usually set by the LSB level

Clearly, the asynchronous conversion takes advantage of the faster comparison cycles, since only one of these $V_{res}[i]$, $\forall i \in [0, N-1]$ will fall within $\pm 1/2$ LSB due to the successive approximation algorithm. The amount of conversion time savings between T_{async} and T_{sync} is a function of the number of bits as well as the profile of $V_{res}[i]$, which depends on the input voltage level. In the extreme case, a 1-bit converter does not benefit from asynchronous processing, since the only comparison cycle is

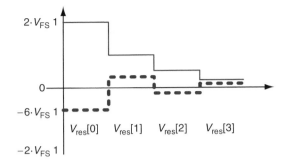

FIGURE 14.4 Best (solid line) and worst (dash line) cases of V_{res} profile.

always limited by the worst-case resolving time. As the number of bits increase, $V_{res}[i]$ will distribute over the full-scale range and thus create time savings. Intuitively, the wider the range of V_{res}, the faster conversion speed it can achieve. With the assistance of numerical analysis of Equation 14.2, the best case scenario is found when input signal is at full swing, i.e., when V_{res} reaches $\pm 1/2 V_{FS}$. When V_{res} alternates its polarity from consecutive comparison cycles, it results in the longest conversion time, as shown in Figure 14.4.

The ratio of T_{async}/T_{sync} of both the best and worst cases is derived as a function of number of bits in order to explore the theoretical performance bound of asynchronous processing. In the best case, $V_{res}[i]$ are simply $V_{FS}/2$, $V_{FS}/4$, $V_{FS}/8$, ..., $V_{FS}/2^N$, assuming a binary successive approximation algorithm. Defining $V_{LSB} = V_{FS}/2^N$, the minimum value of T_{async}/T_{sync} can be expressed as

$$\left.\frac{T_{async}}{T_{sync}}\right|_{min} = \frac{\ln\dfrac{V_{FS}}{V_{LSB}} + \ln\dfrac{V_{FS}}{2V_{LSB}} + \ln\dfrac{V_{FS}}{4V_{LSB}} + \cdots \ln\dfrac{V_{FS}}{2^{N-1}V_{LSB}}}{N\ln\dfrac{V_{FS}}{V_{LSB}/2}}$$

$$= \frac{N^2\ln 2 - \frac{N}{2}(N-1)\ln 2}{N(N+1)\ln 2} = \frac{1}{2} \tag{14.3}$$

In the worst case, the input voltage level that leads to comparison results with alternating polarity can be better understood as the number of bits increases from 2-bit case, assuming V_{res} begins from positive side:

$$\text{2-bit case} \Rightarrow V_{res}[0] - \frac{V_{FS}}{4} < 0$$

$$\text{3-bit case} \Rightarrow V_{res}[0] - \frac{V_{FS}}{4} + \frac{V_{FS}}{8} > 0$$

$$\text{4-bit case} \Rightarrow V_{res}[0] - \frac{V_{FS}}{4} + \frac{V_{FS}}{8} - \frac{V_{FS}}{16} < 0$$

$$\vdots$$

$$\Rightarrow \frac{1}{8}\left(1 + \frac{1}{4} + \frac{1}{16} + \cdots\right) < V_{res}[0] < \frac{1}{4}\left(1 - \frac{1}{4} - \frac{1}{16} + \cdots\right)$$

$$\Rightarrow V_{res}[0] \to \frac{1}{6}V_{FS}, \text{ as number of bit increases} \tag{14.4}$$

Given the derived results from Equation 14.4, the worst-case conversion time occurs when V_{in} is $V_{FS}/3$ or $2V_{FS}/3$ regardless of the number of bits, and therefore the maximum value of T_{async}/T_{sync} is derived as follows:

$$\left. \frac{T_{async}}{T_{sync}} \right|_{max} = \frac{\ln \dfrac{V_{FS}}{1/6 \cdot V_{FS}} + \ln \dfrac{V_{FS}}{1/12 \cdot V_{FS}} + \cdots \ln \dfrac{V_{FS}}{\max\{1/(3 \cdot 2^N) \cdot V_{FS}, V_{LSB}/2\}}}{N \ln \dfrac{V_{FS}}{V_{LSB}/2}}$$

$$= \frac{(N-1)\ln 3 + \ln 2 + \frac{N}{2}(N+1)\ln 2}{N(N+1)\ln 2} \tag{14.5}$$

Note that the ratio of T_{async}/T_{sync} in Equation 14.5 approaches 1/2 as N increases. In conclusion, given the lower and upper bound from Equations 14.3 and 14.5, the maximum resolving time reduction between synchronous and asynchronous cases is twofold. Moreover, the conversion time savings over a synchronous approach increases with higher ADC resolution.

14.4 DESIGN EXAMPLE OF A 6-BIT 600 Ms/s ASYNCHRONOUS ADC

14.4.1 ARCHITECTURE

While there are several possible architectures to incorporate the asynchronous processing concept, the first prototype has utilized only one comparator with a charge redistribution network to achieve a low-complexity implementation similar to an SAR converter. Since the internal comparisons use the same comparator, it does not require special attention to reduce its offset in the analog domain as the global offset can be subtracted in the digital domain. However, the overall conversion speed is slowed down because the comparator must be reset after each comparison cycle. The charge redistribution capacitor network is used to sample the input signal and serves as a DAC for creating and subtracting reference voltages.

Besides asynchronous processing, time interleaving [10] is used to increase the maximum conversion rate over what a single ADC can achieve. Note that there are power and area overheads as the number of parallel converters increases. Therefore, a single asynchronous ADC should be optimized for high speed and small silicon area. In this prototype (Figure 14.5), two ADCs are time interleaved for a doubling of the sample rate over an individual ADC. The two phase (0° and 180°) clocks are provided via on-chip inversion, and used as sampling clocks and reset signals. The high input bandwidth (>4 GHz) of the individual converter achieved here would actually allow additional time interleaving.

There are two critical delay paths in this architecture, which involve signal and timing. For the signal path, each internal comparison result is stored in an SR latch as a buffering stage to the temporary bit caches. For the asynchronous timing path, the comparator's outputs are detected

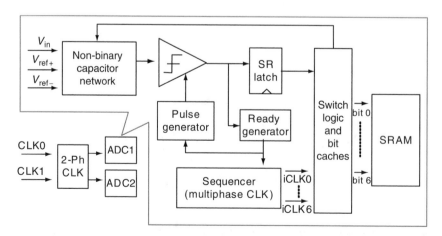

FIGURE 14.5 Simplified block diagrams of the ADC architecture.

by a ready signal generator as a data completion flag of each comparison cycle. This ready signal then drives a sequencer to provide multiple-phase clocks for switching logic and temporary bit caches to store the internal comparison results. A separate pulse generator creates a reset phase for the comparator to avoid any memory effect from the previous comparisons. Note that the ready signal generator, pulse generator, and sequencer are the dedicated digital logic functions to perform asynchronous conversion and they occupy only a small portion of the silicon area.

Finally, the bit streams at the output of the bit caches are designed for high throughput, which raises the difficulty of real-time streaming off-chip. Therefore, a 1K-depth on-chip static random access memory (SRAM) is used to store the converted data, and later readout to off-chip in a much slower rate. The integration of the SRAMs is solely for testing purpose, and occupies the most of the die area.

14.4.2 DYNAMIC COMPARATOR AND READY SIGNAL

The design of the comparator requires special consideration because of the need to generate a data-ready signal. A dynamic comparator that is composed of a preamplifier and regenerative latch is shown in Figure 14.6. The complementary outputs of the comparator are connected to the positive supply during the reset phase and one of the two outputs (Q_p and Q_n) is pulled down to the negative supply during comparison. Therefore, digital logic, which is able to distinguish state $'1''1'$ (reset phase) from state $'1''0'$ or $'0''1'$ (data ready), seems to be sufficient for the ready signal generation. However, one potential issue with asynchronous processing is that the comparator can be in a metastable state when the input is sufficiently small. The time needed for the comparator outputs to fully resolve may take arbitrarily long. As a result, the comparator is designed fast enough such that this only occurs when the input signal is less than 1 LSB, which means the decision does not affect the converter accuracy. In this case, the ready signal generator should still set the flag and the decision result is simply taken from the previous value stored in the SR latch. To achieve this goal, since both outputs (Q_p and Q_n) will drop together to a lower level when the comparator is in metastable state, a simple NAND gate with input threshold above this level is a key solution to the ready signal generation. As Q_p and Q_n are to be in state $'0''0'$ once they drop across the threshold, the NAND gate implemented in complementary CMOS logic will set the flag and continue the remaining conversion process.

There are reset switches in both the preamplifier and latch stages that help to reduce the comparator recovery time during the reset phase. An input offset cancellation is also utilized for the preamplifier stage but is not critical in this ADC architecture as mentioned earlier. Current mirrors between the two stages are useful to reduce charge kickback [11] from the logic level swing of the latch onto the input capacitors preceding the preamplifier. This is especially important since the input capacitor network is pushed to the minimum possible value as will be described shortly.

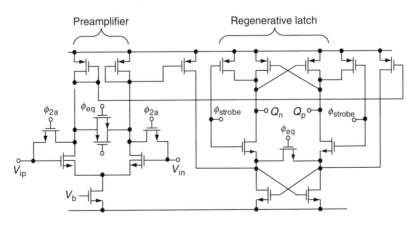

FIGURE 14.6 Dynamic comparator schematic.

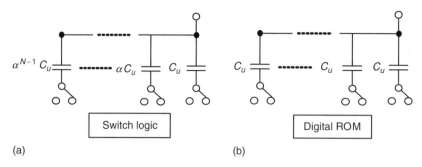

FIGURE 14.7 Conventional implementation of radix creation. (a) Geometrically scaled capacitor array and (b) unitary capacitor array.

14.4.3 Nonbinary Successive Approximation Review

Instead of a binary successive approximation scheme, this ADC adopts redundancy to allow dynamic decision errors for faster conversion speed [12]. In other words, the overlapped search range compensates for wrong decisions made in earlier stages as long as they are within the error tolerance range. This eliminates the constraint of DAC settling accuracy to be less than 1 LSB, and thus helps to reduce the settling time. The equivalent radix is less than 2 and computed as

$$\text{Radix} = 2^{\left(N_{\text{bit}}/\left(N_{\text{bit}}+N_{\text{rdn}}\right)\right)} \tag{14.6}$$

where
N_{bit} is the target bit resolution
N_{rdn} is the redundant bit

In this prototype, 1 extra redundant bit is used for the target of 6-bit resolution. Therefore, the equivalent radix is about 1.81 and results in about 50% reduction in DAC settling time while sacrificing just 15% conversion time due to the extra comparison cycle.

In terms of implementation, there are two basic approaches as shown in Figure 14.7. The geometrically scaled capacitor array makes use of a parallel bank of capacitors ratioed from 1, α, α^2, ..., α^{N-1}, as shown in Figure 14.7a. The advantage of this approach is the low complexity of the switching logic, since only one capacitor will be switched at each comparison cycle. The propagation delay and power consumption through switching logic are expected to be lower. However, in this prototype, the ratio α is a noninteger, which significantly increases layout complexity and matching difficulty [13] for a full array. On the other hand, a unitary capacitor array (Figure 14.7b) can be used to avoid this noninteger matching issue. The nonbinary code words are stored in a digital read-only memory (ROM). However, the propagation delay through a digital ROM is much larger compared to the previous case due to the longer logic depth. Moreover, the total input capacitance of both schemes is on the order of 2^N times unit capacitance. Even for a 6-bit case, the total capacitance can be on the order of picofarads with just tens of femtofarads unit capacitance, which is set by matching and parasitic considerations. This causes additional power consumption as well as the difficulty to maintain a high input bandwidth.

14.4.4 Series Nonbinary Capacitive Ladder

Another approach was therefore taken to create an arbitrary radix, i.e., in effect an analog ROM. This approach uses a ladder structure of a nonbinary capacitor array, which allows a significant reduction in the input capacitance with relaxed matching and layout requirements.

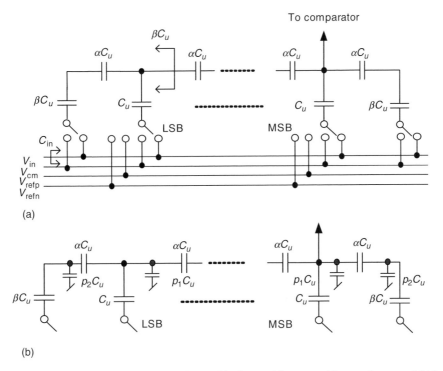

FIGURE 14.8 Series nonbinary capacitive ladder. (a) Ideal case without parasitic capacitance and (b) inclusion of parasitic capacitance.

Three different sizes of capacitors with ratios $1:\alpha:\beta$ are used to build the ladder and are shown in Figure 14.8a. The approach is to have the equivalent capacitance at every internal node be identical, i.e., βC_u. Therefore, the charge redistribution from one section to the adjacent one always sees the capacitive divider between αC_u and βC_u. This division ratio will determine the radix of the SA algorithm. Based on the above observations, the design equation of this ladder is derived as

$$\begin{cases} \beta = 1 + \alpha \| \beta \\ \text{Radix} = 1 + \dfrac{\beta}{\alpha} \end{cases} \tag{14.7}$$

where operator $\|$ is defined as $x \| y = xy/(x + y)$.

Due to the series connection of the capacitors, the equivalent capacitance is decreased, which reduces the DAC settling time and the total input capacitance. The traditional trade-off between matching property and total input capacitance is removed since it does not depend on reducing the unit capacitance size. The total input capacitance of the proposed ladder is no longer dependent on the number of ADC bits, and is calculated as

$$C_{\text{in}} = [1 + 2(\alpha \| \beta)] C_u \tag{14.8}$$

One potential issue with this ladder structure is the vulnerability to the parasitic capacitance due to interconnects or the capacitor itself, especially when the capacitor is implemented as a low-cost metal-oxide-metal (MOM) finger capacitor available in a standard digital CMOS process instead of a higher quality metal-insulator-metal capacitor. The extra capacitance introduced at the floating nodes can change the effective radix value if the parasitic capacitance is not negligible to the capacitors in the ladder. In this prototype, the unit capacitance is set at the minimum possible value and MOM capacitors have nonnegligible fringing capacitances, which necessitates a new design equation including the parasitic capacitance, p_1C_u and p_2C_u denoted in Figure 14.8b.

$$\begin{cases} \beta = 1 + \alpha \| \beta' + p_1 \\ \text{Radix} = 1 + (\beta'/\alpha) \\ \beta' = \beta + p_2 \end{cases} \tag{14.9}$$

By solving Equations 14.7 and 14.9, one can show that the new ratios (α_{mod} and β_{mod}) should be modified according to the following relations with the original ones (α_{org} and β_{org}):

$$\begin{cases} \alpha_{mod} & = (1 + p_1)\alpha_{org} \\ \beta_{mod} & = (1 + p_1)\beta_{org} - p_2 \end{cases} \tag{14.10}$$

In this design, a standard capacitor model with typical electronic design automation (EDA) extraction tools was used to estimate the parasitic capacitances and found accurate enough for the desired ADC resolution. Combining the proposed ladder structure in a passive bottom plate sampling network, the input bandwidth achieves >4GHz because of the small total input capacitance of 90 fF.

14.4.5 DIGITAL CALIBRATION SCHEME

As the systematic error is accounted for in the modified design equation, the ADC is still vulnerable to the random error, such as capacitor mismatch and parasitic variation. These random errors can change the effective radix from MSB to LSB bit, and thus reduce the linearity of the ADC. Similar to the gain error of a residue amplifier in a pipeline ADC, the digital combining weights need correction by estimating the real gain [14]. In this prototype, a foreground digital calibration scheme is developed to correct the combining weights and currently implemented off-chip. The approach was to inject a known input signal to the ADC, and use the converted outputs with initial combining weights to reconstruct the input signal. By using the reconstructed signal as the reference for an least mean square (LMS) loop, the combining weights can be adapted to the real values. Alternatively, the combining weights can be directly calculated through matrix operation using the orthogonality principle. The reason for using an LMS loop is to reduce the algorithmic complexity to enable a potential on-chip integration.

A ramp signal that spans over the full swing range is injected as the known signal as shown in Figure 14.9a. The reconstruction of the reference signal is done through a best linear curve fitting of the ADC outputs with initial guess of the combining weights. Next, the same ADC output code words are fed into an adaptive finite impulse response (FIR) filter to converge the real combining weights. The simulation results showed that the quantization error can be improved after calibration using several hundred samples. Alternatively, a sine wave of a certain frequency and full swing amplitude can be used as the prior information as illustrated in Figure 14.9b. By using an fast Fourier transform (FFT) processor, the sine wave is reconstructed by extracting its amplitude, phase, and offset at the fundamental frequency. The benefit of using a sine wave rather than a ramp is the potentially easier on-chip implementation with high linearity so that the digital calibration can be turned into an on-chip self-calibration scheme and extended to higher ADC resolution.

14.4.6 VARIABLE DUTY-CYCLED CLOCK

There are two criteria for the global sampling clock: (a) it needs to have a variable duty cycle for testing purposes by allowing the time between tracking and conversion phases to be varied to explore the ADC performance and (b) it should have very low jitter if radio frequency (RF) subsampling is to be used. In fact, the root mean squared (RMS) jitter should be on the order of picoseconds using worst-case case analysis to support the subsampling capability.

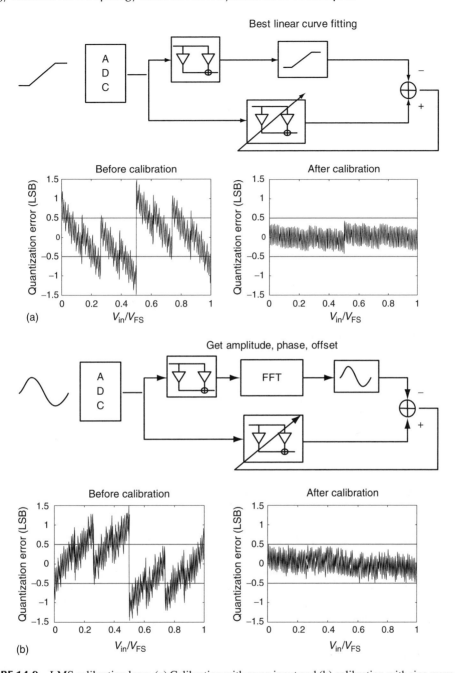

FIGURE 14.9 LMS calibration loop. (a) Calibration with ramp input and (b) calibration with sine wave input.

The clock generation is illustrated in Figure 14.10 and uses two sinusoidal waves that are generated off-chip with a tunable phase skew in between. The waveforms are then regenerated on-chip and combined with an AND gate. The phase skew determines the duty cycle of the clock source. Another 180° phase-shifted clock is achieved by simply inverting the two sinusoidal waves and going through the same combination logic. Special attention was paid in both logic and layout levels to ensure the exact 180° phase shift. Any phase imbalance causes extra distortion or required additional calibration. Finally, the clock jitter is minimized by careful layout, a dedicated power domain, and a clean clock source with extra bandpass filtering. In addition, the edge rate of the sampling clock

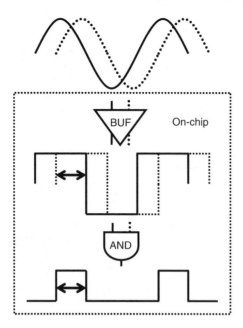

FIGURE 14.10 Variable duty-cycled clock generation.

should be high enough to reduce the jitter, which results in extra power dissipation in the large-sized buffers. The jitter due to intrinsic noise of the logic gates is analyzed and simulated to ensure that it is well below the specification.

14.4.7 HIGH-SPEED DIGITAL LOGIC

The speed of the entire asynchronous and switching logic is also critical to the speed of conversion rate. Therefore, all the digital logic in the critical path is custom dynamic logic and optimized using logical efforts [15] as well as careful layout. The dynamic logic uses a weak keeper transistor to avoid charge leakage and enhance noise invulnerability. Moreover, the dynamic registers are designed for minimal clock loading as these are driven by the asynchronous logic. Note that the pulse duration is adjusted by a variable MOS resistor operated deep in the triode region in order to explore the trade-off between conversion speed and dynamic error. The less critical digital blocks such as bit caches, and SRAM controller are made from standard cells provided by the foundry to save design time. Nevertheless, the timing constraint between ADC and testing SRAM is still tight, which requires careful design of the interface circuitry.

14.4.8 MEASURED RESULTS

The prototype ADC was fabricated in a 1.2 V 0.13 μm six-metal one-poly digital CMOS process [16]. Chip on board packaging was used on two versions of printed circuit board (PCB) designs to measure ADC performance below and above Nyquist frequency (above Nyquist to investigate the use of sub-sampling). A photomicrograph is shown in Figure 14.11. The total chip size measures $1.7 \times 1.4 \, \text{mm}^2$, while each ADC occupies only $250 \times 240 \, \mu\text{m}^2$, which reduces the overhead of time interleaving.

The static performance is characterized through differential nonlinearity (DNL) and integral nonlinearity (INL) measurement. As shown in Figure 14.12, DNL and INL improve from over 1 LSB to within half LSB after combining weights calibration as described in Section 14.4.5. It is equivalent to 2 dB signal-to-noise distortion (SNDR) improvement, which implies that the random error at 6-bit level is not significant. The dynamic performance measurements (Figure 14.13a) show

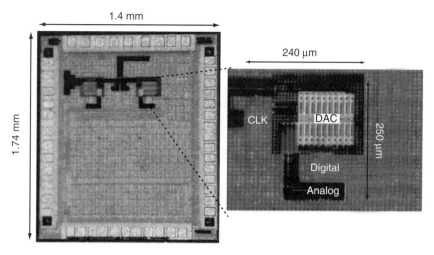

FIGURE 14.11 DNL and INL before and after combining weights calibration.

that the ENOB of a single ADC scales from 5.3 bit at 300 MS/s to 3.7 bit at 500 MS/s, demonstrating the straightforward trade-off between ENOB and conversion rate, which is inherent to the proposed ADC architecture. In Figure 14.13b, the dynamic performance is further explored using RF input above Nyquist range from 3 to 5 GHz, showing that the SNDR remains above 30 dB even with an input frequency over 4 GHz.

Figure 14.13c shows the performance of time interleaving two of the ADCs to achieve 600 MS/s sampling rate at twice the power and area. Off-chip digital subtraction of each ADC offset removes spurious tones improving the SNDR by 0.7 dB. Note that there is little reduction of SNDR at lower frequency, but as the input frequency increases above 300 MHz, the clock skew between paths yields

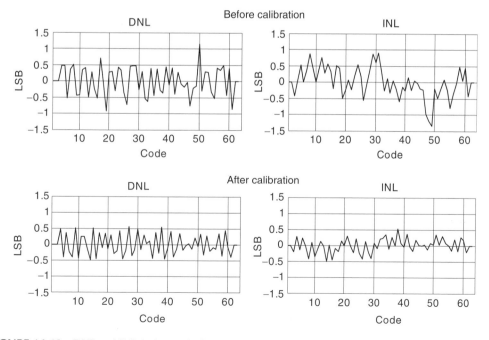

FIGURE 14.12 DNL and INL before and after combining weights calibration.

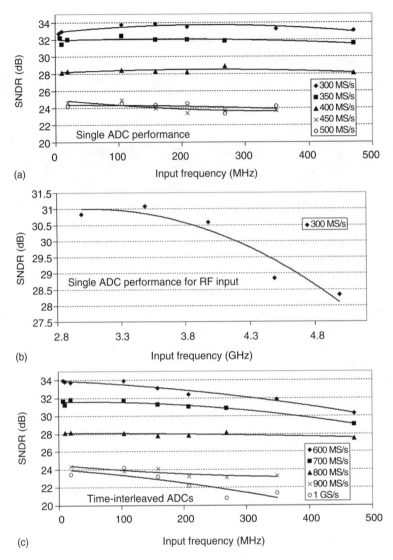

FIGURE 14.13 Measured SNDR versus f_s and f_{in} for single ADC (a) below and (b) above Nyquist frequency; for time-interleaved one (c).

a several dB SNDR reduction. To prove this, the clock skew is extracted through Hilbert transformer and then compensated by digital interpolation. The measured results show that SFDR improves 11 dB and SNDR improves 2 dB. The total power consumption excluding SRAM and IO pads consumes 5.3 mW while analog, digital, and clock section consume 1.2, 3.2, and 0.9 mW, respectively. The performance of the chip is summarized in Table 14.1.

14.5 SCALING TREND OF AN ASYNCHRONOUS ADC

As a medium-resolution ADC is not limited by KT/C noise, it generally benefits from technology scaling. This also applies to the proposed asynchronous ADC architecture, since most circuits are open-loop and digitally operated while only limited by capacitor matching accuracy. A first-order power and speed analysis for technology scaling is explored assuming a constant field scaling, i.e., dimension of transistors and supply voltage scales down by $1/S$. The conversion time is mainly

TABLE 14.1

Performance Summary (25°C)

Technology	0.13 μm 6M1P digital CMOS		
Package	Chip on board		
Resolution	6 bit		
Sampling rate	300–500 MS/s for single ADC		
	(600 MS/s-GS/s for time-interleaved one)		
Supply voltage	1.2 V		
Input 3 dB bandwidth	>4 GHz		
Peak SNDR	34 dB at 600 MS/s		
FOM	0.22 pJ/conversion step		
	Analog	1.2 mW	
Power	Digital	3.2 mW	Total power 5.3 mW
	Clock	0.9 mW	

dominated by the signal tracking time, comparator speed, and digital propagation delay. The value of the capacitor array is assumed to be fixed to preserve the matching property, and the on-resistance of a MOS switch is deliberately scaled down by fixing W. Therefore, the tracking time constant ($R_{on}C_s$), comparator bandwidth ($\propto f_T$ of a transistor), and digital gate delay (CV/I) all scale down by $1/S$ when velocity saturation is included. From a power perspective, if the overdrive voltage and W/L are assumed fixed, the analog power scales down as supply voltage. The digital switching power (fCV^2) scales down by $1/S^2$ [17], while that of clock network scales less due to the relatively larger sized sampling switch. Table 14.2 summarizes the scaling trend, which predicts the figure-of-merit (FOM) defined in Equation 14.11 improves at least $1/S^2$ and thus becomes even more attractive as the technology is scaled:

$$\text{FOM} = \frac{\text{Power}}{2^{\text{ENOB}}f_s} \tag{14.11}$$

As an example, the FOM achieved in the asynchronous ADC design described in Section 14.4 about 0.22 pJ per conversion step even when input is above the Nyquist frequency as it was designed for subsampling RF signal.

Figure 14.14 shows a common usage of ADC topologies in terms of resolution and sampling rate. Traditionally, flash-type ADCs, including subranging and folding converters, dominate over the high-speed and medium-resolution regime, while consuming redundant power. To save the extra power consumption, an SAR converter is a better solution, however, it is normally used for high to medium resolution while limited to speeds of tens of MHz to KHz range. By using the asynchronous processing and scaled technology, the SA algorithm implementation has been increased to hundreds of MHz with the potential of even improving more in the future. The proposed architecture can be easily extended to higher ADC resolution before being limited by KT/C noise. In fact, the FOM

TABLE 14.2

Technology Scaling on the Proposed ADC Architecture

T_{track}	T_{comp}	T_{dig}
$RC \sim 1/S$	$1/f_T \sim 1/S$	$RC \sim 1/S$
P_{analog}	P_{clk}	P_{dig}
$IV \sim 1/S$	$fCV^2 \sim [1/S - 1/S^2]$	$fCV^2 \sim 1/S^2$

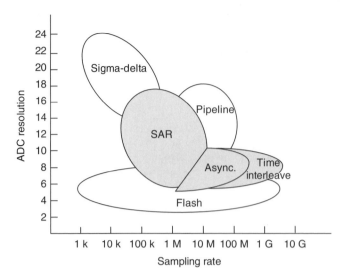

FIGURE 14.14 Future role of the SA architecture.

improves as the ADC bits increases, since the speed scales down proportionally while quantization levels (2^{ENOB}) scale up exponentially. Finally, the greater than 4 GHz input bandwidth allows the potential to time-interleave close to 8 GHz sampling rate for a Nyquist ADC.

14.6 APPLICATIONS OF THE PROPOSED ADC TOPOLOGY

The proposed asynchronous ADC with nonbinary successive approximation topology provides a path heading toward high power efficiency, sampling speed, and bandwidth without the need of a higher-than-sampling-rate clock. These advantageous features potentially enable new system possibilities. Once the ADC is capable of sampling RF signals with higher sampling rate, one system possibility is to move it closer to the antenna and thus reduce the analog complexity, i.e., a mostly digital receiver topology. As an example, a subsampling radio architecture [18] has been proposed to integrate the frequency translation function into a subsampling ADC for a low-complexity implementation of 3–10 GHz UWB systems, which is based on the impulse radio approach. Many recent published UWB system solutions, including both orthogonal frequency division multiplexing (OFDM) and direct-sequence spread spectrum (DSSS) approaches, have adopted a direct-conversion architecture, as shown in Figure 14.15a. A major challenge of this approach is that the overall complexity still remains high, which means a dramatic cost and power reduction from the current wireless solutions is unlikely. For example, the transmitter of a wideband OFDM radio requires high-speed DAC, up-conversion mixers, oscillators, and power amplifier with linearity and peak-to-average ratio constraints because of the multicarrier transmission. On the other hand, an impulse radio simply uses a pulser to drive the antenna, and radiates a passband pulse shaped by the response of the wideband antenna and any possible bandpass filter, as shown in Figure 14.15b. The hardware elimination of mixers and local oscillators (LO) for mixing and the reduced linearity requirement imply lower complexity implementations of a transmitter.

On the receiver side, the direct-conversion architecture utilizes two paths (I and Q) of LO, frequency synthesizer, and mixer to down-convert the passband signal to baseband prior to the ADCs. According to the published literature, these extra analog circuit blocks for frequency translation can contribute significant portion of the total power consumption. Alternatively, the frequency translation can be done in the sampling process using subsampling as a part of ADC. This results in only one receiving path and dramatically reduces the component count compared to a direct-conversion architecture. The remaining analog blocks prior to the ADC are amplifiers

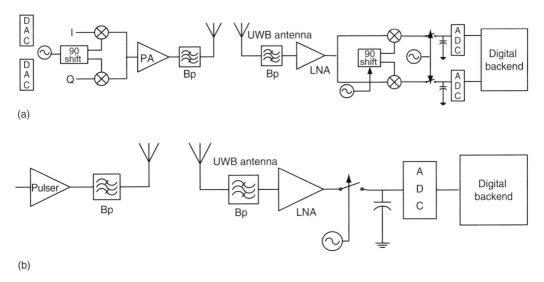

(a)

(b)

FIGURE 14.15 Radio architectures. (a) Direct-conversion radio architecture and (b) subsampling radio architecture.

and bandpass filters. The sampled data are processed by a digital-matched filter in order to reach the matched filter bound for optimal detection [19]. The proposed system avoids wideband analog processing by utilizing complex signal processing to the digital backend, which results in a more efficient solution. Moreover, the single analog receiving path eliminates the analog I/Q mismatch problem caused by the variability of IC fabrication. The proposed architecture is suitable for low-cost data communications as well as fine timing extractions for locationing or positioning applications. The ADC design example introduced in Section 14.4 is well suited in this transceiver architecture.

Another favorable feature of an asynchronous ADC is its easy extension for time-interleaved topology, which is proven to be an efficient way of scaling up conversion rate beyond a single ADC can ever achieve. Recent studies [20,21] are good examples of advancing high conversion rate with a conventionally conceived as lower-conversion-rate ADC architecture. In the future, time-interleaving will keep playing a critical role of pushing higher sampling speed for any given technology and ADC architecture. The asynchronous ADC architecture eliminates the need for internal clocks required by conventional SAR and pipeline ADC, and thus dramatically relaxes the clocking complexity for parallelism. Another potential limitation of time-interleaved ADCs is the input bandwidth constraints since the capacitor load increases proportionally. By using the proposed capacitive ladder network, the capacitive loading of each sampling network is minimized and thus allocate margin for more parallel ADCs. Note that the multiphase sampling clock still needs to be distributed across the ADC array, which will eventually limit the time-interleaved ADC performance and cost.

Certainly, the most important feature of the proposed asynchronous techniques is to improve the power efficiency of the ADC. A recently reported study [22] has further advanced it based on the described asynchronous technique with a charge-based DAC in 90 nm CMOS technology. The sampling rate of the ADC is scalable up to 50 MS/s. FOM reports 65 fJ per conversion-step measured at low input frequency with 20 MS/s although ENOB drops close to Nyquist frequency due to its limited input bandwidth. This confirms that the technology scaling trend favors the asynchronous ADC topology.

Finally, to get a big picture of how asynchronous techniques have helped advancing the ADC power efficiency, the FOM is compared among the recent 6–8 bit, high-speed (>10 MS/s) ADCs

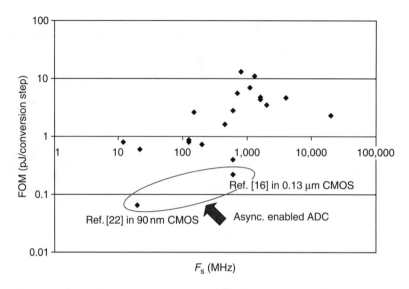

FIGURE 14.16   FOM comparisons with recent >10 MHz, 6–8 bit ADCs from ISSCC 00–05.

in ISSCC 2000–2007, as shown in Figure 14.16. It has successfully pushed the ADCs of this category into the sub-pJ per conversion-step regime. In the future, more judicious ways of combining asynchronous techniques with conventional ADC topologies can be expected, and thus enable more interesting system solutions.

REFERENCES

1. Authorization and use of software defined radios, Federal Communications Commission, First Report and Order, ET Docket 00–47, September 2001.
2. In the matter of facilitating opportunities for flexible, efficient, and reliable spectrum use employing cognitive radio technologies, Federal Communications Commission, Notice of Proposed Rule Making and Order, December 2003.
3. E. Buracchini, The software radio concept, *Commun. Magazine IEEE*, 38(9), 138–143, 2000.
4. I. O'Donnell, M. Chen, S. Wang, and R. Brodersen, An integrated, low power, ultra-wideband transceiver architecture for low-rate, indoor wireless systems, *IEEE CAS Workshop on Wireless Communications and Networking*, September 2002.
5. J. Corcoran, K. Poulton, and T. Hornak, A 1 GHz 6b ADC system, *IEEE International Solid-State Circuits Conference (ISSCC) Digest of Technical Papers*, pp. 102–103, February 1987.
6. P. Xiao, K. Jenkins, M. Soyuer, H. Ainspan, J. Burgharts, M. Hyun Shin Dolan, and D. Harane, A 4 b 8 GSample/s A/D converter in SiGe bipolar technology, *IEEE International Solid-State Circuits Conference (ISSCC) Digest of Technical Papers*, pp. 124–125, February 1997.
7. B. Murmann and B. Boser, A 12-bit 75-MS/s pipelined ADC using open-loop residue amplification, *IEEE J. Solid-State Circuits*, 38(12), 2040–2050, December 2003.
8. G. Promitzer, 12-bit low-power fully differential switched capacitor noncalibrating successive approximation ADC with 1 MS/s, *IEEE J. Solid-State Circuits*, 36, 1138–1143, 2001.
9. H. J. M. Veendrick, The behavior of flip-flops used as synchronizers and prediction of their failure rate, *IEEE J. Solid-State Circuits*, SC-15, 169–176, 1980.
10. W. Black and D. Hodges, Time interleaved converter arrays, *IEEE J. Solid-State Circuits*, 15(6), 1022–1029, 1980.
11. K. Bult and A. Buchwald, An embedded 240-mW 10-b 50 MS/s CMOS ADC in 1-mm^2, *IEEE J. Solid-State Circuits*, 32, 1887–1895, 1997.

12. F. Kuttner, A 1.2V 10b 20MSample/s non-binary successive approximation ADC in 0.13μ CMOS, *IEEE International Solid-State Circuits Conference (ISSCC) Digest of Technical Papers*, vol. 1, pp. 176–177, February 2004.
13. A. Hastings, *The Art of Analog Layout*. Englewood Cliffs, NJ: Prentice Hall, 2001.
14. A. N. Karanicolas, H. Lee, and K. L. Bacrania, A 15-b 1-MSample/s digitally self-calibrated pipeline ADC, *IEEE J. Solid-State Circuits*, 28, 1207–1215, 1993.
15. I. Sutherland, R. Sproull, and D. Harris, *Logical Effort: Designing Fast CMOS Circuits*. San Francisco, CA: Morgan Kaufmann, 1999.
16. M. S. W. Chen and R. W. Brodersen, A 6b 600 MS/s 5.3mw asynchronous ADC in 0.13μm CMOS, *IEEE International Solid-State Circuits Conference (ISSCC) Digest of Technical Papers*, vol. 49, pp. 574–575, February 2006.
17. J. M. Rabaey, A. Chandrakasan, and B. Nikolic, *Digital Integrated Circuits: A Design Perspective*. Upper Saddle River, NJ: Pearson Education, 2003.
18. M. S. W. Chen and R. W. Brodersen, A subsampling UWB radio architecture by analytic signalling, *Proceedings ICASSP*, vol. 4, pp. 533–536, May 2004.
19. J. G. Proakis, *Digital Communications*. New York: McGraw Hill, 1995.
20. K. Poulton, R. Neff, B. Setterberg, B. Wuppermann, T. Kopley, R. Jewett, J. Pernillo, and C. Tan, A 20 GS/s 8 b ADC with a 1 MB memory in 0.18 μm CMOS, *IEEE International Solid-State Circuits Conference (ISSCC) Digest of Technical Papers*, vol. 1, pp. 318–496, February 2003.
21. D. Draxelmayr, A 6b 600 MHz 10mw ADC array in digital 90 nm CMOS, *IEEE International Solid-State Circuits Conference (ISSCC) Digest of Technical Papers*, vol. 1, pp. 264–527, February 2004.
22. J. Craninckx and G. van der Plas, A 65 fJ/conversion-step 0-to-50 MS/s 0-to-0.7 mW 9b charge-sharing SAR ADC in 90nm digital CMOS, *IEEE International Solid-State Circuits Conference (ISSCC) Digest of Technical Papers*, pp. 246–600, February 2007.

15 High-Frequency Filters for Data Communication Applications

Manisha Gambhir, Vijay Dhanasekaran, Jose Silva-Martinez, and Edgar Sánchez-Sinencio

CONTENTS

15.1 INTRODUCTION

With the advent of high-speed communication systems and wireless technology, there has been an increased demand for high-frequency circuits. High data rate communications systems require baseband circuitry suitable to cover wide signal bandwidths (BWs) up to few GHz. Currently, most of the integrated circuit solutions consist of analog-to-digital converters to digitize the incoming signals. Any digitizing architecture requires front-end filtering to limit the signal BW and prevent aliasing. Thus an analog filter is a universal part on any front-end chain. Wideband integrated filters are particularly needed for high data rate communication systems, digital versatile disk (DVD), and disk drive read channel systems and antialiasing filters for direct conversion architectures.

The scope of this chapter is limited to understanding and design of wideband integrated filters for data communication applications. Typical requirements for such applications entail lowpass

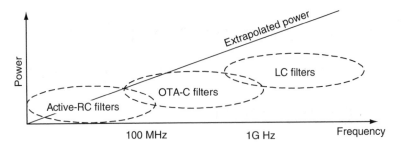

FIGURE 15.1   Filter topologies for various BW.

filtering with BW ranging from hundreds of megahertz to few gigahertz range. For such frequency range, the choice of filters is limited to transconductor-C (operational transconductance amplifier (OTA)-C/G_m–C) and inductance, capacitance (LC) filters. Hence, this chapter would exclusively focus on these two types. The approach adapted here is to evolve from generalized concepts and prototypes to specialized filters for specific applications.

With increase in BW of operation, theoretically a linear increase in power consumption is expected for a given architecture. However, prudent choice of architecture for higher BWs can result in better normalized power efficiency as depicted in Figure 15.1 and later elaborated upon in this chapter.

Complementary metal oxide semiconductor (CMOS) wideband OTA-C filters have been reported with BW up to 550 MHz [1,2] while the wideband LC filters have been reported for BWs greater than 800 MHz [3]. Apart from wide BW, high-gain around resonance frequencies may be required for equalization of data pulses. Most of the equalizing filters reported so far have confined to the BW of 43–200 MHz [4–8] with equalizing gain of around 14 dB. The difficulty lies in achieving high boost gains for a wideband structure with a reasonable power budget. In this chapter, we also present two equalizing "boost" filters with BWs of 330 MHz and 1 GHz and designed for high-frequency boost gain of 24 dB [9,10].

Section 15.2 covers the fundamental concepts in filter design such as impedance and frequency scaling while introducing "power–noise product" as the measure for the efficiency of the filters. Sections 15.3 and 15.4 present basic prototypes and practical design techniques involving G_m–C and LC filter topologies. Section 15.5 exclusively covers filters for equalizing applications. As discussed earlier, the key challenge here is the implementation of significant high-frequency gain using efficient topologies. This section also introduces some seminal ideas about comparison of OTA-C and LC topologies that can help the reader to make prudent choices for different specifications.

15.2   FILTER DESIGN: SCALING FUNDAMENTALS

Before venturing into design of specific filters, we revisit the fundamentals on filter design.

Impedance and frequency are covered here with regard to generalized filter networks. We also develop the concept of power–noise product as an important measure of the filter efficiency.

15.2.1   IMPEDANCE SCALING

The transfer function of an electrical network remains unaltered if the impedance of all the elements is scaled by the same factor (α). In case an active element like transistor is used, the transconductance must be scaled by $1/\alpha$ to retain the transfer function.

Consider, for example, the G_m–C integrator shown in Figure 15.2. Transconductor G_m can be thought as an ideal voltage-controlled current source. The OTA output current, proportional to the input voltage, is carried through the integrating capacitor C to yield the transfer function $V_{out}/V_{in} = G_m/sC$. The transfer function of the integrator (G_m/sC) remains unaltered if both G_m and C are scaled

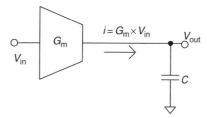

FIGURE 15.2 A G_m–C integrator.

by the factor $1/\alpha$. How does one decide the absolute value of G_m and C since any scaled version will do equally well? This is where the noise performance of the circuit comes into play. The integrated input-referred noise (thermal) of the circuit in the BW 0 to $G_m/(2\pi C)$ is given by $8/3 * kT\gamma/(2\pi C)$ where k is Boltzmann's constant, T is absolute temperature in Kelvin, and γ (usually >1) is the noise factor of the transconductor. The elements G_m and C are typically scaled such that the total noise given by $8/3kT\gamma/C$ meets the noise specification. This is a very important principle that lets the designers "scale" a circuit according to the noise performance required by the circuit.

15.2.2 PARALLEL CIRCUITS

Using parallel circuits (Figure 15.3) is equivalent to impedance scaling. N circuits in parallel is equivalent to $\alpha = 1/N$. It is easy to see this in the example of G_m–C integrator shown in Figure 15.3 where all the nodes of the circuit (input and output) are shorted together. This is even true for more complex networks with multiple internal nodes. Since all the circuits connected in parallel are identical, all the internal nodes should experience the same voltage swing. This means that all the nodes can be shorted together without affecting the functionality of the circuit. This is equivalent to putting all the N elements of the network in parallel, which in turn scales the impedance by the factor $1/N$.

15.2.3 FREQUENCY SCALING

If a transfer function $T2(s)$ can be derived from $T1(s)$ such that $T2(s/\beta) = T1(s)$, then $T2(s)$ is said to be the "frequency scaled" version of $T1(s)$, where β is the frequency scaling factor. In case of the integrator example in Figure 15.2, the transfer function $T1(s) = G_m/sC$ is changed to $T2(s) = \beta G_m/sC$ if the integrator is frequency scaled by $\beta(s \rightarrow s/\beta)$. In case of the resistance,

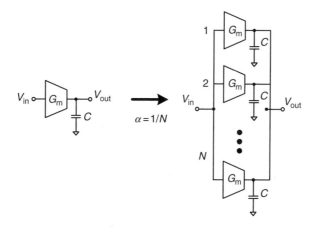

FIGURE 15.3 N-parallel G_m–C integrators.

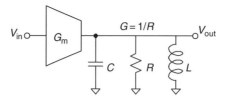

FIGURE 15.4 RLC resonator.

inductance, capacitance (RLC) resonator shown in Figure 15.4, the transfer function $T3(s)$ and its frequency-scaled counterpart $T3'(s/\beta)$ are given by

$$T3(s) = \frac{G_{\mathrm{m}}s/C}{s^2 + s\frac{G}{C} + \frac{1}{LC}}; \quad T3'(s/\beta) = \frac{G_{\mathrm{m}}\beta s/C}{s^2 + s\frac{\beta G}{C} + \frac{\beta^2}{LC}} \tag{15.1}$$

15.2.4 Frequency Scaling Under Constant Noise

$T2(s) = \beta G_{\mathrm{m}}/sC$ can be achieved by either raising G_{m} by β times or by dropping C by β times. Since the integrated noise depends only on C, the G_{m} must be scaled to keep the same noise performance [11]. For RLC resonator, there are two ways to achieve frequency scaling of β: (a) scale both L and C by $1/\beta$ and keep G and G_{m} constant and (b) scale L by $1/\beta^2$, G and G_{m} by β and keep C constant. It can be shown that the method B provides constant noise frequency scaling.

15.2.5 Power–Noise Product

From the discussions regarding impedance scaling, it can be seen that the input referred integrated thermal noise is directly proportional to the scaling factor α. Also, the power dissipated by the transconductor G_{m} is inversely proportional to α. Thus, the quantity power × noise (number of OTAs and capacitors scale equally) is independent of the impedance scale factor α. Therefore, it can be construed that power–noise product is an important figure-of-merit (FOM) for a circuit. It helps assess the topology independent of the impedance levels of the network [12]. The concept of directly trading power with the noise performance is almost universal to all electronic circuits (as long as the noise is thermal noise). It can also be applied to amplifiers, filters, oscillators, and even digital circuits (where the jitter performance is directly traded with power dissipation).

For constant noise frequency scaling, the transconductor G_{m} has to be scaled by a factor β while the capacitor remains invariant. This means that power dissipation is directly proportional to β. Hence, the ratio of power–noise product to the frequency of operation serves as a good FOM to compare different filter topologies. For example, G_{m}–C filters realizing the same transfer function but designed to meet different noise specifications and frequency of operation can be compared for power efficiency using this FOM.

15.3 G_{m}–C TOPOLOGIES

G_{m}–C topologies also known as OTA-C are one of the most popular topologies of the wideband filters with moderate dynamic range. Typical applications of such filters are hard disk read channels, video filtering, and certain wireless systems.

This class of filters obtains its name from the fact that open-loop transconductors driving a capacitor constitute the basic building block. In the active signal processing block, the transconductor does not employ any local feedback mechanism. Thus, naturally the linearity performance of such filters, limited by the transconductors, is moderate. However, absence of any local feedback enables the high-frequency operation.

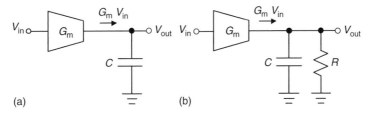

FIGURE 15.5 (a) Lossless G_m–C integrator and (b) lossy G_m–C integrator.

The basic cell is an integrator composed of an OTA loaded by a capacitor. The high frequency of operation of an OTA is possible because the voltage-to-current conversion is carried out by the input stage and internal node impedances are low. Since the OTA operates in open loop, it is able to operate at higher frequencies than active RC, switched capacitor and other techniques that use local feedback. While very linear capacitors are available in current CMOS technologies, the bottleneck is the design of efficient OTAs. An ideal OTA is an infinite BW voltage-controlled current source, with an infinite input and output impedance. G_m–C (or OTA-C) filter techniques are revised in this section and the effects of OTA limitations on filter performance are discussed.

15.3.1 Architectural Considerations

The most commonly used G_m–C architectures belong to the state-variable filters. These topologies are based on integrators, hence it is advisable to fully understand the properties of the basic cell, and to learn some fundamental rules for the manipulation of these blocks to efficiently design intricate filters. The lossless integrator is shown in Figure 15.5a. The voltage gain is given by the ratio of the transconductance gain and the load admittance ($A_v = G_m/sC$); the equivalent integrator's time constant is denoted as C/G_m–C. The lossy integrator is depicted in Figure 15.5b and is obtained by adding a grounded resistor in parallel with the load capacitor. Its voltage transfer function can be found using conventional circuit analysis techniques as:

$$\frac{V_{\text{out}}(s)}{V_{\text{in}}(s)} = \frac{G_m}{sC + \frac{1}{R}} = \frac{G_m R}{1 + sRC} \tag{15.2}$$

In this expression, $s(= j\omega)$ is the frequency variable. Notice that the transfer function of the lossless integrator can be found from Equation 15.2, making $R = \infty$. Since the input impedance of the transconductance stage is very large, integrators can be directly connected to one another to design more complex filter structures. A simple state-variable biquadratic (second-order) section is obtained if two integrators are connected in a loop, as depicted in Figure 15.6. Notice that the two back-to-back

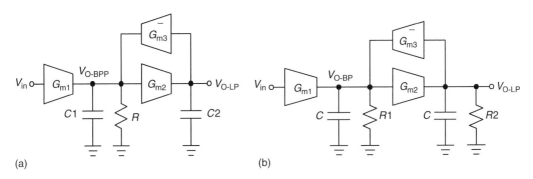

FIGURE 15.6 Second order G_m–C topologies: (a) terminated at the input and (b) double terminated.

transconductors (G_{m2} and G_{m3}) along with grounded capacitor C2 emulate a grounded inductor [13]. This class of biquads is also known as gyrator-based structures.

For Figure 15.6a, the lowpass and bandpass transfer functions are

$$H_{LP}(s) = \frac{V_{o-LP}(s)}{V_{in}(s)} = \frac{\frac{G_{m1}G_{m2}}{C1C2}}{s^2 + \frac{1}{RC1}s + \frac{G_{m2}G_{m3}}{C1C2}} = \frac{\left(\frac{G_{m1}}{G_{m3}}\right)\left(\omega_o^2\right)}{s^2 + \frac{\omega_o}{Q}s + \omega_o^2} \qquad (15.3)$$

and

$$H_{BP}(s) = \frac{V_{o-BP}(s)}{V_{in}(s)} = \frac{\left(\frac{G_{m1}}{C1}\right)s}{s^2 + \frac{1}{RC1}s + \frac{G_{m2}G_{m3}}{C1C2}} = \frac{(G_{m1}R)\left(\frac{\omega_o}{Q}\right)s}{s^2 + \frac{\omega_o}{Q}s + \omega_o^2} \qquad (15.4)$$

The relevant filter parameters can be identified from these expressions. For the lowpass filter, the squared resonance frequency ($\omega_o^2 = G_{m2}G_{m3}/C1C2$) is controlled by the product of the G_m/C gain factor of the two integrators in the loop. The filter's BW (BW $= \omega_o/Q = 1/RC1$) is determined by the R-C product lumped at the lossy node. The lowpass filter's direct current (DC) gain can be obtained evaluating Equation 15.3 at $s = 0$, leading to $A_{DC} = G_{m1}/G_{m3}$. The key advantage of the state-variable-based biquadratic filter is its flexibility since the DC gain, resonance frequency (ω_0), and quality factor (Q) can be controlled independently with G_{m1}, G_{m2}, and G_{m3}, and R, respectively. These properties make this topology very attractive when programmable filters are required. In the case of the bandpass output, the gain peaks at $\omega = \omega_o$, yielding $A_{peak} = G_{m1}R$. Notice that at the resonant frequency $\omega = \omega_o$, the gain at the two outputs are related by

$$\left|\frac{V_{o-LP}}{V_{o-BP}}\right|_{\omega=\omega_o} = \frac{\frac{G_{m2}}{C2}}{\omega_o} = \sqrt{\frac{G_{m2}}{G_{m1}}\frac{C1}{C2}} \qquad (15.5)$$

Thus, to maintain the similar peak gain at both the outputs of biquadratic section (of Figure 15.6a), it is desirable to design the filter such that $V_{o-LP}|\omega = \omega_o = V_{o-BP}|\omega = \omega_o$. In most practical cases, the design conditions $G_{m1} = G_{m2}$ and $C1 = C2$ make the layout process modular and repeatable. A nice feature of this topology is that both lowpass and bandpass outputs are easily available. These outputs can be easily manipulated to generate equalizing zeros.

The double-terminated topology shown in Figure 15.6b is based on an LC ladder prototype; usually ladder-based structures have lesser passband sensitivity to tolerances on the components, making the topology more tolerant to process–voltage–temperature variations. The transfer functions at nodes V_{o-LP} and V_{o-BP} terminals are

$$\frac{V_{o-LP}(s)}{V_{in}(s)} = \frac{\frac{G_{m1}G_{m2}}{C1C2}}{s^2 + \left(\frac{1}{R1C1} + \frac{1}{R2C2}\right)s + \frac{G_{m2}G_{m3}R1R2+1}{R1R2C1C2}} \qquad (15.6)$$

and

$$\frac{V_{o-BP}(s)}{V_{in}(s)} = \frac{\frac{G_{m1}}{C1}\left(s + \frac{1}{R1C1}\right)}{s^2 + \left(\frac{1}{R1C1} + \frac{1}{R2C2}\right)s + \frac{G_{m2}G_{m3}R1R2+1}{R1R2C1C2}} \qquad (15.7)$$

This topology presents two major differences with the previous biquad: both ω_o and BW are function of all passive elements, hence adjusting $R1$ and/or $R2$ for Q control has some impact on the resonant frequency as well. Notice that V_{o-BP} is not a true bandpass output as the zero is located at $\omega = 1/R1C1$ instead of at $\omega = 0$, resulting in nonzero DC gain; Equation 15.7 is often referred as resonator. This topology shows reduced sensitivity to variations on the nominal values in comparison with Figure 15.6a.

15.3.2 BUILDING BLOCKS AND NONIDEALITIES

A suitable architecture for low-voltage application is the OTA based on a common-source configuration shown in Figure 15.7a; it consists of two transistors only. If a long channel transistor is used to make the V-to-I conversion, the drain current is given by the following quadratic equation [14]:

$$I_D \cong \frac{\mu_n C_{OX}}{2} \frac{W}{L} (V_{GS} - V_T)^2 \cong \frac{G_m (V_{GS} - V_T)}{2} = \frac{G_m V_{DSAT}}{2} \tag{15.8}$$

where

    G_m is the small-signal transconductance

    $V_{DSAT} = V_{GS} - V_T$, μ_n, and C_{OX} are the overdrive voltage, mobility of the carriers in the channel, and the oxide capacitance per unit area, respectively

If a small signal is applied to the gate of the single-ended topology ($v_{GS} = V_{GS} + v_{gs}$ with V_{GS} and v_{gs}, the DC and alternating current [AC] input signal, respectively), the small-signal transconductance can be formally defined as the variation of the drain current normalized to small variations at the gate; hence G_m can be written as:

$$G_m = \left.\frac{\partial i_d}{\partial v_{GS}}\right|_Q \cong \mu_n C_{OX} \frac{W}{L} V_{DSAT} \cong \sqrt{2\mu_n C_{OX} \frac{W}{L} I_D} \tag{15.9}$$

The voltage-to-current conversion is nonlinear in nature due to the nonlinear characteristics of the transistor. Even if second-order effects are ignored, it is evident from Equation 15.8 that i_d presents a current component proportional v_{in}^2. For a sinusoidal input voltage v_{in}, it can be shown after some mathematical manipulations that current output presents three components: a DC component, the desirable term proportional to the input signal, and an undesirable component at two times the frequency of the fundamental component, usually termed second-order nonlinearity. In addition, this configuration is very sensitive to the undesirable voltage fluctuations or supply noise present at the source of the transistors.

To improve immunity toward supply noise and other external perturbations, most practical designs employ differential topologies. The simplest fully differential transconductor is shown in Figure 15.7b. In these topologies, the differential signal is referred to an AC small-signal ground, which is isolated from the analog ground usually through a current source. Furthermore, the current source injects noise with same amplitude and phase to the two differential arms, rendering the noise

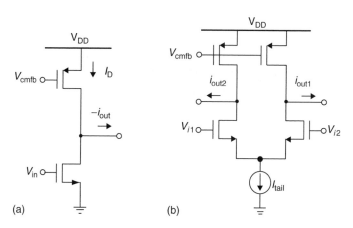

FIGURE 15.7 Basic transconductance topologies: (a) single-ended structure and (b) fully differential topology.

common-mode in nature. It can be seen that differential structures are not so sensitive to common-mode inputs; if the parasitic impedances attached to the common-source are ignored, it can be shown that the single-ended output current for the differential transconductor shown in Figure 15.7b is

$$i_{\text{out1}} = \frac{G_{\text{m1}}(v_{\text{i1}} - v_{\text{i2}})}{2}\left(1 + \sqrt{1 - \left(\frac{v_{\text{i1}} - v_{\text{i2}}}{2V_{\text{DSAT1}}}\right)^2}\right)$$ (15.10)

where G_{m1} is the transconductance of M1. Because of the symmetry of the circuit, differential current is $i_{\text{out1}} - i_{\text{out2}} = 2i_{\text{out1}}$. Assuming that the differential input signal $v_{\text{i1}} - v_{\text{i2}} < 2\ V_{\text{DSAT1}}$, Equation 15.10 can be expanded in terms of Taylor coefficients. If a sinusoidal input signal is used, it can be shown that the ratio of the amplitude of the component at three times the fundamental frequency to the amplitude of the fundamental component is

$$\text{HD3} \cong \frac{1}{32}\left(\frac{V_{\text{in}}}{V_{\text{DSAT}}}\right)^2$$ (15.11)

where V_{in} is the amplitude of the input signal. For a moderate third harmonic distortion (HD3) requirement of $-40\,\text{dB}$ (<0.01); the amplitude of the input signal must be limited to $V_{\text{in}} < 0.56\ V_{\text{DSAT}}$. Notice that high linearity for a given input signal amplitude demands large overdrive voltage (V_{DSAT}). Another effect of the third-order nonlinearity is that a component proportional to V_{in}^3 is also present at the fundamental frequency; this component makes the overall transconductance gain voltage–dependent

$$G_{\text{m}'} = \mu_n C_{\text{OX}}\frac{W}{L}V_{\text{DSAT}}\left(1 - (3 \times \text{HD3})\,V_{\text{in}}^2\right) = \mu_n C_{\text{OX}}\frac{W}{L}V_{\text{DSAT}}\left[1 - (\text{IM3} \times V_{\text{in}}^2)\right]$$

$$= G_{\text{m}}\left[1 - (\text{IM3} \times V_{\text{in}}^2)\right]$$ (15.12)

where
  IM3 $= 3$HD3 is the third-order intermodulation distortion
  $G_{\text{m}'}$ is the transconductance gain accounting harmonic effects

IM3 reduces the transconductance gain when large input signals are applied, and thus affects the performance of the entire filter. For instance the filter's resonance frequency, $\omega_o (= \sqrt{G_{\text{m2}}G_{\text{m3}}/C1C2})$, reduces by a relative fraction of IM3 $\times\ V_{\text{in}}^2$ in presence of a large input signal.

Apart from linearity, noise performance of the transconductor needs to be considered carefully. Active devices generate noise due to the random fluctuations of the carriers; a detailed analysis on the nature of the noise sources present in the transistors can be found in Ref. [15]. Since this chapter deals with high-frequency applications, noise is dominated by the thermal component (white noise density). Section 15.2 summarized importance of noise performance vis-à-vis power for active circuits. In this section, we examine it in particular reference of transconductors. For a single transistor, the gate-referred spectral noise density (expressed as noise power in a unit frequency BW) is given by

$$v_n^2 \cong \left(\frac{4kT}{G_{\text{m}}}\right)\ (V^2/\text{Hz})$$ (15.13)

where
  kT is the thermal energy
  k is the Boltzmann's constant
  T is the absolute temperature

If the noise contribution of all the transistors is accounted and referred to an equivalent input noise source, it can be demonstrated that the equivalent input-referred noise density of the differential pair–based OTA depicted in Figure 15.7b is

$$v_{n\text{OTA}}^2 \cong \left(\frac{8kT}{G_{m1}}\right)\left(1 + \frac{G_{m2}}{G_{m1}}\right) = \left(\frac{8kT}{G_{m1}}\right)(1 + \text{NF}) \qquad (15.14)$$

where parameter NF accounts for the noise contribution of additional transistors in an OTA. Equivalent input referred noise for the entire filter can be computed by accounting for noise generated by all lossy and active elements and referring them back to the input port. For example, equivalent input-referred noise density for the singly terminated biquadratic filter (Figure 15.6a) shown earlier can be written as:

$$v_{\text{eq\_in}}^2 \cong \left(\frac{8kT}{G_{m1}}\right)(1+\text{NF}_1) + \frac{4kTR_1}{G_{m1}^2 R_1^2} + \left|\frac{1 + sR_1C_1}{G_{m1}R_1}\right|^2 \left(\frac{8kT}{G_{m2}}\right)(1+\text{NF}_2) + \left(\frac{G_{m3}}{G_{m1}}\right)^2 \left(\frac{8kT}{G_{m3}}\right)(1+\text{NF}_3)$$

$$(15.15)$$

where NF_i corresponds to the noise contribution of additional transistors in G_{mi}. This chapter limits itself to discussing basic transconductor structures and their properties. Other variants of transconductors may be employed based on the particular application. To enhance transconductance/tail current ratio, complementary differential pair–based transconductor [16] may be used but only at cost of increased input parasitic and reduced headroom. A simple inverter-based transconductor [17] may be highly suited for high-frequency low-voltage application but with susceptibility to supply noise. For further readings on transconductors suitable for high-frequency operation, readers can follow [18,19].

15.3.3 Design Procedure

Actual design of a G_m–C filter is usually an iterative procedure undertaken to satisfy various specifications. Parameters such as in-band and stop-band attenuation, phase ripple, pass-band, and stop-band edge constitute filtering specification. Based on these specifications filter's type, order and exact transfer function can be derived. Well-known techniques and pre-evaluated tables are available in literature to arrive at this step [20]. Once the filter has been specified by its transfer function, parameters such as signal swing, noise, and linearity determine design specifications of the filter. The usual aim is to design the filter for the performance specified with minimum power. It is easy to notice that swing, noise, and linearity specifications are quite interrelated and closely tied with the specification of the entire analog chain in a system. Thus the design specifications may be derived by careful optimization of system performance, power optimization for the filter block in general, and system in whole.

To illustrate the design procedure, let us consider the design of the single-terminated G_m–C biquadratic section. Assuming that ω_o, Q, IM3, peak input signal and noise specifications are given. The unknowns in this case are transistor dimensions (W and L), bias current, and G_m and C values for minimum power consumption. Simplifying assumption of using equal capacitance value and equal integrator time constants are made such that: $C_1 = C_2$, $G_{m2} = G_{m3}$. Also, for unity DC gain: $G_{m1} = G_{m2}$. Using similar-sized components simplifies layout but may not be optimum in all cases. Based on above assumptions, simplified filter design equations are

$$\omega_o = \frac{G_{m1}}{C_1}; \quad BW = \frac{\omega_o}{Q} = \frac{1}{RC_1} \qquad (15.16)$$

If $G_{m1} = G_{m2} = G_{m3}$, then the expression for noise density can be further simplified to

$$v_{eq\_in}^2 \cong \left(\frac{8kT}{G_{m1}}\right)(1 + \text{NF}_1)\left\{2 + \left|\frac{1 + s/\text{BW}}{Q}\right|^2 + \frac{1}{2Q^2}\right\} \qquad (15.17)$$

Thus, G_{m1} is selected to be sufficiently large to meet the required noise specifications. As discussed earlier, this trades-off directly with the power consumption and the silicon area. Finally appropriate overdrive voltage (V_{dsat}) is selected based on linearity requirements, supply headroom, and signal swing. Taking into account accumulated nonlinearity of the three OTAs, filter's IM3 can be roughly expressed as the addition of the power of the harmonic distortion components, leading to

$$\text{IM3} \cong \frac{9}{32}\left(\frac{V_{in}}{V_{DSAT}}\right)^2 \qquad (15.18)$$

where the input signal is $V_{in}\sin(\omega t)$. The above expression assumes that the distortion components generated by each OTA are not correlated and total power of the IM3 is addition of the distortion power components generated by each of the three transconductors. Although this expression is not precise, it still provides a good starting point while keeping the design equations wieldy.

Having evaluated G_m value (based on noise) and V_{DSAT} (based on linearity), transconductor design parameters and tail current can be fixed:

$$I_D \cong \frac{G_m V_{DSAT}}{2}; \quad \text{or} \quad G_m = \frac{I_{tail}}{V_{DSAT}} \qquad (15.19)$$

Considering Equation 15.18 in conjunction with Equation 15.19, G_m/I_{tail} is given by

$$\frac{G_m}{I_{tail}} = \frac{\sqrt{\frac{32}{9}\text{IM3}}}{V_{in}} \qquad (15.20)$$

Since the tail current defines the OTA current consumption, the power per OTA is $I_{tail}V_{DD}$. In this example, all OTAs are similar, leading to a total current consumption given by $NI_{tail}V_{DD}$. The trade-offs involving power consumption are better appreciated if we combine the previous results in a single expression as

$$\text{Power} = \left(\frac{3V_{DD}V_{in}}{\sqrt{\frac{32}{9}\text{IM3}}}\right)G_m \qquad (15.21)$$

While the term inside parentheses is directly given by filter specifications, G_m is dictated singularly through noise specifications; hence power consumption is entirely dictated by filter's specification. Evaluating R and C based on computed G_m value and relation (Equation 15.16) completes the design of this filter.

Nonidealities of the transconductor such as finite output impedance and parasitic capacitances may have significant impact on filter's accuracy. Due to the lack of space, such issues are not discussed here. Readers should refer to Refs. [13,21,22] for a detailed discussion about these issues.

15.4 LC TOPOLOGIES

On-chip inductors are routinely used for narrowband RF circuits. However, their use in broadband filtering has been limited. Robust and low-sensitivity filters can be constructed using passive ladder structures. Passive filters are composed by inductors, capacitors, and resistors requiring zero static power. Another remarkable advantage of passive ladder filter is its optimum noise performance. Since

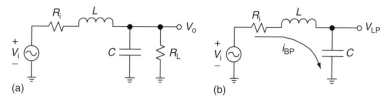

FIGURE 15.8 (a) Second-order doubly terminated ladder filter and (b) simplified biquadratic filter.

inductors and capacitors are noiseless elements, only resistive terminations contribute to the filter's noise. In this section, basics of LC topologies are considered with the example of second-order RLC filters and its various properties.

15.4.1 ARCHITECTURAL CONSIDERATIONS

The basic voltage driven doubly terminated LC ladder filter is depicted in Figure 15.8. An evident advantage of passive RLC filters is its zero power consumption due to the lack of active elements [13]. Linearity is limited by the passive elements, which in most technologies can be fabricated to be quite linear. However, limited range of inductors that can be realized on-chip limits the application of such circuit for BW above 500 MHz. The filter's transfer function of Figure 15.8a can be derived using conventional circuit analysis techniques as:

$$\frac{v_o}{v_{in}} = \frac{\frac{1}{LC}}{s^2 + \left(\frac{1}{R_L C} + \frac{R_i}{L}\right)s + \frac{1+\frac{R_i}{R_L}}{LC}} \tag{15.22}$$

The fundamental design equations are

$$\omega_o^2 = \frac{1+\frac{R_i}{R_L}}{LC}; \quad BW = \frac{\omega_o}{Q} = \frac{1}{R_L C} + \frac{R_L}{L}; \quad A_{DC} = \frac{1}{1+\frac{R_i}{R_L}} \tag{15.23}$$

Note that the relevant filter parameters such as resonance frequency (ω_o), BW, and dc gain (A_{DC}) cannot be determined independent of each other. In other words, filter parameters are determined by several components with mutual interdependence, making them more tolerant to variations in single element.

This, however, makes these topologies unsuitable for applications requiring programmability. For example, if the inductor or capacitor is used to program filter's cut-off frequency, the BW is affected as well. Another relevant limitation of this topology is the lack of bandpass output, which may often be required for certain applications as we shall learn in the later parts of this chapter. The only noise contribution is due to the termination resistors as the practical inductor and capacitors are quasinoise-free elements.

A flexible topology, with similar properties as the biquadratic filter discussed in previous section, is obtained if R_L is removed as shown in Figure 15.8b. The design equations can be easily obtained from Equation 15.23, making $R_L = \infty$, and repeated here for convenience

$$\omega_o^2 = \frac{1}{LC}; \quad BW = \frac{R_i}{L}; \quad A_{DC} = 1 \tag{15.24}$$

Notice that ω_o and BW can now be adjusted independently through C and R_i, respectively; DC gain is unity in this topology. Noise contribution of only R_i is relevant and input referred noise density

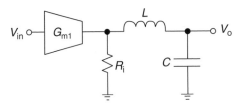

FIGURE 15.9 OTA-RLC filter.

is $4kTR_i$. Also notice that the current flowing through the network is bandpass in nature. It can be shown that

$$\frac{i_{bp}}{v_{in}} = \frac{\frac{s}{L}}{s^2 + \left(\frac{R_i}{L}\right)s + \frac{1}{LC}} \tag{15.25}$$

An efficient equalizing filter solution that exploits the bandpass nature of the current flowing through such RLC network and the lowpass voltage across the load capacitor is presented in the last section of this chapter.

15.4.2 Design Considerations

Since both input and output impedances are finite, it can be construed that interfacing passive LC filters presents certain difficulties. Hence, active buffers may be employed, which in turn add to the overall noise, nonlinearity, and power consumption. An input-buffered RLC filter suitable for high-frequency applications can be obtained from the previous topologies applying the Norton-Thevenin equivalent. The resulting topology is shown in Figure 15.9.

Although this topology maintains the properties of the original passive ladder filter, the G_m stage adds noise and limits the linearity performance of the filter. If the noise contribution of R_i is represented as current, it is not difficult to reflect this noise component to the input, leading to the following expression for the overall input-referred noise density:

$$v_{eq\_in}^2 \cong \left(\frac{8kT}{G_{m1}}\right)(1 + NF_1) + \left(\frac{4kT}{G_{m1}}\right)\left(\frac{1}{G_{m1}R_i}\right) = \left(\frac{8kT}{G_{m1}}\right)(1.5 + NF_1) \tag{15.26}$$

It is assumed in this equation that filter's DC gain is unity; $G_{m1}R_i = 1$. If compared with the G_m–C topology (Equation 15.17), it is evident that this topology presents smaller noise level. IM3 is also lower since there is only one nonlinear block. The most relevant trade-off is the one associated with noise–power-distortion in the OTA, which can be related based on the following result:

$$\text{Power} = \left(\frac{V_{DD} \times V_{in}}{\sqrt{32 \times IM3/3}}\right)G_{m1} \tag{15.27}$$

Thus, the noise is entirely determined by G_{m1} and R_i. Equation 15.27 assumes that V_{DSAT} has been obtained in accordance with the linearity specifications as shown in subsection 15.3.3. Also, observe that inductor may use significant silicon area, limiting this approach to RF applications.

15.5 EQUALIZING FILTERS

With the background of G_m–C and LC topologies, specific design examples can be examined. The specific class of filters we analyze in this section are the equalizing filters. Equalizing filters or the boost filters realize a lowpass filtering function with magnitude peaking around the resonance frequency.

Analog front ends especially for disk drive systems and DVD applications suffer from channel losses and intersymbol interference at high frequency, which degrades the performance of the system.

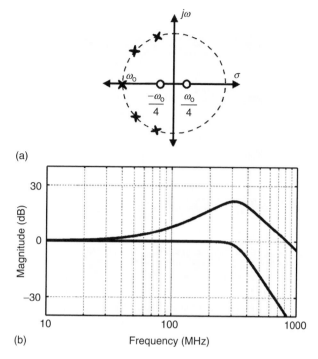

(a)

(b)

FIGURE 15.10 (a) S-plane location of poles and zeros for 24 dB boost setting and (b) magnitude response for fifth-order Butterworth filter with 0 and 24 dB boost. (From Gambhir, M., Dhanasekaran, V., Silva-Martinez, J., and Sánchez-Sinencio, E., *IEEE Trans. Circuits Syst. I, Reg. Papers*, 54, 458, 2007. With permission.)

In order to compensate for these losses and effectively slim the data pulses, boost filters are commonly employed in such systems. Boost filters provide the necessary lowpass filtering before the digitization along with a programmable high-frequency gain for equalization. The need to support higher data rate implies wider BW. For high speed, high-density data systems, it is desirable to have maximum boost gain up to 24 dB [9].

Choice between an LC and G_m–C realization is directly related to the frequency of operation and to some extent area and noise constrains. Focusing on a fifth-order Butterworth filter with two equalizing zeros, we shall examine G_m–C and LC realization working at 330 MHz and 1.1 GHz, respectively. Later in this section, we shall also compare the two topologies for their suitability.

Figure 15.10 shows the pole zero constellations for the fifth-order Butterworth boost filter with two equalizing zeros. The two real zeros are placed symmetrically across $j\omega$ axis such that their cumulative phase contribution is negligible. The location of zeros needs to be programmable from ∞ to $\omega_0/4$, which corresponds to boost gain of 0 to 24 dB at ω_0.

One of the efficient schemes for realizing zeros for equalization gain has been proposed in Ref. [9]. In this scheme, two real axis zeros are realized by combining bandpass and lowpass signals. The conceptual diagram for such architecture is shown in Figure 15.11. The proposed architecture splits the boost gain between two stages. The cascaded representation of the transfer function is given by

$$H_{\text{boost}}(s) = \frac{\omega_o s \sqrt{K} + \omega_o^2}{s^2 + \frac{\omega_o}{Q_1}s + \omega_o^2} \cdot \frac{\omega_o s \sqrt{K} - \omega_o^2}{s^2 + \frac{\omega_o}{Q_2}s + \omega_o^2} \cdot \frac{\omega_o}{s + \omega_o} \qquad (15.28)$$

Here, Q_1 and Q_2 refer to the quality factor of biquad 1 and 2 and their values are 0.618 and 1.618, respectively (for the Butterworth realization). K determines the placement of zeros and its value ranges from 0 to 16 for 0 to 24 dB high-frequency boost.

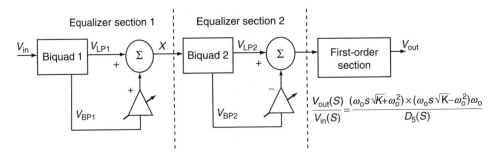

FIGURE 15.11 Scheme used to realize equalization gain ($D_5(S)$ represents fifth-order Butterworth poles). (From Dhanasekaran, V., Gambhir, M., Silva-Martinez, J., and Sánchez-Sinencio, E., *IEEE J. Solid-State Circuits*, 12, 2411, 2007. With permission.)

15.5.1 Boost Filter Architecture Using G_m–C Techniques

The implementation of the architecture shown in Figure 15.11 involves realization of bandpass and lowpass signals as well as analog summers. While the bandpass and lowpass voltage signals are readily available in a G_m–C biquad discussed in Section 15.3, the analog summation is most easily realized by adding signals in current mode. Conceptual realization (using integrators and weighted summers) of this scheme is shown in Figure 15.12. First four integrators (INT1–4) and two summers (S1–2) can be realized using cascade of two standard biquads. $V_{LP1,2}$ and $V_{BP1,2}$ in Figure 15.12 correspond to the lowpass and bandpass nodes of the biquads. $\sqrt{K}$ gain path can be implemented using a variable transconductor from bandpass node of each biquad to next available summing node. Figure 15.13 shows the detailed OTA-C implementation (shown single-ended for clarity) of the proposed architecture.

Implementation of fifth-order filter without additional summers requires the first-order section to be the last one to provide a summing node for current signals generated in biquad 2. The low Q biquad has higher input capacitance than the other biquad. It is power efficient to keep it at the input of the filter since the preceding driver can be easily designed to push the input pole location to high frequency. Given these factors, the order of the sections is optimum if biquad 1 is chosen to be the low Q section and the first-order to be the last one.

For implementing the 330 MHz boost filter using above architecture, the complementary CMOS transconductor [16] is chosen as the primary cell. The fully programmable boost OTA is realized using a Gilbert cell–based analog multiplier. Such a transconductor provides wide programmability and a nearly constant input and output loading conditions across boost control.

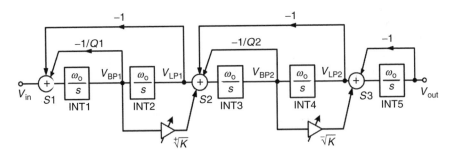

FIGURE 15.12 Conceptual illustration of proposed boost filter architecture. (From Gambhir, M., Dhanasekaran, V., Silva-Martinez, J., and Sánchez-Sinencio, E., *IEEE Trans. Circuits Syst. I, Reg. Papers*, 54, 458, 2007. With permission.)

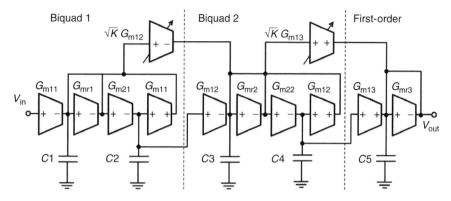

FIGURE 15.13 OTA-C implementation of proposed boost filter (single-ended version).

The prototype for OTA-C realization of the boost filter was implemented using 0.35 μm CMOS technology [9]. The choice of optimum architecture and efficient transconductors result in a power-efficient design. The filter consumes current of 13 mA from ±1.65 V supply. The −3 dB BW measured with 0 dB boost setting is 330 MHz and the maximum achievable boost is about 24 dB. Figure 15.14 shows the measured magnitude response for 0–24 dB boost conditions. Filter's linearity performance was measured around the highest frequency of interest using two-tone intermodulation tests. Third-order intermodulation product was −41 dB with respect to the test tones. The signal-to-noise ratio (SNR) is measured to be 49 dB under no boost condition.

15.5.2 BOOST FILTER ARCHITECTURE USING LC TECHNIQUES

To implement the equalizer sections shown in Figure 15.11, both second-order lowpass and bandpass signals are required. While concurrent availability of bandpass and lowpass signals in a G_m–C biquad

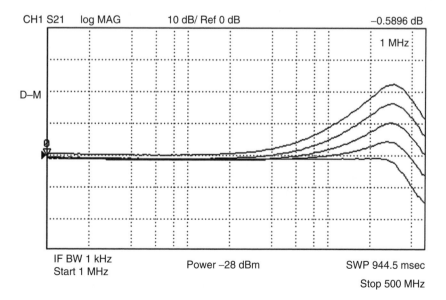

FIGURE 15.14 Measured transfer function of the OTA-C filter for varying boost gains (0–24 dB). (From Gambhir, M., Dhanasekaran, V., Silva-Martinez, J., and Sánchez-Sinencio, E., *IEEE Trans. Circuits Syst. I, Reg. Papers*, 54, 458, 2007. With permission.)

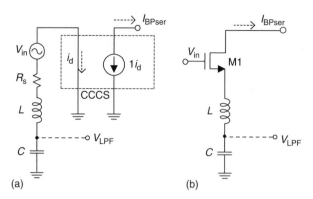

FIGURE 15.15   (a) Series resonator prototype and (b) transistor implementation of (a). (From Dhanasekaran, V., Gambhir, M., Silva-Martinez, J., and Sánchez-Sinencio, E., *IEEE J. Solid-State Circuits*, 12, 2411, 2007. With permission.)

renders the implementation of the boost architecture in Figure 15.11 simple, the LC implementation calls for certain creative modifications. The traditional parallel LC section used to realize a bandpass function is not amenable to generate lowpass signals. However, a series resonator prototype, such, as one shown in Figure 15.15a, generates both bandpass and lowpass signals, albeit in different domains. The current flowing through the series LC resonator (I_{BPser}) is bandpass in nature, while the capacitive element C integrates this current to generate lowpass voltage signal (V_{LPF}).

An active implementation of such series resonator prototype that uses just one transistor is shown in Figure 15.15b. Transistor M1 serves for multiple operations: generates the bandpass current, acts as a buffer for the input, and provides termination for the series resonator. The fact that the Butterworth transfer function demands a low-Q value further validates the choice of LC prototype, with M1 acting as the intentional loss for the resonator.

For actual implementation of the fifth-order Butterworth filter with equalizing zeros, the two series-resonator-based LC equalizer sections are cascaded. A simplified single-ended version of the complete LC filter that realizes fifth-order Butterworth function is shown in Figure 15.16. Currents from transistors M1 and M3 (I_{BPF}) are required to be variable for programmability of equalization gain. This is achieved by variable gain Gilbert cell–based current attenuators A_1 and A_2

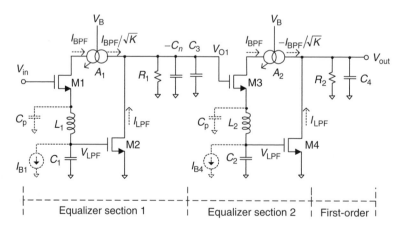

FIGURE 15.16   Simplified schematic (single-ended) of the fifth-order Butterworth filter. (From Dhanasekaran, V., Gambhir, M., Silva-Martinez, J., and Sánchez-Sinencio, E., *IEEE J. Solid-State Circuits*, 12, 2411, 2007. With permission.)

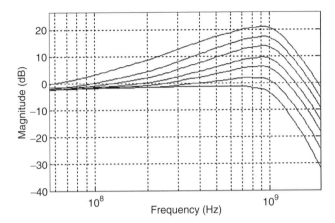

FIGURE 15.17 AC magnitude across 0–23.6 dB boost measured for 1.1 GHz LC boost filter. (From Dhanasekaran, V., Gambhir, M., Silva-Martinez, J., and Sánchez-Sinencio, E., *IEEE J. Solid-State Circuits*, 12, 2411, 2007. With permission.)

controlled through V_B. The real pole at the first biquad output is pushed to 3 GHz by using a negative capacitor $-C_n$ (similar to one proposed in Ref. [23]), which is designed to counter the parasitic and common-mode detector capacitance at the output node of the first biquad (C_3). Ignoring the parasitic capacitance C_p and using node equations at V_{O1} and V_{out}, the complete transfer function $H(s) = V_{out}(s)/V_{in}(s)$ can be written as:

$$H(s) = \frac{(sC_1 + g_{m2})}{s^2 L_1 C_1 + s\frac{C_1}{g_{m1}} + 1} \cdot \frac{(sC_2 + g_{m4})}{s^2 L_2 C_2 + s\frac{C_2}{g_{m3}} + 1} \cdot \frac{1}{s(C_3 - C_n)R_1} \cdot \frac{1}{1 + sC_4 R_2} \qquad (15.29)$$

The prototype for 1 GHz equalizing filter [10] was fabricated using 0.18 μm standard CMOS technology. Thick metal-6 layer is used for inductors. Experimental magnitude plots, thus obtained, are shown in Figure 15.17. A maximum boost gain of 23.6 dB is achieved. The filter displays −3 dB frequency (under 0 dB boost) of 1.15 GHz. For a two-tone input (250 mV p-p with tones at 925 and 975 MHz), third-order intermodulation distortion (IM3) of −48 dB is observed at 0 dB boost setting (shown in Figure 15.18). Measurement results also show an SNR of 47 dB.

15.5.3 G_m–C versus LC Structures

Appropriate choice of filter topology: G_m–C or LC is crucial for a design optimized for power and (or) area. While, for low and very high BW architecture choices are G_m–C and LC, respectively, careful analyses need to be performed to weigh relative merits of these topologies for BWs in mid-frequency range. It can be seen that the active elements used to emulate an inductor in a G_m–C resonator section would make these filters noisier (or less power efficient) than their LC counterpart. However, area constrains in realizing passive inductors for filter in few hundreds of MHz range rule out the LC prototypes for such frequency range. Thus the choice of topology for a high-frequency filter: LC or G_m–C depends heavily on frequency of operation, SNR requirement, and area constrains. We shall hereby analyze these dependencies in context of equalizing LC and OTA-C biquad structures discussed above. The analysis has been kept generic enough to be extrapolated to other LC and OTA-C topologies.

While comparing the power efficiency of the LC and OTA-C prototype, we will use the *power–noise product*, which has been well elaborated upon in Section 15.2. For simplicity, we shall consider a single equalizing section (instead of the whole filter). Considering the LC "Equalizer Section 1" shown in Figure 15.16, noise at node V_{o1} (output of first equalizing section) is evaluated. The noise

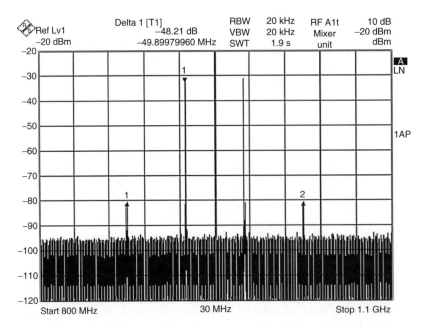

FIGURE 15.18 Measured intermodulation distortion for 1.1 GHz LC boost filter. (From Dhanasekaran, V., Gambhir, M., Silva-Martinez, J., and Sánchez-Sinencio, E., *IEEE J. Solid-State Circuits*, 12, 2411, 2007. With permission.)

of active elements (M1, M2, R_1, and I_{B1}) is expressed in terms of $V^2_{nG_{m2}}$ (input referred noise density of M2). A gain of 4 is assumed for the bandpass path (corresponding to a boost gain = 12 dB per section, thus $g_{m1} = 4QG_{m2}$) $R_1 = 1/G_{m2}$ is assumed to ensure 0 dB low-frequency gain. Expressions for noise density due to lowpass path (V^2_{nLPF}) and bandpass path (V^2_{nBPF}) of the series-LC equalizer section (at node V_{o1}) are thus derived as:

$$V^2_{nLPF-LC} = V^2_{nG_{m2}} \left| \frac{2\omega_o}{s + 2\omega_o} \right|^2 \left\{ \frac{1}{4Q} \left| \frac{\omega_o^2}{D(s)} \right|^2 + \frac{1}{4Q} \left| \frac{\omega_o(Qs + \omega_o)}{D(s)} \right|^2 + 2 \right\} \tag{15.30}$$

$$V^2_{nBPF-LC} = 4V^2_{nG_{m2}} \left| \frac{2\omega_o}{s + 2\omega_o} \right|^2 \left\{ \frac{1}{Q} \left| \frac{s\omega_o}{D(s)} \right|^2 + Q \left| \frac{\omega_o^2}{D(s)} \right|^2 \right\} \tag{15.31}$$

where
$D(s) = s^2 + \omega_o s/Q + \omega_o^2$
Q is the quality factor of the biquad

The terms within the curly braces in Equation 15.30 correspond to the noise contribution of M1, I_{B1}, and M2 and R_L in that order and the terms within the curly braces in Equation 15.31 correspond to the noise contribution of M1 and I_{B1} in that order. The total power consumed by the LC biquad can be expressed as: $P_{LC} = (2 + 4Q)P_{G_{m2}}$ (as $g_{m_1} = 4Qg_{m_2}$ and $R_1 = 1/g_{m_2}$), where $P_{G_{m2}}$ is the power consumption of g_{m2}.

Expressions for $V_{nLPF-GmC}^2$, $V_{nBPF-GmC}^2$ (noise density due to lowpass and bandpass path of the G_m-C equalizer section: Biquad 1 in Figure 15.13), and P_{GmC} (total power consumed by G_m-C equalizer section) can also be derived in a similar manner [10]. Finally, the relative power efficiency of the LC equalizer section (η) is defined as the ratio of integrated power–noise product of G_m-C equalizer section to that of the LC one and is given by

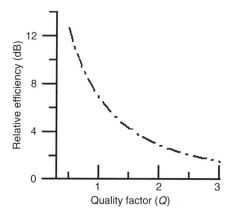

FIGURE 15.19 Relative power efficiency (for LC and G_m–C equalizing filter sections) versus quality factor.

$$\eta = \frac{P_{G_m C} \int_0^{\omega_0} (V^2_{nLPF-G_m C} + V^2_{nBPF-G_m C}) d\omega}{P_{LC} \int_0^{\omega_0} (V^2_{nLPF-LC} + V^2_{nBPF-LC}) d\omega} \tag{15.32}$$

To directly compare the power efficiency of LC series resonator based equalizing topology to the G_m–C one; η is plotted as a function of Q in Figure 15.19. It is evident from the plot that the implemented LC series resonator based biquads (with $Q = 0.618$ and $Q = 1.618$) are on an average about 7.3 times more power efficient than G_m–C ones.

Though LC filters in general are more efficient than the G_m–C ones, their area required especially at low frequencies and SNR can be prohibitively large.

To formulate the area relations for LC and G_m–C filters as function of SNR and frequency, we depend on the impedance and frequency scaling concepts discussed in Section 15.2. Let A_{Co}, A_{gmo} and A_L be the total area taken by capacitors, transistors, and inductors of an LC equalizer section, respectively, for SNR_o (47 dB) and cut-off frequency f_o (1.1 GHz). C, L, and g_m values can then be projected as a function of SNR and f by applying impedance scaling and frequency scaling (for constant noise) respectively. Capacitor, transistor, and inductor area scales roughly by the same factor as C, g_m, and L, respectively. Thus area of scaled LC filter (as a function of SNR and cut-off frequency) can be expressed as:

$$\text{Area}_{LC}(\text{SNR}, f) = A_{Co} \frac{\text{SNR}}{\text{SNR}_o} + A_{gmo} \frac{\text{SNR}f}{\text{SNR}_o f_o} + A_L \left(L_o \frac{\text{SNR}_o}{\text{SNR}} \left(\frac{f_o}{f} \right)^2 \right) \tag{15.33}$$

If $\eta_{1,2}$ represents the value of η obtained for the two biquads (with $Q_1 = 0.618$ and $Q_2 = 1.618$), area of a corresponding G_m–C filter (as a function of SNR and f) can be expressed in terms of A_{Co} and A_{gmo} as:

$$\text{Area}_{G_m C}(\text{SNR}, f) = \frac{1}{2} \sum_{i=1}^{2} \left[A_{Co} \frac{\eta_i (2 + 4Q_i)}{2(7 + 1/Q_i)} \frac{\text{SNR}}{\text{SNR}_o} + A_{gmo} \eta_i \frac{\text{SNR}f}{\text{SNR}_o f_o} \right] \tag{15.34}$$

In 0.18 μm technology, Area$_{LC}$ (46 dB, 1.1 GHz) = 630 K μm^2 [10], which is about twice of Area$_{G_m C}$ (47 dB, 1.1 GHz). However, from Equations 15.33 and 15.34 it is projected that Area$_{G_m C}$ would outrun Area$_{LC}$ beyond certain f for a given SNR and beyond certain SNR for a given f. Figure 15.20 captures this trend by plotting both the areas in K-μm^2 across SNR and f. For instance, at $f = 2$ GHz, Area$_{G_m C}$ equals Area$_{LC}$ for SNR of 44 dB and progressively more area for higher SNR. This trend suggests that the LC biquad can achieve much better power efficiency without area penalty at sufficiently high frequencies or SNR and vice versa.

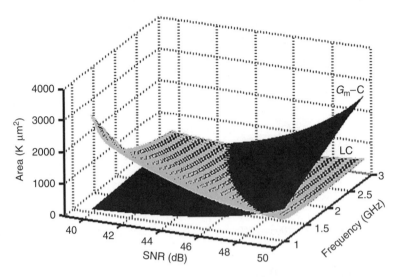

FIGURE 15.20 Area comparison for G_m–C and LC equalizer section. (From Dhanasekaran, V., Gambhir, M., Silva-Martinez, J., and Sánchez-Sinencio, E., *IEEE J. Solid-State Circuits*, 12, 2411, 2007. With permission.)

Note that the above relations (Equations 15.33 and 15.34) are very generic and can be extended to any LC and OTA-C filter with corresponding substitutions for A_{Co}, A_{gmo} and A_L and η.

15.6 CONCLUSIONS

Increasing demand of high data rate systems has driven the rapid evolution of wideband integrated filters and equalizers. This chapter examined G_m–C and LC topologies for this particular class of filters.

The initial sections of this chapter present the fundamentals of filter design along with a generic overview of G_m–C and LC topologies with intent to familiarize the readers with the most commonly employed structures. Also discussed are the typical transconductors for wideband filters along with their design trade-offs. Simple top-down design procedures are provided evolving from filter specifications to block level and transistor design.

Special focus of this chapter was on design considerations for equalizing filters. With two specific examples of G_m–C and LC equalizing filters, architectural and design details are discussed and measurement results are presented for the prototypes. G_m–C and LC filters are examined for power efficiency and area requirements. Finally, it leaves the readers with specific tools to analyze and directly compare relative merits of the two topologies to help them make prudent design choices based on the specifications.

REFERENCES

1. P. Pandey, J. Silva-Martinez, and X. Liu, A 500 MHz OTA-C 4th order lowpass filter with class AB CMFB in 0.35 µm CMOS technology, *Proc. IEEE Custom Integr. Circuits Conf.* (CICC), Orlando, FL, pp. 57–60, October 2004.
2. S. Pavan and Y.P. Tsividis, Widely programmable high-frequency continuous time filters in digital CMOS technology, *IEEE J. Solid-State Circuits*, 35(4), 503–511, 2000.
3. N.M. Nguyen and R.G. Meyer, Si IC-compatible inductors and LC passive filters, *IEEE J. Solid-State Circuits*, 25(4), 1028–1031, 1990.
4. S. Dosho, T. Morie, and H. Fujiyama, A 200 MHz seventh-order equiripple continuous-time filter by design of nonlinearity suppression in 0.25 µm CMOS process, *IEEE J. Solid-State Circuits*, 37(5), 559–565, 2002.

5. F. Rezzi, I. Bietti, M. Cazzaniga, and R. Castello, A 70 mW seventh-order filter with 7–50 MHz cutoff frequency and programmable boost and group delay equalization, *IEEE J. Solid-State Circuits*, 32(12), 1987–1999, 1997.

6. G. Bollati, S. Marchese, M. Demicheli, and R. Castello, An eighth-order CMOS low-pass filter with 30–120 MHz tuning range and programmable boost, *IEEE J. Solid-State Circuits*, 36(7), 1056–1066, 2001.

7. I. Mehr and D.R. Welland, A CMOS continuous-time Gm-C filter for PRML read channel applications at 150 Mb/s and beyond, *IEEE J. Solid-State Circuits*, 32(4), 499–513, 1997.

8. J.E.C. Brown, P. J. Hurst, B.C. Rothenberg, and S.H.A. Lewis, CMOS adaptive continuous-time forward equalizer, LPF, and RAM-DFE for magnetic recording, *IEEE J. Solid-State Circuits*, 34(2), 162–169, 1999.

9. M. Gambhir, V. Dhanasekaran, J. Silva-Martinez, and E. Sánchez-Sinencio, Low power architecture and circuit techniques for high boost wideband Gm-C filters, *IEEE Trans. Circuits Syst. I, Reg. Pap.*, 54(3), 458–468, 2007.

10. V. Dhanasekaran, M. Gambhir, J. Silva-Martinez, and E. Sánchez-Sinencio, A 1.1 GHz 5th order active-LC Butterworth filter with 23 dB equalizing gain, *IEEE J. Solid-State Circuits*, 42(12), 2411–2420, 2007.

11. S. Pavan and Y. Tsividis, Time scaled electrical networks—Properties and applications in the design of programmable analog filters, *IEEE Trans. Circuits Syst. II*, 47(2), 161–165, 2000.

12. G. Efthivoulidis, L. Toth, and Y.P. Tsividis, Noise in Gm-C filters, *IEEE Trans. Circuits Syst. II*, 45(3), 295–302, 1998.

13. R. Schaumann and M. E. Van Valkenburg, *Design of Analog Filters*. New York: Oxford University Press, 2003.

14. B. Razavi, *Design of Analog CMOS Integrated Circuits*. New York: McGraw-Hill, 2000.

15. P. R. Gray and R. G. Meyer, *Analysis and Design of Analog Integrated Circuits*. New York: Wiley, 1984.

16. J. Silva-Martinez, J. Adut, J.M. Rocha-Perez, M. Robinson, and S. Rokhsaz, A 60 mW 200 MHz continuous-time seventh order linear phase filter with on-chip automatic tuning system, *IEEE J. Solid-State Circuits*, 38(2), 216–225, 2003.

17. B. Nauta, A CMOS transconductance-C filter technique for very high frequencies, *IEEE J. Solid-State Circuits*, 27(2), 142–153, 1992.

18. E.A.M. Klumperink and B. Nauta, Systematic comparison of HF CMOS transconductors, *IEEE Trans. Circuits Syst. II*, 50(10), 728–741, 2003.

19. M. Chen, A.N. Mohieldin, and J. Silva-Martinez, Linearized OTAs for high-frequency continuous-time filters: A comparative study, *Midwest Symp. on Circuits and Syst.*, pp. 111–149, August 2002.

20. A.I. Zverev, *Handbook of Filter Synthesis*. New York: Wiley, 1967.

21. H. Khorramabadi and P.R. Gray, High-frequency CMOS continuous-time filters, *IEEE J. Solid-State Circuits*, SC-19(6), 939–948, 1984.

22. E. Sánchez-Sinencio and J. Silva-Martinez, CMOS transconductance amplifiers, architectures and active filters: A tutorial, *Proc. IEE Circuits, Devices, Syst.*, 147, 3–12, 2000.

23. B. Jung, R. Harjani, A wide tuning range VCO using capacitive source degeneration, *Proc. ISCAS*, pp. IV145–IV148, Vancouver, May 2004.

16 Continuous Time $\Sigma\Delta$ Modulators

Robert Sobot

CONTENTS

Despite its long history, apparent simplicity, and elegance, a sigma–delta ($\Sigma\Delta$) modulator remains one of the most elusive circuits in microelectronics. So far, no researcher has presented a comprehensive analysis and model that completely and accurately describes all aspects of the circuit's operation. Rather, circuit developments have resulted from work carried out by a large number of contributors, with only incremental advances in inner circuit operation knowledge. In this chapter, the analysis and design methodology for a center frequency tunable continuous time bandpass (CT BP) $\Sigma\Delta$ modulator, in the high-frequency (HF) signal range, is presented. In particular, some of outstanding problems related to this particular group of modulators, for example, loop delay compensation, center frequency tunability, and higher order stability, are addressed. It appears that $\Sigma\Delta$ modulators have the potential to include a tunable frequency element, combined with frequency mixing, analog/digital (A/D) conversion, and a filtering function, in a single device. Clearly, there is enough motivation to invest more research efforts in this area.

16.1 STATE OF THE ART

The principles of delta-modulation were first introduced in patent by C.C. Cutler, filed in 1954 and granted in 1960 [4]. Since that time, numerous papers, designs, books, and chapters have been published on the topic [18,23]. Today, a large number of circuit designs have emerged from this classical feedback loop concept, known as the sigma–delta ($\Sigma\Delta$) modulator. Regardless of the specific design, all $\Sigma\Delta$ modulator loops contain at least the following functional blocks: summing circuit at the input, loop filter, sampled A/D converter at the output, and digital/analog (D/A) converter in the feedback loop.

For the purpose of our discussion, the following classifications are arbitrarily introduced. Conceptually, the family of $\Sigma\Delta$ circuits can be divided into two groups: discrete time (DT) and continuous time (CT). This categorization is based solely on the time-domain characteristics of the $\Sigma\Delta$ loop filter. If the filter was designed in the z-domain, i.e., implemented as either a switched capacitor (SC) or switched current (SI) filter, then the $\Sigma\Delta$ modulator is referred to as a DT $\Sigma\Delta$ modulator. Similarly, if the filter was designed in the s-domain, then the $\Sigma\Delta$ modulator is designated as a CT $\Sigma\Delta$ modulator. The next level of classification is based on the filter type: lowpass (LP) or bandpass (BP) $\Sigma\Delta$ modulator. Regardless of the filter type, a connection can be made to either a 1 bit or multibit A/D converter. The general assumption is that the feedback D/A converter has the same number of bits as the A/D converter. Lastly, based on the relationship between the input signal frequency and the A/D sampling frequency, oversampled and undersampled $\Sigma\Delta$ modulators are defined.

16.1.1 SURVEY OF PUBLISHED DESIGNS

With the exception of a few milestone works, referenced for completeness, this section reviews CT BP $\Sigma\Delta$ modulator designs, which were supported by the experimental data. Note that a large number of publications related to z-domain and s-domain (simulation results only) designs are beyond the scope of this survey.

In the late 1980s, contributions to the fundamental understanding of bandpass $\Sigma\Delta$ modulation operation have been very significant. In 1989, Schreier and Snelgrove [29] published one of the first theoretical works dedicated to the bandpass transfer function. Several papers by Jantzi et al. [14,15] contributed a working methodology for calculating the noise transfer function (NTF) and signal transfer function (STF) for a $\Sigma\Delta$ loop.

The first CT BP $\Sigma\Delta$ modulator was presented in 1990 by Dressler [6], in which the lowpass filter inside $\Sigma\Delta$ loop was intuitively replaced with a resonator to achieve the bandpass effect. A summary of the experimental data for this second-order, discrete implementation design is as follows: signal-to-noise ratio (SNR) = 55 dB, center frequency f_0 = 2.5 MHz, bandwidth (BW) = 80 kHz, and clock frequency f_{clk} = 10 MHz.

Further advances were made in 1991 by Thurston et al. [38], who suggested a method for determining the analytical expression for a bandpass loop filter transfer function, $H(s)$, in the s-domain. Their research demonstrated the first application of the impulse-invariant transformation technique for performing the domain transformation $H(z) \rightarrow H(s)$, in order to solve for the CT BP $\Sigma\Delta$ loop filter transfer function. This concept, driven by emerging cellular networks, was proved through experiment, thus paving the way for future CT BP $\Sigma\Delta$ research. A summary of the experimental data for this second-order, discrete implementation design is as follows: SNR = 50 dB, f_0 = 7.5 MHz, BW = 100 kHz, and f_{clk} = 10 MHz. Also in the same year, while not strictly in the CT category, Horrocks [12] proposed a tunable center frequency, second-order z-domain bandpass filter.

In May 1992, Jantzi et al. [16] published the first monolithic implementation of a BP $\Sigma\Delta$ modulator. A summary of the fourth-order, SC implementation design is as follows: 3 μm CMOS technology, SNR $= 63$ dB, $f_0 = 455$ kHz, BW $= 10$ kHz, and $f_{clk} = 1.82$ MHz. Then in June, Tröster et al. [39] published the first monolithic implementation of a CT BP ΣΔ modulator. With cellular network application in mind, they used a binary complementary metal oxide semiconductor (BiCMOS) A/D array. A summary of the second-order monolithic implementation, with an external LC resonator is as follows: 1.2 μm/7 GHz BiCMOS A/D array technology, SNR $= 55$ dB, $f_0 = 6.5$ MHz, BW $= 200$ kHz, and $f_{clk} = 26$ MHz.

In 1995, Shoaei [30] contributed an analytical methodology for s-domain CT BP ΣΔ modulator analysis and design, covering both second- and fourth-order modulators. In addition, an alternative method for $H(s)$ synthesis, for the quarter of the sampling frequency ($f_s/4$) CT BP ΣΔ modulator, was introduced.

A second-order tunable g_m–C-based CT BP ΣΔ modulator, with programmable center frequency $f_0 = 24.4$ MHz/62.5 MHz, was developed by Raghavan et al. in 1997 [27]. Taking advantage of the high clock speed, i.e., high oversampling ratio (OSR), in fast InGaAs technology, an SNR $= 92$ dB was achieved. The reported data for this implementation were: InGaAs technology, area 750×750 μm^2, power dissipation DC$_{pwr} = 1.4$ W, SNR $= 92$ dB, $f_0 = 55.6$ MHz, BW $= 366$ kHz, and $f_{clk} = 4$ GHz. Also published in 1997, was a fourth-order modulator design developed by Jayaraman et al. [17]. He investigated the application of a power amplifier (PA) in combination with a ΣΔ modulator. A summary of the reported data is as follows: GaAs technology, SNR $= 63$ dB, $f_0 = 800$ MHz, BW $= 200$ kHz, and $f_{clk} = 3.2$ GHz.

In 1998, Gao and Snelgrove [8] developed a second-order design, claiming it to be the first design in the GHz range including on-chip LC components. This design exhibited: 0.5 μm bipolar junction technology (BJT), SNR $= 57$ dB, $f_0 = 950$ MHz, BW $= 200$ kHz, and $f_{clk} = 3.8$ GHz. Currently, this design has the highest reported center frequency. In the same time, Gao et al. [7] reported a fourth-order design with the following data: $V_{CC} = 5.0$ V, DC$_{pwr} = 350$ mW, area 0.85×1.46 mm^2, $f_{clk} = 4.0$ GHz, BW $= 4$ MHz, $f_0 = 1$ GHz, SNR $= 53$ dB.

The first working design of a sixth-order CT BP ΣΔ modulator was developed by van Engelen et al. in 1999 [40]. Published data for this implementation are as follows: 0.5 μm CMOS technology, SNDR $= 67$ dB, $f_0 = 10.7$ MHz, BW $= 200$ kHz, $f_{clk} = 30/80$ MHz. During the same year, Tao and Khoury [37] demonstrated their downconversion frequency translation approach, using a second-order modulator as the downconverter. This design used discrete inductors with an SC integrator. A summary of the reported data is as follows: 0.35 μm CMOS technology, SNDR $= 54$ dB, $f_0 = 100$ MHz, BW $= 200$ kHz, and $f_{clk} = 400$ MHz.

A comprehensive study on the excess loop delay, with a novel approach to compensation, was published in 2000 by Maurino and Mole [22]. A summary of the experimental data for this fourth-order design is as follows: SiGe technology, SNR $= 68$ dB, $f_0 = 200$ MHz, BW $= 200$ kHz, and $f_{clk} = 800$ MHz. The following year, Raghavan et al. [26] developed the second generation of their InGaAs design, a fourth-order CT BP ΣΔ modulator. They reported the following data for their implementation: power supply $V_{CC} = \pm 5$ V, power dissipation DC$_{pwr} = 3.2$ W, area 3.3×1.7 mm^2, SNR $= 75.8$ dB, $f_0 = 180$ MHz, BW $= 1$ MHz, and $f_{clk} = 4$ GHz.

One of the most accomplished designs so far was developed in 2002 by Schreier et al. [28], in which a robust industrial mixed-signal approach was used to design their commercially available chip. Their sixth-order CT design used external inductors for the first resonator, RC (with trimming capacitors tolerant to within 1%) for the second resonator, and SC for the third resonator. They reported the following data for their implementation: 0.35 μm BiCMOS technology, power supply $V_{CC} = 2.7$ V, power dissipation DC$_{pwr} = 50$ mW, area 5.0 mm^2, tunable range 10/300 MHz, BW $= 333$ kHz, SNR $= 81$ dB, and $f_{clk} = 3/32$ MHz. In the same journal issue, the group lead by Henkel published their design of fourth-order CT design in 0.65 μm BiCMOS technology [10]. Their results demonstrated that if quadrature CT ΣΔ modulation is used, the antialiasing intermediate frequency (IF) bandpass filter can be eliminated. Furthermore, use of a CT polyphase filter eliminates the problem of an I and Q channel mismatch, while reducing the sensitivity of the circuit to excess loop delays. They reported the following data: power supply $V_{CC} = 2.7$ V, power dissipation $DC_{pwr} = 21.8$ mW, area 2.1×2.9 mm^2, $f_0 = 1$ MHz, BW $= 1$ MHz, SNDR $= 56.2$ dB, and $f_{clk} = 25/100$ MHz.

The third generation of the original design proposed by the Raghavan group was published in 2004 by Cosand et al. [3]. A summary of the fourth-order CT is as follows: InGaAs technology, power supply $V_{CC} = \pm 5$ V, power dissipation $DC_{pwr} = 3.5$ W, area 3.3×2.6 mm^2, tunable center frequency range $f_0 = 140/210$ MHz, BW $= 1$ MHz, SNDR $= 78$ dB, and $f_{clk} = 4$ GHz.

In 2005, the author introduced the concept of CT BP $\Sigma \Delta$ modulators with fractional delays [32,35]. This novel fourth-order CT BP concept was confirmed by experimental data as follows: SiGe technology, power supply $V_{CC} = 3.3$ V, power dissipation $DC_{pwr} = 1.0/1.2$ W, total chip area is 2.3×2.3 mm^2, tunable center frequency range $f_0 = 185/289$ MHz, BW $= 20$ MHz, maximal SNR $= 50$ dB, and $f_{clk} = 0.6/1.2$ GHz.

16.1.2   CURRENT DESIGN TRENDS

Although there are numerous publications available concerning CT BP $\Sigma \Delta$ modulators, on an average, only one successful design per year has been reported over the last 15 years, illustrating that casual attempts at modulator designs are very likely to fail. This leaves room for further improvement on both theoretical and practical sides of the problem.

From the overview presented in Section 16.1.1, it would appear that the major development directions relate to increasing both the center frequency and SNR. However, there are many other desirable circuit features that could be emphasized, such as wider bandwidth, higher filter order, lower power requirements, and tunability of the center frequency. Researchers have used all means available to achieve these goals; advances in theoretical work, fast technologies, and higher OSRs are the most commonly exploited methods. Typical limiting constraints are power consumption and stability. Higher order topologies may deliver higher SNR values, however, they are inherently unstable. Conversely, designing modulators that are tunable to any frequency relative to the sampling clock, i.e., f_s/n type as opposed to the $f_s/4$ type, creates the potential for better utilization of the modulator functionality (frequency translation) and a reduction in power consumption through the use of lower clock frequencies. While the numerical methods generally used to design the loop filter function do produce valid results, analytical methods may contribute more to the fundamental understanding of the problem.

Due to the simplistic methods commonly used to define the design space (and often incompatible reporting format), the CT BP $\Sigma \Delta$ center frequency f_0 and maximal SNR, a fair comparison of various designs proved to be difficult. The designs exhibit various levels of performance depending on their specific combination of modulator order N, number of slicing levels in the A/D and D/A blocks (BIT), BW, and OSR. From reported data of designs listed in this section, one possible figure-of-merit center frequency could be shown versus logarithm of the center frequency f_0 (Figure 16.1) in which the horizontal axis is logarithm of the center frequency f_0 and the vertical axis is the figure-of-merit center frequency.

In this case, the figure-of-merit center frequency (CF) is defined as

$$CF = \frac{\log f_0}{\log \frac{SNR \times BW}{N \times OSR \times BIT}}$$

It is interesting to note that, by this criteria, all designs appear to be bounded in this log–log space with the following limits CF_L:

$$CF_L = \begin{cases} -0.78 \log f_0 + 10.5 \\ -0.78 \log f_0 + 9.2 \end{cases} \qquad (16.1)$$

which are shown as dashed lines in Figure 16.1.

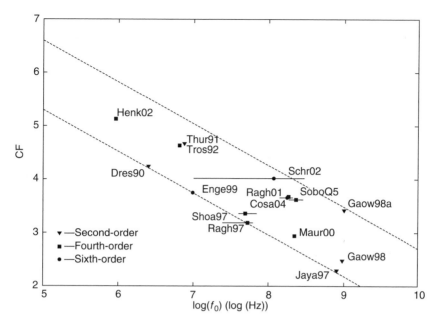

FIGURE 16.1   Figure-of-merit CF for CT BP ΣΔ modulator designs.

16.2   BASIC PRINCIPLES OF ΣΔ MODULATION

Although the basic principles are intuitively simple, a detailed analysis of a ΣΔ modulator is still under development; a complete closed-form analytical model covering all aspects of ΣΔ modulation does not currently exist. As a result, researchers are forced to develop their own semiempirical solutions. This section gives a brief introduction of ΣΔ modulation; a more detailed introduction can be found, for example, in Refs. [18,23,32].

16.2.1   QUANTIZATION OF A CONTINUOUS SIGNAL

Inherently, digital logic circuits process only two levels of an input signal, while analog logic circuits operate with continuous signals. Therefore, a quantizer is required to provide an interface between these two signal domains, i.e., it converts the continuous signal into an n-level discrete step function.

When the maximal input signal amplitude is mapped onto $2N$ output levels, where N is any integer greater than zero ($N = 1, 2, \ldots$), the system is referred to as an N-bit quantizer. In the special case where $N = 1$, the gain G is arbitrarily set. In general, unity gain approximation is used, which loosely states that the 1 bit quantizer has $G = 1$. Therefore, it is inherently linear as $f(x)$ passes through only two points, selected arbitrarily; this is the main reason for its widespread use.

16.2.2   QUANTIZATION NOISE

Quantization errors occur even in ideal quantizers and subsequently in all A/D converters. The error signal's root mean square (rms) and average values can be determined by applying a stochastic approach used for general input signals [18] with these two commonly made assumptions: (1) the quantization error signal is random, with its amplitude always bounded by $\pm\Delta/2$; i.e., only the nonoverloaded quantizer is considered and (2) the input signal is sampled at a sufficiently high sampling frequency, $f_s = 1/T$.

It was shown in the literature [18] that the rms value of the quantization noise is not a function of the sampling frequency, i.e., $\xi_{rms}^2 = \Delta^2/12$. Finally, the definition of the OSR is given as OSR $\equiv$ $f_s/2f_0$. Furthermore, it is assumed that the load impedance is $Z_0 = 1$, which conveniently allows the rms value to be calculated as $\sqrt{\text{power}}$. In addition, by doubling the OSR, the quantization noise power decreases by 3 dB. Thus, an increase in f_s alone is not an efficient way to improve the SNR.

16.2.3 SNR OF AN IDEAL A/D CONVERTER

The SNR of a sinusoidal is given by $\text{SNR}_{\max} = (6.02N + 1.76)$ dB, which is 1.76 dB more power than, for example, the corresponding sawtooth signal. This equation is a well-known result, generally used to estimate N relative to $\text{SNR}_{\max}$ for an A/D converter [18].

In a linear model of $\Sigma\Delta$ loop, the quantizer has been replaced by a summing node, where $e[n]$ is a Gaussian error signal. The following assumptions are made regarding the quantization error signal: (a) the error signal is a stationary random process; (b) the error signal and input signals are not correlated; (c) the error signal has a white noise profile; and (d) the error signal has a uniform probability distribution. These assumptions are referred to as the soft version of the white noise approximation [23,24]. This approach will be used to describe the noise-shaping operational principles. Furthermore, the white noise model allows for a precise calculation of the quantization error's rms, for a large variety of inputs and nonloading systems. The main drawback of this approach is the inability to predict idle tones and noise patterns; it also does not account for excess noise due to overload [23].

A more general form of the SNR equation considers the initial estimate of the SNR relative to the OSR, the NTF order m, and number of bits n, when determining the $\Sigma\Delta$ architecture capable of achieving the required system specifications:

$$\text{SNR}_{\max} \equiv 10\log\left(\frac{P_s}{P_n}\right) = 20\log\left[(2^n - 1)\sqrt{\frac{3}{2}}\frac{\sqrt{2m+1}}{\pi^m}\text{OSR}^{\left(m+\frac{1}{2}\right)}\right] \qquad (16.2)$$

This formula illustrates the trade-offs between complexity ($\Sigma\Delta$ loop order) and clock speed (OSR) for a given $\text{SNR}_{\max}$. There are many different configurations that will achieve an $\text{SNR}_{\max} \approx 110$ dB, such as a second-order loop with OSR = 256, a third-order loop with OSR = 64, or a fourth-order loop with OSR = 32. In addition, by doubling the OSR, different gains in SNR can be achieved; for example, 3 dB for a zero-order shaping loop, 9 dB for a first-order shaping loop, and 15 dB for second-order shaping loop. For a detailed mathematical analysis of the quantization error and liner modeling, the readers are advised to refer to the literature [1,18,23,24].

16.3 FRACTIONAL $\Sigma\Delta$ MODULATORS

This section focuses on the development of an analytical procedure for obtaining a closed-form transfer function for the loop filter of a tunable CT BP $\Sigma\Delta$ modulator. Analytical models for tunable fourth-order CT BP $\Sigma\Delta$ loop, both zero-delay and fractional-delay, are presented. The developed procedure is extended to introduce a new category of CT BP $\Sigma\Delta$ modulators with fractional delays ($f\Sigma\Delta$) [35].

16.3.1 THEORETICAL BACKGROUND

One challenge in designing a CT BP $\Sigma\Delta$ modulator is the lack of an s-domain analytical expression for the tunable loop filter transfer function. Hence, the first BP $\Sigma\Delta$ designs were achieved simply by following intuition and replacing the LP $\Sigma\Delta$ loop filter with a BP loop filter [12,38]. As a result, the designs were nonoptimal for the given loop order.

Shortly after, an analytical form of the second- and fourth-order $f_s/4$ CT BP loop transfer functions was published by Shoaei [30]. These functions were derived through the application of the impulse-invariant transformation. The current trend is to create a tunable, higher order CT BP $\Sigma\Delta$ modulator for RF frequencies [31,33].

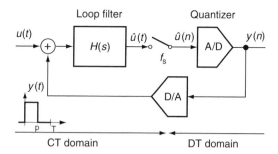

FIGURE 16.2 Block diagram of a CT $\Sigma \Delta$ modulator. (From Sobot, R., Stapleton, S., and Syrzycki, M., *IEEE Trans. Circuits Syst.-I*, 53, 264, 2006.)

The control and compensation of the excess loop delay is another important property that is considered during CT BP $\Sigma \Delta$ design [2,11]. The analytical procedure outlined in this chapter enables the proposal of a novel $\Sigma \Delta$ loop architecture, where the traditional CT BP loop filter function is replaced by a filter function with fractional delays. The architecture exploits the fact that the overall $\Sigma \Delta$ loop timing (including the excess loop delay) must be equal to a fractional multiple of the sampling period, which is matched by the CT loop filter transfer function.

A block diagram of a CT $\Sigma \Delta$ modulator loop is shown in Figure 16.2. The CT domain signals $u(t)$ and $y(t)$ are added before being processed by the CT loop filter $H(s)$. The output signal from the filter $\hat{u}(t)$ is sampled by a switch at a frequency of f_s. The DT signal $\hat{u}(n)$ is then quantized by the A/D block. The feedback loop consists of a D/A converter, which is usually modeled with a zero-order-hold (ZOH) function. The output of the ZOH is the CT pulse function $y(t)$. Clearly, the overall $\Sigma \Delta$ loop gain is a DT function. Therefore, an exact correspondence between the CT filter $H(s)$ and its equivalent DT filter $H(z)$ can be derived only at the sampling points.

The mathematical equivalence between the $H(z)$ and $H(s)$ functions can exist only at the sampling points. In other words, at the sampling points, the inverse z-transform of $H(z)$ must be equivalent to the inverse Laplace transform of $H(s)$.

16.3.2 $\Sigma \Delta$ LOOP FILTER TRANSFER FUNCTION

The underlying assumption in most of the z-domain $\Sigma \Delta$ loop expressions is that the sampling frequency is four times higher than the signal frequency. The equivalent description is that the complex signal frequency is at the $\theta = \pi/2$ angle. Indeed, the $\Sigma \Delta$ loop transfer functions within this category are referred to as the $f_s/4$ functions. The natural starting point for determining tunable versions of the $\Sigma \Delta$ transfer functions is the $f_s/4$ case, as the z-plane location of the input signal frequency is at the midpoint between the DC frequency ($\theta = 0$) and half the sampling frequency ($\theta = \pi$). This symmetry enables the relative change of the input signal frequency to the sampling frequency equally in both directions [35].

16.3.3 ZERO-DELAY CT BP $\Sigma \Delta$ MODULATORS

The analytical methodology outlined in Ref. [35] resulted in a closed-form solution of a CT BP $\Sigma \Delta$ functions which look, for example, as

$$H_0(s) = \frac{\frac{H_{03}}{T} s^3 + \frac{H_{02}}{T^2} s^2 + \frac{H_{01}}{T^3} s + \frac{H_{00}}{T^4}}{\left[s^2 + \left(\frac{\theta - \delta}{T} \right)^2 \right] \left[s^2 + \left(\frac{\theta + \delta}{T} \right)^2 \right]} \tag{16.3}$$

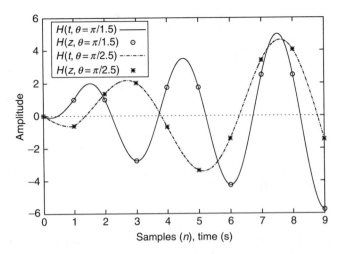

FIGURE 16.3 Impulse response $h(t)$ and corresponding tunable $h[n]$. (From Sobot, R., Stapleton, S., and Syrzycki, M., *IEEE Trans. Circuits Syst.-I*, 53, 264, 2006.)

where

$$H_{03} = -\frac{\theta}{2}\frac{\sin\theta}{(1-\cos\theta)} - \frac{1}{4}$$

$$H_{02} = \frac{\theta^2}{4}\frac{(3-4\cos\theta)}{(1-\cos\theta)} - \frac{\theta}{2}\frac{\sin\theta}{(1-\cos\theta)}$$

$$H_{01} = -\frac{\theta^3}{2}\frac{\sin\theta}{(1-\cos\theta)} + \frac{\theta^2}{4}$$

$$H_{00} = \frac{\theta^4}{4}\frac{3-4\cos\theta}{(1-\cos\theta)}$$

and δ is the separation angle between the pole pairs and the center notch frequency [33]. For $\delta = 0$, the two pairs of poles collapse into a double pole.

The pulse-invariance equivalence corresponding to the tunable s-domain function (16.3) is shown in Figure 16.3, for $\theta = \pi/1.5$, $\pi/2.5$. Ordinarily, the first two samples of this function are equal to zero in the case of the $f_s/4$ modulator [30]. A change in the tuning parameter θ causes the second sample to move away on either side, while the rest of the samples follow accordingly. The plot confirms the impulse-invariance equivalence of the s-domain function with its counterpart in the z-domain.

For the results presented in this section, OSR $= f_s/(2\,\mathrm{BW}) = 100$, which means the linear model (16.2) predicts $\mathrm{SNR}_{max} \approx 88.9\,\mathrm{dB}$. As a comparison, the Ardalan bound [1] is more conservative than the linear model and predicts $\mathrm{SNR}_{max} \approx 77.6\,\mathrm{dB}$, while the actual simulations show a value of $\mathrm{SNR}_{max} \approx 86\,\mathrm{dB}$.

The AC simulation results for NTF and STF related to (16.3) are shown in Figure 16.4. The tuning parameter θ controls the position of the notch in the NTF response, as well as the centering of the STF passband response. The commonly cited criteria for $\Sigma\Delta$ loop stability, the Lee's rule [21], are further constrained by most researchers to a gain equal to 1.6 in order to allow for process variations.

16.3.4 Delayed CT BP ΣΔ Modulators

The equations shown so far describe zero-delay $\Sigma\Delta$ loops. The zero-delay term refers to $\Sigma\Delta$ architectures without the feedback latch. In other words, the loop transfer function is completely determined

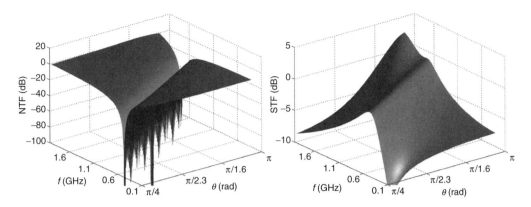

FIGURE 16.4 AC simulation of NTF and STF for fourth-order tunable CT BP ΣΔ model. (From Sobot, R., Stapleton, S., and Syrzycki, M., *IEEE Trans. Circuits Syst.-I*, 53, 264, 2006.)

by the loop filter transfer function, $H_0(z)$ or $H_0(s)$. To emphasize the importance of the zero-delay formulation of the loop function, the following statement should be kept in mind. The overall loop delay around any CT ΣΔ loop must be equivalent to the zero-delay formulation. In other words, any delay introduced inside the loop path, due to circuitry propagation times, must be subtracted from the loop filter transfer function. This situation is the main cause of problems related to the loop delays encountered in CT ΣΔ modulators. Analytical methodologies inherited from z-domain mathematics allow for the calculation of the loop filter functions with an integer number of clock delays. Usually, a latch is used to introduce a one-clock period delay in the feedback part of the loop. Consequently, the loop filter must be modified to exclude the one-clock delay from its own transfer function, so that the overall ΣΔ loop transfer function is still equivalent to the zero-delay scheme.

Instead of creating the delayed versions of the loop filter in the z-domain, let us start from the s-domain equations. The Laplace transformation of (16.3) creates the time-domain equivalent expression of $H_0(s)$. The newly created time-domain function is then shifted in the time domain by fractional multiples of the clock period T. Finally, the application of the inverse Laplace transformation produces the delayed s-domain version of $H_0(s)$.

The importance of being able to design a fractional multiple-clock delay CT ΣΔ loop version becomes more evident at high frequencies. The high-frequency input signals are associated with even higher frequencies of the sampling clocks. To make the situation even worse, the parasitic layout capacitances become more dominant factors at high frequencies. It is easy to visualize a situation where the overall loop delay is longer than one-clock period due to parasitic capacitances alone [32]. The main argument of this section is that instead of trying to cancel the unavoidable loop delay, it should be incorporated in the design.

16.4 CT BP $f\Sigma\Delta$ MODULATOR DESIGN

The preceding sections introduced the theoretical background for CT BP $f\Sigma\Delta$ modulators. In this section, a transistor level design of a fourth-order tunable CT BP $f\Sigma\Delta$ modulator in 0.5 μm/f_T = 47 GHz SiGe technology is presented.

The modulator is a fully differential g_m–C based circuit that demonstrates a practical implementation of the analytical methodology used in designing the fractional CT loop transfer functions $H(s)$ presented in Section 16.3. A loop filter transfer function is designed to compensate for the unavoidable loop delay.

16.4.1 DESIGN ARCHITECTURE

A diagram of the top level modulator architecture, based on the analytical model published in Ref. [35], is shown in Figure 16.5. The CT loop filter transfer function $H(s)$ and summing circuit are

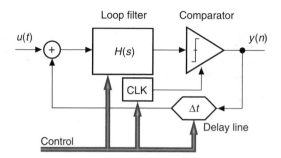

FIGURE 16.5 Block diagram for a CT $f\Sigma\Delta$ modulator. (From Sobot, R., Stapleton, S., and Syrzycki, M., *Proceedings of IEEE Canadian Conference on Electrical and Computer Engineering*, 2007, pp. 530–534, April 2007.)

both g_m–C based structures, while the design employs a 1 bit comparator to serve as the quantizer function.

Conventionally, in order to correctly synthesize both the NTF and STF of the modulator, $H(s)$ is calculated to have a delay around the loop of either zero or an integer multiple of the clock cycle. Using this approach, the unavoidable parasitic loop delays must be dealt with separately [11]. However, in order to correctly calculate the fractional delay $H(s)$, the time delay required by the filter should be set to the loop delay value; this results in the proper synthesis of both the NTF and STF. The same analytical model enables the calculation of the center frequency within the range f_s/n, where n is now a positive real number instead of an integer. The last two statements outline the main features of the design methodology presented in this chapter.

16.4.2 g_m–C Resonator

A discussion of several practical architectures, whose transfer functions have a one-to-one mapping corresponding to (16.3), is given in Ref. [30]. In this work, the g_m–C biquad resonator based architecture shown in Figure 16.6 was used [36] (each of the g_m blocks is tuned with its own biasing current and is not shown here for brevity). There are two resonators built around g_{r1}, g_{f1}, C_1, C_2 and g_{r2}, g_{f2}, C_3, C_4 structures. Circuit blocks g_{t1} and g_{t2} are required to fine tune the rather poor Q-factor of g_m–C-based resonators. The transfer function of the resonator in Figure 16.6 is

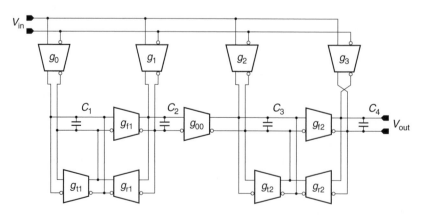

FIGURE 16.6 Block diagram of a fourth-order g_m–C resonator. (From Sobot, R., Stapleton, S., and Syrzycki, M., *Proceedings of IEEE Canadian Conference on Electrical and Computer Engineering*, 2007, pp. 530–534, April 2007.)

$$H(s) = \frac{V_{\text{out}}}{V_{\text{in}}} = \frac{\frac{g_3}{C_4}s^3 + \frac{g_2 g_{f2}}{C_3 C_4}s^2 + \left(\frac{g_1 g_{00} g_{f2}}{C_2 C_3 C_4} + \frac{g_3 g_{f1} g_{r1}}{C_1 C_2 C_4}\right)s + \frac{g_0 g_{f1} g_{00} g_{f2} + g_2 g_{f1} g_{f2} g_{r1}}{C_1 C_2 C_3 C_4}}{\left(s^2 + \frac{g_{f1} g_{r1}}{C_1 C_2}\right)\left(s^2 + \frac{g_{f2} g_{r2}}{C_3 C_4}\right)} \quad (16.4)$$

The ideal analytical coefficients are mapped into g_m values by equating the numerator and denominator coefficients in (16.3), with their respective counterparts in (16.4), where capacitors are set to $C = C_1 = C_2 = C_3 = C_4 = 26\,\text{pF}$.

16.4.3 Design of the g_m Stage

The schematic for the g_m stage used throughout the design is shown in Figure 16.7 [36]. The circuit topology is based on the multi-tanh architecture [9]. Transistors Q_3–Q_4 make the first asymmetric differential pair while transistors Q_5–Q_6 make the second. A size ratio of 4 was used for the Q_3–Q_4 and Q_5–Q_6 transistor pairs, which created an offset of $\pm 55\,\text{mV}$ relative to the input voltage. Their combined gain was approximately 30% larger than the gain of the individual input stages; the linear range was extended relative to the single differential pair.

16.4.4 Comparator

A traditional latch–slave configuration for a high-speed comparator [13] is shown in Figure 16.8 [36]. In accordance with the requirements of (16.3) transfer function, the NRZ pulse shape is used. The signal input stage consists of a differential preamplifier, which performs the following three functions: (1) it amplifies weak signals at the loop filter output; (2) it shifts the common mode voltage; and (3) it buffers possible kickback signals, which are reflected back from subsequent latch and slave stages. The input buffer stage consists of the common-collector voltage buffers Q_3 and Q_4, along with the passive loads R_3 and R_4. Similar buffering stages are inserted between the preamplifier and

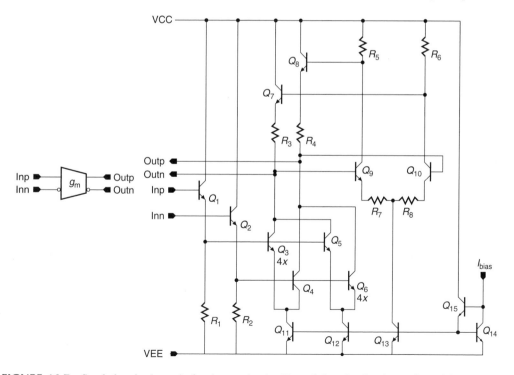

FIGURE 16.7 Symbol and schematic for the g_m circuit. (From Sobot, R., Stapleton, S., and Syrzycki, M., *Proceedings of IEEE Canadian Conference on Electrical and Computer Engineering*, 2007, pp. 530–534, April 2007.)

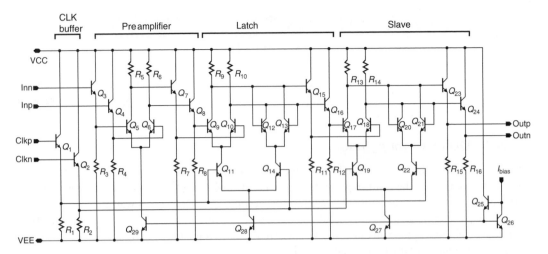

FIGURE 16.8 A detailed schematic of the NRZ comparator. (From Sobot, R., Stapleton, S., and Syrzycki, M., *Proceedings of IEEE Canadian Conference on Electrical and Computer Engineering*, 2007, pp. 530–534, April 2007.)

the latch, between the latch and slave stages, and at the output of the comparator. At the input stage, common-collector voltage buffers Q_1–Q_2 and passive load R_1–R_2 are used for the differential clock signal. Differential pairs Q_{11}–Q_{14} and Q_{19}–Q_{22} serve as the current-switching elements for the latch and slave stages, respectively. Biasing currents for the differential pairs are provided by the current mirror, which consists of transistors Q_{25}–Q_{29}. This circuit was simulated with clock frequencies of up to 5 GHz. However, the implemented $f\Sigma\Delta$ circuit works at much lower speeds, which greatly relaxes the design margins of the implemented comparator circuit.

16.4.5 POSTLAYOUT SIMULATION

With the transistor level design of the $f\Sigma\Delta$ modulator finished and the final layout completed, the last stage of the design process was to perform a postlayout simulation. The results of the postlayout simulations for two different tuning positions, $f_s/4$ and $f_s/5$ configurations, are shown in Figure 16.9.

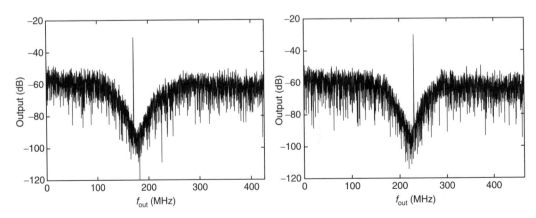

FIGURE 16.9 Simulated postlayout $f\Sigma\Delta$ response, using 8192 clock cycles and an FFT of 65,536, for (left) an $f_s/4$ configuration and (right) an $f_s/5$ configuration. (From Sobot, R., PhD thesis, Simon Fraser University, 2005.)

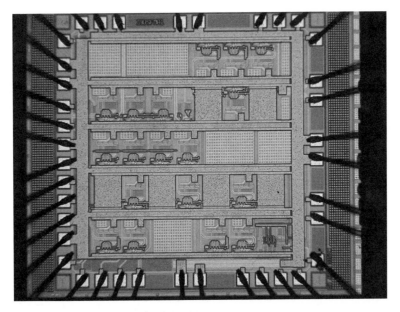

FIGURE 16.10 Microphotograph of the $f\Sigma\Delta$ chip. (From Sobot, R., Stapleton, S., and Syrzycki, M., *Proceedings of IEEE Canadian Conference on Electrical and Computer Engineering*, 2007, pp. 530–534, April 2007.)

16.5 PROTOTYPE CHIP TESTING

This section describes the prototype SiGe chip, the details of the test procedure, and the measured results confirming the validity of the proposed methodology. The test chip was manufactured in $0.5\,\mu m/f_T = 47\,GHz$ SiGe technology, and a microphotograph of the chip is shown in Figure 16.10. The total size of the layout, including the bonding pads and on-chip test voltage controlled oscillator (VCO) circuit, is $2.3 \times 2.3\,mm^2$. The chip was bonded to a high-frequency 80-pin ceramic package and soldered onto a test board [36]. An operating frequency range of up to 4.8 GHz was specified for the package and test board combination. The overall design assumed a 50 Ω environment; therefore, all on-board high-frequency differential lines were laid out as 50 Ω microstrip transmission lines.

16.5.1 EXPERIMENTAL RESULTS

As the chip is essentially a CT filter design, a method of estimating the process variations is required. In order to establish correlation between the simulation results and the measurement results, a simple g_m replica-based on-chip VCO was used. The control currents (i.e., biasing currents), obtained using the aforementioned procedure, yielded an NTF, which was very close to a correct $f\Sigma\Delta$ frequency plot.

Although the currently implemented resonator can be tuned to a much wider frequency range, in order to have the overall $f\Sigma\Delta$ tuned to the same range, some of the control current values (used to control the numerator coefficients in (16.4)) would have to be larger than 20 mA. Thus, the specification list in Table 16.1 includes both achieved tunability range of the $f\Sigma\Delta$, as well as the resonator by itself. The experimental results shown in Figure 16.11 demonstrate the operation of the $f\Sigma\Delta$ modulator for one of the signal frequencies within the maximal tuning range of 185 to 289 MHz. The application of the delayed versions to the $f\Sigma\Delta$ transfer function revealed that the parasitic loop delay was less than 0.1 T at the given sampling frequency [36].

16.6 CT BP $f\Sigma\Delta$ APPLICATIONS

In this section, a couple of possible applications for tunable CT BP $f\Sigma\Delta$ modulators are proposed. First, a new architecture for IF to RF conversion is presented. The architecture is based on a tunable

TABLE 16.1

Measured Specifications for the $f\Sigma\Delta$

Technology	BiCMOS
	0.5 μm SiGe
	$f_T = 47$ GHz
Total die size	2.3 mm × 2.3 mm
Power supply	3.3 V ± 10%
Power consumption	1.0/1.2 W (DC total)
Output HF power	40 mW
Operational bandwidth	20 MHz
Sampling frequency	$f_s = 0.6/1.2$ GHz
Maximal SNR	50 dB
(resBW = 20 kHz)	
Maximal DR	47 dB
$f\Sigma\Delta$ Tunability range	185–289 MHz
Resonator range	60–295 MHz

Source: From R. Sobot, S. Stapleton, and M. Syrzycki. Fractional sigma–delta modulator in SiGe. In *Proceedings of IEEE Canadian Conference on Electrical and Computer Engineering*, 2007, pp. 530–534, April 2007.

CT BP $f\Sigma\Delta$ modulator in combination with Manchester coder and decoder [34]. Second, a combination of a PA and CT BP $f\Sigma\Delta$ modulator in either open or closed loop configuration is presented. The ability to accomodate for the PA propagation delay during the loop filter design phase, and therefore include the PA inside the loop, opens up the possibility for further integration of a transmitting path [19].

16.6.1 FRACTIONAL DELAY $\Sigma\Delta$ UPCONVERTER

A software-defined radio is one of the most active research areas in modern circuit design. Specifically, an efficient way to continuously program the input/output RF converter needs to be

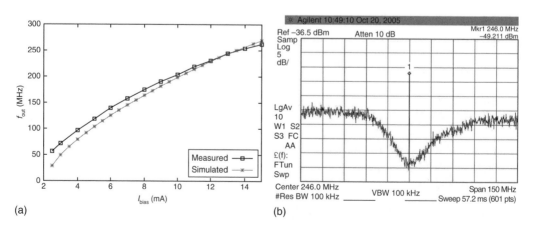

(a)   (b)

FIGURE 16.11 (a) A comparison of measured and simulated control currents for the internal resonator. (From Sobot, R., PhD thesis, Simon Fraser University, 2005.) (b) Measured frequency response for input signal at 246 MHz. (From Sobot, R., Stapleton, S., and Syrzycki, M., *Proceedings of IEEE Canadian Conference Electrical Computer Engineering*, 2007, pp. 530–534, 2007.)

developed. A number of researches have found that a pulse-width-modulated (PWM) signal has potential for efficient RF transmission [25]. More recently, ΣΔ modulators were used for supporting both downconversion of an RF signal [37] and upconversion [17,20] of an IF signal. However, there are still several unsolved problems related to application of ΣΔ modulators for an IF signal upconversion.

First, carrier frequencies required by modern communication systems are on the order of multiples of GHz. The implication is that the ΣΔ circuits, if used as in the references above, have to be sampled at least twice the signal rate, most often four times. Subsequently, application of the traditional z-domain ΣΔ modulators is severely limited. The CT class of modulators is somewhat better suited for the task. However, the loop filter has to be implemented either with passive LC circuits, which is not suitable for IC implementation, or with on-chip low-Q inductors, which necessitates a Q-enhancement circuit. Alternatively, a g_m–C filter implementation can be used with a cost of reduced working frequency and increased nonlinearity.

Second, a ΣΔ modulator by itself is a frequency multiplier, which generates predictable images of the input signal. Even though that SNRs of the images and the input signal are close, the image power levels decline rather quickly.

Third, the loop delay associated with CT ΣΔ modulators presents additional difficulty especially when additional circuit blocks, such as a modulator and a PA, are added inside the loop.

The circuit architecture presented in this section addresses all three types of the problems discussed above. Further, it demonstrates a potentially big advantage of using tunable CT BP fΣΔ modulators, in combination with an encoding technique, for IF signal upconversion. In Figure 16.12, a block diagram of a fractional delay ΣΔ upconverter is shown. The first novelty of the circuit is the addition of two alien blocks inside the loop, namely Manchester encoder and decoder. If a PA is included as the last block of the encoder then $me[n]$ signal is ready to enter the antenna-interfacing circuit. The second novelty is in application of CT BP fΣΔ loop filter [35] to compensate for the additional delays introduced by the alien circuits.

The important advantage of architecture in Figure 16.12 is that the fΣΔ upconverter allows the loop filter to operate at a much lower variable IF frequency than the RF output frequency. This advantage, aside from relaxing the tuning problem, translates into greatly reduced circuit implementation requirements. Also, the fΣΔ may be tuned so that IF and the half sampling frequency $f_s/2$ are closer, with further reduction in circuit complexity, fΣΔ loop stability requirements, and power. Limiting factor to this approach is the loop stability which must obey the Lee's rule [21].

A mixed-signal simulator, within a standard analog IC design environment, was used to perform verification of the proposed topology. There are a number of possible combinations for the sampling frequency f_s and $N f_s$, which produced the desired RF image of the input signal. The simulated IF frequency of 100 MHz and the sampling frequency used in this example are $f_s = 448$ MHz and $N = 5$, shown in Figure 16.13 from DC to $f_s/2$. The frequency spectrum of upconverted WCDMA RF output signal $y[n]$ is shown in Figure 16.13 from DC to Nf_s. The small zoom-in window shows

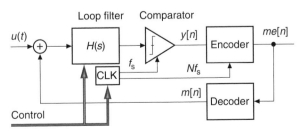

FIGURE 16.12 Block diagram of a fractional delay ΣΔ upconverter. (From Sobot, R., Stapleton, S., and Syrzycki, M., *IEE Electron. Lett.*, 41, 15, November 2005.)

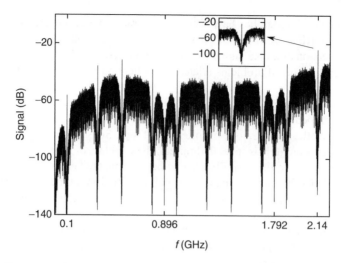

FIGURE 16.13 Frequency–domain plot of a WCDMA 2.14 GHz RF signal, *me*[*n*]. (From Sobot, R., Stapleton, S., and Syrzycki, M., *IEE Electron. Lett.*, 41, 16, November 2005.)

frequency spectrum around the $f_{RF} = 2.24\,\text{GHz} - 100\,\text{MHz} = 2.14\,\text{GHz}$. Finally, the time delay of the alien circuits is estimated to be $T/8$, which is accounted for during the design of the $H(s)$ transfer function. The Manchester modulation suppresses the DC component while enhancing the RF signal.

16.6.2 $f\Sigma\Delta$-Based PA

An architecture for power amplification with high power efficiency is presented. The simulation of a class-S amplifier in combination with CT BP $f\Sigma\Delta$ modulator shows power-added efficiency (PAE) of 40.1% and SNR = 60 dB for a two-tone signal spaced at 4.64 MHz. A WCDMA signal with peak-to-average ratio (PAR) of 8.7 dB demonstrates PAE = 16.6% and SNR = 42 dB.

A working version of CT BP $f\Sigma\Delta$ modulator, which may include encoding upconverter, is used with a CMOS class-S PA designed in 0.5 μm/$f_T = 47$ GHz BiCMOS SiGe technology (Figure 16.14), while Figure 16.15 shows a block diagram of a fractional $\Sigma\Delta$ class-S amplifier. Depending upon the frequency range of operation, an encoding upconverter [3] can be added as the output stage of the modulator. A continuous time signal $u(t)$ is converted by the CT BP $f\Sigma\Delta$ modulator into an NRZ PWM signal $y[n]$, which is then used to drive class-S PA. Out-of-band switching noise is removed with the BP filter so that RF(t) signal is created and delivered into the antenna.

A mixed-signal simulator, within a standard analog IC design environment, was used to perform verification of the proposed topology. For purposes of this experiment, the CT BP $f\Sigma\Delta$ modulator is sampled with sampling frequency $f_s = 800$ MHz and NTF notch was set at 181 MHz, i.e., $f_s/4.42$.

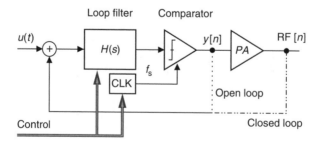

FIGURE 16.14 Block diagram of a power amplifier with $f\Sigma\Delta$ modulator. (From Sobot, R., PhD thesis, Simon Fraser University, 2005.)

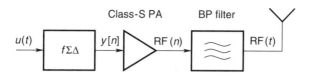

FIGURE 16.15 Block diagram of a fractional $\Sigma\Delta$ class-S PA. (From Sobot, R., PhD thesis, Simon Fraser University, 2005.)

First, two-tone test is performed with two signals centered around $f_0 = 181$ MHz with separation of ±2.32 MHz from the center. This configuration achieved SNR = 60 dB within BW = 10 MHz, where the third-order intermodulation product (IM3) is at 41 dB below the signal level. PAE is found to be 40.1%. Second, WCDMA signal centered at the same frequency f_0, with a PAR of 8.7 dB, SNR = 42 dB within BW = 10 MHz. In this case, PAE is 16.6%. The currently reported PAE numbers for class-AB amplifiers are in the range of 30%–44% [5].

16.7 SUMMARY

In this chapter, the basics of $\Sigma\Delta$ modulation were first introduced, followed by the development of an analytical model for a CT BP $f\Sigma\Delta$ modulator transfer function. Then, a mixed-signal behavioral modeling technique was presented, which facilitated the transistor level design of the modulator prototype. Details of the tuning coefficient mapping from the analytical model to the behavioral model and from the behavioral model to the physical model were shown. A short review of the experimental results was followed by a discussion of possible applications that would benefit from the properties of the $f\Sigma\Delta$ modulators.

REFERENCES

1. S.H. Ardalan and J.J. Paulos. An analysis of nonlinear behavior in delta–sigma modulators. *IEEE Transactions on Circuits and Systems*, CAS-34:593–603, June 1987.
2. J.A. Cherry and W.M. Snelgrove. Excess loop delay in continuous-time delta–sigma modulators. *IEEE Transactions on Circuits and Systems II: Analog and Digital Signal Processing*, 46(4):376–389, April 1999.
3. A.E. Cosand, J.F. Jensen, H.C. Sun, and C.H. Fields. IF-sampling fourth-order bandpass $\Delta\Sigma$ modulator for digital receiver applications. *IEEE Journal of Solid-State Circuits*, 39(10):1633–1639, October 2004.
4. C.C. Cutler. Transmission system employing quantization. U.S. Patent 2,927,962, March 1960.
5. Y. Ding and R. Harjani. A high-efficiency CMOS +22 dBm linear power amplifier. *IEEE Journal of Solid-State Circuits*, 40(9):1895–1900, September 2005.
6. H.J. Dressler. Interpolative bandpass A/D conversion—Experimental results. *IEE Electronics Letters*, 26(20):1652–1653, September 1990.
7. W. Gao, J.A. Cherry, and W.M. Snelgrove. A 4 GHz fourth-order SiGe HBT bandpass $\Delta\Sigma$ modulator. Symposium on VLSI Circuits, 1998. *Digest of Technical Papers*, pp. 174–175, June 1998.
8. W. Gao and W.M. Snelgrove. A 950-MHz IF second order integrated LC bandpass delta–sigma modulator. *IEEE Journal of Solid-State Circuits*, 33(5):723–732, May 1998.
9. B. Gilbert. The multi-tanh principle: A tutorial overview. *IEEE Journal of Solid-State Circuits*, 33(1):2–17, January 1998.
10. F. Henkel, U. Langmann, A. Hanke, S. Heinen, and E. Wagner. A 1-MHz-bandwidth second-order continuous-time quadrature bandpass sigma–delta modulator for low-IF radio receivers. *IEEE Journal of Solid-State Circuits*, 37(12):1628–1635, December 2002.
11. L. Hernandez and S. Paton. Continuous time sigma–delta modulators with transmission line resonators and improved jitter and excess loop delay performance. *IEEE International Symposium on Circuits and Systems, ISCAS 2003*, pp. 25–28, May 2003.
12. D.H. Horrocks. A second-order oversampled sigma–delta modulator for bandpass signals. *Proceedings of the IEEE International Symposium on Circuits and Systems 1991, ISCAS '91*, pp. 1653–1656, June 1991.

13. M. Hotta, K. Maio, N. Yokozawaand, T. Watanabe, and S. Ueda. A 150-mW, 8-bit video-frequency A/D converter. *IEEE Journal of Solid-State Circuits*, 21(2):318–323, April 1986.

14. S.A. Jantzi, R. Schreier, and W.M. Snelgrove. A bandpass $\Delta\Sigma$ A/D convertor for a digital AM receiver. *Proceedings of the International Conference on Analogue to Digital and Digital to Analogue Conversion*, pp. 75–80, September 1991.

15. S.A. Jantzi, R. Schreier, and W.M. Snelgrove. Bandpass sigma–delta analog-to-digital conversion. *IEEE Transactions on Circuits and Systems*, 38(11):1406–1409, November 1991.

16. S.A. Jantzi, W.M. Snelgrove, and P.F. Ferguson. A fourth-order bandpass sigma–delta modulator. *Proceedings of the IEEE Custom Integrated Circuits Conference*, pp. 16.5.1–16.5.4, May 1992.

17. A. Jayaraman, P. Asbeck, S. Beccue, K.C. Wang, and K. Nary. Bandpass delta–sigma modulator with 800 MHz center frequency. *Gallium Arsenide Integrated Circuit (GaAs IC) Symposium*, pp. 15–17, October 1997.

18. D.A. Johns and K. Martin. *Analog Integrated Circuit Design*. John Wiley, New York, 1997.

19. T. Johnson, R. Sobot, and S. Stapleton. CMOS RF class D amplifier with a fractional sigma–delta bandpass modulator. *Microelectronics Journal*, 38(3):439–446, March 2007.

20. J. Keyzer, J. Hinrichs, A. Metzger, M. Iwamoto, I. Galton, and P. Asbeck. Digital generation of RF signals for wireless communications with band-pass delta–sigma modulation. *Microwave Symposium Digest, 2001 IEEE MTT-S International*, Vol. 3, pp. 2127–2130, May 2001.

21. W.L. Lee. A novel higher order interpolative modulator topology for high resolution oversampling A/D converters. Master's thesis, Massachusetts Institute of Technology, Cambridge, MA, 1987.

22. R. Maurino and P. Mole. A 200-MHz IF 11-bit fourth-order bandpass $\Delta\Sigma$ ADC in SiGe. *IEEE Journal of Solid-State Circuits*, 35(7):959–967, July 2000.

23. S. Norsworthy, R. Schreier, and G. Temes. *Delta–Sigma Data Convertors: Theory, Design and Simulation*. IEEE Press, Washington, DC, 1997.

24. A.V. Oppenhiem and R.W. Schafer. *Digital-Time Signal Processing*. Prentice-Hall, Englewood Cliffs, NJ, 1999.

25. F.H. Raab. Radio frequency pulsewidth modulation. *IEEE Transactions on Communications*, 21:958–966, August 1973.

26. G. Raghavan, J.F. Jensen, J. Laskowski, M. Kardos, M.G. Case, M. Sokolich, and S. Thomas III. Architecture, design, and test of continuous-time tunable intermediate-frequency bandpass delta–sigma modulators. *IEEE Journal of Solid-State Circuits*, 36(1):5–13, January 2001.

27. G. Raghavan, J.F. Jensen, R.H. Walden, and W.P. Possey. A bandpass $\Sigma\Delta$ modulator with 92 dB SNR and center frequency continuously programmable from 0 to 70 MHz. *IEEE International Solid-State Circuits Conference*, pp. 214–215, February 1997.

28. R. Schreier, J. Lloyd, L. Singer, D. Paterson, M. Timko, M. Hensley, G. Patterson, K. Behel, J. Zhou, and W.J. Martin. A 50 mW bandpass $\Sigma\Delta$ ADC with 333 kHz BW and 90dB DR. *IEEE International Solid-State Circuits Conference, ISSCC 2002*, 1:216–217, April 2002.

29. R. Schreier and M. Snelgrove. Bandpass sigma–delta modulaiton. *IEE Electronics Letters*, 25(23): 1560–1561, November 1989.

30. O. Shoaei. Continuous-time delta–sigma A/D converters for high speed applications. PhD thesis, Carleton University, Ottawa, Canada, 1995.

31. O. Shoaei and W.M. Snelgrove. Design and implementation of a tunable 40MHz–70MHz Gm–C bandpass $\Delta\Sigma$ modulator. *IEEE Transactions on Circuits and Systems II: Analog and Digital Signal Processing*, 44(7):521–530, July 1997.

32. R. Sobot. Design methodology for continuous-time bandpass sigma–delta modulators. PhD thesis, Simon Fraser University, Burnaby, Canada, 2005.

33. R. Sobot, S. Stapleton, and M. Syrzycki. Tunable center frequency continuous time bandpass $\Delta\Sigma$ modulators for wireless transceivers. *Proceedings of IEEE Vehicular Technology Conference 2004-Fall, VTC2004*-Fall, Vol. 7, pp. 4956–4960, September 2004.

34. R. Sobot, S. Stapleton, and M. Syrzycki. A fractional delay $\Sigma\Delta$ upconverter. *IEE Electronics Letters*, 41(23):15–16, November 2005.

35. R. Sobot, S. Stapleton, and M. Syrzycki. Tunable continuous time bandpass sigma–delta modulators with fractional delays. *IEEE Transactions on Circuits and Systems-I: Regular Papers*, 53(2):264–273, February 2006.

36. R. Sobot, S. Stapleton, and M. Syrzycki. Fractional sigma–delta modulator in SiGe. *Proceedings of IEEE Canadian Conference on Electrical and Computer Engineering*, 2007, pp. 530–534, April 2007.
37. H. Tao and J.M. Khoury. A 400-Ms/s frequency translating bandpass sigma–delta modulator. *IEEE Journal of Solid-State Circuits*, 34(12):1741–1752, December 1999.
38. A.M. Thurston, M.J. Hawksford, and T.H. Pearce. Bandpass implementation of the sigma–delta A–D conversion technique. *International Conference on Analogue to Digital and Digital to Analogue Conversion*, 1991, pp. 81–86, September 1991.
39. G. Tröster, H.J. Dressler, J. Arndt, H.J. Golberg, W. Schardein, K. Schopp, P. Sieber, A. Wedel, and E. Zocher. An interpolative bandpass converter on a 1.2 μm BiCMOS analog/digital array. *Symposium on VLSI Circuits, 1992. Digest of Technical Papers*, pp. 102–103, June 1992.
40. J.A.E.P. van Engelen, R.J. van de Plassche, E. Stikvoort, and A.G. Venes. A sixth-order continuous time bandpass sigma–delta modulator for digital radio IF. *IEEE Journal of Solid-State Circuits*, 34(12):1753–1764, December 1999.

Part IV

Circuits for Communications

17 Low-Voltage Nanometer-Scale CMOS RF Front-End Block Design Employing Magnetic Feedback Techniques

Yannis Papananos, Georgios Vitzilaios, and Gerasimos Theodoratos

CONTENTS

17.1 INTRODUCTION

A broad range of modern high-volume consumer applications require the availability of low-power and wireless microsystems. These systems, among others, should conciliate low-voltage, low-power operation with reduced overall dimensions and versatility. In addition, recent advances in complementary metal oxide semiconductor (CMOS) processes lead to nanometer-scale transistor structures, suitable for high-frequency operation [1–4]. Design in nanometer scale allows for the supply voltage of the digital circuitry to be reduced to or below 1 V, which imposes many challenges in the design of the analog part of the system. The main reason is that classical design topologies become impractical when the lowest possible power supply is desired [1]. The abovementioned reasons and conditions set the framework for the design and implementation of up-to-date wireless systems.

It is thus obvious that low-noise amplifiers (LNA) and mixers, being critical blocks in virtually every modern communication system, are affected by the aforementioned constraints, and the introduction of new design topologies becomes important. In this chapter, we will present design approaches suitable for low-voltage operation and we will present techniques to improve circuit performance that may operate with power supplies below 1 V, focusing on the linearity improvement of both LNA and mixer designs. The chapter examines design limitations for low-voltage supply and shows how magnetic feedback can be used for improving the reverse isolation and linearity of a low-voltage LNA and for enhancing the noise and linearity performance of a mixer operating at 1 V.

17.2 ADVANCES IN LOW-VOLTAGE CMOS DESIGN

Before examining any particular circuit, we will briefly examine some of the different approaches used by the designers, both in industry and academia, in order to have functional designs in power supply levels that can be as low as 0.8–0.5 V. The first approach relies on the recent advances in CMOS technology: the minimum width of the MOS transistor is reduced in the nanometer scale, and technologies with minimum feature size of 40–30 nm have already emerged. These new devices offer, among others, two advantages. The threshold voltage is greatly reduced, so that the designer can "stack" more transistors in a reduced power supply and due to reduced parasitics, high-frequency operation is facilitated. These characteristics allow for well-known topologies, which require a transistor stack, for example, cascode LNAs, to remain functional. Such designs are frequently referred in the literature [3]. This approach relies on technology advances and depends on technology maturity, and does not aim at design limits, which require the lowest possible device stack.

An alternative approach, which we will follow in this chapter, depends on a different strategy. The designer utilizes recent advances in technology but aims for topologies that require the minimum number of stacked devices, in order to reach the lowest possible limit of the power supply. For this reason, new techniques like magnetic feedback [1] or forward body biasing [4] have been introduced to solve problems that arise from lack of cascode devices. The magnetic feedback approach will be thoroughly examined.

17.3 SINGLE-TRANSISTOR LNA TOPOLOGIES

17.3.1 LIMITATIONS OF SINGLE-TRANSISTOR LNAs

One of the most recognizable circuits in radiofrequency (RF) LNA design is the cascode LNA topology, presented in Figure 17.1a, which has been used in the past decades in the vast majority of the commercial RF front-ends. The reason is that the topology is robust, reliable, and relatively simple to design. The main advantage of the topology is the fact that the gain is concentrated in a "common gate" stage (transistor M_2), minimizing the voltage swing at the drain terminal of transistor M_1. This effectively means that since the voltage swing at the drain terminal of M_1 is low, the signal leakage through parasitic paths to the gate terminal will be small and thus the input and output nodes of the topology are isolated.

Unfortunately, when the lowest possible supply voltage is targeted, the topology becomes nonoptimal, since the cascode transistor occupies voltage headroom in order for the transistor to operate in saturation and the LNA may enter compression. If the power supply is the limiting factor, then the designer has to rely on single-transistor LNA topologies. In its simplest form, such an amplifier is presented in Figure 17.1b, which is effectively a single-transistor amplifier with tuned load (inductor L_2 and capacitor C_1) and inductive degeneration (inductor L_1). In this simple form, it is almost certain

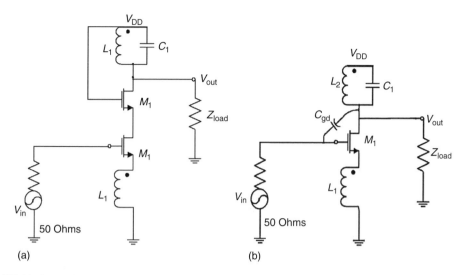

FIGURE 17.1 (a) Simple cascode LNA and (b) single-transistor LNA. (From Vitzilaios, G., Popanamos, Y., Theodoratos, G., and Vasilopoulos, A., IEEE TCAS-II, 53, 971, 2006. With permission.)

that the amplifier is not operational. The reason is that the reverse isolation of the topology is greatly deteriorated in comparison to its cascode counterpart.

Examining the circuit more closely, we can easily see that the gate-drain capacitance of transistor M_1 (C_{gd}) is creating a noninverting signal path between the input and output nodes of the amplifier. For more applications, the inverse signal flow will make the amplifier unable to meet the reverse isolation specs. In addition, this capacitance is reducing the amplifier gain and bandwidth (this phenomenon is also enhanced due to the Miller effect). It is thus important to introduce ways to negate the capacitance effects.

17.3.2 REVERSE ISOLATION ENHANCEMENT UTILIZING MAGNETIC FEEDBACK

The limited reverse isolation of the amplifier of Figure 17.1b can be effectively increased by magnetic feedback, which can be implemented by employing integrated transformers. Transformers can be used to provide magnetic feedback that can be modeled as shown in Figure 17.2. The nature of the feedback depends on the direction of the currents I_1 and I_2. In Figure 17.2, the transformer provides positive feedback at the input. The mutual conductance M is given by $M = k\sqrt{L_1L_2}$, where k is the coupling coefficient. In Figure 17.3a, the elegant design approach of Cassan and Long [1] is presented. The LNA has a single-transistor amplifier core, as shown in Figure 17.3b, but now the load and degeneration inductor form a transformer that can be used to form a negative feedback

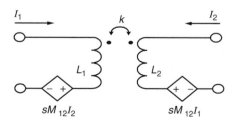

FIGURE 17.2 Application of feedback through a transformer. (From Vitzilaios, G., Popanamos, Y., Theodoratos, G., and Vasilopoulos, A., IEEE TCAS-II, 53, 971, 2006. With permission.)

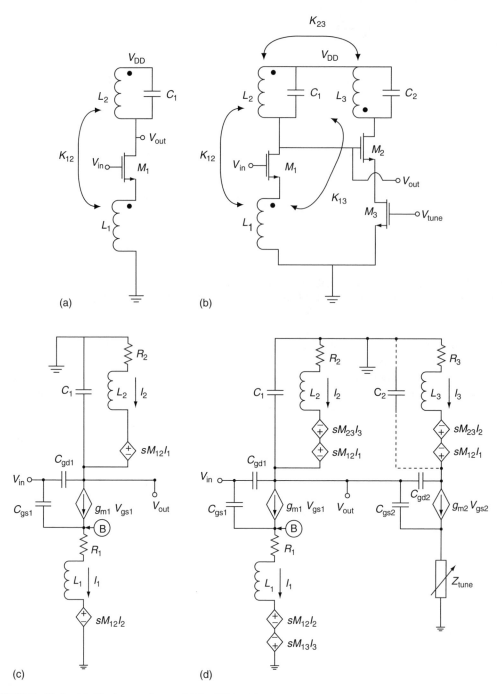

FIGURE 17.3 (a) Single-transformer LNA (STL) [1], (b) triple transformer LNA (TTL), [2] (c) equivalent small-signal model of the STL, and (d) equivalent small-signal model of the TTL. (From Vitzilaios, G., Popanamos, Y., Theodoratos, G., Vasilopoulos, A., IEEE TCAS-II, 53, 971, 2006. With permission.)

loop, whose beneficial effects are used for reverse isolation enhancement. The analysis of the loop is complex and has to rely on signal flow charts. Instead, we will proceed through a qualitative analysis to explain how the negative feedback loop has beneficial effects with regard to the reverse isolation enhancement.

Examining the small-signal equivalent circuit of Figure 17.3c, we can see that the current-dependent voltage source $sM_{12}I_2$, which models the feedback action of the transformer, is applied at the source terminal of the amplifying transistor. Since the current I_2 is a function of the output voltage V_{out} and by considering the polarity of the dependent source, we can assume that there will be a signal path, which is a function of V_{out} to the source terminal that can be designed in such a manner to neutralize the passive feedforward (between nodes V_{in} and V_{out}) due to the parasitic capacitance C_{gd} and thus enhance the reverse isolation of the amplifier. The neutralizing signal is a function of the transformer parameters (inductance values and coupling coefficient) and so the transformer may be designed in such a manner to achieve the neutralizing condition. Such a design has been successfully fabricated and measured, and the reader is refered to Ref. [1] for the full mathematical analysis.

An alternative method of utilizing magnetic feedback to enhance the reverse isolation of the LNA is shown in Figure 17.3b. The topology is referred to as a triple transformer LNA (TTL), since we now have three magnetic feedback loops to perform various functions. The neutralization principle is the same as described above, but this topology allows for higher gain and is more suitable to drive on-chip loads. The proposed LNA employs multiple inductive transformer feedback to introduce negative and positive feedback at specific nodes of the topology. This permits for both gain and reverse isolation to be set at high values and be almost independent of each other. The use of multiple feedback relaxes the values of the inductor-coupling coefficients and offers design flexibility regarding the values of the individual inductors.

17.3.3 TRIPLE TRANSFORMER LNA ANALYSIS

Before analyzing the topology, we will characterize the devices that are used. Transistor M_1 is the amplifying transistor, inductors L_1 and L_2 are the degeneration and load inductors, respectively, and C_1 represents the equivalent load of the following mixer input stage. Inductor L_3, capacitance C_2, and transistors M_2 and M_3 form a "compensating" stage [5]. The three-inductor transformer is used to provide feedback, electronically controlled by V_{tune}. The coupling coefficient k_{23} represents positive feedback at the output node, which is used to increase the output impedance and thus the gain of the amplifier. Coupling coefficients k_{12} and k_{13} represent negative feedback at the source of the amplifying transistor, which in combination with the inductive degeneration action of inductor L_1 are used to increase the reverse isolation of the amplifier. It should be noted that as positive and negative feedback is applied on different nodes, there is no mutual cancellation. The small-signal equivalent model is depicted in Figure 17.3d, where transistor M_3, operating in the triode region of operation, is represented by the variable load Z_{tune}.

$$\begin{bmatrix} \dfrac{g_{m2}}{1+g_{m2}Z_{tune}} & -1 & 0 & 0 & 0 \\ 1 & -sM_{23} & R_2+sL_2 & sM_{12} & 0 \\ s(C_{gd1}+C_1) & 0 & -1 & 0 & -g_{m1} \\ 0 & sM_{13} & -sM_{12} & -(sL_1+R_1) & 1 \\ 0 & 0 & 0 & 1 & g_{m1}+sC_{gs1} \end{bmatrix} \begin{bmatrix} V_{out} \\ I_3 \\ I_2 \\ I_1 \\ V_B \end{bmatrix} = \begin{bmatrix} 0 \\ 0 \\ V_{in}(sC_{gd1}-g_{m1}) \\ 0 \\ V_{in}(g_{m1}+sC_{gs1}) \end{bmatrix}$$

(17.1)

As stated above, the mathematical analysis of the multiple feedback loops is tedious. For the sake of completeness though, we provide a set of equations that can be solved to provide the transfer function of the amplifier, given in Equation 17.1 and referred to as the small-signal model of Figure 17.3d.

The parasitic capacitances of transistor M_2 are omitted for reasons of simplicity and capacitor C_2, which is used for high-frequency loop stability (to be examined later) is neglected in a first-order analysis.

The small-signal model of Figure 17.3d reveals the negative feedback that is used for the reverse isolation enhancement, represented by the current-controlled voltage sources $sM_{12}I_2$ and $sM_{13}I_3$.

More precisely, examining the voltage at node B of Figure 17.3b and 17.3d and taking into account the relative phase of the currents, the source of the amplifying transistor is at a voltage level:

$$v_{source} = I_1R_1 + I_1sL_1 + sM_{12}L_1 + sM_{13}I_3 \tag{17.2}$$

$$v_{source} = I_1R_1 + I_1sL_1 + sM_{12}I_2 \tag{17.3}$$

for the TTL (Equation 17.2) and STL (Equation 17.3), respectively. This voltage is increased with respect to the voltage of inductive degeneration alone, which represents negative feedback. This is the feedback that is used for C_{gd} neutralization and reverse isolation enhancement. In the TTL, the required feedback is attained by the combined effect of two feedback loops ($sM_{12}I_2$ and $sM_{13}I_3$), allowing the designer to optimize the coupling coefficients and the compensating current I_3 in order to simultaneously satisfy the reverse isolation and gain specifications of the design.

One of the most important characteristics of the TTL is the fact that it provides a positive feedback loop that effectively increases the output impedance and the gain of the amplifier. In order to prove this statement, we will examine the LNA response at the output node. For a single-transistor amplifier with no feedback, the input impedance of the load-resonant tank $L_2 // C_1$ can be found from the small-signal equivalent circuit in Figure 17.3c, assuming that the current-controlled voltage sources $sM_{12}I_2$ and $sM_{12}I_1$ are zero. By inspection

$$\frac{v_{out}}{i_2} = \frac{R_2 + sL_2}{s^2L_2C_1 + sR_2C_1 + 1} \tag{17.4}$$

When feedback is in place, Equation 17.4 becomes

$$\frac{v_{out}}{i_2} = \frac{R_2 + sL_2}{s^2L_2C_1 + s(R_2C_1 - \alpha M_{23}I_3 + \beta M_{12}I_1) + 1} \tag{17.5}$$

$$\frac{v_{out}}{i_2} = \frac{R_2 + sL_2}{s^2L_2C_1 + s(R_2C_1 + \gamma M_{12}I_1) + 1} \tag{17.6}$$

in the cases of the TTL (Equation 17.5) and STL (Equation 17.6), respectively. In Equation 17.5, α and β are coefficients relating I_3 and I_1 to the output voltage. The exact values of α and β can be calculated by solving Equation 17.1. In Equation 17.6, γ relates I_1 to the output voltage.

The examination of the second term in the denominator of the abovementioned equations reveals that the feedback represented by the $\beta M_{12}I_1$ and $\gamma M_{12}I_1$ terms is an unwanted effect. It reduces the quality factor of the load inductor, limiting the gain and the output impedance of the topology and has no contribution to the reverse isolation enhancement. In the TTL, the feedback represented by the $-\alpha M_{23}I_3$ term can be designed to totally cancel $\beta M_{12}I_1$. If necessary, the feedback can be designed so that $\alpha M_{23}I_3 > \beta M_{12}I_1$. This represents Q-enhancement of the load inductor and provided that it is a low-Q device, the gain can be enhanced with acceptable bandwidth. The gain enhancement is controlled by the current I_3, set by V_{tune} at the gate of transistor M_3. It should be noted that using magnetic coupling to enhance the gain has minimal effect on the topology's noise figure (NF). This is due to the fact that the thermal noise of the parasitic resistance R_3 of the compensating circuit inductor L_3 is coupled to the output with gain <1, since $k_{23} < 1$. Given that the value of R_3 is inherently small, the effect will be minimal. In addition, as the compensating circuit's current is low (1.2 mA) and transistor's M_3 (operating at the triode region) channel length is short, the noise contributions of M_2 and M_3 when coupled to the output are small and thus the compensating stage noise contribution is kept low (4% of the total output noise).

17.3.3.1 Stability Issues

In order to ensure the stability of the circuit, from Equation 17.5 we derive that the circuit is stable when

$$\alpha M_{23}I_3 \leq R_2 C_1 + \beta M_{12}I_1 \qquad (17.7)$$

Since the Q-enhancement (achieved when $\alpha M_{23}I_3 > \beta M_{12}I_1$) is electronically tunable, it is possible to satisfy Equation 17.7 under component parameter, supply voltage, and temperature variations. In order to take into account the nonlinear nature and the frequency dependence of the feedback loop, an additional stabilization method was adopted, as described in Ref. [5]. The technique relies on resonating the inductor of the compensating circuit with a parallel capacitor C_2, as shown in Figure 17.3d. This is done in order to reduce the small-signal compensating current flowing into inductor L_3 at high frequencies, and to reduce the feedback at these frequencies, thus stabilizing the circuit. The stabilization method has an effect on the L_2 resonating frequency [5], which is given by

$$f_{\text{res}} \approx \frac{1}{\sqrt{L_1(C_1 + C_2) + C_1 C_2 R_1 R_2}} \qquad (17.8)$$

17.3.3.2 Multiple Transformer Layout

The design and layout of a triple integrated transformer is a challenging task, in order to achieve optimum high-frequency performance with minimum silicon footprint. The proposed transformer layout is depicted in Figure 17.4. This figure represents the footprint of the transformers used in the differential version of the LNA. It is obvious that the structure is extremely area-efficient, as it incorporates six inductors in the effective area of one differential inductor. The design was made

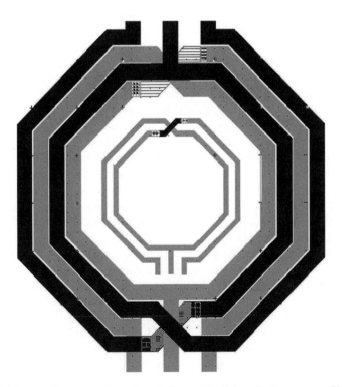

FIGURE 17.4 Triple transformer topology layout. (From Vitzilaios, G., Popanamos, Y., Theodoratos, G., Vasilopoulos, A., IEEE TCAS-II, 53, 971, 2006. With permission.)

easier by the fact that the coupling coefficients k_{13} and k_{12} are relatively small (approximately 0.3). This way, inductor L_1 could be placed inside inductors L_2 and L_3 without overlapping, facilitating a high-frequency design.

17.3.3.3 Typical Performance

The designs of the STL and the TTL were simulated using device files from the 0.13 μm IBM CMOS 8-metal technology. A full-set typical QFN package parasitics was used, while the transformer parameters were extracted using EM software. The circuits were loaded using extracted parameters of a mixer designed for wireless local area network (WLAN) applications. The results presented refer to conjugate-matched designs to a 50 Ω input. The amplifying transistor, the load, and degeneration inductor and the amplifying stage bias currents are the same in both cases. Only the coupling coefficients and the compensating current I_3 are treated as design variables.

Figure 17.5 illustrates a comparison simulation of the TTL and the STL gain response, in the case where the amplifiers achieve the same reverse isolation. The TTL gain enhancement is 2 dB, while the matching network gain is 1 dB less. If necessary, the compensating stage may provide an extra 1.5 dB gain, with 1 dB reverse isolation deterioration. In the same figure, a comparison of the reverse isolation and NF of the STL and TTL is presented, under the same gain conditions. The TTL reverse isolation performance is better by 3.5 dB, and the NF deterioration is minor (0.06 dB10) due to the reasons explained in Section 17.3.3.

17.3.4 ALTERNATIVE METHOD FOR REVERSE ISOLATION ENHANCEMENT

A third method for LNA reverse isolation enhancement that also relies on magnetic feedback is presented in Figure 17.6. The LNA consists of the amplifying transistor M_1, the load inductor L_2, and the degeneration inductor L_1, while resonance is achieved by capacitor C_1. The reverse isolation of such a topology is not adequate for most wireless standards. The reverse isolation may be greatly improved by introducing a neutralization branch, comprising inductor L_3 and transistor M_2. It must be noted that the gate and source terminals of M_2 are short-circuited. The drain terminal of M_2 is connected to the gate terminal of M_1, while V_b biases the device.

The principle of operation can be explained by examining the schematic of Figure 17.7. The neutralization effect is based on the cross-coupled capacitors C_n, as shown in Figure 17.7 [1]. The method relies on additional signal paths (comprising neutralization capacitors C_n) so that the net signal flow through the capacitance C_{gd} of the amplifying transistors is zero. Since the drain voltages of the differential pair are 180° phase-shifted, the current through C_N is in opposite polarity with the current flowing through C_{gd}, and provided that $C_{gd} = C_N$, the two currents are equal in magnitude. Under these conditions, neutralization is achieved. The technique requires differential drive and precise matching of C_{gd} and C_N. If not, stability problems may occur, since the feedback is positive if $C_{gd} < C_N$. Thus, neutralization is only partial in order to keep the circuit stable under process, temperature, and power supply variations. In the topology of Figure 17.6, a differential drive is not necessary, since the 180° out-of-phase signal is created by utilizing the appropriate direction of the currents flowing through inductors L_3 and L_2. A simple first-order analysis can be done by examining the small-signal equivalent model of the LNA shown in Figure 17.8. In this figure, the voltage-controlled current source (VCCS) representing the transistor M_2 is omitted since the gate and source terminals of the device are short-circuited, while the VCVS representing the feedback from inductor L_3 to L_2 is omitted since the AC signal flowing through M_2 is small compared to the signal flowing at the output branch of the amplifier. The current flowing through the capacitance C_{gd1} is given by

$$i = (V_{out} - V_{in})sC_{gd1} \qquad (17.9)$$

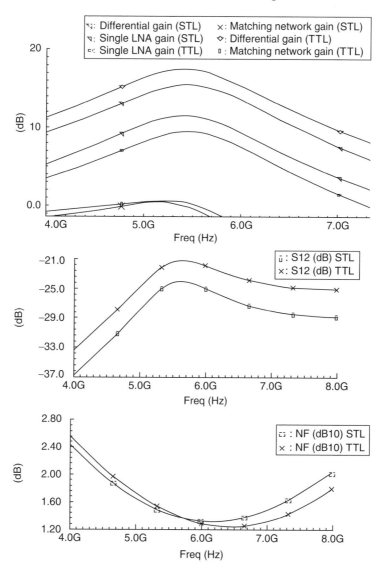

FIGURE 17.5 Simulated gain, reverse isolation, and NF response of the triple transformer LNA and the STL. (From Vitzilaios, G., Popanamos, Y., Theodoratos, G., Vasilopoulos, A., IEEE TCAS-II, 53, 971, 2006. With permission.)

while the current flowing through the capacitance C_{gd2} is

$$i' = (V_A - V_{in})sC_{gd2} = \left(sM_{23}I + i'sL_3 - V_{in}\right)sC_{gd2}$$

$$= \left(sk\sqrt{L_2L_3}I + i'sL_3 - V_{in}\right)sC_{gd2} = \left[\frac{sk\sqrt{L_2L_3}I - V_{in}}{1 - s^2L_3C_{gd2}}\right]sC_{gd2} \qquad (17.10)$$

The currents i and i' have the opposite phase, and so the neutralization effect can be achieved. In order to find the neutralization condition, the oversimplified model of Figure 17.8 is not adequate, and one must consider a high-frequency transistor model. The model must include the extrinsic and intrinsic transistor capacitances. The fact that the short-channel transistors M_1 and M_2 are operating in saturation and cutoff, respectively, must be considered, since this implies that even if the transistors

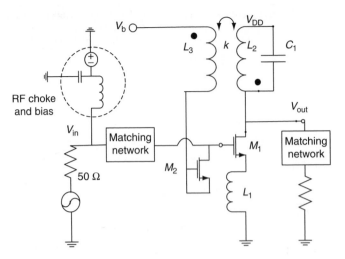

FIGURE 17.6 Reverse isolation enhancement method utilizing an active device. (From Vitzilaios, G. and Papananos, Y., IEEE TCAS-II, 2007. Accepted for publication. With permission.)

M_1 and M_2 are of the same type and size, their corresponding C_{gd} will not be the same [6]. In general

$$C_{gd(1,2)} = C_{gdint(1,2)} + C_{gdext(1,2)} \qquad (17.11)$$

where
   $C_{int()}$ corresponds to the intrinsic capacitance
   $C_{ext()}$ corresponds to the extrinsic capacitance

   The extrinsic gate-drain capacitance is given by [6]

$$C_{gdext(1,2)} = C_{gsext(1,2)} = WC_o'' \qquad (17.12)$$

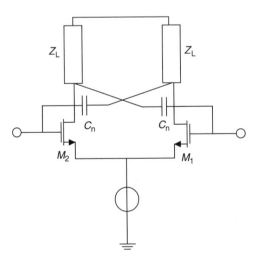

FIGURE 17.7 Principle of operation of the reverse isolation enhancement method. (From Vitzilaios, G. and Papananos, Y., IEEE TCAS-II, 2007. Accepted for publication. With permission.)

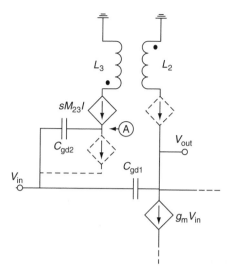

FIGURE 17.8 Small-signal equivalent model of the LNA. (From Vitzilaios, G. and Papananos, Y., IEEE TCAS-II, 2007. Accepted for publication. With permission.)

where C_o'' is a capacitance per unit width of the device, including overlap, fringing, and top contributions [6]. For transistors M_1 and M_2, the extrinsic part, which is geometry-dependent, is matched since the devices are identical. For the intrinsic part, shown in Figure 17.9, the assumption that $C_{gdint1} = 0$ in saturation and strong inversion does not hold, since short-channel devices are considered [6]. This part is not matched with the present setting, since for transistor $M_2 C_{gdint2} = 0$(cutoff) but for transistor $M_1 C_{gdint1} \neq 0$.

Regarding the intrinsic part of M_1, as stated in [6]

$$C_{gdint1} \approx C_{inner\_fringing} = W \frac{2}{\pi} \varepsilon_{Si} \ln \left[1 + \frac{d_j}{t_{ox}} \sin \left(\frac{2}{\pi} \frac{\varepsilon_{ox}}{\varepsilon_{Si}} \right) \right] \qquad (17.13)$$

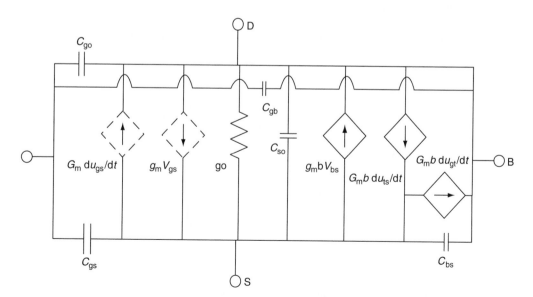

FIGURE 17.9 Intrinsic capacitances in a MOS device. (From Vitzilaios, G. and Papananos, Y., IEEE TCAS-II, 2007. Accepted for publication. With permission.)

where

d_j is the junction depth

ε_{Si} and ε_{ox} are the permittivity of silicon SiO_2, respectively

t_{ox} is the oxide thickness

In order to achieve full neutralization, the designer can use the degree of freedom provided by the integrated transformer since the current flowing through C_{gd2} is a function of k. Appropriate scaling of k would provide the appropriate condition in order for full neutralization to occur, which will arise when the net signal flowing through capacitances C_{gd1} and C_{gd2} sums up to zero at the amplifier input.

Unfortunately, the value of k is limited by layout considerations and cannot be effectively used to fully eliminate the parasitic capacitance. The proposed solution provides an additional mechanism in order to achieve full capacitance cancellation. This is done through the V_b voltage shown in Figure 17.6. Depending on the value of this voltage, and due to the fact that the device is symmetric, the drain and source terminals are interchanged and the device is biased just above weak inversion, in the verge of moderate inversion. In this region of operation, the neutralizing extrinsic capacitance is C_{gsext2}, which, according to Equation 17.12, equals C_{gdext2}. In addition, $g_{m2 \text{ moderate inversion}} \rightarrow 0$ and most importantly $C_{gsint2} \neq 0$. More precisely, using the general capacitance model in [6]:

$$C_{gsint2} = W_2 L_2 C_{ox}{}' \left[\frac{3}{2} + \frac{f(u)}{u} \right]^{-1}$$

$$u = \frac{I_{DS}}{I_Z} \tag{17.14}$$

$$f(u) = \frac{1}{2}\left(\sqrt{1+4u} + 1 \right)$$

where

I_z is the characteristic current in moderate inversion

$C_{ox}{}'$ is the oxide capacitance per unit area

It is thus possible to calculate and set the voltage V_b (and thus set I_{DS}) and achieve the full neutralizing condition. To facilitate calculations, an ideal transformer is assumed and thus the condition would be that the full parasitic capacitances of M_1 and M_2 are the same. If the devices are identical, the condition reduces to

$$C_{gd1} = C_{gs2} \xrightarrow{C_{gdext1}=C_{gsext2}} C_{gdint1} = C_{gdint2} \Rightarrow$$

$$W_2 L_2 C_{ox}{}' \left[\frac{3}{2} + \frac{f(u)}{u} \right]^{-1} = W_1 \frac{2}{\pi} \varepsilon_{Si} \ln\left[1 + \frac{d_j}{t_{ox}} \sin\left(\frac{2}{\pi} \frac{\varepsilon_{ox}}{\varepsilon_{Si}} \right) \right] \tag{17.15}$$

$$g_{m2(\text{verge\_of})\text{moderate\_inversion}} \rightarrow 0$$

Solving for I_{DS}

$$I_{DS} = \frac{\left(\dfrac{W_2 L_2 C'_{ox}}{W_1 \frac{2}{\pi}\varepsilon_{Si}\ln\left[1+\frac{d_j}{t_{ox}}\sin\left(\frac{2}{\pi}\frac{\varepsilon_{ox}}{\varepsilon_{Si}}\right)\right]} + \frac{5}{2} \right)}{\left[\dfrac{W_2 L_2 C_{ox}{}'}{W_1 \frac{2}{\pi}\varepsilon_{Si}\ln\left[1+\frac{d_j}{t_{ox}}\sin\left(\frac{2}{\pi}\frac{\varepsilon_{ox}}{\varepsilon_{Si}}\right)\right]} - \frac{3}{2} \right]^2} I_Z \tag{17.16}$$

In order to demonstrate the effect, a graph of the gain (S_{21}) and reverse isolation (S_{12}) of the LNA as a function of the voltage V_b is plotted in Figure 17.10a. As it is obvious from the S_{21} graph, there is a local minimum where the full capacitance neutralization occurs.

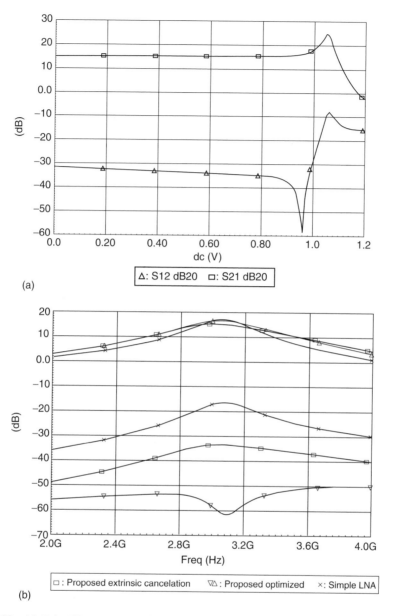

(a)

(b)

FIGURE 17.10 (a) Gain (S21) and reverse isolation (S12) of the LNA as a function of V_b. (b) Gain (S21) and reverse isolation (S12) of the LNA (i) without the proposed technique, (ii) with extrinsic cancellation only, and (iii) with optimum V_b value. (From Vitzilaios, G. and Papananos, Y., IEEE TCAS-II, 2007. Accepted for publication. With permission.)

In Figure 17.10b, a comparison simulation graph is presented that shows the behavior of the amplifier of the gain and reverse isolation of the amplifier when no cancellation occurs, when the extrinsic part of the capacitance is canceled, and when full cancellation is applied. In the last case, it becomes obvious that the effect is significantly better.

At this point, two comments must be made: first, the abovementioned graphs are assuming that the feedback path that is deteriorating the reverse isolation behavior incorporates only the parasitic capacitance C_{gd}, which is not true in a real design, since there are other parasitic signal paths that deteriorate the behavior. Secondly, since the actual cancellation procedure is actually application

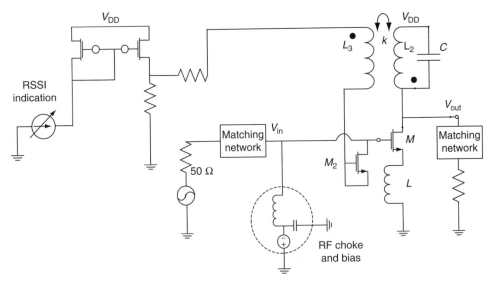

FIGURE 17.11 The proposed method with RSSI bias control. (From Vitzilaios, G. and Papananos, Y., IEEE TCAS-II, 2007. Accepted for publication. With permission.)

of positive feedback, utilizing intrinsic part cancellation may inherently create stability problems depending on the technology variations. In that case, we need to rely on a receiver signal strength indicator (RSSI) that will sense the gain expansion when we are reaching the local minimum as indicated in Figure 17.10a. The RSSI signal can then be used to control the value of the bias voltage of the compensating transistor within limits to avoid possible instability problems. A possible schematic diagram of such a topology is presented in Figure 17.11. This complicates the design, but is necessary if the application requires very good reverse isolation behavior. It should be noted here that, for most applications, canceling out the extrinsic part only is providing adequate isolation, but the method could inherently provide a complete cancellation solution.

17.4 LNA LINEARIZATION UTILIZING MAGNETIC FEEDBACK

In this section, we will examine how magnetic feedback can be used in order to increase the linearity of an integrated LNA, while still focusing on low-voltage applications. Highly linear circuits are of paramount importance in RF systems as nonlinearity causes many problems, including harmonic generation, gain compression, desensitization, blocking, cross modulation, and intermodulation [7].

Several linearization techniques have been introduced, especially in power amplifier (PA) design [9]. However, despite their explicit improvement in linearity, their complex structure (usually based on sophisticated feedback schemes) does not allow them to be used in LNA design where low-noise requirements are of prime importance. Regarding LNA linearization, an elegant LNA linearization technique is the derivative superposition method [7,11]. Predistortion methods have also been reported. For example, in Ref. [8], adaptive gate biasing is proposed, while in Ref. [9], a shunt field-effect transistor (FET) predistortion branch is proposed for PA linearization. In this design, the third-order derivative of the predistortion branch transfer function is used to partially cancel the IMD3 response generated by the main amplifier. A significant improvement in the IIP3 value is reported at the expense of reduced gain in the passband.

The technique that we will present utilizes the third-order derivative of the highly nonlinear, combined transfer function of a three-transistor network predistortion branch for partial cancellation of the IMD3 of the main amplifier. In addition, a magnetic feedback method is used to achieve maximum linearity for a wide range of input power values. The technique aligns the predistortion signal vector with the IMD3 signal vector generated by the main amplifier leading to vector cancellation and optimum linearity performance. Furthermore, magnetic feedback is used to shift the "sweet spot" position introduced by the predistortion branch in order to ensure linearity improvement for a wide range of input power values. The technique improves linearity without significant gain and NF degradation, while both the predistortion signal phase and the sweet spot position are electronically tunable.

17.4.1 PREDISTORTER TOPOLOGY

17.4.1.1 Shunt Transistor Predistorter

A shunt transistor predistorter (STP) [9] is presented in Figure 17.12a. For demonstration purposes, we utilize the STL amplifier presented in the previous section. The drain current i_d can be characterized by a Taylor series expansion of the gate voltage around the bias point. Considering mild nonlinearities, the output current is approximated by the following equation [6]:

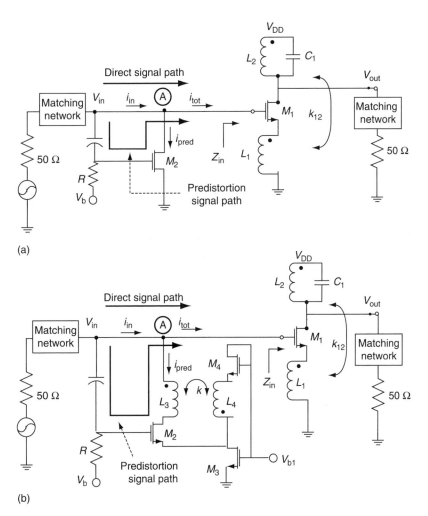

FIGURE 17.12   (a) 1-V LNA with the STP and (b) 1-V LNA with the proposed predistorter (PP).

$$i_d(v_g) = \frac{dI_d}{dV_g}\bigg|_{V_G} v_g + \frac{dI_d^2}{2dV_g^2}\bigg|_{V_G} v_g^2 + \frac{dI_d^3}{3dV_g^3}\bigg|_{V_G} v_g^3 = -\frac{nI_{dss}}{V_p}\left(1 - \frac{V_g}{V_p}\right)^{n-1}\bigg|_{V_G} v_g$$

$$-\frac{n(n-1)I_{dss}}{2V_p^2}\left(1 - \frac{V_g}{V_p}\right)^{n-2}\bigg|_{V_G} v_g^2 - \frac{n(n-1)(n-2)I_{dss}}{6V_p^3}\left(1 - \frac{V_g}{V_p}\right)^{n-3}\bigg|_{V_G} v_g^3$$

$$= g_m^{(1)}v_g + g_m^{(2)}v_g^2 + g_m^{(3)}v_g^3 \tag{17.17}$$

In this equation, I_d and V_g are the large signal drain current and gate voltage, v_g and i_d are incremental gate voltage and drain current, respectively, around the quiescent bias point (I_D, V_G), I_{dss} is the value of I_d for zero V_g, V_p is the transistor pinch-off voltage, n ranges from 1.5 to 2.5, and $g_m^{(1)}$ indicates the ith-order derivative of I_d with respect to V_g. Under appropriate biasing, M_2 is used to generate the opposite polarity IMD3 signal with respect to that of M_1, and is utilized for LNA linearization by IMD3 product cancellation [9].

The topology achieves a large increase in linearity performance with a significant decrease in power gain [9]. It may suffer from phase delay problems at high frequencies, and this negates the linearization effect. Finally, the linearization depends on the $g_m^{(3)}$ value, which is degraded around the optimum bias point, leading to linearity degradation. The abovementioned drawbacks are addressed by an alternative predistorter, to be presented in the following section.

17.4.1.2 Proposed Predistortion Scheme

The proposed predistortion circuit is presented in Figure 17.12b. It consists of transistors M_2, M_3, and M_4. Transistor M_4 is biased near the subthreshold region by V_{b1} and provides a highly nonlinear load to transistor M_3. The output node of M_3 is connected to the source node of M_2 (biased near the subthreshold region by V_b), degenerating the device. Neglecting for the moment the transformer comprising L_3 and L_4, a first-order analysis that is applicable in both the STP [9] and the PP is presented next. Application of Kirchoff's current law at node A provides the total input current flowing towards the LNA:

$$i_{tot} = i_{in} - i_{pred} \tag{17.18}$$

where i_{pred} is the drain current of M_2. Assuming that the value of resistor R is large, v_{in} will appear at the gate terminal of M_2, and so i_{pred} may be Taylor expanded to give

$$i_{pred} = i_{d2}(v_{in}) = g_{m2}^{(1)}v_{in} + g_{m2}^{(2)}v_{in}^2 + g_{m2}^{(3)}v_{in}^3 \tag{17.19}$$

The total output current (i_{d1}) of the LNA is found by adding the current contribution of the direct signal path and the contribution of the predistortion signal path. The direct signal path provides a current given by the following equation:

$$i_{d1(direct\_path)}(v_{in}) = g_{m1}^{(1)}v_{in} + g_{m1}^{(2)}v_{in}^2 + g_{m1}^{(3)}v_{in}^3 \tag{17.20}$$

The predistortion signal contribution is found by transforming the current signal to an equivalent voltage at the gate of M_1.

$$v_{pred} = i_{d2}(v_{in})Z_{in} = \left(g_{m2}^{(1)}v_{in} + g_{m2}^{(2)}v_{in}^2 + g_{m2}^{(3)}v_{in}^3\right)Z_{in} \tag{17.21}$$

where Z_{in} is the effective input impedance of the LNA. This signal is then amplified by M_1 to provide the output current contribution of the predistortion signal.

$$i_{d1(predistortion\_path)}(v_{pred}) = g_{m1}^{(1)}v_{pred} + g_{m1}^{(2)}v_{pred}^2 + g_{m1}^{(3)}v_{pred}^3 \tag{17.22}$$

Substituting Equation 17.21 into Equation 17.22 and adding Equation 17.20, we get the transfer characteristic equation of the LNA with predistortion. Analyzing the first- and third-order product terms

$$i_{d1} = \left(i_{d1(direct\_path)} + i_{d1(predistortion\_path)}\right) \approx g_{m1}^{(1)}\left(v_{in} - v_{pred}\right) + g_{m1}^{(3)}\left(v_{in}^{3} - v_{pred}^{3}\right)$$

$$\approx \left(g_{m1}^{(1)} - g_{m1}^{(1)}\left(g_{m2}^{(1)}Z_{in}\right)\right)v_{in} + \left(g_{m1}^{(3)} - g_{m1}^{(1)}\left(g_{m2}^{(3)}Z_{in}\right)\right)v_{in}^{3} \qquad (17.23)$$

Equation 17.23 indicates that optimum performance will arise when

$$\left(g_{m1}^{(1)}\left(g_{m2}^{(1)}Z_{in}\left(v_{in}\right)\right)\right) \to 0 \Rightarrow g_{m2}^{(1)} \to 0 \qquad (17.24)$$

$$\left(g_{m1}^{(3)}v_{in}^{3} = g_{m1}^{(1)}\left(g_{m2}^{(3)}Z_{in}\left(v_{in}\right)^{3}\right)\right) \qquad (17.25)$$

Under condition of Equation 17.24, minimum degradation on the amplifier gain and NF will be achieved, since the first-order signal of the predistortion branch that subtracts from the amplifier gain will tend to zero and so will be the current-dependent noise contribution of M_2. Under condition of Equation 17.25, maximum IMD3 cancellation will occur. (It should be noted that since the right-hand term of condition Equation 17.25 is complex, phase control is required, as will be explained later on.) In Figure 17.13, the first-, second-, and third-order derivatives of the M_2 drain current of the

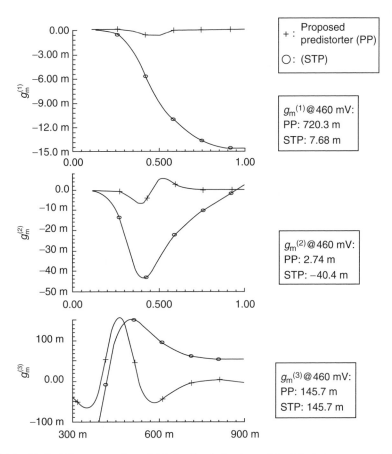

FIGURE 17.13 Typical first-, second- and third-order derivatives of the M_2 drain current of the PP and the STP with respect to the M_2 gate voltage.

two topologies with respect to the gate voltage of M_2 are compared. With the proposed topology, the $g_{m2}^{(1)}$ value is kept small due to the degeneration action of M_3, which leads to M_2 gain compression and thus to $g_{m2}^{(1)}$ slope reduction. In the STP, the M_2 source is grounded. The $g_{m2}^{(3)}$ term is high and is set equal in both cases due to the nonlinear feedback action of transistor M_4, which increases the nonlinearity of transistor M_3. The combined action of M_3 and M_4 provides a nonlinear voltage at the source terminal of M_2 that increases its nonlinear characteristic. In addition, the transformer directly couples the M_3, M_4 nonlinearities at the output node of the predistorter. This indicates that the proposed topology allows for $g_{m2}^{(1)}$ to be reduced by an order of magnitude compared to the STP, in order to satisfy the condition of Equation 17.24, while the condition of Equation 17.25 is satisfied by the same amount by both topologies.

However, both topologies suffer from a serious limitation: At high frequencies, where the signal quantities must be treated as vectors that are subject to phase shifts, Equation 17.18 must be rewritten in vector form:

$$\vec{I}_{tot} = \vec{I}_{in} - \vec{I}_{pred} \qquad (17.26)$$

When the relative phase of these vectors deviates from the ideal value of 0°, the predistorter effect is reduced and is negated for a phase difference of 90°. In addition, the M_2 bias voltage is increased with increasing input power due to the self-bias effect, since the initial biasing is near the subthreshold region. The self-biasing limits the linearization effect due to $g_{m2}^{(3)}$ degradation away from the optimum bias point, as shown in Figure 17.13.

The solution for the abovementioned problems is addressed by the two techniques presented below. First, the predistorter generates a sweet spot, that is, an improvement of the linearity performance in a narrow range of input power values due to a transistor gain expansion. This occurs for an input power of −14.6 dBm as shown in Figure 17.14. In the proposed topology, it is created by the transistor M_2 compression, which is ensured to occur for low input power values due to the following effect: The feedback action from M_4 increases the output voltage at the drain node of M_3 in a fast rate when the input power is increased. This results in M_2 compression due to increased voltage at the source terminal. The compression of M_2 occurs for a gate bias voltage around 510 mV ($g_{m2}^{(1)}$ graph in Figure 17.13). Unfortunately, the sweet spot improves the linearity of the topology for

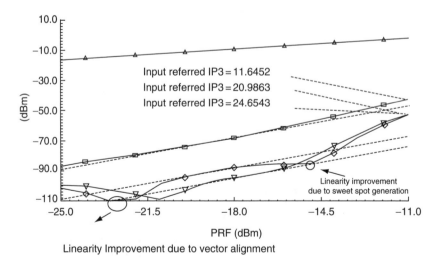

FIGURE 17.14   Comparison IIP$_3$ graphs of the predistortion procedure.

only a narrow range of input power. The sweet spot position may be changed by the transformer comprising inductors L_3 and L_4. For the examination of the feedback, we will use the transformer model used in the previous paragraphs.

The transformer feedback in inductor L_4 is used to increase the voltage at the source of M_2 at an even faster rate, leading to the compression of M_2 for lower input power values and, therefore, moving the sweet spot to a user-defined power level.

The second technique relies on the fact that the transformer may be used for manipulating the relative angle between the vectors in Equation 17.26, in order to satisfy condition of Equation 17.25 and ensure maximum linearity by vector cancellation. In integrated transformers, the in-band phase difference between the currents and between the induced voltages in the primary and secondary inductors of a transformer deviates by a significant amount from the ideal 0° difference in the case of positive feedback. This imperfection is used in the proposed method to the designer's advantage.

The above statement can be demonstrated by examining the phase characteristics of integrated transformers. A simplified model is used and is indented for demonstration purposes only, since the actual model used for simulations was extracted through EM software. Figure 17.15a represents the application of a signal to a loaded integrated inductor. In Figure 17.15b, the situation where the inductor value is large is depicted. If $L \gg R$, then the current and voltage vectors are orthogonal. In integrated inductors, the inductance values are limited in the nH range and the real part of the load R cannot be neglected. The corresponding signals are depicted in Figure 17.15c. It is obvious that there is a phase shift of the current vector, indicated by the angle φ due to the fact that the value of X_R becomes comparable to X_L. In Figure 17.15d, the application of a signal to a lossy transformer model is shown. The current and voltage vectors in the L_2 branch are depicted in Figure 17.15e. As stated above, there is a finite angle φ between the vectors. The input signal is induced in the L_1 branch through the current-dependent voltage source, $sM_{12}i_{in}$, providing a voltage orthogonal to I_{in}. The nonzero value of R' in Figure 17.15d will result in a phase difference φ' between $sM_{12}i_{in}$ and the

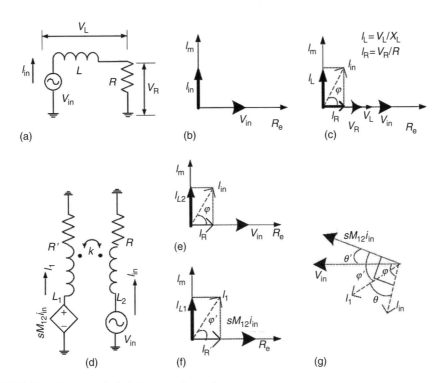

FIGURE 17.15 Vector analysis in integrated transformers.

resulting current I_1 as shown in Figure 17.15f. The current and voltage vectors are placed together under the appropriate phase relations in the same graph (Figure 17.15g). It is evident that there is a deviation from the ideal case of the ideal 0° phase difference by θ and θ' in the voltage and current vectors, respectively. The coupling coefficient k only affects the $sM_{12}i_{in}$ magnitude, and changing it will not alter the values of θ and θ'.

Figure 17.16 represents a simplified version of the predistortion procedure when the transformer is in place. The current source $I_{(M_2)}$ represents the predistortion current at the drain source of M_2, while Z_{load} represents the load the main amplifier presents to M_2. V_{pred}' is the modified predistortion signal at the gate terminal of M_1 and I_4 is the nonlinear current flowing in inductor L_4. When the inductors are coupled (Figure 17.16a), the corresponding vectors are shown in Figure 17.16b. The phase difference between $I_{(M_2)}$ and I_4 is represented by the angle θ' and between the induced voltages V_3 and V_4 by the angle θ. Both angles are exaggerated for demonstration purposes.

According to the previous analysis, angles θ' and θ are constant, different from 0°, set by the L_3, L_4 values and are not a function of the transformer coupling coefficient. It should be noted that V_3 represents the predistortion signal V_{pred} when the coupling coefficient is zero. In order to find the modified predistortion signal V_{pred}', the vector summation of the voltages V_3 and V_4 of Figure 17.16b must be performed to give

$$\vec{V}_{pred'} = (j\omega L_3)\vec{I}_{(M_2)} + (j\omega M_{34})\vec{I}_4 = (j\omega L_3)\vec{I}_{(M_2)} + \left(j\omega k\sqrt{L_3 L_4}\right)\vec{I}_4 \qquad (17.27)$$

The magnitude and phase of $\vec{V}_{pred'}$ are given in in the following equations:

$$\text{mag } V_{pred'} = \sqrt{\left[\left|i_{(M_2)}L_3\right| + |i_4 M_{34}| \cos\theta\right]^2 + \left[|i_4 M_{34}| \sin\theta\right]^2} \qquad (17.28)$$

$$\text{phase } V_{pred'} = \varphi = \arctan\left(\frac{[|i_4 M_{34}| \sin\theta]}{\left[\left|i_{(M_2)}L_3\right| + |i_4 M_{34}| \cos\theta\right]}\right) \qquad (17.29)$$

This vector is thus phase-shifted by an angle φ with respect to the case where there is no coupling (represented by vector V_3). This angle is controllable by manipulating the vector V_4, either through the value of k or by changing i_4, as it is evident from Equation 17.27.

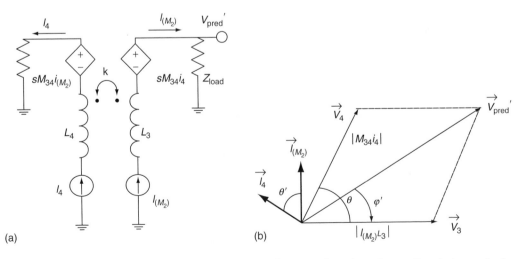

(a)   (b)

FIGURE 17.16 (a) Simplified schematic of the predistorter when there is coupling between L_3, L_4. (b) Corresponding current and voltage vectors.

This indicates that the designer can shift the initial phase angle of the predistorter signal in order for condition of Equation 17.25 to be satisfied. It can be commented that as it is stated in Ref. [5], condition of Equation 17.25 is not providing optimum linearity due to the M_1 second-order products that appear at the LNA output. Thus, the phase should be further shifted in order to partially cancel the second-order terms according to the condition described in Ref. [11]. This is represented in Figure 17.14, where optimum vector positioning provides the local minimum in third-order product response for an input power of $-23\,\mathrm{dBm}$. The linearity response is optimized by bringing this minimum and the sweet spot closer together, making the linearity improvement range wider and centering it at the desired range of input power values. The optimization procedure is graphically represented in Figure 17.14. In this figure, the two distinct areas of linearity improvement due to vector cancellation and sweet spot creation are indicated, when the sweet spot position is not optimized. When the transformer is used to move the sweet spot position at lower input power, it is evident that the input power range where linearity improvement occurs may be maximized. Since the coupling coefficient value is restricted by the amount of feedback required for bringing the two minima close, the values of L_3 and L_4 must be chosen such that the initial angle θ brings the shifted vector in the vicinity of maximum cancellation. The predistortion branch may be electronically tuned by a voltage (manipulating current i_4 in Equations 17.27 through 17.29). The phase-shifting procedure is not independent of the sweet spot position shifting, so it is necessary to include two electronically controlled biases as shown in Figure 17.12b. This allows for maximum linearity to be simultaneously achieved by both optimum phase difference and sweet spot position.

To summarize, the design principle of the proposed predistorter relies on three factors: First, a highly nonlinear predistortion branch is used; this provides an adequate $g_{m2}^{(3)}$ term, while the $g_{m2}^{(1)}$ term is small, avoiding gain and NF deterioration. Second, the transformer-based method maximizes efficiency by ensuring vector cancellation. Finally, the predistortion branch introduces a sweet spot, whose position is changed by the transformer in order to achieve maximum linearity in a wide range of input power values. The use of two control voltages ensures linearity improvement under component parameter deviations, especially in the inductance values. This is done by shaping the third-order product response according to the conditions provided by the fabricated circuit, utilizing each one of the three linearity improvement factors in the optimum way.

17.4.1.3 Typical Performance

A comparison study of the LNA without predistortion and the proposed design was made using device files from a $0.13\,\mu m$ CMOS technology. A set of typical QFN package parasitics was used and the integrated transformer parameters were extracted using EM software. In all simulations, appropriate, frequency-dependent models based on the S-parameters provided by the EM simulations were used. All simulation results refer to conjugate-matched designs to a $50\,\Omega$ input and a typical $50\,\Omega$ load. The main amplifier was kept the same in all cases and only minor modifications in the matching networks were made when the different predistorters were used. Figure 17.17 illustrates a comparison simulation of the power gain response (S_{21}), reverse isolation (S_{12}), NF, and the IIP_3 performance of the topologies. The operation frequency was set at $5\,\mathrm{GHz}$. Simulation results show a 1 dB gain loss and 0.44 dB NF deterioration of the proposed topology compared to the LNA without predistortion, which are attributed to the nonzero $g_{m2}^{(1)}$ value of the topology. The reverse isolation performance is 1 dB better and the IIP_3 value is increased by 10.3 dB for a range of -26 to $-17\,\mathrm{dBm}$ of input power.

17.5 LOW-VOLTAGE MIXER DESIGN

Mixers are essential building blocks of virtually every RF system. A poor mixer performance, especially in terms of nonlinear distortion, greatly aggravates the dynamic range of the overall system. Mixers can be passive or active. The latter category, in the form of the well-known Gilbert cell [12], has prevailed in the majority of applications as it entails advantages such as large conversion gain,

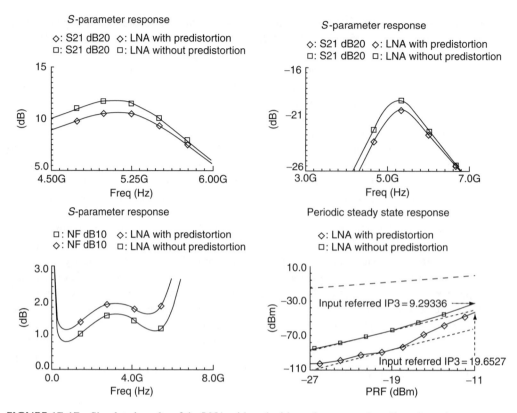

FIGURE 17.17 Simulated results of the LNA with and without the proposed predistortion scheme.

high port-to-port isolation, and good noise performance. The price to pay is deterioration in linearity performance. In modern communication systems, as MOS devices shrink into deep submicron dimensions and the supply voltage scales down to levels at or below 1 V, the problems caused in a system by the nonlinear characteristics of the devices are intensified. Hence, the mixer's optimization is vital in meeting the stringent requirements of modern wireless radio communication systems. In mixers, the complex nature of the distortion-producing mechanisms [13,14] makes its linearity analysis and optimization a difficult task.

In this section, we will present an approach for mixer design suitable for low-voltage operation. The mixer topology utilizes the "inductive-resonance" parasitic capacitance neutralization method [15,16], accompanied by an adaptive biasing scheme in order to achieve simultaneous optimization for gain, noise, and linearity performance. We will also address one of the most important problems in integrated mixers, namely the limited local oscillator (LO) to RF isolation, while preserving all the advantages of the topologies referred in the bibliography [16]. Finally, we will introduce a methodology for second harmonic signal injection [17–19] that significantly enhances the linearity performance, making it comparable with the passive mixer counterparts. The operating principle of the technique is to take advantage of the nonlinearities of the devices to internally cancel the intermodulation distortion. The proposed technique exploits the beneficial interaction of the nonlinearities of the input and the switching stage of a Gilbert cell mixer that results in an improved overall linearity performance [13].

17.5.1 CONVENTIONAL MIXER TOPOLOGY

The conventional, fully balanced Gilbert cell mixer topology is presented in Figure 17.18. The input stage of the mixer consists of two grounded transistors and does not use a classical differential stage

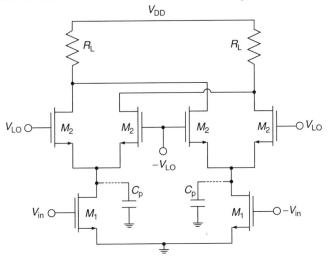

FIGURE 17.18 Conventional mixer topology.

with a current tail. This topology is more suitable for low-voltage designs and it also achieves better linearity performance [14].

The parasitic capacitance C_p at the source nodes of the switching transistors is responsible for the deterioration of the overall mixer performance. Intermodulation analysis of the conventional mixer [13,14] reveals that this parasitic capacitance greatly affects the linearity of the mixer, while it is responsible for the presence of significant amount of noise, generated by transistors M_2, to the output. Additionally, C_p causes leakage of the RF current (produced by the input stage transistors M_1) to ground and thus degrades both mixer's conversion gain and NF.

In Ref. [16], a capacitance neutralization technique is utilized to improve the mixer performance. The technique relies on placing an integrated inductor L at the sources of switching transistors (inductive-resonance technique). The inductor value is appropriately chosen in order to resonate with the parasitic capacitance present at this node, at the RF frequency of interest. This technique increases the conversion gain of the mixer while simultaneously improving both noise and linearity performance [16].

17.5.2 ALTERNATIVE MIXER TOPOLOGY

The mixer to be presented preserves all the advantages of the one previously reported [16] and additionally offers great LO to RF isolation and enhanced linearity performance. In Figure 17.19, the proposed mixer topology is depicted. As it can be seen, the input and switching stages are magnetically coupled with an integrated transformer and thus can be biased independently, facilitating the simultaneous optimization of both stages.

Since the large LO signal appears as common signal at the source nodes of the switching transistors, the transformer prevents this signal from appearing at the drains of transistors M_2 except in the case where large circuit mismatches are present. However, this mixer topology permits a cascode input stage (even for the restricted 1 V supply) which, combined with the transformer, provides high isolation for the input transistors M_1 even in the case of significant circuit mismatches.

It has to be noted that the presence of the integrated transformer has the same purpose and effect as the inductor in Ref. [16]. In Figure 17.20, the AC equivalent of the circuit around the transformer is depicted, where L_1, L_2 are the individual self-inductances of the transformer's inductors, k is the coupling coefficient of the transformer, and C_1, C_2 are the parasitic capacitances at the primary and secondary inductor of the transformer, respectively. The transconductance G_M is referred to the real part of the impedance presented at the source nodes of the switching transistors. After simple

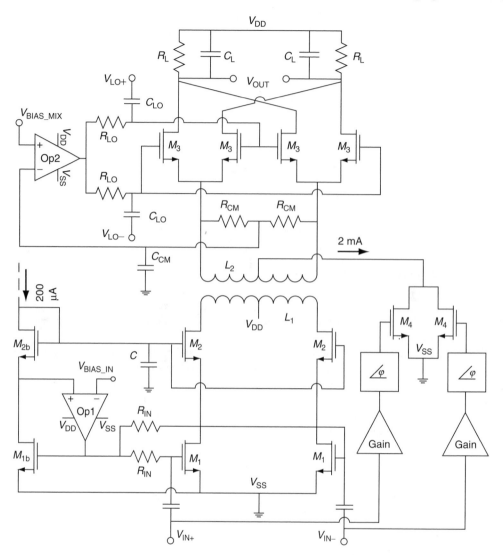

FIGURE 17.19 Mixer topology with integrated transformer.

mathematical calculations, it can be shown that the output current, I_{out}, at the secondary of the transformer reaches its maximum value when

$$s^2 (C_1 L_1 + C_2 L_2) + 1 = 0 \qquad (17.30)$$

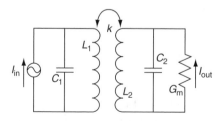

FIGURE 17.20 AC transformer equivalent.

The same equation describes the resonance condition providing the exact frequency for which the parasitic capacitance C_2 is neutralized with respect to a signal connected at the secondary of the transformer. It is thus clear that Equation 17.30 is equivalent to a "resonance" expression that occurs in the case where a simple integrated inductor exists [15,16]. So, the proposed mixer topology has same advantages and similar performance with the previous one reported, especially as far as noise is concerned.

Additionally, Equation 17.30 indicates that the RF power is not grounded via parasitic paths and is transferred with small losses to the secondary of the transformer. More specifically, assuming that Equation 17.30 is valid and that ideal coupling ($k = 1$) is achieved, the current gain of the transformer is given by

$$\frac{I_{\text{out}}}{I_{\text{in}}} = \sqrt{\frac{L_1}{L_2}} \tag{17.31}$$

Equations 17.30 and 17.31 indicate that the presence of the transformer not only prevents RF current, produced by transistors M_1, from leaking into parasitic paths, but also can passively enhance the AC current that will drive the switching stage. Although the ratio of Equation 17.31 can be sufficiently large, an ideal coupling condition cannot be achieved in practice and thus the current gain is restricted. Therefore, the physical layout of the integrated transformer becomes very critical.

At this point, a comment on the transistor sizes and the transformer parameter selection should be made. A medium (3–6) Q factor for the inductor elements of the integrated transformer is desirable in order to achieve a broadband resonance, which unavoidably affects the transistor dimensions and the choice of the transformer's coils. In order to establish a fast switching procedure, the switching transistors must have small length and significantly large width. However, by increasing the switching transistors width, the parasitic capacitor C_2 is proportionally increased, leading to a narrower resonance bandwidth. These are contradicting requirements and so a compromise for the transistor sizes must be made in order to achieve a relatively broadband design that preserves a good mixer performance under process, voltage, and temperature variations.

The mixer of Figure 17.17 is dynamically biased at both the input and the switching stages. Adaptive biasing of the input stage is ensured with the use of an operational amplifier (Op1). The DC current of the input stage is accurately determined by the current mirror consisting of transistors M_1, M_2, M_{1b}, and M_{2b}. Since the subcircuit comprising transistors M_{1b} and M_{2b} is a replica of the one consisting of M_1, M_2, the operational amplifier forces the DC voltages at the drains of the input transistors (M_1) to be locked at the enforced value $V_{\text{BIAS\_IN}}$. Simulations showed that the appropriate choice for the value of this voltage is very critical especially for the linearity performance of the mixer.

The configuration presented in Figure 17.19 is also responsible for the biasing of the switching stage, since the drain current of the transistors M_4 of Figure 17.19 is injected at the center tap of the secondary of the transformer. One important problem in low-voltage mixers is that the bias current of the switching stage increases as the LO amplitude rises. This is due to the fact that the source nodes of the switching transistors, which are also the drain nodes of the biasing transistors, follow the LO signal, present at their gates, for the positive part of the period and thus the DC voltage of these nodes increases with the LO amplitude. In order to overcome this particular problem, dynamic biasing is also used in the switching stage. Resistors R_{CM} and capacitor C_{CM} are used as a lowpass filter in order to sense the DC at the critical nodes and Op_2 is used to lock these nodes to an appropriate DC value ($V_{\text{BIAS\_MIX}}$) by dynamically biasing the gates of the switching transistor. Thus, the DC bias current of the switching stage remains constant in process, voltage, and temperature variations.

Transistors M_4 are also used for the harmonic injection that is used for mixer linearization. In the next section, an intermodulation distortion analysis of the proposed mixer topology will be presented in order to validate its excellent linearity performance.

17.5.3 LINEARIZATION TECHNIQUE

In the proposed mixer, second-order harmonics are produced from the configuration at the right-hand side of Figure 17.19. The circuitry consists of the transistors M_4, a gain, and a phase-shift stage. The implementation of those stages is shown in Figure 17.21. The combination of R_{PH} and C_{PH} form an R-C tank, which is responsible for the required phase shift while the differential stage is responsible for the overall gain of this configuration. The output of the topology of Figure 17.21 is connected directly to the gates of the transistors M_4. Inductor L is used in order to resonate the parasitic capacitances that appear at these nodes at the frequency of interest. Since transistors M_4 are driven by differential signals, their combined drain current comprises of only even-order harmonics. Odd-order harmonics have opposite phase and thus they are cancelled out.

The produced harmonics are injected, in current form, through the center tap of the secondary of the transformer and, due to the circuit symmetry, the signal splits into equal parts and flows into the switching stage. A first-order theoretical analysis of the linearity performance can be based on Volterra series theory. This mathematical tool can provide equations that indicate how the nonlinearities of the mixer subcells can interact in a way to improve overall linearity performance [13]. In the following analysis, we will show how the linearity behavior of the input, the second harmonic injection, and the switching stage can be combined in a profitable manner. The AC output voltage can be written as a time-varying Volterra series as follows:

$$V_{out}(t) = \sum_{n=1}^{\infty} \int_{-\infty}^{+\infty} \int \cdots \int B_n(t, \omega_1, \ldots, \omega_n) \cdot \prod_{k=1}^{n} I_{sw}(\omega_k) \cdot e^{j\omega_k t} d\omega_k \qquad (17.32)$$

where

B_n are the frequency domain time-varying Volterra kernels
$I_{sw}(\omega)$ is the Fourier-transformed AC current that flows into the switching stage

The time-varying behavior of the Volterra kernels is attributed to the presence of the large LO signal and thus they are periodic functions with the same period as the LO signal. The switching stage current, although slightly affected by the LO presence in general, can be represented, in a first-order approximation, by a time-invariant Volterra series as follows:

$$I_{sw}(t) = \sum_{n=1}^{\infty} \int_{-\infty}^{+\infty} \int \cdots \int A_n(\omega_1, \ldots, \omega_n) \cdot \prod_{k=1}^{n} V_{in}(\omega_k) e^{j\omega_k t} d\omega_k \qquad (17.33)$$

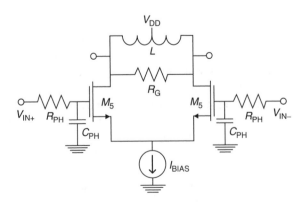

FIGURE 17.21   Gain and phase shift of the second harmonic injection scheme.

The convolution of Equations 17.32 and 17.33 gives the final relation between output and input voltages:

$$V_{\text{out}}(t) = \sum_{n=1}^{\infty} \int\int_{-\infty}^{+\infty}\int C_n(t, \omega_1, \ldots, \omega_n) \prod_{k=1}^{n} V_{\text{in}}(\omega_k)e^{j\omega_k t}\mathrm{d}\omega_k \qquad (17.34)$$

For the calculation of the third-order intermodulation distortion in a downconversion mixer, the first negative harmonics of $C_1(t, \omega_{\text{in}})$ and $C_3(t, \omega_{\text{in}}, \omega_{\text{in}}, -\omega_{\text{in}})$ are necessary. First- and third-order Volterra kernels are given by

$$C_1(t, \omega_a) = B_1(t, \omega_a)A_1(\omega_a) \qquad (17.35)$$

$$C_3(t, \omega_a, \omega_b, \omega_c) = B_1(t, \omega_a + \omega_b + \omega_c) \cdot A_3(\omega_a, \omega_b, \omega_c)$$

$$+ \frac{2}{3} \cdot \begin{bmatrix} B_2(t, \omega_a, \omega_b + \omega_c) \cdot A_1(\omega_a) \cdot A_2(\omega_b, \omega_c) + \\ B_2(t, \omega_b, \omega_a + \omega_c) \cdot A_1(\omega_b) \cdot A_2(\omega_a, \omega_c) + \\ B_2(t, \omega_c, \omega_a + \omega_b) \cdot A_1(\omega_c) \cdot A_2(\omega_a, \omega_b) \end{bmatrix}$$

$$+ B_3(t, \omega_a, \omega_b, \omega_c) \cdot A_1(\omega_a) \cdot A_1(\omega_b) \cdot A_1(\omega_c) \qquad (17.36)$$

The configuration at the right-hand side of Figure 17.19 is responsible for the second harmonic injection into the switching stage. It is obvious from Equation 17.33, which represents the switching stage current, that the even-order coefficients A_n incorporate the products of the second harmonic injection circuit. Thus, appropriate choice of gain and phase shift of this circuit can lead to minimization of the nonlinear terms of Equation 17.36. By decreasing this third-order kernel, the overall linearity performance of the mixer is significantly enhanced.

17.5.4 TYPICAL PERFORMANCE

The design has been simulated using a standard 0.13 μm CMOS technology with BSIM3v3 MOSFET model files. In Figure 17.22, the third-order intermodulation distortion in terms of amplitude

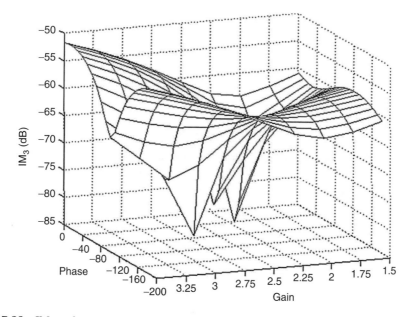

FIGURE 17.22 IM_3 performance versus gain and phase shift. Simulations were carried out for an RF and LO frequency of 5 and 5.05 GHz, respectively, two RF input tones of −24 dBm each, and differential LO amplitude of 600 mV.

difference of first- and third-order products (IM_3) of a two-tone test is depicted versus both gain and phase shift. From this graph, it is obvious that appropriate choice of the design parameters of the configuration of Figure 17.19 will lead to a large linearity improvement.

It is clear from Figure 17.22 that the best linearity performance is achieved for the combination of gain and phase shift that can be extracted from the graph and thus the circuit of Figure 17.19 has been designed appropriately.

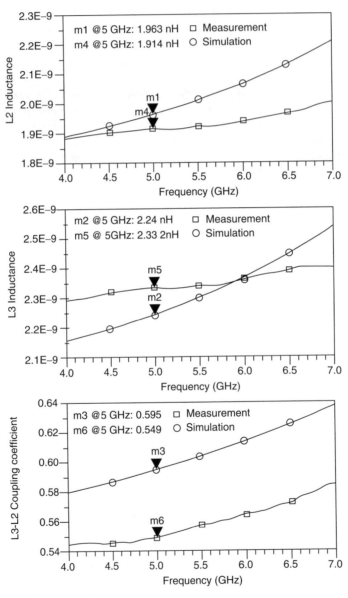

FIGURE 17.23 Comparison of the simulated and measured values of the characteristics of the triple transformer.

17.6 INTEGRATED TRANSFORMER DESIGN METHODOLOGY

In this section, we will describe briefly a design flow procedure in order to implement integrated transformers. The first step is to create a parametric cell, implementing the geometry of the inductive device: a differential inductor, a transformer, a balun or for example, the triple transformer structure of Figure 17.4. The parametric cell design provides the flexibility to employ different geometries of the structure and, therefore, implement a range of inductance values and coupling factors in the case of transformers. The procedure implemented is technology agnostic so the design can be easily adapted to different CMOS or BiCMOS technologies with different metal layers and provide a Design Rule Check-free layout.

Once the layout is in place, EM simulations provide a full set of S parameters for the particular structure. The tools used in our case are Agilent's Momentum and Ansoft's HFSS. The S-parameter set is the basis for the calculation of the electrical performance of the integrated inductive elements: it can be either used as is in an N-port configuration for small-signal electrical simulations or it can be used as a basis to derive a detailed lumped-element model for the particular structure. The lumped-element model is then introduced to the final schematic of the LNA or mixer structure for small or large signal simulations.

As an example of the above procedure, we present the triple transformer of Figure 17.4. This structure was fabricated and measured using a probe station and network analyzer. We then compared EM simulation and hardware measurement results. The results revealed a difference smaller than 5% (Figure 17.23). This kind of difference is acceptable due to both simulation and measurement errors. Therefore, the derived lumped-element models can be trusted since they predict the behavior of the element precisely: from the design point of view, this is very important since we can now smoothly introduce the custom transformer cells in our standard design flow.

17.7 CONCLUSIONS

In this chapter, integrated RF front-end components (LNAs and mixers) suitable for low-power design are provided. We have used magnetic feedback through integrated transformers as a main design tool, and we have shown that it can be a useful tool for a variety of operations. More precisely, we examined how we can use transformers for reverse isolation enhancement, linearity, and noise improvement and LO to RF isolation. Recent advances in integrated transformer modeling allow for accurate modeling and extraction with adequate precision of the critical parameters (inductance values, quality factor, coupling coefficient). This effectively means that integrated transformers can become versatile, reliable design instruments that may facilitate low-voltage design and will be increasingly used in commercial products.

REFERENCES

1. D. J. Cassan and J. R. Long, A 1-V transformer feedback low-noise amplifier for 5 GHz wireless LAN in 0.18-μm CMOS, *IEEE J. Solid-State Circuits*, 38(3), 427–435, 2003.
2. A. R. Shahani, D. K. Shaeffer, and T. H. Lee, A 12 mW wide dynamic range CMOS front-end for a portable GPS receiver, *IEEE J. Solid-State Circuits*, 32(12), 2061–2070, 1997.
3. S. Lee, J. Bergervoet, K. S. Harish, D. Leenaerts, R. Roovers, R. van de Beek, and G. van der Weide, A broadband receive chain in 65 nm CMOS, *Solid-State Circuits Conference, 2007, ISSCC 2007, Digest of Technical Papers. IEEE International*, pp. 418–612, 11–15 February 2007.
4. D. -K. Wu, R. Huang, and Y. -Y. Wang, A low-voltage and low-power CMOS LNA using forward-body-bias NMOS at 5 GHz, *Solid-State and Integrated Circuit Technology, 2006. ICSICT '06, 8th International Conference*, pp. 1658–1660, 2006.
5. B. Georgescu, H. Pekau, J. Haslett, and J. McRory, Tunable coupled inductor Q-enhancement for parallel resonant *LC* tanks, *IEEE Trans.Circuits Syst. II, Analog Digit. Signal Process.*, 50(10), 705–713, 2003.
6. Y. Tsividis, *Operation and Modeling of the MOS Transistor*, McGraw-Hill Series in Electrical Engineering, McGraw-Hill, New York, 1987.

7. T. W. Kim, B. Kim, and K. Lee, Highly linear receiver front-end adopting MOSFET transconductance linearization by multiple gated transistors, *IEEE J. Solid-State Circuits*, 39(1), 223–229, 2004.

8. V. Aparin, G. Brown, and L. E. Larson, Linearization of CMOS LNA's via optimum gate biasing, *Proc. IEEE ISCAS IV*, 748–75, 2004.

9. M. G. Kim, C. H. Kim, and H. K. Yu, An FET-level linearization method using a pre-distortion branch FET, *IEEE Microw. Guided Wave Lett.*, 9(6), 233–235, 1999.

10. G. Vitzilaios, Y. Popananos, G. Theodoratos, and A. Vasilopoulos, A 1-V 5.5 GHz CMOS LNA with multiple magnetic feedback, *IEEE TCAS-II*, 53(9), 971–975, 2006.

11. V. Aparin and L. E. Larson, Modified derivative superposition method for linearizing FET low-noise-amplifiers, *IEEE Trans. Microw. Theory Tech.*, 53(2), 571–581, 2005.

12. B. Gilbert, A precise four-quadrant multiplier with subnanosecond response, *IEEE J. Solid-State Circuits*, 3, 365–373, 1968.

13. M. T. Terrovitis and R. G. Meyer, Intermodulation distortion in current-commutating CMOS mixers, *IEEE JSSC*, 35, 1461–1473, 2000.

14. G. Theodoratos, A. Vasilopoulos, G. Vitzilaios, and Y. Papananos, Calculating distortion in active CMOS mixers using Volterra series, *IEEE ISCAS*, 2249–2252, May 2006.

15. H. Sjoland, A. Karimi-Sanjaani, and A. A. Abidi, A merged CMOS LNA and mixer for a WCDMA receiver, *IEEE JSSC*, 38, 1045–1050, 2003.

16. T. A. Phan, C. W. Kim, Y. A. Shim, and S. G. Lee, A high performance CMOS direct down conversion mixer for UWB system, *IEICE Trans. Electron.*, E88-C, 2316–2321, 2005.

17. C. S. Aitchison, M. Mbabele, M. R. Moazzam, D. Budimir, and F. Ali, Improvement of third-order inter-modulation product of RF and microwave amplifiers by injection, *IEEE Trans. Microwave Theory Tech.*, 49, 1148–1154, 2001.

18. C. W. Fan and K. K. M. Cheng, Theoretical and experimental study of amplifier linearization based on harmonic and baseband signal injection technique, *IEEE Trans. Microwave Theory Tech.*, 50, 1801–1806, 2002.

19. S. Kusunoki, K. Kawakami, and T. Hatsugai, Load-impedance and bias-network dependence of power amplifier with second harmonic injection, *IEEE Trans. Microwave Theory Tech.*, 52, 2169–2176, 2004.

20. G. Vitzilaios and Y. Popananos, A magnetic feedback method for low-voltage CMOS LNA reverse-isolation enchancement, *IEEE TCAS-II*, 2007.

18 InGaP-HBT Power Amplifiers

Kazuya Yamamoto

CONTENTS

18.1 INTRODUCTION

InGaP/GaAs heterojunction bipolar transistor (InGaP-HBT) power amplifiers (PAs) are widely used in global system for mobile communications and code division multiple access (CDMA) handsets and wireless local area network metropolitan area network terminals [1–8], because the InGaP-HBTs possess high power density and high reliability [9,10] with single voltage operation and excellent reproducibility, leading to low cost and high yield. Since advanced wireless communications systems such as wideband/narrowband CDMA and orthogonal frequency division multiplexing (OFDM) systems use nonconstant envelope signals with relatively high peak-to-average power ratios, the systems strongly require not only high-efficiency operation but also high linear operation for the PAs.

A typical block diagram of a CDMA PA and its peripheral circuits is illustrated in Figure 18.1. The PA amplifies the modulated signal from the Si-RF large-scale integration (LSI) up to a specific output power level, and then transmits it to the antenna port through the isolator, duplexer, and antenna switch. Figure 18.2 shows an example of the original spectra and their regrowth of CDMA-modulated signals, where Figure 18.2a shows the input signal (RFin) of the PA and Figure 18.2b shows the spectral regrowth at the PA output (RFout) caused by the distortion of the PA. Since this regrowth involves signal quality degradation and may give unwanted signal interference to adjacent channels, the regrowth levels (signal distortion levels) are strictly restricted by air-interface specifications of each system. These distortion levels are often characterized as adjacent channel power ratio (ACPR) or adjacent channel leakage power ratio (ACLR). As another current technology trend in addition to low distortion characteristics, there are strong requirements for smaller and thinner package size at a low cost. In the linear PA design, therefore, it is essential to realize low distortion characteristics with simple circuit topology suited for smaller size.

This section introduces comprehensive design examples of a CDMA PA while focusing on the relationship between the distortion characteristics, bias circuits, and output matching conditions.

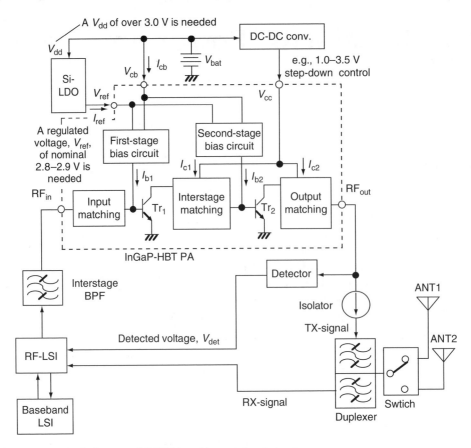

FIGURE 18.1   Block diagram of HBT PA and its peripheral circuits.

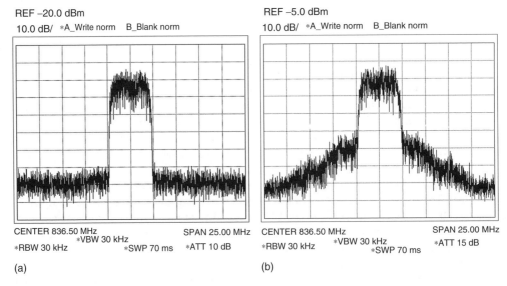

FIGURE 18.2   Spectral examples: (a) W-CDMA-modulated input signal at RF_{in} and (b) W-CDMA output signal at RF_{out}.

In addition, design and measurement results for a switchable-path PA and a low-reference-voltage operation PA are demonstrated as recent circuit design technologies.

18.2 POWER AMPLIFIER DESIGN

In the HBT PAs, bias circuits play the important role of distortion characteristics as well as temperature dependence of quiescent current. This is much different from usual field-effect transistor (FET) amplifier design, because CDMA HBT PAs consume base current on the order of several microamperes to several milliamperes. In contrast, the FET amplifiers hardly consume gate current, and hence the FET amplifier allows us to use very simple bias circuit such as a resistive divider. This section describes the bias circuit design and the relationship between the bias circuits, distortion characteristics, output matching conditions while introducing circuit simulations useful for practical design.

18.2.1 BASIC BIAS CIRCUIT TOPOLOGY

Typical input–output characteristics of a two-stage HBT PA—for example, which is depicted in Figure 18.1—are shown in Figure 18.3. Operating collector currents, I_{c1} and I_{c2}, vary greatly with the increase in output power. With regard to I_{c2} of the second stage collector current, I_{c2} varies from quiescent current of about 40 up to 380 mA at a target output power of 28 dBm. Taking into account a peak-to-average power ratio of about 3.5 dB in the 3GPP W-CDMA specification, the bias circuit also needs to supply the second power stage, Tr_2, with base current corresponding to a peak collector current of about 600 mA. In addition, the bias circuit should have less temperature dependence for quiescent current in order to suppress power gain variation over temperature.

There are two basic bias circuit topologies, which can satisfy these requirements, as shown in Figure 18.4. One is a current-mirror-based topology with a β-helper and the other is an emitter-follower-based topology. In the figure, the emitter finger numbers of the power stage, the emitter follower, and the β-helper are determined to attain the target output power. Note that both the topologies need a reference voltage (V_{ref}), which is independent of battery voltage variation. The voltage, V_{ref}, is usually generated from an Si low-voltage drop-out (LDO) regulator, as depicted

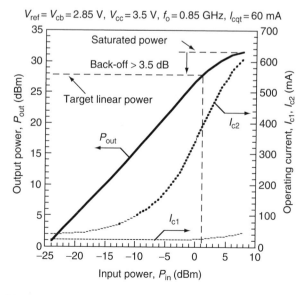

FIGURE 18.3 Typical input–output characteristics for a CDMA PA.

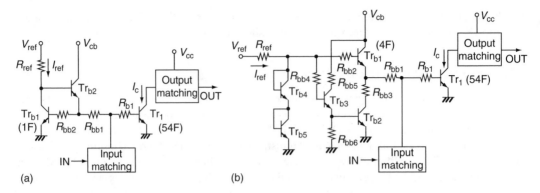

FIGURE 18.4 Circuit schematic examples: (a) current-mirror-based bias circuit and (b) emitter-follower-based bias circuit.

in Figure 18.1. Figure 18.5 compares the simulated temperature dependence of quiescent current between the emitter-follower-based and current-mirror-based topologies. The two topologies can provide temperature insensitivity characteristics for quiescent current. As shown in the figure, the emitter-follower-based topology basically has the advantage of consuming less reference current, I_{ref}, over the current-mirror-based one, because DC current-gain, β, is usually larger than the current-mirror ratio. This section, therefore, focuses on the emitter-follower-based bias circuit design.

Regarding temperature dependence of quiescent current in the emitter-follower-based topology, there are two kinds of typical control methods: base side control (Figure 18.6a) and emitter side control (Figure 18.6b). It is to be noted that in the emitter side control, two diode-connected transistors, Tr_{b2} and Tr_{b3}, are used for reducing control sensitivity of I_{a2}. As shown in the simulation results of Figure 18.7, appropriate temperature-dependent current sources, I_{a1} and I_{a2}, can give almost no temperature-dependent characteristics of quiescent current for the two methods. An actual design

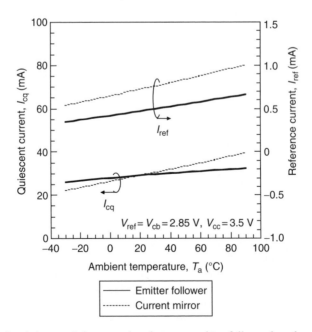

FIGURE 18.5 Simulated characteristic comparison between emitter-follower-based and current-mirror-based bias circuits.

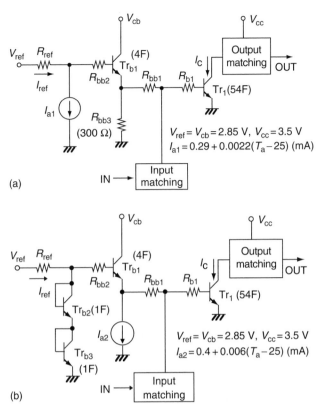

FIGURE 18.6 Circuit schematics for realizing temperature independence of quiescent collector current: (a) base side control scheme and (b) emitter side control scheme.

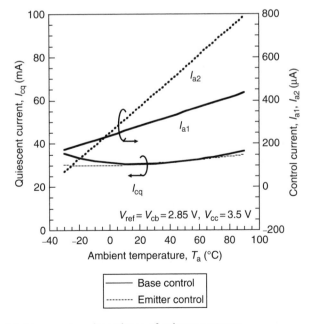

FIGURE 18.7 Simulated temperature dependence of quiescent current.

example of Figure 18.6b is the circuit schematic previously shown in Figure 18.4b. In the circuit, Tr_{b2} and Tr_{b3} work as a temperature-dependent current source [11].

18.2.2 BIAS DRIVE AND AM-AM/AM-PM CHARACTERISTICS

As described in Section 18.1, distortion characteristics such as ACPR or ACLR are one of the important factors to characterize linear PAs for CDMA systems. In the design, however, it is time-consuming and not comprehensive for circuit designers to directly simulate such distortion characteristics. Instead, AM-AM/AM-PM characteristics at fundamental frequency are often used for predicting ACPR or ACLR, because the characteristics are basically comprehensive and such ACPR or ACLR characteristics can be calculated on the basis of the AM-AM/AM-PM characteristics [12,13]. Here, we do not forget that fundamental-frequency-based AM-AM/AM-PM characteristics do not include the contribution of envelope and harmonics to distortion [14–17]. To put it briefly, in the linear PA design, it is essential to realize flat AM-AM/AM-PM characteristics over a wider output power range.

Before describing the relationship between bias circuit and distortion characteristics, let us consider the basic relationship between bias drive and AM-AM/AM-PM characteristics. Figure 18.8 shows the schematics of voltage- and current-drive power stages, and their simulated output transfer

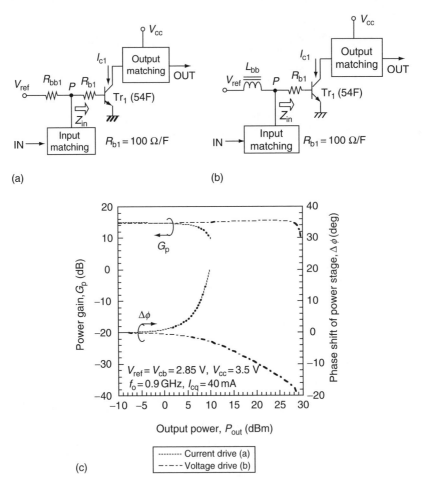

FIGURE 18.8 Circuit schematics for (a) current-drive mode and (b) voltage-drive mode, and (c) simulated output characteristic comparison for current- and voltage-drive power stages.

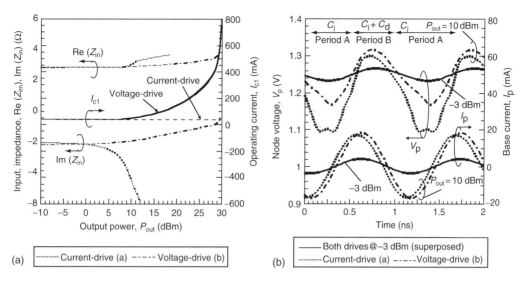

FIGURE 18.9 (a) Simulated input impedance for current- and voltage-drive stages, and (b) simulated node voltage and current waveforms.

characteristics. In Figure 18.8c, gain versus output power represents AM-AM characteristics, and phase shift versus output power represents AM-PM characteristics. We can see that the phase shift between current and voltage drives is opposite to each other. Regarding the gain behavior, the current drive gives gain compression characteristics together with lead-phase characteristics. In contrast, while keeping lag-phase characteristics, the voltage drive offers weak gain expansion characteristics until strong gain compression is observed.

The simulation results of the input impedance (Z_{in}) shown in Figure 18.8a and 18.8b are plotted in Figure 18.9a together with operating current. Figure 18.9b shows the simulated voltage- and current-waveforms at the node, P, in that case. As shown in Figure 18.9a, in the case of current-drive, as the output power increases, the imaginary part of Z_{in}, which corresponds to the reciprocal of the input capacitance, increases in the minus direction. On the other hand, in the case of voltage-drive, as the output power increases, the imaginary part of Z_{in} increases in the plus direction. Figure 18.9b helps understand the input capacitance variation. In the case of the current drive, the peak base voltage (V_p) increases greatly in the turn-off direction during the period (A) with the increase in output power, while in the case of the voltage drive, the increase of V_p in the turn-off direction is very small with the increase of output power. In contrast, the turn-on period (B) of the voltage drive becomes longer with the increase of output power. Junction capacitance is dominant during the period (A), while diffusion capacitance is dominant during the period (B). In addition, the junction capacitance decreases with the decrease of the base voltage. As a result, the input capacitance rapidly decreases in the current-drive mode, while the capacitance gradually increases in the voltage-drive mode as shown in Figure 18.9a. This rapid decrease in input capacitance causes impedance mismatch, thus resulting in the power gain decrease in the current-drive mode, as shown in Figure 18.8c. In contrast, in the voltage-drive mode, the gradual gain increase is observed as shown in the figure, because the operating current, I_{c1}, increases rapidly with the increase in input power, as shown in Figure 18.9a.

To give analytical verification of the above behavior, we have used a well-known hybrid-π-type model as shown in Figure 18.10. The extracted parameters for each finger are listed in the table, where the HBT used for the extraction was fabricated using in-house InGaP HBT processes [18]. Using these parameters, we can implement the following small-signal-based analysis. In Figure 18.10, S_{21} of the HBT block between P and Q is expressed by

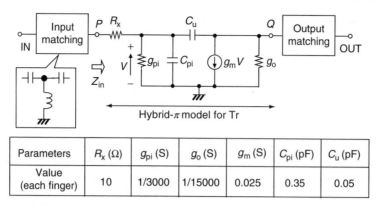

FIGURE 18.10   Small-signal equivalent circuit model and its parameter table.

$$S_{21} \approx \frac{-2g_m}{g_{pi} + j\omega \left(C_{pi} + C_u\right)}$$

$$= \frac{2g_m}{\sqrt{g_{pi}^2 + \omega^2 \left(C_{pi} + C_u\right)^2}} \exp\left(-j\left\{\pi + \tan^{-1}\left[\frac{\omega\left(C_{pi} + C_u\right)}{g_{pi}}\right]\right\}\right), \qquad (18.1)$$

where the following approximation was used: R_x is negligible and $g_m \gg \omega C_{pi}$. Equation 18.1 indicates that the increase in C_{pi} gives the delay of phase shift and the decrease gives the lead as shown in Figure 18.8c. Thus, the analytical formula and simulation presented in this section can clearly explain the difference in the gain and phase shift behavior between the current- and voltage-drive modes.

18.2.3 Bias Circuits and AM-AM/AM-PM Characteristics

This section describes the detailed relationship between the bias circuit, output matching, and AM-AM/AM-PM characteristics. Understanding the relationship is most useful for the actual design of linear PAs for use in CDMA and OFDM systems.

First, let us consider the emitter-follower-based bias circuit and its power stage for the second stage, which operates in the voltage-drive mode, as shown in Figure 18.11. Taking into account the peak-to-average power ratio of about 3.5 dB, output matching is set to power matching so as to deliver a saturated output power of more than 31 dBm, where the emitter finger numbers for the power stage and emitter follower were set at 54 and 8. The quiescent current, reference voltage, and bias supply voltage were set at typical values (of PA products) of 30 mA, 2.85 V, and 2.85 V. Figure 18.12 compares the simulated AM-AM/AM-PM characteristics between the ideal voltage drive (Figure 18.8b) and the emitter-follower-based drive without base feed resistance, R_{bb1} (Figure 18.11) under the same output matching condition. The figure indicates that the emitter-follower-based bias circuit without R_{bb1} works as a closely ideal voltage drive mode. However, the gain expansion and phase shift of the two seem relatively large, and some way of suppressing them is often necessary for practical design.

The simulated dependence of R_{bb1} is shown in Figure 18.13. Figure 18.13a indicates that higher feed resistance (R_{bb1}) is effective in suppressing the gain expansion and phase shift, although overloaded resistance ($R_{bb1} = 50\ \Omega$) creates a gain dip in the middle power range. As can be seen in Figure 18.13b, the operating current decreases with the increase of R_{bb1}, thereby suppressing the gain expansion. The behavior of voltage and current drives mentioned earlier helps understand the effect of R_{bb1}, because loading R_{bb1} makes the bias mode gradually change from voltage drive to current drive.

Figure 18.14 compares the simulated output characteristics between the bias circuit with the decoupling capacitor for the reference node, C_{ref}, and that without C_{ref}. As clearly shown in the

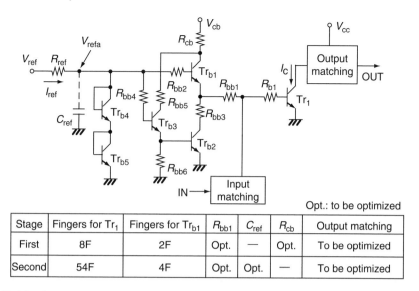

FIGURE 18.11 Circuit schematic of emitter-follower-based bias circuit used for first- and second-stage behavior.

Stage	Fingers for Tr_1	Fingers for Tr_{b1}	R_{bb1}	C_{ref}	R_{cb}	Output matching
First	8F	2F	Opt.	—	Opt.	To be optimized
Second	54F	4F	Opt.	Opt.	—	To be optimized

Figure 18.14b and 18.14c, the decoupling capacitor, C_{ref}, suppresses the node voltage variation of V_{refa} greatly, thus resulting in the improvement of linear power levels. The reason is that the decrease in DC base voltage, V_{b1}, with C_{ref} is smaller than without C_{ref}. However, because the use of C_{ref} enhances the gain expansion and phase shift, we need an appropriate selection of R_{bb1} and an optimum output matching condition in order to deliver flat AM-AM/AM-PM characteristics.

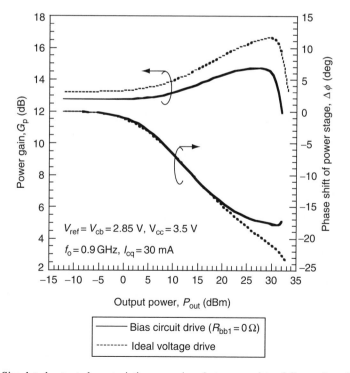

FIGURE 18.12 Simulated output characteristic comparison between emitter-follower-based bias circuit drive and ideal voltage drive.

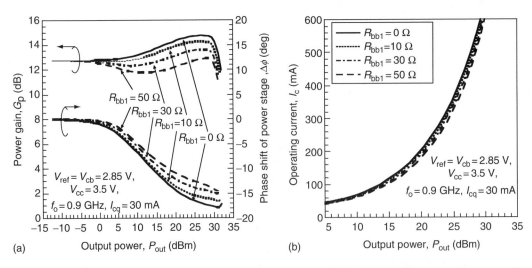

FIGURE 18.13 Simulated bias feed resistance dependence: (a) AM-AM/AM-PM characteristics and (b) operating current.

Next, we consider the first-stage design of the bias circuit and the matching condition. In the linear PAs, the first stage usually takes charge of gain- and phase-shift compensation for the second stage in addition to the role of a driver stage. The gain expansion and lag-phase shift should be, therefore, compensated using the inverse characteristics of the first stage. Figure 18.15 shows the simulated AM-AM/AM-PM characteristics for the first stage, where in the figure the characteristics with $R_{bb1} = 50\ \Omega$ are compared with that without R_{bb1} under the same power-matching condition. The emitter-finger numbers for the first stage and its emitter follower are listed in Figure 18.11. As can be seen in Figure 18.15, the gain expansion can be suppressed using R_{bb1}, although there is a relatively large gain dip. We can predict that from the viewpoint of the previous basic relationship, this gain dip probably occurs in the transition range between current- and voltage-drive modes. As

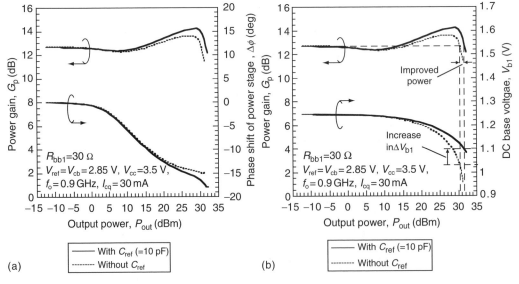

FIGURE 18.14 Simulated output characteristic comparison between bias circuit with C_{ref} and that without C_{ref}: (a) AM-AM/AM-PM characteristics, (b) gain and DC base voltage of power stage

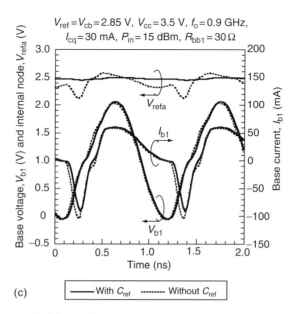

FIGURE 18.14 (**Continued**) (c) internal node voltage and current waveforms of V_{b1}, V_{refa}, and I_{b1}.

described earlier, the dip is mainly due to the input impedance variation, especially input capacitance variation of the power HBT.

Figure 18.16 shows the simulated output characteristic comparison between the power-matching and gain-matching load conditions. In this figure, the characteristics of the bias circuit with the collector load resistance, R_{cb}, are also plotted for comparison. The gain-matching load condition basically provides gain compression and lead-phase characteristics like the current drive mode, because the increase in the operating current is suppressed as shown in Figure 18.16b. In other words, the behavior involving this suppression is considered as an analogy with constant current

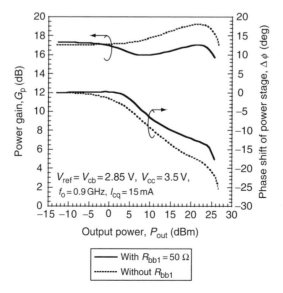

FIGURE 18.15 Simulated AM-AM/AM-PM characteristic comparison for first stage between the bias circuit with Rbb1 and that without Rbb1 under the same power matching condition.

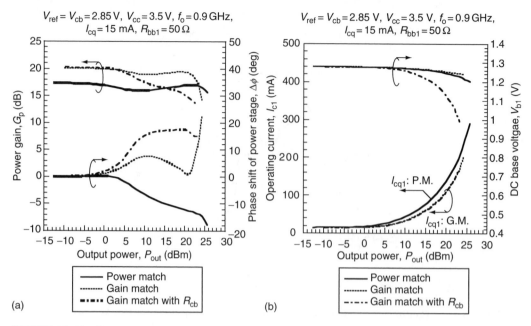

FIGURE 18.16 Simulated characteristic comparison for first stage between power- and gain-matching conditions: (a) AM-AM/AM-PM characteristics and (b) operating current and DC base voltage. In the figure, the characteristics of additional collector resistance, R_{cb}, are also plotted for the comparison.

drive. In addition, the use of R_{cb} limits the output current (I_{b1}) of the emitter-follower (Tr$_{b1}$) while involving the decrease in the collector node voltage of Tr$_{b1}$. Under the gain-matching condition, therefore, the use of R_{cb} gives stronger gain compression and more lead-phase shift, as shown in Figure 18.16b.

Based on the design flow described here, we can design two-stage PA for use in CDMA systems. Figure 18.17 shows the simulated and measured overall output characteristics of the two-stage amplifier for 0.85 GHz-band (Band V) W-CDMA applications. In Figure 18.17a, the distortion characteristics of ACLR1(5 MHz-offset) are calculated on the basis of the simulated AM-AM/AM-PM characteristics. Good flatness of the overall gain and phase shift is obtained, thanks to appropriate compensation for relatively large gain and phase shift of the second stage. As shown in Figure 18.17b, the calculated power gain and ACLR1 agree well with the measured ones. Under the high-speed data packet access (HSDPA)-compliant W-CDMA test condition of 3.4 V, measurement reveals that the fabricated PA module delivers an output power of 28 dBm, a power gain of 28.5 dB, and power-added efficiency (PAE) of more than 40% while keeping good ACLR1 characteristics of less than −40 dBc.

18.3 RECENT POWER AMPLIFIER TECHNOLOGIES

This section introduces two recent circuit technologies for CDMA PAs. Recent studies on CDMA PAs are categorized into the following three main techniques:

(1) Low-power enhanced efficiency (LPEE) technique [19]
(2) Isolator-less technique [20–24]
(3) Low-reference-voltage operation technique

The isolator-less PAs using the isolator-less technique are required for removing relatively expensive and high-package-height isolators from a front-end block. As illustrated in Figure 18.1, the isolator

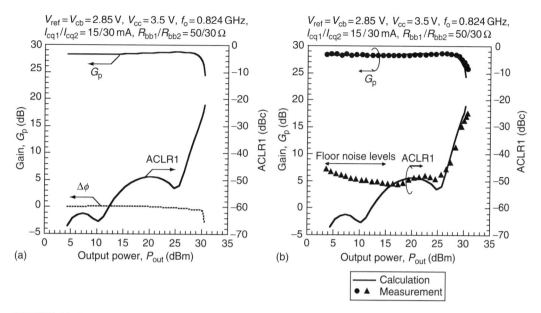

FIGURE 18.17 (a) Simulated AM-AM/AM-PM characteristics and AM-AM/AM-PM-based calculation of ACLR1 for Band V two-stage PA and (b) gain and ACLR characteristic comparison between calculation and measurement.

follows the PA and prevents the PA from being affected by load mismatching at an antenna terminal, because a usual PA cannot satisfy distortion specifications easily under strong mismatching load conditions. The balanced amplifier approach is one of the most promising ways to realize isolator-less PAs [20,21], although some other approaches have been reported up-to-date [22–24]. A detailed description about (1) and (3) will be provided later in the chapter.

18.3.1 Parallel Amplifier Approach

According to the probability distribution function (PDF) of CDMA systems, the probability of a full output power state (e.g., more than 27 dBm) of a PA is much lower than that of a low output power state (e.g., less than 8 dBm) [19]. It is, therefore, important to operate the PA with higher efficiency in the low-power state as well as in the high power state. This is the reason why the LPEE technique has been focused on the CDMA PA design recently. The LPEE technique basically covers a load modulation (switching load) approach [25–30], a parallel amplifier approach [31–34] including stage-bypass approaches [35], a DC-DC converter approach [36–39] including polar modulation [40], and a dynamic/adaptive bias control approach [41–46]. Of these approaches, the use of a DC-DC buck converter is the most popular, because collector supply voltage control using the DC-DC converter can reduce power dissipation of a conventional PA at low output power levels without degrading significant distortion characteristics. On the other hand, strong requirements for smaller on-board area occupation and lower cost have promoted the parallel amplifier approach, which does not need the DC-DC converter. In this section, a switchable-path PA is presented as an example of the parallel amplifier approach, and its performance is compared with a conventional PA with an external DC-DC buck converter.

The switchable-path PA configuration is illustrated in Figure 18.18 [47]. The PA consists of two-stage amplifier paths in parallel and a CMOS bias controller. The CMOS bias controller generates a reference voltage and plays the role of the digital interface [48] with a baseband LSI. For example, as shown in the figure, when V_{mod1} is "low (high power mode)", the main-path PA turns on and the subpath PA turns off. In contrast, when V_{mod1} is "high (mid/low power mode)," the main-path PA

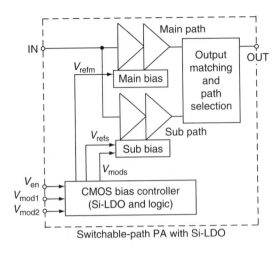

Switchable-path PA with Si-LDO

V_{en}	V_{mod1}	V_{mod2}	V_{refm}	V_{refs}	V_{mods}	Power mode
L	D/C	D/C	L	L	—	Shutdown
H	L	L	H	L	L	High-power mode
H	H	L	L	H	L	Mid-power mode
H	H	H	L	H	H	Low-power mode

FIGURE 18.18   Block diagram of switchable-path PA with Si-LDO and its operation table.

turns off and the subpath PA turns on. In addition, when V_{mod2} becomes "high," the quiescent current of the subpath PA steps down and the PA is switched to low-power mode. Thus, 2 bit digital mode selection is used for switching both the amplifier chain and quiescent current.

In order to reduce the power consumption in each power mode, the transistor size (emitter finger number) in the main-path was optimized for high-power operation, while the size in the subpath was optimized for mid- and low-power operations. Since a two-stage amplifier configuration is adopted for both the main- and subpaths as shown in Figure 18.18, gain variation between the high-power and mid- and low-power modes is expected to be kept small, compared to that of conventional stage-bypass-type PAs [35,48]. Actually, the mid- and low-power mode gain of the switchable-path PA is 5–10 dB higher than that of bypass-type PA [35,48], which may lead to lower RFIC output power and lower total current dissipation of the transmitter block in the transmit state.

Typical RF performance of the switchable-path MMIC PA module fabricated using in-house HBT processes [18,49] was measured under the HSDPA-modulated up-link signal test condition ($\beta_c = \beta_d = \beta_{hs} = 12/15/19.2$). The HBT MMIC PA module, which was fabricated using a glass epoxy substrate, is as small as 4.0 mm × 4.0 mm × 1.1 mm in size. Figure 18.19 shows the measured output characteristics at a supply voltage of 3.4 V and an operating frequency of 1.95 GHz (Band I). The PA delivers PAE of 40%/24%/7% and gain of 27 dB/24 dB/20 dB at output power of 27 dBm/16 dBm/8 dBm in the high-/mid-/low-power modes, respectively, while keeping ACLR1 (±5 MHz offset) less than −40 dBc. Each PAE is as high as that of a PA with a DC-DC converter, and the maximum gain variation between all power modes is about 7 dB. The PA also delivers very flat gain in each power mode and keeps the steady gain discrepancy between the power modes. As can be seen in Figure 18.19b, PAE in the high-power mode decreases to 8% at a 16 dBm output power. However, switching from the high-power mode to the mid-power mode gives much PAE improvement from 8% to 24%. In a low output power range of less than 8 dBm, reduction in the quiescent current is effective in realizing lower power consumption, because the PA in the low-power mode operates with a sufficient margin for specifications of ACLR1.

In order to verify the effectiveness of the switchable-path PA, let us compare the PDF-based current consumption. The PDF-based current is often used as the most important figure-of-merit in

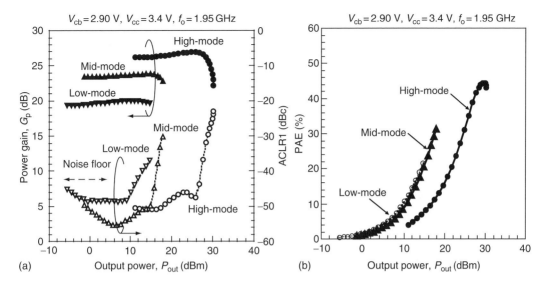

FIGURE 18.19 Measurement results for Band I swithcable-path PA: (a) power gain and ACLR1 characteristics, and (b) PAE.

CDMA PAs. Figure 18.20a shows the measurement-based estimation of total collector current (I_{ct}) and supply voltage (V_{cc}) versus PA's output power (P_{out-PA}) characteristics. In the figure, please note the assumption that a conventional PA, that is a standard 3 mm × 3 mm PA product, operates with a DC-DC converter and the switchable-path PA operates without a DC-DC converter. Typical efficiency versus output current (I_{out}) characteristics of a DC-DC converter are plotted in Figure 18.20b, where output voltage (V_{cc}) is used as a parameter. Figure 18.20c shows the estimated PDF on output current (I_{ct}) for the two amplifier topologies. We can see that in the case of a delicate supply voltage control (e.g., small step control of 1.0–3.5 V with 0.5 V step), the conventional PA with a DC-DC converter is capable of operating with lower current consumption in the high output power range of more than 16 dBm. Based on Figure 18.20c, we can calculate the PDF-based total current consumption. The calculated current consumptions of Figure 18.20c are 26.1 mA for the PA with a DC-DC converter and 28.0 mA for the switchable-path PA without a DC-DC converter.

Thus, the switchable-path PA allows handset designers to remove a DC-DC converter by the sacrifice of small current consumption.

18.3.2 LOW-REFERENCE-VOLTAGE OPERATION POWER AMPLIFIER

As is well known, CDMA PAs and their peripheral circuits often face relentless pressure for low power consumption so as to make battery life of the handsets longer, because the CDMA PAs consume quiescent current continuously. In addition to low power consumption, the proliferation of multiband phones also gives the CDMA PAs strong pressure for simplifying the integration of the PAs into the phones. With regard to the simplification, the reduction in reference voltage (V_{ref}) of the bias circuits in the PA is one of the most effective ways to promote the simplification, because reducing a V_{ref} to 2.5 V or less allows the PA to share power supplies with an Si-RF IC or a baseband LSI, thereby reducing the number of power supplies available in a phone. It is, however, not easy to reduce the V_{ref} of the HBT PAs basically. The reason for it is that as depicted in Figures 18.1 and 18.4, the PAs often use on-chip emitter-follower-based bias circuits which operate with a V_{ref} of higher than 2 V_{be}, or 2.6 V in order to realize low quiescent current and high power operation at the same time [50–54]. Therefore, the PA and its peripheral circuits such as an Si LDO regulator usually require a final battery voltage of 2.85 V or higher.

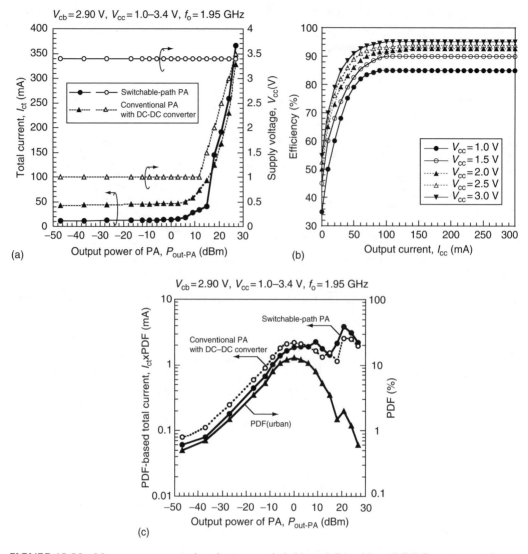

FIGURE 18.20 Measurement comparison between switchable-path PA without DC-DC converter and conventional single-path PA with DC-DC converter: (a) total collector current and supply voltage, (b) typical efficiency of DC-DC converter, and (c) PDF-based total collector current.

On the other hand, a few advanced manufacturers have recently reported on bipolar field-effect transistor (BiFET) technologies, which enable the integration of HBTs and FETs on the same chip at a low cost [55–59]. Successful use of these technologies allows the V_{ref} operation of less than 2 V_{be} because a source-follower-based bias circuit is available for the bias circuits instead of an emitter-follower-based one [57]. However, these technologies need extra mask steps, thereby leading to the increase in the production cost. In contrast to the BiFET, some recent researches on reduction of built-in potential of the HBT have been reported [60,61]. Although lowering the potential in the PA is very effective in realizing low-V_{ref} operation, currently all HBT manufacturers do not have such fabrication process technologies.

Regarding circuit design techniques for low-V_{ref} operation, some important works on HBT bias circuits have been reported to date [2,62]. These works imply the possibility of 2.6–2.7 V low-V_{ref} operation. In these studies, however, the 2 V_{be} problem remains. Therefore, basically these bias

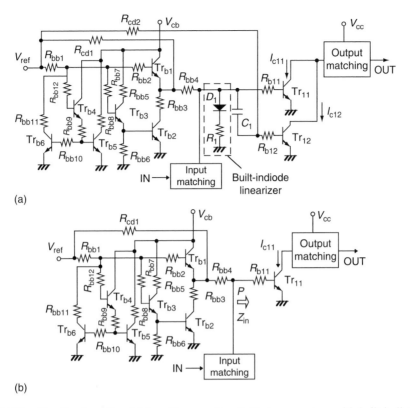

FIGURE 18.21 Circuit schematics: (a) low-V_{ref} operation divided power stage with built-in linearizer and (b) nondivided, voltage- and current-drive power stage.

circuits cannot feed sufficient quiescent current for their power stages under a condition of 2.5 V or less without any supplementary schemes.

This section presents an example of circuit design solutions to realize low-V_{ref} operation CDMA PAs. Figure 18.21a shows the circuit schematic for the power stage and bias circuit operable with a low-V_{ref} of 2.5 V or less [18]. The schematic for the nondivided-type power stage only with voltage and current drive feeding is also shown in Figure 18.21b for the comparison. The configuration of Figure 18.21a features an AC-coupled divided power stage configuration with a built-in diode linearizer, as shown in the figure. In Figure 18.21a, the high resistance, R_{cd1}, is inserted between the V_{ref} terminal and the power stage, Tr_{11}, for current drive. This current injection path (R_{cd1}) feeds easily adequate quiescent current to the power stage even under a low-V_{ref} condition where the emitter follower (Tr_{b1}) cannot work sufficiently. Tr_{12} is the additional AC-coupled power stage and R_{cd2} is the current injection resistance for Tr_{12}. The shunt-type diode linearizer (D_1) [63] is added together with a resistor (R_1) for insertion loss adjustment. The divided power stage and the built-in diode linearizer are the key blocks in realizing smooth output characteristics over a wide temperature and output power range, as shown in the next paragraph. The feature of this configuration is two different kinds of bias feeding to the AC-coupled power stages: one stage is operated with both voltage- and current-drives, while the other stage is operated with only current drive. This aspect is much different from that reported in Ref. [64]. A high-temperature compensation block (Tr_{b4} to Tr_{b6}) is added instead of two diode-connected transistors as depicted in Figure 18.4b.

The principle of operation is as follows. As shown in Figure 18.16b, the base bias voltage of the HBT power stage usually decreases as input power increases. This behavior allows the bias state of the power stage in Figure 18.21a and 18.21b to be changed smoothly from the current drive

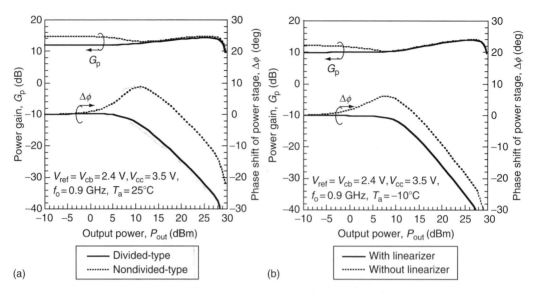

FIGURE 18.22 Simulated output characteristic comparison (a) between divided- and non-divided-type power stage at $T_a = 25°C$, and (b) between divided-type power stage with linearizer and that without linearizer at $T_a = -10°C$.

mode to the voltage drive mode with the increase of input power. The additional power stage (Tr_{12}) and the diode linearizer (D_1) help realize further smooth transfer characteristics. Therefore, we can predict that the power stage will deliver smooth output transfer characteristics even under a low-V_{ref} condition of 2.5 V or less.

Figure 18.22 shows the simulated results of the power stages shown in Figure 18.21a and 18.21b. Figure 18.22a compares the simulated output transfer characteristics between the divided-type power stage and the nondivided type one at room temperature. In the simulation of Figure 18.22a, the diode linearizer was not included so as to verify the effectiveness of the divided power stage configuration. The schematic for the divided-type power stage is accordingly the same as the schematic without the linearizer of Figure 18.21a. Figure 18.22b compares the simulated characteristics between the divided-type power stage of Figure 18.21a with the linearizer and that without the linearizer at a low temperature of $-10°C$. Figure 18.22a shows that the divided-type power stage configuration can give smooth gain and monotonic phase shift, while the nondivided power stage has a gain dip and a nonmonotonic phase shift. The unwanted gain dip and nonmonotonic phase shift is caused by the transition from current drive to voltage drive, because Figure 18.22a implies that the dip and phase shift are represented by the combination of the characteristics for only voltage drive and those for only current drive, as previously shown in Figure 18.8c. As can be seen in Figure 18.22b, the divided power stage with the linearizer provides successful suppression of the gain dip and phase shift even under the $-10°C$ low-temperature condition when the gain dip and nonmonotonic phase shift tend to take place. The first stage was designed so as to compensate for the gain expansion and lag-phase shift of the second stage, as explained previously.

The block diagram, module photograph, and MMIC die micrograph of the fabricated HBT PA module are depicted in Figure 18.23. The PA module is a two-stage amplifier including 50 Ω input and output matching, and is assembled on a 4 × 4 mm glass epoxy substrate. The GaAs MMIC consists of two power stages and their bias circuits, and its die size is as small as 0.76 mm².

Measurement was done using a 900 MHz J-CDMA (Japan-narrowband CDMA: IS-95B compliant) test-set. Supply voltage conditions were as follows: V_{c1} and V_{c2} were 3.5 V, and both V_{ref} and V_{cb} were 2.4 V. Figure 18.24a shows the measured output characteristics at room temperature. The PA exhibits a 27.5 dBm output power, a 26.5 dB power gain, and a 40% PAE with ACPR1(885 kHz

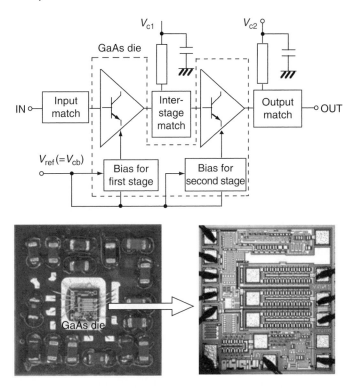

FIGURE 18.23 Low-V_{ref} operation PA module: block diagram, module photo, and GaAs HBT MMIC die mirograph.

offset) of -50 dBc. Figure 18.24b compares the measured output characteristics between the PA with the linearizer and that without the linearizer at a low temperature of $-10°C$. As previously predicted by simulation, the PA with the linearizer has almost no gain dip and degraded ACPR in the middle power level. In contrast, in the PA without the linearizer, a relatively large gain dip and its corresponding ACPR degradation are observed. This experimental result validates the effectiveness of the bias and power stage configuration shown in Figure 18.21a. The measured temperature

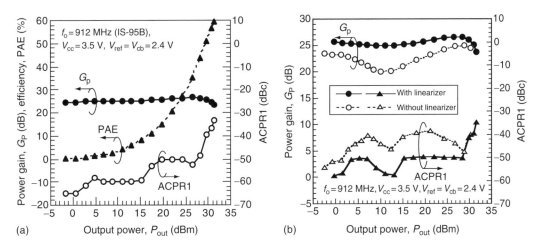

FIGURE 18.24 Measured output characteristics under J-CDMA modulation test: (a) $T_a = 25°C$ and (b) comparison at $T_a = -10°C$ between the PA with linearizer and that without linearizer.

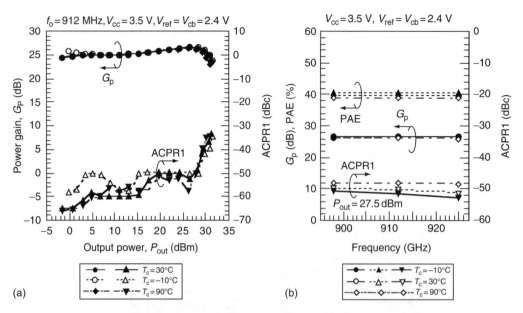

FIGURE 18.25 Measured output characteristics over a wide temperature range under J-CDMA modulation test: (a) input-output power characteristics and (b) frequency response at an output power of 27.5 dBm.

dependence of output characteristics under the J-CDMA modulation test is shown in Figure 18.25a. Over a wide temperature range from $-10°C$ to $90°C$, the PA exhibits a smooth gain variation and sufficiently low ACPR characteristics due to the built-in linearizer, as expected. Figure 18.25b shows the measured frequency response at a 27.5 dBm J-CDMA output power as a parameter of temperature. Over a $-10°C$ to $90°C$ temperature range and a 898–925 MHz frequency range, the PA achieves more than 26 dB power gain, more than 39% PAE, and ACPR of less than -48 dBc. These measurements indicate that the fabricated PA module satisfies the specifications required for J-CDMA PAs [50,52] and has RF performance comparable to currently available products [50,52,53].

18.4   CONCLUSIONS

The chapter provides comprehensive design points for InGaP-HBT MMIC linear PAs while focusing on the relationship between distortion characteristics, bias circuits, and output matching conditions. It demonstrates the design and fabrication for the switchable-path PA and the low-reference-voltage operation PA as two examples of recent CDMA PA technologies. The author expects that the circuit design techniques presented here will help circuit designers understand linear PA design, and lead to further evolution of low-cost, small-size CDMA handset terminals.

REFERENCES

1. T. Hirayama, N. Matsuo, M. Fujii, and H. Hida, PAE enhancement by intermodulation cancellation in an InGaP/GaAs HBT two-stage power amplifier MMIC for W-CDMA, in *IEEE GaAs IC Dig.*, pp. 75–78, 2001.
2. M. Yahagihara, M. Ishii, M. Nishijima, and T. Tanaka, InGaP/GaAs power HBT MMIC for W-CDMA, in *IEICE Microw. Workshop Dig.*, pp. 217–220, 2001.
3. K. Kobayashi, T. Iwai, H. Itoh, N. Miyazawa, Y. Sano, S. Ohara, and K. Joshin, 0.03-cc super-thin HBT-MMIC power amplifier module with novel polyimide film substrate for W-CDMA mobile handsets, in *Proc. 32nd European Microw. Conf.*, pp. 199–202, 2002.
4. Y.-W. Kim, K.-C. Han, S.-Y. Hong, and J.-H. Shin, A 45% PAE/18 mA quiescent current CDMA PAM with a dynamic bias control circuit, in *IEEE RFIC-S Dig.*, pp. 365–368, 2004.

5. Y. Yang, K. Choi, and K. P. Weller, DC boosting effect of active bias circuits and its optimization for class-AB InGaP-GaAs HBT power amplifiers, *IEEE Trans. MTT*, 52(5), 1455–1463, 2004.

6. S. Xu, D. Frey, T. Chen, A. Prejs, M. Anderson, J. Miller, T. Arell, M. Singh, R. Lertpiriyapong, A. Parish, R. Rob, E. Demarest, A. Kini, and J. Ryan, Design and development of compact CDMA/WCDMA power amplifier module for high yield low cost manufacturing, in *IEEE CSIC-S Dig.*, pp. 49–52, 2004.

7. Y. Aoki, K. Kunihiro, T. Miyazaki, T. Hirayama, and H. Hida, A 20 mA quiescent current two-stage W-CDMA power amplifier using anti-phase intermodulation distortion, in *IEEE RFIC-S Dig.*, pp. 357–360, 2004.

8. Y. Yang, Power amplifier with low average current and compact output matching network, *IEEE Microw. Wireless Components Lett.*, 15(11), 763–765, 2005.

9. N. Pan, R. E. Welser, C. R. Lutz, J. Elliot, and J. P. Rodrigues, Reliability of AlGaAs and InGaP heterojunction bipolar transistors, *IEICE Trans. Electron.*, E82-C(11), 1886–1894, 1999.

10. W. Liu, S. K. Fan, T. Henderson, and D. Davito, Temperature dependences of current gains in GaInP/GaAs and AlGaAs/GaAs heterojunction bipolar transistors, *IEEE Trans. Electron. Devices*, 40(7), 1351–1353, 1993.

11. M. Moriwaki, Y. Yamamoto, and K. Maemura, U.S. patent (US 2004/0251967 A1) pending.

12. K. G. Gard, H. M. Gutierrez, and M. B. Steer, Characterization of spectral regrowth in microwave amplifiers based on the nonlinear transformation of a complex Gaussian process, *IEEE Trans. MTT*, 47(7), 1059–1069, 1999.

13. F. Zavosh, M. Thomas, C. Thron, T. Hall, D. Artusi, D. Anderson, D. Ngo, and D. Runton, Digital predistortion techniques for RF power amplifiers with CDMA applications, *Microwave J.*, 22–50, 1999.

14. J. H. Kim, J. H. Jeong, S. M. Kim, C. S. Park, and K. C. Lee, Prediction of error vector magnitude using AM/AM, AM/PM distortion of RF power amplifier for high order modulation OFDM system, in *IEEE MTT-S Dig.*, 2005.

15. S. Yamanouchi, K. Kunihiro, and H. Hida, OFDM error vector magnitude distortion analysis, *IEICE Trans. Electron.*, E89-C(12), 1836–1842, 2006.

16. H. Kawasaki, T. Ohgihara, and Y. Murakami, An investigation of IM3 distortion in relation to bypass capacitor of GaAs MMIC's, in *IEEE MMWMC-S Dig.*, pp. 119–122, 1996.

17. S. Goto, T. Kunii, T. Oue, K. Izawa, A. Inoue, M. Kohno, T. Oku, and T. Ishikawa, A low distortion 25 W class-F power amplifier using internally harmonic tuned FET architecture for 3.5 GHz OFDM applications, in *IEEE IMS Dig.*, pp. 1538–1541, 2006.

18. K. Yamamoto, T. Moriwaki, T. Otsuka, H. Ogawa, K. Maemura, and T. Shimura, A CDMA InGaP/GaAs-HBT MMIC power amplifier module operating with a low reference voltage of 2.4 V, *IEEE J. SSC*, 42(6), 1282–1290, 2007.

19. D. A. Teeter, E. T. Spears, H. D. Bui, H. Jiang, and D. Widay, Average current reduction in (W)CDMA power amplifiers, in *IEEE RFIC-S Dig.*, pp. 429–432, 2006.

20. G. Zhang, S. Chang, and A. Wang, WCDMA PCS handset front end module, in *IEEE IMS Dig.*, pp. 304–307, 2006.

21. G. Zhang, S. Chang, and Z. Alon, A high performance balanced power amplifier and its integration into a front-end module at PCS band, in *IEEE RFIC-S Dig.*, pp. 251–254, 2007.

22. A. van Bezooijen, C. Chanlo, and A. H. M. van Roermund, Adaptively preserving power amplifier linearity under antenna mismatch, in *IEEE MTT-S Dig.*, pp. 1515–1518, 2004.

23. A. Keerti and A. Pham, Dynamic output phase to adaptively improve the linearity of power amplifier under antenna mismatch, in *IEEE RFIC-S Dig.*, pp. 675–678, 2005.

24. G. Berretta, D. Cristaudo, and S. Scaccianoce, CDMA2000 PCS/Cell SiGe HBT load insensitive power amplifiers, in *IEEE RFIC-S Dig.*, pp. 601–604, 2005.

25. S. Kim, J. Lee, J. Shin, and B. Kim, CDMA handset power amplifier with a switched output matching circuit for low/high power mode operations, in *IEEE MTT-S Dig.*, pp. 1523–1526, 2004.

26. S. Kim, K. Lee, P. J. Zampardi, and B. Kim, CDMA handset power amplifier with diode load modulator, in *IEEE MTT-S Dig.*, 2005.

27. J. Nam, J.-H. Shin, and B. Kim, A handset power amplifier with high efficiency at a low level using load-modulation technique, *IEEE Trans. MTT*, 53(8), 2639–2644, 2005.

28. T. Kato, K. Yamaguchi, and Y. Kuriyama, A 4 mm-square 1.9 GHz Doherty power amplifier module for mobile terminals, in *Proc. IEEE Asia-Pacific Microw. Conf.*, 2005.

29. F. Lepine, R. Jos, and H. Zirath, A load modulated high efficiency power amplifier, in *Proc. European Microw. Conf.*, Manchester, pp. 411–414, 2006.

30. T. Apel, Y. -L. Tang, and O. Berger, Switched Doherty power amplifiers for CDMA and WCDMA, in *IEEE RFIC-S Dig.*, pp. 259–262, 2007.

31. J. H. Kim, Y. S. Noh, and C. S. Park, An InGaP-GaAs HBT MMIC smart power amplifier for W-CDMA mobile handsets, *IEEE J. SSC*, 38(6), 905–910, 2003.

32. J. H. Kim, K. Y. Kim, Y. H. Choi, and C. S. Park, A power efficient W-CDMA smart power amplifier with emitter area adjusted for output power levels, in *IEEE MTT-S Dig.*, pp. 1165–1168, 2004.

33. T. Tanoue, M. Ohnishi, and H. Matsumoto, Switch-less-impedance-matching type W-CDMA power amplifier with improved efficiency and linearity under low power operation, in *IEEE MTT-S Dig.*, 2005.

34. G. Hau, C. Caron, J. Turpel, and B. MacDonald, A 20 mA quiescent current 40% PAE WCDMA HBT power amplifier module with reduced current consumption under backoff power operation, in *IEEE RFIC-S Dig.*, pp. 243–246, 2005.

35. K. Kawakami, S. Kusunoki, T. Kobayashi, M. Hashizume, M. Shimada, T. Hatsugai, T. Koimori, and O. Kozakai, A switch-type power amplifier and its application to a CDMA cellphone, in *Proc. European Microw. Conf.*, Manchester, pp. 348–351, 2006. (LPEE: DC-DC conversion including polar modulation.)

36. I. A. Rippke, J. S. Duster, and K. T. Kornegay, A single-chip variable supply voltage power amplifier, in *IEEE RFIC-S Dig.*, pp. 255–258, 2005.

37. J. Lee, J. Potts, and E. Spears, DC/DC converter controlled power amplifier module for WCDMA applications, in *IEEE RFIC-S Dig.*, pp. 77–80, 2006.

38. K. Kawakami, S. Kusunoki, T. Kobayashi, M. Hashizume, M. Shimada, T. Hatsugai, T. Koimori, and O. Kozakai, A switch-type power amplifier and its application to a CDMA cellphone, in *Proc. European Microw. Conf.*, Manchester, pp. 348–351, 2006.

39. G. Hau, J. Turpel, J. Garrett, and H. Golladay, A WCDMA HBT power amplifier module with integrated Si DC power management IC for current reduction under backoff operation, in *IEEE RFIC-S Dig.*, pp. 75–78, 2007.

40. K. Kunihiro, K. Takahashi, S. Yamanouchi, T. Hirayama, H. Hida, and S. Tanaka, A polar transmitter using a linear-assisted delta-modulation envelope-amplifier for WCDMA applications, in *Proc. European Microw. Conf.* (Manchester), pp. 137–140, 2006.

41. J. Deng, P. Gudem, L. E. Larson, and P. M. Asbeck, A high-efficiency SiGe BiCMOS WCDMA power amplifier with dynamic current biasing for improved average efficiency, in *IEEE RFIC-S Dig.*, pp. 361–364, 2004.

42. Y. S. Noh and C. S. Park, An intelligent power amplifier MMIC using a new adaptive bias control circuit for W-CDMA applications, *IEEE J. SSC*, 39(6), 967–970, 2004.

43. J. Nam, Y. Kim, J.-H. Shin, and B. Kim, A high-efficiency SiGe BiCMOS WCDMA power amplifier with dynamic current biasing for improved average efficiency, in *Proc. 34th European Microw. Conf.*, Amsterdam, pp. 329–332, 2004.

44. Y. Yang, High efficiency CDMA power amplifier with dynamic current control circuits, in *IEEE CSIC-S Dig.*, pp. 53–56, 2004.

45. H.-T. Kim, K.-H. Lee, H.-K. Choi, J.-Y. Choi, K.-H. Lee, J.-P. Kim, G.-H. Tyu, Y.-J. Jeon, C.-S. Han, K. Kim, and K. Lee, High efficiency and linear dual chain power amplifier without/with automatic bias current control for CDMA handset applications, in *Proc. European Microw. Conf.*, Amsterdam, pp. 337–340, 2004.

46. J. Deng. R. Gudem, L. E. Larson, D. Kimball, and P. M. Asbeck, A SiGe PA with dual dynamic bias control and memoryless digital predistortion for WCDMA handset applications, in *IEEE RFIC-S Dig.*, pp. 247–250, 2005.

47. T. Shimura, K. Yamamoto, M. Miyashita, K. Maemura, and M. Komaru, InGaP HBT MMIC power amplifiers for L-to-S band wireless applications, in *IEICE Microw. Workshop Dig.*, pp. 225–230, November 2007.

48. AWT6273R, HELP3™ Cellular/WCDMA 3.4 V/29 dBm linear power amplifier module, in Anadigics data sheet.

49. K. Yamamoto, M. Miyashita, T. Moriwaki, S. Suzuki, N. Ogawa, and T. Shimura, A 0/20 dB step linearized attenuator with GaAs-HBT compatible, ac-coupled, stack type base-collector diode switches, in *IEEE IMS Dig.*, pp. 1693–1696, 2006.

50. RF2162, 3 V 900 MHz linear power amplifier, in RF Micro Devices data sheet.

51. AWT6112, Cellular dual mode AMPS/CDMA 3.4 V/28 dBm linear power amplifier module, in Anadigics data sheet.
52. CXG1178K, JCDMA power amplifier module, in Sony semiconductor data sheet.
53. CX77144, Power amplifier module for CDMA (887–925 MHz), in Skyworks data sheet.
54. AWT6271, HELPTM Cellular/WCDMA 3.4 V/28 dBm linear power amplifier module, in Anadigics data sheet.
55. O. Krutko, K. Xie, M. Shokrani, A. Gupta, and B. Gedzberg, Structures and methods for fabricating integrated HBT/FET's at competitive cost, U.S. Patent US 2005/0184310 A1, August 2005.
56. C. J. Wei, Y. Zhu, C. Cismaru, A. Klimashov, and Y. A. Tkachenko, Four terminal GaAs-InGaP bifet DC model for wireless application, in *Proc. IEEE APMC*, pp. 4–7, 2005.
57. A. G. Metzger, P. J. Zampardi, R. Ramanathan, and K. Weller, Drivers and applications for an InGaP/GaAs merged HBT-FET (BiFET) technology, in *IEEE Topical Workshops on Power Amplifiers for Wireless Communications Dig.*, January 2006.
58. A. Gupta, B. Peatman, M. Shokrani, W. Krystek, and T. Arell, InGaP-Plus™—A major advance in GaAs HBT technology, in *IEEE CSIC-S Dig.*, pp. 179–182, 2006.
59. A. G. Metzger, P. J. Zampardi, M. Sun, J. Li, C. Cismaru, L. Rushing, R. Ramanathan, and K. Weller, An InGaP/GaAs merged HBT-FET (BiFET) technology and applications to the design of handset power amplifiers, in *IEEE CSIC-S Dig.*, pp. 175–178, 2006.
60. L. Rushing, P. Zampardi, and M. Sun, Reliability evaluation of InGaAsN for PA handset applications, in *Proc. CS MANTECH Conf.*, New Orleans, pp. 57–60, 2005.
61. P. J. Zampardi, M. Sun, L. Rushing, K. Nellis, K. Choi, J. C. Li, and R. Welser, Demonstration of a low V_{ref} PA based on InGaAsN technology, in *IEEE Topical Workshop on Power Amplifier for Wireless Communications Dig.*, 2006.
62. E. Järvinen, S. Kalajo, and M. Matilainen, Bias circuits for GaAs HBT power amplifiers, in *IEEE MTT-S Dig.*, pp. 507–510, 2001.
63. K. Yamauchi, K. Mori, M. Nakayama, Y. Mitsui, and T. Takagi, A microwave miniaturized linearizer using a parallel diode with a bias feed resistance, *IEEE Trans. MTT*, 45(12), 2431–2435, 1997.
64. S. Shinjo, K. Mori, H. Ueda, A. Ohta, H. Seki, N. Suematsu, and T. Takagi, A low quiescent current CV/CC parallel operation HBT power amplifier for W-CDMA terminals, *IEICE Trans. Electron.*, E86-C(8), 1444–1450, 2003.

19 ΔΣ Digital-RF Modulation for Adaptive Wideband Systems

Albert Jerng and Charles G. Sodini

CONTENTS

Next-generation wireless systems aim to provide high data rates on the order of 1 Gb/s in order to support demand for high-speed mobile Internet applications. In addition to increasing channel bandwidths, wireless systems are employing techniques such as OFDM and modulation schemes such as 64-QAM to pack more bits per hertz. The system choices lead to higher signal-to-noise ratio (SNR) requirements and higher peak-to-average power ratios (PAPR) in the signals. Thus, higher dynamic range is required in the circuits.

19.1 INTRODUCTION

This research introduces a new transmitter architecture that targets high data rate wideband systems [1]. ΔΣ digital–RF modulation efficiently modulates a radiofrequency (RF) carrier with very wide bandwidths by replacing high dynamic range analog circuits with high-speed digital circuits. This technique enables power and area savings in the implementation of a wideband transmitter as complementary metal oxide semiconductor (CMOS) transistors continue scaling. Additionally, a wideband digital–RF modulator can be software-defined to transmit multiple frequency channels, with variable bandwidths and modulation schemes within the band. The multitude of wireless standards requires distinct radio designs for each set of specifications. A wideband, programmable RF modulator with high dynamic range is the key building block for a universal transmitter targeting future high data rate systems.

The conventional *I-Q* modulator consists of a digital-to-analog converter (DAC), analog filter, and analog mixer. The DAC and analog filter become more difficult to design as the bandwidth and dynamic range requirements of the transmitter increase. At high frequencies, timing errors and nonlinear capacitances limit DAC dynamic range [2], rather than static DC errors. Power consumption in the analog reconstruction filter increases proportional to the signal bandwidth for a constant dynamic range [3]. The scaling of CMOS transistors and supply voltages creates further challenges from the standpoint of dynamic range. A mismatch between *I* and *Q* paths and DC offsets cause modulator image and local oscillator (LO) leakage signals, respectively. Closed loop phase locked loop (PLL) modulation directly modulates the voltage controlled oscillator (VCO) without requiring a DAC or analog filter [4]. However, the data bandwidth is limited by the relatively narrow PLL loop bandwidth required to suppress synthesizer phase noise. It is unsuitable for wideband applications with bandwidths on the order of 100 MHz.

Oversampling $\Delta\Sigma$ concepts [5] can be applied to create a digitally controlled vector modulator that provides a continuous range of output phase and amplitude values. In Figure 19.1, filtered *I,Q* digital data are oversampled and converted into 1 bit output streams by digital $\Delta\Sigma$ modulators. The phase shifter needs to either pass the LO signal or invert it, and can be realized trivially by a differential current steering switch in CMOS. While the quadrature LO signals being modulated toggle between only two phases, their sum represents a continuous range of phase/amplitude modulation based on the duty cycle of the oversampled $\Delta\Sigma$ bit-stream. The modulation of the RF carrier is correctly encoded but obscured by a large amount of high-frequency quantization noise. An RF bandpass filter removes the outband quantization noise and reconstructs the modulated RF signal. The concept can be extended to multibit $\Delta\Sigma$ modulators using binary- or unary-weighted LO current-steering cells.

This architecture replaces the DAC, analog reconstruction filter, and analog mixer with a high-speed $\Delta\Sigma$ modulator, a digital-RF converter (DRFC) based on current-steering switches, and a passive RF bandpass fitter (BPF). Both baseband and RF inputs to the DRFC are fully switching digital signals and no distortion results from signal clipping. Thus, analog design issues such as noise/linearity trade-offs, DC offsets, and *I-Q* matching are eliminated. Unlike the conventional *I-Q* modulator, the digital-RF modulator benefits from digital CMOS scaling since the power and area of the high-speed $\Delta\Sigma$ modulators will decrease as channel lengths and supply voltages are reduced. The DRFC building block combines the functionality of a DAC and mixer, and enables greater integration. Passive RF filtering is attractive because it has high dynamic range and consumes no power. In digital process scaling, there has also been a trend of increasing levels of metallization and lower resistance routing. As a result, on-chip inductors with higher *Q* can be built using lower loss metals that are farther away from the substrate. Higher inductor *Q* allows the design of sharper, more selective passive bandpass filters, improving the quantization noise suppression of a $\Delta\Sigma$ digital-RF modulator.

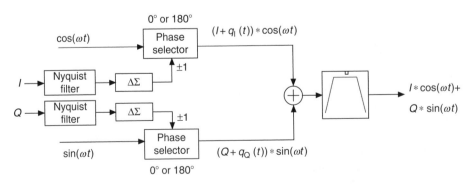

FIGURE 19.1 Digital $\Delta\Sigma$ RF modulator. (From Jerng, A. and Sodini, C., *IEEE J. Solid-State Circuits*, 42, 1710, 2007. With permission.)

19.2   DIGITAL–RF CONVERSION

An RF digital–analog converter (RF DAC) was introduced in Ref. [6]. In general, the output of a DAC contains the desired analog signal as well as its images around each multiple of the DAC clock frequency. The RF DAC uses one of these high-frequency clock images as an RF output. A sine-wave at the desired clock image frequency modulates the DC bias voltage of the DAC current source. This increases the clock image power by mixing the DAC impulse response to the clock image frequency. One drawback is that the RF DAC outputs substantial energy at other frequencies, including its primary output near DC. The DRFC in Ref. [7] uses a balanced version of the RF DAC unit cell. The balanced design is a more efficient RF modulator because the low-frequency response around DC is rejected and the RF output is now the primary output. The DRFC unit cell is identical in structure to a Gilbert-cell mixer. The difference is that the digital baseband inputs drive the top pair of current-steering switches, multiplying a balanced RF carrier signal by ±1, based on the digital data. In contrast, the Gilbert-cell mixer's bottom differential pair is driven by an analog baseband signal and must be linearized.

The DRFC performs a mixing operation between the digital baseband signal and the local oscillator signal to produce a modulated RF output. It merges the functions of the DAC and a mixer, while eliminating the analog filtering between the two. However, the frequency spectrum of the digital signal repeats itself at all multiples of the sampling rate or clock frequency. These clock images are upconverted by the DRFC without any filtering besides the sinc response associated with the zero-order hold in the digital–RF interface. When using a digital ΔΣ modulator, high-frequency-shaped quantization noise is upconverted without any filtering. In either case, an RF bandpass filter is required at the output of the DRFC. In the previous work [6,7], substantial off-chip filtering is required to eliminate high-frequency clock images and quantization noise and produce a clean transmit spectrum.

The fundamental difficulty with direct digital–RF conversion is the transmission of spurious emissions outside the signal band that are difficult to filter at RF frequencies. One approach to the filtering problem involves embedding a semidigital finite impulse response (FIR) reconstruction filter in the digital–RF interface. In Ref. [8], the 1 bit output of a ΔΣ modulator goes through a six-tap digital delay line. Each output of the delay line is applied to the switch input of an RF DAC cell whose current source is weighted with the appropriate FIR filter coefficient. The drawback of this approach is that a large number of taps are needed to implement an FIR filter with reasonable attenuation. For example, in Ref. [9], a 128-tap delay line realizes the equivalent transfer function of a second-order analog filter with −20 dB/decade slope in the stopband. For high RF output frequencies and wide baseband bandwidths with high sampling rates, a large number of delay taps and weighted current sources in the digital–RF interface will consume a large amount of power. In Ref. [8] with a six-tap FIR filter, the RF output spectrum at 1 GHz contains a significant amount of out-of-band quantization noise. The magnitude of this noise is approximately −35 dBc at a frequency offset of 15 MHz from a 1 GHz single-tone output.

Our design integrates a high-Q passive LC bandpass filter into the load of the digital–RF conversion circuit. Figure 19.2 shows a circuit schematic of our quadrature DRFC with load filter. This realization integrates both the DRFC and RF BPF under a single supply. In the unit cells, quadrature phases of an RF carrier are applied to differential pairs biased by tail current sources. The output currents of the differential pairs are routed through differential current-steering switches controlled by the I and Q digital ΔΣ modulator output bits. The resulting output currents from each unit cell are summed and then filtered by a passive LC network that also performs I–V conversion. The filter does not consume any additional voltage headroom due to the inductor, and acts as a tuned load to provide high gain for the DRFC.

Passive LC filtering at RF is attractive because it provides high dynamic range with no power consumption. At multi-GHz RF frequencies, LC filters are relatively small in terms of die area compared to analog RC-based filters. As signal bandwidths increase, active analog filters consume more power for a given dynamic range, while on-chip passive LC bandpass filters become more

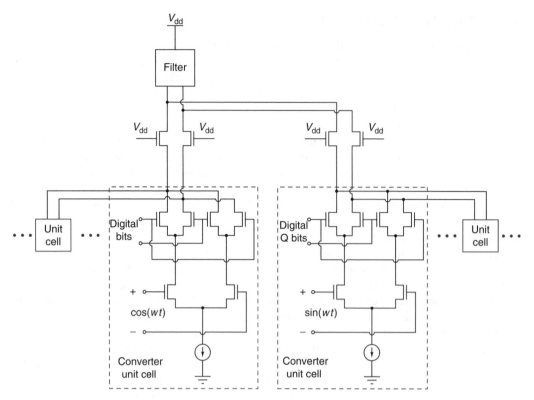

FIGURE 19.2 Quadrature DRFC core. (From Jerng, A. and Sodini, C., *IEEE J. Solid-State Circuits*, 42, 1710, 2007. With permission.)

feasible due to a reduction in the required Q. Thus, direct digital–RF conversion becomes attractive for wideband systems. However, the steepness of a passive LC filter's roll-off is limited by the finite Q of on-chip passives, and the feasibility of this approach depends on the RF filtering requirements.

The selectivity requirements for the BPF depend on the location and magnitude of the spurious signals. Oversampling $\Delta\Sigma$ modulation pushes the clock images farther away, and also reduces the number of unit cells required by the converter. This reduces the power consumption and area of the DRFC, and minimizes routing parasitics in a high-frequency converter. The spurious signals are dominated by the shaped out-of-band quantization noise, whose magnitude can be engineered through design of the $\Delta\Sigma$ noise transfer function (NTF).

19.3 $\Delta\Sigma$ SYSTEM ARCHITECTURE

A prototype $\Delta\Sigma$ digital–RF modulator, with block diagram shown in Figure 19.3, was designed to achieve greater than 1 Gb/s data rate using OFDM modulation with 1 MHz subcarrier spacing and 200 MHz RF bandwidth centered at 5.25 GHz. By using M-ary QAM modulation up to 256-QAM on each subcarrier, a maximum data rate of 1.6 Gb/s can be transmitted. Assuming a required SNR of 30 dB for 256-QAM and a PAPR of 15 dB for the 200 subcarrier OFDM signal, the modulator needs to provide an SNR > 45 dB.

In designing the $\Delta\Sigma$ RF modulator, the oversampling clock frequency and $\Delta\Sigma$ NTF must be carefully chosen to maximize the in-band SNR, minimize out-of-band spurious signals, and prevent aliasing of clock images and quantization noise. The implemented RF modulator utilizes a second-order, 3 bit $\Delta\Sigma$ modulator clocked at 2.625 GS/s. With a baseband *IQ* bandwidth of 200 MHz, the oversampling ratio for the modulator is approximately 13. The LO frequency is 5.25 GHz and the RF reconstruction filter is a fourth-order LC bandpass filter implemented using coupled resonators. The

ΔΣ Digital–RF modulator

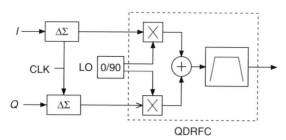

FIGURE 19.3 ΔΣ digital–RF modulator. (From Jerng, A. and Sodini, C., *IEEE J. Solid-State Circuits*, 42, 1710, 2007. With permission.)

modulator achieves a simulated in-band SNR of 52 dB over the 200 MHz passband. Design details regarding the choice of system parameters and the high-speed ΔΣ modulator design are provided in Ref. [1]. The next section will provide design details for the autotuned fourth-order LC bandpass filter.

19.4 LC BANDPASS FILTER DESIGN

The design of a passive LC bandpass filter involves several challenges. In order to attain a sharp roll-off, the on-chip passives used in the filter must have high Q. The finite Q of on-chip inductors typically limits the overall Q to the range of 10–20, depending on the parameters of the process. With high Q and a narrow passband, any variations in capacitance or inductance over process and temperature will cause a shift in the filter center frequency and a large amplitude loss in the fixed RF bandwidth of the system. Meanwhile, noise and spurious signals at out-of-band frequencies may fall in the shifted passband of the filter. A practical realization must include an automatic control loop to stabilize the filter center frequency over process and temperature variations. In order to attain higher resonator Q, active Q-enhancement can be added but will be accompanied by a penalty in power consumption and dynamic range.

A conventional bandpass design method is to take a lowpass prototype ladder filter and perform a lowpass to bandpass transformation by placing a capacitor in series with all inductors and an inductor in parallel with all capacitors. The resulting ladder contains too many inductors, occupying large die area. A narrowband approximation to the bandpass ladder filter can be realized with shunt LC resonator sections that are capacitively coupled [10]. This topology minimizes the number of inductors required in the filter. Further area reduction is achieved by converting the topology into its differential form, as shown in Figure 19.4. Symmetric differentially wound inductors take up less area than two equivalent single-ended inductors. In addition, wasteful spacing between inductors is eliminated, allowing a more compact layout. The capacitor area is reduced by a factor of 4 in the differential resonator implementation.

The coupled resonator design methodology follows in a manner analogous to conventional ladder design using filter look-up tables [10]. Based on the normalized resonator quality factor defined as

$$q_o = \frac{\Delta f}{f_m} Q_o \qquad (19.1)$$

normalized coefficients of coupling k, and normalized source and load q values are tabulated for coupled ladder lowpass prototypes. In Equation 19.1, Δf is the filter bandwidth, f_m is the filter center frequency, and Q_o is the unloaded resonator Q.

The un-normalized bandpass parameters are given by [10]

$$K_{i,k} = k_{i,k} \frac{\Delta f}{f_m} \qquad (19.2)$$

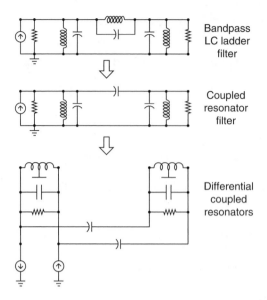

Bandpass
LC ladder
filter

Coupled
resonator
filter

Differential
coupled
resonators

FIGURE 19.4   Filter topology.

$$Q_i = q_i \frac{f_m}{\Delta f} \tag{19.3}$$

Choosing an inductance value L, the filter component values can be calculated using [10]

$$\frac{1}{2\pi \sqrt{LC_N}} = f_m \tag{19.4}$$

where
$\quad C_N$ are the nodal capacitances with all other nodes shorted to ground
$\quad Cc_{i,k}$ are the coupling capacitances between nodes i and k and are equal to $K_{i,k}C_N$

The ith resonator will consist of an inductance L and a capacitance $C = C_N - Cc_{i-1} - Cc_{i+1}$. The source and load resistances can be found from the un-normalized Q_i using [10]

$$R_i = \omega L Q_i \tag{19.5}$$

The loss represented by the finite Q of the resonator can be approximated with a resistor R_p in parallel with the inductor. If we make the approximation that R_p is constant over the narrow bandwidth of the filter, then the physical source and load resistors required by the design can be calculated using

$$R_{S,L} = \frac{R_i R_p}{R_p - R_i} \tag{19.6}$$

For a particular filter order, there is a minimum resonator quality factor Q_o required to realize the filter's transfer function. The minimum resonator Q required for a fourth-order Bessel BPF at 5.25 GHz with bandwidth 260 MHz is

$$Q = \frac{f_m}{\Delta f}(q_o) = \frac{5.25e9}{260e6}(1.297) = 26.2 \tag{19.7}$$

In a given filter type, i.e., Chebyshev, or Bessel, higher order filters provide sharper selectivity, but require pole locations with higher Q's.

19.4.1 AREA CONSIDERATIONS

A straightforward way to reduce filter area is to minimize the required order of the filter and thus the number of resonators. A Chebyshev filter has the sharpest attenuation characteristics and can be used to minimize the required order. However, the Chebyshev response will also require a higher resonator Q to realize the desired pole locations. In a given process, there is generally a design space for inductors that trades-off area for Q [11]. The area of the inductor increases to maximize Q. Because of the trade-off between area and Q, a fourth-order Chebyshev BPF with two resonators may not necessarily be smaller in area than a sixth-order Bessel BPF, whose three resonators each require lower Q.

19.4.2 IMPEDANCE CONSIDERATIONS

According to Equation 19.5, the equivalent resistance of the ith resonator is proportional to both L and Q_i. Higher equivalent resistance at resonance is advantageous because for a given desired output voltage swing, less current is required in the DRFC driving the filter. Higher Q filter designs and larger valued inductors can save power in the DRFC.

19.4.3 RESONATOR DESIGN

A high-Q passive LC filter must be tunable. This is most readily accomplished by incorporating a varactor as the resonator's capacitance. A large ratio between the tunable capacitance and fixed capacitance in the resonator maximizes tuning range. The fixed capacitance is made up of parasitic routing capacitances and loading capacitances on the filter nodes. A larger inductor value reduces the overall capacitance required at resonance, and causes the fixed parasitic capacitances to be a greater percentage of the total capacitance. The inductance must be chosen small enough to insure that the tuning range is greater than the expected center frequency variation.

Varactor design is influenced by Q and linearity. Q is typically limited by the on-chip inductor, although at higher frequencies the varactor Q can become significant since

$$Q_{\text{var}} = \frac{1}{\omega C R_\text{s}} \tag{19.8}$$

The overall resonator Q can be expressed as

$$Q_{\text{res}} = \left(\frac{1}{Q_{\text{ind}}} + \frac{1}{Q_{\text{var}}} \right)^{-1} \tag{19.9}$$

The varactor can also cause signal distortion through its nonlinear C–V characteristic. The voltage across the varactor varies as a function of the input signal driving the filter. This creates a signal-dependent capacitance in the filter that will result in distortion. The magnitude of the distortion products can be calculated by first writing an equation for the tank capacitance, C, as a function of the input signal, V, using a power series expansion:

$$C(V) = C_0 + C_1 V + C_2 V^2 + C_3 V^3 + C_4 V^4 + \cdots \tag{19.10}$$

Figure 19.5 depicts a typical differential resonator design consisting of two inductors, a varactor in series with a fixed capacitor, and a resistor representing the overall resonator loss. The input to the resonator is a current-mode sine wave at the resonance frequency, ω_o. The resulting output voltage

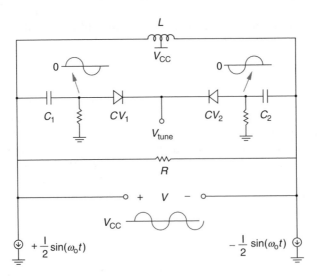

FIGURE 19.5   Differential resonator with nonlinear $C(V)$. (From Jerng, A. and Sodini, C., *IEEE J. Solid-State Circuits*, 42, 1710, 2007. With permission.)

will consist of a sine wave at ω_o as well as harmonics due to the nonlinear capacitance. The output voltage, V, can be expressed as

$$V = IZ = I\left(R\|j\omega L\|\frac{1}{j\omega C(V)}\right) \tag{19.11}$$

The solution to Equation 19.11 is not staightforward because C is a function of V, which in turn is a function of C. The analysis can be greatly simplified by assuming that V only contains frequencies of the original input current signal. This assumption is valid because the distortion products will generally be much smaller than the fundamental signals and will not influence $C(V)$. Using this assumption, one can derive an expression for the resonator current as a function of the resonator voltage to determine the level of distortion products:

$$I = VY = V\left(\frac{1}{R} + j\omega C(V) - \frac{j}{\omega L}\right) \tag{19.12}$$

The relevant distortion products to consider are those that will fall into the passband of the filter. Harmonics of ω_o will be at much higher frequencies and be filtered. When two tones at closely spaced frequencies ω_1 and ω_2 undergo third-order nonlinearity, distortion known as IM3 products will appear as tones at frequencies $2\omega_1 - \omega_2$ and $2\omega_2 - \omega_1$. When the tone spacing is small compared to the bandwidth of the filter, the IM3 products will appear in-band.

In the general case, one will substitute $V = A\sin(\omega_1 t) + A\sin(\omega_2 t)$ into Equation 19.12 and find the ratio between the coefficients of the fundamental currents and the IM3 currents. Since both the fundamental and IM3 frequencies are in the passband of the filter, the actual output voltage can be calculated as the current times the impedance at resonance, R. The ratio between fundamental and IM3 voltages can be used to calculate the output IP3 voltage (OIP3) of the filter.

In a differential implementation, as shown in Figure 19.5, the tank capacitance $C(V)$ will be an even function of the differential tank voltage. In other words, $C(+A) = C(-A)$ due to the symmetry of the circuit. Note that the differential capacitance $C(V)$ in Figure 19.5 is the series combination of $C1$, $C2$, $CV1$, and $CV2$. Since $C(V)$ is an even function, only the even powers of V in Equation 19.10 are required. We can now substitute Equation 19.10 and $V = A\sin(\omega_1 t) + A\sin(\omega_2 t)$

into Equation (19.12). The following equation for I can be written where we have only used the even powers of $C(V)$ up to 2. It is also assumed that $\omega_1 \simeq \omega_2 \simeq \omega_o$.

$$I = [A\sin(\omega_1 t) + A\sin(\omega_2 t)]\left[\frac{1}{R} - \frac{j}{\omega_o L} + j\omega_o C_o + j\omega_o C_2(A\sin(\omega_1 t) + A\sin(\omega_2 t))^2\right] \quad (19.13)$$

Near resonance, $-(j/\omega_o L)$ and $j\omega_o C_o$ will approximately cancel. The relevant terms from the multiplication in Equation 19.13 for IM3 calculations are then

$$I = \frac{A}{R}\sin(\omega_1 t) + \frac{A}{R}\sin(\omega_2 t) + \frac{3}{4}j\omega_o C_2 A^3\left[\sin(2\omega_1 - \omega_2)t + \sin(2\omega_2 - \omega_1)t\right] \quad (19.14)$$

Given the voltage magnitude, A, of the two tones, the power ratio between the IM3 tones and the fundamental tones is calculated to be

$$IM3(\text{dBc}) = 20\log 10\left|\frac{3\omega_o C_2 R A^2}{4}\right| \quad (19.15)$$

The IM3 depends on the second-order coefficient, C_2, of the power series expansion of $C(V)$, and the effective resistance R of the tank at resonance. A higher tank Q will have higher R and result in worse IM3 performance. This indicates a trade-off between filter selectivity and filter distortion in tunable filters. A higher C_2 also causes worse distortion. In general, reducing the tuning range of the filter will lower C_2. Thus, there is also a trade-off between tuning range and distortion in tunable filters.

19.4.4 PROTOTYPE 5.25 GHz FILTER DESIGN

A passive LC bandpass filter centered at 5.25 GHz was designed using the above design procedure with $q_o = 1.297$. The filter order is limited to a four-order Bessel bandpass due to the Q of the on-chip inductors. A schematic of the filter is shown in Figure 19.6. A three-turn differential inductor was designed and optimized for Q using the EM simulator Sonnet. Simulated differential inductance and Q were 2.2 nH and 26 at 5.25 GHz. The metal width and spacing was 8 and 4 μm, respectively. A M1 shield was placed underneath the inductor to reduce substrate losses. PN-junction varactors were used for the resonator load capacitances. The varactor capacitance varies from 0.2 to 0.46 pF when the tuning voltage across the varactor ranges from 2.2 to 0.3 V. Metal-insulator-Metal (MiM)

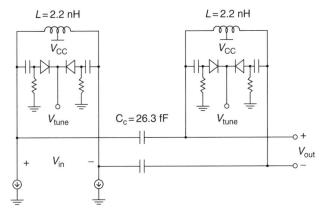

FIGURE 19.6 LC BPF schematic. (From Jerng, A. and Sodini, C., *IEEE J. Solid-State Circuits*, 42, 1710, 2007. With permission.)

1.2 pF capacitors in series with the varactors serve two purposes. First, they linearize the $C-V$ characteristics of the varactor and minimize distortion. Second, they allow the varactor to be configured with its cathode at the virtual ground point of the differential resonator. The parasitic diode from n- to substrate is then at a virtual ground, preventing it from degrading the resonator Q. The series MiM caps do, however, reduce the filter tuning range. The filter is designed to tune from 4.8 to 5.6 GHz, corresponding to a tuning range of $\pm 8\%$. Parallel plate capacitors using the top two metal layers were utilized to implement the small 26.3 fF coupling capacitors. Minimizing resistance in the layout connections to the varactors and inductors was critical for maintaining a high quality factor in the resonator.

The differential tank capacitance $C(V)$ for the 5.25 GHz filter in Figure 19.6 was found through simulations that included extracted layout parasitics. The actual tuning range with parasitics was approximately 500 MHz. Using MATLAB®, $C(V)$ at $V_{tune} = 2$ V was fit to a polynomial expression with coefficients $C_0 = 416.06$ fF and $C_2 = 1.1739$ fF. Using Equation 19.15 and assuming a maximum expected differential peak voltage of 0.6 V in the filter, the IM3 products are calculated to be -46.5 dBc with 0.3 V differential output for each of the two fundamental tones. Circuit simulations in SpectreRF showed the IM3 products to be -51 dBc for the same conditions. Simulations show that for a larger tuning range of 1 GHz, the IM3 products increase to -25 dBc.

19.4.5 AUTOMATIC TUNING LOOP

Automatic frequency tuning can be implemented by configuring a replica resonator or the filter itself as a VCO and locking it to a separate reference frequency in a PLL [12]. These PLL tuning systems are costly in terms of die area and circuit complexity. This design adapts a tuning technique used in baseband filters [13] for use at RF frequencies.

The tuning scheme takes advantage of the fact that the phase difference between filter input and filter output is 90° at the center frequency. According to Equation 19.4, there is a resonant condition between L and C_N at the filter center frequency, where $C_N = C_p + C_c$. This condition can be written using admittances as

$$\frac{1}{j\omega_o L} + j\omega_o C_N = \frac{1}{j\omega_o L} + j\omega_o C_p + j\omega_o C_c = 0 \qquad (19.16)$$

The admittance of each resonator, consisting of L and C_p, at the filter center frequency is then

$$Y_{resonator} = \frac{1}{j\omega_o L} + j\omega_o C_p = -j\omega_o C_c \qquad (19.17)$$

The impedance of each resonator at the filter center frequency is thus

$$Z_{resonator} = \frac{j}{\omega_o C_c} \qquad (19.18)$$

By modelling each resonator as an impedance of jX as in Figure 19.7, where $X = 1/\omega_o C_c$, one can derive the transfer function and the input to output phase relationship of the coupled resonator filter at the filter center frequency:

$$\frac{V_{out}}{V_{in}} = \frac{jR}{X} \qquad (19.19)$$

$$\frac{V_{out}}{I_{in}} = \frac{V_{in}}{I_{in}} \times \frac{V_{out}}{V_{in}} = \frac{RX^2}{R^2 + X^2} \times \frac{jR}{X} = \frac{jR^2 X}{R^2 + X^2} \qquad (19.20)$$

From Equation 19.19, it can be seen that the filter output will lead the filter input by 90°. As the resonator Q decreases, $R^2 \ll X^2$, and the insertion loss of the filter will increase proportional to R^2 or Q^2 since $Q = R/w_o L$.

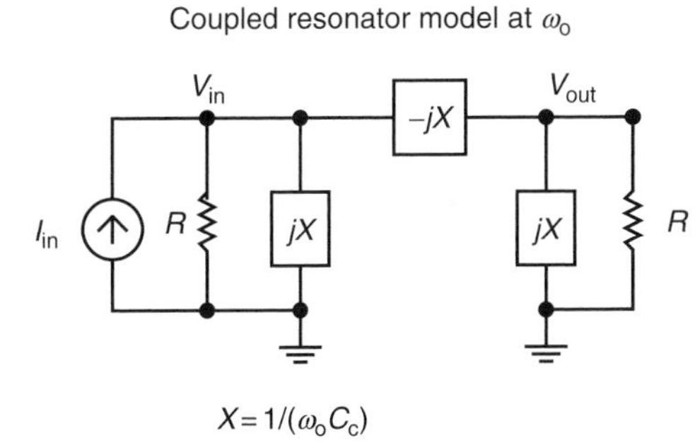

Coupled resonator model at ω_0

$$X = 1/(\omega_0 C_c)$$

FIGURE 19.7 Coupled resonator model.

Figure 19.8 shows a simplified block diagram of the self-tuning loop using single-ended signals. All circuits are implemented differentially and all signals are taken differentially except for the op-amp output. The filter input and output are lightly coupled through small capacitors to a high-frequency phase detector. The differential outputs of the phase detector are applied to a differential-input, single-ended-output op-amp that drives the control voltage of the varactors in the resonators. The feedback loop forces zero differential voltage between the phase detector outputs, which corresponds to the condition of 90° phase difference between the phase detector inputs. Since the filter will always be centered at the system LO frequency, the 5.25 GHz LO signal driving the DRFC can be used to calibrate the filter. The filter does not need to be reconfigured as an oscillator. Self-tuning avoids matching issues, and adds minimal additional circuitry. Most importantly, the filter is calibrated in its actual circuit implementation within the integrated DRFC, including all parasitic effects of the circuit and layout.

19.4.6 DIGITAL TUNING LOOP

The automatic tuning loop can be slightly modified to implement an all-digital tuning loop. This was not implemented, but will be discussed to show how a couple of simple changes can improve the tuning loop with respect to varactor nonlinearity. In Figure 19.9, the op-amp can be replaced with a comparator, and the analog varactor can be replaced with a bank of digitally switchable capacitors. The output of the comparator can be applied to a digital state machine, which will increment or decrement the number of capacitors being switched in, starting from a predetermined initial condition. When the comparator output bit switches polarity, the tuning loop is locked and the digital state machine outputs are held constant. The advantage of the digital tuning loop is that the resonator capacitance will remain fixed as a large signal is applied to the filter. Thus, the distortion due to nonlinear varactor capacitance will be eliminated.

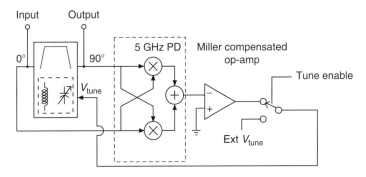

FIGURE 19.8 Tuning loop block diagram. (From Jerng, A. and Sodini, C., *IEEE J. Solid-State Circuits*, 42, 1710, 2007. With permission.)

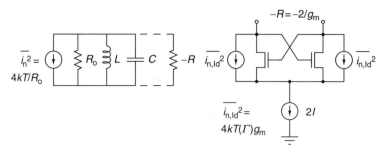

FIGURE 19.9 Measured filter with and without Q-enhancement.

19.4.7 Q-Enhancement

Passive tunable-LC filter performance is limited by the finite Q of on-chip inductors and varactors. One solution is to enhance Q by adding a negative resistance circuit in parallel with the LC resonator [14,15]. A simple model of this is shown in Figure 19.10 using a mixed oxide semiconductor (MOS) cross-coupled differential pair to generate the negative resistance. The issue with this approach is that noise and nonlinearities from the active circuit limit the achievable dynamic range as the Q-enhancement factor increases [16]. This is particularly difficult for receive filter design where the dynamic range constraint at the low noise amplifier (LNA) is severe. For a transmit filter, the signal level is much larger. Analysis can be done to relate dynamic range to the Q-enhancement factor and bias current of the negative resistance circuit.

The Q and effective impedance at resonance with Q-enhancement can be found from Figure 19.10 as follows.

$$Q_o = \frac{R_o}{\omega L} \tag{19.21}$$

$$Q_{enh} = \frac{R_o \| - R}{\omega L} = \left(\frac{R_o}{\omega L}\right)\left(\frac{-R}{R_o - R}\right) = Q_o\left(\frac{1}{1 - (R_o/R)}\right) \tag{19.22}$$

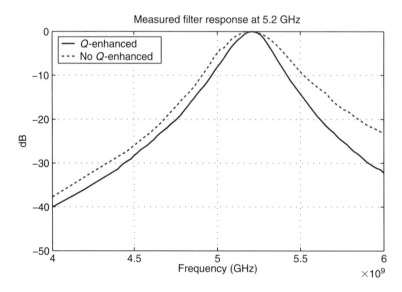

FIGURE 19.10 Q-enhanced resonator model.

If we define

$$E = \frac{1}{1 - (R_o/R)} \qquad (19.23)$$

then

$$R_{enh} = R_o E \qquad (19.24)$$

Lowering R increases the Q-enhancement factor E. Since $R = 2/g_m$ for the cross-coupled MOS pair, a larger E also implies larger g_m, which increases the noise power spectral density of the MOS drain current noise sources. Thus, Q-enhancement increases the noise power at the resonator nodes both through the increased R_{enh} as well as through an increase in g_m. The maximum linear signal swing at the resonator nodes with the negative resistance circuit can be approximated as being proportional to the gate overdrive of the MOS transistors $V_{gs} - V_t$. For short channel devices, the following equations hold [17]:

$$I = W v_{sat} C_{ox} (V_{gs} - V_t) \qquad (19.25)$$

$$g_m = W v_{sat} C_{ox} \qquad (19.26)$$

$$(V_{gs} - V_t) = \frac{I}{g_m} \qquad (19.27)$$

Equation 19.27 indicates that an increase in g_m must be accompanied by an increase in current to maintain the same linear signal swing. Accordingly, the current in the negative resistance circuit must increase as Q is enhanced to maintain constant linearity. From Figure 19.10, we can find a general expression for the approximate dynamic range of the Q-enhanced LC filter:

$$DR = \frac{Max\ Power}{Noise\ Power} \qquad (19.28)$$

The maximum linear power at the resonator output is defined as

$$Max\ Power \approx \frac{\left(\frac{1}{4}\sqrt{2}(V_{gs} - V_t)\right)^2}{2R_{enh}} = \frac{I^2}{16 g_m^{\ 2} R_o E} \qquad (19.29)$$

In Equation 19.29, a voltage swing that yields IM3 products that are $<-60\,dBc$ is chosen as the maximum linear voltage swing. The value $\frac{1}{4}\sqrt{2}(V_{gs} - V_t)$ is found from simulations. The noise power at the resonator output is equal to

$$Noise\ Power = \overline{i_{n,R_o}^{\ 2}} R_{enh} \Delta f + \overline{i_{n,I_d}^{\ 2}} R_{enh} \Delta f \qquad (19.30)$$

$$\overline{i_{n,R_o}^{\ 2}} R_{enh} = \frac{4kT}{R_o} R_o E = 4kTE \qquad (19.31)$$

$$\overline{2 i_{n,I_d}^{\ 2}} R_{enh} = (2)4kT\gamma g_m R_o E \qquad (19.32)$$

$$Noise\ Power = 4kTE[1 + 2\gamma g_m R_o]\Delta f \qquad (19.33)$$

In Equations 19.32 and 19.33, γ represents the channel noise coefficient [18] that is equal to 2/3 in long-channel devices. Measurements have shown that γ is higher than 2/3 for short channel devices [19]. Substituting Equations 19.29 and 19.33 into Equation 19.28,

$$DR = \frac{I^2}{(16)4kT g_m^{\ 2} R_o E^2 [1 + 2\gamma g_m R_o]} \Delta f \qquad (19.34)$$

TABLE 19.1

Current versus Q-Enhancement for

DR = 60 dB Over 200 MHz BW

E	g_m (mS)	$V_{gs} - V_t$	I (mA)
1.1	0.125	0.4	0.05
1.2	0.23	0.5	0.12
1.3	0.32	0.6	0.19
1.4	0.39	0.7	0.28
1.5	0.46	0.8	0.37
1.6	0.52	0.89	0.46
1.7	0.57	0.98	0.55
1.8	0.61	1.065	0.65
1.9	0.65	1.15	0.75
2	0.69	1.24	0.85
2.5	0.83	1.67	1.4
3	0.92	2.09	1.9
3.5	0.985	2.51	2.5
4	1.03	2.93	3
4.5	1.07	3.35	3.6
5	1.1	3.77	4.16

Using Equation 19.34, one can calculate the required current I versus enhancement factor E, given an initial Q_o and a desired dynamic range. Table 19.1 lists the required I, g_m, and $V_{gs} - V_t$ for a cross-coupled pair negative resistance circuit assuming an initial Q_o of 20, a noise bandwidth of 200 MHz, $\gamma = 2$, and a dynamic range requirement of 60 dB. Also, a center frequency of 5.25 GHz and a 2.2 nH differential inductance in the resonator is assumed. According to Table 19.1, a Q-enhancement factor of 2 is achievable with a total current of 1.7 mA, yielding a resonator Q of 40. For enhancement factors greater than 2, voltage headroom becomes an issue due to the high required $V_{gs} - V_t$ to maintain a dynamic range >60 dB. The corresponding currents also become quite large given that each resonator requires its own negative resistance circuit. The negative resistance circuit based on a cross-coupled pair can be biased with a constant-g_m current [20], making the negative resistance fairly stable over process and temperature. For moderate Q-enhancement factors, the Q-enhanced resonator will not oscillate. By factoring the expected Q variation of the resonator into the filter design, it is likely that no explicit Q-tuning loop will be required.

Measured results for a prototype 5.25 GHz LC bandpass filter design with integrated automatic tuning loop are given in Ref. [1]. Figure 19.9 compares the measured results for the filter without Q-enhancement [1] to a version of the filter with Q-enhancement subject to dynamic range constraints. The Q-enhancement factor was designed to be 1.8 using 1.5 mA current in cross-coupled differential pairs loading each resonator. The Q-enhanced filter has a -3 dB bandwidth of 210 MHz and provides 14 dB attenuation at ± 300 MHz offset from 5.2 GHz.

19.5 EXPERIMENTAL RESULTS

A prototype IC implemented in 0.13 µm CMOS integrates the RF bandpass filter and autotuning circuitry with two high-speed digital $\Delta\Sigma$ modulators, a quadrature digital-IF upconverter, a quadrature DRFC, and a quadrature LO path with polyphase filter and LO-limiting buffers. A block diagram is shown in Figure 19.11. The integrated circuit (IC) was packaged in an 88-QFN package with exposed ground paddle. Attention is paid in the layout to isolating the high-speed digital circuitry from the DRFC converter circuit. Separate power and ground are provided to the digital circuits and RF circuits. The digital supplies are further subdivided, with separate power and ground

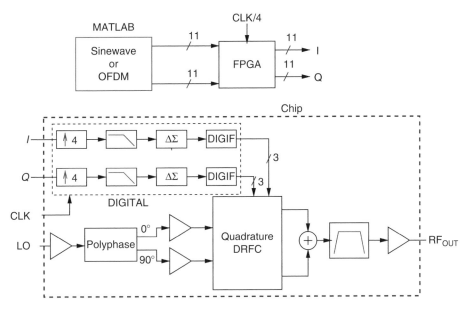

FIGURE 19.11   Test chip block diagram. (From Jerng, A. and Sodini, C., *IEEE J. Solid-State Circuits*, 42, 1710, 2007. With permission.)

assigned to the clock circuitry. Guard rings are placed around the digital block, the DRFC cells, and the LC bandpass filter. Each of the guard rings are tied to separate dedicated ground pads. Space is intentionally left between the DRFC/BPF block and the digital block to minimize substrate coupling. In addition, a blocking (BFMOAT) layer is drawn on the empty space between blocks. This layer blocks P-well and N-well implants, creating regions with higher substrate resistance.

The integrated RF modulator achieves a measured signal-to-noise distortion ratio (SNDR) of 49 dB over a 200 MHz RF bandwidth centered at 5.25 GHz. The bandpass filter is effective in attenuating out-of-band quantization noise from the second-order, 3 bit ΔΣ modulator, allowing the transmitter to meet Federal Communications Commission (FCC) spectral limits outside the 5.15–5.35 GHz UN-II band. The error vector magnitude (EVM), when transmitting a 256-QAM OFDM baseband signal with 100 MHz I,Q bandwidth, was measured to be 30 dB. Assuming 64-QAM modulation across the 200 MHz RF bandwidth, the prototype RF modulator is capable of transmitting a data rate of 1.2 Gb/s. Figure 19.12 plots the output spectrum at 5.25 GHz with a 12 MHz digital baseband sine input. The reader may refer to Ref. [1] for additional figures and measurement results.

The power consumption and die area of the entire modulator are summarized in Table 19.2. All blocks use a supply voltage of 1.5 V except for the DRFC core. A die photo is shown in Figure 19.13. The modulator consumes 187 mW and occupies a die area of 0.72 mm². A general figure-of-merit (FOM) characterizing the energy/bit efficiency of a modulator is (power consumption/data rate). For our modulator,

$$\frac{\text{Power}}{\text{Data rate}} = \frac{187e - 3}{1.2e9} = 0.16 \ \frac{\text{nJ}}{\text{bit}} \tag{19.35}$$

Since two-thirds of the power is consumed in the digital processing, this FOM can be expected to improve with digital process scaling.

19.6   APPLICATIONS FOR WIDEBAND DIGITAL–RF MODULATION

As CMOS technology continues to scale, and wireless system bandwidths increase, direct digital–RF modulation will become more attractive. One driver toward increased bandwidths is the desire

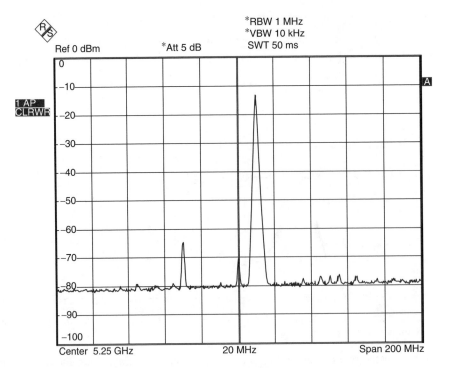

FIGURE 19.12   Measured output spectrum: Digital–IF, 12 MHz sine-wave input. (From Jerng, A. and Sodini, C., *IEEE J. Solid-State Circuits*, 42, 1710, 2007. With permission.)

for higher data rates. Ultrawide band (UWB) currently specifies a channel bandwidth of 528 MHz. Recent allocations of spectrum in the 60 GHz region provide over 5 GHz of contiguous bandwidth available for use. Extremely wide bandwidths will require the use of new RF architectures to minimize power consumption.

A UWB system contains channels occupying frequencies between 3.1 and 10.6 GHz [21]. Each channel uses 128 OFDM subcarriers and the targeted EVM is −20 dB [21]. Assuming a PAPR of

TABLE 19.2

Digital–RF Modulator Power Consumption/Die Area

	Power Consumption (mW)	Die Area
Digital block	120	0.16 mm²
QDRFC core	10 (2.5 V)	0.03 mm²
QDRFC data/clock drivers	33	Included in core
LO polyphase/buffers	20	0.21 mm²
BPF tuning circuitry	4	0.12 mm²
BPF	0	0.2 mm²
Total	187	0.72 mm²

Source: (From Jerng, A. and Sodini, C., *IEEE J. Solid-State Circuits*, 42, 1710, 2007. With permission.)

FIGURE 19.13 Die photo. (From Jerng, A. and Sodini, C., *IEEE J. Solid-State Circuits*, 42, 1710, 2007. With permission.)

~15 dB, an SNR of 35 dB is required. Table 19.3 summarizes the system parameters for a proposed ΔΣ digital–RF modulator transmitting a UWB channel centered at 3.96 GHz. The simulated SNR over the 500 MHz channel bandwidth is 40 dB. The wide fractional bandwidth of UWB enables on-chip integration of an RF bandpass reconstruction filter. However, because the UWB system contains channels anywhere from 3.1 to 10.6 GHz, the challenge in such an implementation is to tune the RF bandpass filter over this wide frequency range. One solution to this problem is to implement the digital–RF modulator at a fixed IF frequency, and then mix to the desired RF frequency using a variable LO.

TABLE 19.3

UWB System Example (Band 1, Channel 2 at 3.96 GHz)

LO frequency	3.96 GHz
Clock frequency	3.96 GHz
Order of ΔΣ NTF	2nd
Quantizer resolution	3
BPF center frequency	3.96 GHz
BPF BW	500 MHz
BPF type	Butterworth
BPF order	3
BPF Q requirement	23
Simulated in-band SNR (500 MHz BW)	40 dB

A second driver pushing bandwidths higher for existing systems is power amplifier (PA) linearity and efficiency. The use of OFDM and its high PAPR has created large interest in finding power efficient PA linearization schemes. Examples of recent approaches include polar modulation, out-phasing, and PA predistortion. Each of these techniques requires a much wider bandwidth in the IQ baseband signals. In both polar modulation and out-phasing, a constant envelope phase-modulated RF signal is generated as input to a switching class PA with higher efficiency. For OFDM systems such as WLAN or WiMax with maximum channel bandwidths of 20 MHz, the phase modulation cannot be implemented via closed-loop PLL modulation without severe phase noise degradation. $\Delta\Sigma$ digital–RF modulation can be used to implement an accurate, efficient wideband phase modulator for use in proposed PA linearization schemes.

Another application for this work is in software-defined radios that use adaptive transmitters. By modulating an entire band of spectrum with sufficient dynamic range, a wideband digital–RF modulator can transmit any number of channels within the band using varying bandwidths and modulation schemes. In this way, the transmitter can be configured to utilize and share spectrum in a certain band on an adaptive basis, depending upon channel conditions, spectrum availability, and the presence of interferers from other systems. One benefit of this approach is that a fixed frequency PLL can be used to generate the LO signal for the transmitter. Frequency tuning over the band is provided by the wideband digital baseband signal.

Finally, wideband digital–RF modulation is not limited to the use of $\Delta\Sigma$ modulation. An over-sampling converter without noise shaping can also be used. In order to achieve the same dynamic range as a $\Delta\Sigma$ modulator, an oversampling converter without noise shaping will require more bits, consuming more power and area in the digital–RF interface. The RF filter will need to attenuate clock images, but will not need to filter out-of-band-shaped quantization noise.

19.7 CONCLUSION

The $\Delta\Sigma$ digital–RF modulator is a power- and area-efficient modulator for achieving high data rates in wideband systems. Measured results demonstrate a 1.2 Gb/s data rate over 200 MHz RF bandwidth centered at 5.25 GHz. Spurs associated with direct digital–RF conversion have been eliminated through the integration of a high-Q, self-tuned RF bandpass filter. The $\Delta\Sigma$ digital–RF modulator is amenable to digital CMOS scaling and software radio.

REFERENCES

1. A. Jerng and C. Sodini, A wideband $\Delta\Sigma$ digital–RF modulator for high data rate transmitters, *IEEE J. Solid-State Circuits*, 42(8), 1710–1722, 2007.
2. W. Schofield, D. Mercer, and L. S. Onge, A 16b 400MS/s DAC with < -160dBm/Hz noise power spectral density, *Int. Solid-State Circuits Conf.*, 2003, pp. 126–127.
3. F. Rezzi, I. Bietti, M. Cazzaniga, and R. Castello, A 70-mW seventh-order filter with 7–50 MHz cutoff frequency and programmable boost and group delay equalization, *IEEE J. Solid-State Circuits*, 32(12), 1987–1998, 1997.
4. M. Perrott, T. Tewksbury, and C. Sodini, A 27-mW CMOS fractional-N synthesizer using digital compensation for 2.5-Mb/s GFSK modulation, *IEEE J. Solid-State Circuits*, 32(12), 2048–2060, 1997.
5. J. Candy and G. Temes, *Oversampling Delta-Sigma Data Converters*. IEEE Press, Washington, DC, 1992.
6. S. Luschas, R. Schreier, and H. Lee, Radio frequency digital-to-analog converter, *IEEE J. Solid-State Circuits*, 39(9), 1462–1467, 2004.
7. P. Eloranta and P. Seppinen, Direct-digital RF modulator IC in 0.13 μm CMOS for wide-band multi-radio applications, *ISSCC, 2005*, pp. 532–533.
8. S. M. Taleie, T. Copani, B. Bakkaloglu, and S. Kiaei, A bandpass delta–sigma RF-DAC with embedded FIR reconstruction filter, *ISSCC, 2006*, pp. 578–579.
9. D. Su and B. Wooley, A CMOS oversampling D/A converter with a current-mode semidigital reconstruction filter, *IEEE J. Solid-State Circuits*, 28(12), 1224–1233, 1993.
10. A. Zverev, *Handbook of Filter Synthesis*. John Wiley, New York, 1967.

11. F. Rotella, D. Howard, M. Racanelli, and P. Zampardi, Characterizing and optimizing high Q inductors for RFIC design in silicon processes, *IEEE Radio Freq. Integr. Circuits Symp.*, 2003, pp. 339–342.
12. X. He and W. Kuhn, A 2.5 GHz low power, high dynamic range, self-tuned Q enhanced LC filter in SOI, *IEEE J. Solid-State Circuits*, 40(8), 1618–1628, 2005.
13. H. Khorramabadi and P. Gray, High-frequency CMOS continuous-time filters, *IEEE J. Solid-State Circuits*, SC-19(6), 939–948, 1984.
14. R. Duncan, K. Martin, and A. Sedra, A Q-enhanced active-RLC bandpass filter, *IEEE Trans. Circuits Syst. II*, 44(5), 341–347, 1997.
15. T. Soorapanth and S. Wong, A 0-dB IL 2140 $\pm$ 30 MHz bandpass filter utilizing Q-enhanced spiral inductors in standard CMOS, *IEEE J. Solid-State Circuits*, 37(5), 579–586, 2002.
16. W. Kuhn, D. Nobbe, D. Kelly, and A. Orsborn, Dynamic range performance of on-chip RF bandpass filters, *IEEE Trans. Circuits Syst. II*, 50(10), 685–694, 2003.
17. D. Hodges, H. Jackson, and R. Saleh, *Analysis and Design of Digital Integrated Circuits*, 3rd ed. McGraw Hill, New York, 2004.
18. A. van der Ziel, Thermal noise in field effect transistors, *Proc. IEEE*, pp. 1801–1812, August 1962.
19. A. Abidi, High-frequency noise measurements on FETs with small dimensions, *IEEE Trans. Electron Devices*, ED-33(11), 1801–1805, 1986.
20. J. Steininger, Understanding wide-band MOS transistors, *IEEE J. Solid-State Circuits*, 6(3), 26–31, 1990.
21. ECMA-368, High Rate Ultra Wideband PHY and MAC Standard, www.ecma-international.org, Accessed on June 2006, December 2005.

20 Mitigation of CMOS Device Variability in Digital RF Processor

Khurram Waheed and Robert Bogdan Staszewski

CONTENTS

20.1 INTRODUCTION

The recent market demand for ultralow-cost cell phones by billions of first-time users in the developing countries has spurred development of single-chip second generation (2G) radios. Modern 2G radios integrate a radiofrequency (RF) transceiver with a digital baseband (DBB) processor in scaled complementary metal oxide semiconductors (CMOS). The ultimate goal of a phone-on-a-chip has not yet been realized due to various integration issues of low-voltage CMOS with 2 W RF power amplifiers, 20 V battery chargers, and receiver bandpass RF surface acoustic wave (SAW) filters. Hence, the integration at the RF and DBB level still provides the lowest cost solution, even though it has repeatedly proven to be a complex technological challenge [1].

To further drive the cost down, transition to a nanoscale digital CMOS technology (feature size ≤ 100 nm) with no mask adders is necessary. However, the RF and analog coexistence with larger-scale digital circuitry in nanoscale digital CMOS presents numerous issues. Many of these problems are mitigated by transforming the RF functionality into an all-digital architecture of a frequency synthesizer and transmitter [2] or a digitally intensive discrete-time architecture of a receiver [3].

Despite these recent architectural advances, the core RF circuits still experience some of the conventional RF system issues, such as device parameter spread and mismatch, performance variability due to environmental conditions, and parasitic coupling. Integration of analog/RF circuits with digital processors brings tremendous benefits of using freely available but powerful digital logic and memory to assist in calibration, compensation, linearization, predistortion, built-in self-test (BIST), etc.

Digital RF processing techniques are focused on using digitally intensive signal processing methods in RF to deliver the ever-increasing levels of wireless terminal functionality in a shrinking form factor. Techniques developed at Texas Instruments for the Digital RF Processor (DRP) platform transform the RF functionality into digital or digitally intensive implementation such that it reaps all the well-known benefits of digital design and automation flow.

Figure 20.1 highlights the 2nd generation of DRP architecture. At the heart of the transceiver lies the all-digital PLL (ADPLL) [2], generating local oscillator (LO) and almost all other clocks. The ADPLL-based transmitter employs the polar architecture with all-digital phase/frequency and amplitude modulation paths. The receiver [3] employs a discrete-time architecture in which the RF signal is directly sampled and processed using analog and digital signal processing (DSP) techniques. The antenna RF input signal is amplified and converted into the current domain by a low-noise transconductance amplifier (LNTA). The RF current is then directly sampled or mixed to zero-IF or very low IF in the charge domain. The signal is then filtered and converted into the digital domain for further conditioning. A digitally controlled crystal oscillator (DCXO) [4] generates a high-quality basestation–synchronized frequency reference such that the transmitted carrier frequencies and the received symbol rates are accurate to within 0.1 ppm. A power management system consists of a bandgap generator and multiple low drop-out (LDO) linear regulators to supply voltage to various radio subsystems as well as to provide good noise isolation between them. An RF-BIST [5] executes an autonomous transceiver performance and compliance testing of the global system for mobile (GSM) standard [6]. The script processor (SCR) [7] handles various TX and RX process calibration, voltage and temperature compensation, sequencing and lower-rate datapath tasks and encapsulates the transceiver complexity in order to present a much simpler software programming model. The data

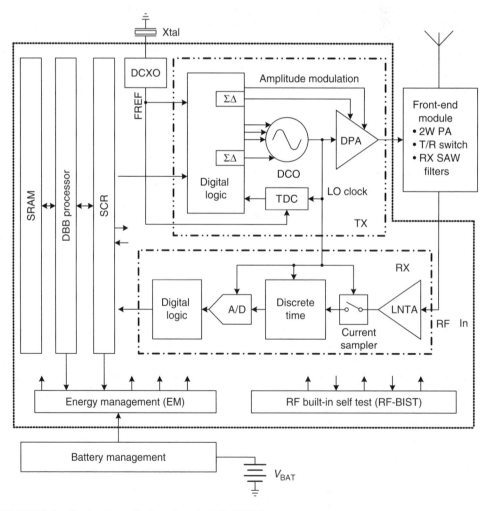

FIGURE 20.1 Single-chip radio based on the 2G of DRP.

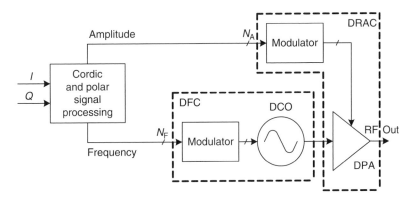

FIGURE 20.2 All-digital RF polar transmitter.

and high-level control is routed from/to DBB processor via data bus router. The transceiver is integrated with the DBB, static random access memory (SRAM) memory in a complete system-on-chip (SoC) solution.

The all-digital RF polar transmitter architecture, which is amenable for nanometer-scale CMOS integration, is shown in a simplified form in Figure 20.2. It exploits the new paradigm by emphasizing (1) fast switching characteristics or high f_T of MOS transistors—40 ps and 100 GHz in 90 nm CMOS, and 15 ps and 250 GHz in 45 nm CMOS; (2) high density of digital logic—250 kgates/mm^2 in 90 nm CMOS and 1 Mgates/mm^2 in 45 nm CMOS, and SRAM memory—1 Mb/mm^2 in 90 nm CMOS and 4 Mb/mm^2 in 45 nm CMOS; (3) small device geometries and precise device matching made possible by the fine lithography. The architecture avoids the typical obstacles to RF integration: (1) biasing currents that are commonly used in analog designs; (2) reliance on voltage resolution with ever-decreasing supply voltages and increasing noise and interferer levels; (3) nonstandard devices that are not needed for memory and digital circuits, which constitute majority of the silicon die area of an SoC.

The digital back-end of Figure 20.2 transmitter consists of dense and fast logic to perform sophisticated DSP. Low-cost logic and memory is also used to fix any imperfections of analog devices. The tiny and well-matched devices allow for precise and high-resolution conversions from digital to two analog polar domains: RF frequency/phase and RF amplitude.

The two "DAC" converters, where "A" stands for frequency/phase or RF amplitude analog domain, are best realized using a topology of Figure 20.3. The conversion cell elements are unit-weighted and realized as finest devices that the lithography can create. Further resolution improvement is achieved through high-speed $\Sigma\Delta$ dithering. Consequently, the integer part of the modulator is realized as a binary-to-unit-weighted encoder and the fractional part as a $\Sigma\Delta$ modulator.

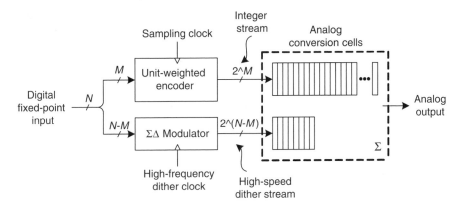

FIGURE 20.3 Generic modulator with uniform array of conversion devices.

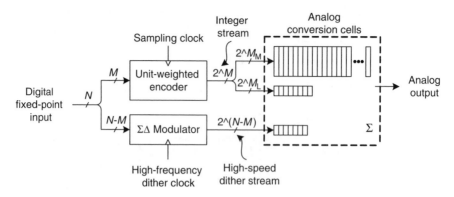

FIGURE 20.4   Generic modulator with segmented conversion devices.

A practical realization of Figure 20.3 converter would result in no more than 8 bits of integer resolution. To break that limitation, Figure 20.4 reveals a modulator structure with segmented arrays of conversion devices in which the ratio of the larger to smaller device weighting is typically a power of two. Thus, an extra few bits of resolution can be achieved.

The binary-to-unit-weighted encoding redundancy in Figures 20.3 and 20.4 (e.g., code 3 could activate *any* three devices, which are not necessarily adjacent) could be further exploited to improve the conversion linearity. Figure 20.5 shows an example of a technique borrowed from the analog data converter field. The dynamic element matching (DEM) is used to perform rotation of active elements at every cycle of the data clock. This way, any device mismatch will be averaged out over the number of participating devices. In this example, the input binary code is 3. Every clock cycle, a different set of three unit-weighted elements is selected. While many selection sequences are possible, the cyclic shift of the active elements as shown is perhaps the simplest to implement and thus most commonly used. If, for example, varactor #4 exhibits particularly large mismatch, then that varactor will be selected proportionally frequently for any code between 1 and 8. For this reason, there will not be a code exhibiting particularly large deviation from the other codes. Compare that to a straight encoding in which only code of 4 will select that problematic varactor.

This chapter deals with the estimation of the random and systematic matching errors of these conversion devices. On new process fabrication technology nodes, delay in the availability of reliable device models, coupled with the challenging RF performance requirements of modern wireless standards, rapidly reduce the acceptable device variability margins. This necessitates statistical analysis of device variability early in the design cycle [16,17]. This task is further complicated by the difficulty in even making reliable laboratory measurements due to the accuracy issues caused by probing noise, minuscule device sizing, dynamic effects, and loading. Some key device characteristics of the digitally-controlled oscillator (DCO) [12], for example, capacitor mismatch, thermal and $1/f$ noise contributions are quite difficult to measure accurately. For small capacitors used in most analog circuits, generally in the range of 0.1–1 pF (picofarad), direct measurement would possess many uncertainties resulting from parasitics in the physical test setup [18,19]. Varactors built now in CMOS are even smaller, typically in atto- to femto-farad range [12]. These varactors also suffer from a relatively low Q-factor and exhibit substantial series resistance due to the n-well material. All these attributes render the characterization of varactors extremely difficult [8,14].

FIGURE 20.5   DEM through cyclic shift of active elements.

It is vital to develop characterization methodologies, which do not penalize the device area budget [1], do not consume extensive testing time, and can be easily ported into the next generations of the CMOS processes with minimal adjustments [13–15]. In this chapter, we present such a characterization technique to estimate the mismatches in transistor switches of an RF digitally controlled pre-power amplifier (DPA) [9] and tuning varactors of the RF DCO [12]. DPA and DCO are vital components of the DRP technology at the heart of the modern GSM/EDGE RF transceiver designs at Texas Instruments. The sections below describe novel techniques used to estimate and arbitrate the device variability in the physical DPA and DCO designs.

20.2 MISMATCHES IN DIGITAL-TO-RF-AMPLITUDE CONVERTER

The direct digital-to-RF-amplitude converter (DRAC) efficiently combines the traditional transmit chain functions of digital/analog (D/A) conversion, filtering, buffering, and mixing or RF output amplitude control into one single circuit. It is based on a DPA for a low-power all-digital GSM/EDGE transmitter proposed in Ref [2]. The DPA [9] (see Figure 20.6) with an efficiency of 17% at 0 dBm output power occupies only 0.005 mm^2 in 90 nm CMOS. The DPA operates as a pseudo-class E RF power amplifier and is driven by a square wave, which is the phase-modulated signal from the ADPLL. An array of N-channel metal oxide semiconductor transistors is used as on/off switches with a certain resistance. The matching network components are chosen to provide a bi-directional current source, second harmonic rejection, switching noise filtering, and critically dampening the switch output. The control logic for each switch comprises an RF digital AND gate whose inputs are the phase-modulated output of the ADPLL and part of the amplitude control word from a digital control block [9].

In order to achieve the tight quantization phase noise (PN) floor necessitated by the cellular modulation schemes, the DPA typically needs to have better than 10 bits of integer resolution. Fractional resolution is achieved by $\Sigma\Delta$ dithering of the least significant bit (LSB) transistors [11]. To achieve such requirements while ensuring monotonic amplitude transfer function characteristics in the presence of device variability necessitates the use of a segmented structure (as proposed in Figure 20.4), i.e., DPA transistors are implemented as $1\times$ and $N\times$ devices, where N is chosen to be 4, 8 or 16, etc. In this work, we will focus primarily on analysis of a DPA structure using $1\times/8\times$ devices. We will refer to the $1\times$ devices as LSB transistors and the $8\times$ devices as most significant bit (MSB) transistors. The performance of such a DPA structure is susceptible to not only the systematic and

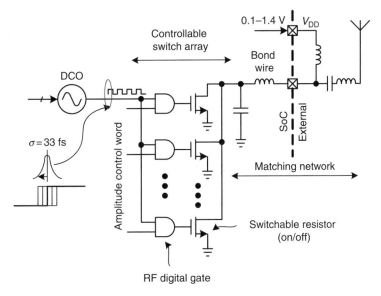

FIGURE 20.6 Circuit schematics of digital to RF amplitude converter.

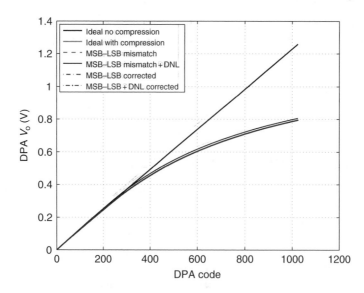

FIGURE 20.7 DPA output voltage versus input digital code under ideal, compressed, and mismatched conditions.

random mismatches in these LSB and MSB devices but also to the systematic sizing ratio mismatch between the MSB/LSB devices. Such a mismatch can invariably occur in spite of following the best circuit layout practices due to the fabrication lithographic process tolerances that impact the MSB and LSB devices differently.

Figure 20.7 shows the DPA output voltage as a function of DPA input digital code in the presence of $+30\%$ mismatch between MSB and LSB devices. Further, MSB and LSB devices have a random variability of 4% and 10% standard deviation, respectively. Note that with MSB devices that are larger than N times $1\times$ devices, the output power will be slightly higher (see Figure 20.8b) than in the absence of such a mismatch and vice versa.

Figure 20.8a shows the DPA output power as a function of the code, while Figure 20.8b provides a zoomed-in view near max codes. It can be seen that the DPA exhibits significant compression (c.a. 4 dB) at max output.

Figure 20.9 presents the voltage step size of the DPA as a function of the input digital code. It is evident that the DPA step size is not uniform across the input digital code, rather the impact of compression on the differential voltage step size of the DPA is clear.

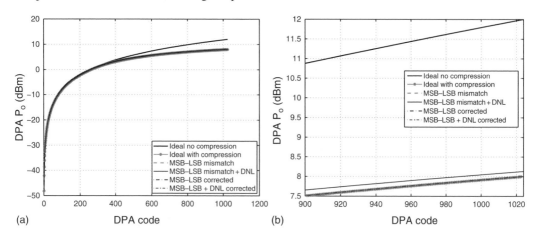

FIGURE 20.8 Impact of device mismatch and compression on DPA output power. (a) DPA output power in dBm and (b) Zoomed in view of DPA output power in dBm.

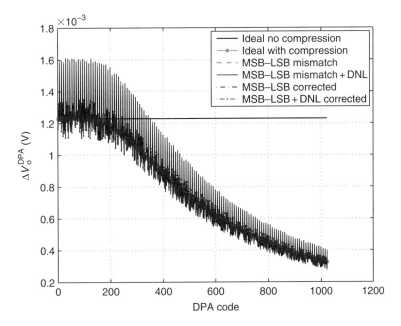

FIGURE 20.9 DPA differential step size as a function of digital input code.

At lower codes (see Figure 20.10a), the DPA step size seems somewhat periodic because of the parity between LSB and MSB step sizes. However, refer to the zoomed-in version shown in Figure 20.10b, the MSB–LSB mismatch also gets compressed at higher DPA input codes. This nonlinear behavior is process-, voltage-, and temperature-dependent and can be further aggravated if, for example, the drive strengths of the MSB and LSB are not designed carefully. This level of differential nonlinearity (DNL) is clearly not acceptable and severely degrades the fidelity of

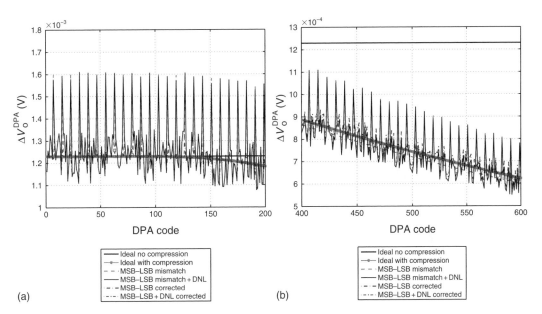

FIGURE 20.10 Impact of device mismatch and compression on DPA differential step size. (a) DPA DNL in linear region (lower input codes) and (b) DPA DNL in compression region (mid input codes).

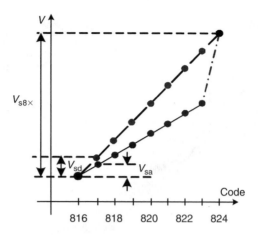

FIGURE 20.11 Code-to-voltage DPA slope variation due to MSB–LSB ratio mismatch.

amplitude modulation. Such a periodic pattern emanating from MSB–LSB transistors not only degrades the close-in performance of the polar transmitter due to spectral regrowth but also has the potential to create spurious content in the complex modulated envelope output from the polar transmitter.

To better illustrate the instantaneous impact of the MSB–LSB segmentation ratio mismatch, Figure 20.11 shows a zoomed version of the mismatch at high DPA codes. The $8\times$ step size is $Vs_{8\times}$, and the desired $1\times$ step size V_{sd} is exactly one-eighth of $Vs_{8\times}$. The actual physical $1\times$ step size on the other hand is V_{sa} (which is smaller than V_{sd} as shown in the example figure). If the number of 1x steps is increased monotonically, it will result in a straight line that has a different slope and tends to diverge from the desired DPA slope. Note that the common starting point of the two lines is the DPA output voltage for the 102nd MSB device (or code 816). Let us assume that the value of the eight unit-weighted LSB bits at a certain instant of time is N. Then if we connect the bits to the DPA $1\times$ transistors directly, the equivalent voltage output will be $N \times V_{sa}$. But if we scale N by the ratio V_{sd}/V_{sa}, with fractional precision using $\Sigma\Delta$ dithering, before it is passed on to the DPA, then the DPA output voltage will be $N \times V_{sd}/V_{sa} \times V_{sa} = N \times V_{sd}$, which is the desired linear output voltage. Therefore, if the MSB/LSB ratio mismatch is known precisely, the amplitude modulation input digital code to the DPA can be predistorted to compensate for this artifact. However, due to compression and the presence of the device-level random variations in the MSB–LSB transistors, this MSB/LSB ratio mismatch cannot be characterized reliably in a straightforward manner. However, we have determined that such an estimate can be predicted accurately using power spectral density (PSD) of the differential DPA steps.

The normalized PSD of the differential DPA steps is shown in Figure 20.12 for two different ratios between MSB and LSB devices. Assuming DPA codes are swept at the normalized frequency of f_s, then the LSB to MSB transitions will be hit f_s/N times. This fact manifests itself in the shown single-sided spectrum as $N/2$ spectral peaks, N being the number of LSB transistors that ideally equal an MSB transistor. The single-sided spectrum has been plotted as $10 \times \log(\Delta V_o^{DPA})$. Further, the ratio mismatch, r_{ML}, between MSB and LSB transistors can be estimated by the relative level of the DNL spectral peaks caused by the ratio mismatch, i.e.,

$$r_{ML} = N10^{(P_1-3)/20} \tag{20.1}$$

where P_1 is the level of the first harmonic peak due to the presence of the device ratio mismatch: 3-dB is subtracted due to the use of a single-sided spectrum. Equation 20.1 provides the correct ratio

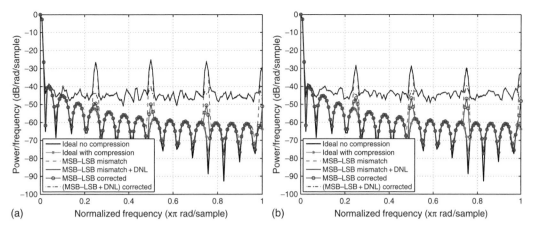

FIGURE 20.12 Spectrum of the DPA differential step size in the presence of MSB/LSB ratio mismatch and device variability. (a) MSB step size 30% larger than corresponding LSB step and (b) MSB step size 20% smaller than corresponding LSB step size.

estimate but it cannot discriminate whether the MSB step is more or less than the $N \times$ LSB steps. This can be estimated using the relation:

$$\text{sign}\left(10 \times \log\left(\max\left(\Delta V_o^{PPA}\right) \cdot \min\left(\Delta V_o^{PPA}\right) / \text{mean}^2\left(\Delta V_o^{PPA}\right)\right)\right) \quad (20.2)$$

Once the ratio is correctly estimated, the ratio can be used to digitally predistort the DPA input codes. This improves the close-in spectral mask of the EDGE-class TX, which may otherwise fail the stringent spurious limits imposed by the wireless standard as well as coexistence requirements with other radios.

Figure 20.13 shows the close-in modulated Enhanced Data rates for GSM Evolution (EDGE) (8-PSK) spectrum for the first channel in the GSM communication band. In the presence of the MSB–LSB ratio mismatch (see Figure 20.13a), the close-in $\Sigma\Delta$ noise shaping is ruined. This causes the TX to violate the close-in spectral requirements in the corresponding GSM RX band as well as the coexistence requirements with nearby commercial broadcast bands. This is certainly not acceptable. However, if the MSB–LSB segmentation ratio in the DPA is known, a simple circuit as shown in

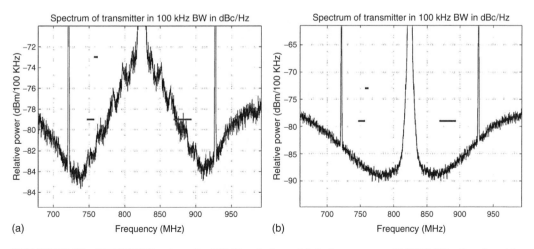

FIGURE 20.13 EDGE TX Spectrum for GSM band, channel 1, in the presence of MSB/LSB ratio mismatch. (a) Close-in spectrum of an EDGE TX in the presence of MSB–LSB ratio mismatch and (b) close-in spectrum of an EDGE TX with digital predistortion for MSB–LSB ratio mismatch.

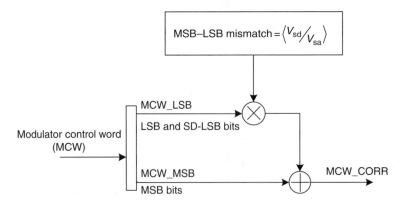

FIGURE 20.14   Schematics of an MSB–LSB ratio mismatch compensation circuit for a DAC modulator.

Figure 20.14 can be used to predistort the input of the DPA. The predistortion is achieved by scaling the LSB contribution in the DPA to counter for the ratio mismatch.

Figure 20.13b shows the EDGE TX close-in spectral emission mask with the systematic MSB–LSB ratio mismatch mitigated using the circuit of Figure 20.14. It is observed that the sigma–delta noise shaping in the DPA is restored and the spectrum at 25 MHz offset from the TX channel improves by approximately 10 dB.

20.3   MISMATCHES IN DIGITALLY CONTROLLED OSCILLATOR

The DCO (see Figure 20.15) has been realized in 90 nm technology optimized for short-channel thin-oxide devices operating as digital switches at only 1.2 V. The DCO is architectured to use a continuous-time, continuous-amplitude analog oscillator core embedded in a digital wrapper to

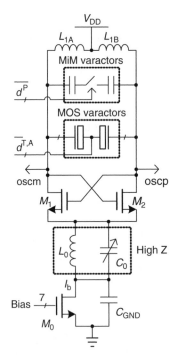

FIGURE 20.15   Circuit schematics of the DCO with three varactor banks.

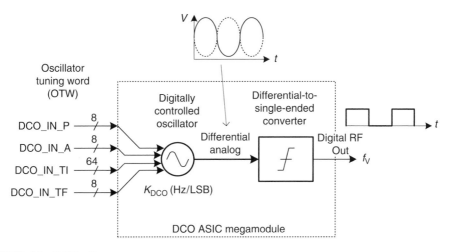

FIGURE 20.16 DCO cell.

realize a DCO [12]. This prevents the analog nature of DCO from propagating to the consequent stages. The frequency output of the LC tank in the DCO can be controlled by either changing the inductance or the capacitance. However, in a monolithic implementation, it is more practical to keep the inductor largely fixed while changing the capacitance of a voltage-controlled device, such as a varactor. Since the digital control of the capacitance is required, the total capacitance is quantized into a number of digitally controlled varactors, which do not necessarily follow the binary-weighted pattern of their capacitance values.

In order to achieve a wider oscillator bandwidth as well as a fine frequency control, the LC tank capacitance has been realized by three different quantization banks of capacitors (see Figure 20.16). Namely, a coarse metal-insulator-metal (MIM)-type process, voltage, and temperature (PVT) calibration bank (PB), a medium-sized inversion-type-MOS acquisition bank (AB), and the fine-grain tracking bank (TB) [2,12]. An example of the relative sizes of PB, AB and TB varactor sizes in a 2G system is shown in Table 20.1.

20.3.1 VARACTOR STRUCTURE IN NANOSCALE CMOS

In the DCO, the finest-grain frequency control is achieved by switching tiny inversion-type CMOS varactors (see Figure 20.17a). For nanoscale processes, the inversion type device is a better varactor candidate due to the well isolation properties in this n-well process as well as featuring more distinctly defined operational regions than the accumulation-type varactor. Unlike varactors from the prior generations of CMOS [8], with a nanoscale varactor, the linear range is quite compressed and has undesirably high gain, which makes the RF oscillator extremely susceptible to noise and operating point shifts. An example C–V curve for a PMOS inversion-type varactor is shown in Figure 20.17b.

TABLE 20.1
DCO Varactor Banks. Δf Is at the High-Band Output

Varactor Bank	Input	Weighting	Step Size (Δf)
PVT	d^{P}	8-bit binary	$\Delta f^{\mathrm{P}} = 4\,\mathrm{MHz}$
Acquisition	d^{A}	64-bit unit	$\Delta f^{\mathrm{A}} = 200\,\mathrm{kHz}$
Tracking int.	d^{T}	64-bit unit	$\Delta f^{\mathrm{T}} = 12\,\mathrm{kHz}$
Tracking fract.	d^{TF}	3-bit unit	$\Delta f^{\mathrm{T}} = 12\,\mathrm{kHz}$

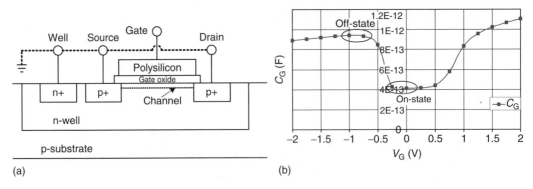

FIGURE 20.17 Positive channel metal oxide semiconductor (PMOS) inversion-type varactor. (a) Structure of an inversion type varactor and (b) gate capacitance versus gate voltage of a measured PMOS varactor with $L = 0.5\,\mu m$, $W = 0.6\,\mu m$, $N = 8$ fingers 122, freq = 2.4 GHz.

For tuning operation, the flat on-state region of the depletion mode and the flat off-state region of the inversion mode are used as two stable binary-controlled operating points. The CMOS varactors employed during the tracking phase are chosen to be the minimum feature size to allow for the finest-grain frequency control physically possible. Further fine-grain control is achieved using fractional high-speed $\Sigma\Delta$ modulators [2,12]. For the DCO TB, the switchable capacitance of the finest differential LSB varactor is on the order of tens of attofarads. Matching of these minimum dimension varactors is important for DCO to ensure monotonic linear tuning characteristics. It is well-known that, as the physical transistor dimensions shrink, mismatch variance may increase as a result of dopant fluctuations and lithographic geometry errors. However, advanced CMOS process lithography today allows for the creation of extremely small-sized but relatively very well-controlled varactors.

20.3.2 FULLY DIGITAL CONTROL OF A CMOS OSCILLATOR

At any instant, the total capacitance contributed by all the three DCO banks: PB, AB, and TB is given by [10]

$$C = C_{0,\text{tot}} + \sum_{k=1}^{N} \overline{d}_k \Delta C_k 2^k \tag{20.3}$$

where
  $C_{0,\text{tot}}$ includes all the parasitic static shunt capacitances due to the varactors in the low-capacitance state (see Figure 20.17)
  N is the total number of varactors in the tuning banks
  $\overline{d}_k$ is the complement of the digital control word for the kth varactor
  ΔC_k is its capacitive contribution

After initial calibration for the PVT and the AB, mostly only TB is active during the DCO tracking operation, therefore (20.3) can be re-expressed as

$$C = C^P + C^A + C_0^T + \sum_{k=1}^{N^T} \overline{d}_k^T \Delta C_k^T \tag{20.4}$$

where
  C^P and C^A are the capacitive contributions of the tracking and ABs respectively
  C_0^T is the sum of the shunt capacitances of the TB varactors

N^{T} is the total number of TB varactors

$\overline{d}_k^{\mathrm{T}}$ is the inverted select bit for the kth TB varactor

ΔC_k^{T} is the switchable capacitance of the kth unit-weighted tracking varactor

A detailed Very High Speed Integrated Circuits (VHSIC), Hardware Discription Language (VHDL) DCO model has been developed, simulated, and calibrated with earlier chips developed for Bluetooth [8] and GSM [20].

TB comprises tiny minimum feature size inversion-type CMOS varactors with capacitance on the order of tens of atto-farads, which can be either switched in or out by the DCO tuning word. Further fine-grain control is achieved using fractional high-speed $\Sigma\Delta$ modulators [11]. Matching of these minimum dimension varactors is important for DCO to ensure monotonic linear tuning characteristics as well as achieve the stringent PN characteristics needed for a cellular class transmitter. It is well-known that, as the physical transistor dimensions shrink, mismatch variance may increase as a result of dopant fluctuations, Refs. [13,16,17] (see and the references therein).

20.3.3 Modeling of Varactor Mismatches

To ensure GSM-compliant transmitter quality, not only it is imperative to use digitally intensive signal processing techniques to compensate for any analog imperfections in the nanoscale RF integration, but it is a requisite to design analog modules in a fashion so as to achieve predictable performance reliably across various lots in volume production. The fabrication varactor mismatch analysis thus is a pivotal step for DCO reliability analysis. For the sake of VHDL modeling, the TB varactors are modeled according to the relation

$$C_{\mathrm{eff}}^{\mathrm{T}} = C_{\mathrm{nom}}^{\mathrm{T}} + C_{\mathrm{sys}}^{\mathrm{T}} + C_{\mathrm{rand}}^{\mathrm{T}} \qquad (20.5)$$

where

$C_{\mathrm{nom}}^{\mathrm{T}}$ is the nominal value of the capacitance

$C_{\mathrm{sys}}^{\mathrm{T}}$ represents the systematic error in the value due to oxide thickness variations and process gradients, etc.

$C_{\mathrm{rand}}^{\mathrm{T}}$ is the random Gaussian distributed error in the physical geometry of the varactors due to the lithographic and other process errors

Due to the physical complexities involved in the estimation of systematic error contributions to the varactor capacitance and the relatively small footprint of varactor arrays, its value was set to zero during the investigations of this work. All the capacitance errors were modeled as Gaussian-distributed with the magnitude of random errors controlled by the standard deviation of the Gaussian density function. To study the effect of mismatches via standard-VHDL modeling, the effect of these capacitance value errors may be introduced in the time-domain model of the DCO [20] as instantaneous frequency deviation. It has been shown in Refs. [10,20] that during the tracking mode of DCO operation, the instantaneous frequency step Δf^{T} is directly proportional to the change in the capacitance value ΔC^{T}, i.e., $\Delta f^{\mathrm{T}}(f) \propto \Delta C^{\mathrm{T}}$. All the random errors were generated at the beginning of a VHDL simulation and were kept constant throughout the simulation. The standard deviation for the mismatches was varied as a percentage of the nominal frequency-dependent DCO gain slope of the TB varactors.

For the purpose of investigation, the analog DCO model with varactor mismatch modeling was integrated into the full-scale RTL and gate-level GSM transmitter models. The intent was to determine the maximum tolerable level of varactor mismatches without violating the GSM target specifications [6]. Several regressions using GSM modulation data were run to determine the quantitative performance degradation caused by the varactor mismatches. The simulated regression results showed that the varactor mismatches did not significantly affect the far-out phase spectrum of the

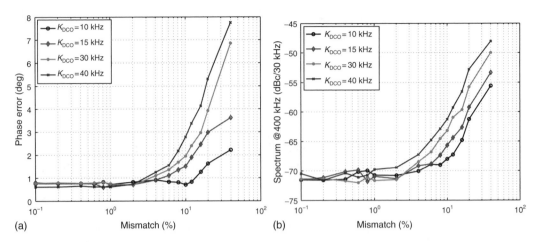

FIGURE 20.18 GSM transmitter output parameters as a function of DCO varactor random Gaussian mismatches. (a) Modulation distortion and (b) modulated spectrum at 400 kHz offset.

transmitter. Rather the main degradation was observed in the close-in spectrum as well as in the root mean squared (RMS) phase error of the demodulated signal. Figure 20.18 shows the degradation of the RMS phase error and PN at 400 kHz offset from the carrier, which is considered the most challenging in the GSM specification. The results are tabulated and have been plotted as a function of the standard deviation of varactor mismatch expressed as a percentage of TB minimum frequency deviation step. Another intuitive observation from the same plots is that a relatively larger mismatch can be tolerated if the TB has a relatively smaller frequency step size.

20.3.4 ASSESSMENT OF PHYSICAL DEVICE MISMATCHES

A novel harmonic technique has been developed to physically assess the amount of mismatch in the GSM transceiver. The technique is based on the notion that the DCO TB with its varactor mismatches can be treated as a synchronous noisy quantizer. The synchronism is due to the clocked feeding of modulation data to the DCO input ports, whereas the noise in the quantizer is due to the physical mismatches in the geometry of nanoscale varactors as a result of fabrication process tolerances. A simple mathematical model is described below. A band-limited modulation signal $m(t)$ is uniformly quantized at the rate $1/T_S$. The nth quantized modulation sample $\hat{m}(n); -\infty < n < \infty$, has a quantization step Δ, given by $\Delta = (m_{max} - m_{min})/2^{B+1} \cong M/2^B$, where m_{max}, m_{min} are the maximum and minimum modulation values, $M = \max(\mathrm{abs}(m_{max}, m_{min}))$ is the maximum (positive or negative) range of the signal and $B + 1$ is the number of bits used in the quantizer codeword. In general, using 2's complement number representation for signed operations, the quantizer fraction is coded as $\hat{m}_B(n) = -a_0 2^0 + a_1 2^{-1} + \cdots + a_B 2^{-B}$, with $|a_i| < 1$ for all i. Therefore $\hat{m}_B(n)$ is always in the range $-1 \le \hat{m}_B(n) < 1$ and the quantized sample is $\hat{m}(n) = M\hat{m}_B(n)$ or $\hat{m}(n) \propto \hat{m}_B(n)$.

For convenience, $\hat{m}(n)$ may be thermometer encoded, which is suitable if each bit in the codeword is used to physically control a frequency control device, such as a varactor in the DCO that is being turned on or off. Defining the quantization to be a functional, say Q, we can express $\hat{m}(n)$ as

$$\hat{m}(n) = Q(m(n)) \tag{20.6}$$

where $m(n)$ is in the range $-(M + \Delta/2) < m(n) < M + \Delta/2$. The quantization error $e(n)$ in the quantized signal is given by $e(n) = m(n) - \hat{m}(n) = m(n) - Q(m(n))$.

For an ideal uniform quantizer, $e(n)$ is uniformly distributed as shown in Figure 20.19a. However, due to the physical limits of fabrication process, the DCO varactors have mismatches and therefore

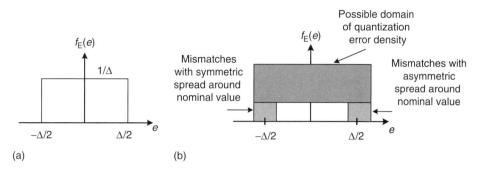

FIGURE 20.19 TB varactor band quantization density functions. (a) Quantization using uniform intervals and (b) quantization in presence of noisy quantization levels.

e has a stochastic density function loosely defined in Figure 20.19b. The density of $f_E(e)$ is scaled to have an integrated area of unity. Observe that, due to noisy quantization levels, the $f_E(e)$ probability density function (pdf) has boundaries, which can have a symmetric or asymmetric spread around the nominal value. Depending on the physics of the fabrication process, the mismatches may or may not have a mean of zero and the spread of the quantization pdf boundary becomes a function of mismatch variance σ_e^2. Consequently, the value of the density function is not a uniform value across the range of e, rather it can take any possible trajectory in the domain shown in Figure 20.19b.

For an ideal quantizer, $e(n)$ is a stationary, uncorrelated, white noise process which is uniformly distributed in the range $-\Delta/2$ to $\Delta/2$, as shown in Figure 20.19a. However, for a real world quantizer, $e(n)$ will not be uniformly distributed due to integral nonlinearities/DNL in the quantizer itself caused by systematic and random device mismatches, layout parasitics, and/or signal characteristics.

For mathematical convenience, we can alternately express the quantization process to be generated by the multiplication of the quantized value with a dirac pulse of duration T_S, i.e.,

$$\hat{s}(t) = \sum_{n=-\infty}^{\infty} \left\lfloor \frac{\text{Sample}(s(t), nT_S) - S_L}{\Delta} \right\rfloor \cdot \delta(t - nT_S) + S_L \tag{20.7}$$

where
T_S is the sampling period
S_L is the lower limit of the quantizer given by $S_L = -S_M - \Delta/2$
$\lfloor \cdot \rfloor$ is the floor operator
Δ is the nominal quantization step as defined above
$\text{Sample}(x, y)$ is the sampling function that samples the signal x at time instant y
$\delta(t - nT_S)$ is a unit-weighted pulse of repetition period T_S

Assuming the sampling rate is significantly larger than the bandwidth of the single-tone signal $s(t)$, i.e., $1/T_S \gg BW(s(t))$, using Taylor series, it is straightforward to conclude that the above process in Equation 20.7 will produce the scaled replicas of the spectrum of the signal $s(t)$ only at the odd multiples of the fundamental signal frequency.

In case the quantizer has noise in its quantization levels, Equation 20.7 needs to be re-written as

$$\hat{s}(t) = \sum_{n=-\infty}^{\infty} \left\lfloor \frac{\text{Sample}(s(t), nT_S) - S_L}{\Delta} \right\rfloor \cdot \delta(t - nT_S) + e(t - nT_S) + S_L \tag{20.8}$$

where $e(t - nT_S)$ is the quantization error-level in the step taken at the sampling instant n, which has a mean of zero (averaged over a large number of quantization levels) and has a variance σ_e^2, which is a measure of the mismatches in the quantization step size. The function $(\delta(t - nT_S) + e(t - nT_S))$ is not a pure unit-weighted Dirac pulse function any more and therefore its spectrum will comprise of even as well as odd harmonics. In this case, the quantization step variance is given by

$$\text{var}(\delta(t - nT_S) + e\,(t + nT_S)) = \text{var}(\delta(t - nT_S)) + \text{var}(e(t + nT_S)) = \sigma_e^2 \qquad (20.9)$$

where $\text{var}(\delta(t - nT_S))$ is a deterministic step with a variance of zero. Thus the mismatch variance is the primary cause of the spectral energy that will appear in the even harmonics of the quantizer spectrum. Note, that in case, the quantizing step function is not ideal, but has finite rise and fall times, then some of the energy in the even harmonics will also be contributed by this waveform distortion. This impairment needs to be appropriately taken into account in the following described harmonic characterization technique.

As a result of the above explained artifacts, a sine-wave, when passed through a noiseless quantizer, produces an output comprising of the fundamental frequency plus the odd harmonics only. However, if the quantizer has noise in the quantization levels, the output will comprise of even as well as odd harmonics. This effect has been captured for a sinusoidal input signal of 100 kHz sampled at 1 GHz in Figure 20.20. The amplitude of these even harmonics and the amount of change that is observed in the odd harmonics has been experimentally verified to be dependent on mismatch variance σ_e^2.

In this work, we have taken advantage of this phenomenon to develop an efficient, low complexity and fast characterization technique to determine the physical mismatches in the TB varactors of the DCO. The spectral amplitude of these even harmonics can be used as a metric to determine the amount of mismatch in the quantization levels. Figure 20.21 presents the results of recording the amplitude of the first three harmonics as a function of the modeled varactor mismatches using a sinewave of 203 kHz for modulation. Figure 20.22 shows a lab measurement using an Agilent™ spectrum analyzer indicating the amplitude of the second harmonic observed by application of a 203 kHz sinewave modulation to the actual GSM transmitter chip.

It can be seen that a -41.26 dB level of second harmonic in Figure 20.22 corresponds to approximately 5% of standard deviation in the varactors constructed in the 90 nm CMOS. Further, lab

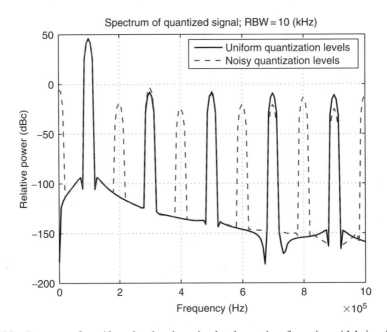

FIGURE 20.20 Spectrum of a uniform-level and a noisy-level quantizer for a sinusoidal signal.

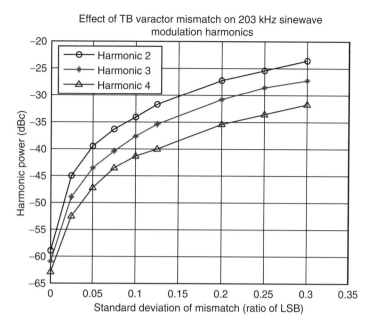

FIGURE 20.21   Harmonic power level of sinewave modulation tones as a function of varactor mismatch.

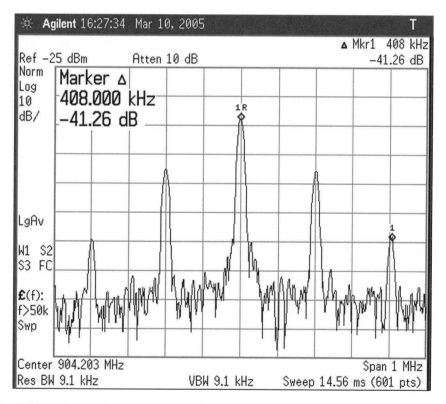

FIGURE 20.22   Lab spectral measurement at DCO output with a 203 kHz sinewave input on a 90 nm GSM transceiver chip.

TABLE 20.2

Lab Performance Measurements on a 90 nm GSM Transceiver Chip

Parameter	Measured Results	Specification
Average 400 kHz modulated spectrum in 30 kHz	−68.9 dB	−60 dB
Average RMS phase error	0.96°	5°

measurements were carried out on several GSM chips to measure the 400 kHz offset PN and the RMS phase error across all GSM channels at various temperatures. The average of lab measurements has been tabulated in Table 20.2. The lab results upon correlation with the characterization plots of Figure 20.18 indicate the varactor mismatches to be smaller than 5%. These characterized values are superior to the statistical mismatch estimates based on the typical 90 nm process data.

20.4 SUMMARY

In this chapter, we have described a DRP-based architecture of an all-digital polar transmitter for the 2G cellular applications. The architecture is based on a dense and fast digital logic to perform DSP, as well as two digital-to-RF converters, namely a DCO and a DPA. The digital logic outputs two high-rate digital sample streams that constitute phase/frequency and amplitude components of a polar coordinate vector representation. The DCO performs conversion from digital code to carrier frequency deviation, whereas DPA performs conversion from digital code to RF amplitude of the carrier. Both of these converters use segmented device banks of varactors (DCO) and MOS transistor switches (DPA), which are amenable to device ratio as well as systematic and random device mismatches. We have also described estimation techniques and suggested compensation algorithms against device variability of the two converters. The proposed techniques allow for fast and accurate mismatch assessment, so that compensation schemes, if needed, may be employed.

REFERENCES

1. A. A. Abidi, RF CMOS comes of age, *IEEE Journal of Solid-State Circuits*, 39(4), 549–561, 2004.
2. R. B. Staszewski, J. Wallberg, S. Rezeq, C.-M. Hung, O. Eliezer, S. Vemulapalli, C. Fernando, K. Maggio, R. Staszewski, N. Barton, M.-C. Lee, P. Cruise, M. Entezari, K. Muhammad, and D. Leipold, All-digital PLL and transmitter for mobile phones, *IEEE Journal of Solid-State Circuits*, 40(12), 2469–2482, 2005.
3. K. Muhammad, Y.-C. Ho, T. Mayhugh, C.-M. Hung, T. Jung, I. Elahi, C. Lin, I. Deng, C. Fernando, J. Wallberg, S. Vemulapalli, S. Larson, T. Murphy, D. Leipold, P. Cruise, J. Jaehnig, M.-C. Lee, R. B. Staszewski, R. Staszewski, and K. Maggio, The first fully integrated quad-band GSM/GPRS receiver in a 90 nm digital CMOS process, *IEEE Journal of Solid-State Circuits*, 41(8), 1772–1783, 2006.
4. J. (J.-C.) Lin, A low-phase-noise 0.004-ppm/step DCXO with guaranteed monotonicity in the 90-nm CMOS process, *IEEE Journal of Solid-State Circuits*, 40(12), 2726–2734, 2005.
5. R. B. Staszewski, I. Bashir, and O. Eliezer, RF built-in self test of a wireless transmitter, *IEEE Transactions on Circuits and Systems II*, 54(2), 186–190, 2007.
6. 3rd Generation Partnership Project; Technical Specification Group GSM/EDGE Radio Access Network: Radio Transmission and Reception (Release 7), 3GPP 45.005, version 7.0.0 (2005-04).
7. R. Staszewski, T. Jung, R. B. Staszewski, K. Muhammad, D. Leipold, T. Murphy, S. Sabin, J. Wallberg, S. Larson, M. Entezari, J. Fresquez, S. Dondershine, and S. Syed, Software assisted digital RF processor for single-chip GSM radio in 90 nm CMOS, *Proceedings of 2006 IEEE Custom Integrated Circuits Conference*, pp. 81–84, September 2005.

8. R. B. Staszewski, C.-M. Hung, D. Leipold, and P. T. Balsara, A first multi-gigahertz digitally controlled oscillator for wireless applications, *IEEE Transactions on Microwave Theory and Techniques*, 51(11), 2154–2164, 2003.

9. P. Cruise, C.-M. Hung, R. B. Staszewski, O. Eliezer, S. Rezeq, D. Leipold, and K. Maggio, A digital-to-RF-amplitude converter for GSM/GPRS/EDGE in 90-nm digital CMOS, *Proceedings of 2005 IEEE Radio Frequency Integrated Circuits (RFIC) Symposium*, pp. 21–24, June 2005.

10. R. B. Staszewski, C. Fernando, and P. T. Balsara, Event-driven simulation and modeling of phase noise of an RF oscillator, *IEEE Trans. on Circuits and Systems I*, 52(4), 723–733, 2005.

11. R. B. Staszewski, S. Rezeq, C.-M. Hung, P. Cruise, and J. Wallberg, Sigma-delta noise shaping for digital-to-frequency and digital-to-RF-amplitude conversion, *Proceedings of Fifth International Workshop on SoC for Real-Time Applications*, Banff, Canada, pp. 154–159, July 2005.

12. C.-M. Hung, R. B. Staszewski, N. Barton, M.-C. Lee, and D. Leipold, A digitally controlled oscillator system for SAW-less transmitters in cellular handsets, *IEEE Journal of Solid-State Circuits*, 41(5), 1160–1170, 2006.

13. K. Waheed and R. B. Staszewski, Harmonic characterization of mismatches in deep submicron varactors for a digitally controlled RF oscillator, *Proceedings of IEEE Midwest Symposium on Circuits and Systems*, session B3L-G, pp. 951–954, August 2005.

14. K. Waheed and R. B. Staszewski, Characterization of deep-submicron varactor mismatches in a digitally controlled oscillator, *Proceedings of 2005 IEEE Custom Integrated Circuits Conference*, sec. 18-3, pp. 605–608, September 2005.

15. K. Waheed and R. B. Staszewski, Digital RF processing techniques for device mismatch tolerant transmitters in nanometer-scale CMOS, *Proceedings of 2007 IEEE International Symposium on Circuits and Systems*, sec. 19.3, pp. 1253–1256, May 2007.

16. H. Masuda, S. Ohkawa, A. Kurokawa, and M. Aoki, Challenge: Variability characterization and modeling for 65- to 90-nm processes, *Proceedings of 2005 IEEE Custom Integrated Circuits Conference*, pp. 593–599, September 2005.

17. P. R. Kinget, Device mismatch and tradeoffs in the design of analog circuits, *IEEE Journal of Solid-State Circuits*, 40(6), 1212–1224, 2005.

18. K. El-Sankary and M. Sawan, A digital blind background capacitor mismatch calibration technique for pipelined AD, *IEEE Transactions on Circuits and Systems II*, 51(10), 507–510, 2004.

19. Y. Chiu, Inherently linear capacitor error-averaging techniques for pipelined A/D conversion, *IEEE Transactions on Circuits and Systems II*, 47(3), 229–232, 2000.

20. K. Waheed and R. B. Staszewski, Time-domain behavioral modeling of a multigigahertz digital RF oscillator using VHDL, *IEEE Midwest Symposium on Circuits and Systems*, Cincinnati, OH, August 7–10, 2005.

21 Front-End Circuits for Multi-Gb/s Chip-to-Chip Links

Anthony Chan Carusone

CONTENTS

21.1 INTRODUCTION

To avoid becoming a bottleneck in large electronic systems, the bit rates that must be communicated over chip-to-chip communication links are increasing exponentially. Traditionally, the data are communicated over printed circuit board (PCB) traces, which have accommodated increasing bit rates with simple circuits and advancing process technologies. However, links operating at speeds on the order of 1 GHz begin to encounter the inherent bandwidth limitations of the interconnect.

One approach to overcome the bandwidth limitations of chip-to-chip links is to use alternative interconnect technologies. This may include the use of low-loss materials for PCB fabrication, optical [1], or even wireless [2] interconnect technologies. However, the cheapest method for increasing bit rates over existing chip-to-chip links remains low-power electronic signal processing integrated into transceivers at either end, whenever this is possible.

Well-known techniques are being used for this purpose including transmit equalization [3–5], receive-side equalization [6–8], decision feedback equalization [4,9,10], and multilevel modulation [3,4,8]. They are used for both serial [3,8,11] and parallel [9,12,13] chip-to-chip links at bit rates from 1–10 Gb/s. Crosstalk is generally identified as a major source of interference in these links [8,12,14]. Crosstalk cancelation is possible only when the aggressors are on-chip [8,12], often at the expense of increased power, complexity, and decreased signal swing when low supply voltages offer little headroom.

This chapter will identify major impairments to multi-Gbps chip-to-chip links, describe techniques to model them, and survey the methods being used to mitigate those impairments and enable future links.

First, typical chip-to-chip links are described and categorized in Section 21.2. Then, models applicable to multi-Gbps chip-to-chip links are developed in Section 21.3. In Section 21.4, equalization is described, including its implementation at the transmitter and the receiver. Section 21.5 compares different modulation schemes. A common criterion for selecting an appropriate modulation scheme is examined and found to be oversimplified. Finally, a summary and conclusions are presented in Section 21.6.

21.2   APPLICATIONS

In this chapter, we are concerned with links carrying traffic at data rates from 1 to 100 Gbps (measured per channel in the case of a parallel bus). At these rates, interconnect of even just a few centimeters in length has a propagation delay that is appreciable compared to a bit period. Hence, PCB traces are designed to have a specific characteristic impedance, and termination is provided at both transmitter and receiver.

State-of-the-art fully differential chip-to-chip links remain more than one order of magnitude faster than their single-ended counterparts. This more than offsets the twofold increase in the number of pins and traces required to accommodate them. Hence, this chapter will focus on fully differential links, which are used almost invariably for high-performance chip-to-chip applications.

The most popular PCB trace geometries for implementing controlled-impedance chip-to-chip transmission lines are the microstrip and stripline, both shown in Figure 21.1. The microstrip may be of a lower cost than the stripline since it occupies only two layers on the surface of the PCB. However, the electromagnetic fields in the stripline are completely confined to a uniform dielectric medium, making it less sensitive to environmental variations than the microstrip whose electrical characteristics may, for example, vary depending on the coating applied to the surface of the PCB.

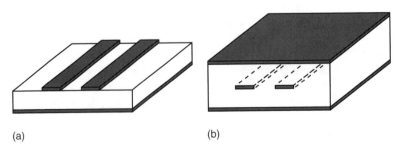

(a)   (b)

FIGURE 21.1   Popular PCB trace geometries for high-performance chip-to-chip communication: (a) a microstrip transmission line; (b) a stripline transmission line.

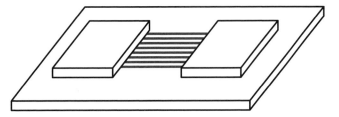

FIGURE 21.2   A parallel chip-to-chip bus.

The vast majority of high-performance chip-to-chip links may be broadly categorized into two groups:

- *Parallel chip-to-chip buses*: Several channels in parallel provide a high-throughput communication link between two integrated circuits (ICs) on the same PCB.
- *Serial links over backplanes*: These links comprise only a single electrical connection over which all data traffic is aggregated. They are generally longer and traverse more discontinuities than parallel buses such as connectors.

The nature and relative severity of the various channel impairments differ significantly between the two groups.

21.2.1   PARALLEL BUSES

A typical parallel chip-to-chip bus is depicted in Figure 21.2. It is typically characterized by multiple interconnects densely routed in parallel between ICs and punctuated by relatively few discontinuities such as connectors. Such links are required, for example, between a central processing unit and a peripheral bridge or coprocessor, between processor and memory, or between high-speed data converters and digital signal processors. The number of parallel connections varies widely depending on the application, of course, from 4 to 128 or more.

Transceivers for these applications must consume low power and low area. All of the parallel data streams are synchronized to the same clock, thereby permitting any power or area overhead associated with clocking to be amortized over the width of the bus. The most stringent channel impairments are crosstalk from neighboring transceivers, and in the case of very high-speed links, intersymbol interference (ISI).

At present, consumer products include parallel bus links at data rates of 5 Gbps [15] and rates of up to 20 Gbps per transceiver pair have been demonstrated in experimental platforms [16].

21.2.2   SERIAL LINKS OVER BACKPLANES

Several important applications require a high-speed link between ICs on separate PCBs. For example, the links connecting high-speed physical layer transceivers and routers inside network switches, or between daughter cards in a high-performance parallel-processing computer. Wide parallel buses connecting PCBs require large connectors, which are expensive and physically large. An obvious alternative is to aggregate the traffic from the entire bus onto a single serial link. An example of such a link is shown in Figure 21.3 in which the transmitter and the receiver are separated by two connectors and three lengths of PCB trace. The "backplane" generally refers to the intermediate PCB, carrying traffic between the two daughter cards, but similar design challenges arise in applications in which transmitter and receiver are separated by only one connector and two lengths of PCB trace.

A common requirement in serial links is to provide electrical isolation between the transmitter and the receiver. A series-connected AC-coupling capacitor is generally provided for this purpose.

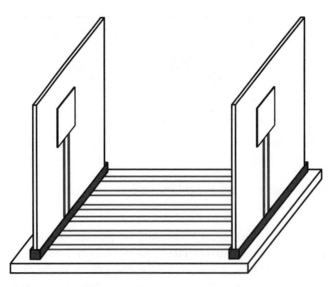

FIGURE 21.3 A serial link over a backplane.

AC-coupling allows the transmitter and receiver to operate with different common-mode voltages, easing interoperability and providing an extra degree of freedom in the design of transceivers. However, the discontinuity presented by the coupling capacitor, including the vias, which may be required to access the surface of the PCB, represents yet another discontinuity in the link. It also introduces baseline wander in the presence of long strings of consecutive identical bits, and is therefore usually accompanied by a DC-balanced line code such as 8b10b coding [17]. Clearly, in comparison with the short parallel buses described above, backplane serial links face significantly tougher bandwidth limitations due to the discontinuities.

21.3 LINK MODELING

This section briefly reviews the modeling of point-to-point chip-to-chip links. A complex chip-to-chip link, such as the serial link depicted in Figure 21.3, may comprise several sections of stripline or microstrip transmission line with mismatched terminations and IC packaging at either end and punctuated by vias and connectors along the way. The frequency response of a link is very sensitive to the discontinuities, so they must be included in link models.

The PCB traces and discontinuities are linear two-port networks. The approach taken in this section is to obtain a linear two-port description of each transmission line segment and discontinuity and then cascade them to generate a model of the entire link. Unfortunately, the two-port descriptions of various portions of the link are often specified in different formats. Hence, all two-port descriptions must be converted into a common form before combining them to construct the complete link model. Transmission matrices are a particularly useful two-port description since they are easily cascaded.

Section 21.3.1 provides some background on electrical two-port networks and transmission matrices. Section 21.3.2 then provides a scalable two-port description of PCB interconnect including skin effect and dielectric losses. Section 21.3.3 follows with examples of complete links modeled in this way.

21.3.1 LINEAR TWO-PORT NETWORKS

A general linear two-port electrical network is shown in Figure 21.4. The quantities $V_1(j\omega)$, $I_1(j\omega)$, $V_2(j\omega)$, and $I_2(j\omega)$ are complex-valued phasor representations of the voltages and

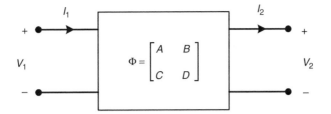

FIGURE 21.4 A general linear two-port electrical network and its transmission matrix.

currents at ports 1 and 2 of the network, respectively, at angular frequency $\omega = 2\pi f$. A transmission matrix, $\Phi(j\omega)$ is 2×2 with frequency-dependent complex-valued entries $A(j\omega)$, $B(j\omega)$, $C(j\omega)$, and $D(j\omega)$, that relate the port voltages and currents:

$$\begin{bmatrix} V_1(j\omega) \\ I_1(j\omega) \end{bmatrix} = \begin{bmatrix} A(j\omega) & B(j\omega) \\ C(j\omega) & D(j\omega) \end{bmatrix} \begin{bmatrix} V_2(j\omega) \\ I_2(j\omega) \end{bmatrix} = \Phi(j\omega) \begin{bmatrix} V_2(j\omega) \\ I_2(j\omega) \end{bmatrix} \tag{21.1}$$

The matrix Φ provides a complete description of a linear electrical two-port network. Electrical two-ports may otherwise be described by

- Impedance matrices,
- Admittance matrices, and
- Scattering parameters

Any of these may be straightforwardly transformed into transmission matrices via the expressions provided in the appendix.

The transmission matrix of series-connected two-ports is simply the product of the transmission matrices of the constituent two-ports. Figure 21.5 shows a simple example of two two-port networks connected in series having transmission matrices $\Phi_1(j\omega)$ and $\Phi_2(j\omega)$, respectively. The voltage and current phasors at the far ends of the series connection may be written in terms of $\Phi_1(j\omega)$ and $\Phi_2(j\omega)$, resulting in a transmission matrix describing the entire link, $\Phi(j\omega)$:

$$\begin{bmatrix} V_1(j\omega) \\ I_1(j\omega) \end{bmatrix} = \Phi_1(j\omega) \begin{bmatrix} V_2(j\omega) \\ I_2(j\omega) \end{bmatrix} = \Phi_1(j\omega) \Phi_2(j\omega) \begin{bmatrix} V_3(j\omega) \\ I_3(j\omega) \end{bmatrix} \equiv \Phi \begin{bmatrix} V_3(j\omega) \\ I_3(j\omega) \end{bmatrix} \tag{21.2}$$

This result is straightforwardly generalized, so that the transmission matrix of any number of two-ports connected in series is given by the product of the constituent two-ports:

$$\Phi(j\omega) = \prod_k \Phi_k(j\omega) \tag{21.3}$$

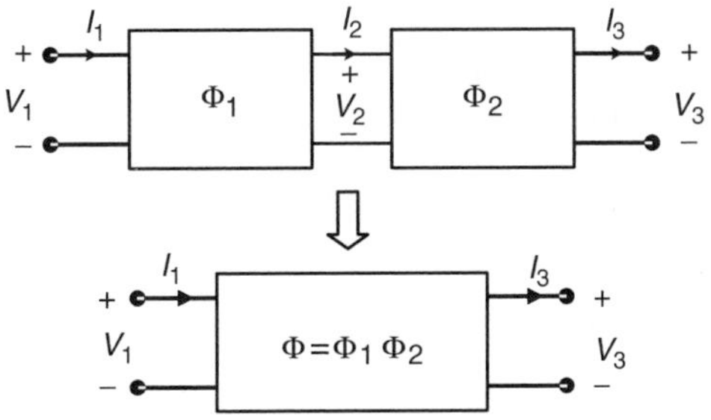

FIGURE 21.5 A series connection of linear two-port networks.

A transmission matrix for the entire link may be built in this way from transmission matrices for the terminations at either end, discontinuities such as vias and connectors, and the intervening sections of lossy PCB trace. This will be done in Section 21.3.3.

21.3.2 THE TRANSMISSION LINE

Whereas the discontinuities of a link are complex structures best characterized using either electromagnetic simulation or direct measurement, PCB traces for high-performance chip-to-chip links employ transmission line geometries for which scalable and accurate models are available. This section describes such a scalable model and develops the corresponding transmission matrices. The advantages of such scalable models (as opposed to direct measurement) are manifold. Parameters such as the length of the link, the location of discontinuities, and the loss tangent can be easily swept to evaluate the performance and robustness of a given transceiver in the complete variety of channels likely to be encountered in the field. Furthermore, it provides an exact result with no measurement noise. The best approach is to use measurement results from a few test cases to refine a scalable model of the link.

For high-performance chip-to-chip communication, wired interconnects are transmission lines, which may be modeled using a quasistatic transverse electromagnetic (TEM) approximation. In general, the resulting model comprises frequency-dependent values of resistance, inductance, conductance, and capacitance per unit length, as shown in Figure 21.6.

The frequency dependence of R, L, G, and C can be described in terms of a few parameters to model skin effect and dielectric losses [18].

The skin effect is modeled by a complex-valued frequency dependence in R, which includes the conductors' DC loss per unit length (R_0, often negligible for short traces) and a skin effect constant (R_S):

$$R(f) = R_0 + \sqrt{f}(1+j)R_s \tag{21.4}$$

In combination with a frequency-independent value for series inductance (L_0), this results in a loss that increases at 10 dB/decade.

Conduction through the dielectric is generally negligible, $G_0 = 0$. But, alternating electromagnetic fields cause heating and, hence, loss in the dielectric that increases with frequency. This is modeled using a complex-valued frequency-dependant per-unit-length capacitance:

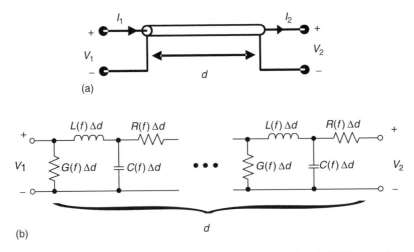

(a)

(b)

FIGURE 21.6 (a) A general transmission line of length d and (b) a quasistatic TEM approximate model.

TABLE 21.1

Model Parameter Values Typical of a 100 Ω Differential Impedance Interconnect on an FR4 PCB

R_0 (Ω/m)	≈ 0
R_S (Ω/m$\sqrt{\text{Hz}}$)	$8.7 \cdot 10^{-9}$
L_0 (nH/m)	817
G_0 (Ω^{-1}/m)	0
C_0 (pF/m)	148
θ_0	0.022
f_0 (GHz)	10

$$C(f) = C_0 \left(\frac{jf}{f_0} \right)^{-2\theta/\pi} \tag{21.5}$$

where

C_0 is the transmission line's capacitance per unit length at low frequencies

θ and f_0 are parameters that provide for linearly increasing (20 dB/decade) losses at very high frequencies

Since dielectric losses impose strict bandwidth limitations on chip-to-chip communication, there is strong motivation to consider low-loss materials such as teflon for this purpose. However, the epoxy laminate FR4 remains a popular dielectric material in spite of its relatively high loss due to its mechanical robustness and low cost. Once the frequency-dependent values of R, L, G, and C are known, they can be converted to transmission matrices using formulas in the appendix.

21.3.3 COMPLETE LINKS

The transmission matrix of PCB traces can be combined with those of discontinuities using Equation (21.3) to make a complete link model. A set of MATLAB® functions useful for combining two-port descriptions of various linear networks into a single transmission matrix and then performing the required inverse fast Fourier transform to obtain a time-domain representation is available for free from my Web site.* An example using the functions to create a chip-to-chip channel model is also provided. These MATLAB functions were used for the examples in this section.

The interconnect model is a differential microstrip with 100 Ω matched terminations on an FR4 dielectric. Model parameter values can be obtained from analytical expressions for a known trace geometry, electromagnetic simulations, and/or fitting to real measured results. The values used here are typical of a 100 Ω differential impedance microstrip on an FR4 PCB and are given in Table 21.1.

Figure 21.7 shows the influence of a simple package model on the frequency and transient responses of a link between two ICs on the same PCB. A 5 cm link has a 3 dB bandwidth exceeding 20 GHz without including IC packaging, but the 3 dB bandwidth drops to just over 2.5 GHz with the inclusion of a simple package model. The reflection due to the package's discontinuity is clearly visible in the step response and there is a large ripple in the frequency response.

Over longer channels, the effects of PCB trace losses become more significant than the effects of discontinuities. For example, introducing package models into the 20 cm link reduces the 3 dB bandwidth by less than 25%, from 2.2 to 1.7 GHz. This is sensible since the reflections introduced by the packaging must traverse a longer length of PCB trace before appearing at the receiver and,

---

* www.eecg.utoronto.ca/~tcc/MATLABABCD.zip

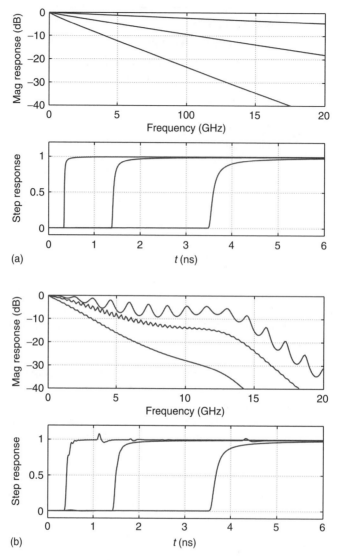

FIGURE 21.7 Frequency and step responses of 5, 20, and 50 cm chip-to-chip links on FR4 dielectric (a) without and (b) with simple IC package models included at either end.

hence, are attenuated by the trace losses. Hardly any reflections are visible in the 50 cm transient response. However, the skin effect and dielectric losses are large, contributing almost 25 dB loss at 10 GHz.

In Figure 21.7, the only discontinuities are the IC packages at either end of the link. This is the case in, for instance, point-to-point parallel links between two ICs on the same PCB. However, serial links over a backplane have more discontinuities, most notably the connectors between the motherboard and daughter cards. Measured responses for two real backplanes are plotted in Figure 21.8. Both channels have the same daughter cards with 10 cm of interconnect each and the same backplane connectors; one channel has a 10 cm length of interconnect on the motherboard, and the other has 40 cm of interconnect on the motherboard. Even though they are very different lengths, both channels have similar responses because the loss and reflections due to connectors are the dominant effect.

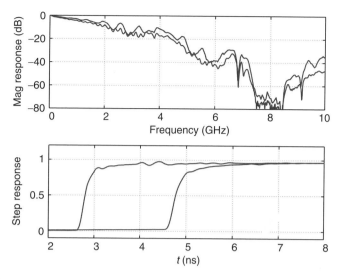

FIGURE 21.8 Measured frequency and step responses of two real backplane links comprising 10 and 40 cm motherboards and two 10 cm daughter cards.

21.4 EQUALIZATION

It is well established that equalization is necessary for multi-Gb/s communication over many practical chip-to-chip links. A wide variety of filters have been employed at both the transmitter and the receiver for this purpose.

Figure 21.9 presents a system model for the linear portion of a chip-to-chip link. It comprises a transmit filter with frequency response $B(\omega)$ and impulse response $b(t)$ operating on a transmit signal $u(t)$, a channel with frequency response $H(\omega)$ and impulse response $h(t)$, and a receiver with frequency response $C(\omega)$ and impulse response $c(t)$. Equalization may be incorporated into the transmitter $B(\omega)$ and/or the receiver $C(\omega)$.

Recently, both transmitter and receiver filters have been combined with decision feedback to equalize long or challenging links [19–21]. With such a wide variety of equalizer architectures in use, there is great interest in comparing their performance to arrive at the "optimal" design for a given link.

21.4.1 TRANSMIT EQUALIZATION

Practical considerations often limit the peak output voltage of the transmitter to a constant L. Hence, the transmit filter must satisfy the following constraint for all possible inputs $u(t)$:

$$|(u \times b)(t)| = \left| \int u(t - t') b(t') dt' \right| \leq L \tag{21.6}$$

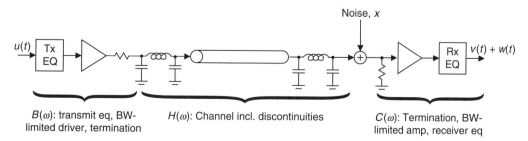

FIGURE 21.9 System model of the linear portions of a chip-to-chip link.

It is always possible to choose an input $u(t)$ for which $\left|\int u(t - t')b(t')dt'\right| = \int |u(t - t')||b(t')|dt'$. So, to ensure that (21.6) is satisfied for any input $u(t)$, the following expression must also hold:

$$\int \left|u\left(t - t'\right)\right| \left|b\left(t'\right)\right| dt' \leq L \tag{21.7}$$

Without loss of generality, we may assume that the input to the transmit filter is normalized to have a signal swing within the range $|u(t)| \leq 1$. Therefore, the transmit filter must satisfy the following equation:

$$\int \left|b\left(t'\right)\right| dt' \leq L \tag{21.8}$$

Multiplying by $|e^{j\omega t'}| = 1$,

$$\int \left|b\left(t'\right)\right| dt' = \int \left|b\left(t'\right)\right| \left|e^{j\omega t'}\right| dt' \tag{21.9}$$

$$\geq \left|\int b\left(t'\right) e^{j\omega t'} dt'\right| = |B\left(\omega\right)| \tag{21.10}$$

Combining Equations (21.8) and (21.10) gives

$$|B(\omega)| \leq L \tag{21.11}$$

Equation 21.11 must hold for all ω if the transmit waveform's peaks are to be limited in amplitude to the constant L. The simplest transmitter satisfying (21.11) is an amplifier with constant gain L. If any spectral shaping is to be introduced at the transmitter to compensate for frequency-dependent channel losses, it is done by reducing the gain at some frequencies below L, in accordance with (21.11). Hence, peak-constrained transmit equalization is often referred to as "de-emphasis" since it equalizes the channel response by attenuating the low-frequency portion of the transmit signal spectrum below the level attainable with a simple transmitter having a constant gain L.

The filter $B(\omega)$ in many chip-to-chip links is a finite impulse response (FIR) filter. A typical implementation is illustrated in Figure 21.10. A conventional transmitter without de-emphasis (Figure 21.10a) may be straightforwardly modified to provide de-emphasis as shown in Figure 21.10b by partitioning the output driver into segments (12 in this case) and driving some segments with

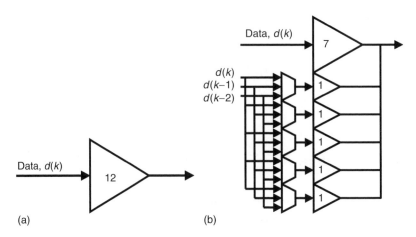

FIGURE 21.10 (a) A conventional transmitter without de-emphasis. (b) A modification to provide de-emphasis by partitioning the output driver into 12 segments.

delayed versions of the transmitted data. Much of the transmitter's power budget is consumed by the output drivers, which are typically required to provide a signal swing of several hundred millivolts to 50 Ω loads, in which case the over-head of the digital multiplexors in Figure 21.10b is negligible. The circuitry required to generate the delayed data signals $d(k-1)$ and $d(k-2)$ is not shown in Figure 21.10, but at worst comprises one or more latches clocked at the data rate.

It is clear that the peak transmitted voltage in Figure 21.10b remains the same regardless of the multiplexor settings (which define the FIR filter response) since the maximum current that can be delivered to the load always remains equal to the sum of all 12 output driver segments. Hence, this implementation of transmit equalization automatically maintains a constant peak-transmitted voltage while providing a flexible and digitally programmable frequency response. The overhead associated with this implementation, compared to a transmitter with no equalization, comprises mainly the digital multiplexors whose power consumption should be much smaller than that of the output stages anyway.

A disadvantage of transmit equalization, if the filter parameters are to be made adaptive, is that a mechanism must be provided for communicating information about the received waveform back to the transmitter. There is also another more subtle disadvantage in terms of performance in the performance of the link in the presence of noise, which is discussed in the next subsection.

21.4.2 RECEIVE EQUALIZATION

Forward equalization at the receiver may be performed by either a discrete-time filter [6–8] or a continuous-time filter [19,20,22] preceding the decision device. Unlike a de-emphasis filter, which processes digital inputs, a linear receive equalizer must be capable of handling the full dynamic range of signals that can appear at the receiver front-end, generally making it a more challenging circuit design.

For the link dominated by independent noise in Figure 21.9, the signal at the output of the receiver front-end comprises the desired signal from the far end transmitter, $v(t)$, plus a noise component, $w(t)$, where

$$u(t) = (u * b * h * c)(t) \qquad (21.12)$$

and

$$w(t) = (x * c)(t) \qquad (21.13)$$

Assume that for a given channel, the transmitter and receiver must combine to provide a desired frequency response, $KE(\omega)$, which equalizes the channel. Therefore

$$B(\omega)C(\omega) = KE(\omega) \qquad (21.14)$$

Without loss of generality, the frequency-dependent term $E(\omega)$ satisfies $|E(\omega)| \le 1$ and the gain constant K normalizes the received signal amplitude. The transmitter and receiver impulse responses must, therefore, satisfy

$$(b * c)(t) = Ke(t) \qquad (21.15)$$

where $e(t)$ is the inverse Fourier transform of $E(\omega)$.

Combining (21.12) and (21.15), it becomes clear that using either transmit de-emphasis, a receive-side linear equalizer, or any combination of the two satisfying (21.14) will yield the exact same noiseless received signal:

$$v(t) = K(ueh)(t) \qquad (21.16)$$

This is not surprising since the channel and equalizer are ideally linear systems, so the order in which they appear has no effect on the end-to-end pulse response. The noise, however, only passes through the receiver, and so has a power spectral density given by

$$S_w(\omega) = S_x(\omega)|C(\omega)|^2 \qquad (21.17)$$

where $S_a(\omega)$ is the power spectral density of $a(t)$. Therefore, to minimize the noise power, it is necessary to minimize the magnitude response of the receiver (thereby attenuating the noise) while still satisfying (21.14) to equalize the channel. If the transmitter is peak-constrained to an amplitude of L, it must satisfy (21.11). So, subject to (21.11) and (21.14), the noise power at the receiver (21.17) is minimized as follows:

$$|B(\omega)| = L \tag{21.18}$$

$$|C(\omega)| = \frac{K}{L}|E(\omega)| \tag{21.19}$$

So, when the transmitter is peak-constrained, signal-to-noise ratio (SNR) at the receiver is maximized by using a transmitter with a flat frequency response (hence, no de-emphasis) and performing all of the equalization at the receiver. In many high-speed serial links, the dominant noise is actually the superposition of many independent noise sources and is well approximated by a bounded Gaussian distribution [14]. In this case, maximizing the SNR also minimizes the bit error rate.

In summary, assuming ideal implementations of transmit and receive equalization are possible and putting aside their complexity and power consumption, receive equalization offers better performance than peak-constrained transmit equalization (de-emphasis). Nevertheless, transmit equalization is often preferred for its simpler implementation. Furthermore, the performance advantages of receive equalization can become very small in parallel links, as discussed in the next section.

21.4.3 CROSSTALK

In a parallel bus of chip-to-chip links, the primary source of noise is often self-crosstalk from neighboring channels. Since, as shown in Figure 21.11, the primary noise source is influenced by the transmitter design, very little performance advantage is offered by receive equalization.

Specifically, let $u_2(t)$ equal the signal being communicated over the link while $u_1(t)$ is an independent aggressor. If, as before, the combination of transmitter and receiver is required to have a frequency response of $KE(\omega)$ (21.14), then any combination of transmit- and/or receive-side equalization will yield the same noiseless received signal:

$$u(t) = (u_2 * b * h * c)(t)$$
$$= K(u_1 * e * g)(t) \tag{21.20}$$

However, unlike independent noise links, the aggressors in self-cross talk links are communicating data using a similar transmitter, $B(\omega)$. Therefore, regardless of where the equalization is performed, the noise at the output of the receiver front end will be exactly the same:

$$w(t) = (u_1 bgc)(t)$$
$$= K(u_1 eg)(t) \tag{21.21}$$

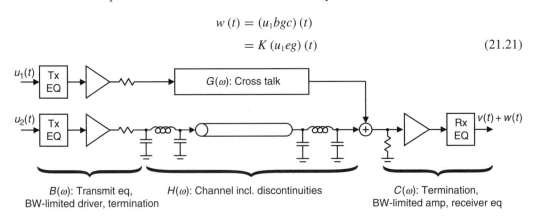

FIGURE 21.11 System models for a parallel chip-to-chip link limited by self-crosstalk.

Equations 21.20 and 21.21 show that the performance of a link limited by self-crosstalk is the same using transmit de-emphasis, receiver equalization, or any combination of the two. This is intuitive since the transmitter and receiver appear in both the signal path and the dominant noise path. The result is also applicable in the presence of transmitter and receiver jitter since these also affect both the signal and noise paths.

In practice, parallel links are subject to both independent and self-crosstalk noise. Hence, receive equalization still offers some performance advantage, albeit very small when the noise is dominated by self-crosstalk.

21.5 MODULATION

Simple binary nonreturn-to-zero signaling is still commonly used for most chip-to-chip interfaces, primarily due to its straightforward hardware implementation. However, baseband pulse amplitude modulation (PAM) with four [3,23], six [24], and eight levels [11] has been suggested. A common question is: when should multilevel modulation be used instead of binary signaling?

As in all band-limited channels, bandwidth-efficient signaling schemes are required to extend the data rates over chip-to-chip interconnect significantly beyond $2\times$ the channel bandwidth. In this context, what *precisely* is the "channel bandwidth?" This question will be considered for serial and parallel links separately.

21.5.1 SERIAL LINKS

Using 4-PAM halves the operating frequency compared with binary signaling, resulting in reduced channel losses. However, the level spacing is also reduced, by one-third or 9.5 dB. So, 4-PAM is thought to be advantageous whenever the reduced level spacing is compensated for by improved signal transmission and lower noise.

A common approach is to quantify the improvement in signal transmission and the reduction in noise obtained by going to 4-PAM using the ratio of the noise spectrum to the channel response (noise-to-channel ratio, NCR):

$$\text{NCR}(\omega) = \frac{S_x(\omega)}{|H(\omega)|^2} \tag{21.22}$$

If one is comparing binary and 4-PAM for communication at a data rate of f_b, then the simple rule states that 4-PAM is beneficial whenever

$$20 \log_{10} \left(\frac{\text{NCR}(\pi f_b)}{\text{NCR}(\pi f_b/2)} \right) > 9.5 \text{dB} \tag{21.23}$$

Assuming NCR decreases linearly with increasing frequency on a log–log plot, Equation 21.23 indicates that 4-PAM signaling is preferable to binary signaling as long as the slope of the NCR curve exceeds 9.5 dB per octave (31.5 dB per decade) at one-half the data rate [4,8]. The argument is straightforwardly extended to conclude that M-PAM is preferable to N-PAM whenever

$$20 \log_{10} \left(\frac{\text{NCR}\left(2\pi f_b/\log_2 N\right)}{\text{NCR}\left(2\pi f_b/\log_2 M\right)} \right) > 20 \log_{10} \left(\frac{(M-1)}{(N-1)} \right) \text{dB} \tag{21.24}$$

Unfortunately, the preceding analysis is somewhat oversimplified, as demonstrated by simulations in, for example, Refs. [14,25]. Its greatest shortcoming is that it fails to recognize the broadband nature of baseband communication signals. "Spot" measures of NCR at $f_b/2$ and $f_b/4$ are insufficient to characterize the performance of binary and 4-PAM links. Since baseband binary signals have considerable spectral content at low frequencies, they can be very reliable even when the channel response is quite poor at one-half the bit rate. Hence, the straightforward criteria described above unfairly favor 4-PAM signals.

Advocates of 4-PAM signaling also often correctly point out that less high-frequency boost is required to equalize a 4-PAM signal than a binary signal at the same data rate since the binary spectrum extends to higher frequencies where channel losses increase. However, due to their reduced level spacing, 4-PAM signals are more sensitive to residual ISI. Hence, in practice, 4-PAM signals may actually be more difficult to equalize since greater accuracy is required in defining the equalizer response.

21.5.2 Parallel Buses

In parallel bus chip-to-chip links dominated by self-crosstalk, deciding which modulation to use is complicated by the fact that both the signal and noise are influenced by the transmitted spectrum.

This problem has been heavily researched for communication over digital subscriber loops where crosstalk from multiple subscribers is a dominant source of noise [26]. Under these circumstances, it has been shown that the transmitted spectrum that maximizes the total capacity of all links is given by the solution to a quadratic equation:

$$|G(\omega)|^2 \left(|G(\omega)|^2 + |H(\omega)|^2\right) S_u^2(\omega) + S_x(\omega) \left(2|G(\omega)|^2 + |H(\omega)|^2\right)S_u(\omega)$$
$$+S_x(\omega)\left(S_x(\omega) - A|H(\omega)|^2\right) = 0 \qquad (21.25)$$

In (21.25), the noise power includes both crosstalk from channels with the same transmitted spectrum, $S_u(\omega)|G(\omega)|^2$, and other noise, $S_x(\omega)$.

The results presented in Ref. [27] extend this analysis to consider peak-power-limited PAM communication, which is more applicable to chip-to-chip links than the Gaussian distributions considered in Ref. [26]. In both cases, the analysis indicates that the transmit spectrum should be mostly restricted to frequencies where $|H(\omega)|^2/|G(\omega)|^2 \gtrsim 1$. The procedure then is to choose a modulation method so that the Nyquist rate is below the frequency at which $|H(\omega)|^2/|G(\omega)|^2 = 1$.

A particularly interesting result of this analysis is that multilevel modulation is generally unnecessary for purely far-end cross talk (FEXT)–limited channels. Because the cross talk is also attenuated by the channel's through response, it is impossible to have $|H(\omega)|^2/|G(\omega)|^2 \gtrsim 1$ at any frequency.

21.6 CONCLUSION

Modern integrated chip-to-chip communication transceivers include various combinations of transmitter de-emphasis, receiver equalization, and multilevel modulation. These alternatives were analyzed and their performance compared.

Transmit- and receive-side linear equalization were found to offer equivalent performance when crosstalk from synchronized links with similar transmitters are the dominant outside interference. However, receive-side linear equalization is preferable when other (independent) noise sources are dominant. These conclusions are also applicable to the linear portion of a decision feedback equalizer. Ideal circuit implementations were assumed to compare the achievable performance using both approaches. In practice, receive-side equalizers are a more difficult circuit design since they must process the full dynamic range of the received signal whereas transmit equalizers accept quantized (often binary digital) inputs. However, receive-side equalizers are more easily adapted online when no reverse channel is present.

A comparison of binary and multilevel modulation was also presented. The common practice of using the slope of a channel's NCR curve to predict the relative performance of binary and 4-PAM signaling was shown to be somewhat oversimplified. It unfairly favors multilevel modulation. Furthermore, multilevel modulation was found to be generally undesirable for purely FEXT-limited channels. The problem of finding the optimal transmit spectrum for a given chip-to-chip link remains an open question requiring further study.

Of course, as the data rates over a given channel is increased, eventually multilevel signaling becomes necessary. Consider the migration of voice-band modems to more bandwidth-efficient modulation schemes throughout the 1980s and 1990s. However, bandwidth-efficient modulation demands more sophisticated equalization than binary signaling. For example, a 4-PAM transmitter in [28] employed 13 taps of de-emphasis to compensate for 14.5 dB loss at one-half the symbol rate. The cost associated with the additional silicon area and power consumption of such sophisticated equalization could, instead, be applied to lowering the channel loss by employing a low-loss dielectric [29].

21.A   APPENDIX A

Transmission matrices provide a description of linear electrical two-port networks that are particularly useful for modeling complex chip-to-chip links. In this appendix, mathematical transformations are provided from several other electrical two-ports network descriptions to transmission matrices.

21.A.1   IMPEDANCE MATRICES

Referring to Figure 21.4, impedance matrices relate the terminal voltages and currents of a two port as follows:

$$\begin{bmatrix} V_1 \\ V_2 \end{bmatrix} = \begin{bmatrix} Z_{11} & Z_{12} \\ Z_{21} & Z_{22} \end{bmatrix} \begin{bmatrix} I_1 \\ I_2 \end{bmatrix} \tag{21.A.1}$$

A transmission matrix describing the same two-port is given by

$$\Phi = \frac{1}{Z_{21}} \begin{bmatrix} Z_{11} & \Delta Z \\ 1 & Z_{22} \end{bmatrix} \tag{21.A.2}$$

where

$$\Delta Z = Z_{11}Z_{22} - Z_{12}Z_{21} \tag{21.A.3}$$

21.A.2   ADMITTANCE MATRICES

Referring to Figure 21.4, admittance matrices relate the terminal voltages and currents of a two port as follows:

$$\begin{bmatrix} I_1 \\ I_2 \end{bmatrix} = \begin{bmatrix} Y_{11} & Y_{12} \\ Y_{21} & Y_{22} \end{bmatrix} \begin{bmatrix} V_1 \\ V_2 \end{bmatrix} \tag{21.A.4}$$

A transmission matrix describing the same two-port is given by

$$\Phi = -\frac{1}{Y_{21}} \begin{bmatrix} Y_{22} & 1 \\ \Delta Y & Y_{11} \end{bmatrix} \tag{21.A.5}$$

where

$$\Delta Y = Y_{11}Y_{22} - Y_{12}Y_{21} \tag{21.A.6}$$

21.A.3   SCATTERING PARAMETERS

Scattering parameters are a linear two-port network description often obtained from laboratory measurements on a network analyzer. The four frequency-dependent complex-valued scattering parameters S_{11}, S_{12}, S_{21}, and S_{22} are related to the entries of an equivalent transmission matrix as follows:

$$A = \frac{(1 + S_{11})(1 - S_{22}) + S_{12}S_{21}}{2S_{21}} \tag{21.A.7}$$

$$B = Z_0 \frac{(1 + S_{11})(1 + S_{22}) - S_{12}S_{21}}{2S_{21}} \tag{21.A.8}$$

$$C = \frac{1}{Z_0} \frac{(1 - S_{11})(1 - S_{22}) - S_{12}S_{21}}{2S_{21}} \qquad (21.A.9)$$

$$D = \frac{(1 - S_{11})(1 + S_{22}) + S_{12}S_{21}}{2S_{21}} \qquad (21.A.10)$$

21.A.4 RLGC VALUES

Per-unit-length values of series resistance R, series inductance L, shunt conductance G, and shunt capacitance C are sometimes derivable from a knowledge of the physical geometry and material properties of a transmission line. Once available, these may be transformed into the transmission matrix of a transmission line segment of length d:

$$\Phi = \begin{bmatrix} \cosh(\gamma d) & Z_0 \sinh(\gamma d) \\ \frac{1}{Z_0} \sinh(\gamma d) & \cosh(\gamma d) \end{bmatrix} \qquad (21.A.11)$$

where the characteristic impedance is

$$Z_0 = \sqrt{\frac{(R + j\omega L)}{(G + j\omega C)}} \qquad (21.A.12)$$

and the propagation constant is

$$\gamma = \sqrt{(R + j\omega L)(G + j\omega C)} \qquad (21.A.13)$$

ACKNOWLEDGMENTS

The author is sincerely thankful and appreciates Mike Bichan for providing the measurement results of backplane channels and Amer Samarah for his help with the figures and appendix.

REFERENCES

1. D. V. Plant and A. G. Kirk, Optical interconnects at the chip and board level: Challenges and solutions, *Proc. IEEE*, 88(6), 806–818, 2000.
2. M.-C. F. Chang, V. P. Roychowdhury, L. Zhang, H. Shin, and Y. Qian, RF/wireless interconnect for inter- and intra-chip communications, *Proc. IEEE*, 89(4), 456–466, 2001.
3. R. Farjad-Rad, C. K. Yang, M. Horowitz, and T. H. Lee, A 0.3-μm CMOS 8-Gb/s 4-PAM serial link transceiver, *IEEE J. Solid-State Circuits*, 35(5), 757–764, 2000.
4. J. Zerbe, C. Werner, V. Stojanovic, F. Chen, J. Wei, G. Tsang, D. Kim, W. Stonecypher, A. Ho, T. Thrush, R. Kollipara, M. Horowitz, and K. Donnelly, Equalization and clock recovery for a 2.5–10 Gb/s 2-PAM/4-PAM backplane transceiver cell, *IEEE J. Solid-State Circuits*, 38(12), 2121–2130, 2003.
5. A. Martin, B. Casper, J. Kennedy, J. Jaussi, and R. Mooney, 8 Gb/s differential simultaneous bidirectional link with 4 mv 9 ps waveform capture diagnostic capability, in *IEEE Int. Solid-State Circuits Conf.*, February 2003, pp. 78–79.
6. J.-Y. Sim, Y.-S. Sohn, H.-J. Park, C.-H. Kim, and S.-I. Cho, 840 Mb/s CMOS de-multiplexed equalizing transceiver for DRAM-to-processor communication, in *Symp. VLSI Circuits*, June 1999, pp. 23–24.
7. J. E. Jaussi, G. Balamurugan, D. R. Johnson, B. K. Casper, A. Martin, J. T. Kennedy, N. Shanbhag, and R. Mooney, An 8 Gb/s source-synchronous I/O link with adaptive receiver equalization, offset cancellation and clock deskew, *IEEE J. Solid-State Circuits*, 40(1), 80–88, 2005.
8. Y. Hur, M. Maeng, C. Chun, F. Bien, H. Kim, S. Chandramouli, E. Gebara, and J. Laskar, Equalization and near-end cross talk (NEXT) noise cancellation for 20-Gb/s 4-PAM backplane serial I/O interconnections, *IEEE Trans. Microwave Theory and Techniques*, 53(1), 246–255, 2005.
9. Y.-S. Sohn, S.-J. Bae, H.-J. Park, and S.-I. Cho, A 2.2 Gbps CMOS look-ahead DFE receiver for multidrop channel with pin-to-pin time skew compensation, in *Custom Integr. Circuits Conf.*, September 2003, pp. 473–476.

10. R. Payne, B. Bhakta, S. Ramaswamy, S. Wu, J. Powers, P. Landman, U. Erdogan, A.-L. Yee, R. Gu, L. Wu, Y. Xie, B. Parthasarathy, K. Brouse, W. Mohammed, K. Heragu, V. Gupta, L. Dyson, and W. Lee, A 6.25Gb/s binary adaptive DFE with first post-cursor tap cancellation for serial backplane communications, in *IEEE Int. Solid-State Circuits Conf.*, February 2005, pp. 68–69.

11. D. J. Foley and M. P. Flynn, A low-power 8-PAM serial transceiver in 0.5-μm digital CMOS, *IEEE J. Solid-State Circuits*, 37(3), 310–316, 2003.

12. J. L. Zerbe, P. S. Chau, C. W. Werner, W. F. Stonecypher, H. J. Liaw, G. J. Yeh, T. P. Thrush, S. C. Best, and K. S. Donnelly, A 2 Gb/s/pin 4-PAM parallel bus interface with transmit cross talk cancellation, equalization, and integrating receivers, in *IEEE Int. Solid-State Circuits Conf.*, February 2001, pp. 66–67.

13. K. Chang, S. Pamarti, K. Kaviani, E. Alon, S. Shi, T. J. Chin, J. Shen, G. Yip, C. Madden, R. Schmitt, C. Yuan, F. Assaderaghi, and M. Horowitz, Clocking and circuit design for a parallel I/O on a first-generation CELL processor, in *IEEE Int. Solid-State Circuits Conf.*, February 2005, pp. 526–527.

14. V. Stojanovic and M. Horowitz, Modeling and analysis of high-speed links, in *Custom Integr. Circuits Conf.*, September 2003, pp. 589–594.

15. E. Guizzo and H. Goldstein, Expressway to your skull, *IEEE Spectr.*, 43(8), 34–39, 2006.

16. J. E. Jaussi, B. K. Casper, M. Mansuri, F. O'Mahony, K. Canagasaby, J. Kennedy, and R. Mooney, A 20Gb/s embedded clock transceiver in 90 nm CMOS, in *IEEE Int. Solid-State Circuits Conf.*, February 2006, pp. 340–341.

17. A. X. Widmer and P. A. Franaszek, A dc-balanced, partitioned-block 8B/10B transmission code, *IBM J. Res. Devel.*, 27(5), 440–451, 1983.

18. H. Johnson and M. Graham, *High-Speed Signal Propagation: Advanced Black Magic*. Englewood Cliffs, NJ: Prentice Hall, 2003.

19. N. Krishnapura, Q. Barazande-Pour, M. Chaudhry, J. Khoury, K. Lakshmikumar, and A. Aggarwal, A 5Gb/s NRZ transceiver with adaptive equalization for backplane transmission, in *IEEE Int. Solid-State Circuits Conf.*, February 2005, pp. 60–61.

20. M. Sorna, T. Beukema, K. Selander, S. Zier, B. Ji, P. Murfet, J. Mason, W. Rhee, H. Ainspan, and B. Parker, A 6.4Gb/s CMOS SerDes core with feedforward and decision-feedback equalization, in *IEEE Int. Solid-State Circuits Conf.*, February 2005, pp. 62–63.

21. K. Krishna, D. A. Yokoyama-Martin, S. Wolfer, C. Jones, M. Loikkanen, J. Parker, R. Segelken, J. L. Sonntag, J. Stonick, S. Titus, and D. Weinlader, A 0.6 to 9.6 Gb/s binary backplane transceiver core in 0.13 μm CMOS, in *IEEE Int. Solid-State Circuits Conf.*, February 2005, pp. 64–65.

22. S. Gondi, J. Lee, D. Takeuchi, and B. Razavi, A 10 Gb/s CMOS adaptive equalizer for backplane applications, in *IEEE Int. Solid-State Circuits Conf.*, February 2005, pp. 328–329.

23. J. T. Stonick, G.-Y. Wei, J. L. Sonntag, and D. K. Weinlader, An adaptive PAM-4 5 Gb/s backplane transceiver in 0.25 μm CMOS, *IEEE J. Solid-State Circuits*, 436–443, 2003.

24. K. Farzan and D. A. Johns, A low-complexity power-efficient signaling scheme for chip-to-chip communication, in *IEEE Int. Symp. Circuits Syst.*, Vol. 5, pp. 77–80, May 2003.

25. B. K. Casper, M. Haycock, and R. Mooney, An accurate and efficient analysis method for Gb/s chip-to-chip signaling schemes, in *Symp. VLSI Circuits*, June 2002, pp. 54–57.

26. I. Kalet and S. Shamai, On the capacity of a twisted-wire pair: Gaussian model, *IEEE Trans. Commun.*, 38(3), 379–383, 1990.

27. S. Shamai, On the capacity of a twisted-wire pair: Peak-power constraint, *IEEE Trans. Commun.*, 38(3), 368–378, 1990.

28. A. Amirkhany, A. Abbasafar, J. Savoj, M. Jeeradit, B. Garlipp, V. Stojanovic, and V. Horowitz, A 24 Gb/s software programmable multi-channel transmitter, in *Symp. VLSI Circuits*, June 2007, pp. 38–39.

29. B. Casper, G. Balamurugan, J. E. Jaussi, J. Kennedy, M. Mansuri, F. O'Mahony, and R. Mooney, Future microprocessor interfaces: Analysis, design and optimization, in *Custom Integr. Circuits Conf.*, September 2007, pp. 479–486.

22 Clock and Data Recovery Circuits

Jafar Savoj

CONTENTS

22.1   INTRODUCTION

The raw speed and low cost of modern complementary metal oxide semiconductor (CMOS) technologies has fueled the evolution of integrated circuits (ICs) into large systems-on-chip (SOC). The increase in the aggregate clocking speed and the number of supported features in an SOC require high-speed data transceivers at the input/output (IO) ports, operating at speeds as high as few tens of gigabits per second.

Traditional implementations of high-speed transceivers utilized various techniques to achieve maximum operating rate and highest signal purity despite the speed limitations of available fabrication processes. These techniques included using external components, additional mask layers through the fabrication process, and choice of high-power logic families. Other implementations benefited from exotic fabrication technologies that provided higher raw speeds but did not achieve low-cost and low-power dissipation. Low yield of devices fabricated in such processes did not allow the integration of the entire system and made the circuits only appropriate for use in multichip systems.

Modern designs benefit from much higher process speeds and target full integration of transceivers in very large scale integration (VLSI) technologies. Such implementations consider critical issues including small area and power consumption of the core, as well as provisions to allow the use of low-cost package and printed circuit board. On-chip signal processing techniques are devised to reverse the dispersive effects of these materials stemming from poor impedance control of the lines.

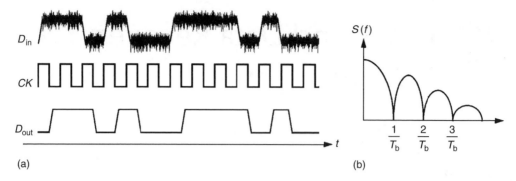

FIGURE 22.1   (a) Role of the CDR circuit and (b) spectrum of random data.

The design of the clock and data recovery (CDR) circuit is the most challenging task in implementation of any wireline receiver. To fulfill stringent requirements for power consumption, integrability, and ease of portability across process nodes, new architectures have transitioned from fully analog into hybrid and digital structures. In this chapter, we study the evolution of CDR architectures and describe some of their key implementations.

The task of the CDR circuit is to regenerate the data arriving at the receiver as shown in Figure 22.1a. The received signal experiences attenuation and distortion from the communication channel and is further corrupted by deterministic and random sources of disturbance such as crosstalk, reflection, and supply noise. The regeneration circuit inside the receiver is synchronized with the incoming data and captures the data at the instant of maximum eye opening, typically at the center of the bit period. The sampling clock signal should achieve high purity (in terms of timing jitter) because the purity of the regenerated signal can never exceed that of the sampling clock signal. Therefore, clock recovery focuses on synchronizing the sampling clock with the data, positioning the sampling edge of the clock at the center of the data eye, and reducing the timing jitter of the clock signal to the smallest possible.

As shown in Figure 22.1b, the power spectral density function of random data lacks frequency components at the data rate and its integer multiples. Therefore, the synchronizing clock signal cannot be directly derived from the data. Clock recovery techniques provide solutions to extract the synchronization information from the data.

In this chapter, we provide an overview of common CDR architectures and describe the implementation of their key building blocks. Next, we focus on the design of high-performance LC oscillators for CDR circuits. Finally, we study the means of characterizing jitter for CDR circuits.

22.2   INJECTION-LOCKED CDR CIRCUIT

Edge detection of random data transforms its spectrum to contain a frequency component at the frequency of data rate. Edge detection can be implemented either in the form of differentiation and rectification, or multiplication of the signal by a delayed version of itself. Bandpass filtering of the resulting signal can extract the synchronizing frequency [1]. However, the center frequency to bandwidth ratio of such filters is very high and beyond the practical limits for the integration of filters in VLSI technologies.

The major architectures suitable for implementing integrated CDR circuits were originally deployed for synchronization in television receivers. These receivers relied on injection locking of the received synchronization signal into a scanning oscillator to produce a scanning signal at the receiver [2]. Figure 22.2 depicts the utilization of injection locking in CDR circuits based on this concept. The output of the edge detector, formed with a delay element and a multiplier, is injected into a voltage-controlled oscillator (VCO). The VCO output locks to the burst of pulses arriving

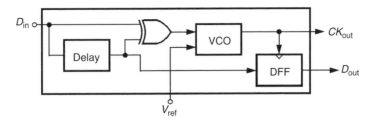

FIGURE 22.2 Injection-locked CDR circuit. (From Nogawa, M., Nishimura, K., Kimura, S., Yoshida, T., Kawamura, T., Togashi, M., Kumozaki, K., and Ohtomo, Y., *ISSCC Digest of Technical Papers*, 228, February 2005. With permission.)

at its input and produces an output clock signal with the same frequency as the input. Due to its injection locking nature, this circuit acquires lock, instantaneously upon the arrival of pulses at the input and loses lock when the pulses disappear. As a result, this architecture is suitable for burst-mode communications [3]. The recovered clock is used to retime the data inside a D-flip-flop (DFF). This structure can only lock to the incoming pulses if the frequency of the free-running VCO is close to that of the incoming pulses. A reference voltage (V_{ref}) generated by an auxiliary loop is used to drive the VCO towards its locking range. We will later describe means of producing the reference voltage in the context of aided frequency acquisition.

22.3 PHASE-LOCKED CDR CIRCUIT

With relative poor immunity of injection-locked circuits to noise, an automatic frequency and phase-controlled synchronizing system that relied on phase locking of a local scanning oscillator to the horizontal and vertical pulses in a television receiver was introduced in Ref. [2]. This structure resulted in improvements in television service despite severe noise conditions and expanded the coverage area. This mechanism for phase synchronization still remains a common approach for clock recovery in wireline communication systems. The original discoveries for high immunity of these circuits to noise are applicable to CDR circuits and play an important role in the jitter performance of wireline transceivers.

Figure 22.3 shows the architecture of a phase-locked CDR circuit. A phase detector (PD) produces an output that is proportional to the phase difference between the input data and the VCO (we will later see that this information can be as simple as a logic value to indicate whether the clock sampling edge is early or late). The combination of a voltage-to-current ($V–I$) converter and a lowpass filter (LPF), formed at minimum by a resistor and a capacitor in series, provides a control voltage required to sustain the VCO oscillation at the desired phase and frequency. Finally, a decision circuit is used to retime the data using the recovered clock signal. A thorough study into the dynamics of phase-locked systems can be found in Ref. [4]. CDR circuits are mostly followed by a demultiplexer that converts the serial recovered data into parallel lines.

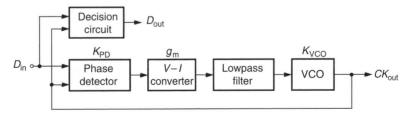

FIGURE 22.3 Phase-locked CDR circuit.

In order to achieve minimum bit error rate (BER), either the input data or the recovered clock has to be delayed so that the sampling edge of the recovered clock falls in the middle of the data eye, a condition that is hard to guarantee by design despite process and temperature variations. To achieve the optimum sampling instant, some PD topologies not only provide the phase error information, but also automatically retime the data. We refer to this concept as automatic retiming and only focus on this type of PDs in this chapter.

22.3.1 FULL-RATE CDR

In a full-rate CDR circuit, the frequency of the recovered clock equals the incoming data rate, allowing a clock rising (or falling) edge to appear in the middle of every bit. Single-edge retiming avoids degradation of the quality of detection arising from the duty cycle mismatch of the clock signal. We categorize the full-rate circuits based on the type of PD used inside them.

22.3.1.1 Linear PD

Hogge [5] developed a circuit consisting of a cascade of two flip-flops and two XOR gates. The circuit produces two pulses at every transition of input data sequence (Figure 22.4). One of the pulses has a width equal to the phase difference between the data edge and the clock edge (Error), and the other pulse has a constant width equal to one-half of a bit period (Reference). The difference of these two pulses is integrated by the $V - I$ converter (or a charge pump for lower speed implementations) and the LPF. The characterization of the integrated value versus phase error over one bit period indicates a linear relationship. The slope of the output is an indication of the PD gain (K_{PD}). Both inputs to the Reference XOR and one input to the Error XOR experience a CK-Q delay inside the flip-flops from the instant that the clock signal is asserted. The input data arriving at the first XOR gate are delayed to account for this effect.

When the integral of the difference falls to zero, the loop locks, indicating that the phase difference between the clock and the data has reached one-half of a bit period and the clock samples the data in the middle of the eye. Interestingly, D_{out} is the retimed version of the input in lock condition.

The Hogge PD was commonly used in CDR implementations until late 1990s due to its simplicity and predictable characteristic over process and temperature. With the advancement of digital CDR solutions, binary PDs have enjoyed more popularity in recent years. However, there have been efforts to utilize the linear characteristic of Hogge PD in hybrid analog/digital solutions [6].

22.3.1.2 Binary PD

As the operation speed of CDR circuits increases, binary PDs that rely on samples of the data to provide the phase error information in the form of early or late signals, and achieve higher immunity to nonidealities of analog process become more popular. Alexander PD [7] is perhaps the most

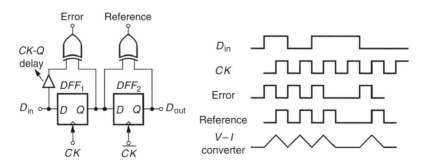

FIGURE 22.4   Hogge PD. (From Hogge, C., *IEEE J. Lightwave Technol.*, LT-3, 1312, 1985. With permission.)

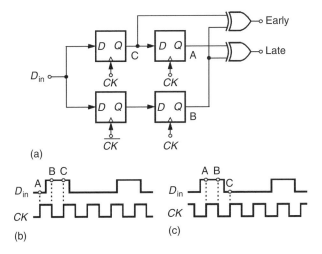

FIGURE 22.5 (a) Alexander PD, (b) late clock, and (c) early clock.

commonly used binary PD. As shown in Figure 22.5a, the operation of the circuit is based on obtaining three consecutive samples of the data, one at the edge and two from the previous and the current bits. Similarity of the edge sample to the current bit is an indication of a late clock signal (Figure 22.5b), and its similarity to the previous bit is an indication of early clock (Figure 22.5c). Three identical samples indicate that no data transition has occurred and the PD output is tristated. This feature prohibits the PD from producing a continuous early or late signal when experiencing no transitions due to a long sequence of identical bits.

The comparison of data samples is performed using two XOR gates. The edge sample is obtained with the falling edge of the clock and is retimed with the rising edge. This retiming aligns B with A and C. Similar to the Hogge PD, the difference of early and late signals is integrated by the V–I converter and the LPF. The second data sample (A) is usually used as the retimed data because the cascade of two flip-flops further improves the quality of the retimed data.

In lock condition, the recovered clock dithers around data transition, producing early and late pulses such that their average effect over a time interval on the order of the inverse of the loop bandwidth is zero. In practice, metastability of flip-flops transforms the early/late characteristic of the PD into a linear characteristic with high gain.

The gain of the Alexander PD is further reduced due to the jitter of the sampling clock. In Ref. [8], the actual characteristic of the PD is predicted by convolving the probability density function (PDF) of the clock jitter and the characteristic derived from the metastability analysis of the flip-flops.

22.3.2 SUBRATE CDR

A CDR circuit is typically followed by a demultiplexer that converts the serial data stream into parallel lines at lower rate. Data demultiplexing requires low-frequency clock signals, produced by dividing the frequency of the recovered clock. If clock recovery could be performed using clock signals at a subrate of the input data, power consumption could be reduced due to lower speed clocking and removal of high-power frequency dividers from the circuit. Furthermore, such schemes enable operation at higher speeds for any CMOS process node.

22.3.2.1 Linear PD

Linear PDs can be implemented to operate with subrate clock signals to relax speed constraints. A half-rate linear architecture only requires differential phases of the clock, a unique feature distinguishing this structure from a binary scheme requiring quadrature phases to operate. The circuit

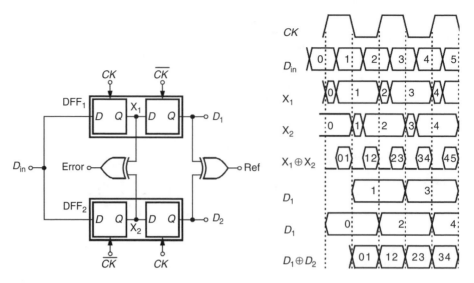

FIGURE 22.6   Half-rate linear PD. (From Savoj, J. and Razavi, B., *IEEE J. Solid-State Circuits*, 36, 761, 2001. With permission.)

consists of four latches and two XOR gates as shown in Figure 22.6 [9]. Data is applied to the inputs of two sets of cascaded latches, each constituting a flip-flop that retimes the data. Since the flip-flops are driven by a half-rate clock, the two output sequences are the demultiplexed waveforms of the input sequence if the clock samples the data in the middle of the bit period.

The output of each latch tracks its input for half a clock period and holds the value for the other half, resulting in the waveforms shown in Figure 22.6 for X_1 and X_2. The XOR of these two pulses consists of pulses that appear when a data transition occurs. The width of these pulses is determined by the clock edge on one side and data transition on the other, indicating that the average value of the pulse sequence contains information about the phase difference. However, the random nature of the data and the periodic behavior of the clock make this average value pattern-dependent. Therefore, a reference signal must also be generated whose average conveys this dependence.

The two waveforms D_1 and D_2 contain the samples of the data at the rising and falling edges of the clock. Thus, the XOR of D_1 and D_2 contains pulses as wide as half the clock period for every data transition, serving as the reference signal. The height of *Error* signal should be scaled by a factor of 2 with respect to *Ref*, since *Error* is half as wide as *Ref* in lock condition.

22.3.2.2   Binary PD

An Alexander PD obtains the edge and bit samples using the rising and falling edges of a full-rate clock and is shown in Figure 22.7a. If half-rate clock phases are utilized, the same samples can be obtained using the in-phase and quadrature clock signals. Similarly, a quarter-rate PD relies on half-quadrature clock phases to obtain similar samples. Since mismatch between multiple phases of the clock degrades the performance of subrate circuits, utilization of these circuits is usually limited to half-rate and quarter-rate operation.

Figure 22.7b depicts a half-rate implementation of the Alexander PD [10]. The two data samples are obtained using the rising and falling edges of the in-phase clock. The quadrature clock signal obtains the edge sample. Similar to the full-rate approach, two XOR gates compare these samples to derive early and late information. A modification to this circuit includes introduction of an extra

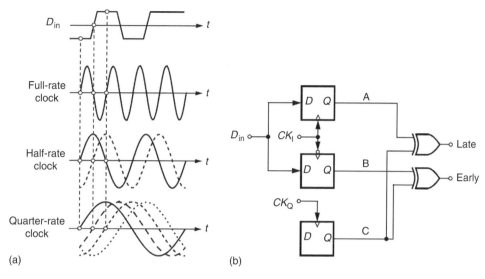

FIGURE 22.7   (a) Subrate sampling and (b) half-rate binary PD.

flip-flop, retiming C with CK_I, in order to provide alignment of the edge sample with data samples to improve the performance of the circuit.

22.3.3 AIDED FREQUENCY ACQUISITION

The oscillator used inside the CDR circuit is designed to have a wide tuning range to provide reliable oscillation at the frequency of interest despite process and temperature variations. CDR loops are usually designed for a small closed-loop bandwidth to allow higher suppression of input jitter. The CDR circuit has a capture range on the order of its loop bandwidth. As a result, if the initial VCO frequency, at the moment of startup, is outside the CDR capture range, the circuit may never acquire phase lock. Also, any sudden disturbance of the clock phase might force the loop out of lock, even if it has initially acquired lock. Therefore, phase-locked CDR circuits must utilize frequency acquisition schemes for stable operation. Acquisition circuits usually rely on a reference signal, used for clock synthesis inside the transmitter. Other schemes depend on samples of data itself for frequency acquisition. These topologies are usually referred to as reference-less frequency acquisition circuits. The study of such circuits is beyond the scope of this chapter. Here, we consider circuits that utilize a reference signal for frequency acquisition.

The dual-loop architecture shown in Figure 22.8 [11] relies on the wide capture range of a phase-frequency detector (PFD) operating with periodic signals [12] and consists of two loops sharing the charge pump (V–I converter), the LPF, and the VCO. A multiplexer selects the mode of operation by passing the output from either the PFD of loop I or the PD of loop II. When the circuit starts, loop I acquires lock to the reference signal, producing a VCO output frequency of Nf_{ref}. The difference between this frequency and the data rate is usually very small and within the acquisition range of the CDR circuit. Consequently, when loop II is enabled, the CDR circuit can acquire lock to the incoming data. The lock detector, designed in the form of a digital counter, determines the instant of transition.

This topology provides acquisition with very small overhead in hardware and area because the building blocks can be shared between both the loops. In a real implementation, the loop parameters should be chosen based on the requirements for loop II with attention to the stability of loop I.

Along with aided frequency acquisition, dual-loop architectures can provide flexibilities in the design of the CDR circuit. For example, the circuit proposed in Ref. [13] can be used to reduce of

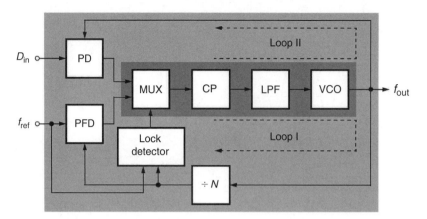

FIGURE 22.8 Dual-loop lock acquisition. (From Gutierrez, G., Kong, S., and Coy, B., *Proceedings of Custom Integrated Circuits Conference*, pp. 575–578, May 1998. With permission.)

VCO gain (K_{VCO}) in a CDR circuit and lower its timing jitter, since a smaller VCO gain translates the noise on the control line of the VCO into smaller jitter.

As shown in Figure 22.9, the structure consists of two loops utilizing identical VCOs. The control voltage of the VCO in the primary loop is split into coarse and fine lines. In this implementation, the coarse voltage is provided by the auxiliary loop. When the auxiliary loop acquires lock, it produces a control voltage required to drive the main loop into its capture range. This voltage is heavily filtered to avoid excess jitter in the primary VCO. The auxiliary loop can theoretically be turned off to save power when phase lock to random data has been achieved. However, this circuit should power up instantaneously in case of a sudden loss of lock. A similar auxiliary loop can be implemented in an injection-locked CDR topology to drive the VCO within the capture range of the circuit.

22.4 PHASE-ROTATING CDR CIRCUIT

Traditionally, high-speed transceivers have utilized separate VCOs for the transmitter and the receiver due to a slight frequency difference between the received and transmitted data. This difference stems from frequency mismatch of the far-end and local crystal oscillators.

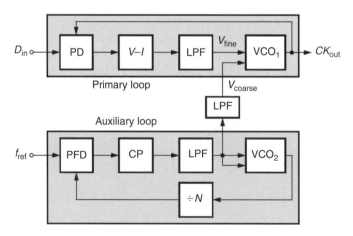

FIGURE 22.9 Lock acquisition and K_{VCO} reduction. (From Scheytt, J.C., Hanke, G., and Langmann, U., *ISSCC Digest of Technical Papers*, 348, February 1999. With permission.)

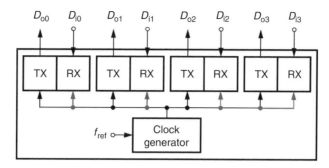

FIGURE 22.10   Clocking for multiple transceivers.

The increase in the aggregate data rate processed by an SOC can be accommodated using parallel streams of high-speed data. For example, 10 Gb/s attachment unit interface (XAUI) and Quad OC-192 standards achieve total operating rates of 10 and 40 Gb/s using four transceivers in parallel, each operating at 2.5 and 10 Gb/s, respectively.

The large-area overhead of LC oscillators, required for high-frequency signal generation and lower jitter, is more pronounced if each transceiver has separate VCOs for the transmitter and the receiver. Therefore, means of sharing oscillators among all transceivers must be sought.

Figure 22.10 demonstrates a configuration where the output of a single clock generator is routed to four transmitters. Since all four transmitters are synchronized, a shared clocking structure reduces the area. The introduction of resonant clocking schemes has significantly reduced the overall power consumption of such clocking networks [14].

The variation of the incoming phase (and possibly frequency) among the received signals can only be resolved using separate CDR circuits for each receiver. To save power and area, we must consider means of utilizing the clock, produced by the clock generator as the base signal and rotating its phase to track the edges of the incoming data sequences.

Figure 22.11 depicts the structure of a generic phase-rotating architecture. Multiple phases of a clock signal produced by the synthesizer enter the phase rotator whose output clock phase is steered by the signals produced by the controller. In order to rotate the output phase along all four quadrants (full 360°), both polarities of the in-phase and quadrature clock signal (CK_I and CK_Q) must be utilized, indicating that differential signaling must be used.

Phase rotation in any quadrant can be accomplished with two orthogonal phases. The control signals for the phase rotator can be either analog or digital. Digital operation is based on rotating the output phase in fine steps such that the input data edge at any point in time is bound between clock phases selected by consecutive digital codes.

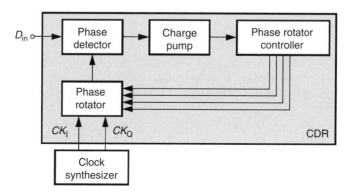

FIGURE 22.11   Phase-rotating CDR circuit.

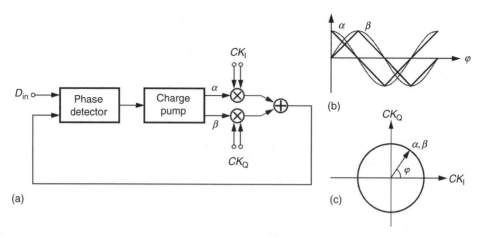

FIGURE 22.12 (a) CDR with analog phase rotator, (b) coefficients, and (c) recovered clock. (From Yang, F., O'Neill, J.H., Inglis, D., and Othmer, J., *IEEE J. Solid-State Circuits*, 37, 1813, 2002. With permission.)

The PD and the charge pump (V–I converter) used in such structures are similar to the ones used in phase-locked circuits. In a digital approach, a binary PD is preferred and the charge pump is replaced by a digital filter. We will study digital rotation in the context of digital CDR circuits.

Figure 22.12a represents an analog implementation of the phase rotator [15,16]. In this circuit, the rotator consists of two multipliers and an adder, and linearly combines the input signals in the form of $\alpha CK_I + \beta CK_Q$. Phase lock can be achieved if α and β represent cosine and sine of the instantaneous phase difference between the incoming data and the synthesizer frequency (Figure 22.12b). Note that if α and β are differential and can achieve both positive and negative values, phase rotation can be performed across all four quadrants (Figure 22.12c).

The design of the circuit can be simplified if α and β are approximated with triangular waveform as shown in Figure 22.12b. In Ref. [15], α and β signals are the differential outputs of a charge pump circuit and only possess positive values. As a result, 360° phase rotation across all four quadrants is accomplished by replacing a clock signal by its complement when the rotated vector passes the boundary between two quadrants. A boundary select circuit changes the polarity of the clock signal when its coefficient reaches zero. In Ref. [16], the circuit is modified to produce α and β signals that can assume both positive and negative values, hence alleviating the requirement for clock substitution and smoothing the transition across the quadrant boundaries.

Figure 22.13 shows a low-voltage implementation of the analog phase rotator. The circuit consists of two Gilbert multipliers whose outputs are summed. In order to provide reliable low-supply operation, the phase coefficients are mirrored.

22.5 FULLY DIGITAL CDR CIRCUIT

The transition from an analog to a digital CDR architecture can be explained using the digital representation of the system, derived by substituting loop parameters in s-domain with those in z-domain [17]. Based on the approximation of e^{-sT} with $1 - sT$, s can be replaced in z-domain:

$$z^{-1} = e^{-sT} \Rightarrow s \approx \frac{1 - z^{-1}}{T} \tag{22.1}$$

As a result, the combination of the analog VCO, V–I converter, and the LPF can be redefined as

$$\left(\frac{K_{\text{VCO}}}{s}\right)\left(K_{\text{PD}}R + \frac{K_{\text{PD}}}{sC}\right) \rightarrow \left(\frac{K_{\text{DPC}}}{1 - z^{-1}}\right)\left(K_\varphi + \frac{K_{\text{f}}}{1 - z^{-1}}\right)z^{-N} \tag{22.2}$$

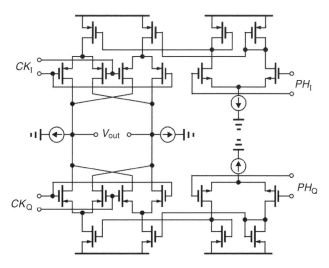

FIGURE 22.13 Analog phase rotator. (From Yang, F., O'Neill, J.H., Inglis, D., and Othmer, J., *IEEE J. Solid-State Circuits*, 37, 1813, 2002. With permission.)

Figure 22.14 depicts the system-level representation of a digital CDR based on the transformation of Equation 22.2. In this system, K_{DPC} is the gain of the digital-to-phase converter (DPC), a circuit that selects one of the m equally spaced clock phases according to its input digital code, and passes that to its output. The additional term given by z^{-N} represents the latency around the loop.

Figure 22.15 shows the actual implementation of the CDR circuit [17]. The PD usually has a binary subrate structure to relax the speed constraints of the digital logic. The resulting early/late information are either summed or processed in a majority vote circuit in order to reduce the width of the parallel error bus.

The frequency integrator compensates for the difference between the clock frequency and the data rate. It should have enough bits to cover the maximum frequency difference between the reference clock and the data. The gain stages are realized by shift registers. In order to reduce the volume of computation in each cycle and reduce power consumption, only the top bits of the phase and frequency integrators are passed to the proceeding stage. The discarded bits are a source of dither. The number of these bits is chosen so that the dithering of the frequency integrator is not a major source of jitter.

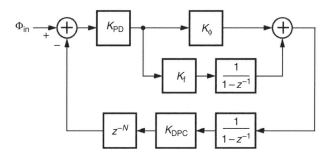

FIGURE 22.14 Digital representation of a CDR circuit. (From Sonntag, J.L. and Stonick, J., *IEEE J. Solid-State Circuits*, 41, 1867, 2006. With permission.)

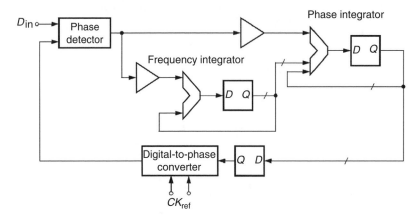

FIGURE 22.15   Digital CDR circuit.

22.6   LC OSCILLATORS IN CDR CIRCUITS

The performance of an oscillator has a strong impact on the output jitter of the closed-loop CDR circuit. In Ref. [18], it was shown that when an oscillator is placed inside a loop with unity gain bandwidth of f_u, jitter accumulates with the square root of time and saturates thereafter. The saturated value of jitter was calculated in Ref. [19] and represented as

$$\Delta t_{CDR} = \sqrt{\frac{f_0}{4\pi f_u}} \Delta t_{VCO} \tag{22.3}$$

where
   Δt_{CDR} is the total CDR jitter
   f_0 is the VCO center frequency
   Δt_{VCO} is the cycle-to-cycle jitter of the open-loop VCO

Equation 22.3 indicates that for applications with a fixed loop bandwidth, the most effective way to reduce the output jitter is reduction of the open-loop VCO jitter. In Ref. [20], it was also shown that the phase noise due to white noise sources and the cycle-to-cycle jitter are related, indicating that VCOs should be designed with the smallest possible phase noise.

VCO phase noise, regardless of circuit topology, is a strong function of its power consumption. In general, an increase by a factor of N in power consumption is equivalent to adding the output of N identical oscillators in phase. This structure achieves an increase in the total output power by N^2, while the uncorrelated noise power only increases by N, concluding that when power is increased by N phase noise decreases by the same factor [21]. As a result, all figures-of-merit defined to describe the performance of oscillators are normalized to power consumption.

Despite using an area-consuming integrated inductor, LC oscillators achieve the lowest phase noise compared to other VCO topologies. Unlike ring oscillators, whose frequency is a function of inverter or differential pair delay and highly dependent on the technology, the oscillation frequency of an LC oscillator is determined solely from the values of the integrated inductor and the total output capacitance and can be very high as long as the cross-coupled devices provide gain. LC oscillators therefore provide the highest oscillation frequency in any particular technology. Together, these features make topologies of Figure 22.16 attractive for implementation in CDR circuits.

The oscillators of Figure 22.16 usually use an N-channel metal oxide semiconductor (NMOS) varactor to achieve frequency tunability. Formed by placing an NMOS inside an N-well

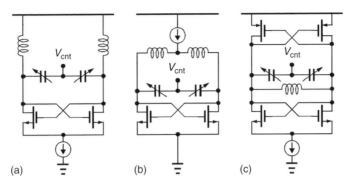

FIGURE 22.16 Conventional LC oscillators: (a) bottom current, (b) top current, and (c) complementary.

(Figure 22.17a), the structure achieves the maximum capacitance in accumulation and the minimum capacitance in depletion (Figure 22.17b). In order to achieve the maximum capacitance variability (maximum VCO tuning range), the gate-source voltage of the varactors should possess both positive and negative values.

In an LC VCO, gates of the varactors are usually connected to the output nodes and the control voltage produced by the preceding LPF is connected to its source/drain. The control voltage can typically swing close to supply and ground rails. Therefore, to achieve the maximum tuning range, the common-mode level of the VCO output should be close to mid-supply. The structure of Figure 22.16a can be replaced with structures of Figure 22.16b and 22.16c to resolve this issue.

The complementary VCO structure of Figure 22.16c is shown to achieve the lowest phase noise for a given power consumption. This structure provides a higher g_m for a given power consumption because the current flows through both the NMOS and the P-Channel metal oxide semiconductor (PMOS) cross-coupled pairs. Also, if the NMOS and PMOS switching devices are sized so that the VCO has identical rise and fall times, its output achieves single-ended symmetry, a phenomenon that is believed to reduce the output phase noise [22].

The study of phase noise in oscillators has been an active research topic in recent years. In Ref. [23], the major sources of phase noise in LC oscillators were identified to be the resonator noise, current-source noise, differential pair noise, and AM-to-PM conversion in the varactors.

Resonator noise can be reduced by designing inductors that have higher quality factor (Q). However, it is believed that as the frequency of oscillation increases, varactors, and not the inductors, contribute to reduction of the Q of the tank. Alternative tuning techniques can be sought to alleviate the impact of low-Q varactors [24].

The current-source noise translates up and down in frequency due to the switching of the cross-coupled pair and enters the resonator. As a result, components of tail current noise at dc and $2f_0$ can contribute to the output phase noise. Low-frequency noise and $2f_0$ device noise appear as amplitude noise and phase noise, respectively. An extra inductor and the capacitor can form high impedance at $2f_0$ and block the noise from entering the resonator as shown in Figure 22.18 [25].

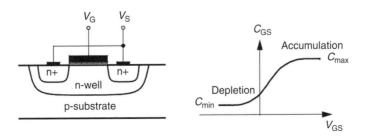

FIGURE 22.17 MOS varactors.

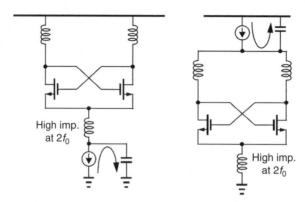

FIGURE 22.18 Noise filtering in LC oscillators. (From Hegazi, E., Sjoland, H., and Abidi, A.A., *IEEE J. Solid-State Circuits*, 36, 1921, 2001. With permission.)

AM-to-PM conversion represents the transformation of current-source amplitude noise into phase noise due to voltage tunability of the varactor capacitance. The varactor *C–V* characteristic achieves its highest slope in mid-range and its smallest slope at the two boundaries. This characteristic poses a significant design challenge because oscillators are nominally designed to operate in the center of the tuning range, where they experience the highest AM-to-PM conversion and exhibit the highest phase noise. The problem is more pronounced if the number of uncorrelated noise sources in the oscillator increases. Examples include quadrature and multiphase oscillators.

This problem can be circumvented by reducing K_{VCO}. However, this is not trivial since the VCO must have sufficient tuning range, up to a few gigahertz, running from a low supply voltage. VCO calibration is an effective technique for K_{VCO} reduction [26]. The coarse output capacitance is selected from of a bank of linear capacitors (such as fringe metal capacitors) and the fine capacitance is provided by a varactor as shown in Figure 22.19.

When the system powers up, a calibration algorithm switches the metal capacitors and drives the VCO frequency to a range that can be covered with the fine varactors. Despite having a large tuning range, K_{VCO} of this oscillator is significantly reduced because the fine varactor constitutes a small fraction of the total output capacitance. Further studies into the operation and optimization of VCOs can be found in Ref. [27].

22.7 JITTER IN CDR CIRCUITS

Jitter in CDR circuits is characterized with three parameters. Jitter generation is a measure of the jitter produced by the circuit itself. The major sources of this jitter are the phase noise of the VCO,

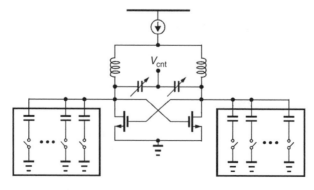

FIGURE 22.19 K_{VCO} reduction with calibration.

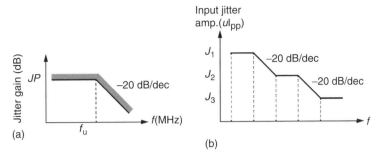

FIGURE 22.20 (a) Jitter transfer and (b) jitter tolerance masks.

and supply and substrate noise. Utilization of LC oscillators and reduction of K_{VCO} are effective means of lowering this type of jitter.

Jitter transfer describes the closed-loop CDR circuit in terms of its unity-gain bandwidth and maximum allowed peaking. The latter is a provision to avoid excessive peaking of jitter when multiple CDR circuits are cascaded in a long-haul wireline link. The characterization of transfer function is performed by modulating the edge of the input data signal with a low-frequency sinusoid and monitoring the response of the CDR circuit to this modulation. The ratio of the magnitude of the modulation frequency appearing in the recovered clock to that in the input data is an indication of jitter transfer magnitude at the frequency of modulation. Repetition of this process over various modulation frequencies produces the jitter transfer characteristic of the CDR circuit (Figure 22.20a) [28].

Jitter tolerance is an indication of the maximum input jitter that a CDR circuit can tolerate before it erroneously detects a bit. Figure 22.20b shows a jitter tolerance mask. This mask demonstrates the maximum amplitude of phase modulation that can be corrected by the loop as a function of phase modulation frequency.

Satisfaction of both jitter tolerance and transfer specifications using a single CDR loop is not trivial. Wider loop bandwidth improves trackability of the input jitter but lowers output jitter suppression. For example, a 10 Gb/s optical system requires a transfer bandwidth of 160 kHz and a tolerance bandwidth of about 4 MHz. This problem can be solved using a cleanup phase-locked loop (PLL) (Figure 22.21).

In this circuit, the main CDR loop is designed to have a wide bandwidth for jitter tolerance. The output of this loop is the input to another PLL with a much narrower bandwidth. The cascade

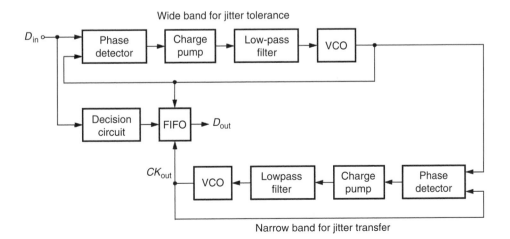

FIGURE 22.21 Cleanup PLL.

TABLE 22.1
State-of-the-Art CDR Circuits

Reference	Speed (Gb/s)	Process	Architecture	Power Dissipation (mW)	Bit Error Rate	Area (mm²)
[30]	40	65 nm (SOI)	Quarter rate	72	$<10^{-12}$	0.03
[31]	25	90 nm	Half rate	172	$<10^{-12}$	0.07
[32]	16	90 nm	Subrate	35	$<10^{-10}$	0.05
[33]	10	90 nm	Phase rotating	300[a]	$<10^{-9}$	0.43
[34]	6.25	0.13 μm	Digital	152[a]	$<10^{-15}$	0.56[a]
[35]	1.6	0.13 μm	Digital	12	$<10^{-12}$	0.1
[36]	1.5	90 nm	Digital	1`	NA	0.023
[37]	33.8	90 nm	Injection locked	98	NA	0.64
[38]	20	90 nm	Injection locked	175	$<10^{-9}$	0.96

[a] Quoted for the entire transceiver.

of these two loops achieves the narrow bandwidth, required for jitter transfer. In order to achieve the same low bandwidth for the recovered data, the circuit uses a first in, first out (FIFO), in which write and read clock signals are provided by the main CDR and the cleanup PLL, respectively. This block removes the additional jitter on the recovered data arising from retiming with the clock signal generated by the wide bandwidth loop [29].

22.8   CONCLUSIONS

Designing the CDR circuit is the most challenging task in the implementation of a wireline receiver. With the advancement in CMOS processes, analog architectures have transitioned into hybrid and digital solutions. New techniques have enabled sharing of a single clock generator for multiple transceivers, integrated into the same chip, and have resulted in significant reduction of power consumption and chip area.

LC oscillators, despite consuming larger area compared to their ring counterparts, achieve much higher speeds of oscillation and superior timing jitter. Novel design methodologies have enabled their utilization in the design of high-speed CDR circuits. Also, circuit and architecture level innovations are used to fulfill stringent requirements for generation, transfer, and tolerance of jitter in modern integrated CDR circuits. Table 22.1 summarizes the key features of some of the recently published CDR circuits. These prototypes are considered to be state-of-the-art at the time of this publication.

REFERENCES

1. A. W. Buchwald, Design of integrated fiber-optic receivers using heterojunction bipolar transistors, PhD Thesis, University of California, Los Angeles, CA, January 1993.
2. K. R. Wendt and G. L. Fredendall, Automatic frequency and phase control of synchronization in television receivers, *Proceedings of the IRE*, pp. 7–15, January 1943.
3. M. Nogawa, K. Nishimura, S. Kimura, T. Yoshida, T. Kawamura, M. Togashi, K. Kumozaki, and Y. Ohtomo, A 10 Gb/s burst-mode CDR IC in 0.13 μm CMOS, *ISSCC Digest of Technical Papers*, 228–229, February 2005.
4. B. Razavi (ed.), *Monolithic Phase-Locked Loops and Clock Recovery Circuits*, IEEE Press, Piscataway, NJ 1996.
5. C. Hogge, A self-correcting clock recovery circuit, *IEEE Journal of Lightwave Technology*, LT-3, 1312–1314, 1985.
6. M. H. Perrott, Y. Huang, R. T. Baird, B. W. Garlepp, L. Zhang, and J. P. Hein, A 2.5 Gb/s multi-rate 0.25 μm CMOS CDR utilizing a hybrid analog/digital loop filter, *ISSCC Digest of Technical Papers*, 328–329, 2006.

 7. J. D. H. Alexander, Clock recovery from random binary data, *Electronics Letters*, 11, 541–542, 1975.
 8. J. Lee, K. S. Kundert, and B. Razavi, Analysis and modeling of bang-bang clock and data recovery circuits, *IEEE Journal of Solid-State Circuits*, 39, 1571–1580, 2004.
 9. J. Savoj and B. Razavi, A 10 Gb/s CMOS clock and data recovery circuit with a half-rate linear phase detector, *IEEE Journal of Solid-State Circuits*, 36, 761–768, 2001.
10. M. Rau, T. Oberst, R. Lares, A. Rothermel, R. Schweer, and N. Menoux, Clock/data recovery PLL using half-frequency clock, *IEEE Journal of Solid-State Circuits*, 32, 1156–1159, 1997.
11. G. Gutierrez, S. Kong, and B. Coy, 2.488 Gb/s silicon bipolar clock and data recovery IC for SONET (OC-48), *Proceedings of Custom Integrated Circuits Conference*, pp. 575–578, May 1998.
12. C. A. Sharpe, A 3-state phase detector can improve your next PLL design, *EDN*, pp. 55–59, September 1976.
13. J. C. Scheytt, G. Hanke, and U. Langmann, A 0.155, 0.622, and 2.488 Gb/s automatic bit rate selecting clock and data recovery IC for bit rate transparent SDH systems, *ISSCC Digest of Technical Papers*, 348–349, February 1999.
14. R. Palmer, J. Poulton, W. J. Dally, J. Eyles, A. M. Fuller, T. Greer, M. Horowitz, M. Kellam, F. Quan, and F. Zarkeshvari, A 14 mW 6.25 Gb/s transceiver in 90 nm CMOS for serial chip-to-chip communications, *ISSCC Digest of Technical Papers*, 440–441, February 2007.
15. T. H. Lee, K. S. Donnelly, J. T. C. Ho, J. Zerbe, M. G. Johnson, and T. Ishikawa, A 2.5 V CMOS delay-locked loop for an 18 MBit, 500 megabyte/s DRAM, *IEEE Journal of Solid-State Circuits*, 29, 1491–1496, 1994.
16. F. Yang, J. H. O'Neill, D. Inglis, and J. Othmer, A CMOS low-power multiple 2.5–3.125 Gb/s serial link macrocell for high IO bandwidth network ICs, *IEEE Journal of Solid-State Circuits*, 37, 1813–1821, 2002.
17. J. L. Sonntag and J. Stonick, A digital clock and data recovery architecture for multi-gigabit/s binary links, *IEEE Journal of Solid-State Circuits*, 41, 1867–1875, 2006.
18. J. A. McNeil, Jitter in ring oscillators, *IEEE Journal of Solid-State Circuits*, 32, 870–879, 1997.
19. B. Razavi, *Design of Integrated Circuits for Optical Communications*, New York: McGraw Hill, 2003.
20. F. Hertzel and B. Razavi, A study of oscillator jitter due to supply and substrate noise, *IEEE Transactions on Circuits and Systems II: Analog and Digital Signal Processing*, 46, 56–62, 1999.
21. B. Razavi, A study of phase noise in CMOS oscillators, *IEEE Journal of Solid-State Circuits*, 31, 331–343, 1996.
22. A. Hajimiri and T. H. Lee, Design issues in CMOS differential LC oscillators, *IEEE Journal of Solid-State Circuits*, 34, 717–724, 1999.
23. J. J. Rael and A. A. Abidi, Physical processes of phase noise in differential LC oscillators, *Proceedings of Custom Integrated Circuits Conference*, pp. 569–572, May 2000.
24. A. Maxim, A varactor-less 10 GHz CMOS LC-VCO for optical communications transceiver SOCs using caged inductors, *Proceedings of Custom Integrated Circuits Conference*, pp. 663–670, September 2006.
25. E. Hegazi, H. Sjoland, and A. A. Abidi, A filtering technique to lower LC oscillator phase noise, *IEEE Journal of Solid-State Circuits*, 36, 1921–1930, 2001.
26. H. Darabi, S. Khorram, E. Chien, M. Pan, S. Wu, S. Moloudi, J. C. Leete, J. J. Rael, M. Syed, R. Lee, B. Ibrahim, M. Rofougaran, and A. Rofougaran, A 2.4 GHz CMOS transceiver for Bluetooth, *ISSCC Digest of Technical Papers*, 200–201, February 2001.
27. D. Ham and A. Hajimiri, Concepts and methods in optimization of integrated LC VCOs, *IEEE Journal of Solid-State Circuits*, 36, 896–909, 2001.
28. L. M. DeVito, A versatile clock recovery architecture and monolithic implementation, *Monolithic Phase-Locked Loops and Clock Recovery Circuits, Theory and Design*, B. Razavi (ed.), IEEE Press, New York, 1996 [invited paper].
29. L. DeVito, Clock recovery and data retiming, MEAD Workshop on IC Design for Optical Communication Systems, Monterey, March 2001.
30. T. Toifl, C. Menolfi, P. Buchmann, C. Hagleitner, M. Kossel, T. Morf, J. Weiss, and M. Schmatz, A 72 mW 0.03 mm^2 inductorless 40 Gb/s CDR in 65 nm SOI CMOS, *ISSCC Digest of Technical Papers*, 226–227, February 2007.
31. C. Kromer, G. Sialm, C. Menolfi, M. Schmatz, F. Ellinger, and H. Jackel, A 25 Gb/s CDR in 90 nm CMOS for high-density interconnects, *ISSCC Digest of Technical Papers*, 326–327, February 2006.
32. S. Palermo, A. E. Neyestanak, and M. Horowitz, A 90 nm CMOS 16 Gb/s transceiver for optical interconnects, *ISSCC Digest of Technical Papers*, 44–45, February 2007.

33. M. Meghelli, S. Rylov, J. Bulzacchelli, W. Rhee, A Rylyakov, H. Ainspan, B. Parker, M. Beakes, A. Chung, T. Beukema, P. Pepeljugoski, L. Shan, Y. Kwark, S. Gowda, and D. Friedman, A 10 Gb/s 5-tap-DFE/4-tap-FFE transceiver in 90 nm CMOS, *ISSCC Digest of Technical Papers*, 80–81, February 2006.

34. K. Krishna, D. Yokoyama-Martin, S. Wolfer, C. Jones, M. Loikkanen, J. Parker, R. Segelken, J. Sonntag, J. Stonick, S. Titus, and D. Weinlader, A 0.6 to 9.6 Gb/s binary backplane transceiver core in 0.13 mm CMOS, *ISSCC Digest of Technical Papers*, 64–65, February 2005.

35. P. K. Hanumolu, M. G. Kim, G.-Y. Wei, and U.-K. Moon, A 1.6 Gbps digital clock and data recovery circuit, *Proceedings of Custom Integrated Circuits Conference*, pp. 603–606, September 2006.

36. K.-H. Chao, P.-Y. Wang, and T.-H. Hsu, 0.0234 mm^2/1 mW DCO based clock/data recovery for Gbit/s applications, *Digest of Symposium on VLSI Circuits*, pp. 132–133, June 2007.

37. L.-C. Cho, C. Lee, and S.-I. Liu, A 33.6–33.8 Gb/s burst-mode CDR in 90 nm CMOS, *ISSCC Digest of Technical Papers*, 48–49, February 2007.

38. J. Lee, A 20 Gb/s burst-mode CDR circuit using injection-locking technique, *ISSCC Digest of Technical Papers*, 46–47, February 2007.

Part V

Circuits for Imaging and Sensing

23 Low Power CMOS Imager Circuits

Alexander Fish and Orly Yadid-Pecht

CONTENTS

23.1 INTRODUCTION

Wide utilization of portable battery-operated devices in modern multimedia applications triggered a demand for high-density ultra low-power image sensor [1,2]. Active pixel sensor (APS), implemented in a standard complementary metal oxide semiconductor (CMOS) technology, became a very attractive solution for these applications, rivaling traditional charge coupled devices (CCDs). Offering significant advantages in terms of low power, low voltage, and monolithic integration [3–7], CMOS technology allows for the fabrication of so called "smart" image sensors. The term "smart" sensor relates to the ability of the imager to integrate analog and digital signal processing onto the same substrate with the sensor and its digital interface. This signal processing can perform functions of real-time object tracking [8–17], motion detection [18,19], image compression [20,21], and many others [22–24]. Low-power smart image sensors are very useful in a variety of applications, such as space, automotive, medical, security, industrial, and others. Today's state-of-the-art CMOS imagers, used in portable applications, like cellular videophones, consume only 2–50 mW of power. However, the requirements of the next generation of portable devices from image sensors are expected to be more stringent in terms of power dissipation, consuming less than 1 mW [2].

There are many ways to reduce power dissipation in CMOS image sensors. Usually, the target is achieved by continuous technology scaling and aggressive supply voltage reduction [1,25]. However, in contrast to conventional digital circuits that achieve reduced area and better performance in state-of-the-art submicron processes, CMOS image sensors are more sensitive to technology scaling, requiring a stable, well-characterized technology [5]. In addition to reduced photo collection efficiency and increased dark current, technology scaling and supply voltage reduction significantly affect the output swing of the sensor, producing a serious impact on the signal-to-noise ratio (SNR) and dynamic range (DR) of the imager [1,3,26].

Many papers on low-power high-performance CMOS image sensors have been published in the literature [1,2,29–33,43]. Some of these papers focused on low-power "camera-on-a-chip" development for portable and space applications, where power dissipation is the main demand [2,29–33]. The term "camera-on-a-chip" was first introduced by Fossum [32] and it relates to the detector array with on-chip timing, bias generation, control and signal chain electronics, including analog-to-digital conversion. Research on low-power camera-on-a-chip using CMOS imagers began in 1995 by NASA at the Jet Propulsion Laboratory [32] and in 1999, the first large format, high-quality, fully digital, fully programmable, and low-power camera-on-a-chip was presented by Pain et al. [30]. Fabricated in 0.5 μm CMOS technology, this camera provided 512 × 512 resolution with pixels size of 11.6 μm × 11.6 μm and 8 mW of power dissipation with a voltage supply of 3.3 V. In 2003, Cho et al. [2] presented a 176 × 144 resolution camera-on-a-chip, dissipating only 550 μm of power and operating from a 1.5 V supply. In this design, low-power sensor design methodology was the first consideration at all design levels. It included low-voltage operation, output chain optimization, custom designed timing, and control block, utilizing a low-power successive approximation analog-to-digital converter (ADC), power management, minimizing the number of circuit blocks and others.

Looking for additional ways to reduce power, many researchers attempt to utilize alternative technologies for image sensor implementations [34–38]. One of these alternatives is silicon-on-insulator (SOI) technology, offering ultra low-power figures mainly because of reduced parasitic capacitances and reduced leakage currents due to lower area parasitic junctions. SOI is very popular in low-power digital circuit designs and potentially can be useful for low-power image sensor designs [39]. However, because of their too thin top silicon film (usually bellow 2000 Å), conventional vertical photodiode and photogate are not suitable for implementation in SOI. To overcome this barrier, a hybrid bulk/SOI pixel structure was proposed [40]. In this implementation, the photodiode was fabricated on the substrate of the SOI wafer after etching away the buried oxide. All other transistors, including the reset transistor and the in-pixel amplifying transistors, were implemented on the top SOI thin film. This way, the performance of this imager is similar to the performance of a standard sensor fabricated in a bulk technology, while achieving higher speed and lower power dissipation.

Since today's image systems contain very complex mixed signal designs, their power reduction is not a trivial task. Usually, it involves various low-power design methods, commonly used in low-power, low-voltage analog and digital circuit designs. The goal of this chapter is to present where power is dissipated in CMOS imagers and to figure out possible solutions to reduce this power dissipation.

First, we present sources of power dissipation in CMOS image sensors with reference to smart image sensor architectures. Then, considerations for power reduction in state-of-the-art smart CMOS image sensors are discussed, showing practical examples, where the mentioned techniques can be useful. We show that power dissipation can be reduced at all design levels—technology, device, circuit, logic, architecture, algorithm, and system integration. The interaction between the different design levels and their influence on general power dissipation is also presented.

23.2 SMART CMOS IMAGE SENSOR ARCHITECTURE

Figure 23.1 shows the general architecture of a smart CMOS APS-based image sensor. The core of this architecture is a camera-on-a-chip, consisting of a pixel array, Y-addressing circuitry with a row driver, X-addressing circuitry with a column driver, analog front end (AFE), ADC, digital timing and control block, bandgap reference, and a clock generator. Optional analog and digital

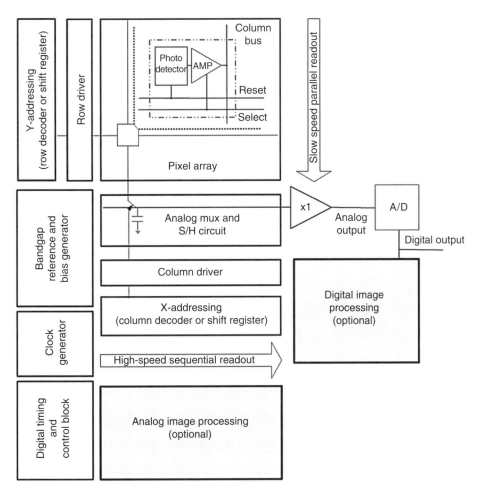

FIGURE 23.1 General architecture of a smart CMOS APS-based image sensor.

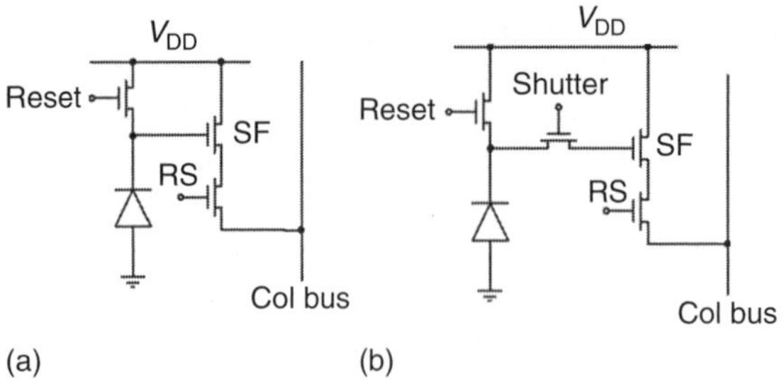

FIGURE 23.2 (a) Basic photodiode APS pixel and (b) scheme of a commonly used global shutter (snapshot) pixel. (From Yadid-Pecht, O. and Etienne-Cummings, R., *CMOS Imagers: From Phototransduction to Image Processing*, Kluwer Academic, Dordrecht, 2004. With permission.)

processing blocks upgrade the camera-on-a-chip core to a smart imager, and they are used to perform additional functions that can vary from one design to another depending on the application and system requirements. Note that, in contrast to the camera-on-a-chip components, the placement of analog and digital processing blocks is symbolic and does not represent their real location on the die. A brief description of the main imager building blocks (APS pixel array, scanning circuitry, AFE, ADC, reference, digital timing, and image processing circuits) is presented herein.

23.2.1 Active Pixel Sensor Pixel Array

The imager pixel array consists of N by M active pixels. The most popular is the basic photodiode APS pixel, employing a photodiode and a readout circuit of three transistors: a photodiode reset transistor (Reset), a row select (RS) transistor, and a source-follower (SF) transistor. The scheme of this pixel is shown in Figure 23.2a and its detailed description can be found in Ref. [4]. The in-pixel amplification is performed using the SF amplifier implemented by the in-pixel SF transistor and the source-follower bias (SFB) transistor, located at the bottom of the column and shared for all pixels in that column. The Enable transistor in Figure 23.3 is used as a digital switch to, when necessary, switch off current through the SFB transistor. Many types of photodetectors and pixels can be found in the literature. This includes a p-i-n photodiode, photogate, and pinned photodiode-based pixels, operating either in rolling shutter or in global shutter (snapshot) readout modes. The difference between these modes is that in the rolling shutter approach, the start and end of the light collection for each row is slightly delayed from the previous row, leading to image distortion when there is relative motion between the imager and the scene. On the other hand, the global shutter technique

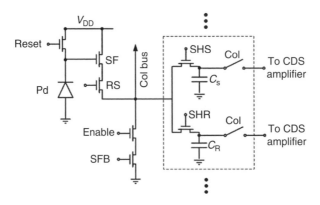

FIGURE 23.3 Column-parallel signal and reset S/H circuits.

uses a memory element inside each pixel and provides capabilities similar to a mechanical shutter: it allows simultaneous integration of the entire pixel array and then stops the exposure while the image data is read out. The in-pixel amplification in the global shutter sensor can be performed by an SF amplifier identical to that in a rolling shutter pixel, shown in Figure 23.2a. The full transistor scheme of a commonly used photodiode-based global shutter pixel is shown in Figure 23.2b and its detailed description can be found in Ref. [4].

The operation of both rolling shutter and global shutter pixels can be divided into two main stages: reset and phototransduction. During the reset stage, the photodiode capacitance is charged to a reset voltage by turning on the Reset transistor. In the rolling shutter mode, this reset voltage is read out to the corresponding sample-and-hold (S/H) circuits in a column-correlated double sampling (CDS) circuit by turning on the in-pixel RS and column-shared SHR transistors (see Figure 23.3). The main purpose of the CDS circuit is to eliminate fixed pattern noise (FPN) caused by random variations in the threshold voltage of the reset and pixel amplifier transistors, variations in the photodetector geometry, and variations in the dark current. There are many ways to perform CDS, but almost all solutions, in addition to S/H, utilize an amplifier located either at the bottom of each column or outside the array that is shared by all the columns in the array. During the phototransduction stage, the photodiode capacitor is discharged through a constant integration time at a rate approximately proportional to the incident illumination. Therefore, a bright pixel produces a low analog signal voltage and a background pixel gives a high signal voltage. This voltage is read out to the corresponding S/H by enabling the RS transistor of the pixel and SHS transistor. In the rolling shutter mode, the CDS circuit outputs the difference between the reset voltage level and the photovoltage level. In the global shutter sensors, the FPN elimination is more complicated, but also possible.

Note that, although the presented pixels are commonly used in today's cameras-on-a-chip, many smart imagers employ more complicated pixels. Some of them perform analog image processing tasks at the pixel level. A very good example for these imagers is a neuromorphic sensor, in which each pixel consists of a photodetector and local circuitry, performing spatiotemporal computations on the analog brightness signal [10–13]. Another example is an imager, in which A/D conversion is performed in the pixel level.

23.2.2 SCANNING CIRCUITRY

Unlike CCD image sensors, CMOS imagers use digital memory style readout, usually employing Y-addressing and X-addressing to control the readout of output signals through the analog amplifiers and allow access to the required pixel. The array of pixels is accessed in the row-wise fashion using the Y-addressing circuitry. All pixels in the row are read out into column analog read out circuits in parallel and then are sequentially read out using the X-addressing circuitry (see Figure 23.1).

Generally, the Y-addressing and X-addressing circuitry can be implemented either using digital decoders or shift registers. The most acceptable solution is to use shift registers, because this solution reduces power dissipation and the number of global buses, compared to imagers, where decoders are used. An additional important role of shift registers in CMOS imagers is the definition of regions (windows) of interest. A serial selection of regions of interest and their subsequent processing can greatly facilitate the computation complexity and significantly reduce power dissipation. This issue is discussed in detail in the next sections.

23.2.3 ANALOG FRONT END

As mentioned, all pixels in a selected row are processed simultaneously and sampled onto S/H circuits at the bottom of their respective columns. Due to this column-parallel process, for the array, having M columns, AFE circuitry usually consists of the $2M$ S/H circuits, M size analog multiplexer, controlled by the X-addressing circuitry, one or M amplifiers to perform CDS, and one or more analog variable gain amplifiers (VGAs). These VGAs are usually used for color processing and for signal amplification for further analog processing or analog-to-digital conversion. The number of amplifiers required to perform the CDS functionality depends on the chosen CDS architecture and is equal to M in case the subtraction is done separately for each column. The consideration for using one or M amplifiers closely depends on the ADC type used, as described in Section 23.2.4. The choice of an AFE configuration depends on many factors, including the type of sensor being used, DR, resolution, speed, noise, and power requirements. The considerations regarding making appropriate AFE choices for imaging applications can be found in Ref. [42].

23.2.4 ANALOG-TO-DIGITAL CONVERSION

ADC is an inherent part of state-of-the-art smart image sensors. There are many considerations for on-chip ADC implementations. First, the ADC must support video rate data. This rate depends on the array size and on the specific application. For example, for $1 M$ size imagers operating at 60 frames/s, 60 Msamples are required. Second, the ADC must have at least 8 bit resolution with low integral nonlinearity (INL) and differential nonlinearity (DNL). And finally, the ADC should be efficient by means of power and area; it should not influence image quality by introducing noise through substrate coupling or other crosstalk mechanisms [3]. There are three general approaches to implementing sensor array ADC:

- *Pixel-level ADC*, where every pixel has its own converter [43,44]. This approach allows parallel operation of all ADCs in the APS array, so a very low speed ADC is suitable. Using one ADC per pixel has additional advantages, such as higher SNR and simpler design.
- *Column-level ADC*, where an array of ADCs is placed at the bottom of the APS array and each ADC is dedicated to one or more columns of the APS array [45,46]. All these ADCs are operated in parallel, so a low-to-medium-speed ADC design can be used depending on the sensor array size. The disadvantages of this approach are the necessity to fit each ADC within the pixel pitch (i.e., the column width) and the possible problem of mismatch among the converters on different columns.
- *Chip-level ADC*, where a single ADC circuit serves the whole APS array [47,48]. This method requires a very high-speed ADC, especially if a very large array is used. The architecture shown in Figure 23.1 utilizes this approach for ADC implementation.

The chosen approach for ADC implementation strongly influences the type of CDS circuits that can be used. In the case of chip level ADC, the CDS amplifier can be implemented either in each column or one CDS amplifier can serve the whole array. In the case of column-level ADC, only the column-level CDS amplifiers can be implemented, requiring N amplifiers to perform the CDS.

Many techniques for ADC were presented over the past few years [49]. However, only a few of them were implemented in conjunction with CMOS image sensors. The most popular ADCs for imager applications are based on four fundamental architectures: successive approximation, sigma–delta, single-slope, and pipelined architecture. The pixel-level approach is suitable for simple sigma–delta, single slope, and modifications of successive approximation ADCs. Single-slope, sigma–delta, and successive approximation ADCs can be implemented as column-level ADCs. Finally, successive approximation and pipelined ADCs are suitable for the chip-level approach. In general, it is not straightforward to figure out what is the preferable technique and approach for ADC implementation. This depends mainly on the array size, required resolution, readout rate, and area and power limitations.

23.2.5 REFERENCE CIRCUITS

Bandgap reference and current generators are used to produce on-chip analog voltage and current references for building blocks like amplifiers, ADCs, digital clock generators, and others. It is very important to design high-precision and temperature-independent references, especially in high-resolution state-of-the-art image sensors, where the temperature of the dye can vary by many tens of degrees.

23.2.6 DIGITAL TIMING

Digital timing, clock generators, and control blocks aim to control the whole system operation. Their implementation on the chip level decreases the number of required I/O pads and thus reduces system power dissipation. Synchronized by the generated clock, the digital timing and control block produces the proper sequencing of the row address, column address, ADC timing, and the synchronization pulses creation for the pixel data going off-chip. In addition, it controls the synchronization between the imager and the analog and digital processing.

23.2.7 IMAGE PROCESSING

Although these blocks are optional, they play a very important role in today's smart image sensors. Conventional vision systems are put at a disadvantage by the separation between a camera for "seeing" the world, and a computer or digital signal processor (DSP) for "figuring out" what is seen. In these systems, all information from the camera is transferred to the computer for further processing. The amount of processing circuitry and wiring necessary to process this information completely in parallel is prohibitive. In all engineered systems, such computational resources are rarely available and are costly in terms of power, space, and reliability. In contrast to a conventional camera-on-a-chip, which only captures the image and transfers it for further processing, smart image sensors reduce the computational cost of the processing stages interfaced to it by carrying out an extensive amount of computation at the focal plane itself (analog and digital image processing blocks in Figure 23.1), and transmitting only the result of this computation. Both analog and digital processing can be performed either in the pixel or in the array periphery. There are advantages and disadvantages for both methods. In-pixel digital image processing is very rare because it requires pixel-level ADC implementation and results in a very poor fill factor and large pixel size. *In-pixel analog image processing* is very popular, especially in the field of neuromorphic vision chips. In these chips, in-pixel computations are fully parallel and distributed, since the information is processed according to the locally sensed signals and data from the pixel neighbors. Some neuromorphic visual sensors operate in the subthreshold region and therefore have very low-power dissipation. The in-pixel analog image processing implementation usually results in increased pixel size, but allows more efficient computation. Other applications employing in-pixel analog processing are tracking chips, wide DR sensors, motion and edge detection chips, compression chips, etc. The periphery analog processing approach assumes that analog processing is performed in the array periphery without penalty on the imager's spatial resolution and it is usually done in a column-parallel manner. While this approach has computational limitations compared to in-pixel analog processing, it provides several significant benefits, like area saving, reuse of the processing circuits for better marching, and, potentially, more pixels. *Periphery digital processing* is the most standard and usually simpler choice. It is performed following the A/D conversion, utilizes standard existing techniques for digital processing, and is most often done on the chip level. The main disadvantage of this approach is its inefficiency by means of area occupied and power dissipated. Note that all mentioned techniques can be mixed and applied together on one chip to achieve better results. Examples of such imagers are presented in Section 23.4.

23.3 SOURCES OF POWER DISSIPATION IN IMAGE SENSORS

The contribution of different image sensor components to overall power dissipation may vary significantly from system to system. For example, pixel array power dissipation can vary from a number of microwatts for a small array employing three-transistor APS architecture to hundreds of milliwatts for large-format smart imagers employing in-pixel analog or digital processing. Another example is the scanning circuitry that dissipates hundreds of nanowatts to several microwatts and thus can be neglected in very small arrays operating at very low rate per frame. On the other hand, it becomes more important in large-format arrays, operating at hundreds of frames per second, dissipating power varying from hundreds of microwatts to several milliwatts. Therefore, it is essential to understand the sources of power dissipation in image sensors in order to estimate the influence of each component on overall power dissipation. This section describes sources of power dissipation in smart CMOS image sensors with reference to the general imager architecture, presented in Section 23.2. Theoretical analysis of the power dissipation of the sensor array, scanning circuitry, and AFE is discussed herein in detail and then demonstrated through practical examples in Section 23.4.

23.3.1   SENSOR ARRAY POWER DISSIPATION

Power dissipation of a conventional three-transistor APS array operating in the rolling shutter regime is given by

$$P_{\text{Array}} = F_{\text{R}} NM \left(E_{\text{reset}} + E_{\text{SF}} \right) + N \times M \times P_{\text{leak}} \tag{23.1}$$

where

F_{R}	is a frame rate
N	is the number of rows
M	is the number of columns
E_{reset}	is the energy required for pixel reset
E_{SF}	is the energy dissipated by the in-pixel SF during one frame
P_{leak}	is the power dissipated due to the reset and the RS transistors leakage

The E_{reset} component can be expressed by

$$E_{\text{reset}} = V_{\text{DD}} V_{\text{Reset}} C_{\text{Pd}} \tag{23.2}$$

where

     V_{Reset} is the reached reset voltage on the photodiode

     C_{Pd} is capacitance that relates to the photodiode capacitance and to the capacitance at the gate of the SF transistor (see Figure 23.2a)

In the case where PMOS is used as a reset transistor, the V_{Reset} voltage value is approximately equal to V_{DD}. During the reset stage, half of the energy drawn from the power supply ($E_{\text{reset}}/2$) is stored on the photodiode capacitance C_{Pd}, while half of the energy is dissipated in the reset transistor. During the phototransduction stage, the energy stored on C_{Pd} is dissipated by the photogeneration mechanism.

The in-pixel SF amplifier is used to read out an analog signal to the column S/H capacitors. As mentioned, the SFB transistor is located at the bottom of the column and is shared for all pixels in that column. The E_{SF} component can be approximated by

$$E_{\text{SF}} = V_{\text{DD}} V_{\text{S}} \left(C_{\text{S}} + C_{\text{col}} \right) + V_{\text{DD}} V_{\text{R}} \left(C_{\text{R}} + C_{\text{col}} \right) + I_{\text{LN}} V_{\text{DSF\_Sav}} T_{\text{CS\_charge}}$$
$$+ I_{\text{LN}} V_{\text{DSF\_Rav}} T_{\text{CR\_charge}} + I_{\text{LN}} \left(V_{\text{DD}} - V_{\text{S}} \right) \alpha_{\text{S}} + I_{\text{LN}} \left(V_{\text{DD}} - V_{\text{R}} \right) \alpha_{\text{R}} \tag{23.3}$$

where C_{S} is the signal S/H capacitor (see Figure 23.3), V_{S} is the maximum value on the C_{S} capacitor, C_{R} is the reset S/H capacitor, V_{R} is the maximum value on the C_{R} capacitor, C_{col} is the capacitance of the column line, I_{LN} is the bias current supplied by the SFB transistor, $T_{\text{CS\_charge}}$ and $T_{\text{CR\_charge}}$ are the time required to charge the C_{S} and C_{R} capacitors, respectively, $V_{\text{DSF\_Sav}}$ and $V_{\text{DSF\_Rav}}$ are the average drain-source voltages on the SF transistor during the charge of the C_{S} and C_{R} capacitors, respectively, α_{S} and α_{R} are the parameters defined as

$$\alpha_{\text{S}} = T_{\text{S\_read\_out}} - T_{\text{CS\_charge}}; \quad \alpha_{\text{R}} = T_{\text{R\_read\_out}} - T_{\text{CR\_charge}} \tag{23.4}$$

where $T_{S\_read\_out}$ and $T_{R\_read\_out}$ are periods when the RS transistor is open to allow signal and reset readout, respectively. These parameters should be as small as possible, i.e., allowing the source-follower operation only for a very short time, required for proper signal and reset sampling. In most imagers, the signal and reset output chains are symmetric, resulting in $\alpha_S = \alpha_R = \alpha_{S/R}$, $C_R = C_S = C_{S/R}$, $T_{CS\_charge} = T_{CR\_charge} = T_{CS/R\_charge}$. Assuming the worst case where $V_S = V_R = V_{S/R}$ and $V_{DSF\_Sav} = V_{DSF\_Rav} = V_{DSF\_S/Rav}$, Equation 23.3 can be simplified to

$$E_{SF} = 2V_{DD}V_{S/R}\left(C_{S/R} + C_{col}\right) + 2I_{LN}V_{DSF\_S/Rav}T_{CS/R\_charge} + 2I_{LN}\left(V_{DD} - V_{S/R}\right)\alpha_{S/R} \tag{23.5}$$

The first and the second terms of Equation 23.5 relate to the power dissipation that is necessary for proper readout operation, while the third one should be reduced as much as possible. It should be observed that the C_{col} capacitance of the column line becomes more and more important in large-format imagers. Note that the presented analysis assumes the worst case, where the C_{Pd}, C_S, and C_R capacitors should be charged from zero each frame. In practice, the array power dissipation will be smaller mainly due to reduced first term of Equation 23.5.

Finally, the third power component P_{leak} is caused by the leakage currents through the reset transistor. This leakage becomes more important when the imager employs a PMOS as a reset transistor and is implemented using modern submicron technologies (0.18 µm and below). It consists of a number of components: subthreshold current, source-drain band-to-band tunneling (BTBT), and gate leakage [50]. Since currently image sensor implementation in advanced processes below 0.13 µm is avoided due to low sensor performance, the subthreshold component is still dominant. However, in future imagers, designed in 90 and 65 nm processes, the BTBT and gate leakages will have a more significant impact on the sensor's power dissipation and imager performance. The subthreshold current of a MOS device, taking into account weak inversion, drain-induced barrier lowering (DIBL), and the body effect, is given by [50]

$$I_{sub\_th} = I_0 \times e^{\left(\frac{V_{GS} - V_{TH}}{nv_t}\right)} \times \left(1 - e^{\frac{-V_{DS}}{v_t}}\right) \times e^{\frac{\eta V_{DS}}{nv_t}} \tag{23.6}$$

where

$$I_0 = \mu_0 C_{ox} \frac{W}{L}(n-1)v_t^2 \tag{23.7}$$

where

V_{TH} is threshold voltage
$v_t = kT/q$ is thermal voltage
η is DIBL coefficient
n is subthreshold swing coefficient of the transistor
μ_0 is zero bias mobility
C_{ox} is gate oxide capacitance

Because of the dependence on the reset transistor drain–source voltage, the subthreshold current varies with pixel illumination intensity, producing an impact not only on power dissipation, but also on image quality.

Power dissipation of the global shutter sensor array can be analyzed in the same way and it is very similar to power dissipation of the rolling shutter imager. The main difference is in higher peak power dissipation of global shutter sensors caused by simultaneous operation of all pixels in the array. For example, peak energy dissipation due to the reset stage of a conventional N by M global shutter imager is given by

$$E_{peak\_reset} = N \times M \times V_{DD}V_{Reset}C_{Pd} \tag{23.8}$$

But the problem gets worse in the case of a more complicated imager, operated in the global shutter regime. In this imager, a number of analog/digital circuits should operate simultaneously in all pixels of the array, causing very high peak power.

Although an analysis of a simple pixel was presented, it can be easily extended to any kind of smart pixel. For pixels employing in-pixel analog and/or digital processing, the pixel power dissipation can be expressed as

$$P_{\text{Array}} = F_R N M \left(E_{\text{reset}} + E_{\text{read\_out}} + E_{\text{analog}} + E_{\text{digital}} \right) + N M P_{\text{leakage}} \quad (23.9)$$

where
E_{analog} and E_{digital} are energy dissipation components dissipated by in-pixel analog and/or digital processing during one frame

$E_{\text{read\_out}}$ is the energy dissipated for signal readout from the pixel (in a conventional pixel it is equal to E_{SF})

P_{leakage} is the power dissipated due to the in-pixel leakage currents.

Power dissipation of different analog and digital circuits is discussed in the following subsections. Note that some imagers operate in continuous mode and do not require a reset, resulting in $P_{\text{reset}} = 0$.

23.3.2 POWER DISSIPATION IN SCANNING CIRCUITRY

Similarly to conventional digital circuits, there are three major sources of power dissipation in imager scanning circuitry:

$$P_{\text{digital}} = p_t \left(C_L V V_{\text{DD}} f_{\text{clk}} \right) + I_{\text{SC}} V_{\text{DD}} + P_{\text{leakage}} \quad (23.10)$$

The first term represents the switching component power, where C_L is the loading capacitance, f_{clk} is the clock frequency, and p_t is the probability that a power-consuming transition occurs (the activity factor). In most cases, when the scanning circuitry is implemented using the CMOS design technique, the voltage swing V is the same as the supply voltage V_{DD}, however, in some logic circuits, such as single-gate pass-transistor implementations, the voltage swing on some internal modes may be slightly less [51,52]. The second term is due to the direct-path short circuit current I_{SC}, which arises when both NMOS and PMOS transistors are simultaneously active, conducting current directly from supply to ground [53]. Finally, the third term is the leakage component P_{leakage} which mechanism is similar to that of P_{leak} described in Section 23.3.1.

Usually, the dominant term in a "well-designed" digital circuit is the switching component. The picture can be different when discussing image sensor scanning circuitry. Due to a relatively low frequency and a small activity factor, the switching component is less dominant than in the standard high-performance digital circuits, while the leakage current becomes more important, especially in the modern submicron technologies with reduced threshold voltage. For example, one cell of row shift register operates at 30 kHz frequency with activity factor of 0.001 for the 1 Mpixels array operating at 30 frames/s. A column shift register cell of the same array has 30 MHz frequency with the same activity factor of 0.001. In some smart imagers, where the shift register is used for active windows definition, the frequency and the activity factor can be much smaller than in standard imagers (see the tracking imager example in Section 23.4). Compared to state-of-the-art digital circuits operating at gigahertz frequencies and having much higher activity, the switching component is less dominant in sensors scanning circuitry. In summary, we can define the activity factor and the operation frequency for each cell in the row and in the column circuitries as

$$p_{t\_\text{row}} = 1/N; \quad p_{t\_\text{column}} = 1/M \quad (23.11)$$

$$f_{\text{row}} = F_R N; \quad f_{\text{column}} = F_R N M \quad (23.12)$$

Thus, Equation 23.10 can be rewritten separately for row and column circuitries, designed using CMOS design technique:

$$P_{\text{row\_scanning}} = C_L V_{\text{DD}}^2 F_R + I_{\text{SC}} V_{\text{DD}} + P_{\text{leakage}} \quad (23.13)$$

$$P_{\text{column\_scanning}} = C_L V_{\text{DD}}^2 F_R N + I_{\text{SC}} V_{\text{DD}} + P_{\text{leakage}} \quad (23.14)$$

where $P_{\text{row\_scanning}}$ and $P_{\text{column\_scanning}}$ are associated with the power dissipation of each cell in the row and in the column circuitries, respectively. The C_L capacitance relates to the load capacitance of each cell in the row and in the column circuitries. For example, the load capacitance that is seen by the output cells of the row driver corresponds to the capacitance of the reset and the RS transistor gates and to the capacitance of the row interconnect.

23.3.3   ANALOG FRONT END POWER DISSIPATION

Power dissipation of AFE strongly depends on the chosen imager and AFE architectures. However, the following general expression can be written to describe AFE power dissipation:

$$P_{\text{AFE}} = P_{\text{SF\_bias}} + P_{\text{CDS\_amp}} + P_{\text{gain\_amp}} \tag{23.15}$$

where

$P_{\text{SF\_bias}}$ is the power dissipated in the column SFB transistor
$P_{\text{CDS\_amp}}$ and $P_{\text{gain\_amp}}$ are power components dissipated by the CDS and gain amplifiers, respectively

The $P_{\text{SF\_bias}}$ depends on the column SFB current I_{LN} flowing through the SFB transistor (see Figure 23.3) and on the voltage on the transistor SFB that varies according to the reset and signal values of all pixels in the column. For the case of a conventional three-transistor APS, where the voltage on SFB transistors can reach maximum value of $V_{\text{DD}} - 2\,V_{\text{TH}}$, the worst case for $P_{\text{SF\_bias}}$ component can be approximated as

$$P_{\text{SF\_bias}} = N2I_{\text{LN}}(V_{\text{DD}} - 2V_{\text{TH}})T_{\text{read\_out}}F_R M \tag{23.16}$$

where $T_{\text{read\_out}} = T_{\text{S\_read\_out}} = T_{\text{R\_read\_out}}$, assuming that signal readout time is equal to reset readout time.

It is more difficult to present an analysis for the $P_{\text{CDS\_amp}}$ and $P_{\text{gain\_amp}}$ components, since they are strongly architecture- and circuit-dependent. For example, in the case of column-parallel CDS, M CDS amplifiers are required, compared to the case of one global CDS amplifier. Although low-power analog amplifier design is out of the scope of this chapter, the fundamental limits for the analog circuit's minimum power consumption are presented. These limits can be also useful in Section 23.3.7, where a comparison between power dissipation of analog and digital image processing is discussed. According to the analysis presented in Ref. [54], the minimum power necessary to realize a single pole, assuming an ideal 100% current efficient transconductor, meaning that all the current pulled from the supply voltage is used to charge the integrating capacitor, and assuming a circuit that can handle rail-to-rail signal voltages, is given by

$$P_{\text{min}} = 8kTf\,\text{SNR} \tag{23.17}$$

where

f is the frequency of operation (required bandwidth)
SNR is the required SNR

This absolute limit is very steep, since it requires a factor of 10 power increase for every 10 dB of SNR improvement. In a similar way, the minimum power required for a single-stage common-source voltage amplifier of gain A_v (operated in weak inversion) can be defined as

$$P_{\text{min\_amp}} = 8nkTfA_v\text{SNR} \tag{23.18}$$

where n is the slope factor [54]. As can be seen, the minimum power for an amplifier is nA_v times larger that the absolute minimum given by Equation 23.18. Note that this power dissipation does not include the power spent in bias circuitry. Moreover, it does not include additional sources of noise, like flicker noise in the devices, noise coming from the power supply, or generated on the chip by other blocks of the imager.

In conclusion, the most important result of this is that the power dissipation linearly depends on the required SNR and the required bandwidth, meaning that the frame rate, array size, and required precision are the main factors affecting AFE power dissipation.

23.3.4 POWER DISSIPATION IN ANALOG-TO-DIGITAL CONVERSION

ADC power dissipation can provide a high contribution to the overall power dissipated by the imager [3], making low-power ADC design a very important issue in CMOS sensors design. Generally, the trend in ADCs is toward higher speed and higher resolution at a reduced power level and supply voltage. Lower supply voltage implies a smaller input voltage range, and hence greater susceptibility to noise from all potential sources: power supplies, references, digital signals, etc. A variety of ADCs are used in image sensors and multiple possible approaches of their implementations (see Section 23.2.4) make it very difficult to perform a general analysis of ADC power dissipation.

However, when quantifying the efficiency of ADC implementation, power dissipation can be represented in terms of the required resolution and sample rate. Power dissipation is a nonlinear function of the resolution and sample rates and a large number of figures-of-merit exist to relate all these parameters. The most widely accepted measure of the effectiveness of the design is the energy per conversion step, defined as

$$E_{\text{conv}} = \frac{P_{\text{D}}}{2^N f_{\text{S}}} \tag{23.19}$$

where
   P_{D} is ADC power dissipation
   N is the resolution
   f_{S} is the sample rate

The effectiveness of ADC strongly depends on the technology, ADC architecture, and circuits used. Successive approximation (SA) and sigma–delta are usually reported as power-efficient architectures for CMOS image sensors implementation [2,28]. However, because the decimation takes a lot of power in sigma–delta ADC, the SA seems to be the best solution for low-power ADC implementation, optimizing the resolution, area, and sampling rate. Good examples of column-parallel and chip-level SA low-power implementations can be found in Refs. [30] and [2], respectively.

23.3.5 POWER DISSIPATION IN BANDGAP REFERENCE AND CURRENT GENERATORS

The power dissipated in bias circuitry is wasted and, in principle, should be minimized. However, inadequate bias schemes may increase the noise and therefore require a proportional increase in power. For example, a bias current would be noisier if it was obtained by multiplying a smaller current. In many large-format imagers, power dissipation of the bandgap and current generators can be neglected. Usually, the maximum power dissipation of a few tens of microwatts is achieved.

23.3.6 POWER DISSIPATION IN DIGITAL TIMING AND CONTROL BLOCK

Digital timing and control blocks are digital circuits, and therefore the power dissipation can be analyzed in a similar way as was done in Section 23.3.2. The difference is that, unlike the scanning circuitry, the switching component cannot be easily defined and therefore the ratios between the switching and leakage components are more dependent upon the architecture of the imager. In any case, the operation of the digital timing and control block is still at frequencies much lower than standard digital circuits, making the leakage power dissipation component very important.

23.3.7 POWER DISSIPATION IN IMAGE PROCESSING

Power dissipation of analog and digital processing mainly depends on the complexity of the required processing. In some smart imagers, this component can be negligible while in other designs, it may even exceed 90% of the overall power dissipation [55]. When designing a smart sensor, the main question is whether to perform the required functionality in the analog or in the digital domains. Of course, this is not a simple question, but at least a partial answer can be found by observing the analysis presented in Ref. [54]. In this analysis, the minimum power for an analog system was compared to that of a digital system. In a digital system, each elementary operation requires a certain number m of binary gate transition cycles, each of which dissipates an amount of energy E_{tr}. Thus, the minimum power can be presented by

$$P_{min\_dig} = mfE_{tr} \tag{23.20}$$

where f is the signal bandwidth. The number m of transitions is only proportional to some power α of the number of bits N, and therefore power consumption is only weakly dependent on SNR (essentially logarithmically):

$$m \cong N^\alpha \approx \left[\log(\text{SNR})\right]^\alpha \tag{23.21}$$

A comparison with an analog system is obtained by estimating the number of gate transitions that are required to compute each period of the signal, which for a single pole digital filter can be estimated approximately to

$$m \cong 50N^2 \tag{23.22}$$

Immunity to thermal noise imposes an absolute minimum energy per transition E_{trmin} estimated to $8kT$, which provides the absolute minimum power limit. However, in real life, $E_{trmin} = C_L V_{DD}^2$ is forced to a much higher value by the need to recharge the load capacitance to the supply voltage. Figure 23.4 shows the summary of this comparison. It can be easily seen that for systems where small SNR is acceptable, the analog circuits may consume much less power than the digital. However, for systems requiring large SNR, the analog solution is most inefficient power wise.

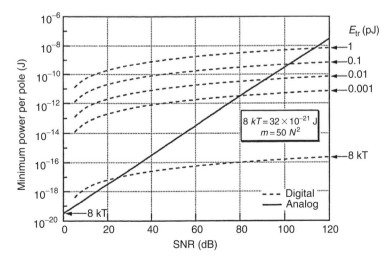

FIGURE 23.4 Minimum power for analog and digital circuits. (From Enz, C.C. and Vittoz, E.A., In Tutorial for 1996 International Symposium on Circuits and Systems, R. Cavin and W. Liu (Eds.), IEEE Service Center, Piscataway, CA, pp. 79–133, 1996. With permission.)

23.4 POWER REDUCTION IN SMART CMOS IMAGE SENSORS

Low-power sensor design methodology can be considered at all levels—technology, circuit and logic, architecture, algorithm, and system integration. This approach was first proposed by digital circuit designers [56] and was then utilized in low-power camera-on-a-chip design by Cho et al. [2]. In this section, we discuss power reduction at mentioned design levels and show the interactions between these levels. In addition, we provide practical examples that demonstrate how the discussed considerations can be used in practical designs.

Figure 23.5 presents a view of all possible levels where power consumption can be reduced. The interaction between the levels is shown using arrows. If two levels do not interact, it means that power dissipation reduction at each one of these levels can be done independently, without affecting considerations for power reduction at the other level. In cases where the levels are signed as interacted, the applied techniques at one of the levels can affect considerations for power reduction at the other level. For example, the technology and the algorithm levels do not interact with each other, but they are the basis for the device and architecture levels, respectively. The power dissipation at the device level strongly depends on the chosen technology, while the power dissipation at the architecture level depends on the chosen algorithm. Of course, there is no influence of the technology used on the optimization at the algorithm and architecture levels, meaning that these tasks can be done independently. Power reduction at the circuit and logic levels strongly depends both on the chosen technology and available devices and on the imager architecture. Final system integration interfaces between all design levels.

23.4.1 POWER REDUCTION AT THE TECHNOLOGY AND DEVICE LEVELS

23.4.1.1 General Considerations

The right choice of technology process is very important for the successful fabrication of low-power sensors. Similarly to digital circuits, CMOS imagers benefit from technology scaling by reducing pixel size, increasing resolution, and integrating more analog and digital circuits on the same chip with the sensor. However, low-power sensor implementation in advanced submicron technology below 0.18 μm is very problematic, mainly because of reduced photoresponsivity, dark current, and increased leakages. Of course, a conventional digital process can be modified to address image sensor requirements [5], but as designers we cannot change a process and instead have to choose from a variety of existing technologies available in the market. When choosing the preferable technology, the trade-offs are price, required resolution, required frame rate, image quality, and expected power dissipation. For example, for some applications, where the low-power dissipation is the main

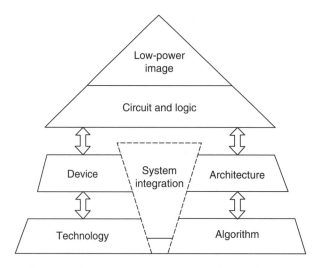

FIGURE 23.5 View of all possible levels, where power consumption can be reduced.

demand, while image quality and price are less important, alternative technologies, such as silicon-on-sapphire (SOS), can be utilized. SOS is one of the existing kinds of SOI technology, where sapphire-aluminum oxide (Al_2O_3) is used as the substrate. SOS technology has a very low-power figure mainly because of reduced parasitic capacitances and reduced leakage currents due to lower area parasitic junctions. SOI/SOS are very popular in low-power digital circuit designs and can be potentially useful for low-power image sensor design. It has been previously shown that providing the option of backside illumination due to the transparent substrate of SOS wafers can be a good alternative for the design of low-power CMOS image sensors. A number of practical examples of SOS low-power imagers can be found in Refs. [35–38].

In any case, the chosen technology must support the design techniques that should be utilized at the circuit and logic design levels, allowing the use of all relevant devices and connectivity required for the implementation of these techniques. It is recommended that the chosen technology will have the following features: (a) the option to implement transistors with various threshold voltages, (b) the possibility for separate body biasing of both NMOS and PMOS transistors, (c) the ability to operate at low voltage, while also having high threshold, high V_{DD} transistors, (d) the option to implement thick gate transistors for gate leakage reduction, (e) a large number of available metals for interconnect capacitance and crosstalk reduction, and finally (f) reduced capacitances.

The next level, where power can be reduced, is the device level. Considerations for power reduction at the device level are strongly dependent on the chosen technology, and the available layers and devices. Herein, we present an example of a design where power dissipation was optimized at the device level.

23.4.1.2   Practical Example—Self-Powered Active Pixel Sensors

In this example, a self-powered CMOS active pixel sensor (SPS) architecture for power reduction at the device level is presented. The concept of SPS was firstly presented in Ref. [65] and it employs novel power generation photodiode (PGPd) as additional photosense elements for self-power generation. The SPS pixel architecture allows the employment of incidental light energy to produce power for the APS reset operation, in-pixel amplifier, and part of the AFE circuitry, hence reducing the sensor power dissipation from the regular power supply.

Figure 23.6a shows an example of a possible implementation of the basic SPS structure in conjunction with a 3-T photodiode CMOS APS employing a PMOS reset transistor. According to this design, the V_{DD}' line is common to all pixels and is connected to the PGPd elements, which can be located inside each pixel or alternatively, in the sensor periphery. The SPS operates as follows: under illumination conditions, the voltage at node V_{DD}' is increased to V_{DD} by the photocurrent, which discharges the PGPd space-charge capacitance. This photoelectric process is comparable to the regular photodiode (Pd) operation, where the voltage across the junction decays proportionally to the illumination level. Note that the PGPd is connected in an open-circuit configuration as long as the reset and the RS transistors are off, which is similar to the Pd integration mode. The photoelectric charge, induced by the PGPd element, is used for Pd reset during the reset stage and for readout operation. Figure 23.6b depicts a possible SPS implementation in a standard n-well CMOS process. The PGPd element is formed by the p+-N-well junction, while the Pd is implemented by the n+-Psub junction. It is very important to examine the efficiency of the SPS technique by comparing the power generated by the PGPd elements to the power dissipated by the APS array.

The photocurrent produced in the PGPd may be considered as a current source and is given as follows:

$$I_{gen} = q\Phi\eta A \tag{23.23}$$

where
   q is the electron charge
   Φ is the photon flux
   η is the quantum efficiency
   A is the photosensitive area of the PGPd

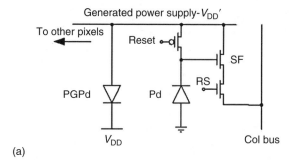

(a)

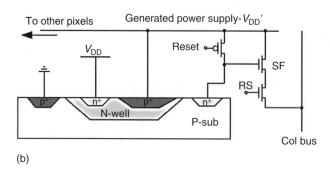

(b)

FIGURE 23.6 (a) Possible implementation of SPS and (b) SPS implementation in a standard N-well CMOS process. (From Fish, A., Hamami, S., and Yadid-Pecht, O., *IEEE Trans. Circuits Systems II*, 53, 131, 2006. With permission.)

Therefore, the total power produced by all PGPds on the die is given by

$$P_{\mathrm{gen}} = I_{\mathrm{gen}} V_{\mathrm{DD}}{}' \approx I_{\mathrm{gen}} V_{\mathrm{DD}} = q \Phi \eta A V_{\mathrm{DD}} \tag{23.24}$$

As can be seen from (23.24), the power generated by the PGPd elements linearly depends on the area of all PGPd devices. In this analysis, we will answer the following question: what is the PGPd area required to supply enough power for the sensor array and SFB transistor operation. From Section 23.3.1, the power P_{diss} dissipated by the sensor array and SFB transistor is given by

$$P_{\mathrm{diss}} = F_{\mathrm{R}} NM \left(E_{\mathrm{reset}} + E_{\mathrm{SF}} \right) + NMP_{\mathrm{leakage}} + P_{\mathrm{SF\_bias}} \tag{23.25}$$

To allow an efficient operation of the SPS under the defined requirements, P_{gen} should be larger than P_{diss}, i.e.,

$$P_{\mathrm{gen}} \geq P_{\mathrm{diss}} \tag{23.26}$$

Therefore, the requirements for minimum area of the PGPd elements can be written as

$$A \geq \frac{F_R \times N \times M \times (E_{\mathrm{reset}} + E_{\mathrm{SF}}) + N \times M \times P_{\mathrm{leakage}} + P_{\mathrm{SF\_bias}}}{q \Phi \eta V_{\mathrm{DD}}} \tag{23.27}$$

An illustrative example of the SPS structure in 0.18 μm process is depicted herein. The 64 × 64 sensor array operating at 30 frames per second at 1.8 V is assumed. The parameter values listed in this section are typical to a 0.18 μm process and will be utilized to illustrate the relevant expressions.

The E_{reset} component is calculated using (23.2) and is equal to

$$E_{\text{reset}} = V_{\text{DD}} V_{\text{Reset}} C_{\text{Pd}} = 0.032 \text{ pJ} \qquad (23.28)$$

assuming $V_{\text{Reset}} = 1.2\,\text{V}$ and $C_{\text{Pd}} = 15\text{fF}$.

The E_{SF} component is calculated using (23.5) and is equal to

$$E_{\text{SF}} = 2V_{\text{DD}} V_{\text{S/R}} \left(C_{\text{S/R}} + C_{\text{col}} \right) + 2I_{\text{LN}} V_{\text{DSF\_S/Rav}} T_{\text{CS/R\_charge}} = 1.26 \text{ pJ} + 0.53 \text{ pJ} = 1.79 \text{ pJ} \quad (23.29)$$

assuming $V_{\text{S/R}} = 0.7\,\text{V}$, $V_{\text{DSF\_S/Rav}} = 1.4\,\text{V}$, $C_{\text{S/R}} = 0.4\,\text{pF}$, $C_{\text{col}} = 0.1\,\text{pF}$, $\alpha_{\text{S/R}} = 0$, and $T_{\text{read\_out}} = 250\,\text{ns}$.

The $P_{\text{SF\_bias}}$ component is given by (23.16) and is equal to

$$P_{\text{SF\_bias}} = N2I_{\text{LN}}(V_{\text{DD}} - 2V_{\text{TH}})T_{\text{read\_out}} F_{\text{R}} M = 32.3 \text{ nW} \qquad (23.30)$$

assuming $I_{\text{LN}} = 0.75\,\mu\text{A}$, $N = M = 64$, and $F_{\text{R}} = 30$.

The generated photocurrent density is calculated using

$$J_{\text{ph}} = q\Phi\eta \cong 2\mu \left[\text{A/cm}^2 \right] \qquad (23.31)$$

where
$\eta = 0.3$ for typical p$^+$/n-well junction
$\Phi = 4 \times 10^{13}\,\text{ph/cm}^2\text{s}$

Note that the room light flux and clear sky flux correspond to $\Phi = 4 \times 10^{12}$ and $\Phi = 4 \times 10^{15}$, respectively.

Thus, neglecting E_{reset}, the minimum required area of the PGPd elements can be calculated using (23.27):

$$A \geq \frac{30 \times 64 \times 64 \times 1.79 \times 10^{-12} + 64 \times 64 \times 10^{-12} + 32.3 \times 10^{-9}}{3.6 \times 10^{-6}} = 7.12 \text{ mm}^2 \qquad (23.32)$$

assuming $P_{\text{leakage}} = 1\,\text{pA}$.

According to (23.32), the PGPd active area required for a proper operation for 64×64 sensor array and SFB transistor at the given illumination level and without need of a conventional power supply is $7.12\,\text{mm}^2$. This can be significantly reduced in some applications (e.g., space applications), where sometimes only small regions of interests are processed. In these applications, the required area of the PGPd element is reduced by factor of M, shown in the following expression:

$$M = \frac{N_{\text{row}} N_{\text{col}}}{XY} \qquad (23.33)$$

where X and Y are the number of rows and columns in the region of interest, respectively. For example, for the region of 10×15 the required area is only $0.26\,\text{mm}^2$.

A test chip, consisting of a 7×7 SPS array, was implemented in a standard $0.18\,\mu\text{m}$ CMOS technology and successfully tested in the laboratory. A detailed quantitative and qualitative analysis was carried out and successfully demonstrated that the proposed SPS architecture allows generation of electric power for the whole APS array and signal readout circuitry operation. More details on the SPS technique can be found in Ref. [65]. However, a relatively low PGPd element efficiency requires a very large area for SPS technique implementation, resulting in increased costs. Thus, the SPS architecture was found as most suitable for imagers, where low power is the main demand.

23.4.2 Power Reduction at the Algorithm and Architecture Levels

23.4.2.1 General Considerations

Power reduction through the algorithm selection is based on minimizing the number of operations and hence the number of required hardware resources. In conventional cameras-on-a-chips, there are not so many options to reduce power dissipation through the algorithm level. Usually these imagers are limited by rolling or global shutter operation requirements and are expected to provide full resolution analog/digital grayscale/color images at the required frame rate. In smart image sensors, the situation is different. Many smart imagers are not required to provide a full resolution image at real time; however, they are required to perform real-time image processing. In contrast to a standard camera-on-a-chip, the choice of a suitable algorithm in smart sensors is the main key for low-power dissipation. Power dissipation reduction at the algorithm level can be summarized in the following points:

- Reduction of an operational activity of different processing blocks
- Definition and activation of windows of interest, consisting of the required features, while cutting off the remaining pixels of the array and periphery circuitry, responsible for their operation
- Image transfer for further processing only in the cases where it is essential—some tracking systems, where only the coordinates of the tracked targets should be transferred for further processing can be a good example
- Computation of the required functions based on the computation results from the previous frame; in many cases, there is no need to perform calculations from scratch as updated results from previous frames can be used instead, if an update is required
- Taking the correct decision when to perform the computation in an analog or in a digital mode
- Performing compute-intensive and power-hungry functions off-chip, if possible

The design at the architecture level is strongly dependent on the decisions made at the algorithm level. However, optimization at the architecture level can influence the algorithm itself and even change it. Two main directions for power reduction at the architecture level can be identified:

1. General optimization at the architecture level that does not depend on the algorithm and does not influence the algorithm choice
2. Algorithm-dependent architecture optimization that feeds back to the algorithm level

The first direction may include the following optimizations:

- The use of power management and temporarily turn-off of unused blocks
- The use of shift registers for readout and for window definition
- The reduction of the number of global busses
- The use of parallelism, column-parallel image processing, pipelining, etc.

The second direction includes the optimization of the number, kind, and placement of circuit blocks. In some cases, the proposed algorithm cannot be implemented using power-efficient architectures and thus should be changed in order to achieve the required task. This mostly happens because each sensor array dictates the physical pitch and the directions for data readout. Examples of power optimization at the algorithm and the architecture levels are shown in the following practical example.

23.4.2.2 Practical Example—Sensor for Multiple Targets Detection and Tracking

Visual tracking of salient targets in the field of view (FOV) is a very important operation in machine vision, star tracking, and navigation applications. The same task is also performed by biological systems. To accomplish this real-time operation, both in engineering and biological systems, a large amount of information is to be processed in parallel. This parallel processing is a very complicated task that demands huge computational resources. Thus, a serial selection of regions of interest and their subsequent processing can greatly facilitate the computational complexity. This phenomenon is known as visual attention in biological systems.

In this section, a sensor for multiple targets detection and tracking is presented. The employed algorithm and the imager architecture were optimized by utilizing the spotlight model of attention [57], allowing for an efficient implementation of a real-time tracking sensor. The spotlight model assumes that only a single region of interest is processed at a certain time point and supposes smooth movement to other regions of interest. The described sensor allows parallel computations and is distributed, but on the other hand, most of the image processing is performed in the array periphery, achieving image quality and high spatial resolution. The sensor aims to output the coordinates of all tracking targets in real time. Similar to biological systems, which are limited in their computational resources, engineering applications are constrained with low-power dissipation. Thus, maximum efforts have been taken to reduce power consumption in the proposed sensor. Figure 23.7 shows a general architecture of the tracking sensor. Note that only a brief architecture description is presented here, and more information on the discussed sensor can be found in Ref. [57].

The sensor has two modes of operation: target acquisition and target tracking. In the acquisition mode, the sensor goals are to locate the three most salient targets in the FOV and calculate their centroid coordinates. Also, the sensor should be able to eliminate bad pixels and other small, bright artifacts in the processed image. These aims are achieved in the following way: all neighboring pixels are connected by resistors (implemented by transistors), creating the resistive network. The pixels voltage becomes smaller if pixels are more exposed and the local minimum of the voltage distribution can be regarded as the centroid of the target corresponding to the exposed area. This minimum can be found using an analog loser-take-all (LTA) circuit [58]. Note that, to achieve a higher spatial resolution, only a Y-direction resistive network was included in the array. The distribution in the X-direction is applied in the array periphery (via the LTA circuit).

At the first stage of the acquisition mode, all pixels of the whole image are activated. The global minimum, corresponding to the brightest target, is located using a one-dimensional LTA circuit. To achieve this purpose, the whole image is scanned row by row (using the readout row shift register), finding the local minimum in each row. Then the row local minima are input to the same LTA circuit again and the global minimum is computed. Once the first brightest target is found, the system defines a small-size programmable window, with the center located at the target centroid coordinates. While finding the second bright target in the FOV, all pixels of the first window consisting of the brightest target found during the first search are deactivated. This way, the bright pixels of the first target do not influence the result of the second search. The third target is found in the same way. As a result, at the end of the acquisition mode, all centroid coordinates of the three most salient targets in the FOV are stored in the memory and three small windows around these coordinates are defined. The window definition is performed using two digital shift registers. Thus, six shift registers are required to define three different windows. The acquisition mode control block is responsible for defining and positioning these active windows.

Once the sensor has acquired the three most salient targets, the tracking mode is initiated. The predefined windows serve as a spotlight in biological systems, such that only the regions inside the windows are processed. At a certain time, only one window is processed, so the system needs to switch between the windows during the tracking. In contrast to biological systems, these "spotlights" attend only to the regions predefined in the acquisition mode. Thus, even if new, more salient objects appear during the tracking, attention to the chosen regions is not influenced. Because the sensor is

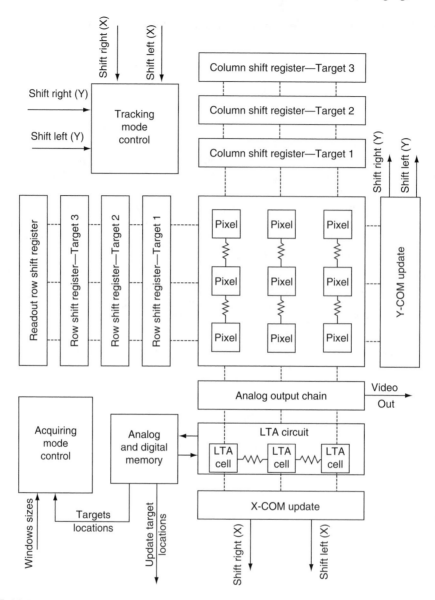

FIGURE 23.7   General architecture of the tracking sensor. (From Fish, A., Spivakovsky, A., Golberg, A., and Yadid-Pecht, O., IEEE ICECS, pp. 543–546, Tel-Aviv, Israel, December 2004. With permission.)

in tracking mode most of the time, it is very important to achieve very low-power dissipation in this mode. In the proposed system, this is accomplished in two ways:

- Only pixels of the three active windows and the circuitry responsible for proper centroid detection and pixels readout are active. The remaining circuits (including most pixels of the array) are disconnected from the power supply.
- The circuit does not calculate the new centroid coordinates. A simple analog circuit (COM update block in Figure 23.7) checks if the new centroid location differs from the centroid location of the previous frame. In the case that no difference is found, the circuit does not need to perform any action, significantly reducing system power dissipation. This principle suits the general idea of "no movement–no action." If the target changes its position, the

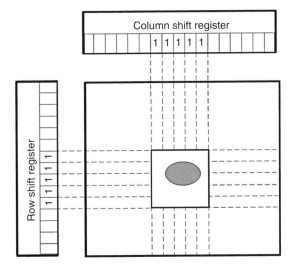

FIGURE 23.8 Window definition using two shift registers. (From Fish, A., Spivakovsky, A., Golberg, A., and Yadid-Pecht, O., IEEE ICECS, pp. 543–546, Tel-Aviv, Israel, December 2004. With permission.)

"shift left" or "shift right" (both for x and y) signals are produced by the COM update blocks. These signals are used as inputs to the tracking mode control block and afterward the appropriate shift register performs movement to the right direction, correcting the location of the window.

Figure 23.8 shows how an active window definition is performed using two shift registers. This windows definition method allows switching from one target of interest to another without any need to access the memory and load the new target coordinates. This way the power dissipation of the scanning circuitry can be significantly reduced, optimizing the leakage component of its power dissipation [66].

23.4.3 POWER REDUCTION AT THE CIRCUIT AND LOGIC LEVELS

Power reduction through the circuit and logic levels can be divided into four parts: leakage current, low-voltage operation, power-efficient digital design, and optimization of sensor output chains. We will examine each part independently and later describe a practical example of a wide DR snapshot APS that utilizes these techniques.

23.4.3.1 Leakage Control

Power reduction by leakage current control can be accomplished by the following actions:

- Differentiation between the "active" and "sleep" modes of the certain circuit by insertion of a "sleep" transistor—insertion of NMOS transistor in series with the SFB transistor (see Figure 23.3) can be a good example.
- Reduction of the leakage current using the "stacked scheme"—by stacking two off transistors, the subthreshold current is reduced significantly compared to a single off device due to simultaneous reductions in gate-source, body bias, and drain-source voltages [59].
- Multiple V_{TH} design—applying different leakage control techniques that usually are not used in image sensors design, like multithreshold-voltage CMOS (MTCMOS), variable threshold CMOS (VTCMOS), dynamic V_{TH} design, cutoff CMOS (SC-CMOS), and others. The MTCMOS technique utilizes both high- and low-threshold voltage MOS transistors in

a single chip [60], the VTCMOS technique applies reverse back gate bias to cutoff leakage current in standby mode by increasing the magnitude of the threshold voltage [61], and the dynamic threshold method adjusts the active leakage power based on the desired frequency of operation and in the case of SC-CMOS a low-VTH cutoff NMOS or PMOS is added to a circuit in-series with the rest of the logic. In the case of an NMOS insertion, the gate is overdriven by a negative voltage in standby mode to minimize the leakage [62].

23.4.3.2 Low-Voltage Operation

Reduction of the power supply voltage is a key element in low-power CMOS imagers. However, the design of a low-voltage CMOS sensor involves several well-known challenges. Employing multiple voltage supplies can relax the problem. The idea of multiple V_{DD} grows up from the dual-V_{DD} approach of digital circuit design, where the gates of the noncritical paths have the reduced supply voltage V_{DDL}, while those on the critical paths have V_{DDH}. This results in reducing the power without degrading the entire circuit performance. Similarly to digital circuits, in image sensors high V_{DD} can be used in the critical places, where V_{DD} reduction has a significant effect, while low-voltage supply can be useful in other places.

23.4.3.3 Power-Efficient Digital Control Circuitry

The custom design of the particular blocks in digital circuitry using nonstandard design techniques, like pass-transistor-logic (PTL), may significantly improve power dissipation, but the design cost would increase. Custom digital design can be very useful when an in-pixel digital processing approach is used.

23.4.3.4 Optimization of Sensor Output Chains

A standard three-transistor APS pixel includes only one analog output and two digital control inputs. The advanced smart sensor usually has more complicated structures. In addition to an increased number of analog outputs and digital control inputs, it can include digital outputs generated by the pixel. The demand for this increased number of pixel inputs and outputs, the desire for a high fill factor, and the requirement for low-voltage operation result in the necessity to optimize analog and digital sensor output chains for area efficient low-power, low-voltage operation. Possible output chain optimizations can include the following:

- Optimizing for increased DR with low-voltage operation—this includes utilizing boot-strapped reset pulse, designing column analog readout such that only unipolar NMOS (or PMOS) switches are used, etc.
- Applying the same output chain for analog and digital signals to reduce the area and power in the case of complicated pixels.
- Using nonstandard output chains such as the complementary active pixel sensor (CAPS) method and the active column sensor (ACS) technique. The CAPS architecture removes the threshold voltage influence on the available output swing, and was presented by Xu in 2002 [27]. Using this method 1 V V_{DD} APS was successfully designed in 0.25 μm CMOS process. Having a 12 μm × 10 μm pixel size with a fill factor of 30%, this image sensor dissipated only 18 mW of power. According to the ACS method, firstly introduced in Refs. [63,64], a unity gain amplifier (UGA) was implemented instead of a conventional source-follower. UGA common transistors were implemented at the columns of the array and only the differential input transistor is situated inside the pixel. This way the fill factor and pixel properties were not affected, while a true UGA with increased output swing and better linearity was achieved. Both CAPS and ACS techniques allow operation at reduced supply voltage.

23.4.3.5 Practical Example—Wide Dynamic Range Snapshot APS

In this example, the design challenge was to achieve low-power dissipation of the pixel at the given algorithm and imager architecture. The described imager allows capturing of fast moving objects in the FOV and provides wide DR by applying adaptive exposure time to each pixel, according to the local illumination intensity level. The algorithm for DR expansion, used in the presented sensor, is based on the algorithm recently implemented in a rolling shutter APS [67], but is applied to a snapshot APS here [68]. According to this algorithm, the total integration time is subdivided into several integration times, which are progressively shorter. At the beginning of the frame, all pixels in the imager are reset. Then the photodiode output of each pixel is compared with an appropriate threshold, at certain time points, by enabling the in-pixel comparator. This operation is applied simultaneously to all pixels in the array to ensure snapshot operation of the imager. If any of the checks determine that the pixel will saturate at the end of the current integration time, then the pixel is reset again and is allowed to start integrating light again, but for a shorter period of time. The binary information regarding whether the reset was applied or not is saved locally in the pixel memory, and is transmitted during the next integration subperiod to the external digital storage in the upper part of the sensor, associated with each pixel. This reading enables proper scaling of the value being read out and enables the pixel value to be described in a floating-point representation. Figure 23.9 shows the implementation of a single pixel; the pixel operation is fully described in Ref. [67].

The presented circuit operates as follows: at the beginning of the frame the pixel is reset by applying "global reset" = "0" and "cond reset" = "0." This way the internal line "reset" is equal

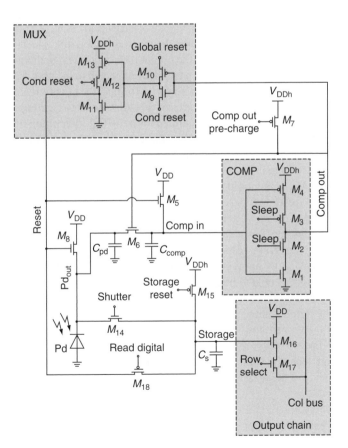

FIGURE 23.9 Implementation of a single pixel of the WDR global shutter APS. (From Fish, A., Belenky, A., and Yadid-Pecht, O., *IEEE Trans. Circuits Systems II*, 52, 729, 2005. With permission.)

to "1_h" (1.8 V) regardless of the internal line "Comp out" value, charging the photodiode (Cpd) and internal line "Comp_in" (C_{comp}) to $V_{reset} = V_{DDh} - V_{thN} \sim V_{DD}$. (Note that there is a difference between "1" that equals $V_{DD} = 1.2$ V and "1_h" that equals $V_{DDh} = 1.8$ V.) At the same time, the internal line "Comp out" is precharged to "1_h" by negative pulse of "Comp out precharge" = "0." The reset phase is stopped by applying "Global Reset" = "1" and "Cond Reset" = "1" and the photodiode starts discharging according to the energy of incident light. At this stage, the total capacitance connected to the photodiode is given by $C'_{pd} = C_{pd} + C_{comp}$. At the first time point $T_1 = T_{INT} - T_{INT}/X^1$ (see Equation 23.25), the output of the photodiode (voltage on C'_{pd}) is compared with an appropriate threshold, associated with the switching threshold voltage of the inverter. This comparison is performed by enabling the inverter operation ("Sleep" = "1_h" and "not(Sleep)" = "0").

If "Pd out" < threshold, meaning that the pixel will saturate at the end of the integration time, then "Comp out" = "1_h" (determined by the inverter). At the same time, "Cond Reset" falls to "0" by applying a short negative pulse, causing M_9 and M_{10} to operate as standard inverters and enabling the operation of the inverter, consisting of M_{11}, M_{12}, and M_{13}. As a result (for "Comp out" = "1_h"), the photodiode is reset again. The binary information regarding whether the reset was applied or not is saved locally in the storage capacitor (C_s) by "Read digital" = "0" at the time when "Cond Reset" = "0" and is transmitted during the next subintegration period to the external digital storage in the upper part of the sensor array, associated with the certain pixel, to enable proper scaling of the value read. The readout of this digital signal is performed through the regular output chain, used for analog signal readout, by allowing "Row Select" = "1."

If "Pd out" > threshold, meaning that the pixel will not saturate at the end of the integration time, then "Comp out" = "0" (determined by the inverter). In this case, the photodiode is not reset ("Cond Reset" = "0," "Comp out" = "0" => "Reset" = "0") and transistor M_6 is turned off, separating C_{pd} and C_{comp}. Once the comparison is stopped by turning the inverter back to the "sleep" mode and applying "Global Reset" = "1" and "Cond Reset" = "1," the photodiode continues discharging, according to the energy of incident light. This time, the total capacitance connected to the photodiode is given only by C_{pd}. At the following time points when comparison is performed, the C_{pd} is already disconnected from the C_{comp}, causing the voltage saved on C_{comp} (greater than threshold) to be compared to the threshold produced by the inverter. Thus, no resets are applied until the full integration time is finished.

At the end of the full integration time T_{INT}, the capacitor C_{pd} is connected to the capacitor C_{comp} by applying "Comp out precharge" = "0" and the final photodiode voltage on the capacitor $C'_{pd} = C_{pd} + C_{comp}$ is determined by the charge transfer between C_{pd} and C_{comp}. This way this final voltage is independent M_6, be it closed or open during the last integration period. Note that if Comp out" was equal to "1_h" (reset was performed at the last time point), the capacitor C_{pd} would have already been connected to the capacitor C_{comp}.

The next stage is the transfer of the charge accumulated in the photodiode capacitor C'_{pd} to a storage capacitor C_s by applying "Shutter" = "1_h." Before this charge transfer occurs, the storage capacitor C_s is reset to V_{DDh} by applying "Storage Reset" = "0." Once this charge transfer has been completed, the photodiode is able to begin a new frame exposure, and the charge just transferred to the in-pixel memory is held there until it is read out at its assigned time in a row-by-row readout sequence through the output chain.

As was already mentioned, low supply voltage is a main key for power consumption reduction. On the other hand, power voltage reduction impacts the sensor output swing. In this design, a dual power supply voltage technique was used in order to reduce power dissipation, while achieving an acceptable output swing of the output chain. In addition, several techniques for power reduction were used in this design:

- Inclusion of leakage current controls using the stacked scheme design (applied in the comparator)
- Insertion of the circuits into "sleep" mode when not operating (applied on the Mux, the inverter and the comparator)

- Usage of a simple digital inverter as an analog comparator
- Usage of a nonstandard low-power Mux. Usually, the usage of this circuit is avoided because of threshold voltage drop [67]. In this snapshot pixel, the regular state of the MUX is "Global Reset" = "1" and "Cond Reset" = "1" and, if "Comp out"="1" leakage would not exist
- Usage of the same output chain for analog and digital signals

A test chip of a 32×32 array has been implemented in a standard $0.35\,\mu m$ CMOS technology, achieving power dissipation of only $18.95\,nW$ (single pixel) for an 8 bit (equivalent to DR$\sim$107 dB) expansion at room light and $29\,nW$ at a high illumination level equivalent to clear sky. The test chip was successfully tested in the laboratory and showed to be fully functional comparing to our previous pixel design, implementing the same functionality, the presented low-power imager achieved a power dissipation reduction of more than 70%.

23.5 CONCLUSIONS

In this chapter, we have briefly considered power reduction in state-of-the-art smart CMOS image sensors. Sources of power dissipation in image sensors were presented and explained with reference to smart image sensor architecture. An approach for ultra low-power smart CMOS sensor design at different design levels was presented. Accordingly, the presented approach power reduction of all imager components can be reduced at technology, device, circuit, logic, architecture, algorithm, and system integration levels. Practical examples of low-power smart CMOS image sensors recently proposed by the authors were briefly described. Although we could not provide more detailed explanations on existing low-power design techniques due to the limited space allowed, we hope we have succeeded in presenting some basic concepts in low-power imager design that can be useful to beginners in the area of image sensors design.

REFERENCES

1. C. Xu, W. Zhang, W. Ki, and M. Chan, A 1.0 V VDD CMOS active-pixel sensor with complementary pixel architecture and pulsewidth modulation fabricated with a 0.25 μm CMOS process, *IEEE Journal of Solid State Circuits*, 37(12), 1853–1859, 2002.
2. K. Cho, A. I. Krymski, and E.R. Fossum, A 1.5 V 550 μW 176 × 144 autonomous CMOS active pixel image sensor, *IEEE Transactions on Electronic Devices* (Special Issue on Image Sensors), 50, 96–105, 2003.
3. E. Fossum, CMOS image sensors: Electronic camera-on-a-chip, *IEEE Transactions on Electronic Devices*, 44, 1689, 1997.
4. O. Yadid-Pecht and R. Etienne-Cummings, *CMOS Imagers: From Phototransduction to Image Processing*, Kluwer Academic, Dordrecht, 2004.
5. A. El Gamal and H. Eltoukhy, CMOS image sensors, *IEEE Circuits and Devices Magazine*, May/June, 6–20, 2005.
6. E.R. Fossum, Active pixel sensors: Are CCD's dinosaurs? *Proceedings of SPIE, Charged-Coupled Devices and Solid State Optical Sensors III*, vol. 1900, pp. 30–39, 1993.
7. M. Bigas, E. Cabruja, J. Forest, and J. Salvi, Review of CMOS image sensors, *Microelectronics Journal*, 37(5), 433–451, 2006.
8. O. Yadid-Pecht, B. Pain, C. Staller, C. Clark, and E. Fossum, CMOS active pixel sensor star tracker with regional electronic shutter, *IEEE Journal of Solid State Circuits*, 32(2), 285–288, 1997.
9. A. Fish, D. Turchin, and O. Yadid-Pecht, An APS with 2-dimensional winner-take-all selection employing adaptive spatial filtering and false alarm reduction, *IEEE Transactions on Electronic Devices* (Special Issue on Image Sensors), 50(1), 159–165, 2003.
10. V. Brajovic and T. Kanade, Computational sensor for visual tracking with attention, *IEEE Journal of Solid-State Circuits*, 33(8), 1998.
11. T. Horiuchi and E. Niebur, Conjunction search using a 1-D, analog VLSI-based attentional search/tracking chip, *Conference for Advanced Research in VLSI*, D.S. Wills and S.P. DeWeerth (Eds.), pp. 276–290. IEEE Computer Society, Los Alamitos, CA, 1999.

12. G. Indiveri, Neuromorphic analog VLSI sensor for visual tracking: Circuits and application examples. *IEEE Transactions on Circuits and Systems II*, 46(11), 1337–1347, 1999.

13. C.S. Wilson, T.G. Morris, and P. DeWeerth, A two-dimensional, object-based analog VLSI visual attention system, *Twentieth Anniversary Conference on Advanced Research in VLSI*, vol. 20. pp. 291–308, IEEE Computer Society Press, Los Alamitos, CA, 1999.

14. M. Clapp and R. Etienne-Cummings, A dual pixel-type imager for imaging and motion centroid localozation, *Proceedings of ISCAS'01*, Sydney, Australia, May 2001.

15. N. Mei Yu, T. Shibata, and T. Ohmi, A real-time center-of-mass tracker circuit implemented by neuron MOS technology, *IEEE Transactions on Circuits and Systems II*, 45(4), 495–503, 1998.

16. R.C. Meitzler, K. Strohbehn, and A.G. Andreou, A silicon retina for 2-D position and motion computation, *Proceedings of ISCAS'95*, New York, 3, 2096–2099, 1995.

17. T. Komuro, I. Ishii, M. Ishikawa, and A. Yoshida, A digital vision chip specialized for high-speed target tracking, *IEEE Transactions on Electron Devices* (Special Issue on Image Sensors), 50(1), 191–199, 2003.

18. A. Simoni, G. Torelli, F. Maloberti, A. Sartori, S.E. Plevridis, and A.N. Birbas, A single-chip optical sensor with analog memory for motion detection, *IEEE Journal of Solid-State Circuits*, 30(7), 800–806, 1995.

19. M. Clapp and R. Etienne-Cummings, Dual pixel array for imaging, motion detection and centroid tracking, *IEEE Sensors Journal*, 2(6), 529–548, 2002.

20. S. Kawahito, M. Yoshida, M. Sasaki, K. Umehara, D. Miyazaki, Y. Tadokoro, K. Murata, S. Doushou, and A. Matsuzawa, A CMOS image sensor with analog two-dimensional DCT-based compression circuits for one-chip cameras, *IEEE Journal of Solid-State Circuits*, 32(12), 2023–2029, 1997.

21. K. Aizawa, H. Ohno, Y. Egi, T.Hamamoto, M. Hatory, H. Maruyama, and J. Yamazaki, On sensor image compression, *IEEE Transactions on Circuits and Systems for Video Technology*, 7(3), 543–548, 1997.

22. B.E. Shi, A low power orientation selective vision sensor, *IEEE Transactions on Circuits and Systems–II: Analog and Digital Signal Processing*, 47(5), 435–440, 2000.

23. E. Culurciello, R. Etienne-Cummings, and K. Boahen, Arbitrated address-event representation digital image sensor, *Electronics Letters*, 37(24), 1443–1445, 2001.

24. A.G. Andreou, R.C. Meitzler, K. Strohbehn, and K.A. Boahen, Analog VLSI neuromorphic image acquisition and pre-processing systems, *Neural Networks*, 8(7–8), 1323–1347, 1995.

25. H.S. Wong, Technology and device scaling consideration for CMOS imagers, *IEEE Transactions on Electronic Devices*, 43(12), 2131–2142, 1996.

26. H.S. Wong, CMOS image sensors—recent advances and device scale considerations, *IEDM Technical Digest.*, 201–204, 1997.

27. C. Xu, W. Ki, and M. Chan, A low-voltage CMOS complementary active pixel sensor (CAPS) fabricated using a 0.25 μm CMOS technology, *IEEE Transactions on Electron Devices* (Special Issue on Image Sensors), 23(7), 398–400, 2002.

28. L.G. McIlrath, A low-power low-noise ultrawide-dynamic-range CMOS imager with pixel-parallel A/D conversion, *IEEE Journal of Solid-State Circuits*, 36(5), 846–853, 2001.

29. K. Yoon, C. Kim, B. Lee, and D. Lee, Single-chip CMOS image sensor for mobile applications, *IEEE Journal of Solid-State Circuits*, 37, 1839–1845, 2002.

30. B. Pain, G. Yang, B. Olson, T. Shaw, M. Ortiz, J. Heynssen, C. Wrigley, and C. Ho, A low-power digital camera-on-a-chip implemented in CMOS active pixel approach, 12th International Conference on VLSI Design—VLSI for the Information Appliance, Goa, India, January 1999.

31. R.H. Nixon, S.E. Kenemy, B. Pain, C.O. Staller, and E.R. Fossum, 256 × 256 CMOS active pixel sensor camera-on-a-chip, *IEEE Journal of Solid-State Circuits*, 21(12), 2046–2050, 1996.

32. E. Fossum, Low power camera-on-a-chip using CMOS active pixel sensor technology, *IEEE Symposium on Low Power Electronics*, pp. 74–77, 1995.

33. C.C. Liebe, L. Alkalai, G. Domingo, B. Hancock, D. Hunter, J. Mellstrom, I. Ruiz, C. Sepulveda, and B. Pain, Micro APS based star tracker, *Proceedings of IEEE Aeroconf*, vol. 5, pp. 2285–2300, 2002.

34. C. Shen, C. Xu, R. Huang, W. Zhang, P.K. Ko, and M. Chan, A new active pixel sensor (APS) architecture on SOI substrate for low voltage operation, *Proceedings of 9th International Symposium on IC Technology, Systems and Applications*, Singapore, pp. 275–278, September 3–5, 2001.

35. E. Culurciello and A.G. Andreou, A 16 × 16 silicon on sapphire CMS photosensor array with a digital interface for adaptive wavefront correction, *Proceedings of ISCAS*, Vancouver, May 2004.

36. C. Xu, C. Shen, P.K. Ko, and M. Chan, An active pixel sensor (APS) based on high gain CMOS compatible lateral bipolar transistor (LBT) on SOS substrate with backside illumination, *IEEE Proceedings on Sensors*, 2, 1283–1286, 2003.

37. A. Apsel, E. Culurciello, A. Andreou, and K. Aliberti, Thin film pin photodiodes for optoelectronic silicon on sapphire CMOS, *Proceedings of the 2003 IEEE International Symposium on Circuits and Systems*, Bangkok, pp. 908–911, May 2003.

38. A. Fish, E. Avner, and O. Yadid-Pecht, Low-power global/rolling shutter image sensors in silicon on sapphire technology, *Proceedings of IEEE ISCAS*, Kobe, Japan, May 2005.

39. D. Antoniadis, SOI CMOS for low-power systems, in *Low Power CMOS Design*, A. Chandrakasan (Ed.), IEEE Press, Reading, MA, 1997.

40. C. Xu, W. Zhang, and M. Chan, A low voltage hybrid bulk/SOI CMOS active pixel image sensor, *IEEE Electronic Device Letters*, 22, 248–250, 2001.

41. E.Y. Chou, A.J. Budrys, and K.M. Cham, Low power salient integration mode image sensor with a low voltage mixed-signal readout architecture, *International Symposium on Low Power Electronics and Design*, Monterey, CA, August 10–12, 1998.

42. K. Buckley, Selecting an analog front-end for imaging applications, *Analog Dialogue*, 34(6), 1–5, 2000.

43. D.X.D. Yang, A. El Gamal, B. Fowler, and H. Tian, A 640 × 512 CMOS image sensor with ultra wide dynamic range floating point pixel level ADC, *IEEE ISSCC*, WA 17.5, 1999.

44. B. Pain, S. Mendis, R. Scober, R. Nixon, and E. Fossum, Low-power low-noise analog circuits for on-focal-plane signal processing of infrared sensors, *Proceedings of SPIE*, Volume 1946, Infrared Detectors and Instrumentation, Albert M. Fowler (ed.), 365–374, October 1993.

45. A. Dickinson, S. Mendis, D. Inglis, K. Azadet, and E. Fossum, CMOS digital camera with parallel analog-to digital conversion architecture, IEEE Workshop on Charge Coupled Devices and Advanced Image Sensors, April 1995.

46. A. Krymski and N. Tu, A 9-V/Lux-s 5000-frames/s 512 × 512 CMOS sensor, *IEEE Transactions on Electronic Devices*, 50, 136–143, 2003.

47. S. Smith, J. Hurwitz, M. Torrie, D. Baxter, A. Holmes, M. Panaghiston, R. Henderson, A. Murray, S. Anderson, and P. Denyer, A single-chip 306 × 244-pixel CMOS NTSC video camera, *ISSCC Digest of Technical Papers*, pp. 170–171, February 1998.

48. M. Loinaz, K. Singh, A. Blanksby, D. Inglis, K. Azadet, and B. Acland, A 200 mW 3.3 V CMOS color camera IC producing 352 × 288 24b video at 30frames/s, *ISSCC Digest of Technical Papers*, pp. 186–169, February 1998.

49. Analog Devices, Mixed-signal and DSP design techniques, Technical report.

50. V. De, Y. Ye, A. Keshavarzi, S. Narendra, J. Kao, D. Somasekhar, R. Nair and S. Borkar, Techniques for leakage power reduction, in *Design of High-Performance Microprocessor Circuits*, A. Chandrakasan, W.J. Bowhill, and F. Fox (Eds.), Wiley-IEEE Press, pp. 46–62, September 2000.

51. A.P. Chandrakasan and R.W. Brodersen, Minimizing power consumption in digital CMOS circuits, *Proceedings of IEEE*, 83, 498–523, 1995.

52. K. Yano, K.T. Yamanaka, T. Nishida, M. Saito, K. Shimohigashi and A. Shimizu, A 3.8 ns CMOS 16 × 16 multiplier using complementary pass transistor logic, *IEEE Journal of Solid-State Circuits*, 25, 388–395, 1990.

53. H.J.M. Veendrick, Short-circuit dissipation of static CMOS circuitry and its impact on the design of buffer circuits, *IEEE Journal of Solid-State Circuits*, SC-19, 468–473, 1984.

54. C.C. Enz and E.A. Vittoz, CMOS low-power analog circuit design, in *Emerging Technologies, Tutorial for 1996 International Symposium on Circuits and Systems*, R. Cavin and W. Liu (Eds.), IEEE Service Center, Piscataway, CA, pp. 79–133, 1996.

55. P. Dubek and P.J. Hicks, A general-purpose processor-per-pixel analog SIMD vision chip, *IEEE Transactions on Circuits and Systems-I: Regular Papers*, 52(1), pp. 13–20, 2005.

56. J. Rabay and M. Pedram, *Low Power Design Methodologies*, Kluwer Academic, Dordrecht, 1996.

57. A. Fish, A. Spivakovsky, A. Golberg, and O. Yadid-Pecht, VLSI sensor for multiple targets detection and tracking, IEEE ICECS, Tel-Aviv, Israel, December 2004, CD ROM.

58. A. Fish, V. Milrud, and O. Yadid-Pecht, High speed and high resolution current loser-take-all circuit of o(N) complexity, *IEEE ICECS*, December 2004, CD ROM.

59. S. Narendra, V. De, D. Antoniadis, A. Chandrakasan and S. Borkar, Scaling of stack effect and its application for leakage reduction, *ISLPED 2001*, pp. 195–200, 2001.

60. S. Mutoh, T. Douseki, Y. Matsuya, T. Aoki, S. Shigematsu and J. Yamada, 1-V power supply high-speed digital circuit technology with multithreshold-voltage CMOS, *IEEE Journal of Solid-State Circuits*, 847–854, 1995.

61. T. Kuroda, T. Fujita, T.S. Mita, T. Nagamatsu, S. Yoshioka, K. Suzuki, F. Sano, M. Norishima, M. Murota, M. Rako, M. Kinugawa, M. Kakumu and T. Sakurai, A 0.9 V, 150 MHz, 10 mW, 4 mm^2, 2 D discrete cosine transform core processor with variable threshold-voltage (V_T) scheme, *IEEE Journal of Solid-State Circuits*, 31(11), 1770–1779, 1996.

62. H. Kawaguchi, K. Nose and T. Sakurai, A super cut-off CMOS (SCCMOS) scheme for 0.5 V supply voltage with picoampere stand-by current, *IEEE Journal of Solid-State Circuits*, 35(10), 1498–1501, 2000.

63. T.L. Vogelsong, J.J. Zarnowski, M. Pace, and T. Zarnowski, Scientific/industrial camera-on-a-chip using active column sensor CMOS imager core, *Proceedings of SPIE (Sensors and Camera Systems for Scientific, Industrial, and Digital Photography Applications)*, 31(11), pp. 102–113, 2000.

64. S. Diller, A. Fish, and O. Yadid-Pecht, Advanced output chains for CMOS image sensors based on an active column sensor approach—a detailed comparison, *Sensors and Actuators: A. Physical*, 116(2), 304–311, 2004.

65. A. Fish, S. Hamami, and O. Yadid-Pecht, CMOS image sensors with self-powered generation capability, *IEEE Transactions on Circuits and Systems II*, 53(11), 131–135, 2006.

66. A. Fish, V. Mosheyev, V. Linkovsky, and O. Yadid-Pecht, Ultra low-power DFF based shift registers design for CMOS image sensors applications, IEEE ICECS, Tel-Aviv, Israel, December 2004, CD ROM.

67. O. Yadid-Pecht and A. Belenky, In-pixel autoexposure CMOS APS, *IEEE Journal of Solid-State Circuits*, 38(8), 1425–1428, 2003.

68. A. Fish, A. Belenky, and O. Yadid-Pecht, Wide dynamic range snapshot APS for ultra low-power applications, *IEEE Transactions on Circuits and Systems II*, 52(11), 729–733, 2005.

24 CMOS Imager Array Design, Operation, and Trends

Mark Jaffe, John Ellis-Monaghan, and Jim Adkisson

CONTENTS

24.1 INTRODUCTION

Since the first complementary mixed oxide semiconductor (CMOS) imagers in the mid-1980s, CMOS imager pixels have evolved from a very simple dynamic random access memory (DRAM)-like architecture along a variety of different paths. The majority of CMOS imager products today employ a four-transistor pixel or one of its simple variants. Pixel architecture evolved first for optical performance, then to accommodate shrinking pixel sizes, and now for higher functionality. This chapter describes the evolution of the CMOS imager pixel circuit, gives a detailed description of the operation of the CMOS imaging pixel, and discusses issues that were encountered as the size of the pixel shrunk.

24.2 PIXEL EVOLUTION

24.2.1 PASSIVE PIXEL

The first CMOS imager pixel was the so-called passive pixel. This employed a row select (RS) line and a column access line through which a reset voltage could be applied and the signal could be

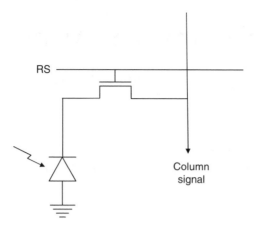

FIGURE 24.1   Passive pixel CMOS imager pixel.

read (Figure 24.1). This pixel performance was limited by the basic charge-sharing operation of getting the signal from the relatively low capacitance photodiode to the much higher capacitance column line through the resistance of the transistor. If a passive pixel were to be employed in a large array, the capacitance on the column line could easily be 1000 times larger than the photodiode capacitance. This would limit the voltage swing to a few millivolts, causing a very poor signal-to-noise ratio, as well as create a very high RC time constant for a read. These issues prevented the passive pixel from achieving a large market presence and drove the development of the active pixel.

24.2.2   ACTIVE 3T PIXEL

The concept of the active pixel is to put an amplifier inside each pixel to drive the output line. This improves both the response time and the voltage swing of the output. The simplest active pixel is the 3T or three-transistor pixel. This pixel uses a single transistor source follower (SF) as the buffer amplifier (Figure 24.2). The three-transistor pixel was the first CMOS imager pixel to achieve commercial success and was widely used in cell phone cameras and a variety of other applications. While the 3T pixel operation was better than the passive pixel and provided reasonable quality images, two fundamental issues limited the performance of 3T active pixels: the pixel has elements that are not sensitive to light and the noise performance is inferior to the mature charge-coupled device (CCD) technologies, which dominated the market. Both of these issues drove pixel architecture and imager semiconductor process changes.

The 3T pixel contains three-transistor gates, with associated source drain diffusions, as well as two metal lines in the row direction: RS and reset, and two metal lines in the column direction: V_{dd} and the data output line. All of these features take up space and therefore limit the fraction of the pixel that contains the photodiode. A figure-of-merit called the fill factor was created to describe the fraction of the pixel area, which was sensitive to light. As the pixel size has decreased, fill factors dropped to less that 50%. This limits the overall sensitivity to incoming light. More advanced semiconductor process lithography has been applied to shrink the feature size of the support features. Also, microlenses on top of each pixel have been developed to focus incident light onto the photosensitive portion of the pixel.

Although the image quality of the 3T pixel was good enough to propel CMOS imagers into the market, the performance was inferior to CCDs, which had dominated the digital imaging market for many years. The major difference was in noise. While noise can be broken down into many categories, the dominant sources in CMOS imagers are (a) fixed pattern noise, which arises from

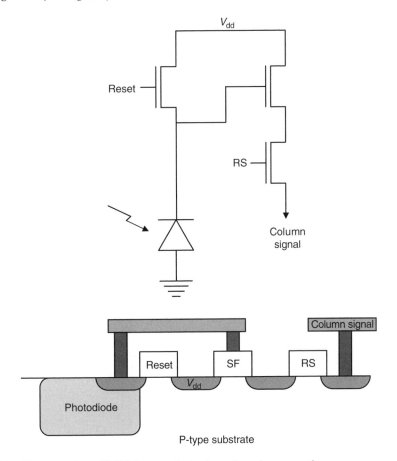

FIGURE 24.2 Three-transistor CMOS imager pixel schematic and cross-section.

differences in the electrical characteristics of the cells, (b) reset noise, which stems from the variation in the reset potential of the photodiode, and (c) thermal noise or dark current that produces electron generation in the absence of light.

Reducing the manufacturing variability of parameters such as transistor thresholds will certainly help reduce fixed pattern noise, but there will be an inevitable distribution in the gains and nominal bias of the read out circuits for each pixel. Double sampling was the obvious tool to use to subtract this effect as much as possible. Virtually all CMOS imagers employ a double sample technique with most imagers sampling a reference level and an image level for each pixel in sequential read operations and subtracting the two readouts with a differential amplifier in the column circuit.

Several techniques have been used to combat the reset noise issue. During the reset of a 3T cell, the voltage on the photodiode rises toward V_{dd}, reducing the drive of the reset transistor. As the voltage approaches a threshold less than the reset gate (RG) potential, the transistor begins to turn off, slowing down the reset. A hard or boosted reset is often employed to combat this effect in which the gate of the reset transistor is driven to a voltage, which is more than a V_{th} above V_{dd}, keeping the reset transistor on during the whole reset and increasing the reset potential of the photo diode to V_{dd} [1]. A feedback reset technique has also been used where the pixel is actively sensed during reset and this sense voltage is used to control either the duration of the reset or the potentials driving the reset [2].

Fundamentally, the structure of the 3T pixel limits the effectiveness of the ability to control the noise. A hard reset or a feedback reset can help reduce the noise of the reset level, but without a full frame of memory, there is no easy way to sample the reset level of the pixel before it started to accumulate charge. Further, the structure requires a metal contact to the photodiode. In order to make a good contact, the doping level must be high, the n-well used to collect charge must come to the surface, and the metal silicide in contact must come into direct contact with the photo diode. All of these factors can significantly increase the thermal generation of electrons. In order to improve the performance of all of these effects, the four-transistor pixel was created.

24.2.3 ACTIVE 4T PIXEL

The 4T pixel (Figure 24.3) adds a transfer gate to the 3T pixel. This seemingly simple addition yields several significant performance advantages that result both from the physical structure as well as the pixel operation. The added transistor allows the photodiode to be low-doped and isolated from the surface by a highly doped pinning layer [3]. There is also no direct metal contact to the photodiode. This physical isolation of the photodiode can significantly reduce the thermal generation. In addition, the low doping levels of the photodiode allows the reset to fully deplete the charge collection well down to zero electrons, virtually eliminating the shot noise associated with the amount of charge in the reset level. Operationally, the pixel is reset by turning on both the transfer and reset signals and read out by transferring charge from the photodiode to a floating diffusion where it drives the gate of the SF. This pixel architecture allows a true correlated double sample. The read sequence

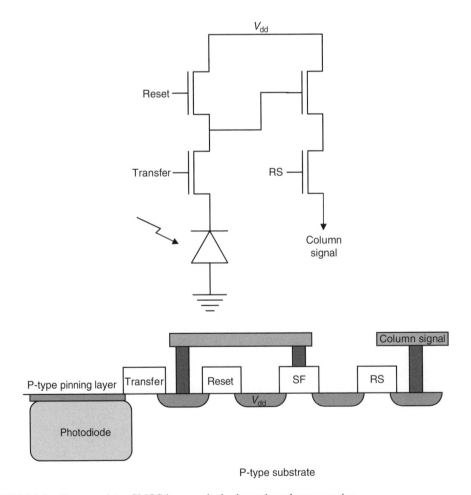

FIGURE 24.3 Four-transistor CMOS imager pixel schematic and cross-section.

is: reset, sample reference, transfer, and sample signal. The two samples are read from the same reset, eliminating much of the temporal noise associated with the reset as well as much of the pixel pattern noise.

A drawback of the 4T pixel is the voltage swing. The voltage on the photodiode in the 3T pixel can swing from ground up to the reset level (V_{dd} or a threshold below). The 4T pixel is forced to adjust to nearly half of this voltage swing. This is because there are two voltage swings: on the photodiode and on the floating diffusion. The photodiode swings from ground when it is "full" of electrons or the electron concentration is equal to the doping level, up to a pinning voltage, which is the point where the photodiode is empty or fully depleted. The pinning voltage can be controlled by the photodiode doping level. All of the electrons in the photodiode must be transferable from the photodiode to the floating diffusion. The floating diffusion, whose voltage is sensed during the pixel read, swings from its reset voltage (V_{dd} or a threshold below) down to a potential when the charge from a full photodiode is transferred. This lower level has to be above the pinning potential or transfer of electrons would stop. Having a $2\times$ lower voltage swing means that the noise has to be more than $2\times$ lower to get equivalent signal-to-noise performance. Typically, 4T pixels exceed this requirement.

24.2.4 GLOBAL SHUTTER AND SPECIALTY PIXELS

As with CCDs, early CMOS imager pixels were both reset and readout in the dark behind a closed mechanical shutter. This was done because image arrays are readout one row at a time and the time it takes to read out an image can be a significant fraction of a second. In medium- to high-light conditions, it often takes longer to read out the image than the exposure time. Figure 24.4a shows the read sequence for a typical CMOS imager with a mechanical shutter. For low-cost cameras, manufacturers desired to eliminate the mechanical shutter since it is expensive to manufacture, adds

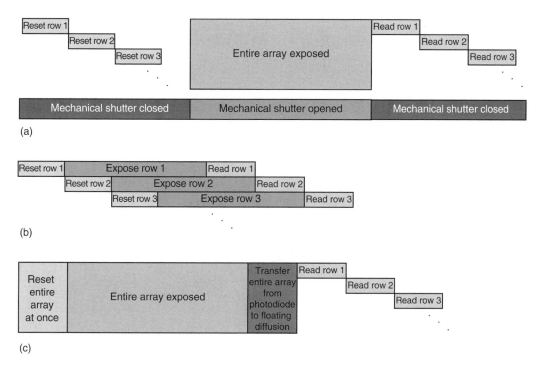

FIGURE 24.4 (a) Standard operation with a mechanical shutter image capture sequence, (b) rolling shutter image capture sequence, and (c) global electronic shutter shutter image capture sequence.

to the size of the camera module, and is difficult to use in a movie mode. The timing sequence used to eliminate the need for a mechanical shutter was called rolling shutter mode. In this mode, the rows of the image are reset in sequence and readout in a rolling sequence (Figure 24.4b). The disadvantage to rolling shutter is that each row is exposed to light for a different time window with the bottom row being exposed significantly later than the top row. When the image is moving, significant motion-related object distortion can occur. An alternative readout scheme known as an electronic global shutter has been created to generate true stop action image without the need for a mechanical shutter.

Electronic global shutter works by resetting the entire array at once, exposing the entire array at once, transferring the contents of all of the photodiodes to the floating diffusions at once, and then reading the array one row at a time (Figure 24.4c) [4]. The difficulty with implementing this technique is that between the time that the charge is transferred to the floating diffusion until it is read out, the floating diffusion must be shielded both from incoming light as well as from electrons in the substrate, which are created by light. Light shields, which are made of opaque materials, usually metal levels, and electron shields, made from implants that create an electric field repelling the electrons, must be added to the process and the physical pixel design to protect the floating diffusion (Figure 24.5). Because the reset is performed globally and so cannot be sampled for each pixel directly before the read, correlated double sample is not possible. A double sample is still

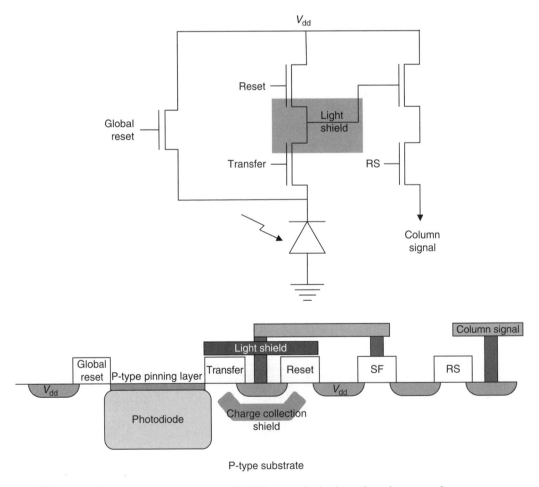

FIGURE 24.5   Five-transistor global shutter CMOS imager pixel schematic and cross-section.

done in the read sequence: sample signal, reset, and sample reference. This sequence, however, is subtracting a reference value from a different reset and thus can remove fixed pattern noise but is not effective at removing the reset noise.

With the evergrowing range of applications for image sensors, an evergrowing selection of pixels have emerged with a large variety of pixel architectures, with many of them aimed at specific applications. Examples include high dynamic range pixels displaying logarithmic or combination linear-logarithmic output response [5], asynchronous detection pixels, which respond to changes in illumination as opposed to light for machine vision applications, in which only the portion of the scene that changes is of interest [6], and pixels utilizing a capacitive transimpedance amplifier to integrate the photocurrent collected [7].

24.3 PIXEL OPERATION

24.3.1 STATIC STATE

CMOS imagers operate in a number of different modes. The most frequent form of operation is called rolling shutter. Referring to Figure 24.6, time T0 describes the static state of the pixel. For the sake of simplicity, we will refer to the charges as electrons. However, there are some vendors who design photodiodes where holes are the primary charge collected and the following discussion would, of course, need to be modified to match this form of operation. In the static pixel state, the pixel is isolated from the column by turning the RS transistor off. The photodiode is also isolated by also keeping the transfer gate off. The floating diffusion is maintained at a higher potential by tying the RG to a high potential. There are several reasons for this configuration. First, the pixel may be illuminated beyond its charge capacity. If this is true, it needs a path for the excess electrons to be carried away from adjacent pixels. Normally, this would travel through the transfer gate. But upon reaching the floating diffusion, you need a path to the pixel V_{dd}, which would be through the reset device. By turning the reset device on (to V_{dd} or to some intermediate voltage less than V_{dd}), this provides a path for excess "bloomed" electrons to be discarded. The RG is also held high for dark current minimization. If the floating diffusion is held high, this floating diffusion becomes a sink for electrons generated in interface states under the transfer gate. If the floating diffusion is allowed to approach the ground, electrons generated in interface states under the transfer gate will be swept to the photodiode and add to the dark current of the pixel.

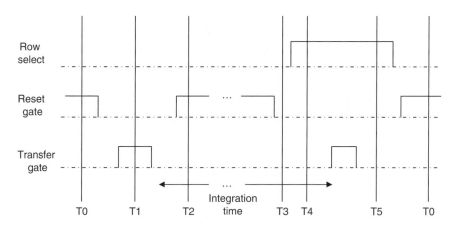

FIGURE 24.6 Timing diagram for 4T pixel operation.

24.3.2 PHOTODIODE RESET

Time T1 shows the setup state to ready the pixel for operation. The first thing needed is to reset the pixel, dumping all the electrons out of the photodiode. This is done by dropping the reset line and raising the transfer gate line. The potential on the floating diffusion must be high enough to draw all the electrons out of the photodiode and onto the floating diffusion.

In Figure 24.7, we plot the cross-section of a portion of our 4T pixel. On the left of the pixel is a heavily doped layer adjacent to the shallow trench isolation (STI) sidewall, which passivates the silicon/silicon dioxide interface with holes. There is also a surface p-type pinning layer, which is electrically connected to the substrate through the STI sidewall. This layer is fairly heavily doped so as to be able to fully deplete through the n-layer when sufficient bias is applied. For both the pinning layer and the STI sidewall layer, the dopant level is high enough to passivate any interface states/traps at the silicon surface with holes [8]. Below the surface pinning layer is the photodiode collection region, a lightly doped n-type implant. The doping of this n-type photodiode is purposely set to a low value so as to fully deplete the diode of all carriers when the transfer gate is raised high and the floating diffusion is at least initially in a high state. The figure also shows the transfer gate adjacent to the photodiode and the floating diffusion on the other side of the transfer gate. This floating diffusion is connected to both the gate of the SF and the source of the RG. In Figure 24.8, we see a simulation of the electron concentration in a cross-section of the pixel though the photodiode. The transfer gate is tied to a high potential as in time T1 (in Figure 24.6) and the electrons are swept out of the photodiode into the inversion layer of the transfer gate and the floating diffusion node. The simulation shows a very small concentration of electrons left in the photodiode but this concentration when applied to photodiode dimensions results in $\ll 1$ electron. After fully depleting the photodiode, the transfer gate is turned off.

The pixel must be fully depleted during reset primarily for noise reasons. In a nonfully depleted pixel, the measured signal will include an additional shot noise term driven by the statistics of measuring a finite number of carriers. Assume that 36 electrons are left behind in the photodiode after reset. Since shot noise is proportional to the square root of the number of carriers, the extra noise will be 6 electrons. Of course, additional shot noise is present simply because of photon counting. If you read 10,000 electrons, you can expect noise on the order of 100 electrons strictly from the signal. The additional 6 electrons added in quadrature from a nonfully depleted reset is insignificant. If the desired photon counting signal results in a 50-electron signal, the noise is nearly doubled by failing to fully reset the pixel. If, however, you are fully depleted, a 10-electron signal will be well above your noise floor of 3–4 electrons.

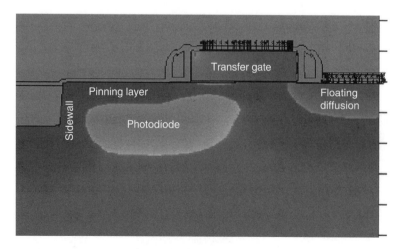

FIGURE 24.7   Doping of cross-section of pinned photodiode.

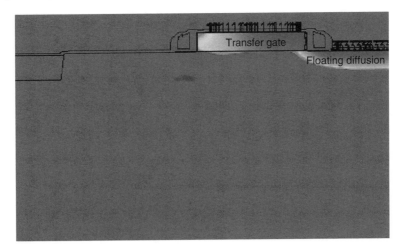

FIGURE 24.8   Electron concentration with transfer gate on and pixel reset.

With the transfer gate off and the photodiode fully depleted, the pixel is ready to accumulate electrons generated by photoillumination. There is some subtlety in the transfer gate off condition. If the transfer gate is held very negative, the gate will accumulate holes at the surface, passivating interface states at the oxide/silicon interface and lowering the dark current generated by the transfer gate [9]. But this costs both power supply (or the circuitry to generate this second power supply) and the loss of some blooming protection. It is now much more difficult for carriers to leak across the transfer gate when the light illumination exceeds the capacity of the photodiode. As an alternative, the transfer gate may be held to a moderate negative value, low enough to see dark current improvement, but not so low as to cause other problems.

24.3.3   PHOTOINTEGRATION

Referring back to Figure 24.6, at time T2, the RG is then brought back up (either to V_{dd} or an intermediate potential) and the photodiode is now ready for photoillumination (in normal operation, the photodiode is always being illuminated, but for purposes of this discussion, we have broken it up into pieces). In Figure 24.9, we see the effect of photoillumination under a strong light source (like

FIGURE 24.9   Electron concentration during blanket photoillumination with transfer gate off.

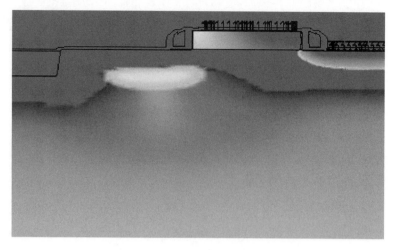

FIGURE 24.10   Electron concentration 4 ns after light illumination ceased.

a flash exposure). Electron–hole pairs are generated in the illuminated region and are collected in the photodiode area. However, these carriers do not stay just where they are generated. The carriers themselves spread out fairly rapidly, moving throughout the pixel area by mutual repulsion, as well as by diffusion processes. Only in areas where there is boron well doping is there a built-in potential high enough to prevent the carriers from diffusing to the floating diffusion and being lost.

Electrons diffusing to an adjacent pixel creates a problem called color crosstalk. For example, electrons generated by red photons might be collected in an adjacent green pixel. Typical remedies include creating electron barriers by adding implants between the pixels or placing an electron sink, such as an n-type substrate, to impede electron motion between the pixels.

At the end of the flash exposure, we now see the effect of the built-in fields in the pixel region. In Figure 24.10, 4 ns after the flash stops, electrons from the surface regions have all been swept away from the surface and stored in the photodiode. This is due to the built-in fields and the high boron concentration at all the surfaces.

In Figure 24.11, we continue to see the evolution of the charge distribution. After 20 ns of flash cessation, most of the electrons have been swept into the photo diffusion with only a fraction of the

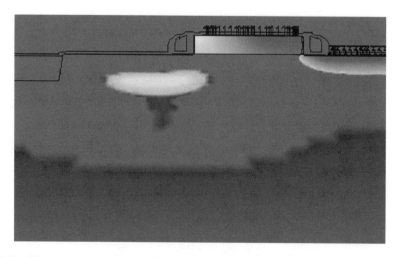

FIGURE 24.11   Electron concentration 20 ns after light illumination ceased.

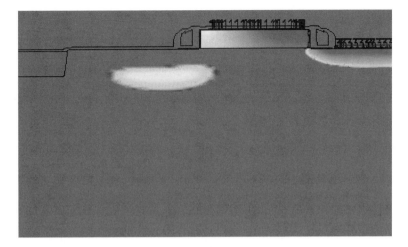

FIGURE 24.12   Electron concentration 60 ns after light illumination ceased.

total concentration left to be gathered. In Figure 24.12, we see that only 60 ns after photoillumination ceased, all the carriers have been collected within the simulation bounds.

24.3.4   CORRELATED DOUBLE SAMPLE READOUT

Now that the photodiode is full of carriers, it is time to read out the charges (see Figure 24.6, T3). To do this, we need to first reset the floating diffusion. As this timing example has the reset tied high during the integration time, we need to bring the RG down. This will float the floating diffusion with a starting value at a high potential. We are now ready to do our correlated double-sampled read. To do correlated double sampling, we first store the value on the floating diffusion. Enabling the RS (time T4) passes the potential through the SF and RS transistors onto the column line and then into a capacitor at the bottom of the column.

Immediately after floating diffusion read, we raise the potential of the transfer gate to a high level. As seen in Figure 24.13, this pulls the electrons from the photodiode to the inversion layer under the transfer gate. Then, when the transfer gate is turned off, the photodiode is emptied of electrons and all these electrons are dumped onto the floating diffusion. The floating diffusion capacitance converts the photodiode charge to a voltage. Ideally you would like this floating diffusion capacitor to be as

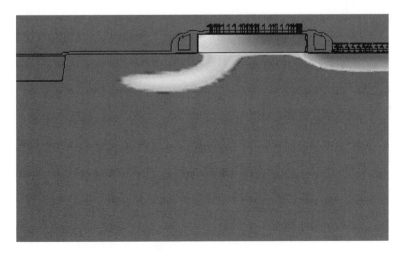

FIGURE 24.13   Electron concentration with illumination off and transfer gate turning on.

small as possible to get the most voltage gain. However, as a portion of this capacitance is junction capacitance and this capacitance is voltage-dependent, the linearity of this amplification can suffer if the capacitance is too small. In addition, lag (which is the incomplete transfer of charges from the photodiode to the floating diffusion) can become a problem if the floating diffusion becomes too small such that the floating diffusion voltage after transfer approaches the pinning potential. If the floating diffusion capacitance is too large, the signal amplification will be too small and the SF noise will dominate the signal at low levels.

At time T5 in Figure 24.6, we read the potential on the floating diffusion through the SF and RS transistors and store this potential in a second capacitor at the bottom of the column. A differential amplifier then measures the signal between the two stored capacitor potentials, which gives us our correlated double sample signal. The correlated double sampling is done for the improved noise immunity gained by measuring the potential of the floating diffusion before and after charge transfer. By disabling the RS and bringing up the RG, we have now completed a cycle and are back to the T0 point in our timing diagram.

24.4 PIXEL SCALING

The scaling of CMOS imager pixel size has been fundamentally driven by the cell phone market, which has been particularly cost-sensitive. The pixel size has had to shrink fast enough to reduce the overall chip size even while the number of pixels in the array has increased from variable gate amplifier to megapixel arrays. The size of the array usually fits a known standard, driven by module and lens standards. This in turn sets the technology—the dimensions of the array are fixed by the optical format and the number of pixels in the array is set by the standard display format. This has set a progression of the pixel size, from 3.3 to 2.7 to 2.2 to 1.75 to 1.4 µm. As seen in Figure 24.14, the scaling in pixel size has accelerated considerably between 2003 and 2007 [10]. The trend toward thin cell phones has also driven pixel size by reducing the space from the lens to the chip. As the space from the lens to the silicon chip shrinks, the angle between the lens and the edge of the array increases. At higher angles, the optical performance degrades. Thus, smaller imaging fields were required, driving even smaller pixels.

The scaling of image sensors is, however, rather different than scaling of a standard CMOS technology [11]. CMOS technology usually benefits from a smaller size with shorter gate delays and

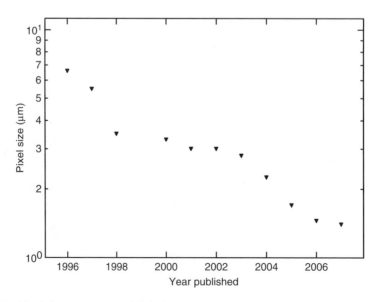

FIGURE 24.14 Pixel size versus year published.

lower capacitance, however, image sensor technology scaling causes fundamental challenges. These challenges can be grouped into three areas: optical transmission, signal capacity, and noise.

24.4.1 Scaling and Optical Transmission

Optical transmission into the photodiode is limited by the metal lines needed to wire the pixel transistors and output the signal. Typically, two to three horizontal wiring channels are needed per pixel row and one or two vertical wiring channels per pixel column. As the pixel shrinks, increasing amount of the available space is taken up by wires and a smaller opening is left for the light to pass to the silicon surface. By using a microlens, much of the metal can be avoided and quantum efficiency of 40% or better can be usually achieved depending on the size of the pixel. As the metal aperture is decreased from 1.2 to 0.6 µm in the tightest dimension, the measured quantum efficiency is reduced from 46% to under 15% (Figure 24.15). Simulations show a similar effect. This has driven the scaling of metal technology from 180 nm in the 3–5 µm pixel generation, 130–150 nm in the 1.75–2.7 µm pixel generations, to 90 nm in the 1.75 µm pixel generations [12–14]. The shrinking aperture is made even more challenging because the chief ray angle of the incoming light near the corner of an array can be as much as 25° from normal incidence. Typically, microlenses are moved slightly across the array to account for the incoming angles. However, as the pixel scales in a given metallization technology node, simple microlenses are incapable of sufficient focusing.

Backside illumination, an emerging technology, offers significant benefit although with corresponding challenges [13,15,16]. In this technique, the imager is built and metallized, then the substrate is thinned, and light imaged from the backside. The main issues with this technique are the ability to thin the backside to a controlled thickness, packaging, electrical crosstalk, and passivation of the backside interface. Creation of a mechanically stable and repeatable recess, while expensive, can be accomplished using bonded silicon-on-insulator wafers. The thickness of the silicon can be set before processing and stopping an etch on the buried oxide film is easy. Techniques from wafer-scale packaging and wafer bonding can be used to solve film stability and cleanliness issues [17]. Since the metallization can now be on the front-side of the wafer, away from the imaging plane, metal lines do not interfere with the light and bump contacts can be made with no concern for opaque defects. Fabrication of color filters and microlenses have been successfully formed on the wafer

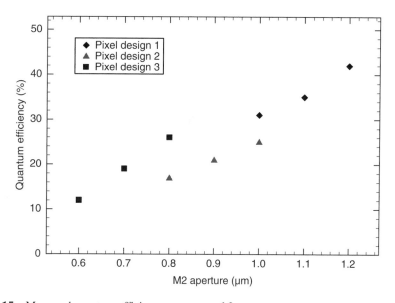

FIGURE 24.15 Measured quantum efficiency versus metal 2 aperture.

backside [16]. The electrical crosstalk and backside interface passivation require additional effort. In the case of a front-side imager, the passivating implants can be done before metallization but late enough to avoid significant thermal diffusion. However, the backside interface is typically not available until after thinning, and well after the metallization is in place. Various methods have been used to passivate the backside, including low-temperature molecular beam epitaxy [15]. This passivation must be effective: blue light will be collected in a short distance near the backside interface and must not be trapped at the interface or allowed to recombine in a heavily doped passivation layer. In addition, for adequate collection, the thickness of the silicon layer must be substantial: on the order of 4 μm or more, while the pixel dimensions are likely to be small, 1.4 μm or less. Without significant tailoring of fields, electrical crosstalk will induce unacceptable hue shifts. For example, electrons created under a blue color filter near the back interface of the wafer must be kept from diffusing into a red pixel. While electrical crosstalk is an increasingly challenging issue for front-side imagers, more options are available. For example, n-type substrates, which can help crosstalk considerably by rejecting electrons generated deep in the silicon from the front, are not available for backside imagers.

As pixel technology scales, the optical performance of even a perfect imaging pixel will degrade. Since most cellphone camera lenses are small, the $f/\#$ is typically 2.8. For images with acceptable blurring, the airy disc resolution metric is probably the most appropriate. With this metric, 3.7 μm is the minimum spacing between pixels for green light. However, for a color imager, the spacing of the repeating Bayer pattern should be used. With this metric, a 1.85 μm pixel technology meets the nonblurring criteria, assuming a nearly ideal lens [18]. Using the more forgiving Sparrow resolution criteria leads to a ~0.9 μm pixel spacing with a Bayer pattern, but with today's nonoptimum lenses, this resolution is unlikely to be acceptable. So, at some point, decreasing pixel size while maintaining pixel count will degrade picture quality. However, that is not to say that subresolution pixels may not be useful [19]. Producing pixels smaller than the resolution limit of the lens is not a hard limit. Below this limit, the image may still appear pleasing, but will have less information content than 1 bit per pixel. Ultimately, the minimum size of a pixel may be set by marketing as opposed to scientific measurements of quality. If an 8 Mpixel camera produces a pleasing image to a consumer, it may sell well even if it contains no more information than a 6 Mpixel camera with a larger pixel.

24.4.2 Scaling and Signal Capacity

Signal capacity is a second key challenge for pixel scaling. The p–n junction that serves as the storage for collected electrons gets smaller with each generation. Assuming that the fraction of total pixel area that is used for the photodiode is constant (fill factor), the total photodiode area reduces by ~1.5× per generation. For a pinned photodiode (4T) detector, the total capacity of the photodiode is the integrated voltage-dependent capacitance times the allowable voltage swing from fully depleted to the minimum voltage that can hold an additional electron. To increase the capacitance with decreasing area will typically require increased fields. This is a challenge as leakage currents increases strongly with field strength. Alternately, the voltage at which the photodiode depletes could be increased, but this too is limited by the need to transfer all the charge during the read, as discussed above. The scaling of the photodiode has thus led to a significant reduction in the total electron capacity of the photodiode, although at a rate slower than the simple 1.5× pixel area reduction [12].

Improving the fill factor has been the key to keeping electron capacity above the pixel area scaling trend. One method uses the entire area underneath the repeating pixel to collect and store light-generated electrons [20]. A second method eliminates wasted space by sharing components of the pixel [21]. This has long been used in the DRAM industry where contacts, isolation trenches, and other features were shared to reduce the size of the DRAM unit cell. The three support transistors in the 4T pixel, the RG, SF, and RS can be shared among a number of photodiodes. Sharing two photodiodes leads to two transfer gates, and the three support field-effect transistors (FETs) for two photodiodes, or 2.5 transistors per photodiode. This sharing scheme was used by some vendors for

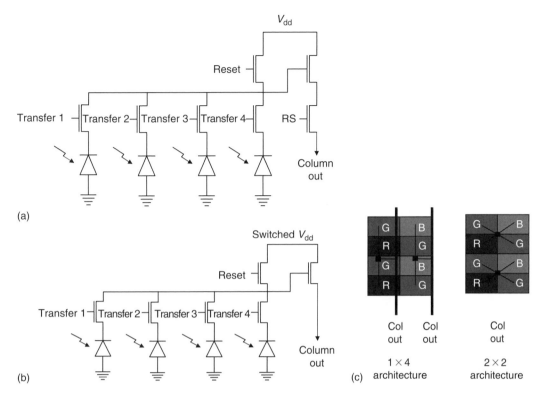

FIGURE 24.16 Shared pixel architectures: (a) 1.75 transistors per pixel; 4 shared photodiodes, (b) 1.5 transistors per pixel, 4 shared photodiodes, and (c) Bayer sharing schematic for 1 × 4 and 2 × 2 architectures.

pixels larger than 2.7 μm [12]. For the 2.2 and 2.7 μm generation, sharing the support electronics among four photodiodes is common, leading to 1.75 transistors per photodiode [22]. A diagram of these circuits is shown in Figure 24.16. While the support electronics must now drive these alternate configurations, the fundamental imager performance is unchanged. A more significant change to pixel operation eliminates the RS transistor by using the V_{dd} line as a row signal [23,24]. In this mode of operation, the V_{dd} signal is at a low potential at all times except when a row is being reset or read out.

Unfortunately, shared pixels now share defects in the support transistors, possibly resulting in yield degradation. A defect that would have caused a single cell fail is now a pair or quad defect. Typically, single-cell fails are correctable, but a fail that affects adjacent pixels in the same color plane usually is not. As shown in Figure 24.16, in a 4 × 1 architecture, floating diffusions and support transistors are shared with two greens and either two reds or two blues, which can result in a nonfixable cluster of 2 adjacent red or blue failures. In a 2 × 2 architecture, there will be two greens, but only a single red and blue pixel. Depending on the correction architecture, the two greens may not be a cluster fail, and therefore correctable.

While sharing transistors between photodiodes can significantly improve available photodiode area, it results in asymmetries relative to a repeating four transistors per photodiode layout [21]. With a Bayer pattern of color filters, each color will have a slightly different metal environment to wire each of the needed transistors to the floating diffusion, control lines, and V_{dd} and output lines. This can lead to hue shifts, particularly at the corners of the array where light is arriving at a significantly higher angle than in the center of the array. An example is shown in Figure 24.17. A second cause of hue shift is crosstalk between pixels due to metallization environment. With an unshared pixel, the metallization is the same. With a shared pixel, the floating diffusions are shared and the metal lines

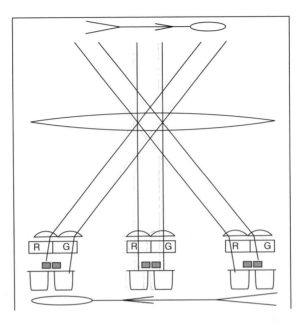

FIGURE 24.17 Asymmetry in mirrored layout leads to hue shift. Metal or transfer gate blocking red and green light on opposite edges of array leading to a greenish head and reddish feet. (From McGrath, R. D., Fujita, H., Guidash, R. M., Kenney, T. J., and Xu, W., 2005 IEEE Workshop on Charge-Coupled Devices and Advanced Image Sensors, Karuizawa, Japan, pp. 9–11, 2005, With permission.)

connecting the floating diffusions to the source follower will be much longer, with subsequently increased coupling. Unless care is taken to equalize the coupling, pixels with different environments may couple to the drive wires differently. Despite these issues, the need to maintain optical and electrical fill factors at a reasonable level have driven the industry to shared architectures.

24.4.3 Pixel Scaling and Noise

As pixel dimensions scale, technology considerations tend to increase noise. However, since the signal is dropping, due to lower electron capacity, the technologist must reduce the noise. One aspect of noise mentioned above, hue shifts, will lead to noise in the final image. Beyond the capacitive coupling and optical hue shifts mentioned above, electrical crosstalk becomes more significant as pixels scale. The photon absorption depth is constant, but the pixel area is shrinking. A second challenge is dark current, which introduces dark current shot noise. Since the leakage requirements are many orders of magnitude more stringent than a DRAM, the technology must have better passivation of interfaces, lower fields, and reduced damage. Clearly, the straightforward implementation of these features does not lend itself well to reduced pixel sizes. Passivation of interfaces, usually with p-type dopant, will smear into the photodiode, reducing the area of the n-type photocollection area. Lower fields typically require smoother, longer profile diffusions, again reducing the area of active collection and storage. Damage can be kept low with low-implant doses, but high-temperature anneals to heal damage are limited by the requirement for limited out-diffusion. While some of the sources of generation are area-dependent and therefore reduce with smaller pixel area, assuming no changes need to be made to maintain capacity, others are more systematic.

One trend in dark current reduction is the removal, where possible, of STI [25]. The STI fabrication tends to roughen the surface due to the etch process, leading to an increased number of states. The curvature of the sidewall will lead to many different planes, as opposed to the well-controlled (100) plane of the surface, which minimizes states. The passivation of the trench with dopants is much more difficult than passivating the surface. The surface is available for dopants at the end of the process, minimizing out-diffusion. In contrast, the bottom of the trench requires either passivation early in the process, where diffusion is a problem, or late in the process where higher doses, alignment tolerances, and multiple implants must be done.

However, as the dark current is reduced below $\sim 30 e^{-}/s$ at $60°C$, other components begin to dominate the signal-to-noise ratio of the imager. A typical approach to reducing the area in standard CMOS would be to scale the voltage, gate oxide, and FET lengths. This is problematic in the CMOS imager, where V_{th} drops across the reset gate and source follower limit the voltage swing on the column. Therefore, V_{dd} has remained relatively constant despite pixel scaling at 2.8–3.3 V. This limits the gate oxide thickness and hence the scaling of the FETs. In addition, since the floating diffusion is sensitive to leakage, particularly in a global shutter mode, the halo implants and increased fields use to scale devices in standard CMOS would have negative effects on noise due to leakage. Smaller widths will make FETs more sensitive to STI edge effects. Smaller lengths move the FETs away from optimum long channel behavior. As the FET width and lengths scale, $1/f$ noise will increase. Correlated double sampling increases the sampling frequency of the source-follower, reducing the effect of $1/f$ noise, but cannot completely eliminate its effect. $1/f$ noise is composed of discrete states. Therefore, depending on the time constant of the trap, some pixels may have transitioned to a different state between the reference read and the signal read of the correlated double sampling output. Being random, this is an uncorrectable effect and leads to "blinking" pixels [26]. Techniques to reduce hot electron effects, such as adding nitrogen to the gate, will tend to increase $1/f$ noise [27].

24.5 CONCLUSIONS

CMOS imagers, which have a large number of applications, have become a mainstream technology. Even though the CMOS imager technologies were cheaper and more adaptable, competition with the prior technology, and CCD image sensors forced advances in the technology to improve the quality of the captured images. Advances in circuit topology, semiconductor process, and operation conditions all were necessary to produce high-quality images. More recently, technology advances have been dominated by the shrinking of the pixel. Significant technical challenges have been faced to produce ever smaller pixels with acceptable image quality, but the industry has consistently innovated and met these challenges and it is now routinely creating pixels as small as 1.75 μm with even smaller pixels in sight.

REFERENCES

1. J. R. Janesick, *Scientific Charge-Coupled Devices*, SPIE Press, Washington, DC, 2001.
2. B. Fowler, M. Godfrey, J. Balicki, and J. Canfield, Low noise readout using active reset for CMOS APS, *Proceedings of SPIE*, 3965, 126, 2000.
3. M. Guidash, T. Lee, P. Lee, D. Sackett, C. Drowley, M. Swenson, L. Arbaugh, R. Hollstein, and S. Donner, A 0.6 μm CMOS pinned photodiode color imager technology, *IEDM Technical Digest*, 927, 1997.
4. M. Wany and G. Israel, CMOS image sensor with NMOS-only global shutter and enhanced responsivity, *IEEE Transactions on Electronic Devices*, 50(1), 57, 2003.
5. E. Fox, J. Heynecek, and D. Dykaar, Wide-dynamic range pixel coupled with linear and logarithmic response and increased signal swing, *Proceedings of the SPIE*, 3965, 4, 2000.
6. P. Lichtsteiner, C. Posch, and T. Delbruck, A 128 × 128 120 dB 3 mW asynchronous vision sensor that responds to relative intensity change, *ISSCC 2006*, p. 208.
7. B. Fowler, J. Balicki, D. How, and M. Godfrey, Low FPN high gain capacitive transimpedance amplifier for low noise CMOS image sensors, *Proceedings of the SPIE*, 4306, 68, 2001.
8. A. Theuwissen, The hole role in solid-state imagers, *IEEE Transactions on Electronic Devices*, 53(12), 2972–2980, 2006.
9. A. Theuwissen, J. Bosiers, and E. Rocks, The hole role, *IEDM Technical Digest*, 799, 2005.

10. L. J. Kozlowski, Noise minimization via deep submicron system-on-chip integration in megapixel CMOS Imaging sensors, *Opto-Electronics Review*, 14(1), 11–18, 2006, and IEDM 2000–2006, ISSCC 2000–2006, IEEE Workshop on CCDs and Advanced Image Sensors 2005 and 2007.

11. J. Nakamura (Ed.), *Image Sensors and Signal Processing for Digital Still Cameras*, CRC Press, Boca Raton, FL, pp. 88–89, 2006.

12. G. Agranov, T. Gilton, R. Mauritzson, U. Boettiger, P. Altice, J. Shah, J. Ladd, X. Fan, F. Brady, J. McKee, C. Hong, X. Li, and I. Patrick, Optical-electrical characteristics of small, sub-4 μm and sub-3 μm pixels for modern CMOS image sensors, *2005 Workshop on CCDs and Advanced Image Sensors*, Karuizawa, Japan, pp. 206–209, June 2005.

13. J. Prima, F. Roy, P. Coudrain, X. Gagnard, J. Segura, Y. Cazaux, D. Herault, N. Virollet, N. Moussy, B. Giffard, and P. Gidon, A 3mega-pixel back-illuminated image sensor in 1T5 architecture with 1.45 μm pixel pitch, *Proceedings of 2007 International Image Sensor Workshop*, pp. 5–8, June 2007.

14. G. Agranov, R. Mauritzson, S. Barna, J. Jiang, A. Dokoutchaev, X. Fan, and X. Li, Super small, sub-2 μm pixels for novel CMOS image sensors, *Proceedings of 2007 International Image Sensor Workshop*, pp. 307–310, June 2007.

15. B. Pain, Fabrication and initial results for a back-illuminated monolithic APS in a mixed SOI/bulk CMOS technology, *Proceedings of 2005 IEEE Workshop on CCD and Advanced Image Sensors*, pp. 102–104, June 2005.

16. S. Iwabuchi, Y. Maruyama, Y. Ohgishi, M. Murmatsu, N. Karaswa, and T. Hirayama, A back-illuminated high-sensitivity small-pixel color CMOS image sensor with flexible layout of metal wiring, *ISSCC Digest of Technical Papers*, 302–303, February 2006.

17. A. Badihi, ShellCase ultrathin chip size package, *Proceedings International Symposium on Advanced Packaging Materials: Processes, Properties and Interfaces*, pp. 236–240, March 14–17, 1999.

18. P. B. Catrysse and B. A. Wandell, Roadmap for CMOS imagers: Moore meets Planck and Sommerfeld, In Sampat, N., DiCarlo, J. M., and Motta, R. J. (Eds.), *Proceedings of the SPIE*, 5678, 1–13, 2005.

19. E. R. Fossum, *Image Sensors and Signal Processing for Digital Still Cameras*, J. Nakamura (Ed.), CRC Press, Boca Raton, FL, 2006.

20. B. Dierickx, G. Meynants, and D. Scheffer, Near-100% fill factor standard CMOS active pixel, *IEEE CCD and AIS Workshop*, Brugge, Belgium, p.1, June 5–7, 1997.

21. R. D. McGrath, H. Fujita, R. M. Guidash, T. J. Kenney, and W. Xu, Shared pixels for CMOS image sensor arrays, *2005 Workshop on CCDs and Advanced Image Sensors*, Karuizawa, Japan, pp. 9–11, June 2005.

22. Mori, 1/4-inch 2-Mpixel MOS image sensor with 1.75 transistors/pixel, *IEEE Journal of Solid-State Circuits*, 39(12), 2426, 2004.

23. Takahashi, A 3.9-μm pixel pitch VGA format 10-b digital output CMOS image sensor with 1.5 transistor/pixel, *IEEE Journal of Solid State Circuits*, 39(12), 2417, 2004.

24. M. Kasano, Y. Inaba, M. Mori, S. Kasuga, T. Murata, and T. Yamaguchi, A 2.0 μm pixel pitch MOS image sensor with an amorphous Si film color filter, *2005 ISSCC Technical Journal*, 348–349, 2005.

25. K. Itonaga, H. Abe, I. Yoshihara, and T. Hirayama, A high-performance and low-noise CMOS image sensor with an expanding photodiode under the isolation oxide, *IEDM Technical Digest*, 791–794, 2005.

26. Y. Degerli, *IEEE Transactions on Electronic Devices*, 47(5), 949–962, 2000.

27. K. W. Chew, K. S. Yeo, S.-F. Chu, Impact of technology scaling on the $1/f$ noise of thin and thick gate oxide deep submicron NMOS transistors, *IEEE Proceedings-Circuits, Devices and Systems*, 151(5), 415–421, 2004.

25 Wide Dynamic Range CMOS Cameras

Steve Collins, Bhaskar Choubey, Hsiu-Yu Cheng, and Stephen Otim

CONTENTS

25.1   INTRODUCTION

Over the past few years, digital cameras have emerged in mainstream consumer applications and largely replaced traditional film cameras. Most of the research effort that has led to this success has been focused upon increasing the number of pixels at the same time reducing costs and power consumption. However, the resulting pixels are not perfect. In particular, these cameras are based upon integrating pixels with a dynamic range of 60 dB or less. Although this is usually sufficient, there are some situations in which the dynamic range is inadequate and as a result details are lost in parts of the image. This is inconvenient for consumers; however, it is preventing cameras from being deployed in important emerging applications such as improving road safety [1].

The problem with extending the dynamic range of existing cameras is that this is difficult to achieve when there is motion in a scene and even for static scenes, a large amount of data is generated. An alternative approach to capturing high dynamic range images is suggested by the response of the human visual system, which is sensitive to relative contrast rather than absolute brightness. When combined with the existence of high dynamic range logarithmic pixels, it has stimulated an interest in pixels with a logarithmic response.

Logarithmic pixels based upon a load transistor operating in weak inversion have been developed by several groups [2–4]. These pixels have the dynamic range required for all the applications that have been identified. However, they are known to suffer from fixed pattern noise. It is shown that this fixed pattern noise can be removed so that the contrast sensitivity of the camera is comparable to the human visual system. Unfortunately, these pixels are vulnerable to temporal noise and can have a very slow response. The temporal response of the pixels can be improved using a reset operation, such as the one used in conventional pixels with a linear response but these modified pixels are still vulnerable to temporal noise. To overcome all these problems, an integrating pixel with a logarithmic response is proposed. These pixels are as fast as conventional integrating pixels and more robust to temporal noise than the previous designs. If the expected performance of these pixels can be confirmed, then they could be employed in a whole range of new applications.

25.2 MOTIVATION

The market for digital cameras has grown rapidly in the past decade and digital cameras now outsell film cameras. This implies that digital cameras are perfect. However, all users have probably been disappointed at times, see for example, the holiday photographs shown in Figure 25.1. In the worst of these two images, on the left, the main subject disappears into the shadows because the majority of the image is brighter than the subject. The photograph on the left is better, however, details such as the skyline of the mountain tops are now missing.

The fundamental problem with the digital images in Figure 25.1 is that the dynamic range of the scene is larger than the input dynamic range of the camera and the camera has been designed to try to image the majority of the scene. The dynamic range of the camera is less than that of the scene because of variations in the brightness of different parts of the scene. The resulting loss of detail is an inconvenience for consumers, however, it is a real problem for other potential large markets. One of these markets is within automobiles where a large dynamic range is required to deal with imaging in tunnels or under bridges with bright roads beyond or with scenes illuminated by headlights [1].

The evolutionary approach to increasing the dynamic range of a camera is to use multiple integration times [5,6] similar to the way in which multiple stops have been used to extend the dynamic range of film cameras [7]. With a 10 bit analog-to-digital converter (ADC), the dynamic range of a camera is 60 dB; this is small compared to the dynamic range of some scenes, which can be as large as 120 dB. Increasing the camera's dynamic range by this amount requires a huge range of integration times. For a static scene, it may be possible, although inconvenient, to extend the integration time by this amount. The problem is that any camera within an automobile must be capable of imaging moving scenes. A high dynamic range camera is therefore required that can image moving scenes.

FIGURE 25.1 Examples of the effect of the limited dynamic range of a digital camera on some holiday photographs.

An interesting example of a high dynamic range imaging system is our own visual system. The human visual system is far too complex to be replicated in a camera. However, the early stages of the human visual system are interesting for two reasons: They suggest an alternative strategy for high dynamic range imaging. Equally importantly, the human visual system is the system through which the images that are acquired are viewed. It will be inefficient to gather and display information that is invisible to the human visual system.

The human visual system has evolved to deal with high dynamic range scenes, such as those that cause problems for cameras. This capability is achieved despite the fact that individual components of the human visual system have a lower dynamic range than a typical camera. An example of this is the response of the photoreceptors in the retina that first convert light into electrical signals. Measurements show that if a photoreceptor is exposing to a short burst of light that is about 100 times brighter than its initial input, its output will saturate. In contrast, a longer exposure to the same light will cause an initial saturated response that gradually reduces to close to its background response. This ability of the photoreceptors to adapt to the existing lighting conditions is one of the features of the human visual system that enables it to image high dynamic range scenes. The overall performance of the human visual system is characterized in psychophysical experiments by measuring the minimum observable local increment in light intensity ΔI when the background intensity is I, sometimes known as the just noticeable difference. Over a large range of high background intensities, it is found that $\Delta I / I$ is a constant [8]. This known as Weber's law and the value of the constant depends on experimental conditions, but is approximately 1%.

It is inefficient to acquire data that are invisible to the human visual system. A response that naturally represents percentage changes in the input is the logarithmic response. In addition, a logarithmic response has an interesting mathematical property; specifically it converts a product of two components to a sum. This property of logarithms has been exploited in several potentially useful image processing algorithms such as homomorphic filtering, bilateral filtering, and some tone-mapping algorithms that reduce the dynamic range of an image [7]. These algorithms are very useful, if not essential, for a system containing high dynamic range pixels because the dynamic range of display devices is similar to, or even smaller, than that of existing digital cameras. This will make it impossible to directly display at least some of the images acquired by any high dynamic range camera. If this problem is solved using any of the algorithms that employ a logarithmic representation of the scene, then it is more efficient to acquire the image using pixels with a logarithmic response.

25.3   LOGARITHMIC PIXELS

Both the sensitivity of the human visual system and the algorithms that may be needed to display a high dynamic range image suggest that it will be more efficient to use a camera with a logarithmic response. Several approaches to obtaining pixels with a logarithmic response have been suggested [2–4]. Of these the most widely investigated and commercially mature approach is to create a pixel in which the photocurrent flows through a load transistor within each pixel. By designing the pixel so that this load device operates in weak inversion, it is possible to create an output voltage that is proportional to the logarithm of the photocurrent. This strategy was employed by Mead to create systems that mimic part of the mammalian retina [9]. Within these systems, the logarithmic response was usually obtained using a pMOS load device. In order to reduce the size of the pixel an nMOS load, transistor is now more commonly used.

A circuit diagram of a pixel with an nMOS load transistor, M_{load}, and a source-follower readout circuit is shown in Figure 25.2. The function of the source-follower readout circuit, formed by M_{sf} and M_{sel} within the pixel together with a shared current source M_{source}, is to selectively connect the pixel to a shared output. In its turn, this shared output is usually connected to a second source-follower, not shown in Figure 25.2, so that each pixel in a two-dimensional array can be selectively connected to a single output channel. This channel often includes an ADC to create a digital output.

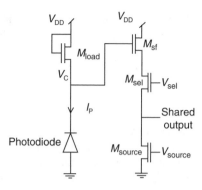

FIGURE 25.2   Schematic circuit diagram of a pixel with a logarithmic response.

Any source-follower circuits between the pixel and the output are designed to function as buffers with a gain close to 1. The functional form of the relationship between the photocurrent and the final output is therefore determined by the load transistor. Ideally, the current flowing through this device is the photocurrent. However, there will be an additional component, which flows in the absence of light and is known as the dark current. The gate-source voltage needed to supply both the photocurrent and the dark current can be calculated using the current–voltage characteristics of the load transistor. A model of these characteristics that covers situations in which the gate-source voltage is both larger and smaller than the threshold voltage of the transistor is the EKV model [10]. In this model, the drain-source current of a mixed oxide semiconductor field-effect transistor (MOSFET) with a gate-source voltage V_{GS} and threshold voltage V_t is given by

$$I_{DS} = I_0 \left[\ln \left(1 + \exp \left((V_{GS} - V_t)/2n\phi_t \right) \right) \right]^2$$

where ϕ_t is the thermal voltage and

$$I_0 = \frac{W}{L} 2n\phi_t^2 \mu C_{ox}$$

is a device- and process-dependent parameter that represents the current flowing through the device when its gate-source voltage is approximately 50 mV larger than its threshold voltage. In effect, this parameter determines the threshold between the device operating in weak inversion and moderate inversion.

Within the pixel, the gate voltage of the transistor is constant and therefore the source voltage, which is also the output voltage of the pixel, varies as a function of the photocurrent. Equating the drain current of the transistor to the sum of the photocurrent, I_{ph}, and the dark current, I_{dark}, leads to [11]

$$V_{out} = V_{DD} - V_t - 2n\phi_t \ln \left(\exp\sqrt{(I_{ph} + I_{dark})/I_0} - 1 \right)$$

Usually the total current flowing through the transistor is smaller than I_0. The response of the pixel can then be represented by the simpler model:

$$V_{out} = V_{DD} - V_t - n\phi_t \ln \left((I_{ph} + I_{dark})/I_0 \right)$$

that is equivalent to the model proposed by Joseph and Collins [12]. If the photocurrent is larger than the dark current, this becomes

$$V_{out} = V_{DD} - V_t - n\phi_t \ln \left(I_{ph}/I_0 \right)$$

which is the desired logarithmic response. The model therefore predicts that the output of the pixel is proportional to the logarithm of the photocurrent as along as the photocurrent is larger than the dark current but smaller than I_0.

Although this model represents the voltage within the pixel, the response of any buffer circuits between this point and the output is linear. These circuits leave the form of the relationship between the photocurrent and the output voltage unchanged. This suggests that the full parametric model of the pixel response is

$$V_{\text{out}} = a' - 2b \ln \left[\exp\sqrt{(dI_{\text{ph}} + c)} - 1 \right]$$

where c represents the effect of the dark current. However, unless a phototransistor is used to amplify the photocurrent, it will be small enough for the pixel's response to be represented by the simpler model:

$$V_{\text{out}} = a' - b \ln \left(dI_{\text{ph}} + c \right)$$

This becomes even simpler when the photocurrent is larger than the dark current

$$V_{\text{out}} = a - b \ln \left(I_{\text{ph}} \right)$$

which is the desired logarithmic response, shown in Figure 25.3. In this case, the dark current is approximately $1\,\text{fA}$ and I_0 is approximately $1\,\mu\text{A}$. As a result, the output voltage of the pixel is proportional to the logarithm of the photocurrent over a range of approximately six decades.

Results such as those in Figure 25.3 show that these pixels have the required dynamic range. A problem with these pixels is that differences between the nominally identical devices within different pixels create a variation in the responses of an array of pixels to a uniform input. These variations, known as fixed pattern noise, degrade the quality of the captured image. Fixed pattern noise occurs in all types of pixels, however, a combination of a comparatively small output voltage range of these pixels and large variations between devices operating in weak inversion means that it is a particularly significant problem in these logarithmic pixels [13]. The effects of fixed pattern noise are clearly seen in the results in Figure 25.4. This figure shows the results of estimating the photocurrents in

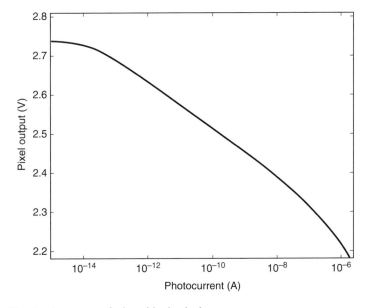

FIGURE 25.3 Simulated response of a logarithmic pixel.

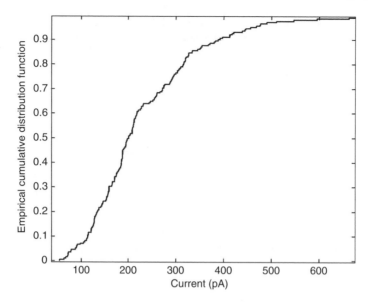

FIGURE 25.4   The measured cumulative distribution of photocurrents estimated assuming that all pixels have an identical response.

200 pixels assuming that all the pixels are identical, in particular that they have the same a and b parameters. These typical results clearly show that the estimated photocurrents vary by an order of magnitude. This level of fixed pattern noise leads to unacceptably poor images.

Various techniques have been proposed to reduce the effects of fixed pattern noise [4,14,15]. The dominant contribution to fixed pattern noise is variation in the parameter a, which adds a variable contribution to the output of each pixel. These variations can be reduced by subtracting the response of a pixel to a uniform input from its response to a scene. Several techniques have been proposed to generate this uniform input, including using the image of a uniform scene, such as a sheet of white paper. The advantage of using an image is that this technique includes all possible sources of variations. Generating and imaging a uniform scene can be inconvenient, but this would not be a significant problem if the image could be acquired and stored as part of the manufacturing process. However, the model for the pixel response shows that its output depends upon temperature-dependent device parameters. Fixed pattern noise correction should therefore be based upon data acquired at the operating temperature.

The problems of using a uniform scene for fixed pattern noise correction can be avoided by occasionally using a transistor as a current sink to draw a current through the load device [16]. In order to ensure the uniformity of this current, these transistors can be shared between pixels with a switch transistor within the pixel, selecting which pixel sources the current at a particular time. Using the response of each pixel to this uniform current, it is possible to correct the pixel response for parameter variations. The results in Figure 25.5 show that after correction for additive (offset) variations the range of estimated photocurrents is reduced to between 182 and 197 pA. The range of photocurrents has therefore been reduced dramatically. If required, the performance of these pixels can be improved further by correcting for variations in both the additive (offset) and multiplicative (gain) parameter. Results in Figure 25.5 show that this more sophisticated correction leads to a further reduction in the range of estimated photocurrents. In this case, the standard deviation in the photocurrent is reduced from 2.7 pA with offset only correction to 1.2 pA with offset and gain correction. Since the mean photocurrent is 190 pA, this means that three standard deviations corresponds to just less than 2%. It could therefore be argued that with averaging over small areas, these pixels could match the contrast sensitivity of the human visual system, which is approximately 1% in bright scenes.

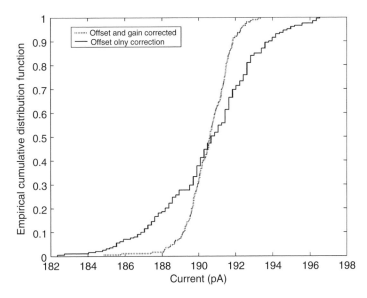

FIGURE 25.5 The measured cumulative distribution function of estimated photocurrents when the outputs from 200 pixels are corrected for either offset or offset and gain variations.

The most commonly known problem, fixed pattern noise, caused by using a load transistor in weak inversion can be solved. However, these pixels have two other significant problems. The first of these problems can be seen in Figure 25.6. This shows both the response of the pixel and the rate of change of the response, given as the change in output when the photocurrent changes by an order of magnitude, which is the b parameter in

$$V_{out} = a + b \log(I_{ph} + I_{dark})$$

Over a wide range of input currents, the typical pixel response changes by 60 mV when in the photocurrent changes by an order of magnitude. A change in photocurrent of 1% therefore corresponds

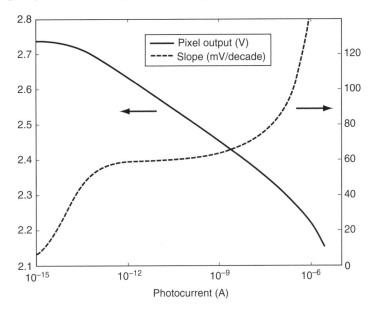

FIGURE 25.6 The output voltage and its rate of change (slope) as a function of the photocurrent.

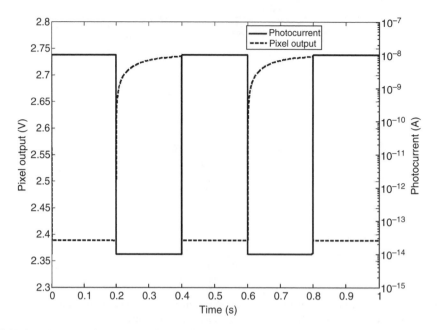

FIGURE 25.7   The dynamic response of a logarithmic pixel, shown on the left-hand axis, to sudden changes in the photocurrent, shown on the right-hand axis.

to a change in output of only 240 μV. This suggests that temporal noise will be a problem. In addition, Figure 25.6 shows that the rate of change of the pixel output reduces as the dark current becomes an increasing proportion of the total current. The pixel output will therefore be particularly vulnerable to noise at low light levels. This problem can be reduced by changing the load device to either two devices in series [17] or a floating-gate device [18] to increase the rate of change of the pixel output. Increasing the rate of change of the response of a logarithmic pixel will make the response less vulnerable to noise. However, it will aggravate the other problem, which is the dynamic response of the pixel.

The problem with the pixel's dynamic response occurs when the photocurrent in the pixel suddenly drops, as might occur when an edge within the scene moves through the pixel position. Any sudden change in input means that the pixel voltage must change, however, the only current available to charge or discharge the pixel capacitance is the difference between the currents flowing through the load transistor and the photodiode. This can be a small fraction of a small current and as a result, the response of the pixel is expected to be slow. In fact, results such as Figure 25.7 show that it can take more than 100 ms for a pixel's response to reach a value close enough to equilibrium to represent the photocurrent to an accuracy of few percent. Since this effect occurs at photocurrents that arise in naturally illuminated scenes, this slow pixel response is a particular problem for applications, such as within automobiles, which demand imaging of moving scenes.

25.4   LOGARITHMIC PIXELS WITH RESET

The reset operation within a conventional integrating pixel creates a response time that is independent of the photocurrent. This suggests that the speed of response of a logarithmic pixel at low light levels can be increased using a reset operation. A schematic circuit diagram for a logarithmic pixel with an additional device, M_{reset}, to allow the pixel voltage to be reset is shown in Figure 25.8. As in the conventional integrating pixels, image acquisition begins when the pixel voltage is first reset to a high value and then isolated. Once the pixel is isolated, the photocurrent discharges the pixel

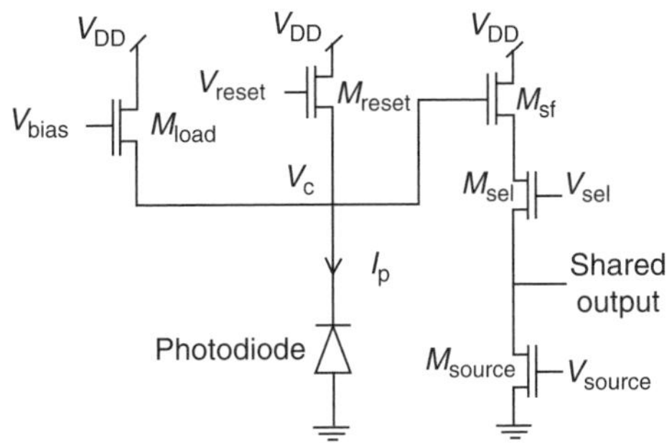

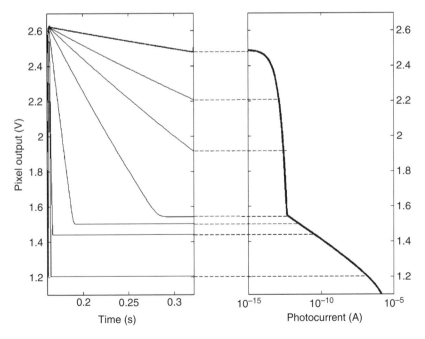

FIGURE 25.8 The circuit diagram for a logarithmic pixel that can be reset.

capacitance and the source voltage of the load transistor decreases. As shown in the left hand part of Figure 25.9, initially this leads to a linear relationship between the output voltage of the pixel and the photocurrent, with the larger photocurrents causing the more rapid changes in the output voltage. However, unlike in the conventional integrating pixel, discharging the capacitance within this pixel decreases the source voltage of a load transistor. If this becomes less than the constant gate voltage of this device, the load transistor will start to supply current. Initially, this current will be too small to balance the photocurrent and this transistor's source voltage will continue to decrease, but at a slower rate. This will lead to an increasing current flowing through the load transistor until the current in this device balances the photocurrent. If this occurs, when the load transistor is operating in weak inversion the result, as shown in the right hand half of Figure 25.9, is a voltage that is proportional to the logarithm of the photocurrent.

If the output of the pixel is sampled at a particular time after integration begins, some photocurrents will be too small to have discharged the pixel enough for the load device to start supplying

FIGURE 25.9 The left hand figure shows the temporal response of a logarithmic pixel with reset to various photocurrents. The left hand figure shows the corresponding output voltages at a particular time after the pixel is isolated.

current. For these currents, the response of the pixel is proportional to the photocurrent. In contrast higher photocurrents will discharge the pixel enough for the load transistor to supply current into the pixel and when this balances the photocurrent the pixel output is proportional to the logarithm of the photocurrent. The value of the photocurrent at which the transition between these two behaviors occurs depends upon several factors including the reset and gate bias voltages. However, for the results in Figure 25.9, the transition occurs at approximately 1 pA.

The description of the operation of the pixel shows that the difference between the current flowing out of the pixel and the current sourced by the load transistor discharges the pixel capacitance C_{PD}. The current flowing out of the pixel, I_P is either dark current or photocurrent, hence

$$I_P = I_{ph} + I_{dark}$$

The difference between this current and the source-drain current flowing through the load device discharges the pixel capacitance

$$I_P = -C_{PD}\frac{dV_C}{dt} + I_0\left[\log\left(1 + \exp((V_{bias} - V_C - V_{T,M_{load}})/2n\phi_t)\right)\right]^2$$

If this capacitance is reset to voltage level V_{reset} and the source-follower buffer can be represented as a gain, G_r, and an offset, O_r, so that

$$V_{out} = G_r V_P - O_r$$

The differential equation can integrated between when the pixel is isolated and an "integration" time t_{int} to give [19]

$$V_{out} = P'\log\left[\exp\left(\frac{V'_{reset} - t_{int}I_pQ'}{P'}\right) + \frac{\left\{1 - \exp\left(\frac{-t_{int}I_pQ'}{P'}\right)\right\}R'}{I_p}\right]$$

where parameters

$$V'_{reset} = G_r V_{reset} - O_r$$

$$P' = G_r n\phi_t$$

$$Q' = G_r/C_{PD}$$

$$R' = 2I_0 \exp\left\{\frac{G_r\left(V_{bias} - V_{T,Mlog}\right) + O_r}{G_r n\phi_t}\right\}$$

have been introduced for convenience. At low photocurrents or short "integration" times

$$\exp\left(\frac{-t_{int}I_pQ'}{P'}\right) \approx 1$$

and the pixel shows the expected linear response

$$V_{out} = V'_{reset} - t_{int}I_pG_r/C_{PD}$$

For higher photocurrents or longer integration times, the arguments of both the exponential functions become negative and so

$$V_{out} = P'\log(R'/I_P) = P'\log(R') - P'\log(I_p)$$

which is the expected logarithmic response.

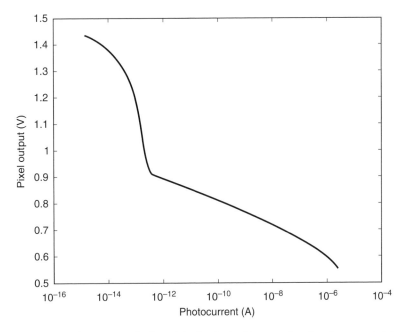

FIGURE 25.10 Measured response of a logarithmic pixel with reset showing the expected transition from a linear to a logarithmic response at higher photocurrent (in this case currents more than several hundred femtoamperes).

The model of the pixel response fits the measured responses, such as the one shown in Figure 25.10, over an input range of 160 dB, including the linear and logarithmic regions of operation and the transition region between these two. In addition, the model correctly predicts the behavior of the pixel to changes in parameters such as the bias voltage and "integration" time. As with all pixels, these pixels also suffer from fixed pattern noise and an important use of the model is to develop a fixed pattern noise correction procedure. Direct use of the relatively complex model requires an iterative process to correctly determine the photocurrent whenever the output corresponds to the transition region. Fixed pattern noise correction is significantly easier if the output corresponds to either the linear or logarithmic regions represented by one of the simpler equations. It is impossible to ensure a pixel output is in one of these regions for a particular "integration" time. However, the transition region is very narrow. This means that by sampling the output twice, it is possible to ensure that one of the two outputs corresponds to one of the two simpler models. One of the simpler equations can then be used to correct the fixed pattern noise.

The logarithmic pixel with reset overcomes some of the problems of the logarithmic pixel based upon a load transistor operating in weak inversion. However, it has a complex response and still has a relatively small rate of change of output voltage over a large input range that is vulnerable to temporal and quantization noise.

25.5 INTEGRATING PIXELS WITH A LOGARITHMIC RESPONSE

The problems that occur in the logarithmic pixel with reset can be overcome by creating integrating pixels with a logarithmic response. An integrating pixel with a logarithmic response can be created by making its effective integration time-dependent upon the photocurrent by disconnecting the input to the source-follower circuit from the photodiode after a time that depends upon the photocurrent [20]. Disconnecting the two parts of the pixel circuit requires a switch device within the pixel. One way in which this switch can be controlled is shown in Figure 25.11. Again this pixel includes a reset

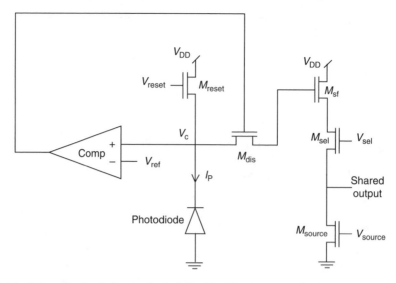

FIGURE 25.11   Schematic circuit diagram for a pixel with a photocurrent dependant effective integration time.

device and a source-follower readout circuit. In addition, it contains a comparator (Comp) and an n-channel device, M_{dis}, between the photodiode and the source-follower input.

As in the conventional integrating pixel, the change in pixel voltage t seconds after integration has started is

$$\Delta V_{\text{c}} = -\frac{I_{\text{ph}}t}{C_{\text{pixel}}}$$

To obtain a logarithmic output, the two parts of the pixel circuit should be disconnected so that the change in pixel voltage that is sampled onto the source-follower to become the pixel output is proportional to the logarithm of the photocurrent. This means that the they should be disconnected at a time t_{dis} when

$$\frac{I_{\text{ph}}t_{\text{dis}}}{C_{\text{pixel}}} = S\ln\left(\frac{I_{\text{ph}}}{I_{\text{ref}}}\right)$$

where

I_{ref} is a user-selected reference current
S is the user-selected pixel gain

The source-follower circuit is isolated from the photodiode when the two inputs to the comparator are equal. If the pixel voltage is reset to a voltage V_{r} by the reset device, then the two inputs to the comparator are equal when

$$V_{\text{ref}}(t) = V_{\text{r}} - \frac{I_{\text{ph}}t}{C_{\text{pixel}}}$$

For this to occur when the change of voltage represents the logarithm of the photocurrent, then this is equivalent to

$$V_{\text{ref}}(t) = V_{\text{r}} - S\ln\left(\frac{I_{\text{ph}}}{I_{\text{ref}}}\right)$$

This last equation can be rearranged to give

$$I_{\text{ph}} = I_{\text{ref}}\exp\left(-(V_{\text{ref}}(t) - V_{\text{reset}})/S\right)$$

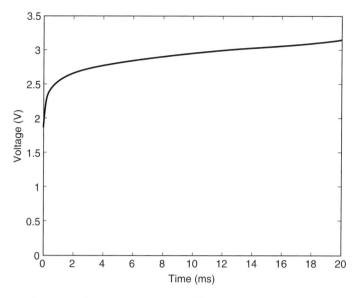

FIGURE 25.12   A typical result obtained using MATLAB® to determine the time-dependant reference voltage needed to achieve a logarithmic response.

Ideally, the aim is to obtain an expression for the reference voltage as a function of time, however, the above equations lead to an expression for time as a function of the reference voltage

$$t = C_{\text{pixel}}((V_{\text{reset}} - V_{\text{ref}}(t)) \exp[(V_{\text{ref}}(t) - V_{\text{reset}})/S])/I_{\text{ref}}$$

This equation has been solved in MATLAB in order to obtain a time-dependent reference voltage. Results such as those in Figure 25.12 show that the reference voltage changes very rapidly in the first fraction of a millisecond before gradually rising to a maximum value. When reference voltages similar to that in Figure 25.12 are used within MATLAB to emulate the operation of the pixel, the results that are obtained, such as those in Figure 25.13, confirm that this reference voltage is expected to lead to a logarithmic response over a wide dynamic range.

Integrating logarithmic pixels have been fabricated using the Austria Microsystems 0.35 μm standard complementary metal oxide semiconductor (CMOS) logic process. The pixel size and fill factor of these prototype pixels were 25 by 25 μm² and 24%, respectively. For all the experiments with this pixel, the time between starting the integration of the photocurrent and sampling the output voltage was set to 20 ms and the voltage was measured using an HP 4155B. The reference voltages used were calculated in MATLAB and then supplied to the pixel using the Agilent Intuilink Waveform Editor and 33250A arbitrary waveform generator.

To demonstrate the functionality of the pixel design, particularly its very wide dynamic range, test pixels were manufactured in which the photodiode was replaced by a MOSFET acting as a voltage-controlled current sink. The gate voltage needed to sink a particular current through the MOSFET was estimated from Cadence Spectre simulation results. The results in Figure 25.14 show that the pixel has a logarithmic response over a dynamic range of 140 dB. Furthermore, the pixel output changes by 228 mV when the photocurrent changes by an order of magnitude. This pixel is therefore less vulnerable to noise than the previous pixels.

The problem with including a comparator within the pixel is that it increases the number of devices and hence the size of each pixel. This problem can be minimized by using a single pMOS transistor, M_{dis} in Figure 25.15, as an imperfect comparator. In this more compact circuit, the reference voltage is connected to the gate of M_{dis}. When the pixel voltage is being reset, the reference voltage is

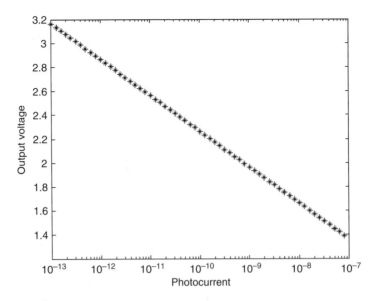

FIGURE 25.13 A typical result that has been obtained using MATLAB to emulate the response of an integrating pixel with a logarithmic response.

held low so that M_{dis} conducts and the input to the source-follower circuit is also reset. Once the pixel is isolated, the pixel capacitance will be discharged by the photocurrent, I_{ph}. At the same time that the pixel is isolated, the reference voltage is increased. Initially, the reference voltage will be less than the pixel voltage and M_{dis} will conduct. However, this device will have a decreasing source voltage and an increasing gate voltage. Eventually, these changing voltages cause the channel resistance of M_{dis} to increase suddenly and it effectively stops conducting. This isolates the gate voltage of the source-follower transistor M_{sf}, which controls the output voltage from the pixel. The effective integration time for the pixel output is therefore determined by the time when M_{dis} stops conducting.

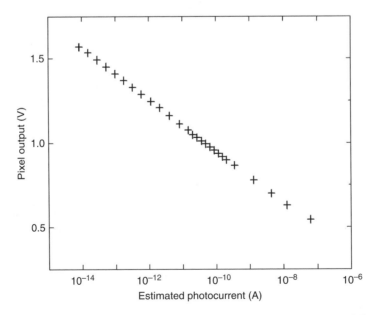

FIGURE 25.14 The typical measured response of an integrating logarithmic pixel containing a comparator.

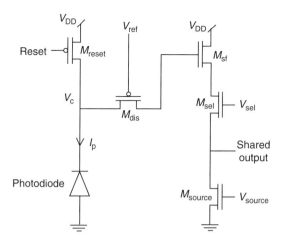

FIGURE 25.15 A schematic circuit diagram for a compact integrating pixel that can have a logarithmic response.

Assuming that M_{dis} acts as an ideal switch, then the compact circuit in Figure 25.15 will have the same response as the circuit in Figure 25.11. The only difference is that the comparator isolates the source-follower when the voltage in the pixel equals the reference voltage. In the case of the more compact pixel, the source-follower is isolated when

$$V_{gate,M_{dis}} - V_{source,M_{dis}} = -V_{th,M_{dis}}$$

Since the reference voltage is connected to the gate of this device and the photodiode is discharging the source, this difference can be taken into account by modifying the reference voltage.

A compact integrating logarithmic pixel has been fabricated using the UMC 0.25 μm, 1P4M, 2.5 V CMOS process to create a pixel with an area of 7.5×7.5 μm^2 and a fill factor of 45%. The optical response of the pixel with a photodiode has been measured using a 150 W quartz tungsten source with neutral density filters to change the illumination intensity. With this equipment, it was possible to create light levels ranging from 0.1 to 10,000 lux. The results in Figure 25.16 show that by changing the reference voltage, it is possible to change the logarithmic response of the same pixel. In particular, the data in this figure show that the change in output voltage per decade change in illumination intensity can be changed from 238 to 341 mV/decade.

The results in Figure 25.16 show that the logarithmic response of the pixel can be controlled by varying the externally generated reference. Alternatively, if the reference voltage is held low enough, the pixel acts as a conventional integrating pixel with a linear response. This integrating mode of operation has been used to determine the dark current in the pixel. To do this, the pixel was placed in the dark and allowed to integrate the resulting dark current for 2 s. The resulting rate of change of the output voltage was found to be 28 mV/s. To convert this to an equivalent illumination level, the experiment was repeated with the pixel illuminated with 0.15 lx, as measured using a Sekonic L-508 light meter. Under these conditions, the rate of change of the pixel voltage was 602 mV/s. This means that the dark current corresponds to 7 mlux.

25.6 THE FUTURE

Over the past few years, there has been a very rapid growth in the market for digital cameras. This success has been achieved because advantages such as convenience easily outweigh any limitations, such as their limited dynamic range. The fact that the dynamic range of a typical camera is smaller than some scenes leads to loss of detail in some areas of a scene, which is an inconvenience for

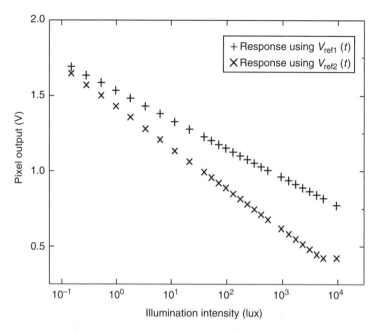

FIGURE 25.16 The measured response of more compact integrating logarithmic pixel.

consumer products, such as digital cameras and camera phones. More seriously it is preventing these cameras from being used in other applications including some that could improve road safety. It is these emerging applications for digital cameras that are currently driving the development of wide dynamic range cameras.

The dynamic range of a conventional digital camera can be increased by acquiring several images with different integration times that are then merged to create a wide dynamic range image. However, this technique can only be applied easily to stationary scenes and it leads to large amounts of data per pixel. Both these problems can be solved using pixels with a logarithmic response that have a wide input dynamic range that is compressed using a nonlinear response that preserves the contrast information that is important to the human visual system. Wide dynamic range logarithmic pixels based upon load transistors operating in weak inversion have been known for many years. However, this type of pixel has several problems that have prevented them from being widely adopted. These problems can be overcome using an integrating pixel with a logarithmic response.

Integrating logarithmic pixels have the potential to capture details in wide dynamic range images of scenes containing motion without increasing the number of bits per pixel. However, this type of pixel has only been proposed very recently and the assessment of its performance is still incomplete. Given the uncertainty with any new idea, this type of pixel is most attractive for emerging applications for digital cameras for which a high dynamic range is essential.

If it can be demonstrated that these pixels have the proposed advantages and there are no insurmountable unforeseen problems, then the improvement in dynamic range may be attractive for existing camera markets, particularly in consumer products. One interesting opportunity when using high dynamic range cameras in these products will be to integrate the cameras with algorithms that compress the dynamic range of the image without degrading the perceptual quality of the image. The result will be a camera that is even easier to use than existing cameras.

Another potential opportunity to employ logarithmic pixels arises because they may help solve a relatively subtle problem. This problem arises because the color of an object in an image depends upon both the reflectance of the objects surface and the spectrum of the illuminant. An important consequence of this problem is that although colour usually forms a key part of a description of an object, it is difficult to use color to recognize objects in a digital image. One possible solution to this problem that has been suggested involves a combination of logarithmic pixels and photodectors with a narrow spectral response [21]. This combination, and the subsequent processing of the captured data, was suggested based upon a theory in which it was assumed that the illuminant could be represented by a black body with a specific temperature and the spectral response of the detector is a delta function. Based upon these assumptions, it was shown that the output from logarithmic pixels could be transformed to remove the variability caused by changing the spectrum of the illuminant. Although it is impossible to create detectors with delta function responses, it appears that the algorithm is useful when the detectors have a narrow spectral response. Materials that appear to have suitable responses have been identified and work has started to integrate them with silicon readout circuits. If this speculative work is successful, then it should be possible to use color in a wide range of applications including automatic visual inspection and machine vision systems.

REFERENCES

1. B. Hoefflinger, *High-Dynamic Range (HDR) Vision*, Springer, Dordrecht, 2007.
2. S. G. Chamberlain and J. P. Lee, Silicon imaging arrays with new photoelements, wide dynamic range and free from blooming, *Proceedings of 1984 Custom Integrated Circuits Conference*, pp. 81–85, 1984.
3. S. Kavadias, B. Dierickx, D. Scheffer, A. Alaerts, D. Uwaerts, and J. Bogaerts, A logarithmic response CMOS image sensor with on-chip calibration, *IEEE J. Solid-State Circuits*, 35(8), 1146–1152, 2000.
4. M. Loose, K. Meier, and J. Schemmel, A self-calibrating single-chip CMOS camera with logarithmic response, *IEEE J. Solid-State Circuits*, 36(4), 586–596, 2001.
5. J. H. Park, M. Mase, S. Kawahito, M. Sasaki, Y. Wakamori, and Y. Ohta, An ultra wide dynamic range CMOS image sensor with a linear response, *SPIE*, 6068, 94–101, 2006.
6. M. Mase, S. Kawahito, M. Sasaki, Y. Wakamori, and M. Furuta, A wide dynamic range CMOS image sensor with multiple exposure-time signal outputs and 12-bit column-parallel cyclic A/D converters, *IEEE J. Solid-State Circuits*, 40(12), 2787–2795, 2005.
7. E. Reinhard, G. Ward, S. Pattanaik, and P. Debevec, *High Dynamic Range Imaging: Acquisition, Display, and Image-based Lighting*, Morgan Kaufmann, San Francisco, CA, 2006.
8. B. A. Wandell, *Foundations of Vision*, Sinauer Associates, Sunderland, MA, 1995.
9. C. Mead, *Analog VLSI and Neural Systems*, Addison-Wesley, Boston, MA, 1989.
10. E. V. C. Enz and F. Krummenacher, An analytical MOS transistor model valid in all regions of operation and dedicated to low voltage and low current applications. *Analogue Integrated Circuits and Signal Processing*, 8(1), 83–114, 1995.
11. S. Otim, D. Joseph, B. Choubey, and S. Collins, Modeling of high dynamic range logarithmic CMOS image sensors, *IEEE Instrumentation and Measurement Technology Conference 2004*, 1, pp. 451–456, 2004.
12. D. Joseph and S. Collins, Modelling, calibration and correction of nonlinear illumination-dependent fixed pattern noise in logarithmic CMOS image sensors, *IEEE Trans Instr. and Meas.*, 51(5), 996–1001, 2002.
13. O. Yadid-Pecht, Wide-dynamic-range sensors, *SPIE J. Opt. Eng.*, 38(10), 1650–1660, 1999.
14. N. Ricquer and B. Dierickx, Active pixel CMOS sensor with on chip nonuniformity correction, *Proceedings of IEEE Workshop CCD/AIS*, 20–21, 1995.
15. L.-W. Lai, C.-H. Lai, and Y.-C. King, A novel logarithmic response CMOS image sensor with high output voltage swing and in-pixel fixed pattern noise reduction, *IEEE Sensors J.*, 4(1), 122–126, 2004.
16. B. Choubey, S. Ayoma, S. Otim, D. Joseph, and S. Collins, An electronic calibration scheme for logarithmic CMOS pixels, *IEEE Sensors*, 6(4), 950–956, 2006.
17. G. Tecchiollo and A. Sartori, Photo-sensitive element for electro-optical sensors U.S. Patent 6891144 B2, 2005.
18. S. Collins, J. Ngole, and G. F. Marshall, A high gain trimmable logarithmic CMOS pixel, *Electronics Letters*, 36(21), 1806, 2000.
19. B. Choubey and S. Collins, Models for pixels with wide dynamic range combined linear and logarithmic response, *IEEE Sensors J.*, 7(7), 1066–1072, 2007.
20. H.-Y. Cheng, B. Choubey, and S. Collins, A high dynamic range integrating pixel with an adaptive logarithmic response, *IEEE Photonics Technol. Lett.*, 19(15), 1169–1171, 2007.
21. G. D. Finlayson and S. D. Hordley, Color constancy at a pixel, *J. Opt. Soc. Am. A*, 18(2), 253–264, 2001.

26 CMOS Focal Plane Spatially Oversampling Computational Image Sensor

Ashkan Olyaei and Roman Genov

CONTENTS

26.1 INTRODUCTION

Many video processing applications employ spatial image transforms such as block-matrix and convolutional transforms. For example, block-matrix transforms such as discrete cosine transform (DCT) or discrete wavelet transform are widely used in various image and video compression algorithm standards. Convolutional transforms are often employed in pattern recognition. These transforms require extensive computational resources for their real-time implementations.

A number of techniques for realizing block-matrix and convolutional transforms in sensory systems have been developed. Dedicated digital signal processors (DSPs) rely on high throughput architectures to compute spatial-weighted sums needed in block-matrix transforms, but require significant area and power resources. At high imager resolutions, the input data rate or memory-processor bandwidth of such processors may limit their sustained throughput [1,2]. An analog-to-digital converter (ADC) to quantize the analog sensory input prior to signal processing is also required for such digital processors.

To overcome these limitations, block-matrix and convolutional transforms have also been implemented in the analog domain directly on the focal plane. Capacitor bank implementations use charge sharing to compute weighted sum and difference [3–5] but may have limited scalability. Current-mode weighted averaging implementations [6] use zero-latency current-mode addition but employ

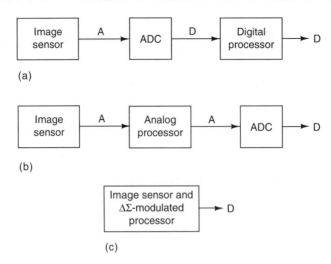

FIGURE 26.1 Video block-matrix and convolutional transforms computing architectures with (a) a digital processor, (b) an analog processor, and (c) the computational image sensor.

multiple matched current mirrors at the expense of increased pixel area. Charge integration and gain-stage voltage summation [7] utilized in variable resolution imaging do not allow for weighted averaging and require additional column-parallel amplifiers. Current-mode vector-matrix multiplication [8] architectures employ floating-gate arrays for block matrix storage and achieve high power efficiency. Kernel-dependent scan-out imager architectures have been shown to reduce memory requirements in focal-plane spatial image processing [9]. A tree-based partitioning algorithm that implements adaptive compression has also been reported [10]. Circuit implementations based on video compression algorithms utilize in-pixel temporal prediction [11–13] and array-based spatial prediction [14] to reduce the amount of transmitted data. All of the aforementioned architectures perform computation in analog very large-scale integration (VLSI) domain and require an extra ADC to provide the digital output.

Analog-digital mixed-domain complementary metal oxide semiconductor (CMOS) imaging and signal processing combine the benefits of the two domains [15]. Analog circuits perform area-efficient and low-power computation directly on the focal plane, eliminating the need for an external processor [16]. The intrinsic parallelism of imaging architectures yields high computational throughput, often beyond that of modern digital processors, allowing to perform complex video processing operations in real time. Digital components provide the output in a convenient digital format, and sustain the accuracy and configurability of such systems.

We present a mixed-signal VLSI implementation of a digital CMOS imager computing block-matrix and convolutional transforms on the focal plane for real-time image processing. The computational image sensor presented here combines image acquisition, signal processing, and quantization in a single compact low-power architecture as shown in Figure 26.1. Our approach combines weighted spatial averaging and oversampling quantization in a single $\Delta\Sigma$-modulated ADC cycle, making focal-plane computing an intrinsic part of the quantization process. The approach yields power dissipation below that of a conventional digital imager while the need for a peripheral DSP is eliminated.

26.2 BLOCK-MATRIX AND CONVOLUTIONAL TRANSFORMS

Block-matrix and convolutional transforms correlate a segment of an image with a spatial kernel, or block matrix, in order to identify statistical redundancies or distinguish particular features depending on the transform type. In the example of image compression, the redundancies are eliminated to reduce data rate. In the case of pattern recognition, the features are employed to form a more precise object description and enhance the classifier performance.

To transform an image I into transformed image T, the kernel or block matrix C is tiled vertically and horizontally across the image. The block-matrix is tiled in an overlapping or nonoverlapping fashion, corresponding to convolutional and block-matrix transforms, respectively. For the case of the block-matrix transform, coefficients of T are obtained by computing the two-dimensional dot product of C and I at each tile location:

$$T_{ij} = \sum_{h=1}^{H} \sum_{v=1}^{V} C_{hv} I_{xy} \tag{26.1}$$

$$x = h + (i-1)H, \quad i = 1, 2, \ldots, \frac{L}{H} \tag{26.2}$$

$$y = v + (j-1)V, \quad j = 1, 2, \ldots, \frac{K}{V} \tag{26.3}$$

where
  $C_{hv} \in \mathbb{Z}$ are the block-matrix coefficients comprising a spatial kernel
  L and K are the image horizontal and vertical sizes, assumed for simplicity to be multiples of
      the kernel dimensions H and V
  h and v are the horizontal and vertical block matrix indices
  i and j are the indices of the block-transformed image

The block-matrix and convolutional transforms are generally computationally expensive. For example, consider an HDTV 1080i imager operating at 30 frames per second (fps). To handle the computational throughput of an 8×8 convolutional transform of the video, an equivalent computational throughput of one 3.5 GHz Pentium processor is needed.

In video compression algorithms, to achieve selective compression of the image, redundant and localized gradient values are filtered out according to a threshold bias, which is based on the required compression ratio and the reconstructed image quality specifications. Another thresholding technique, mainly employed in JPEG image compression, is nonuniform quantization of the block-matrix-transformed image, in which the more significant low-frequency spatial information components are quantized with a higher resolution compared to the less important high-frequency ones.

In pattern recognition systems, to enhance the performance of the classifier, based on a particular feature extraction algorithm, a set of features are extracted from the input images by computing their block-matrix transforms. Both training and classification are then performed on these extracted features.

26.2.1 DISCRETE WAVELET TRANSFORM: HAAR WAVELET EXAMPLE

Two-dimensional Haar wavelet transform is a simple example of a block-matrix transform commonly used in image compression and pattern recognition systems. By extracting horizontal, vertical, and diagonal edges, Haar wavelets register the relationship between intensities among neighboring pixels in different orientations and hence form a "ratio template" [17]. The ratio template is independent of illumination conditions. It truly captures the ratio between various features of an object, which within a class exhibit greater correlation than the absolute intensity values. As a result, a more precise object description is generated at a cost of lower spatial resolution.

The one-dimensional Haar wavelet is composed of the scaling function $\phi(t)$ or the father wavelet, and the wavelet prototype function $\psi(t)$ also known as the the mother wavelet:

$$\phi(t) = \begin{cases} 1 & \text{if } -1 \leq t \leq 1 \\ 0 & \text{otherwise} \end{cases}$$

$$\psi(t) = \begin{cases} 1 & \text{if } 0 < t \le 1 \\ -1 & \text{if } -1 \le t \le 0 \\ 0 & \text{otherwise} \end{cases}$$

The combination of scaling and wavelet prototype functions in two-dimensional space yields the following scalar, horizontal, vertical, and diagonal two-dimensional Haar wavelet functions:

$$\phi(x,y) = \frac{1}{4}\phi(x)\phi(y) \tag{26.4}$$

$$\psi^H(x,y) = \frac{1}{4}\psi(x)\phi(y) \tag{26.5}$$

$$\psi^V(x,y) = \frac{1}{4}\phi(x)\psi(y) \tag{26.6}$$

$$\psi^D(x,y) = \frac{1}{4}\psi(x)\psi(y) \tag{26.7}$$

where the 1/4 coefficient is applied so that the maximum value of the wavelet transform stays within the image intensity range. The equivalent spatial kernels of the scalar and wavelet functions for the first-level Haar wavelet are

$$\Phi_1 = \frac{1}{4}\begin{pmatrix} +1 & +1 \\ +1 & +1 \end{pmatrix} \tag{26.8}$$

$$\Phi_1^H = \frac{1}{4}\begin{pmatrix} +1 & -1 \\ +1 & -1 \end{pmatrix} \tag{26.9}$$

$$\Psi_1^V = \frac{1}{4}\begin{pmatrix} +1 & +1 \\ -1 & -1 \end{pmatrix} \tag{26.10}$$

$$\Psi_1^D = \frac{1}{4}\begin{pmatrix} +1 & -1 \\ -1 & +1 \end{pmatrix} \tag{26.11}$$

Following the block-matrix transform notation in (26.1)–(26.3), the transformation of the image I into the Haar-wavelet-transformed image T is of the linear form:

$$T_{ij} = \sum_{h=1}^{H}\sum_{v=1}^{V} C_{hv} I_{xy} \tag{26.12}$$

$$x = h + (i-1)H, \quad i = 1,2,\ldots,\frac{L}{H} \tag{26.13}$$

$$y = v + (j-1)V, \quad j = 1,2,\ldots,\frac{K}{V} \tag{26.14}$$

$$H = V = 2^b, \quad b = 1,\ldots,B \tag{26.15}$$

where $C_{hv} \in \{-1,+1\}$ are the Haar wavelet coefficients comprising a square spatial kernel in (26.8)–(26.11):

$$C \in \{\Phi_1, \Psi_1^H, \Psi_1^V, \Psi_1^D\} \tag{26.16}$$

and B is the number of levels of Haar transform. B-level Haar wavelet features for $B > 1$ can be obtained by repetitive use of level-one transform, or directly using $3B + 1$ spatially averaging Haar wavelet coefficient kernels of size $H = V = 2^b$, with $b = 1,\ldots,B$.

Figure 26.2 illustrates the spatial kernels of two-dimensional one-, two-, and three-level Haar wavelet transforms. The Haar wavelet transformed images are obtained by computing the block-matrix transforms of the original image with these Haar wavelet kernels. Figure 26.3 depicts the correspondingly computed two-dimensional Haar wavelet transforms of Audrey.

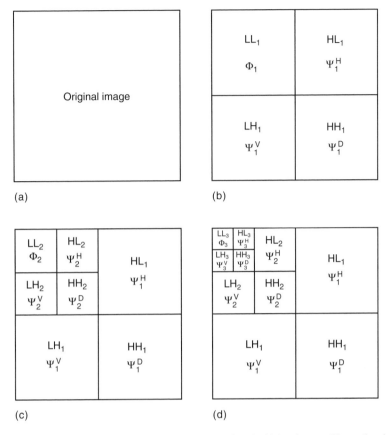

FIGURE 26.2   An illustration of spatial kernels of two-dimensional of (a) an image, (b) one-level, (c) two-level, and (d) three-level Haar wavelet transforms.

26.3   ARCHITECTURE

The block-matrix transform of the form (26.1) can be decomposed as follows:

$$T_{ij} = \sum_{h=1}^{H} \sum_{v=1}^{V} C_{hv} I_{xy} = \sum_{h=1}^{H} T_{ij,h} \tag{26.17}$$

with partial sums

$$T_{ij,h} = \sum_{v=1}^{V} C_{hv} I_{xy} = \sum_{v=1}^{V} |C_{hv}| S_{xy} \tag{26.18}$$

and the sign of C_{hv} factored into the sign-transformed pixel outputs:

$$S_{xy} = \text{sign}(G_{hv}) I_{xy} \tag{26.19}$$

where the notation is consistent with that of (26.2)–(26.3), $C_{hv} = \text{sign}(C_{hv})|C_{hv}|$, and I_{xy} is the output of a pixel at location (x, y).

The proposed mixed-signal VLSI architecture efficiently implements computations (26.19), (26.18), and (26.17), in that order, as depicted in Figure 26.4. Image acquisition and correlated double sampling (CDS) yield offset-compensated pixel output I_{xy} as described in Section 26.4. A switched-capacitor sign unit circuit multiplies pixel output I_{xy} by the sign of a respective kernel

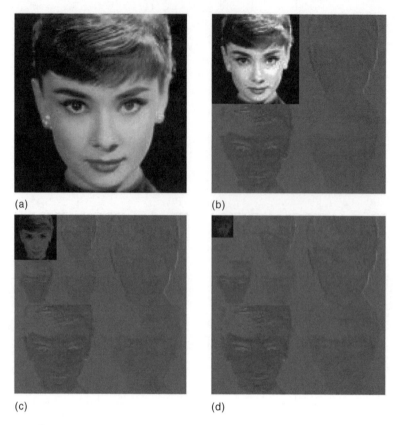

FIGURE 26.3 (a) Audrey and computed Audrey's Haar transforms, (b) one-level, (c) two-level, and (d) three-level transforms.

coefficient via selecting the sampling sequence order yielding a sign-transformed pixel output S_{xy} in (26.19). Section 26.5.1 presents the sign unit circuit implementation. Weighted average of V adjacent pixel outputs in an image column is computed by combining oversampling quantization and selective distributed sampling of the sign-transformed pixel outputs to yield $\hat{T}_{ij,h}$, as discussed in Sections 26.5.2 and 26.5.3. $\hat{T}_{ij,h}$ is the digital representation of $T_{ij,h}$ in (26.18). The switch matrix routes the block-matrix coefficients and their corresponding sign values bit-serially from a ring shift register (*SR*) with a sequence period of V values and spatial period of H columns, synchronously with image readout clock *RowScan* to the oversampling quantizers and sign units, respectively. The operation of the switch matrix is discussed in Section 26.5.4. A simple digital delay and adder loop performs spatial accumulation over H adjacent ADC outputs in the digital domain as they are read out to yield $\hat{T}_{ij}$, which is the digital representation of T_{ij} in (26.17). Section 26.5.5 presents an implementation of this digital accumulation.

26.4 IMAGE ACQUISITION

Image acquisition is performed by the active pixel array, the row control, and the CDS units. The active pixel comprises a resetable n^+-diffusion-p-substrate photodiode, a selectable analog memory, C_{Mem}, and a selectable source follower with shared column-parallel current source biased with *IbiasCol* current as shown in Figure 26.5. The analog memory is implemented as a MOS capacitor for higher density of integration inside the pixel and consequently a larger fill factor.

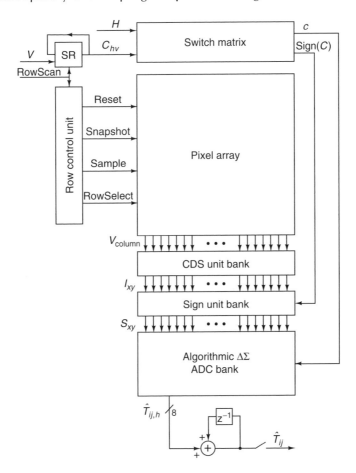

FIGURE 26.4 Top-level architecture of the focal plane spatially oversampling CMOS image compression sensor. Digital accumulation and thresholding (not shown) blocks are implemented off-chip.

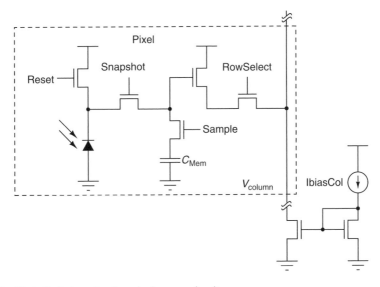

FIGURE 26.5 Photodiode-based active pixel sensor circuit.

The row control unit generates the digital signals *Reset, Snapshot, Sample*, and *RowSelect* controlling the integration and readout phases. There are four modes of operation: snapshot mode, rolling mode, CDS, mode and frame-differencing mode. A switched-capacitor CDS unit circuit suppresses fixed pattern noise by subtracting the reset pixel output from the sensed pixel output. More information on the design of the image acquisition circuits can be found in Ref. [18].

26.5 COMPUTATIONAL QUANTIZATION

This section presents a mixed-signal VLSI implementation of (26.19), (26.18), and (26.17). To simplify notation, in this section we consider Equations 26.18 and 26.19 for a single image column segment (i.e. for given i, h, and x) and the first row of the transformed image (i.e. for $j = 1$ and $y = v$). This simplifies the partial sum in Equation 26.18 to

$$T = \sum_{v=1}^{V} C_v I_v = \sum_{v=1}^{V} |C_v| S_v \qquad (26.20)$$

For a particular image row (i.e. for given v), the sign-transformed pixel output in Equation 26.19 further simplifies to

$$S = \text{sign}(C)I \qquad (26.21)$$

where I is the raw output of a pixel.

26.5.1 SIGN UNIT

The sign unit shown in Figure 26.6a is implemented as a switched-capacitor difference circuit. It applies the sign of the coefficient C to the input by selecting a switched-capacitor sampling sequence order as illustrated in the timing diagram in Figure 26.6b. This directly implements Equation 26.21. The amplifier in the sign unit is the same as in the CDS circuit. For the sake of simplicity, the feedback capacitor reference voltage is shown as ground.

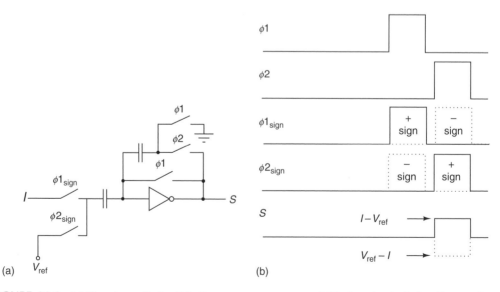

FIGURE 26.6 (a) The sign unit circuit in the sensory processor and (b) sign circuit timing diagram for a continuous range of signed outputs, $V_{\text{ref}} = \min\{I\}$.

26.5.2   ΔΣ-Modulated Multiplying ADC

The spatially compressing image quantizer is implemented as a first-order incremental ΔΣ-modulated ADC extended to an oversampling multiplying ADC [19] as described in this section. The first-order incremental oversampling ADC is depicted in Figure 26.7a. It converts a sequence of analog samples into a digital word representing a quantized version of the average of all samples. It is comprised of a sample-and-hold (S/H) circuit, an integrator, a comparator, and a decimating counter. The rectangular decimation window and initial reset of the accumulator avoid tones in the quantization noise spectrum at DC input that are characteristic of a conventional first-order DS modulator with a low-pass decimation filter. As shown in Figure 26.7b, this architecture can be combined with the sign unit and extended to perform both quantization and signed multiplication of the analog input with a digital word:

$$C = \text{sign}(C)|C| \tag{26.22}$$

with

$$|C| = \sum_{i=0}^{N-1} c[i] \tag{26.23}$$

where $c[i]$ are unsigned unary coefficients of C.

Selective sampling of the sign-transformed pixel output S, controlled by the bit-serial unary sequence $c[i]$, yields an analog sequence $u[i] = Sc[i]$. The first-order modulator converts the sequence $Sc[i]$ into a bit stream $y[i]$ in N cycles, using a 'resetable' (RST) analog integrator:

$$w[0] = 0 \tag{26.24}$$

$$w[i+1] = w[i] + \alpha(Sc[i] - y[i]), \quad i = 0, \dots, N-1 \tag{26.25}$$

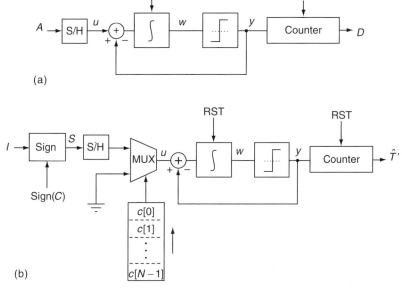

FIGURE 26.7   (a) First-order ΔΣ incremental A/D converter and (b) ΔΣ-modulated multiplying ADC. The sign unit circuit shown in Figure 26.6a also performs the SAH operation. Here the S/H cell is explicitly shown for clarity.

$$w[N+1] = w[N] - \alpha y[N] \qquad (26.26)$$

and a single-bit quantizer:

$$y[0] = -1 \qquad (26.27)$$

$$y[i] = \text{sign}(w[i]), \quad i = 1, \ldots, N \qquad (26.28)$$

where α is the intrinsic gain of the integrator.

A binary counter accumulates the bits $y[i]$ to produce a decimated output. The rectangular decimation window, and initial reset of the integrator, avoid tones in the quantization noise spectrum at DC input that are characteristic of a conventional first-order $\Delta\Sigma$ modulator with low-pass decimation filter [20]. The quantization error (conversion residue) is directly given by the final integrator value $\frac{1}{\alpha}w(N+1)$, as verified by summing (26.25) and (26.26) over i:

$$\sum_{i=0}^{N} y[i] = \sum_{i=0}^{N-1} Sc[i] - \frac{1}{\alpha}w(N+1) \qquad (26.29)$$

where

$$\sum_{i=0}^{N} y[i] = \hat{T}' \qquad (26.30)$$

is the digital output.

This operation yields multiplication of the sign-transformed analog pixel output S with the unsigned digital coefficient $|C|$ defined in (26.23), while a digital output resolution of $\log_2(N)$ bits is warranted:

$$\hat{T}' = |C|S + q' \qquad (26.31)$$

which in combination with (26.21) and (26.22) yields

$$\hat{T}' = CI + q' \qquad (26.32)$$

where

$$|q'| = \left| \frac{1}{\alpha}w[N+1] \right| \qquad (26.33)$$

is the multiplication quantization noise. Higher resolution at a lower oversampling ratio N can be obtained using higher-order incremental conversion [21].

As the input is amplitude-modulated with unary signed coefficients, an error in the amplitude of these coefficients can contribute to the noise. A noise analysis due to this nonideality is given in Ref. [22] where amplitude modulation of an analog sequence with a Hadamard sequence is utilized in the design of a Nyquist-rate $\Delta\Sigma$-modulated ADC. This noise is deemed negligible in this design as a simple analog multiplexer is employed to modulate the input.

26.5.3 $\Delta\Sigma$-Modulated Weighted Averaging ADC

The multiplying ADC architecture in Section 26.5.2 can be extended to perform weighted averaging. Accumulation is achieved by sampling V adjacent pixels in one column, $I_1, I_2 \cdots I_V$, without resetting the integrator or the binary counter. A discrete-time index v is thus introduced.

The architecture of the oversampling weighted averaging ADC is depicted in Figure 26.8. The bit stream $y[vi]$ is now generated for V inputs each sampled N times:

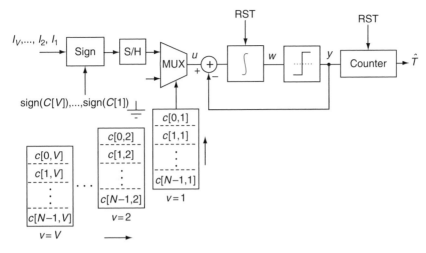

FIGURE 26.8 $\Delta\Sigma$-modulated weighted averaging ADC for $j = 1$ and given i and h. The ADC samples V adjacent pixels in one column, weights each by coefficient C_v, and concurrently quantizes their sum.

$$w[0] = 0 \tag{26.34}$$

$$w[v(i + 1)] = w[vi] + \alpha(S[v]c[i, v] - y[vi]), \quad i = 0, \ldots, N - 1 \tag{26.35}$$
$$v = 1, \ldots, V$$

$$w[V(N + 1)] = w[VN] - \alpha y[VN] \tag{26.36}$$

and a single-bit quantizer:

$$y[0] = -1 \tag{26.37}$$

$$y[vi] = \mathrm{sign}(w[vi]), \quad i = 1, \ldots, N \tag{26.38}$$
$$v = 1, \ldots, V$$

The quantization error, $\frac{1}{\alpha}w[V(N + 1)]$, is obtained similarly:

$$\sum_{v=1}^{V}\sum_{i=0}^{N} y[vi] = \sum_{v=1}^{V}\sum_{i=0}^{N-1} S[v]\, c[i, v] - \frac{1}{\alpha}w[V(N + 1)] \tag{26.39}$$

where

$$\sum_{v=1}^{V}\sum_{i=0}^{N} y[vi] = \hat{T} \tag{26.40}$$

and the notation is consistent with that of (26.24)–(26.29).

This realizes the computation of a weighted sum of sign-transformed pixel outputs $S[v]$ with the unsigned digital coefficients $|C[v]|$, defined in (26.23), with an output resolution of $\log_2(VN)$ bits:

$$\hat{T} = \sum_{v=1}^{V} |C[v]|\, S[v] + q \tag{26.41}$$

which in combination with (26.21) and (26.22) yields

$$\hat{T} = \sum_{v=1}^{V} C[v]\, I[v] + q \tag{26.42}$$

where

$$|q| = \left| \frac{1}{\alpha} w \left[V (N + 1) \right] \right| \qquad (26.43)$$

is the weighted averaging quantization error. We arrive at expression (26.41) for the digital weighted sum, $\hat{T}$, which is a discrete-time equivalent of (26.20).

The optimization of the number of oversampling cycles based on particular block-matrix coefficients can enhance the computational throughput of the architecture. When the maximum absolute value of coefficients in a single row of a block matrix (scaled to all integers), N_V, is less than N, the oversampling computational cycle is stopped on the N_Vth sample and continued on the next row.

Further improvements of the computational throughput of this architecture are achieved by employing an algorithmic $\Delta\Sigma$-modulated ADC [23]. When the maximum of sums of absolute values of coefficients in all columns of a block-matrix (scaled to all integers) is less than VN, the oversampling computational cycle is stopped once all of the coefficients have been fed in. Higher resolution bits are then obtained by subsequent algorithmic residue resampling and extended counting on the residue [23].

26.5.4 SWITCH MATRIX

The switch matrix routes the H different time-dependent block-matrix coefficients and sign signals to L/H groups of adjacent column-parallel ADCs and sign unit circuits, respectively. The block diagram of the switch matrix is shown in Figure 26.9. A total of V pixel rows are sampled while V coefficients are being shifted out from the shift register. The coefficients are looped back to the shift register input to maintain V-row time period. Each kernel coefficient is stored in a binary format of length

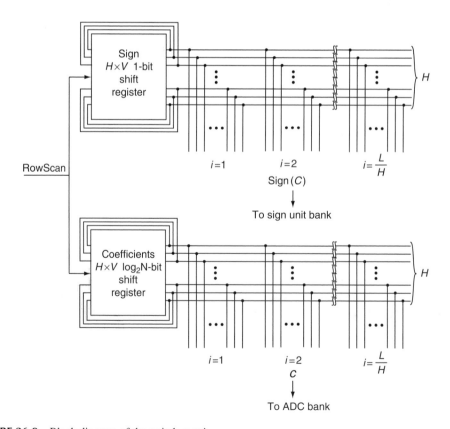

FIGURE 26.9   Block diagram of the switch matrix.

$\log_2(N)$-bits and is digitally oversampled to yield its unary representation of length N bits to match the sampling mechanism of an oversampling ADC and correspondingly weight each pixel output.

26.5.5 DIGITAL ACCUMULATION

Reintroducing the spatial indices i, j, and h back into Equation 26.42 yields the general expression for the columnwise weighted average:

$$\hat{T}_{ij,h} = \sum_{v=1}^{V} C_{hv} I_{xy} + q_{ij,h}$$

where $q_{ij,h}$ is the columnwise weighted averaging quantization noise with standard deviation $\sigma_{ij,h}$.

A simple digital delay and adder loop performs spatial accumulation over H adjacent ADC outputs in the digital domain as they are readout:

$$\hat{T}_{ij} = \sum_{h=1}^{H} \hat{T}_{ij,h} = \sum_{h=1}^{H} \sum_{v=1}^{V} C_{hv} I_{xy} + q_{ij} \qquad (26.44)$$

where q_{ij} is the transformed image quantization noise with standard deviation

$$\sigma_{ij} \approx \sqrt{\sum_{h=1}^{H} \sigma_{ij,h}^2}$$

for i.i.d noise and large H, yielding an additional improvement in SNR. This mixed-signal VLSI computation realizes a block-matrix transform in (26.1) with $H \leq 8$. The switch matrix size scales linearly with H. A maximum of H equal to 8 is chosen here to strike a balance between the switch matrix implementation complexity and the overall functionality. The area overhead of sign unit circuits, switch matrix, and digital accumulator scales linearly with the imager size and becomes small for large K and L. As computing is interleaved with quantization, the extra computational time and thus power dissipation are small compared to those of raw image quantization in a conventional CMOS digital imager.

26.6 COMPARATIVE EXAMPLE

This section compares the presented architecture with a conventional approach where column-parallel algorithmic ADCs performing no computation are employed and an additional peripheral serial digital multiplier and accumulator performs video compression. It is assumed that the kernel is a square matrix of size V with M-bit coefficients and the frame rate is the same in both cases. The comparison is performed for one column only as the two-dimensional computation can be partitioned such that multiplication is performed in the vertical dimension, and only V additions per kernel are performed in the horizontal dimension.

The first-order incremental $\Delta\Sigma$-modulated ADC requires a number of clock cycles exponential with the number of bits of resolution, M. This is a disadvantage compared to the algorithmic ADC, which requires a number of clock cycles proportional to M. On the other hand, the SNR of the spatially oversampling ADC is much higher than that of the algorithmic ADC for the same resolution and the same energy per cycle due to in-pixel and interpixel oversampling and subsequent noise shaping. In thermal noise–limited circuits, power dissipation is linear with SNR. Thus, for the same SNR, power dissipation of the oversampling ADC can be reduced below that of the algorithmic ADC.

The numeric comparison depends on the degree of vertical overlap of kernels in subsequent computations. In the worst case, corresponding to the highest number of computations, the subsequent kernels overlap by $V - 1$ pixels in the vertical dimension. Assuming $V = 8$ and $M = 8$, in the worst case, the power dissipation of the $\Delta\Sigma$-modulated spatially oversampling ADC is 63% of the power dissipation of the algorithmic ADC for the same SNR. In the nominal case, when the kernels do not overlap, the power dissipation of the spatially oversampling ADC is only 8% of the power dissipation of the algorithmic ADC. This assumes multiplication and addition accuracy of 8 bits as necessary for many image compression tasks. In addition, the conventional approach requires a serial digital multiplier and an adder. At HDTV 1020i imager resolution, a computational throughput of several billions of operations per second is required and would need to be delivered by the digital multiplier and adder at the cost of significant additional power dissipation and integration area. In the proposed approach, besides savings in the ADC power dissipation, the need for such a high-throughput DSP is eliminated.

26.7 RESULTS

Experimental results are obtained from a 0.35 μm CMOS prototype containing a 128×128 pixel array and a bank of 128 column-parallel algorithmic $\Delta\Sigma$-modulated ADCs. Figure 26.10 shows the die micrograph of the image compression sensor. Table 26.1 summarizes its electrical and optical characteristics. The values of parameters V and H are programmable in the range of 2 to 8. Any transform in this size range with signed digital coefficients can be computed. Two-dimensional Haar wavelet transform, a block-matrix transform commonly used in image compression [24,25], is

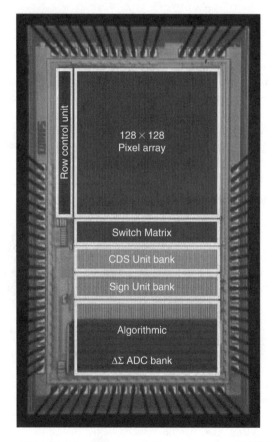

FIGURE 26.10 Die micrograph of the focal plane spatially oversampling CMOS image compression sensor. The integrated 3.1 mm 1.9 mm prototype was fabricated in a 0.35 μm CMOS technology.

TABLE 26.1
Summary of Characteristics

Technology	0.35 µm CMOS
Area	3.1 mm × 1.9 mm
Supply voltage	3.3 V
Array size	128 × 128 pixels
Pixel size	10.45 µm × 10.45 µm
Fill factor	42%
Frame rate	30 fps
Kernel size	2 × 2—8 × 8 programmable
Throughput	4 GMACS in HDTV 1080i DCT
Optical dynamic range	105 dB
Dark current	17.5 fA/pixel
ADC power consumption	4.3 mW
Output resolution	8 bit

chosen here as a simple example to illustrate the functionality of the presented computational imager implementation. The test setup is shown in Figure 26.11.

Figure 26.12a shows an image acquired by the pixel array with 25 ms integration time. The algorithmic $\Delta\Sigma$-modulated ADC performs distributed image sampling and concurrent signed weighted-average quantization, realizing a one-dimensional spatial Haar wavelet transform. Two oversampling phases each of length $N = 32$ clock cycles are interleaved with a single algorithmic residue resampling cycle. Image readout and computational quantization are characterized offline in two sequential steps. A digital delay and adder loop implemented off-chip in digital domain performs spatial accumulation over multiple ADC outputs. This amounts to computing a two-dimensional Haar wavelet transform. Figure 26.12b depicts experimentally measured two-dimensional one-, two-, and three-level Haar wavelet transforms of the original image. Figure 26.12c shows the reconstructed images of the corresponding Haar wavelet transforms. The reconstructed images of one-level Haar transform are compared in Figure 26.13 for various peak SNR and compression ratio.

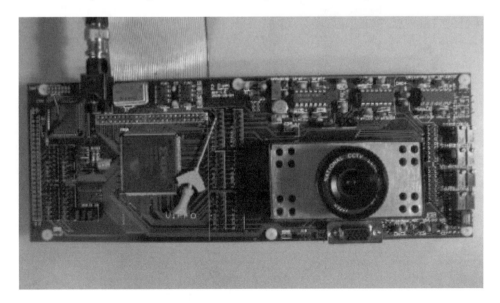

FIGURE 26.11 The printed circuit board for experimental characterization of the 0.35 µm CMOS prototype of the sensory image processor.

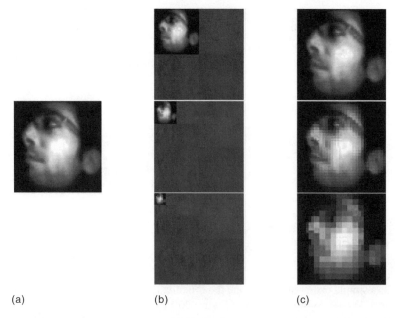

(a)   (b)   (c)

FIGURE 26.12   (a) An image captured by the CMOS image compression sensor at 30 fps. (b) Experimentally recorded one-level (top), two-level (center) and three-level (bottom) Haar wavelet transforms of the image in (a) computed on the CMOS image compression sensor. (c) Reconstructed images for one-level (top), two-level (center), and three-level (bottom) Haar wavelet transforms for the same compression threshold. Compression ratios from top to bottom are 5.33, 20.27, and 41.53.

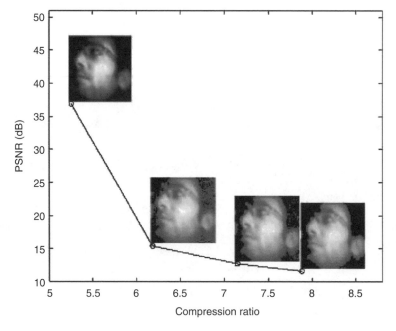

FIGURE 26.13   Reconstructed images obtained by decompression of the experimentally computed one-level transform of the original image (top of Figure 26.12b) for varying compression thresholds.

The horizontal resolution of the imager is limited only by maximum scan-out clock frequency for a given frame rate as is the case in conventional imagers. Area and power dissipation scale linearly with the horizontal imager size. In the vertical dimension, all pixels have to be sampled within the given frame period as set by the programmable spatial kernel with parameters H, V, and coefficients C as well as the imager resolution with parameters L, K, in Equations (26.1–26.3). When computing the DCT using sixty-four 8×8 blocks at 30 fps, the sensory processor is projected to yield a computational throughput of 4 GMACS when scaled to HDTV 1080i resolution. The throughput is based on a conservative quantizer sampling rate of 40 ksps and a pixel integration time of 5 ms. If a higher resolution in the vertical dimension is required, either the integration time has to be reduced or the ADC sampling rate has to increase.

26.8 CONCLUSIONS

We present a mixed-signal VLSI implementation of a digital CMOS imager computing block-matrix transforms on the focal plane for real-time video compression. The approach combines weighted spatial averaging and oversampling quantization in a single algorithmic $\Delta\Sigma$-modulated ADC cycle, making focal-plane computing an intrinsic part of the quantization process. The approach yields power dissipation lower than that of a conventional digital imager while the need for a peripheral DSP is eliminated. The experimental results obtained from a 0.35 µm 128 × 128-pixel CMOS prototype validate the utility of the design for large-scale focal-plane signal processing.

REFERENCES

1. C. Chen, Z. Yang, T. Wang, and L. Chen, A programmable VLSI architecture for 2-D discrete wavelet transform, *IEEE International Symposium on Circuits and Systems (ISCAS'00)*, 1, 619–622, 2000.
2. U. Sjostrom, I. Defilippis, M. Ansorge, and F. Pellandini, Discrete cosine transform chip for real-time video applications, *IEEE International Symposium on Circuits and Systems (ISCAS'90)*, 2, 1620–1623, 1990.
3. S.E. Kemeny, R. Panicacci, B. Pain, L. Matthies, and E.R. Fossum, Multiresolution image sensor, *IEEE Trans. Circuits and Syst. Video Technol.*, 7(4), 575–583, 1997.
4. Q. Luo, and J.G. Harris, A novel integration of on-sensor wavelet compression for a CMOS imager, *IEEE International Symposium on Circuits and Systems (ISCAS'02)*, Scottsdale, AZ, III-325–III-326, May 26–29, 2002.
5. S. Kawahito, M. Yoshida, M. Sasaki, K. Umehara, D. Miyazaki, Y. Tadokoro, K. Murata, S. Doushou, and A. Matsuzawa, A CMOS image sensor with analog two-dimensional DCT-based compression circuits for one-chip cameras, *IEEE J. Solid-State Circuits*, 32(12), 2030–2041, 1997.
6. V. Gruev and R. Etienne-Cummings, Implementation of steerable spatiotemporal image filters on the focal plane, *IEEE Trans. Circuits Syst. II*, 49(4), 233–244, 2002.
7. Z. Zhou, B. Pain, and E.R. Fossum, Frame-transfer CMOS active pixel sensor with pixel binning, *IEEE Trans. Electron. Devices*, 44(10), 1764–1768, 1997.
8. A. Bandyopadhyay, J. Lee, R. Robucci, and P. Hasler, A 80 µW/frame 104 × 128 CMOS imager front end for JPEG Compression, *IEEE International Symposium on Circuits and Systems (ISCAS'05)*, 5318–5321, 2005.
9. E. Artyomov, Y. Rivenson, G. Levi, and O. Yadid-Pecht, Morton (Z) scan based real-time variable resolution CMOS image sensor, *IEEE Trans. Circuits Syst. Video Technol.*, 15(7), 947–952, 2005.
10. E. Artyomov, and O. Yadid-Pecht, Adaptive multiple resolution CMOS active pixel sensor, *IEEE International Symposium on Circuits and Systems (ISCAS'04)*, 4, 836–839, 2004.
11. T. Hamamoto, K. Aizawa, Y. Egi, T. Hamamoto, M. Hatori, H. Maruyama, and J. Yamazaki, On sensor image compression, *IEEE Signal Processing Society Workshop on VLSI Signal Processing*, 61–69, 1995.
12. K. Aizawa, H. Ohno, Y. Egi, M. Hatori, and J. Yamazaki, Image sensor for compression and enhancement, *IEEE Trans. Circuits Syst. Video Technol.*, 7(3), 543–548, 1997.
13. U. Mallik, M. Clapp, E. Choi, G. Cauwenberghs, and R. Etienne-Cummings, Temporal change threshold detection imager, *IEEE International Solid-State Circuits Conference (ISSCC'05)*, 1, 362–363, 2005.
14. D. Leon, S. Balkir, K. Sayood, and M.W. Hoffman, A CMOS imager with pixel prediction for image compression, *IEEE International Symposium on Circuits and Systems (ISCAS'03)*, 4, 776–779, 2003.

15. A. Graupner, J. Schreiter, S. Getzlaff, and R. Schuffny, CMOS image sensor with mixed-signal processor array, *IEEE J. Solid-State Circuits*, 38(6), 948–957, 2003.

16. E.R. Fossum, CMOS image sensor: Electronic camera-on-a-chip, *IEEE Trans. Electron. Devices*, 44, 1689–1698, 1997.

17. M. Oren, C. Papageorgiou, P. Sinha, E. Osuna, and T. Poggio, Pedestrian detection using wavelet templates, *Computer Vision Pattern Recognition* 193–199, 1997.

18. A. Olyaei and R. Genov, Focal-plane spatially oversampling CMOS image compression sensor, *IEEE Trans. Circuits Syst. I*, 49(1), 26–34, 2007.

19. A. Olyaei, and R. Genov, Algorithmic delta-sigma modulated FIR filter, *IEEE International Symposium on Circuits and Systems (ISCAS'06)*, 4407–4410, Kos, Geece, May 21–24, 2006.

20. J. Robert, G.C. Temes, V. Valencic, R. Dessoulavy, and P. Deval, A 16-bit low-voltage CMOS A/D converter, *IEEE J. Solid-State Circuits*, SC-22, 157–163, 1987.

21. O.J.A.P. Nys and E. Dijkstra, On configurable oversampled A/D converters, *IEEE J. Solid-State Circuits*, 28(7), 736–742, 1993.

22. I. Galton and H.T. Jensen, Delta-sigma modulator based A/D conversion without oversampling, *IEEE Trans. Circuits Syst. II*, 42(12), 773–784, 1995.

23. G. Mulliken, F. Adil, G. Cauwenberghs, and R. Genov, Delta-sigma algorithmic analog-to-digital conversion, *IEEE International Symposium Circuits and Systems (ISCAS'02)*, Scottsdale, AZ, May 26–29, 2002.

24. A. Olyaei, and R. Genov, CMOS wavelet compression imager architecture, *IEEE CAS Emerging Technologies Workshop*, St. Petersburg, Russia, June 23–24, 2005.

25. A. Olyaei, and R. Genov, Mixed-signal CMOS Haar wavelet compression imager architecture, *Midwest Symposium on Circuits and Systems (MWSCAS05)*, Cincinnati, OH, August 7–10, 2005.

27 Unified Computer Arithmetic for Handheld GPUs

Byeong-Gyu Nam, Hyejung Kim, and Hoi-Jun Yoo

CONTENTS

27.1 INTRODUCTION

As the mobile electronics advances, handheld devices such as cell phones and personal digital assistants (PDAs) are evolving from text-based applications to various multimedia applications including real-time three-dimensional (3D) computer graphics. Figure 27.1 shows an example 3G system that contains a radiofrequency (RF) frontend, a baseband modem, an application processor for H.264 and 3D computer graphics, and peripherals. For these systems, power- and area-efficient design becomes a very critical issue due to their limited footprints and battery lifetime. The power consumption of each component in these systems can be found in Refs. [1,2] and the power budget allocated for the multimedia application processor is usually less than 300–400 mW [1–3]. This chapter gives an overview of the 3D computer graphics for handheld systems and introduces arithmetic schemes proposed to reduce power and area overheads of the handheld 3D graphics processing units (GPUs), which require various arithmetic operations.

Section 27.2 introduces the handheld 3D computer graphics as a background for this chapter and Section 27.3 discusses several arithmetic schemes used for the handheld GPUs to reduce hardware

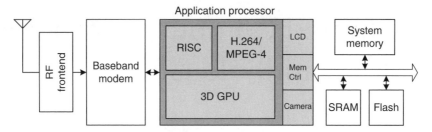

FIGURE 27.1   Example of 3G system architecture.

complexities. Based on these, we present the philosophy of unified arithmetic in Section 27.4 and propose a novel unified arithmetic unit to reduce area and power overheads of handheld graphics systems in Section 27.5. Finally, Section 27.6 summarizes this chapter.

27.2   HANDHELD 3D COMPUTER GRAPHICS

The realization of the real-time 3D graphics on handheld systems has been a challenging issue since the real-time 3D graphics applications inherently require huge computation power while the handheld systems can only provide limited amount of area and power budgets. Thus, the standard embedded graphics application programming interfaces (APIs) support the fixed point or floating point arithmetic with limited dynamic range to be efficiently implemented in software or hardware in the handheld systems. In this section, the basic real-time 3D graphics pipeline is explained and the standard APIs for handheld 3D graphics are introduced.

27.2.1   3D GRAPHICS PIPELINE

The 3D graphics pipeline is mainly composed of geometry and rendering stages. The geometry stage consists of operations like transformation, lighting, and clipping for vertex level processing. With the output of the geometry stage, the rendering stage performs triangle setup, rasterization, texture mapping, depth test, and blending operations for pixel-level processing. This graphics pipeline is illustrated in Figure 27.2.

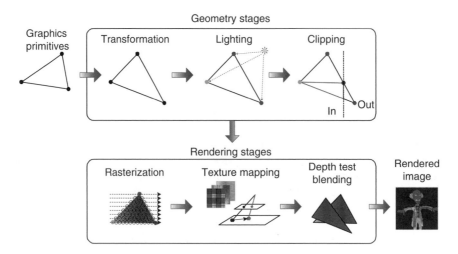

FIGURE 27.2   3D graphics pipeline.

27.2.1.1   Geometry Stage

The main operation of the geometry stage is the coordinate transformation and lighting for input vertices. In graphics pipeline, the input vertices are transformed from object coordinate space into the camera coordinate space using model-view transformation composed of translation, rotation, scaling, and shifting. Using combinations of these operations, the model-view transformation enables animation and reformation of the original model.

In camera coordinate, the lighting operation is carried out determining the intensity of each vertex. The intensity is determined from the interactions between material properties of the vertex and the light sources. The vertex intensity is a composite of ambient, diffuse, and specular lights as shown in Equation 27.1 [4]. The ambient light describes the background light, the diffuse light describes the light scattered equally in all directions, and the specular light describes the light that comes from a particular direction, and it tends to bounce off the surface in a preferred direction.

$$\text{Vertex intensity} = \sum_{i=0}^{n-1} \left(\begin{array}{l} \text{Ambient}_{\text{light}} \times \text{Ambient}_{\text{material}} + \\ (\max\{\boldsymbol{L} \times \boldsymbol{n}, 0\}) \times \text{Diffuse}_{\text{light}} \times \text{Diffuse}_{\text{material}} + \\ (\max\{\boldsymbol{s} \times \boldsymbol{n}, 0\})^{\text{shininess}} \times \text{Specular}_{\text{light}} \times \text{Specular}_{\text{material}} \end{array} \right)_i \qquad (27.1)$$

where

k_c, k_l, and k_q are attenuation coefficients
$\boldsymbol{L}$ is the unit vector of light direction
$\boldsymbol{n}$ is the unit normal vector at vertex
$\boldsymbol{s}$ is the unit normal vector of light vector and view vector

After lighting, each vertex is projected into the 3D screen space through perspective projection. In this space, the vertices outside the view frustum are clipped out and their coordinates are divided by their w-component to transform them into Euclidean space.

27.2.1.2   Rendering Stage

With the output of the geometry stage, the rendering stage performs triangle setup, rasterization, texture mapping, depth testing, and blending operations. The triangle setup operation sets up triangle parameters such as gradients of triangle edge and scan line required for the rasterization operation. The rasterization fills in the triangle with simple linear interpolation based on the parameters from triangle setup.

The texture mapping is applied to enhance the realism of 3D graphics scenes. This operation is to wrap a 3D model object with two-dimensional (2D) texture images, which were obtained by scanning the surface of the original 3D object in real space. Thus, texture mapping can easily represent the surface details such as surface color and roughness. The depth test is performed to remove hidden pixels in a polygon. It compares current depth value with the previous one in the depth buffer to decide whether the current pixel should be discarded or not. The blending is also applied to modulate the pixel colors from lighting and texture mapping or to represent translucent object by blending the new pixel color with the previous one on the 2D screen.

27.2.2   STANDARD APIS FOR HANDHELD 3D GRAPHICS

There are several standard 3D graphics APIs defined for resource-limited handheld systems such as cell phones, PDAs, and portable gaming consoles. These include the OpenGL-ES [5], the Direct3D Mobile [6], and the Java specification request 184 [7]. These are designed to better support the computational capabilities of the handheld devices, which often lack the floating point arithmetic units. Recently, the OpenGL-ES was defined for the embedded systems and there are several versions of its specification now. They were derived from the OpenGL [4], but tailored to the handheld systems

Fixed-function 3D graphics pipeline

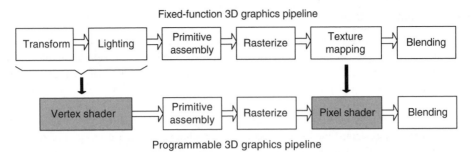

Programmable 3D graphics pipeline

FIGURE 27.3   Programmable graphics pipeline.

by dropping much functionality. The versions 1.0 and 1.1 support fixed point data types and fixed-function graphics pipelines for power- and area-efficient system designs. Even if this approach can reduce area and power overheads of the 3D graphics systems, its graphics effects are quite limited due to the fixed pipeline definition.

The version 2.0 is based on floating point arithmetic and introduces the programmability to the pipeline to allow more flexibility enabling various enhanced graphics effects. The programmable graphics pipeline is composed of programmable vertex shader, programmable pixel shader, and some fixed-function stages. The motivation behind the creation of these user-programmable shaders is the increasing configurability required by the continually evolving graphics APIs and thereby, programmable devices are necessary to support the combinatorial explosion of mode combinations [8].

The shaders substitute certain stages of graphics pipeline by programming of the shader instructions. The vertex shader substitutes the transformation and lighting operations in the geometry stage, while the pixel shader does the texture mapping in the rendering stage. However, the pixel shader is not limited to the texture mapping and extends the graphics pipeline to support per pixel lighting computation. Using this programmability, the vertex and pixel shaders support not only the fixed OpenGL lighting model, but various lighting models to provide more delicate graphics effects. The comparison between traditional fixed-function graphics pipeline and user-programmable pipeline using vertex and pixel shaders is shown in Figure 27.3.

For this programmability, the arithmetic operation set for these shaders should cover the operations that make the OpenGL lighting equation (27.1) and several other functions to support various lighting models. This includes not only additions and subtractions but various complicated vector and elementary functions [9].

27.3   COMPUTER ARITHMETIC FOR HANDHELD GPUs

Since the 3D computer graphics is extremely compute-intensive and data-parallel application, the GPUs adopt highly data-parallel architecture with single instruction multiple data (SIMD) arithmetic units. The GPU architecture is composed of the geometry and rendering pipeline stages as shown in Figure 27.4. It has fixed-function rasterizer, frame buffer operation unit, and two programmable stages incorporating the vertex and pixel shaders in geometry and rendering stages, respectively. The rasterizer interfaces the vertex and pixel shaders by setting up triangle parameters. The frame buffer unit processes several frame buffer operations like depth test and pixel-level blending operations. Basically, the programmable shaders are stream processors and they can substitute certain stages of graphics pipeline by assembly programming of its instructions. They have two main functional units, a four-way vector SIMD unit for vector multiplication and dot product and an elementary function unit for the functions like reciprocal and exponential functions as depicted in Figure 27.5.

In order to reduce computational complexity, the implementation of the low-power GPUs has been adopting various kinds of approximation schemes exploiting the characteristics of the handheld

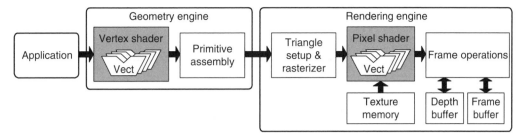

FIGURE 27.4   Overall architecture of GPU.

3D graphics [10]. Since the target applications of the handheld 3D graphics are not the cinematic photorealistic rendering, but the 3D gaming and user interfaces within a limited screen size, the limitations of the human visual system can be exploited more aggressively than for the conventional 3D graphics. Thus a reasonable amount of computation error is allowable without introducing noticeable artifacts. The computer arithmetic schemes used for the low-power GPUs exploiting these properties are discussed in this section.

27.3.1   FLOATING POINT ARITHMETIC

In 3D graphics, real number representation is required to represent the attributes of 3D models and the conventional single-precision floating point number representation is used for this purpose in Ref. [11]. The vertex shader in this work incorporates the vector SIMD unit for vector operations such as vector addition, multiplication, and dot product and the elementary function unit for reciprocal, reciprocal-square-root, logarithm (log2), and exponential (exp2). These units operate at a relatively lower operating frequency of 166 MHz. The vector unit supports single-cycle throughput and latency for the floating point multiplication and single-cycle throughput with four-cycle latency for dot product due to the cascaded four-input floating point adder after four multipliers. The elementary function unit supports single-cycle throughput with seven-cycle latency by interpolating the curves for elementary functions using the second-order minimax approximation proposed in Ref. [12]. The second-order minimax approximation can be performed by evaluating

$$f(x) \cong a_0 + a_1 x + a_2 x^2 \tag{27.2}$$

where the coefficients a_0, a_1, and a_2 are obtained for each approximation subinterval. The architecture for the elementary function unit evaluating this expression is depicted in Figure 27.6. It uses four of

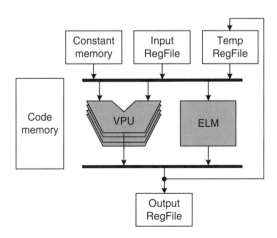

FIGURE 27.5   Block diagram of programmable shaders.

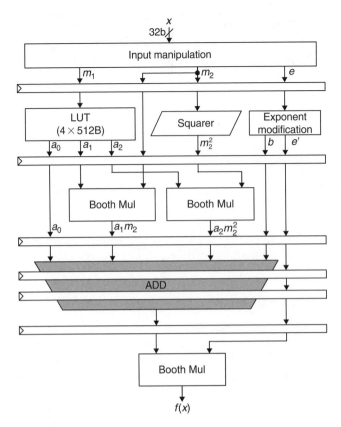

FIGURE 27.6 Overall architecture of elementary function unit. (From Kim, D., Chung, K., Yu, C.-H., Kim, C.-H., Lee, I., Bae, J., Kim, Y.-J., Park, J.-H. Kim, S. Park, Y.-H., Seong, N.-H., Lee, J.-A., Park, J., Oh, S., Jeong, S.-W., Kim, L.-S., *IEEE J. Solid-State Circuits*, 41, 71, 2006.)

512-byte lookup table per each function and the other datapaths like squarer, multiplier, and four-input adder are shared among four elementary function interpolations. The throughput and latency for each operation are listed in Table 27.1.

TABLE 27.1
The Latency/Throughput of Operation Set

Type	Operations	Latency (Cycles)	Throughput (Result/Cycles)
Vector operations	Vector-SIMD addition	2	1
	Vector-SIMD multiplication	1	1
	Vector dot product	4	1
Elementary functions	Reciprocal	7	1
	Reciprocal square root	7	1
	Logarithm (log2)	7	1
	Exponential (exp2)	7	1

Source: From Kim, D., Chung, K., Yu, C.-H., Kim, C.-H., Lee, I., Bae, J., Kim, Y.-J., Park, J.-H. Kim, S. Park, Y.-H., Seong, N.-H., Lee, J.-A., Park, J., Oh, S., Jeong, S.-W., Kim, L.-S., *IEEE J. Solid-State Circuits*, 41, 71, 2006.

TABLE 27.2

The Latency/Throughput of Operation Set

Type	Operations	Latency (Cycles)	Throughput (Result/Cycles)
Vector operations	Vector-SIMD addition	1	1
	Vector-SIMD multiplication	4	1
	Vector-SIMD multiply-and-add (MAD)	4	1
	Vector dot product	5	1/2
	Transform	7	1/4
Elementary functions	Reciprocal	6	1/3
	Reciprocal square root	8	1/5

Source: From Sohn, J.-H., Woo, J.-H., Lee, M.-W., Kim, H., Woo, R., and Yoo, H.-J., *IEEE J. Solid-State Circuits*, 41, 1081, 2006.

27.3.2 FIXED POINT ARITHMETIC

Instead of the floating point numbers, the fixed point number representation is used for the real number representation in Ref. [13]. Since simple integer datapath can be used for the fixed point number representation, fixed point arithmetic unit can operate at a higher clock rate with reduced power and area consumptions.

The vertex shader in this work includes a fixed point four-way 128 bit vector processing unit that is responsible for SIMD addition and multiplication and a 32 bit elementary function unit responsible for reciprocal (RCP) and reciprocal square root (RSQ). It operates at 200 MHz operating frequency and shows single-cycle throughput with four-cycle latency for the vector multiplication and three-cycle throughput of one result every three cycles and six-cycle latency for the RCP. The throughput and latency for supported operations are listed in Table 27.2.

The matrix-vector multiplication of 4×4-matrix with four-element vector, required for object transformation in 3D graphics, is implemented with the first multiply (MUL) and following three multiply-accumulate (MAC) operations, as illustrated in Figure 27.7, to reduce the transformation latency on a unit optimized for the MAC operations rather than the dot product. In order to resolve data dependency between MUL and following MACs, intermediate values of multiplier are bypassed for accumulation as shown in Figure 27.8. In this architecture, two stages of 32×16 Booth multipliers are used instead of a single 32×32 multiplier in order to increase the operating frequency. Therefore, the accumulation is also performed throughout two stages. This four-stage pipelined fixed point multiplier can operate at 30% higher operating frequency with 17% less power consumption than six-stage pipelined single-precision floating point multiplier.

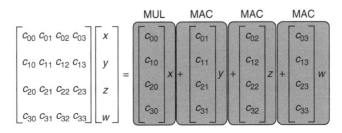

FIGURE 27.7 4-Cycle matrix-vector multiplication.

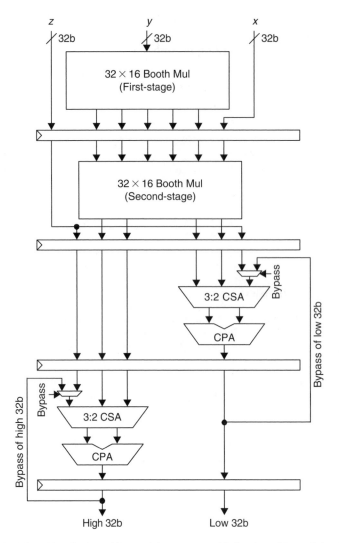

FIGURE 27.8 Multiply-add unit supporting matrix-vector multiplication. (From Sohn, J.-H., Woo, J.-H., Lee, M.-W., Kim, H., Woo, R., and Yoo, H.-J., *IEEE J. Solid-State Circuits*, 41, 1081, 2006.)

The small screens of mobile terminals allow that this 32 bit fixed point representation system can generate scenes with tolerable accuracy loss compared with the floating point number system (FLP). With this approach, 36% energy consumption is reduced on an average.

27.4 PHILOSOPHY OF UNIFIED COMPUTER ARITHMETIC

The arithmetic capabilities of modern computers are surprisingly limited. They do not even support the basic four-function calculations of addition, subtraction, multiplication, and division; their capabilities are usually limited to addition, subtraction, and multiplication. Even the computers with the four-function capability incorporate dedicated arithmetic units like adder, multiplier, and divider instead of a single arithmetic unit, which is capable of all these calculations. This scenario becomes worse when various kinds of arithmetic operations are required in applications like 3D graphics since so many different arithmetic units cannot be incorporated for each operation and thereby software emulations should be adopted, which lead to increased operational latency and power consumption.

For example, compared to the required operations in Ref. [9], the operation sets of the works discussed in previous section are quite limited just including the four-function arithmetic and a few elementary functions like square root, logarithm, and exponential. Moreover, vector and elementary functions are implemented separately on different functional units: vector SIMD and elementary function units. To cope with this issue, the best solution would be that all required arithmetic operations are supported in hardware unit and these are unified into a single arithmetic unit. Historically, the elegance of the unified theories in sciences has attracted people to explore important scientific discoveries, which made significant impacts on many areas of science and technology. Likewise, the unified computer arithmetic can also play an important role in the progress of the science of computer arithmetic.

There have been several works on the unified computer arithmetic. The simplest one can be the unification of addition and subtraction, in which the adder circuit is also used for the subtraction by simply negating the subtrahend. Also there are approaches for more complex operations. One of the early studies is the CORDIC algorithm [14]. It proposed a unified framework to implement various arithmetic operations like trigonometric, hyperbolic, exponential, logarithm, multiplication, division, and square root in a single arithmetic unit in terms of coordinate rotations. However, due to its iterative nature, its performance is quite limited. In Ref. [15], a table-based approximation scheme using rectangular multipliers was proposed for fast function evaluations. It implemented division, logarithm, reciprocal square root, arctangent, sine, and cosine functions in a single arithmetic unit. However, this approach required quite large table size of 1 Mbits. Recently, other table-based function approximation scheme using minimax quadratic interpolation algorithm is proposed in Ref. [16]. It supports reciprocal, square-root, reciprocal square-root, logarithm, exponential, and sine functions with quite small tables of 38 Kb and thereby reduces area overheads. However, all these works mainly focused on the elementary functions and did not take into account the unification of vector and elementary functions in a single arithmetic unit.

27.5 UNIFIED VECTOR AND ELEMENTARY FUNCTION UNIT

In this section, a unified arithmetic unit for various arithmetic operations used in the 3D graphics shaders is proposed Ref. [17]. It unifies vector addition, subtraction, multiplication, division, square root, and dot product with elementary functions like trigonometric functions, power with two variables, and logarithm to arbitary constant bose in a single arithmetic unit. This unit exploits the logarithmic arithmetic to reduce arithmetic complexities of these operations so that it can achieve power- and area-efficient unification.

27.5.1 FIXED POINT HYBRID NUMBER SYSTEM

It is well known that the logarithmic number system (LNS) can simplify various arithmetic operations [18]. However, the addition and the subtraction become more complicated in LNS, where they require nonlinear function evaluations as shown in Table 27.3. A hybrid approach of the LNS and FLP was introduced to solve this problem in Ref. [19], where the addition and subtraction are performed in FLP while other operations are done in the LNS. However, the use of FLP still requires the complicated and power-consuming floating point addition and subtraction, which require denormalization for exponent alignment and normalization of the final result.

In this work, we propose the fixed point hybrid number system (FXP-HNS), exploiting only the strong points of the fixed point number system (FXP) and the LNS: simple addition/subtraction in FXP and other complicated operations in LNS. Although the conversion error between the fixed point and logarithmic numbers carries certain amount of computation error, the FXP-HNS can be useful for the handheld 3D graphics systems that have small screens. Hence, our approach results in simple implementation of complicated arithmetic functions with small size and low power consumption for the handheld systems.

TABLE 27.3

Operations in Ordinary and Logarithmic Arithmetic

Operation	Ordinary Arithmetic	Logarithmic Arithmetic
Multiplication	$x \times y$	$X + Y$
Division	$x \div y$	$X - Y$
Square root	$\sqrt{x}$	$X \gg 1$
Addition	$x + y$	$X + \log_2(1 + 2^{Y-X})$
Subtraction	$x - y$	$X + \log_2(1 - 2^{Y-X})$

Source: From Nam, B.-G., Kim, H., and Yoo, H.-J., *IEEE J. Solid-State Circuits*, 42, 1767, 2007. With permission.

In FXP-HNS, fixed point numbers are represented as 32 bit 2's complement representation with fraction point. It has programmable precision from Q32.0 to Q1.31 to cover the different precision requirements of shader programs. The 32 bit logarithmic numbers in FXP-HNS are represented with 1 bit sign and zero encodings, 6 bit integer, and 24 bit fraction parts. Since the logarithm for zero and negative value is not defined mathematically, the absolute value is used for logarithmic conversion and the zero and sign are encoded into additional bits. The 6 bit integer covers the entire range of the corresponding fixed point number and the remaining 24 bits are used for the fractional part.

27.5.2 NUMBER CONVERTERS

For the conversion between the fixed point and logarithmic numbers in the FXP-HNS, the logarithmic and antilogarithmic converters are proposed. These are based on piecewise linear approximation for the low-power and small-area implementation.

27.5.2.1 Logarithmic Converter

When x is a 32 bit FXP input, the x and its logarithm can be represented as Equations 27.3 and 27.4, respectively:

$$x = 2^k(1 + f) \tag{27.3}$$

$$\log_2 x = k + \log_2(1 + f) \tag{27.4}$$

The k in $\in \{-31, -30, \ldots, 31\}$ is the characteristic of the logarithm and $\log_2(1 + f)$ is the fractional part in the range of $[0, 1)$. In piecewise linear approximation, the nonlinear term $\log_2(1 + f)$ is approximated as

$$\log_2(1 + f) \cong a_i f + b_i \tag{27.5}$$

where a_i and b_i are the approximation coefficients defined for each approximation region i in the range of $1 \leq (1 + f) < 2$. Here, we divide the input $(1 + f)$ into finer subdivisions around the input of 1, since the error increases as the input value gets closer to 1. This results in 24 approximation regions and shows less than 0.8% conversion error. This logarithmic conversion scheme and its error graph are shown in Figure 27.9.

The logarithmic converter based on this scheme is shown in Figure 27.10. The input x is converted into its absolute value through the ABS block. The LOD block computes the characteristic value k' by detecting the leading one bit. It takes the bits following the leading one as the fraction value f. The 14 most significant bits (MSBs) of f are used for addressing the lookup table in the APP block. The APP block approximates the nonlinear fractional part $\log_2(1 + f)$ with $a_i f + b_i$ and its multiplication is implemented by summation of shift terms. Therefore, the APP carries out the summation of five

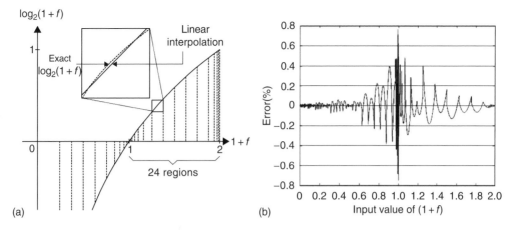

FIGURE 27.9 Proposed logarithmic conversion scheme: (a) logarithmic conversion scheme and (b) error graph. (From Nam, B.-G., Kim, H., and Yoo, H.-J., *IEEE J. Solid-State Circuits*, 42(8), 1767, 2007. With permission.)

terms using the carry save adder (CSA) tree and carry propagation adder (CPA). Since this converter receives input values with variable precision of $Qm.n$, the actual characteristic value k is computed by $k' - n$. The conversion result is obtained by packing the *zero, sign*, characteristic value, and the approximated fraction value into a 32 bit logarithmic number.

27.5.2.2 Antilogarithmic Converter

When X is a logarithmic number, the LNS representation of the X is composed of integer part k and fractional part f. The antilogarithmic conversion of x can be represented as

$$2^X = 2^k 2^f \qquad (27.6)$$

In this equation, 2^k is just a shift operation and the nonlinear term 2^f can be approximated by piecewise linear interpolation as given in the following equation:

$$2^f \cong a_i f + b_i \qquad (27.7)$$

where a_i and b_i are the approximation coefficients defined for each approximation region i in the range of $0 \leq f < 1$. We divide the f into 16 even approximation regions since the error for the antilogarithmic conversion is evenly spread over the entire region. This antilogarithmic conversion scheme shows less than 0.02% conversion error. The antilogarithmic conversion scheme and its error graph are shown in Figure 27.11.

The antilogarithmic converter is shown in Figure 27.12. The input logarithmic number is split into the characteristic value k and fractional value f. The four MSBs of f are used for addressing the lookup table in the APP block. The APP block approximates the nonlinear part 2^f with $a_i f + b_i$ and its multiplication is also implemented by summation of shift terms. The summation of five terms is carried out using the CSA tree and CPA. The enabled *sign* bit negates these terms for negative output. The shift amount is computed by $n - 24 + k$ since the final result should be adjusted to the format with the input precision, i.e. $Qm.n$. Finally, the enabled *zero* bit selects the output of zero.

27.5.3 FXP-HNS Unified Arithmetic Unit

A unified arithmetic unit is presented based on FXP-HNS. It unifies the vector arithmetic operations with elementary functions in a single four-way arithmetic unit. It has fully pipelined architecture with

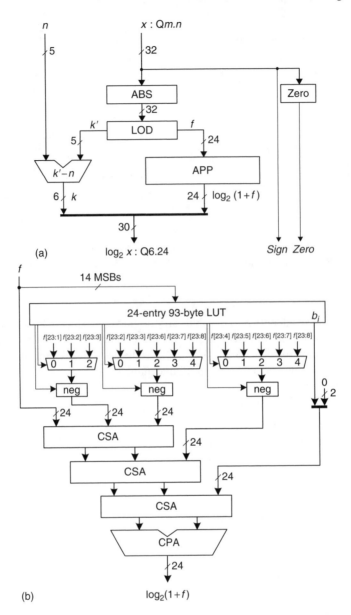

FIGURE 27.10 Block diagram of logarithmic converter: (a) Logarithmic converter and (b) approximation block. (From Nam, B.-G., Kim, H., and Yoo, H.-J., *IEEE J. Solid-State Circuits*, 42, 1767, 2007. With permission.)

four pipeline stages so that it supports single-cycle throughput for all of the supported operations with maximum four-cycle latency. In this architecture, the input operands are converted into the logarithmic domain, where complicated operations are reduced into simple operation, and the results are restored into fixed point domain, where simple additions and subtractions are carried out.

The overall block diagram of the proposed unit is shown in Figure 27.13. The eight 32 bit fixed point input operands, x_i and y_i in each channel, are converted into 32 bit logarithmic number through logarithmic converters at the first stage (LOGC stage). The second stage (LNS stage) contains a programmable CSA tree for a single 30b × 32b multiplier or four-way 30b × 8b multipliers according to target elementary functions. The final summation of the carry and partial sum from the CSA tree

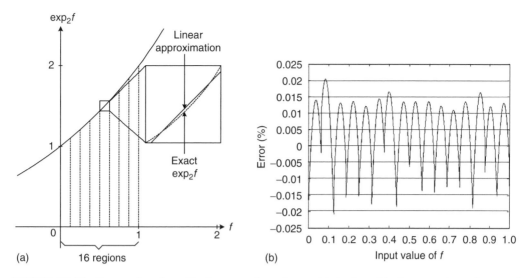

FIGURE 27.11 Proposed antilogarithmic conversion scheme: (a) Antilogarithmic conversion scheme and (b) error graph. (From Nam, B.-G., Kim, H., and Yoo, H.-J., *IEEE J. Solid-State Circuits*, 42, 1767, 2007. With permission.)

can be carried out with the four 38 bit carry propagation adders in the LNS stage. These are also used with four 1 bit shifters for the vector arithmetic operations. The third stage (ALOGC stage) converts the 38 bit computation results from the four adders in logarithmic number into 32 bit fixed point numbers with the given precision of input operand through antilogarithmic converters. The final stage (ADD stage) consists of a CPA tree that can be programmed into a single adder tree with five 32 bit input operands or independent four-way two-input 32 bit adders according to the target operations.

27.5.3.1 Vector Operations

In conventional architecture, only vector-SIMD multiplication, MAD, and vector dot product are implemented for the vector operations [11,13]. In our proposed architecture, the division and square root are also defined as vector-SIMD operations since it is useful for vector-SIMD $x \div \sqrt{y}$ calculation used in vector normalizations required for 3D graphics lighting equation. The multiplication, division, square root, and MAD can be represented by a single generic operation defined as the following expression:

$$(x_i \otimes y_i^p \oplus z_i)_{i \in \{0,1,2,3\}} \quad \text{where } \otimes \in \{\times, \div\}, \oplus \in \{+, -\}, p \in \{0.5, 1\} \tag{27.8}$$

For example, operations such as $x \times y, x \div y, x \times y + z$, and $x \div \sqrt{y} - z, \ldots$ can be represented with this generic operation. This expression can be converted into the expression in FXP-HNS like as in the following equation:

$$(2^{(\log_2 x_i) \oplus (\log_2 y_i \gg q)} \oplus z_i)_{i \in \{0,1,2,3\}} \quad \text{where } \oplus \in \{+, -\}, q \in \{0, 1\} \tag{27.9}$$

According to the expression (27.9), multiplication, division, square, and square root can be converted into addition, subtraction, left shift, and right shift in the logarithmic number domain, respectively. In this architecture, the square root can be computed in conjunction with multiplication or division without any additional cycles. The diagram for this vector-SIMD unit is also shown in Figure 27.13. In this figure, the LNS stage is augmented with a CSA tree for elementary functions. The dot product

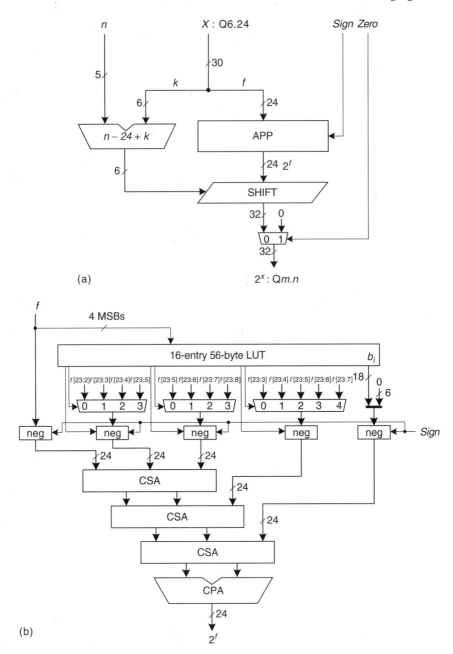

FIGURE 27.12 Block diagram of antilogarithmic converter: (a) antilogarithmic converter and (b) approximation block. (From Nam, B.-G., Kim, H., and Yoo, H.-J., *IEEE J. Solid-State Circuits*, 42, 1767, 2007. With permission.)

in FXP-HNS of expression (27.10) is also implemented using the vector-SIMD unit for vector element multiplication:

$$\sum_{i=0}^{i=3} x_i \times y_i = \sum_{i=0}^{i=3} 2^{\log_2 x_i + \log_2 y_i} \tag{27.10}$$

For the summation of the multiplication results in (27.10), the four-way two-input 32 bit adders used for the vector-SIMD MAD operation are programmed into a single five-input 32 bit adder tree as

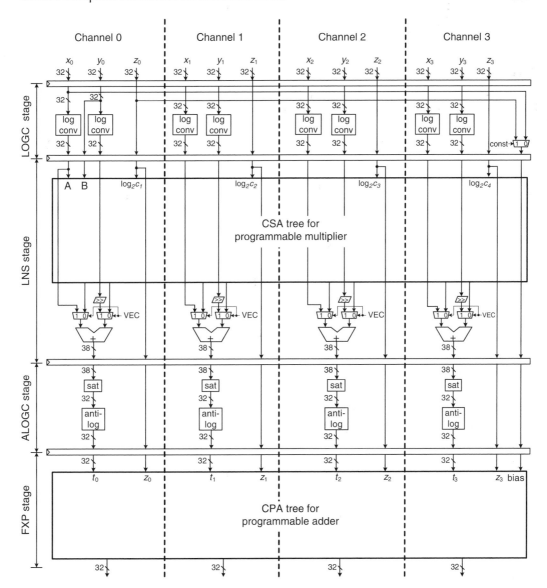

FIGURE 27.13 Unified arithmetic unit in FXP-HNS. (From Nam, B.-G., Kim, H., and Yoo, H.-J., *IEEE J. Solid-State Circuits*, 42, 1767, 2007. With permission.)

shown in Figure 27.14. In this case, the *bias* port is disabled, but it will be used for trigonometric operations.

27.5.3.2 Elementary Functions

The elementary functions such as power of arbitrary exponent, logarithm to arbitrary base, trigonometric, inverse trigonometric, hyperbolic, and inverse hyperbolic functions are unified with vector operations.

The use of Taylor series expansions leads to the unification of trigonometric, hyperbolic, and their inverse functions with the vector operations. For the first five terms of the Taylor series computation, a new generic operation is defined as

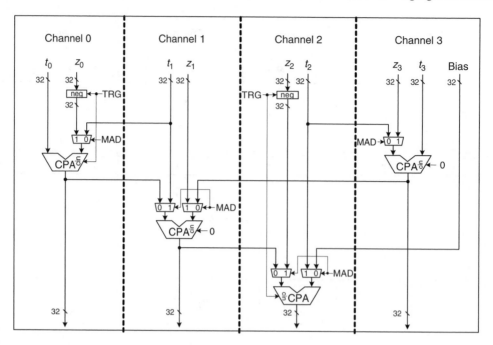

FIGURE 27.14 CPA tree for programmable adder. (From Nam, B.-G., Kim, H., and Yoo, H.-J., *IEEE J. Solid-State Circuits*, 42, 1767, 2007. With permission.)

$$c_0 x^{k_0} \oplus c_1 x^{k_1} \oplus c_2 x^{k_2} \oplus c_3 x^{k_3} \oplus c_4 x^{k_4} \tag{27.11}$$

where

$\oplus \in \{+, -\}$

c_i and k_i are positive real and integer constants, respectively

For example, Taylor series like $\sin x \simeq x - x^3/3! + x^5/5! - x^7/7! + x^9/9!$ and $\cos x \simeq 1 - x^2/2! + x^4/4! - x^6/6! + x^8/8!$ can be represented by this generic operation. The computation of the powers required for this operation can be converted into multiplications in logarithmic number domain by transforming (27.11) into FXP-HNS expression as

$$c_0 x^{k_0} \oplus 2^{(\log_2 c_1 + k_1 \times \log_2 x)} \oplus 2^{(\log_2 c_2 + k_2 \times \log_2 x)} \oplus 2^{(\log_2 c_3 + k_3 \times \log_2 x)} \oplus 2^{(\log_2 c_4 + k_4 \times \log_2 x)} \tag{27.12}$$

where

$\oplus \in \{+, -\}$

$\log_2 c_i$ and k_i are real and positive integer constants, respectively

Each term $2^{(\log_2 c_i + k_i \times \log_2 x)}$ in (27.12) requires an addition and a multiplication in logarithmic number domain with constant $\log_2 c_i$ and x converted into logarithmic number. The first term $c_0 x^{k_0}$ in (27.12) is not converted into the logarithmic number domain since the first term of the Taylor series expansion tends to be just a constant or the input x. For the other four terms, a 30b × 8b radix-4 Booth multiplier is sufficient since the k_i in each term is small integer that can be represented within 8 bits. Therefore, the multiplications required for (27.12) can be computed using four 30b × 8b multipliers. Figure 27.15 shows that each 30b × 8b CSA tree has an extra input port to add the coefficient $\log_2 c_i$, which is presented as an input operand to each channel. The four-way 38 bit CPAs to add the carry and partial sum of Q14.24 from each CSA tree are shared with the four-way adders for the vector arithmetic operations in the LNS stage. The 14 bit integers are saturated into 8 bit values by testing

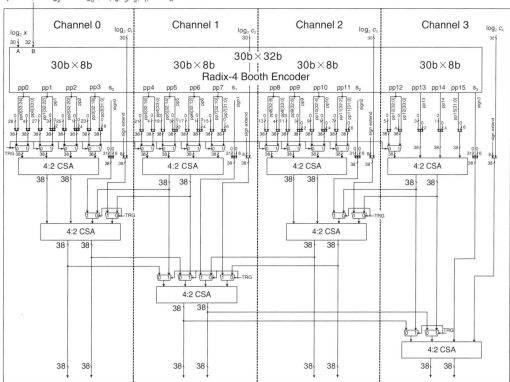

FIGURE 27.15 CSA tree for programmable multiplier. (From Nam, B.-G., Kim, H., and Yoo, H.-J., *IEEE J. Solid-State Circuits*, 42, 1767, 2007. With permission.)

the overflow or underflow conditions. The saturated results of Q8.24 are converted into 32 bit fixed point numbers through antilogarithm converters.

The $\oplus$ operations in (27.12) can be computed with the five-input adder tree shown in Figure 27.14 by programming each CPA into an adder or a subtractor according to the target Taylor series. The first term of the Taylor series is directly fed into this adder tree through the *bias* port. The final result can be obtained from the channel 2 of the ADD stage.

The power function can be converted into the multiplication in logarithmic number domain according to the equation

$$x^y = 2^{(y \times \log_2 x)} \tag{27.13}$$

Therefore, a 32b $\times$ 32b multiplier is required in the LNS stage. As shown in Figure 27.15, this multiplier can be obtained by programming the four 30b $\times$ 8b multipliers used for the trigonometric functions into a single 30b $\times$ 32b multiplier. Since y is in the range of $[0, 128]$ in specular lighting [4], for which the power function is mainly used, it has the format of Q8.24. Therefore, the CSA tree for the 30b $\times$ 32b multiplier computes only 38 MSBs, truncating the 24 least significant bits of fractional part, to make 24 bit fraction value for the antilogarithmic converter. These are added using the 38 bit CPA used for the vector arithmetic operations in channel 3. The multiplication result is saturated to Q8.24 by testing the overflow or underflow conditions. The saturated result is converted into 32 bit fixed point format through the antilogarithmic converter. The final result can be obtained from the ALOG stage of channel 3.

The logarithm to arbitrary constant base can be computed by multiplying a constant with the binary logarithm, according to the equation

$$\log_b x = k \times \log_2 x \quad \text{where } k = 1/\log_2 b \tag{27.14}$$

The multiplication of the constant k can be done using the 30b × 32b multiplier used in power function. The 32 bit constant k of Q8.24 should be given as an input operand. The final result can also be obtained from the channel 3 of the ALOG stage. Figure 27.16 shows the complete block diagram of the proposed arithmetic unit.

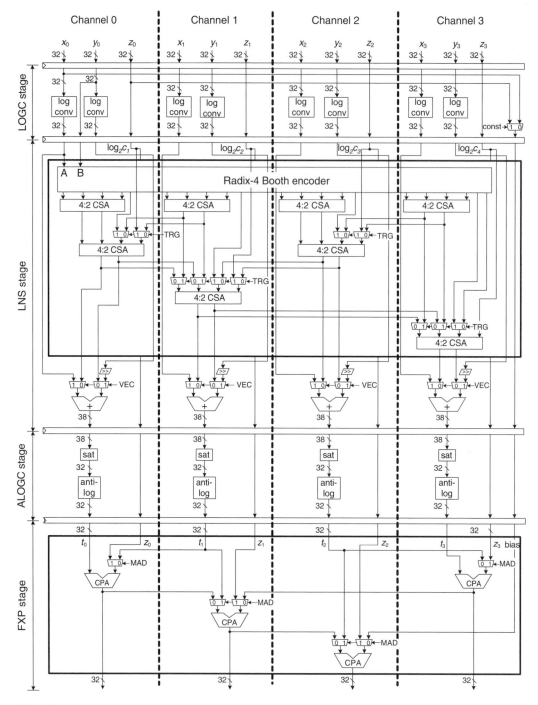

FIGURE 27.16   Complete diagram of the FXP-HNS unified arithmetic unit.

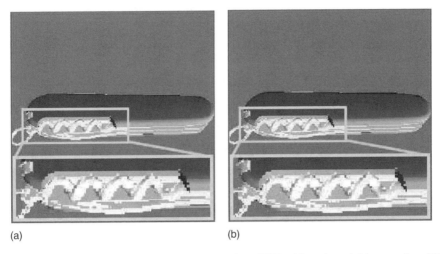

(a)   (b)

FIGURE 27.17 Comparison of TnL effects: (a) scene from FXP arithmetic and (b) scene from FXP-HNS arithmetic. (From Nam, B.-G., Kim, H., and Yoo, H.-J., *IEEE J. Solid-State Circuits*, 42, 1767, 2007. With permission.)

27.5.4 EVALUATIONS

The proposed function unit has been simulated in 3D graphics software environment with FXP and FXP-HNS libraries. The test 3D graphics scenes are shown in Figure 27.17 to show the reliability of the proposed arithmetic unit. The test model in Figure 27.17 consists of 5878 polygons and the test screen size is 512×512. In this figure, the OpenGL TnL operation [4] is tested to show the geometry stage effects. The inbox shows a zoomed image for the accuracy comparison. Unnoticeable error can be found between two images.

A test chip is fabricated by one-poly five-metal 0.18 CMOS technology. A micrograph of 93K gate chip is shown in Figure 27.18, lined out with the regions for each pipeline stage. Its core size is 2.9 mm². The pipeline registers are clock gated by the control of each operation to reduce switching

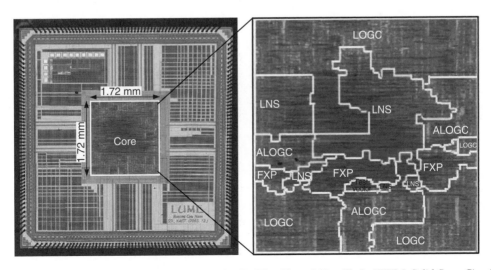

FIGURE 27.18 Chip micrograph. (From Nam, B.-G., Kim, H., and Yoo, H.-J., *IEEE J. Solid-State Circuits*, 42, 1767, 2007. With permission.)

TABLE 27.4

The Power/Latency of Operation Set

Type	Operations	Power (mW)	Latency/Throughput (Cycles)
Vector operations	Vector-SIMD multiply	8.03	3/1
	Vector-SIMD multiply-with-sqrt	8.1	3/1
	Vector-SIMD divide	8.15	3/1
	Vector-SIMD divide-by-sqrt	8.3	3/1
	Vector-SIMD MAD	10.1	4/1
	Vector dot product	9.8	4/1
Elementary functions	Power	10.2	3/1
	Logarithm	11.8	3/1
	Sine	15.2	4/1
	Cosine	15.1	4/1
	Arctangent	15.3	4/1

Source: From Nam, B.-G., Kim, H., and Yoo, H.-J., *IEEE J. Solid-State Circuits*, 42, 1767, 2007. With permission.

power consumption. Table 27.4 summarizes the power consumption and latency of each operation. The pipeline operates at 210 MHz with maximum 15.3 mW power consumption at 1.8 V. The shmoo plot is given in Figure 27.19.

This work is compared with other chip implementations reported in Refs. [11,13]. For the comparison with others, we selected the vector-SIMD multiply-and-add (VMAD) operation, the common operation among the compared works and useful for the 3D geometry transformation. Table 27.5 shows the comparison results.

As the processing speed gets higher, the area and power consumption also increase. Since the area and power consumption are also important factors for handheld devices, we use the following figure-of-merit (FoM) (Equation 27.15) for the comparison with others. It takes into account the area and power consumptions and the accuracy also to evaluate the effect of errors from our chip:

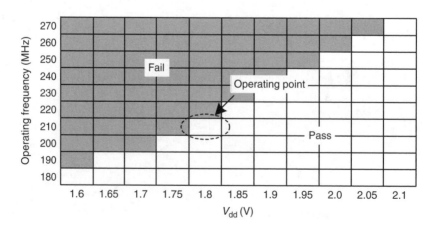

FIGURE 27.19 Shmoo plot. (From Nam, B.-G., Kim, H., and Yoo, H.-J., *IEEE J. Solid-State Circuits*, 42, 1767, 2007. With permission.)

TABLE 27.5

Comparison with Floating Point/Fixed Point Designs

Reference	Performance (VMAD/s)	Accuracy (%)	Power (mW)	Area (mm²)	FoM (VMAD.%/s/mW/mm²)	Technology (μm)
[11]	166M	100	N.A.	N.A.	N.A.	0.13
[13]	200M	100	20.1	3.15	315.88M	0.18
This work	210M	98.25	10.1	372	690.14M	0.18

$$\text{FoM} = \frac{\text{Performance (VMAD/s)} \times \text{Accuracy (\%)}}{\text{Power (mW)} \times \text{Area (mm}^2)} \tag{27.15}$$

According to this FoM, this work shows 2.18 times higher value in spite of the richer operation set supported than others.

27.6 CONCLUSIONS

In the history of computer arithmetic, the unified arithmetic schemes have been playing an important role in the progress of this field. Especially, the unification becomes important for power- and area-efficient designs in the resource-limited handheld systems. Although the handheld 3D graphics require various vector and elementary function computations, previous works on low-power arithmetic for handheld GPUs were based on separated vector and elementary function units.

With the extent of the unified computer arithmetic, we proposed a unified arithmetic unit for various vector and elementary functions required in handheld GPUs. By exploiting the logarithmic arithmetic, the FXP-HNS is newly proposed for power- and area-efficient unification and single-cycle throughput implementation of all supported operations. Novel power- and area-efficient logarithmic and antilogarithmic conversion schemes are proposed for this number system with relatively low conversion errors. Evaluation results show that the proposed FXP-HNS unified arithmetic unit is useful for handheld 3D graphics systems with limited screen size even though it carries little computation errors. Compared with others, our work achieves 2.18 times improvement in terms of the proposed FoM.

Although the application of our work is confined to the 3D computer graphics especially for the handheld systems with limited screen size, it can be extended to more general applications by exploring the schemes to improve the conversion accuracies with reasonable increase of power and area overheads.

REFERENCES

1. M.A. Viredaz and D.A. Wallach, Power evaluation of a handheld computer, *IEEE Micro.*, 66–74, 2003.
2. T. Simunic, L. Benini, P. Glynn, and G. De Micheli, Event-driven power management, *IEEE Trans. Computer-Aided Design Integr. Circuits*, 20(7), 840–857, 2001.
3. R. Woo, S. Choi, J.-H. Sohn, S.-J. Song, and H.-J. Yoo, A 210-mW graphics LSI implementing full 3-D pipeline with 264 Mtexels/s texturing for mobile multimedia applications, *IEEE J. Solid-State Circuits*, 39(2), 358–367, 2004.
4. OpenGL ARB, *OpenGL Programming Guide*, 3rd edn. Addison Wesley, Boston, MA, 1999.
5. Khronos Group, OpenGL-ES 2.0, available at http://www.khronos.org.
6. Microsoft Corporation, *Direct3D Mobile*, available at http://msdn2.microsoft.com/en-us/library/Aa452478.aspx.
7. Java Community Process, *JSR 184*, available at http://www.jcp.org/en/jsr/detail?id= 184.

8. E. Lindholm, M.J. Kilgard, and H. Moreton, A user-programmable vertex engine, *Proc. SIGGRAPH* 2001, pp. 149–158, August 2003.

9. Khronos Group, *The OpenGL-ES Shading Language*, available at http://www.khronos.org.

10. D. Drisu, S. Vassiliadis, S. Cotofana, and P. Liuha, Low cost and latency embedded 3D graphics reciprocation, *Proc. IEEE Int. Symp. Circuits and Systems*, Vancouver, Canada, May 2004.

11. D. Kim, K. Chung, C.-H. Yu, C.-H. Kim, I. Lee, J. Bae, Y.-J. Kim, J.-H. Park, S. Kim, Y.-H. Park, N.-H. Seong, J.-A. Lee, J. Park, S. Oh, S.-W. Jeong, and L.-S. Kim, An SoC with 1.3 Gtexels/s 3-D graphics full pipeline for consumer applications, *IEEE J. Solid-State Circuits*, 41(1), 71–84, 2006.

12. J. Muller, Partially rounded small-order approximations for accurate, hardware-oriented, table-based methods, in *Proc. IEEE Int. Symp. Computer Arithmetic*, Santiago de Compostela, Spain, June 2003.

13. J.-H. Sohn, J.-H. Woo, M.-W. Lee, H. Kim, R. Woo, and H.-J. Yoo, A 155-mW 50-Mvertices/s graphics processor with fixed-point programmable vertex shader for mobile applications, *IEEE J. Solid-State Circuits*, 41(5), 1081–1091, 2006.

14. J.S. Walther, A unified algorithm for elementary functions, *Proc. Spring Joint Computer Conf.*, pp. 379–385, 1971.

15. W.F. Wong and E. Goto, Fast hardware-based algorithms for elementary function computations using rectangular multipliers, *IEEE Trans. Computers*, 43(3), 278–294, 1994.

16. J.-A. Pineiro, S.F. Oberman, J.-M. Muller, and J.D. Bruguera, High-speed function approximation using a minimax quadratic interpolator, *IEEE Trans. Computers*, 54(3), 304–318, 2005.

17. B.-G. Nam, H. Kim, and H.-J. Yoo, A low-power unified arithmetic unit for programmable handheld 3-D graphics systems, *IEEE J. Solid-State Circuits*, 42(8), 1767–1778, 2007.

18. J.N. Mitchell Jr., Computer multiplication and division using binary logarithms, *IRE Trans. Electronic Computers*, 11, 512–517, 1962.

19. F.-S. Lai, and C.-F.E. Wu, A hybrid number system processor with geometric and complex arithmetic capabilities, *IEEE Trans. Computers*, 40(8), 952–962, 1991.

28 Sense Circuits for Integrated Sensors

Sitaraman V. Iyer and Hasnain Lakdawala

CONTENTS

28.1 INTRODUCTION

Over the last two decades, integrated circuit (IC) manufacturing has advanced rapidly, resulting in low-power digital portable devices including music players, cell phones, digital cameras, and the like. Making these personal devices user-interactive and sensitive to the environment requires a combination of real-world sensors and digital computing. The advances in IC manufacturing have not only increased digital computing efficiency substantially but have also enabled integration of sensors with electronics. Sensors integrated with traditional complementary metal oxide semiconductor (CMOS) circuitry are poised to enable the era of "smart" devices that can sense environments and adapt intelligently. There are already several commercially successful integrated sensors such as temperature sensors, hall effect sensors, accelerometers, pressure sensors, microphones, and gyroscopes [1–4].

There are several advantages of integrating sensors with CMOS circuits. Miniaturization and lower costs are possible with tight integration of sensors and circuits. Close proximity enables lower noise and interference and lower power sensing elements. The well-established infrastructure in the electronics design and manufacturing industry can be readily reused to achieve sensor-circuit integration. Combined circuit and sensor designs offer previously unavailable degrees of freedom to the designer to trade-off pros and cons across the sensor-circuit boundary. The combined design also offers unique challenges. Traditional IC manufacturing processes are not necessarily optimized or characterized for sensors. Modeling and simulation tools are often limited to separate domains and do not offer seamless system-level design. Engineers are commonly trained and skilled in either the circuits or the sensors. To meet the challenge of sensor-circuit codesign and to achieve the full potential of integrated sensors, reaching out across disciplines to obtain a broader perspective of issues and techniques is necessary.

This chapter aims to bring together fundamentals from the circuit and sensor design worlds to facilitate cross-domain understanding for both circuit and sensor designers. We will explore underlying common techniques used widely in integrating circuits with sensing elements. Basic concepts of two kinds of sensor systems, the Hall effect sensors and the inertial sensors, their respective circuit designs, and their sense mechanisms will be studied in this chapter. This chapter is intended to provide the reader with a broad foundation to integrated sensors, with insight into the common nonidealities, operation, and design trade-offs. A preferential treatment of CMOS-compatible implementations is also maintained throughout the chapter.

We begin by looking at common challenges and fundamental limitations across different sensor domains. These include sensor and circuit thermal and flicker noise, manufacturing variations and mismatch, and temperature sensitivity. We then look at common architectural techniques used across domains, including the use of dummy sensors, active offset cancellation, use of chopper stabilization, auto-zeroing, and correlated double sampling (CDS). This is followed by individual exploration of two specific sensor systems: Hall effect sensors and inertial sensors covering integrated accelerometers and gyroscopes. The chapter is concluded with a recap of the architecture and circuit techniques applicable across different sensor domains.

28.2 SENSOR AND CIRCUIT NONIDEALITIES

There are a number of constraints that limit the ultimate accuracy of integrated sensors. These include random noise, offsets, manufacturing variations impacting the sensor gain and temperature dependence, to name a few. This section gives a brief overview of some of the common limiting factors.

28.2.1 NOISE

Noise in circuits results from resistor and MOS thermal noise, junction shot noise, and flicker noise in MOS transistors. Sensing elements such as a suspended proof mass in accelerometers also contribute to overall system noise through microscopic random movements, resulting from vibrational energy exchange with the surrounding medium. Random noise can exhibit frequency dependence and is commonly quantified with the noise spectral density that specifies the noise power present per Hertz of the frequency spectrum. Thermal and shot noise tend to have a uniform noise spectral density across a wide frequency spectrum and are referred to as white noise. Flicker noise tends to have increasing content at lower frequency and is typically modeled as having a noise spectral density inversely proportional to the frequency. Thermal noise is a result of the continuous energy exchange between a device and the surrounding ambience through loss mechanisms. Thermal noise in linear resistors can be expressed as an equivalent voltage noise source in series with the resistor with the noise spectral density given as $V_{nr} = \sqrt{4KTR}$. Thus, a 1 kΩ resistor has a voltage noise spectral density of about 4 nV/$\sqrt{\text{Hz}}$. Junction shot noise, present in diodes as well as bipolar transistors, arises due to the quantized nature of the current flowing through the junction. It is expressed as an equivalent noise current source in parallel with the junction and is given as $I_{nj} = \sqrt{2qI_D}$. MOS transistors contain thermal noise due to the channel resistance as well as flicker noise due to charge traps located near the oxide interface. The total input-referred voltage noise spectral density is qualitatively shown in Figure 28.1. At low frequencies, it is dominated by the flicker noise, while at high frequencies it

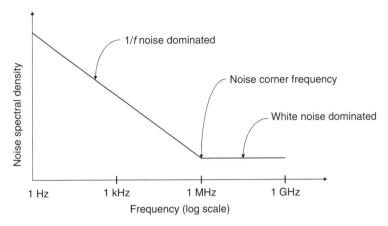

FIGURE 28.1 MOSFET input-referred equivalent voltage noise spectral density.

settles to the thermal noise. The crossover point where the noise starts getting dominated by the thermal noise is commonly referred to as the noise corner frequency. Sensing circuits are usually architectured with the bandwidths directly related to the noise corner frequency as will be seen in subsequent sections.

28.2.2 OFFSET

An analog sensor system is typically expected to produce a linear output in response to an applied physical stimulus such as magnetic field, light, or motion. With zero-input stimulus, the output is expected to be zero. However, this is rarely the case with a raw sensor or with a sensing circuit. Raw sensor outputs almost always have offsets that may be orders of magnitude higher than the resolution of the sensor. Linear amplifiers used as sense circuits also exhibit significant levels of input-referred offset. Offsets can broadly be classified as design- and manufacturing-induced. Design-induced offsets result from design asymmetries in the sensor or the circuit including layout mismatches. These tend to be constant across different dies on the same wafer and across wafers and can be predicted through simulations with accurate modeling of the sensor and circuit layouts. Manufacturing-induced offsets result from manufacturing mismatches and gradients in the sensor or the circuit. Manufacturing-induced offsets can be further classified into systematic and random. Systematic offsets result from gradients and curvatures across a wafer. As a result, sensor parts and transistors that were intended to be matched by design could end up being slightly different in their properties. Random offsets result from random manufacturing variations such as random dopant fluctuations in transistors and dimensional variations in micromachined sensors. In the next section, we will look at common mitigation strategies to reduce offsets to a tolerable level.

28.2.3 GAIN VARIATION

System gain can be defined simply as the ratio of a change in the output voltage to the input physical change. System gain is one of the important specifications that the user would like to be very well controlled. In some critical applications such as acceleration sensors used in airbag deployment, the system gain is used to make potentially life-saving decisions. Real-world manufacturing results in a distribution of parts in terms of gain variation. Sensor properties such as photosensitivity of image sensor pixels and circuit properties such as amplifier gain contribute to overall system gain variation. While absolute properties tend to vary significantly (higher than 50% sometimes), the ratio of similar devices tends to be much better controlled across parts. The lower variance of ratioed devices is exploited extensively across both sensor and circuit designs to achieve lower gain variation.

28.2.4 TEMPERATURE EFFECTS

Almost all physical transducers as well as transistors are sensitive to temperature in multiple ways. Most resistors increase with temperature: thermal noise increases with temperature and transistors become slower at higher temperature. On the sensor side, Hall sensor offsets and stresses in inertial sensors change with temperature. There are a number of techniques that are used to reduce temperature sensitivity as will be seen in the next section.

28.3 TECHNIQUES TO REDUCE OFFSET AND LOW-FREQUENCY NOISE

28.3.1 SENSOR TECHNIQUES

Offsets manifest themselves differently in different sensor domains and are therefore dealt with in different ways. However, there are certain common techniques that can be applied across a wide range of sensor systems. Here we look at three such techniques: differential sensing, use of dummy sensors, and active sensor offset cancellation.

In a single-ended sensor, the input stimulus produces an electrical change (capacitance, resistance, voltage, or current) in one direction only, whereas in a differential sensor, the same input stimulus produces an equal and opposite electrical change in two elements of the sensor. This is explained in Figure 28.2 using a simple resistive divider example. R_{s1} is the sense resistor that changes with the input stimulus and R_f is a fixed resistor in the single-ended sensor. The output voltage is proportional to the difference between R_{s1} and R_f. If there is no effort made to match R_{s1} and R_f nominally, then there will be a static mismatch and a resultant output offset. On the other hand, in the differential sensor, R_{s1} and R_{s2} are both sensitive to the input stimulus but in opposite directions. So, by design, the output offset voltage will be zero. However, manufacturing-induced mismatches between R_{s1} and R_{s2} will still be present and additional steps will have to be taken to reduce the residual offsets. Many surface-machined accelerometers employ a differential sensing mechanism wherein an input acceleration produces a differential capacitance change in two sense capacitors [1].

One way to reduce manufacturing-induced offsets is to use a dummy sensor. Offset cancellation is achieved by using an identical replica of the real sensor that is not sensitive to the input stimulus. This technique has been used in accelerometers [5] and IR bolometers [6]. The offset is canceled to the extent that the real sensor and the dummy sensor are matched. Though this technique incurs a $2\times$ area penalty for the sensor, it can be used effectively in applications where the system cost is not a

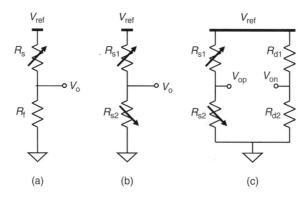

FIGURE 28.2 Examples of sensor topologies using a resistor a sense element: (a) single-ended sensing, offsets produced by design mismatches between R_s and R_f, (b) differential sensing, no design mismatch, manufacturing-induced static mismatch between R_{s1} and R_{s2}, and (c) differential sensing with differential dummy forming a Wheatstone bridge. Eliminates static manufacturing-induced offset also to first order.

strong function of the sensor area and where matching can be achieved with high confidence in high-volume manufacturing (HVM). A fully differential Wheatstone configuration may also be achieved by combining a differential sense scheme with a dummy sensor (Figure 28.2c). Offset variations with temperature (also referred to as bias drifts) may not be well-compensated by the dummy sensor because of the way in which it is desensitized to input stimulus and alternate compensation schemes may be required.

In applications in which the cost of adding a dummy sensor is prohibitive or if the dummy sensor is not expected to match the real sensor, then electronic offset compensation or calibration schemes can be implemented. Capacitive digital-to-analog converter (DAC)-based scheme has been demonstrated for accelerometers in Refs. [7,8]. This method would need the digital offset cancellation code to be stored in fuses or to be redone on power on each time the sensor is powered on.

In some types of sensors such as image and magnetic (Hall effect) sensors, a combination of electronic techniques such as chopper stabilization or CDS and differential sensing is used to significantly reduce the inherent offsets. These two techniques are cornerstones of several electronic sensor systems and will be discussed below.

28.3.2 CHOPPER STABILIZATION

Low-frequency nonidealities such as input offsets and flicker noise directly limit the precision of direct sensor amplifiers because many sensor systems are intended for sensing slowly varying ($<10\,\mathrm{kHz}$) phenomena. One way to separate the sense signal from the low-frequency nonidealities is through chopper stabilization [9,10]. The basic idea is to upconvert, through modulation, the low-frequency sense signal to a higher frequency while maintaining the nonidealities at low frequencies. Following sense signal amplification, demodulation downconverts the amplified signal back to the low-frequency baseband while the low-frequency nonidealities simultaneously get upconverted to the modulation frequency. A time domain view of the chopper stabilization process is shown in Figure 28.3. In this figure, a resistive sensor is used as an example though the concept is applicable to any kind of sensor. The Wheatstone bridge is driven by complementary modulation signals V_{mp} and V_{mn} that result in upconversion of the sense signal. The first-stage amplifier input offsets and low-frequency noise are lumped together into the equivalent input-referred voltage source labeled V_{os}. The output of the first stage shows the superposition of the modulated sense signal and the low-frequency noise plus offset. The demodulator (DEMOD) can be implemented using simple pass-gates. The output of the demodulator block shows the downconverted amplified sense signal superposed on the upconverted noise plus offset. The lowpass filter (LPF) removes the upconverted noise plus offset, resulting in an amplified sense signal.

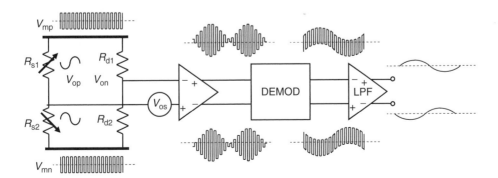

FIGURE 28.3 Chopper stabilization procedure showing upconversion of sense signal, followed by amplification, downconversion of amplified signal, and LPF to eliminate upconverted noise and offsets.

There are several considerations for implementing a chopper-stabilized sensor system. The figure shows only one amplifier stage. Typically, however, the amplification is obtained from multiple stages since it is difficult to obtain adequate gain at high bandwidth from a single stage. With multiple gain stages, the noise and offset of the second and additional stages are not as critical as those of the first stage since those stages see an amplified signal. On the other hand, the dynamic range of the later stages needs to accommodate the larger signal swings. The modulation frequency is typically chosen to be slightly above the $1/f$ noise corner frequency, which is the frequency at which the input-referred noise of the first amplifier stage changes from being $1/f$ noise-dominated to thermal noise–dominated. There is also a trade-off between the bandwidth, power, and noise of the gain stages. The bandwidth of each stage needs to be high enough to allow adequate settling of the chopper-stabilized signals. On the other hand, too high a bandwidth will result in more noise from frequencies that are multiples of the modulation frequency aliasing down to the signal frequencies after demodulation. Higher bandwidths will also result in higher power consumption in the amplifier stages. High-frequency differential noise between V_{mp} and V_{mn} also gets downconverted into the signal band and therefore, the modulation signals should be relatively clean. A more detailed discussion of the design considerations can be found in Ref. [10].

28.3.3 Correlated Double Sampling

An alternate method to separate low-frequency noise and offsets from the sense signal is to use CDS. This is a discrete time method as opposed to chopper stabilization, which is a continuous time method. The basic concept of CDS is illustrated in Figure 28.4a. Basically, this involves taking two samples of the output voltage; the first one with a known common mode voltage (V_{cm}) and the second with the sense signal input (V_{in}). The output voltage (V_{out}) is taken as the difference between the two voltages. As a result, the internal offset of the amplifier does not contribute to the final output (V_{out}). If the two samples are taken close together in time, then this technique will also result in cancellation of the low-frequency noise of the amplifier.

A switched capacitor implementation of CDS is shown in Figure 28.4b. When clock phase ϕ_0 is high, the amplifier is held in unity gain reset mode. At this time, the output V_{out} is $V_{cm} + V_{os}$ where V_{os} is the input-referred offset of the op-amp. The capacitor C_{in} gets charged to the offset voltage V_{os}. After the clock phase ϕ_{0d} goes low and the clock phase ϕ_1 goes high, the sense signal input V_{in} is applied to

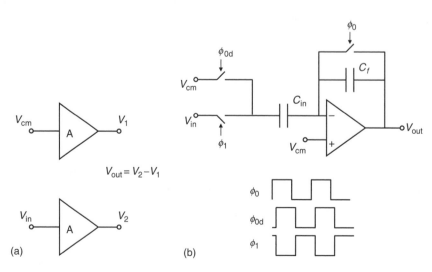

FIGURE 28.4 (a) CDS concept and (b) switched capacitor implementation of CDS.

C_{in}. Because of the negative feedback with capacitor C_f, C_{in} is forced to charge to $V_{cm} + V_{os} - V_{in}$. The resultant additional charge, $C_{in}(V_{cm} - V_{in})$, has to flow through the feedback capacitor and therefore the voltage across C_f changes from 0 to $C_{in}(V_{cm} - V_{in})/C_f$. The resultant output voltage then becomes $V_{cm} + V_{os} + C_{in}(V_{cm} - V_{in})/C_f$. Note that the V_{os} term gets only a unity gain while the input signal (V_{in}) goes through an amplification of C_{in}/C_f. It is also possible to architect the CDS so that the V_{os} terms get completely canceled [10]. Note that there is an inherent assumption that the op-amp has infinite gain. Since this assumption is never true, practically there is a residual offset contribution to the output that is inversely proportional to the DC gain of the op-amp. By choosing the sampling frequency to be above the $1/f$ noise corner frequency of the op-amp, the $1/f$ noise can also be reduced. In addition to the op-amp noise itself, the KT/C noise introduced by the sampling switches also contributes to the output noise. Therefore, the capacitors C_f and C_{in} have to be chosen to be large enough to prevent KT/C from being the dominant noise source. The ratio of the two capacitors determines the amplifier gain. Note that the high-frequency input-referred noise of the op-amp does not get canceled by the CDS and also sees a similar gain as the input sense signal and can easily be the dominant noise contributor to the output. As in the case of the chopper stabilization, the reference voltage V_{cm} contributes directly to the output noise level and should, therefore, be relatively clean. The bandwidth of the op-amp needs to be set high enough to allow the sense signal to settle quickly before the output is sampled by the following stage. At the same time, a CDS system is even more sensitive to noise folding from higher harmonics of the sampling frequency than a chopper-stabilized system. This is because in a chopper-stabilized system, the noise folding from higher harmonics is weighted by the strength of the harmonic in a square wave, i.e., 1/3, 1/5, 1/7, and so on, respectively, for the third, fifth, and seventh harmonics of the modulation frequency assuming a 50% duty cycle modulation voltage. A CDS system, being a sampled data system, experiences full noise folding from all odd harmonics if the op-amp has infinite bandwidth. Finite op-amp bandwidth results in filtering of the higher noise frequencies. Noise at the even harmonics is canceled if the delay between the two samples of the CDS is exactly half of the sampling period. Some noise reduction can also be obtained by choosing a track-and-hold for the output stage rather than a sample-and-hold. With a track-and-hold, the actual sense signal appears at the output for a longer time as opposed to a sampled version [10,11].

28.3.4 Gain Calibration

Gain variation is one of the most common issues encountered across sensor domains. From a product perspective, it is important to guarantee that the sensor gain will be within a specified range, with a smaller range being better. In sense circuits, gain can be well controlled by using ratios of capacitors or resistors to set the gain. In transducers, it is more difficult to ensure a constant transduction gain in HVM. Trimming after test is one option to reduce gain variation, though it is expensive as it significantly increases test and calibration time. A combination of sensor and circuit techniques can also be devised for specific cases. One such implementation in Ref. [8] where the first stage feedback capacitor was also made by the microelectromechanical system (MEMS) techniques used to fabricate the sensor element.

28.3.5 Temperature Compensation

Temperature dependence is an undesired aspect of all sensors other than temperature sensors. Temperature impacts the resistance in Hall effect sensors, the accuracy in IR sensors, photodiode leakage current in image sensors, and introduces stresses in suspended inertial sensors, leading to spurious outputs. The dummy sensor concept used for offset reduction is also useful for temperature compensation in many sensor applications. For more expensive sensors needing very precise temperature calibration, one option is to do the calibration externally with the help of a precision proportional to absolute temperature (PTAT) voltage or current reference [2].

28.4   HALL EFFECT SENSORS

28.4.1   Sensing Mechanism

Integrated magnetic sensors based on the Hall effect have been studied for several decades [12]. The basic concept of the Hall effect is illustrated in Figure 28.5a. When a rectangular plate carrying an electric current (I) is under the influence of a magnetic field (B_z) perpendicular to the plate, an electric field perpendicular to both, the electric current and the magnetic field, is induced. The electric field generates an electric potential difference that is commonly referred to as the Hall voltage (V_H). The Hall voltage is proportional to the applied voltage V to generate the electric current I and to the incident magnetic field (B_z). It also depends on the effective carrier mobility μ_n, the aspect ratio of the plate, and a finite geometry correction factor G.

An equivalent circuit can be devised as shown in Figure 28.5b [12]. It consists of a Wheatstone bridge with the input current being driven between two diagonally opposite terminals and the output Hall voltage being sensed from the other two terminals of the bridge. One of the most severe nonidealities in integrated Hall sensors is the intrinsic sensor offset. The intrinsic offset can be captured in the simple electrical model as resistance mismatches in the Wheatstone bridge as shown in Figure 28.5c [12]. Integrated Hall sensors typically use a n-well as the Hall plate with terminals located at each edge of the n-well. The following section describes the most common nonidealities primarily the Hall sensor offset and offset drift and strategies to cancel that. Additionally, nonlinearity and temperature dependence are other nonidealities that need to be addressed [3].

28.4.2   Nonidealities in Hall Sensors

The intrinsic Hall sensor offsets can be as high as a few millitesla, resulting from factors such as nonuniform doping in the Hall plate, patterning asymmetries, and mechanical stress-induced piezoresistance effects. In comparison, the earth's magnetic field is less than 100 μT. To reduce the overall offset by more than two orders of magnitude, two common strategies for offset reduction are the spinning current method [13] to cancel intrinsic offset of a single sensor and use of multiple orthogonal plates [14] to cancel stress gradient-induced piezoresistive effects. In Ref. [14], a combination of the two techniques is used to achieve 3.5 μT of 3σ offset. The sensitivity of Hall sensors is typically of the order of 1 V/T. Therefore, in addition to reducing the intrinsic sensor offset, very low offset sense circuits are also needed to amplify Hall voltages produced by magnetic fields with strengths

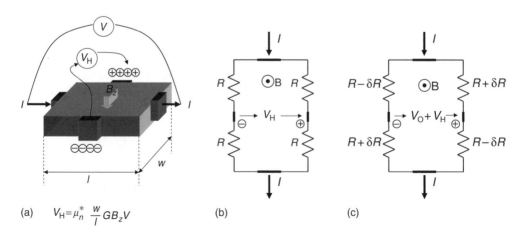

(a) $V_H = \mu_n^* \dfrac{w}{l} G B_z V$ (b)   (c)

FIGURE 28.5   (a) Hall effect concept, (b) equivalent electrical circuit, and (c) equivalent electrical circuit including offset voltage (V_O) resulting from current flow asymmetry.

in the range of 10–100 µT. The spinning current method lends itself naturally to circuit techniques such as chopper stabilization as demonstrated in Refs. [14–16]. A nested chopper technique is used in Ref. [14] wherein the inner (faster) chopping frequency eliminates the offsets of the first amplifier stage and also implements sign changes necessary during the current spinning and the outer (slower) chopping operation eliminates offsets in the subsequent sigma–delta modulator.

For larger range (>1 T) Hall sensors, nonlinearities are an important consideration. Nonlinearities in integrated Hall sensors can be attributed to three factors: the junction field effect (JFE), resulting from modulation of the resistance of the Hall plate due to change in the width of the surrounding depletion layer, resulting from the Hall voltage itself as well as the magnetoresistance effect [12] and due to material and geometric nonlinearities [3]. In Ref. [17], optimal biasing techniques for the substrate (shield) are proposed to reduce JFE induced nonlinearity while the geometric and material nonlinearities are used to cancel each other.

Temperature variations of the sensitivity and the offset also need to be addressed for Hall sensors. There have been several inventions that aim to reduce or eliminate temperature effects in Hall sensors, couple of which are listed in the Refs. [18,19] as being suitable for CMOS implementation. The basic idea underlying many of the inventions is to generate a temperature-compensated reference current or voltage to power the Hall sensor. This is usually accomplished using a combination of bandgap and PTAT reference voltages and currents and resistors with appropriate temperature characteristics. Typically, some amount of the postfabrication trimming of the temperature coefficients of the voltages and currents is also required to account for manufacturing variations.

28.5 INERTIAL SENSORS

28.5.1 ACCELEROMETERS AND GYROSCOPES

Over the past decade or so, integrated inertial sensors including MEMS-based accelerometers and gyroscopes have become widely used in the automobiles and mobile electronics. Single-chip integration of the sense element and the associated electronics was a key enabler for this widespread use. Accelerometer requirements such as bandwidth, linearity, and resolution vary with applications. Automotive applications such as airbag deployment and stability system need higher bandwidths but lower resolution when compared to inertial navigation requirements [20]. Integrated MEMS inertial sensors typically address the low-cost medium resolution market. MEMS inertial sensors can be broadly classified into two types based on the sensor fabrication process: bulk- and surface-micromachined. Surface-micromachined sensors are more easily integrated with CMOS circuitry and will be the focus of the following section.

Surface-micromachined accelerometers commonly use a capacitive sense scheme as shown in Figure 28.6. They consist of a proof mass suspended by means of a suspension spring whose other end is anchored firmly to the chip substrate. The gap between the proof mass and another anchored set of plates forms the sense capacitance. The spring is designed to be compliant in the sense axis and stiffer in other axes. Under external acceleration inertial forces act on the proof mass and displace it, resulting in a sense capacitance change that can then be converted to a voltage change. In order to maximize capacitance change, the proof mass consists of several fingers that are interdigitated with anchored fingers, giving more surface area for the sense capacitance. To linearize the output capacitance change and to make it insensitive to the absolute capacitance value, the sense capacitance is usually configured as a capacitive divider.

Surface-micromachined gyroscopes are more complex than accelerometers primarily because very small displacements (tens of femtometers) have to be measured in the presence of coupling from other motions that are several orders of magnitudes higher [2,21]. Microgyroscopes use an accelerometer internally to measure the Coriolis force–induced displacements. Simplified concept of it vibratory microgyroscopes is shown in Figure 28.7. It consists of an accelerometer nested inside a frame that is itself connected to the anchored substrate by means of a suspension spring. The external

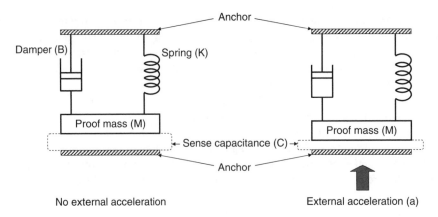

FIGURE 28.6 Surface-micromachined accelerometer concept.

frame is set into oscillations about the *y*-axis through an electrostatic actuator. When attached to a rotating object, the proof mass of the nested accelerometer experiences the Coriolis force, resulting in an acceleration that is proportional to the angular velocity of rotation as well as the instantaneous velocity of *y*-axis oscillations. Sensing this acceleration results in a signal that is proportional to the angular velocity, which is the quantity the microgyroscope aims to sense. In addition to sense circuits needed for the nested accelerometer, usually a closed loop electronic oscillator is built around the external frame to sustain oscillations of the frame along the *y*-axis. The following section discusses the commonly used sense circuit architectures and related trade-offs for microaccelerometers and microgyroscopes.

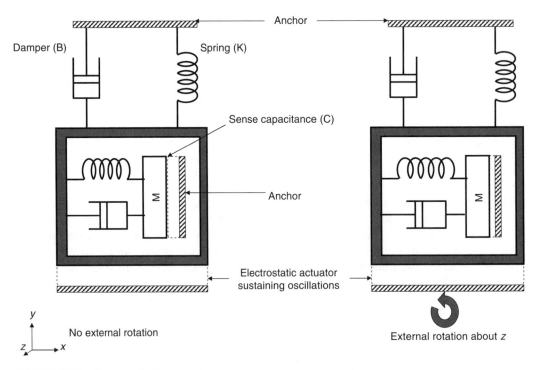

FIGURE 28.7 Concept of vibratory microgyroscope.

28.5.2 Circuit Techniques

Transresistance [21] and transcapacitance amplifiers [7,8,21,22] are commonly used to convert the capacitance change to voltage. As mentioned previously, to eliminate offsets and $1/f$ circuit noise, the sense capacitances are usually modulated at a frequency much higher than the sensor bandwidth. The modulation frequency is typically a few MHz. Simplified single-ended transresistance and differential transcapacitance amplifiers are shown in Figure 28.8a and 28.8b, respectively. Note that differential transresistance and single-ended transcapacitance amplifiers are also feasible. In transresistance amplifiers, the voltage across the sense capacitors is held constant by means of the feedback resistor. Therefore, any change in capacitance results in a charge flow. The differential charge between the two sense capacitors is forced to flow through the feedback resistor and, therefore, causes an output voltage change. The transduction gain is proportional to the ratio of the feedback resistor to the equivalent impedance of the sense capacitance change at the modulation frequency. In the differential transcapacitance amplifier shown, the capacitive negative feedback maintains the differential voltage between the input nodes to be zero, resulting in any differential charge flowing through the feedback capacitors to the op-amp outputs charging them up in the process. The transduction gain is proportional to the ratio of the sense capacitance change to the feedback capacitance. Note that the output voltage is independent of the absolute sense capacitance as well as the parasitic and MOS gate capacitances to first order, if the op-amp has high enough gain. However, the noise gain, i.e. the factor by which the input referred noise of the op-amp is amplified is proportional to the sum of the absolute sense capacitances as well as the parasitic and MOS gate capacitances at the op-amp input. Therefore, it is always preferable to minimize the parasitic capacitance at the input of the op-amps and monolithic integration of circuit and sensor is desirable. While it may be tempting to minimize the input transistor gate capacitance as well, we should recall that a smaller transistor has smaller transconductance and larger input-referred thermal noise and flicker noise. There is, in general, an optimal transistor size range that balances the conflicting design objectives of smaller input-referred noise and smaller noise gain [22,23].

Both switched capacitor-based CDS [8,23] and chopper stabilization [22] techniques have been used for capacitive accelerometers to reduce the impact of offsets and $1/f$ noise. In switched capacitor circuits, the bias for the input nodes of the op-amp is provided by a reset of the feedback capacitors every clock cycle. In chopper-stabilized sense schemes [21,22], the amplifier inputs are biased only once in multiple clock cycles to reduce the noise introduced by the bias conductance. Capacitive sense circuits noise floors have improved steadily over the last decade [20] and sub-10 zF/$\sqrt{\text{Hz}}$ noise floors have been reported [8]. Equivalently, integrated surface-micromachined accelerometers typically have resolutions in the range of 100 µg/$\sqrt{\text{Hz}}$ to 1 mg/$\sqrt{\text{Hz}}$. Higher resolution accelerometers with sigma–delta closed loop operation have been reported [24], relying primarily on a bigger

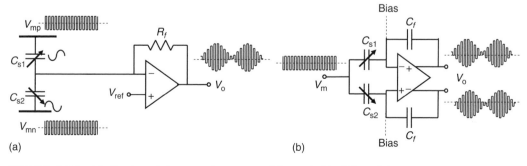

(a)   (b)

FIGURE 28.8 (a) Single-ended transresistance amplifier and (b) differential transcapacitance amplifier (biasing not shown).

proof mass that can be obtained through bulk micromachining combined with a separate sense circuit IC. Digital offset trimming has been demonstrated for accelerometers using a binary-weighted capacitive DAC in Ref. [7] and a ternary capacitive ladder DAC in Ref. [8]. Temperature-induced offset and sensitivity changes are more difficult to compensate. Localized temperature control by heating has been proposed for accelerometers in a standard CMOS process with postprocessing to create mechanical structures [5]. The Analog Devices microgyroscope features a companion PTAT output [21] that can be used for external compensation of the temperature sensitivity. Furthermore, it uses a common-centroid layout with four connected vibratory sensor elements to effectively cancel linear and angular acceleration sensitivity [2]. The microgyroscope is an excellent example of an application that combines circuit and sensor techniques to achieve the design goals.

28.6 CONCLUSIONS

In this chapter, we reviewed common nonidealities in integrated sensors including thermal and flicker noise, offsets, gain variation, and temperature sensitivity. Thermal noise is usually reduced by a careful noise budget analysis incorporating all the noise sources and optimizing the sensor and sense circuits as well as reference voltage generation circuits. Flicker noise can be eliminated by use of chopper modulation or CDS, both of which are well-suited for CMOS circuit implementation. Since sensor bandwidth requirements are typically less than a few kilohertz, these two techniques are used extensively across multiple domains including Hall sensors, image sensors, and inertial sensors. The two techniques can also be used to eliminate circuit offsets. Sensor offsets are usually eliminated by a combination of differential sensing, dummy sensor, or active offset cancellation methods. In Hall sensors, the spinning current method can be viewed as spatial differential sensing where the sign of the offset is reversed in the two polarities while the signal polarity remains the same. Hall sensor systems and microgyroscopes have been built with multiple sense elements that are coupled to cancel out spatial and directional asymmetries and packaging-related stress-induced effects. Active offset cancellation techniques have also been demonstrated in multiple domains and involve digital or analog offset calibration that is done when the sensor is not subjected to external sense stimuli. Temperature effects on offsets and sensitivities are pervasive across sensor domains. Dummy sensor and multiple sense elements can be used to partially cancel out temperature effects. In the circuit domain temperature-dependent voltage and current sources generated using a combination of diode voltage drops and PTAT references can be used for temperature compensation.

While monolithic integration of sensors and circuits opens up possibilities for a myriad of smart devices, in the initial development stages integrated sensors tend to offer lesser raw performance compared to segregated sensor and circuit systems. However, over time with careful co-optimization of the sensor and circuits by leveraging the close proximity of the sensor and circuit to the fullest extent and utilizing circuit techniques to compensate for sensor shortcomings, integrated sensors can provide cost- and area-effective alternatives to traditional sensor systems.

REFERENCES

1. Analog Devices ADXL202 data sheet.
2. J. A. Geen, Very low cost gyroscopes, IEEE Sensors Conference, Irvine, California, October 30–November 3, 2005.
3. R. S. Popovic, *Hall Effect Devices: Magnetic Sensors and Characterization of Semiconductors*, 2nd ed., IOP Publishing, New York, 2003.
4. A. El Gamal and H. Eltoukhy, CMOS image sensors, *IEEE Circuits and Devices Magazine*, 6–20, May/June 2005.
5. H. Lakdawala, G. K. Fedder, Temperature control of CMOS micromachined sensors, *Proc. of 15th IEEE Conf. on Microelectromechanical Systems (MEMS 2002)*, pp. 324–327, January 20–24, 2002.
6. P. Neuzil and T. Mei, A method of suppressing self-heating of bolometers, *IEEE Sensors J.*, 2(4), 207–210, 2004.

7. M. Lemkin and B. E. Boser, A three-axis micromachined accelerometer with a CMOS position-sense interface and digital offset-trim electronics, *IEEE J. Solid-State Circuits*, 34(4), 456–468 1999.
8. S. V. Iyer, H. Lakdawala, R. S. Sinha, E. J. Zacherl, R. T. Unetich, D. M. Gaugel, D. F. Guillou, and L. R. Carley, A 0.5 mm^2 integrated capacitive vibration sensor with sub-10 zF/$\sqrt{Hz}$ noise floor, *Proc. of IEEE Custom Integr. Circuits Conf. (CICC 2005)*, pp. 93–96, September 2005.
9. S. J. Sherman, W. K. Tsang, T. A. Core, R. S. Payne, D. E. Quinn, K. H.-L. Chau, J. A. Farash, and S. K. Baum, A low cost monolithic accelerometer; Product/technology update, *Tech. Dig. IEEE Electron Devices Meeting (IEDM'92)*, pp. 160–161, December 1992.
10. C. C. Enz and G. C. Temes, Circuit techniques for reducing the effects of op-amp imperfections: autozeroing, correlated double sampling, and chopper stabilization, *Proc. IEEE*, 84(11), 1584–1614, 1996.
11. A. Bilotti, G. Monreal, Chopper-stabilized amplifiers with a track-and-hold signal demodulator, *IEEE Trans. Circuits and Systems I: Fundamental Theory and Applications*, 46(4), 490–495, 1999.
12. H. P. Baltes and R. S. Popovic, Integrated semiconductor magnetic field sensors, *Proc. IEEE*, 74, 1107–1132, 1986.
13. P. J. A. Munter, A low-offset spinning-current Hall plate, *Sensors Actuators A (Phys.)*, 22A(1–3), 743–746, 1990.
14. J. C. van der Meer, F. R. Riedijk, E. van Kampen, K. A. A. Makinwa, and J. H. Huijsing, A fully integrated CMOS Hall sensor with a 3.65 µT 3σ offset for compass applications, *Tech. Dig. Int. Solid-State Circuits Conf. (ISSCC)*, pp. 256–247, 2005.
15. A. Bilotti, G. Monreal, and R. Vig, Monolithic magnetic hall sensor using dynamic quadrature offset cancellation, *IEEE. J. Solid-State Circuits.*, 32(6), 829–836, 1997.
16. Y. Hu and W.-R. Yang, CMOS Hall sensor using dynamic quadrature offset cancellation, *Int. Conf. Solid-State and Int. Circuit. Technol.*, pp. 284–286, 2006.
17. Ch. Schott and R. S. Popovic, Linearizing integrated Hall devices, *Tech. Dig. Int. Solid State Sensors and Actuators, 1997 (Transducers '97)* Chicago, vol. 1, pp. 393–396.
18. U. Theus, M. Motz, J. Niendorf, Hall sensor with automatic compensation, U.S. Patent 5260614, November 9, 1993.
19. M. Motz, Temperature compensation circuit for a Hall element, U.S. Patent 6825709, November 30, 2004.
20. N. Yazdi, F. Ayazi, and K. Najafi, Micromachined inertial sensors, *Proc. IEEE*, 86(8), 1640–1659, 1998.
21. J. A. Geen, S. J. Sherman, J. F. Chang, and S. R. Lewis, Single-chip surface micromachined integrated gyroscope with 50°/h Allan deviation, *IEEE J. Solid-State Circuits*, 37(12), 1860–1866, 2002.
22. J. F. Wu, G. K. Fedder, and L. R. Carley, A low-noise low-offset capacitive sensing amplifier for a 50 µg/rt-Hz monolithic CMOS MEMS accelerometer, *IEEE J. Solid-State Circuits*, 39(5), 1860–1866, 2004.
23. M. Lemkin, Micro Accelerometer Design with Digital Feedback Control, PhD thesis, University of California at Berkeley, 1997.
24. H. Kulah, J. Chae, N. Yazdi, K. Najafi, Noise analysis and characterization of a sigma–delta capacitive microaccelerometer, *IEEE J. Solid-State Circuits*, 41(2), 352–361, 2006.

29 Detector Interface Circuits for X-Ray Imaging

Pawel Grybos

CONTENTS

29.1 INTRODUCTION

Nowadays, there is a growing interest in digital position-sensitive x-ray imaging systems for biology, medicine, chemistry, and solid state physics applications. Such systems consist of array of sensors of different shapes (strip or pixel) and readout electronics. Advanced very large system integration (VLSI) electronics and techniques of designing and prototyping integrated circuits (ICs) allow the integration of a large number of channels in a single chip, so that each element of the detector array can be readout by an individual electronic channel. Such solution offers postprocessing capability and good spatial resolution of an image, as well large dynamic range and the possibility of work with high-intensity x-ray radiation. If additionally digital imaging systems are able to extract some energy information about x-ray radiation, it would be very advantageous, especially in medical radiology (contrast improvement and dose reduction), material screening, or diffractometry.

    The fast front-end electronics of a large array of x-ray sensors should amplify and filter small signals from the each sensor element, perform analog-to-digital conversion, and then store the data in digital form on the IC in each channel independently at the same time. Because of the complexity of the multichannel mixed-mode IC, the important problems like power limitation, low level of noise, good matching performance, and crosstalk effects must be solved simultaneously. This chapter discusses the architecture of the detector array and all the abovementioned problems for the readout electronics in the digital x-ray imaging system.

29.2 DETECTOR ARRAY

In various areas of physics, biology, and medicine, imaging methods using position-sensitive detection of x-rays have been developed [1–3]. There are basically two types of position-sensitive detectors:

- Detectors based on charge integration as photographic emulsions, charge-coupled device, etc.
- Detectors that work in the single-photon counting mode

Imaging techniques using detectors of the second type are sometimes called as digital imaging. An advantage of single-photon counting detectors is essentially an infinite dynamic range contrary to the integration-type detectors, which usually have problems with a limited dynamic range and a low contrast of the image. The best known and most frequently used sensors working in the single-photon counting mode are made of semiconductor materials and can have different geometry.

29.2.1 SEMICONDUCTOR MATERIAL FOR DETECTOR

Generally, the sensor medium converts the energy deposited by x-ray photons to electric signals. The choice of the detector material depends on application and the detector should produce the highest possible signal for given energy of x-rays. The material very often used for the low-x-ray energy range is silicon, mainly because of good homogeneity and very well-stabilized technology. A standard 300 μm thick detector converts nearly all 8 keV x-ray, but only ~26.7% of 20 keV x-ray, and ~2% of 60 keV x-ray. Therefore many laboratories are making efforts to produce detectors that are more efficient for high x-ray energy using other semiconductor materials, such as high-purity germanium [4–7] and compound semiconductors like gallium arsenide (GaAs) [8,9], cadmium telluride (CdTe), cadmium zinc telluride (CdZnTe) [10–14], and mercuric iodide [15]. These materials are more suitable for detection of harder x-rays, above 20 keV, because of their higher atomic numbers (see Table 29.1), but there are still many problems to be solved concerning charge trapping, impurity concentration, polarization, contact quality, leakage current, and nonuniformities [2].

29.2.2 DETECTOR GEOMETRY

Position-sensitive semiconductor detector is an array of individual sensor elements and in the single-photon counting mode, each element is readout by an individual electronic channel. In the case of silicon, the detectors are built as matrices of reverse-biased diodes processed on common high-resistivity substrate (1–10 kΩ cm), which is 250–500 μm thick. The sensitive area can be divided

TABLE 29.1
Important Parameters of Semiconductor Material Used for X-Ray Detector

Parameter	Si	Ge	GaAs	CdTe	CdZnTe
Atomic number Z	14	32	32	50	48.5
Energy bandgap (eV)	1.12	0.67	1.43	1.5	1.5–2.2
Energy for electron–hole pair creation (eV)	3.64	2.96	4.27	4.7	~ 4.5
Mobility at room temperature					
Electrons ($cm^2/V\,s$)	1450	4500	8000	1150	1350
Holes ($cm^2/V\,s$)	450	3500	400	110	120

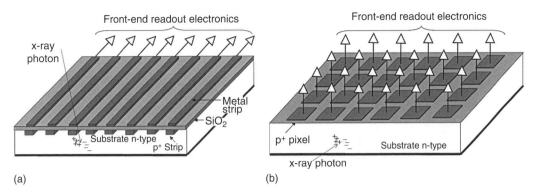

FIGURE 29.1 Simplified view of (a) strip detector and (b) pixel detector.

into individual diodes of the shapes according to the geometrical requirements of the experiment, area to be imaged, and the spatial resolution. The diodes are mostly shaped as strips or pixels (see Figure 29.1).

The charge generated in the detector medium by the x-ray photon is moved by the electric field inside the detector and induces currents in the strip or pixel. Reading these current pulses, one can reconstruct the position of incident particle and obtain a one- or two-dimensional image.

The spatial resolution of such a position-sensitive system is determined by

- Intrinsic spatial resolution of detector resulting from the interaction of photons with the detector material
- Strip/pixel pitch of the detector
- Parameters of the readout electronic system, especially signal-to-noise ratio (SNR)

Keeping in mind that the conversion factor of silicon for generating electron–hole pairs is 3.6 eV/e-h, a single 5 keV photon produces about 1400 electron–hole pairs, which are then collected by the readout electrodes in about 20 ns. In this case, to obtain a reasonable SNR (above 10), the readout electronics should have the equivalent noise charge below 140 el. rms. This is close to the practical limit one can achieve in a complex fast multichannel system working at room temperature.

29.3 REQUIREMENTS AND LIMITATIONS OF THE FRONT-END ELECTRONICS FOR DIGITAL X-RAY IMAGING

A typical signal processing channel is shown in Figure 29.2. A current signal generated in the silicon strip/pixel detector is integrated in a charge-sensitive preamplifier. At the output of the preamplifier, one obtains a voltage step with an amplitude proportional to the total charge generated in the detector.

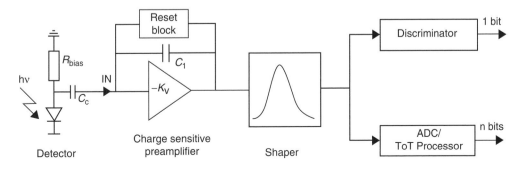

FIGURE 29.2 Principal block of a detector readout system.

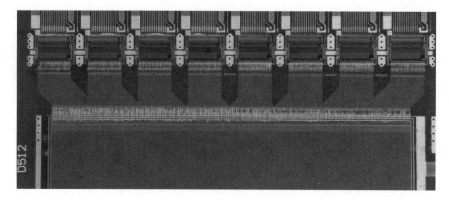

FIGURE 29.3   Fragment of multichip module: 512-strip detector connected to eight 64-channel readout ASICs.

The voltage step is fed to the main amplifier, called a shaper, which provides the pulse shaping according to the timing requirements and the filtration of noise to maximize the SNR.

Further processing of the shaped signal can be done in two possible ways. The first one is based on the so-called binary readout architecture. In this case, the comparator detects the presence of the signal of amplitude above the preset threshold and in response provides 1-bit yes/no information. The second way of processing the shaped signal employs an analog readout architecture, where the amplitude of the signal corresponding to each individual photon is measured, then the information is stored, and used for offline processing.

The number of such front-end channels in readout systems is equal to the number of strips/pixels in the semiconductor detector. In the case of readout electronics for strip detector, multichannel ICs are mostly designed as 32-, 64-, or 128-channel ICs, which are the basis for building larger modules consisting of several hundreds up to a few thousands of readout channels (see Figure 29.3). In the case of readout for pixel detector, a single IC could have up to several tens of thousands of readout channels working independently. For example, Medipix2 IC has 65536 identical front-end channels [15].

The multichannel architecture of readout IC requires that some of the problems must be solved in a different way than in conventional readout electronics and some others are completely new. The main requirements for multichannel ICs used for digital x-ray imaging are as follows:

- Low level of noise of the low-power and high-speed readout electronics
- Multichannel architecture of IC, which imposes limits on the maximum power dissipation and the area of single readout channel
- Minimization of crosstalk effects in the mixed-mode ICs (sensitive analog front-end and fast digital block for further data processing are placed on the same die)
- Good matching performance of analog parameters (gain, offsets, noise) from channel-to-channel.

To the community of electronic VLSI designers, these problems are known to be difficult and suffer from lack of adequate modeling. There are a number of interesting and valuable papers, which describe each of the above aspects separately. In the case of multichannel mixed-mode ICs, all of these aspects are important and they often impose conflicting requirements.

The relationship between different designs and optimization aspects are shown schematically in Figure 29.4. For example, biasing of input stage of front-end electronics with high current is advantageous for the noise optimization, but in multichannel system, it is limited by available power dissipation in a single channel. Some elements of circuit architecture, which are preferred from the point of view of crosstalk minimization (e.g. differential structure in analog blocks), increase the noise. The limitation of a single channel area is in opposition to the low level of $1/f$ noise and to the reduction of mismatch effects.

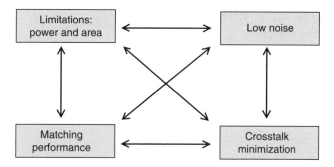

FIGURE 29.4 Relationship between different aspects of multichannel mixed-mode ICs design.

In the following sections, the most important aspects of the general readout channel architecture shown in Figure 29.2 will be discussed, namely, the noise optimization of the low-power complementary metal oxide semiconductor (CMOS) charge-sensitive amplifier (CSA) and requirements for the shaper stage to improve SNR and ensure fast signal processing. Possible analog-to-digital conversion architectures used in multichannel ICs are shown and problems of the crosstalk and matching in mixed-mode multichannel ICs are also discussed.

29.4 CHARGE-SENSITIVE AMPLIFIER

29.4.1 NOISE OPTIMIZATION

The main noise contribution to the detector readout system comes from the detector itself and the CSA. The simplified noise model of a detector and CSA is shown in Figure 29.5. The noise performance of this system can be analyzed using the equivalent input voltage and current noise sources.

In a well-designed CSA, made using CMOS technology, the voltage noise is dominated by the noise of the input MOS transistor and can be expressed as [16,17]

$$\frac{\overline{dv_n^2}}{df} = \frac{8}{3}kT\frac{\gamma}{g_{m1}} + \frac{K_f}{C_{ox}^2}\frac{1}{WL}\frac{1}{f} \tag{29.1}$$

where
  k is the Boltzmann's constant
  T is the temperature
  g_m is the transistor transconductance
  γ ranges from 1/2 (weak inversion) to 2/3 (strong inversion)
  C_{ox} is the gate oxide capacitance per area

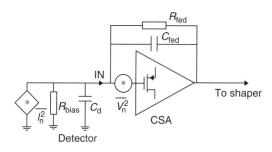

FIGURE 29.5 Simplified noise scheme of a detector and CSA.

W and L are the width and length of the input transistor

K_f is the flicker noise coefficient

f is the frequency

In the above formula excess, noise effects are neglected, because we assume that the charge amplifier in multichannel systems is designed to operate with low current density in the input transistor and at low drain-source voltage [18]. In the shorter form the above equation could be written as

$$\frac{\overline{dv_n^2}}{df} = a + \frac{b}{f} \tag{29.2}$$

where constants a and b are responsible for the thermal and flicker noise, respectively.

A current noise is associated with the detector leakage current I_{DET}, the detector bias resistance R_{bias} (in the case of AC coupling detector only), and the effective feedback resistance R_{fed} in the CSA, so the power spectral density is given by

$$\frac{\overline{di_n^2}}{df} = 2qI_{DET} + \frac{4kT}{R_{bias}} + \frac{4kT}{R_{fed}} \tag{29.3}$$

and it could be rewritten in the shorter form as

$$\frac{\overline{di_n^2}}{df} = c \tag{29.4}$$

In order to find an equivalent noise charge of a detector system, one needs to calculate the noise rms value at the shaper output. Knowing the charge gain of the system, the Equivalent Noise Charge (ENC) can be expressed as [19,20]

$$\text{ENC} = \sqrt{a\frac{F_v C_{in}^2}{T_p} + bF_{vf}C_{in}^2 + cF_i T_p} \tag{29.5}$$

where

C_{in} is the sum of the gate capacitance C_g of the input transistor, the feedback capacitance C_{fed}, and the detector capacitance C_d

T_p is the peaking time at the shaper output (signal reaches the maximum value)

F_v, F_{vf}, and F_i are constants dependent on a filter type

Three immediate observations based on formula (29.5) are important for the optimization of the front-end circuit (see also Figure 29.6):

- Contribution of the input voltage white noise to the ENC is proportional to the total input capacitance and inversely proportional to the square root of the peaking time.
- Contribution of the input current noise is independent of the input capacitance and proportional to the square root of the peaking time.
- Contribution of the input voltage flicker noise to the ENC is proportional to the total input capacitance and independent of the peaking time.

One can now optimize the front-end system taking into account various requirements and constraints implied by a particular application. There are two parameters determined by the detector geometry: detector capacitance and detector leakage current. Both parameters are proportional to the strip/pixel area and depend on the detector fragmentation, which directly influence position-sensitive resolution

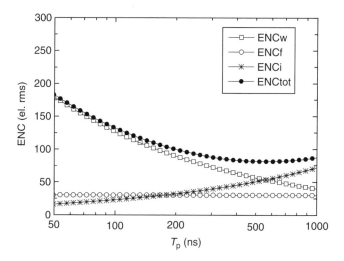

FIGURE 29.6 Contributions of different components of the preamplifier to total ENC versus peaking time.

of the system. Additionally, the detector leakage current is a strong function of temperature, so lowering the temperature is a way to reduce the leakage current and the associated short noise. From formulas (29.3) through (29.5), it is obvious that the values of detector bias resistor R_{bias} and feedback resistor R_{fed} should be relatively high (in the range of tens or hundreds of MΩ) not to contribute to the noise. In the case of R_{fed}, the situation is more complicated, because the high value of this resistor limits the high rate performance of the signal processing chain.

The next parameter, which can be considered as an input to formula (29.5), is the peaking time T_p. For cases without constraint on the peaking time, as for any low rate experiment, one can find an optimum peaking time for other given parameters, voltage, and current noise spectral densities and detector capacitance [20]. However, in many applications, high rate capability is a serious requirement, which has to be taken into account as a limitation for the maximum allowable peaking time.

For a CSA followed by a fast shaper (with short peaking time T_p), the ENC is usually dominated by the voltage noise of the input transistor and is given as

$$\text{ENC} \approx C_{in} \sqrt{\frac{F_v}{T_p} \times 4kT \frac{\gamma}{g_m} + F_{vf} \times \frac{K_f}{C_{ox}WL}} \tag{29.6}$$

For the really short peaking time, if

$$T_p < \frac{F_v}{F_{vf}} \times 4kT \frac{\gamma}{g_m} \times \frac{C_{ox}WL}{K_f} \tag{29.7}$$

the contribution of $1/f$ noise is always smaller than that of the channel thermal noise. In this case, the optimum transistor dimensions can be found in the following way. Under the assumption that the input transistor works in the strong inversion, the C_g may be calculated as

$$C_g = \frac{2}{3}C_{ox}WL + 2C_{ov}W \tag{29.8}$$

where C_{ov} is the overlap gate-diffusion capacitance per channel width unit, equal for drain and source. The transconductance g_m is given by

$$g_m = \sqrt{2\mu C_{ox} \frac{W}{L} I_{DS}} \tag{29.9}$$

The minimization of the ENC, in the case of dominant thermal noise component, leads to the optimum transistor width in the strong inversion W_{SIopt} given by

$$W_{\text{SIopt}} = \frac{C_{\text{det}} + C_{\text{fed}}}{2C_{\text{ox}}L_{\text{min}} + 6C_{\text{ov}}} \qquad (29.10)$$

However, in a modern multichannel CSA, it often happens that the input transistor operates moderate and weak inversion regions, where the abovementioned formula is not valid any more. An adequate variable used to determine the actual inversion level of a MOS transistor operating in saturation is its normalized forward current i_f [21]:

$$i_f = \frac{I_{\text{DS}}}{2n\mu C_{\text{ox}}(W/L)\varphi_{\text{T}}^2} \qquad (29.11)$$

where
  I_{DS} is the drain current
  n is the subthreshold slope factor
  μ is the carrier mobility
  φ_{T} is the thermal voltage

It is commonly assumed, that the weak inversion takes place for $i_f < 0.1$, while the strong inversion region is for $i_f > 10$ [22]. Therefore, MOS transistor has a two decade current transition region called the moderate inversion region, bounded by $i_f = 0.1$ and $i_f = 10$. In a modern multi-channel CSA, the input transistor often operates in this region (see Figure 29.7.) since the current i_f is relatively low because of the following factors:

- High-density front-end systems are designed for a low-power operation, so the I_{DS} current is strongly limited.
- Thermal noise optimization of the CSA input transistor often leads to a high value of g_m and large W/L ratio.
- In modern submicron technologies, smaller channel length L and higher capacitance C_{ox} are available.

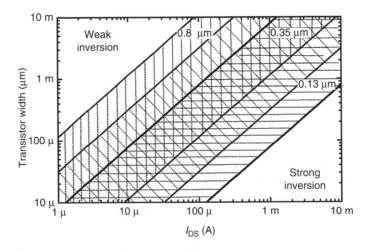

FIGURE 29.7 Limits of the moderate inversion region calculated from Equation 29.1 at $i_f = 0.1$ and at $i_f = 10$. The calculations were performed for three CMOS technology generations 0.13, 0.35, and 0.8 μm, assuming in each case the minimum transistor length available. (From Grybos, P., Idzik, M., and Maj, P., *IEEE Trans. Nucl. Sci.*, 54, 555, 2007. With permission.)

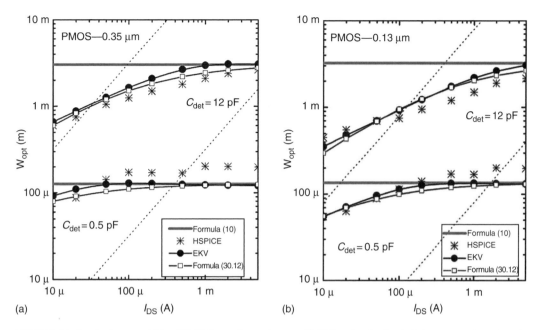

FIGURE 29.8 Optimum width of PMOS transistors for two CMOS technologies (minimum transistor length is used): (a) 0.35 μm and (b) 0.13 μm (dotted lines show the limits of moderate inversion region). (From Grybos, P., Idzik, M., and Maj, P., *IEEE Trans. Nucl. Sci.*, 54, 555, 2007. With permission.)

In that case, the new formula for the optimum transistor width must be used:

$$W_{opt} = W_{SI} \frac{1}{1 + A \left(\frac{W_{SI}}{i_{noW}} \right)^m} \tag{29.12}$$

where

$$i_{noW} = \frac{I_{DS}}{2n\mu C_{ox}(1/L)\varphi_T^2} \tag{29.13}$$

and $m = 0.61$ and $A = 0.25$ are the fitting parameters. This formula is valid for input transistor working in weak, moderate, and strong inversions. The comparisons of the results obtained from above formula with numerical HSPICE simulations (BSIM3 transistor models) and analytical calculation using EKV model are shown in Figure 29.8.

29.4.2 RESET BLOCK

After integration of the current pulse in the preamplifier, the feedback capacitor C_{fed} should be discharged by the reset block (Figure 29.2) during a short period of time to prevent piling up of pulses in the preamplifier. There are two basic techniques used for discharging the feedback capacitance: switch reset and continuous discharge. A switch reset technique is shown schematically in Figure 29.9a. A switching circuit periodically resets the feedback capacitor. Such solution is commonly used in ASICs (Application Specific Integrated Circuit) for the readout of silicon strip detectors in the particle physics experiments on the collider, where, if the signals appear, they appear synchronously in all the channels. The trigger signal for discharging the capacitors is provided by the central clock of the experiment. In x-ray measurements, the signals appear randomly in time and independently in each channel. After receiving a signal from the strip, the circuit has to generate the trigger signal for discharging the capacitor. In order to generate such a trigger signal, one needs to implement a

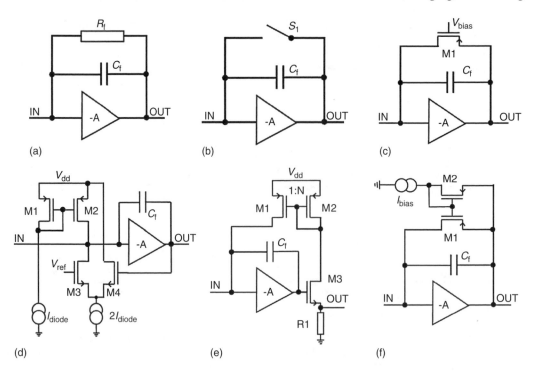

FIGURE 29.9   The most frequently used rest systems in CSA feedback.

threshold discriminator in every channel. The disadvantage of this solution is a possible problem with the switching noise. One could also apply the reset signal to all channels synchronously, after a certain period of time, having in mind the maximum rate of input pulses and the saturation limit of CSA; however, such solution results in additional dead time of the whole system.

Continuous discharging can be completed either by a resistor in parallel to the capacitor or by a controlled current source (see Figure 29.9b through 29.9f). In either case, the discharging component contributes to the parallel noise at the CSA input. In order to limit this noise source, one should use a large value resistor or a low discharging current. Using a physical resistor (Figure 29.9b) is simple, but it is difficult to obtain the large value resistance with a low parasitic capacitance. Instead of a simple resistor, many designers use a feedback metal-oxide-semiconductor field-effect transistor working in triode or saturation region [23–25]. This is the compact solution, with the possibility to control feedback resistance, however, nonlinear effects must be taken into account.

More complex possibilities are shown in Figure 29.9d through 29.9f:

- Discharge system shown in Figure 29.9d uses a differential stage. The baseline recovery after signal integration is achieved by the low-frequency feedback loop that sets the output voltage of the CSA to a reference voltage V_{ref} [26,27]. The effective feedback resistance is equal to $R_{\text{fed}} = 1/g_{m1}$.
- The configuration shown in Figure 29.9e uses a current conveyor feedback [28]. In response to a signal at CSA output, a reference current is produced in a low-value resistor R. This reference current is significantly reduced in the network based on current mirrors and discharges the feedback capacitor.
- The technique shown in Figure 29.9f is also based on a current mirror and uses a current source [29]. If there is no activity, feedback transistor M2 stays in the linear region. When the signal appears, the M2 enters the saturation region and the copy of bias current I_{bias} discharges the capacitor.

There are also two aspects, which should be taken into account while choosing of one of the above options of the reset system:

- The leakage current, which in the case of direct current (DC)-coupled detector, should be automatically accommodated by the CSA; the good candidates are, for example, the solution in Figure 29.9c (with the transistor working in saturation region) and Figure 29.9d.
- The long decay time constant of the preamplifier output signal, which produces the limitations of the pulse rate due to the piling up. Possible solution is the implementation of the pole-zero cancellation circuit, which is relatively easy in the cases shown in Figure 29.9c and 29.9e.

29.5 SHAPER STAGE

The shaper stage after the CSA is added to perform the following tasks:

- Filter the CSA output signal to improve the SNR in the system.
- Add additional gain in the signal processing chain.
- Shorten the pulse duration to reduce the possibility of pile-up pulses.

The choice of the filter type and order (which determines the peaking time) strongly depends on the required energy resolution and high rate operation requirements. There is a wide range of shapers built in hybrid technologies and using components of the shelf. However, very strong additional requirements of multichannel ICs are a low-power budget and a small area occupied by the single channel. Probably for this reason, the simple semi-Gaussian pulse shaper of type CR-(RC)^n, which consists of one RC differentiator and n integrators, becomes one of the most popular in the multichannel ICs. For the RC-(RC)^n filter with the same integrator and differentiator time constants $\tau_i = \tau_d = \tau$, the transfer function is given by [16]

$$H(s) = \left[\frac{s\tau}{1 + s\tau}\right]\left[\frac{A}{1 + s\tau}\right]^n \tag{29.14}$$

where A is the DC gain of the integrator. The noise power spectrum at the CSA output is weighted by the above transfer function and this improves the SNR in the detector system. The constants F_v, F_{vf}, and F_i, which appear in formula (29.5) for the equivalent noise charge, are directly related to the filter type.

The selection of the filter order also modifies the pulse response of the CSA. Assuming an ideal voltage step (equal to Q_{in}/C_{fed}) at the CSA output and taking the transfer function of the filter given by Equation 29.14, one obtains the shaper output signal in the time domain as

$$V_{out} = V_{max}\left(\frac{t}{T_p}\right)^n \exp\left(-\frac{nt}{T_p}\right) \tag{29.15}$$

with

$$V_{max} = \frac{Q_{in}A^n n^n}{C_{fed}n!\exp n} \tag{29.16}$$

where the peaking time $T_p = n\tau$. The family of shaper output pulses for a given time constant τ is shown in Figure 29.10a. Increasing the filter order results in decreasing the signal amplitude, but makes the pulse more symmetrical. Higher filter order is more suitable for high rate application, but to obtain the same peaking time for the higher shaper order, one should shorten the time constant of the filters. The family of the pulses with the same peaking time, but different orders, is shown in

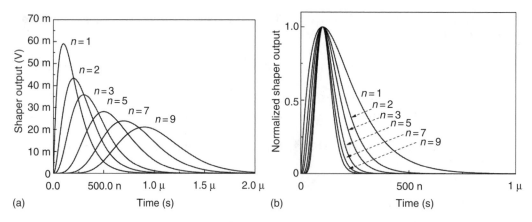

FIGURE 29.10 Shaper output for different filter orders with (a) the same time constant τ and (b) the same peaking time T_p.

Figure 29.10b. For higher order filters, the shaper output pulse returns to the baseline faster, and this directly influences the high rate operation and reduces the probability of pile-up pulses [30].

Higher filter order requires more power and occupies more silicon area. For this reason, many multichannel readout ICs use the shaper RC-(RC)^2, while the readout electronics for pixel detector has a very simple shaper stage. Additionally, for the applications using very short peaking time (high x-ray intensity), the time constants of the CSA have to be taken into account and the CSA and shaper stage must be analyzed together.

29.6   ANALOG-TO-DIGITAL CONVERSION IN MULTICHANNEL IC

Further signal processing at the shaper outputs in the multichannel chip depends strongly on the specific requirements of the foreseen applications. The ideal system should ensure the position and energy measurements. This means that each channel should contain fast high-resolution analog-to-digital converter (ADC) and a big memory buffer. Nowadays the constraints associated with area and power consumption of multichannel ASIC exclude such an approach. Instead, the existing multichannel ASICs for fast digital imaging use intermediate solutions, which reduce the energy resolution or single channel maximum throughput. The most frequently used solutions are as follows (see Figure 29.11):

- One or a few discriminators per single channel [15,30–38]
- A simple low-resolution ADC per single channel [39–41]
- Multiplexing technique with high-resolution ADC [42,43]

The architecture with one or a few amplitude discriminators per single channel is used in various imaging techniques (often employing monoenergetic x-rays), in which it is sufficient to measure spatial distributions of the x-rays of energies above a given threshold or within a given energy window. The existing solutions have even up to nine discriminators per single channel [30]. The counters after the discriminators store the information about the number of signals, which were higher than given threshold level. As each channel works independently, this solution is suitable for very high intensity of x-rays (even more than 1 MHz per single channel). This is a significant advantage in systems comprising hundreds or thousands of channels. Additionally, using this simple approach, by scanning the discriminator threshold level and measuring the integral distribution of pulse amplitudes, one could extract spectroscopic information of x-ray radiation.

If the requirements concerning the resolution of the ADC are not very demanding (5–6 bits), one can implement a simple low-resolution ADC in each channel. One of the proposed solutions is based on what is called time-over-threshold (ToT) method. The idea is shown schematically in

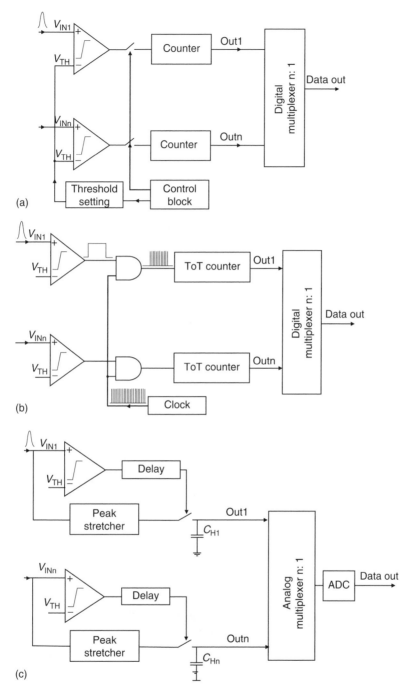

FIGURE 29.11 Three possible schemes of readout at the shaper output: (a) binary architecture, (b) analog architecture employing ToT, and (c) analog architecture employing multiplexing of analog signals.

Figure 29.11b. In the ToT method [39,40], the analog signal from the front-end circuit is applied to a simple threshold discriminator. Signals above the threshold generate a logic pulse at the discriminator output. The duration time of this pulse is measured in a simple way by counting pulses from a clock generator over the period equal to the duration of the discriminator response. The width of the discriminator pulse is a nonlinear function of pulse amplitude and knowing this function for a given pulse shape, one could determine the pulse amplitude at the discriminator input. The basic limitations of ToT processor are the measurement range limited by the discrimination level, low accuracy for small signals just above the threshold, and sensitivity to the time jitter of the discriminator, especially for low amplitudes. The system with ToT processor is relatively simple and because of low power consumption, it can be implemented in every channel. This solution has been implemented in several ASICs for readout of strip/pixel detector in the particle physics experiments. However, in the case of x-ray imaging, one needs to store the data. The possible approach solution is to multiplex all the channels into one serial output or elaborate a scheme of sparse readout [41].

A fully analog readout scheme, employing a standard ADC, is shown schematically in Figure 29.11c. Each channel is equipped with a peak detector (PD) and a sample and hold (S&H) circuit. Because in x-ray applications the input pulses are statistically distributed in time, one needs to implement a threshold discriminator in each channel to generate a trigger signal for the S&H circuit. The analog signals from a certain number of channels are then multiplexed into one ADC, which, in most cases, is an external device, although one can consider integrating it in the front-end ASIC. An obvious limitation of the maximum channel throughput is the multiplexing rate, which is less critical when we reduce the number of channels per ADC and increase the number of ADCs in the system. Another aspect specific for such architecture concerns the control of the multiplexer and the ADC operation. One can either run the multiplexer and the ADC continuously [42], allowing for some probability of the pile-ups in the S&H circuits, or trigger the multiplexer and the ADC upon a signal occurring in the detector [43]. The most commonly used scheme is based on OR gate taking inputs from all the channels. Then each signal occurring in any of the channels triggers the readout sequence. Another trend of the development is to integrate in the front-end ASICs analog memory buffers that can serve as derandomizers and increase the channel throughput in this way (see Chapter 4).

29.7   PHYSICAL IMPLEMENTATION ASPECTS

29.7.1   CROSSTALK IN MIXED-MODE IC

The IC for digital x-ray imaging contains both analog and digital blocks. A major disadvantage of such a mixed-signal IC is the increased interaction (crosstalk) between different parts of the circuit on the same die. Disturbances produced in the fast digital circuit can be transferred to analog-sensitive front-end (see Figure 29.12) through the common power supply lines (supply bounce) or through the common substrate (substrate noise) and significantly reduce the analog chip performance.

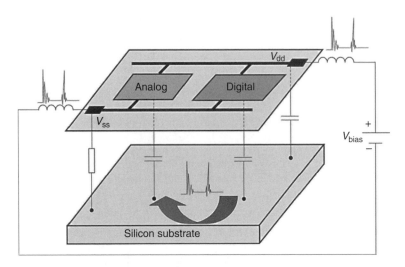

FIGURE 29.12   Disturbance transfer from digital to analog blocks.

TABLE 29.2

Recommendations for Crosstalk Reduction in Mixed-Mode IC

Reduce the Noise Generation in Digital Blocks	Increase the Immunity of Analog Part	Isolation Techniques
Reduce the number of synchronously working digital blocks.	Analog blocks should have a high power supply rejection ratio.	Limit the supply bounce by reducing the parasitic inductance and resistance of connections between the chip and the external world: proper package, multiple bonding pads for power supply bus.
Employ design styles, which generate less disturbances, e.g. current mode logic, source-coupled logic, or low-voltage differential standard (LVDS), especially in I/O blocks.	Fully differential circuits are recommended with high common mode rejection ratio	
Introduce a phase shift between digital clock and analog sampling.	Positive-channel mixed oxide semi-conductor (PMOS) transistors are preferable for the signal handling. In the case of p-type substrate, PMOS transistors are in n-well.	Use separate power supply buses, guard rings, pads, and bonding wires for analog and digital blocks.
Pay special attention to the nodes with large parasitic capacitance to the substrate (large fanout), like buses, I/O drivers, clock distribution networks. The rise and fall times for drivers of these nodes should be as large as the design constraints allow. The voltage swing on these nodes should be minimized.	N-well connected to a clean power supply can be used to shield the transistors from the substrate.	Design carefully the floor planning and the distribution of power supply in the chip. Digital signals should not be routed over the analog portion of the chip or close to the sensitive lines.
	N-channel mixed oxide semi-conductor (NMOS) transistors ones can be used as components of the DC current and it is better to refer them to the substrate, than to the clean supply.	Sensitive and noisy pads must be kept as far as possible from each other due to mutual inductance of the bonding wires or package pins.
	Signals to another subcircuit or to the outside world should be made concurrently with their reference through the dedicated path, rather than rely on the common ground.	Use vertical shielding for large area sensitive devices (even input pads) and for noisy routing channels, like clocks.
	All sensitive high-impedance nodes have to be kept inside the chip, because the chip parasitics are much smaller than those of package or printed circuit board.	Use different decoupling techniques between power supply and the ground: simple capacitors, resistance, inductance, capacitance (RLC) filters, active supply bypass or active guard rings.
	Use minimum required bandwidth for the signal processing.	

Source: Adapted from Refs. [44–60].

There are three main steps to minimize the crosstalk in the mixed-mode IC: (i) to reduce the amount of generated switching noise, (ii) to increase the immunity of the analog part, and (iii) to introduce proper isolation between the analog and the digital part. The detailed recommendations for each of these steps are summarized in Table 29.2.

As an example, the layout of 64 channel mixed-mode readout ASIC, called RX64, which follows the above mentioned rules, is shown in Figure 29.13.

29.7.2 MATCHING PERFORMANCE

The readout IC used in digital x-ray imaging consists of many identical channels, which are supposed to have the same analog parameters like gain, offset, noise, and speed performance. An inherent feature of the VLSI technology is a mismatch that causes time-independent random variations of physical parameters of identically designed devices. The second effect is the asymmetry in the

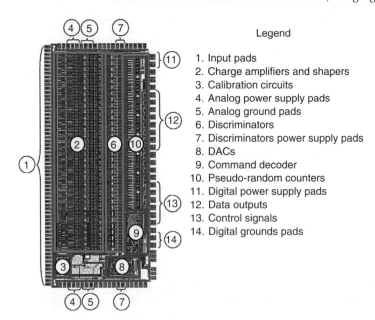

Legend

1. Input pads
2. Charge amplifiers and shapers
3. Calibration circuits
4. Analog power supply pads
5. Analog ground pads
6. Discriminators
7. Discriminators power supply pads
8. DACs
9. Command decoder
10. Pseudo-random counters
11. Digital power supply pads
12. Data outputs
13. Control signals
14. Digital grounds pads

FIGURE 29.13 Layout of 64-channel mixed-mode ASIC RX64. (From Grybos, P., Low noise multichannel integrated circuits in CMOS technology for physics and biology applications, Monograph 117, AGH Uczelniane Wydawnictwa Naukowo-Dydaktyczne, Poland, 2002, pp. 40–76.)

multichannel circuit in the bias condition and layout, which is also a source of additional error in large readout systems.

The problem of the matching must be taken into account at all design stages, starting from the circuit front-end architecture, through the layout drawing and till circuit packaging and bias condition. The most important rules to improve matching are listed below [61–75]:

- Transistors, resistors, and capacitors of larger geometrical dimensions match better but this is contrary to the high-speed and small–chip area requirements. At first, one should find matching sensitive devices in the circuit and then consider a compromise between matching and other design constraints.
- Some circuit architectures are less sensitive to the mismatch effects. For example, the gain of an amplifier should be based on the ratio of two identical devices like capacitors or resistors of the same type. AC coupling between consecutive stages in the signal processing chain eliminates the problem of offset propagation.
- Using the correction digital-to-analog converters (DACs) in each channel separately to tune the gain, offset, etc. is often the only possible solution in the case of very fast signal processing in the front-end electronics. After the calibration measurements, the correction DACs are loaded individually one by one to each channel. This protocol guarantees the cancellation of unwanted parameter spread from channel to channel.
- Evaluation of the matching performance of a design can be done employing Monte Carlo simulation. In this way, one can quickly estimate channel-to-channel matching performance, identify critical devices, and then optimize their dimensions or relevant bias conditions. One should remember that specified in usual way matching parameters, for example for transistors, are valid under certain assumptions (common centroid layout, identical surrounding, no metal crossover, etc.).
- Symmetry improves matching. Symmetry means not only the same layout of matched devices, but also the same temperature, bias condition, parasitic, metal coverage, and the surrounding environment.

- Packaging introduces not only parasitic components (resistance, capacitance, inductance), but also is a source of mechanical stress and heat nonuniformity distribution that adversely affects the matching. It is especially important for multichannel systems, in which several identical chips are mounted together on the same custom-designed multilayer board. So the design of such a board and the process of mounting chips should be handled with special care to limit all the above effects.

29.8 EXAMPLES OF MULTICHANNEL READOUT ICs

29.8.1 DEDIX IC—FAST DIGITAL X-RAY IMAGING

Dual energy digital imaging of x-ray (DEDIX IC) has been developed at the AGH UST Cracow, Poland, as a fast readout ASIC for silicon strip detector [33]. The chip comprises six basic blocks (see Figure 29.14a): 64 analog front-end channels, 2×64 counters with RAM, an input–output block, a control command decoder, control DACs, and a calibration circuit. The ASIC is designed

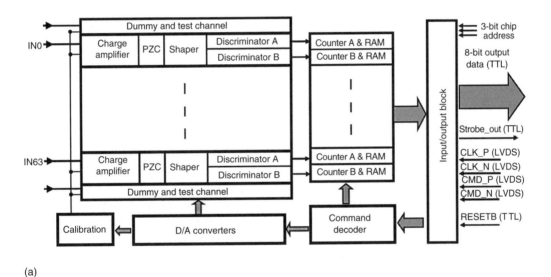

(a)

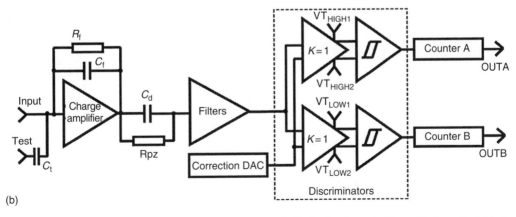

(b)

FIGURE 29.14 Block diagram of (a) DEDIX chip and (b) single channel. (From Grybos, P., Maj, P., Ramello, L., Swientek, K., *IEEE Trans. Nucl. Sci.*, 54, 1207, 2007. With permission.)

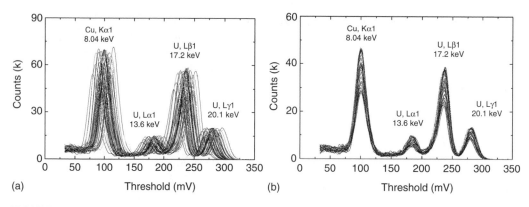

(a)   Threshold (mV)   (b)   Threshold (mV)

FIGURE 29.15 Spectra of Pu-238 radioactive source and Cu Kα line measured with silicon strip detector and DEDIX IC: (a) correction OFF and (b) correction ON. (From Grybos, P., Maj, P., Ramello, L., Swientek, K., *IEEE Trans. Nucl. Sci.*, 54, 1207, 2007. With permission.)

in 0.35 μm austriamicrosystems CMOS process and the layout total area is 3900×5000 μm^2. The block diagram for a single channel is shown in Figure 29.14b. Each channel is built of a CSA with a pole-zero cancellation circuit, a shaper CR-RC^2 with a peaking time of 160 ns, two discriminators, and two independent 20-bit counters.

The threshold voltages for two discriminators in a single channel are set independently, but are common for all 64 channels of the ASIC. Because from the CSA input up to the discriminator inputs, the signal processing chain is DC-coupled, a correction DAC (working in each channel independently) is necessary to minimize the effects of the DC level spread at the discriminator inputs. The examples of the x-ray spectra measured with 64-channel DEDIX chip with the correction circuit ON and OFF are shown in Figure 29.15.

The spectra have been obtained by scanning the discriminator threshold and measuring in the integral distribution of pulses amplitudes. The differences of intensity in different channels are due to small dimensions of x-ray source (different x-ray intensities across the strips of the detector). The effective threshold voltage spread calculated to CSA input is below 7 electron rms.

The DEDIX chip is able to work properly up to the rate of 1 MHz per channel for statistically distributed photons from the x-ray tube, without gain, offsets, and noise degradation. The exemplary results obtained using 8 keV photons from x-ray tube are shown in Figure 29.16. The plots show the

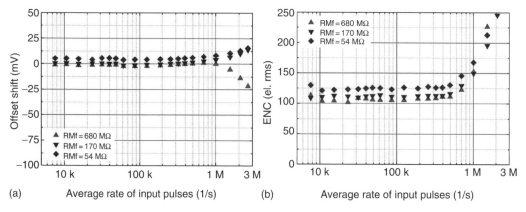

(a)   Average rate of input pulses (1/s)   (b)   Average rate of input pulses (1/s)

FIGURE 29.16 High count rate characteristic of DEDIX chip: (a) DC level shift at discriminator input and (b) noise. (From Grybos, P., Maj, P., Ramello, L., Swientek, K., *IEEE Trans. Nucl. Sci.*, 54, 1207, 2007. With permission.)

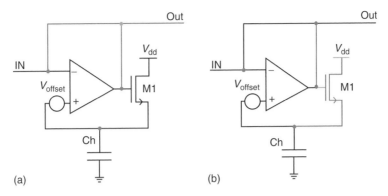

FIGURE 29.17 Simplified scheme of two-phase PD: (a) WRITE phase and (b) READ phase. (From De Geronimo, G., O'Connor, P., Kandasamy, A., *Nuc. Instr. Meth.*, A484, 2002. With permission.)

circuit performance for different settings of CSA feedback resistance R_{Mf}. The chip uses two power supply voltages: 2.2 V (analog parts) and 3 V (discriminators and digital blocks), and consumes about 5 mW per channel. This IC is mainly used in diffractometry measurements.

29.8.2 Peak Detector Derandomizer ASIC

Using fully analog-to-digital conversion per single channel for high rate of input pulses is impractical for most large multichannel systems. One of the solutions is to multiplex analog signal from some channels into one ADC, but Poisson fluctuations of the input pulse rate increase speed requirements both for multiplexer and ADC. To reduce speed requirements for ADC, a group from Instrumentation Division, Brookhaven National Laboratory, Upton, New York, has developed PD derandomizer (PDD) ASIC [76–78]. This circuit works as a data-driven analog first-in, first-out memory buffer between the preamplifiers and the ADC. Amplitudes of signals occurring randomly in time are stored in such a buffer and read out with rate comparable to the average rate of input pulses.

The basis of the chip is two-phase PD, shown schematically in Figure 29.17, which has high absolute accuracy (0.2%) and linearity (0.05%). In the WRITE phase, PMOS transistors M1 load the hold capacitor C_h up to the peak value of input pulse. In the READ phase, it works as unity gain buffer. This PD scheme is offset-free configuration, because in the WRITE–READ cycle, the error of amplifier offset is cancelled.

The PDD ASIC contains 32 input channels and performs simultaneous amplitude and time measurements, together with delivery of the channel address information (Figure 29.18). The chip has

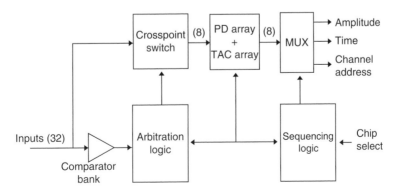

FIGURE 29.18 Block diagram of amplitude and time measurement ASIC. (From O'Connor, P., De Geronimo, G., Kandasamy, A., *IEEE Trans. Nucl. Sci.*, 50, 892, 2003. With permission.)

been fabricated in 0.35 μm CMOS technology. This fully self-triggered ASIC multiplexes 32-shaped input signals into an array of eight PDs (which act as a derandomizing analog memory) with associated time-to-amplitude converters (TAC). Thirty-two comparators monitor the inputs for activity. When the input pulse arrives, the arbitration logic routes the pulse to the next available PD/TAC. The connection is maintained until the peak amplitude is found. This value is stored on the PD hold capacitor until the external ADC is ready to convert this sample.

In comparison with popular techniques, such as a track-and-hold and analog memory, the PDD ASIC enables efficient pulse height measurements at 20 to 300 times higher rates. It can operate at rates up to 300 kHz per channel with 99% efficiency.

29.8.3 MEDIPIX2 IC

Medipix2 [15,79] has been developed at CERN using six-metal CMOS 0.25 μm process. Its readout electronics consists of array of 256 × 256 readout channels and that gives the matrix of 65,536 identical elements in total, with the same geometry as the pixel detector. Each readout pixel occupies an area of 55 × 55 μm, containing 500 transistors and has a static power consumption about 8 μW. A bump bonding technique is used to connect the detector to 20 μm wide octagonal input pads of the readout electronics. The architecture of the single cell is shown in Figure 29.19.

Single pixel cell consists of

- CSA with the differential architecture to reduce the substrate and power supply noises.
- Active structure in the CSA feedback, which provides fast discharge of feedback capacitance and compensation of detector DC leakage current [80].
- Two discriminators with independent threshold settings and independent 3-bit correction DACs. The outputs of the discriminators could be masked in the case of malfunction or excessive noise. The subsequent logic block produces the pulse if the amplitude of the output CSA signal falls within a defined energy window.

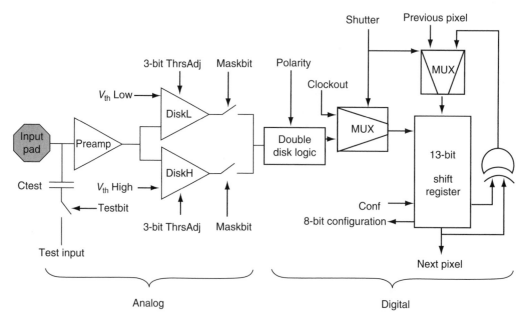

FIGURE 29.19 Single readout pixel cell. (From Llopart, X., Campbell, M., Dinapoli, R., San Segundo, D., Pernigotti, E., *IEEE Trans. Nucl. Sci.*, 49, 2279, 2002. With permission.)

- Shift register, which can operate in two modes depending on the Shutter signals. In one mode, it works as a 13-bit pseudorandom counter and its contents are increased by one by every discriminator pulse. In the second mode, an external clock is applied and it is used to shift the data from pixel to pixel. This mode of operation applies for both loading the value of the configuration DACs in each pixel and reading out the counter contents.

The readout electronics is able to process the signals from the input carriers of both types. The noise performance is 140 el. rms, while the unadjusted threshold variation is around 360 el. rms. The maximum counting rate is about 1 MHz/pixel.

The floor plan of Medipix2 chip is shown in Figure 29.20. Its total area is 14.1×16.1 mm, and the sensitive matrix area of pixels covers 87% of the entire chip (i.e. $1.98 \, \text{cm}^2$). The periphery area, which contains fast shift registers, DACs converters, I/O control logic, and Low-Voltage Differential Signaling (LVDS) drivers and receivers, is located at the bottom part of the chip. Such floor plan enables minimization of the dead area in multichip module.

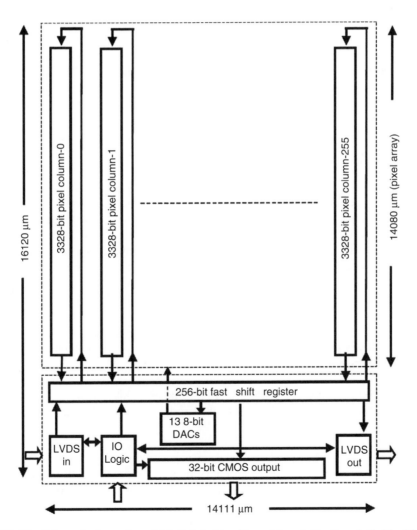

FIGURE 29.20 Schematic floor plan of Medipix2 chip. (From Llopart, X., Campbell, M., Dinapoli, R., San Segundo, D., Pernigotti, E., *IEEE Trans. Nucl. Sci.*, 49, 2279, 2002. With permission.)

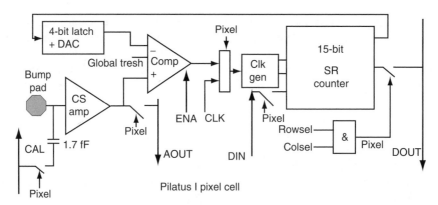

FIGURE 29.21 Schematic view of the pixel unit cell. (From Brönnimann Ch. Eikenbersy, E.F., Henrich, B., Horisberger, R., Huelson, G., Pohl, E., Schmitt, B., Schulze-Briese, C., Suzuki, M., Pomizaki, T., Toyokawa, H., and Wagner, A. *J. Synchrotron Rad.*, 13, 120, 2006. With permission.)

When the pixels are read out or their registers are loaded, the whole pixel matrix is organized in 256 columns (Figure 29.21). The readout of the chip can be performed in serial mode using fast LVDS logic or in parallel via 32-bit CMOS bus. Using the 100 MHz clock, the first mode of the readout takes less than 9 ms, while in the second mode, it is done in 266 μs. The whole chip contains about 33 million transistors.

Medipix2 is very attractive for different applications and it is used in many scientific experiments [78]. Other projects based on Medipix2 IC, which aim to add new functionality or increase chip performance, like Medipix3 [81] and Timepix [82] are under development.

29.8.4 PILATUS IC

At the same time, the similar solution of the readout electronics for pixel detector has been developed for the x-ray measurements at the beamlime of the Swiss Light Source (SLS), as a part of the PILATUS (PIxeL ApparaTUs for the SLS) project [15,83–85]. The goal of this project was to build a hybrid pixel system covering approximately the area of $40 \times 40 \, cm^2$ with 2000×2000 pixels. The first generation PILATUS I chip was designed in 2000 at the Paul Scherrer Institut (PSI), Villigen, Switzerland, using DMILL radiation-tolerant CMOS process (Atmel Temic SA, Nantes, France).

PILATUS I contains an array of $44 \times 78 = 3432$ pixels with a pixel size of $217 \times 217 \, \mu m$. The active area spans $10 \times 17 \, mm^2$. Each pixel contains a low-noise CSA, a single-level comparator with a 4-bit individual threshold adjustment, and 15-bit shift register counter. The noise of bump-bonded chip is 75 el. rms. The total power consumption is 100 μW per pixel. The lowest achievable threshold of the chip is about 3 keV, which allows to measure ^{22}Ti Kα radiation (4.5 keV).

The chip operates in two modes:

- Counting mode, in which all pixels count incoming x-rays photons
- Readout mode, in which the data are readout serially with clock frequency of 10 MHz, within the readout time 6.7 ms

These ICs are used to build the PILATUS module, which is an array of 8×2 chips bump-bonded to silicon pixel sensors. The variation of the threshold settings is an approximately linear function of the comparator voltage. After trimming on the module, which contains about 54,000 pixels (16 chips × 3432 pixels), the spread of the threshold voltage is about 6%, i.e. for the average threshold setting at 1680 el., the threshold dispersion is 112 el. rms.

To improve the yield of good pixel, the chip has been redesigned. The second version of the chip PILATUS II was designed in 2004 at the PSI using the UMC 0.25 μm technology, in which radiation tolerance was achieved by design [81]. The active chip area is 10×17 mm^2 and it consists of an array of 60×97 pixels. Each pixel has a size of 172×172 μm. The single pixel contains similar blocks as the first version, but a comparator has a 6-bit individual threshold adjustment and the counter capacity is 20 bit. The authors claim the count rate up to $\sim$1.5 MHz/pixel/s. The chips mounted in the PILATUS module (an array of 8×2 chips) are readout in parallel within the readout time of about 2 ms. Using these PILATUS modules, PILATUS 6M detector system has been built. This system is composed of 5×12 modules with 2463×2527 pixels and has a total active area of 424×435 mm^2.

REFERENCES

1. Besch H.J. Radiation detectors in medical and biological applications, *Nucl. Instr. Meth.*, A419, 202–216, 1998.
2. Mikulec B. Development of segmented semiconductor arrays for quantum imaging, *Nucl. Instr. Meth.*, A510, 1–23, 2003.
3. De Geronimo G., O'Connor P., Radeka V., and Yu B. Front-end electronics for imaging detectors, *Nucl. Instr. Meth.*, A471, 192–199, 2001.
4. Rossi G., Morse J., Labiche J.-C., Protic D., and Owens A.R. x-ray response of germanium microstrip detectors with energy and position resolution, *Nucl. Instr. Meth.*, A392, 264–268, 1997.
5. Rossi G., Morse J., and Protic D. Energy and position resolution of germanium microstrip detectors at x-ray energies from 15 to 100 keV, *IEEE, Trans. Nucl. Sci.*, 46(3), 765–773, 1999.
6. Vetter K. et al. Three-dimensional position sensitivity in two-dimensionally segmented HP-Ge detectors, *Nucl. Instr. Meth.*, A452, 223–238, 2000.
7. Amman M. and Luke P.N. Three-dimensional position sensing and field shaping in orthogonal-strip germanium gamma-ray detectors, *Nucl. Instr. Meth.*, A452, 155–166, 2000.
8. Chen J., Geppert R., Irsigler R., Ludwig J., Pfister J., Plötze T., Rogalla M., Runge K., Schäfer F., Schmid Th., Söldner-Rembold S., and Webel M. Beam test of GaAs strip detectors, *Nucl. Instr. Meth.*, A369, 62–62, 1996.
9. Smith K.M. GaAs detector status, *Nucl. Instr. Meth.*, A383, 75–80, 1996.
10. Eisen Y., Shor A., and Mardor I. CdTe and CdZnTe gamma ray detectors for medical and industrial imaging systems, *Nucl. Instr. Meth.*, A428, 158–170, 1999.
11. Gostilo V., Ivanov V., Kostenko S., Lisjutin I., Loupilov A., Nenonen S., Sipila H., and Valpas K. Technological aspects of development of pixel and strip detectors based on CdTe and CdZnTe, *Nucl. Instr. Meth.*, A460, 27–34, 2001.
12. Kalemci E. and Matteson J.L. Investigation of charge sharing among electrode strips for a CdZnTe detector, *Nucl. Instr. Meth.*, A460, 527–537, 2002.
13. Chen H., Salah A. Awadalla, Mackenzie J., Redden R., Bindley G., Bolotnikow A.E., Camarada G.S., Carini G., and James R.B. Characterization of traveling heater method (THM) grown $Cd_{0.9}Zn_{0.1}Te$ crystal, *IEEE Trans. Nucl. Sci.*, 54(4), 811–816, 2007.
14. Schieber M., Grybos P., Rodriguez A.E.C., Idzik M., Gaitan J.L., Prino F., Ramello L., Swientek K., and Wiacek P. Evaluation of mercuric iodide ceramic semiconductor detectors, *Nucl. Phys. B*, 61B, 321–329, 1998.
15. Llopart X., Campbell M., Dinapoli R., San Segundo D., and Pernigotti E. Medipix2: a 64-k pixel readout chip with 55 μm square elements working in single photon counting mode, *IEEE Trans. Nucl. Sci.*, 49(5), 2279–2283, 2002.
16. Sansen W. and Chang Z.Y. Limits on low noise performance of detector readout front ends in CMOS technology, *IEEE Trans. Circuits Syst.*, 37(11), 1375–1382, 1990.
17. Manghisoni M., Ratti L., Re V., and Speziali V. Submicron CMOS technologies for low-noise analog front-end circuits. *IEEE Trans. Nucl. Sci.*, 49(4), 1783–1790, 2002.
18. O'Connor P. and De Geronimo G. Prospects for charge sensitive amplifiers in scaled CMOS, *Nucl. Instr. Meth.*, A480, 713–725, 2002.
19. Gatti E. and Manfredi P.F. Processing the signal from solid-state detectors in elementary particle physics, *La Revista del Nuovo Cimento*, 9(1), 1–146, 1986.
20. Chang Z.Y. and Sansen W. Effect of $1/f$ noise on resolution of CMOS analog readout systems for microstrip and pixel detectors, *Nucl. Instr. Meth.*, A305, 553–560, 1991.

21. Enz C., Krummenacher F., and Vitoz E. An analytical MOS transistor model valid in all regions of operation and dedicated to low-voltage and low-current applications, *Analog Integrated Circuits Signal Processing*, 8, 83–114, 1995.

22. Grybos P., Idzik M., and Maj P. Noise optimization of charge amplifier with MOS input transistor working in moderate inversion region for short peaking times, *IEEE Trans. Nucl. Sci.*, 54(3), 555–560, 2007.

23. Beuville E., Borer K., Heijne E., Jarron P., Lisowski B., and Singh S.AMPLEX, a low-noise, low-power analog CMOS signal processor for multielement silicon particle detectors, *NIM A*, 288, 157–167, 1990.

24. Grybos P. Pole-zero cancellation circuit for charge sensitive amplifier with pile-up pulses tracking system, Proceedings of Nuclear Science Symposium, San Diego, CA, October 2006, pp. 226–230.

25. De Geronimo G. and O'Connor P.A. CMOS detector leakage current self-adaptable continuous reset system: Theoretical analysis, *NIM A*, 421, 322–333, 1999.

26. Ludewigt B., Jaklevic J., Kipnis I., Rossington C., and Spieler H. A high rate, low noise, x-ray silicon strip detector system, *IEEE Trans. Nucl. Sci.*, 41(4), 1037–1041, 1994.

27. Vandenbussche J., Leyn L., Van der Plas G., Gielen G., and Sansen W. A fully integrated low-power MOS particle detector front-end for space applications, *IEEE Trans. Nucl. Sci.*, 45(4), 2272–2278, 1998.

28. Sampietro M., Bertuccio G., Fasoli L. Current mirror reset for low-power BiCMOS charge amplifier, *NIM A*, 439, 373–377, 2000.

29. Blanquart L., Mekkaoui A., Bonzom V., and Delpierre P. Pixel analog cells prototypes for ATLAS in DMILL technology, *NIM A*, 395, 313–317, 1997.

30. De Geronimo G., Dragone A., Grosholz J., O'Connor P., and Vernon E. ASIC with multiple energy discriminator for high rate photon counting applications. *IEEE Trans. Nucl. Sci.*, 54(2), 303–312, 2007.

31. Comes G., Loddo F., Hu Y., Kaplon J., Ly F., Turchetta R., Bonvicini V., and Vacchi A. CASTOR: VLSI CMOS mixed analog-digital circuit for low noise multichannel counting applications, *Nucl. Instr. Meth.*, A377, 440–445, 1996.

32. P. Grybos et al. RX64DTH—A fully integrated 64-channel ASIC for a digital x-ray imaging system with energy window selection, *IEEE Trans. Nucl. Sci.*, 52, 839–846, 2005.

33. Grybos P., Maj P., Ramello L., and Swientek K. Measurements of matching and high count rate performance of multichannel ASIC for digital x-ray imaging systems, *IEEE Trans. Nucl. Sci.*, 54(4), 1207–1215, 2007.

34. Lundqvist M., Cederstrom B., Chmill V., Danielsson M., and Hasegawa B. Evaluation of a photon-counting x-ray imaging system, *IEEE Trans. Nucl. Sci.*, 48(4), 1530–1536, 2001.

35. De Geronimo G., O'Connor P., and Grosholz J. A generation of CMOS readout ASICs for CZT detector, *IEEE Trans. Nucl. Sci.*, 47(6), 1857–1876, 2000.

36. Bronnimann Ch. et al. A pixel read-out chip for PILATUS project, *Nucl. Instrum. Methods A*, 465, 235–239, 2001.

37. Lindner M., Blanquart L., Fischer P., Kruger H., and Wermes N. Medical x-ray imaging with energy windowing, *Nucl. Instrum. Methods A*, 465, 229–234, 2001.

38. Seller P. et al. Photon counting hybrid pixel detector for x-ray imaging, *Nucl. Instrum. Methods A*, 455, 715–720, 2000.

39. Becker R., Grillo A., Jacobsen R., Johnson R., Kipnis I., Levi M., Luo L., Manfredi P.F., Nyman M., Re V., Roe N., and Shapiro S. Signal processing in the front-end electronics of BaBar vertex detector, *Nucl. Instr. Meth.*, A377, 459–464, 1996.

40. Manfredi P.F., Leona A., Mandeli E., Perrazo A., and Re V. Noise limits in front-end system based on time-over threshold signal processing, *Nucl. Instr. Meth.*, A439, 361–367, 2000.

41. Feuerstack-Raible M. Overview of microstrip readout chips, *Nucl. Instr. Meth.*, A447, 35–43, 2000.

42. Fiorini C., Longoni A., and Buttler W. Multi-channel implementation of ROTOR amplifier for the readout of silicon drift detectors arrays, *Nuclear Science Symposium*, 2001, San Diego, CA, pp. 143–146.

43. Overdick M., Czermak A., Fisher P., Herzog V., Kjensmo A., Kugelmeier T., Ljunggren K., Nygard E., Pietrzik C., Schwan T., Strand S.-E., Straver J., Weilhammer P., Wermes N., and Yoshioka K. A "bioscope" system using double-sided silicon strip detectors and self-triggering read-out chips, *Nucl. Instr. Meth.*, A392, 173–177, 1997.

44. Blalack T. Design techniques to reduce substrate noise. In: Huijsing J., van de Plassche R., and Sansen W. (eds.), *Analog Circuit Design. Volt Electronics; Mixed-Mode Systems; Low-Noise and RF Power Amplifiers for Telecommunication*, Boston, MA: Kluwer Academic, 1999, pp. 193–217.

45. Su K., Loinaz M.J., Masui S., and Wooley B.A. Experimental results and modeling techniques for substrate noise in mixed-signal integrated circuits, *IEEE J. Solid-State Circuits*, SC-28(4), 420–430, 1993.
46. Gharpurey R. Modeling and analysis of substrate coupling in integrated circuits, Ph.D. dissertation, University of California, Berkeley, CA, 1995.
47. Heijningen M., Compiet J., Wambacq P., Donnay S., Engels M.G.E., and Bolsens I. Analysis and experimental verification of digital substrate noise generation for epi-type substrates, *IEEE J. Solid-State Circuits*, SC-35(7), 1002–1008, 2000.
48. Clement F. Technology impacts on substrate noise. In: Huijsing J., van de Plassche R., and Sansen W. (eds.), *Analog Circuit Design. Volt Electronics; Mixed-Mode Systems; Low-Noise and RF Power Amplifiers for Telecommunication*, Boston, MA: Kluwer Academic, 1999, pp. 173–192.
49. Clement F. Substrate noise coupling analysis in mixed-signal ICs, Presentation from the Workshop on Substrate-Noise Coupling in Mixed-Signal ICs, IMEC, Leuven, Belgium, September 5–6, 2001.
50. Ingels M. and Steyaert M. Design strategies and decoupling techniques for reducing the effects of electrical interference in mixed-mode ICs, *IEEE J. Solid-State Circuits*, SC-32(7), 1136–1141, 1997.
51. Makie-Fukuda K., Kikuchi T., Matsuura T., and Hotta M. Measurement of digital noise in mixed-signal integrated circuits, *IEEE J. Solid-State Circuits*, SC-30(2), 87–92, 1995.
52. Schmerbeck T. Practical aspects in analog and mixed-mode IC design, Laussanne, Switzerland, EPFL Electronics Laboratories Advanced Engineering Course CMOS and BiCMOS IC Design'99: Practical Aspects in Analog and Mix-Mode ICs, 1999.
53. Nauta B. and Hoogzaad G. Substrate bounce in mixed-mode CMOS. In: Huijsing J., van de Plassche R., and Sansen W. (eds.), *Analog Circuit Design. Volt Electronics; Mixed-Mode Systems; Low-Noise and RF Power Amplifiers for Telecommunication*, Boston, MA: Kluwer Academic, 1999, pp. 157–171.
54. Pedder D. Interconnection and packaging of solid-state circuits, *IEEE J. Solid-State Circuits*, SC-24(3), 698–703, 1989.
55. Baker J., Li H., and Boyce D. *CMOS Circuit Design, Layout, and Simulation*, Piscataway, NJ: IEEE Press, 1998.
56. Meyer R. and Mack W. A 1-GHz BiCMOS RF front-end IC, *IEEE J. Solid-State Circuits*, SC-29(3), 350–355, 1994.
57. Pelgrom M., Jeanet Rens A.C., Vertregt M., and Dijkstra M.B. A 25-Ms/s 8-bit CMOS A/D converter for embedded application, *IEEE J. Solid-State Circuits*, SC-29(8), 879–886, 1994.
58. Meyer R. and Mack W. A wideband low-noise variable-gain BiCMOS transimpedance amplifier, *IEEE J. Solid-State Circuits*, SC-29(6), 701–706, 1994.
59. Makie-Fukuda K., Maeda S., Tsukada T., and Matsuura T. Substrate noise reduction using active guard band filters in mixed-signal integrated circuits, *IEICE Tran. Fundamentals*, E80 A(2), 313–320, 1997.
60. P. Grybos, *Low Noise Multichannel Integrated Circuits in CMOS Technology for Physics and Biology Applications*, Monograph 117, AGH Uczelniane Wydawnictwa Naukowo-Dydaktyczne, Cracow, Poland, 2002, pp. 40–76.
61. Plas G., Vandenbussche J., Van den Bosch A., Steyaert M., Sansen W., and Gielen G. MOS transistor mismatch for high accuracy applications, *ProRISC: IEEE Benelux Workshop on Circuits, Systems and Signal Processing*, pp. 529–534, Mierlo, The Netherlands, November 1999.
62. Pelgrom M., Duinmaijer A., and Welbers A. Matching properties of MOS transistors, *IEEE J. Solid-State Circuits*, SC-24(5), 1433–1440, 1989.
63. Shyu J., Temes G., and Krummenacher F. Random error effects in matched MOS capacitors and current sources, *IEEE J. Solid-State Circuits*, SC-19(6), 948–955, 1984.
64. Lakshmikumar K., Hadaway R., and Copeland M. Characterisation and modeling of mismatch in MOS transistors for precision analog design, *IEEE J. Solid-State Circuits*, SC-21(6), 1057–1066, 1986.
65. Bastos J., Steyaert M., Pergoot A., and Sansen W. Mismatch characterisation of submicron MOS transistors, *Analog Integrated Circuits Signal Process*, 12, 95–106, 1997.
66. Serrano-Gotarredona T. and Linares-Barranco B. Systematic width-and-length dependent CMOS transistor mismatch characterisation and simulation, *Analog Integrated Circuits Signal Process*, 21, 271–296, 1999.
67. Laker K. and Sansen W. *Design of Analog Integrated Circuits and Systems*, New York: McGraw-Hill, 1994.
68. Lovett J.S., Welten M., Mathewson A., and Mason B. Optimizing MOS transistor mismatch, *IEEE J. Solid-State Circuits*, SC-33(1), 147–150, 1998.

69. Tuinhout H., Pelgrom M., and Vertreget M. Matching of MOS transistors, EPFL Electronics Laboratories Advanced Engineering Course on Deep Submicron: Modeling and Simulation, Laussanne, Switzerland, October 1998.

70. Plas G., Vandenbussche J., Sansen W., Steyaert M., and Gielen G. A 14-bit intrinsic accuracy Q^2 random walk CMOS DAC, *IEEE J. Solid-State Circuits*, SC-34(12), 1708–1718, 1999.

71. McNutt M., LeMarquis S., and Dunkley J. Systematic capacitance matching errors and corrective layout procedures, *IEEE J. Solid-State Circuits*, SC-29(5), 611–616, 1994.

72. Tuinhout H., Pelgrom M., Penning de Vries R., and Vertegret M. Effects on metal coverage on MOSFET matching, *Technical Digest IEDM'96*, 1996, pp. 735–739.

73. Tuinhout H. and Vertreget M. Test structures for investigation of metal coverage effects on MOSFET matching, *Proceedings IEEE International Conference on Microelectronics Test Structures ICMTS'97*, 1997, pp. 179–183.

74. Unno Y. High-density low-mass hybrid and associated technologies, *Proceedings of 6th Workshop on Electronics for LHC Experiments, CERN 2000–010*, pp.66–76, Cracow, Poland, September 2000.

75. Zhang Q., Liou J.J., McMacken J.R., Thomson J., and Layman P. SPICE modelling and quick estimation of MOSFET mismatch based on BSIM3 model and parametric, *IEEE J. Solid-State Circuits*, SC-36(10), 1592–1595, 2001.

76. De Geronimo G., O'Connor P., and Kandasamy A. Analog CMOS peak detect and hold circuits. Part 2. The two-phase offset-free and derandomizing configuration, *Nuc. Instr. Meth.*, A484, 544–556, 2002.

77. De Geronimo G., Kandasamy A., and O'Connor P. Analop peak detector and derandomizer for high-rate spectroscopy, *IEEE Trans. Nucl. Sci.*, 49(4), 1769–1773, 2002.

78. O'Connor P., De Geronimo G., and Kandasamy A. Amplitude and time measurement ASIC with analog derandomization: First Results. *IEEE Trans. Nucl. Sci.*, 50(4), 892–897, 2003.

79. Campbell M. et al. Readout for a 64 × 64 pixel matrix with 15-b pixel photon counting, *IEEE Trans. Nucl. Sci.*, 45, 751–753, 1998.

80. Krummenacher F. Pixel detectors with local intelligence: An IC designer point of view, *Nuc. Instr. Meth.*, A305, 527–532, 1991.

81. Ballabriga R., Campbell, M., Heijne, E.H.M., Llopart, X., and Tlustos L. The Medipix3 prototype, a pixel readout chip working in single photon counting mode with improved spectrometric performance, *NSS Symposium 2006, Conference Record*, pp. 3557–3561.

82. http://medipix.web.cern.ch

83. http://pilatus.web.psi.ch

84. Brönnimann Ch., Eikenberry E.F., Horisberger R., Hülsen G., Schmitt B., Schulze-Briese C., and Tomizaki T. Continuous sample rotation data collection for protein crystallography with the Pilatus detector, *Nuc. Instr. Meth.*, A510, 24–28, 2003.

85. Brönnimann Ch., Eikenberry E.F., Henrich B., Horisberger R., Hülsen G., Pohl E., Schmitt B., Schulze-Briese C., Suzuki M., Tomizaki T., Toyokawa H., and Wagner A. The Pilatus 1M detector, *J. Synchrotron Rad.*, 13, 120–130, 2006.

30 CMOS Systems and Interfaces for Microgyroscopes

*Ajit Sharma, Mohammad F. Zaman,
and Farrokh Ayazi*

CONTENTS

30.1 INTRODUCTION

A gyroscope is a sensor used to measure angle or velocity of rotation. Spinning wheel gyroscopes that rely on conservation of angular momentum were used for most part of the previous century. However, in recent years, the advent of micromachining technology has made microelectromechanical system (MEMS)-based angular rate sensors increasingly common. Micromachined gyroscopes use vibrating elements to sense rotation and are devoid of any rotating parts or bearings. This makes them ideally suited for inexpensive batch fabrication using planar processes and for potential integration with

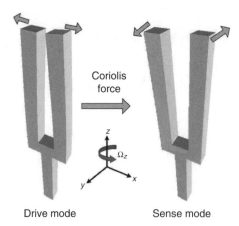

FIGURE 30.1 Demonstration of Coriolis force in a tuning fork subject to rotation. (Original drawing by Ajit Sharma.)

complementary metal oxide semiconductor (CMOS) circuitry [1]. This chapter will focus on systems and CMOS circuits that interface with these micromachined gyroscopes.

30.1.1 PRINCIPLE OF OPERATION

Micromachined gyroscopes are based on the Coriolis effect, where rotation about an axis causes a transfer of energy between two vibratory modes that are orthogonal to the axis of rotation and to each other. Consider the tuning fork shown in Figure 30.1. The tines of the tuning fork are excited into resonance along the x-axis. This mode is referred to as the primary mode or drive mode. When the tuning fork is rotated about the z-axis with angular rate Ω_z, there is an apparent force that acts orthogonally on the tines. This apparent force is referred to as the Coriolis force and causes the tines to deflect along the y-axis with an acceleration equal to $2\vec{v} \times \Omega_z$. The deflection of the tines along the y-axis—referred to as the sense mode—is proportional to the input rotation rate, Ω_z, and is the basis of all vibratory gyroscopes.

30.1.2 APPLICATIONS OF MEMS GYROSCOPES

Micromachined gyroscopes constitute one of the fastest growing segments of the microsensor market. The application domain of these devices is quickly expanding from automotive to consumer and personal navigation systems. A multitude of applications exist in the automotive sector including navigation, antiskid and safety systems, rollover detection, next generation airbag systems, and antilock brake systems [1]. Consumer electronics applications include image stabilization in digital cameras, smart user interfaces in handhelds, gaming, and inertial pointing devices.

Miniature gyroscopes can also be used for navigation purposes. Inertial navigation is the process of determining the position of a body in space by using the measurements provided by accelerometers and gyroscopes installed on the body [2]. Inertial measurement units (IMU) are vital components in aircraft, unmanned aerial vehicles, global positioning system (GPS)-augmented navigation, and personal heading references. Being self-contained, an IMU can perform accurate short-term navigation of a craft/object in the absence of GPS signals [2]. The petroleum industry uses gyroscopic sensors for real-time monitoring and correction of drilling in offshore rigs. Guidance systems and platform stabilization of missiles are but a few of the military applications that require accurate angular rate measurements.

Although bulk-mechanical and optical gyroscope technologies have successfully demonstrated navigation-grade performance, their constituent components are difficult to miniaturize and implement on silicon substrate using conventional CMOS fabrication techniques. Hence MEMS research has extensively focused on the development of vibratory gyroscopes that are capable of approaching subdegree per hour rate resolutions [1].

30.2 PERFORMANCE METRICS

Different parameters are used to specify a high-performance vibratory gyroscope system [3]. These performance metrics are briefly discussed below.

30.2.1 RESOLUTION

The resolution of a gyroscope is the minimum rotation rate that can be distinguished from the noise floor of the system per square root of bandwidth of detection, and is expressed in units of $°/h/\sqrt{Hz}$ (degree per hour per square root of hertz). The overall resolution of the microgyroscope (total noise equivalent rotation (TNEΩ)) is determined by two uncorrelated components: the mechanical (or Brownian) noise equivalent rotation (MNEΩ) and the electronic noise equivalent rotation (ENEΩ):

$$TNE\Omega = \sqrt{MNE\Omega^2 + ENE\Omega^2} \text{ (Units: } °/h) \qquad (30.1)$$

Brownian motion of the structure caused by molecular collisions from surrounding medium represents the mechanical noise component of any vibratory gyroscope [1]. By equating the displacement caused by Brownian motion to the displacement induced by Coriolis acceleration, the mechanical resolution of a vibratory microgyroscope is derived to be

$$MNE\Omega = \frac{1}{2q_{Drive}} \sqrt{\frac{4k_B T}{\omega_o M Q_{EFF}}} \sqrt{BW} \text{ (Units: } °/h) \qquad (30.2)$$

where quantities k_B, T, and BW represent the Boltzmann's constant ($1.38\,e^{-23}$ J/K), operating temperature (K), and measurement bandwidth (Hz), respectively. In the expression, ω_o is the resonant frequency of the sensor, M is the mass, q_{Drive} is the amplitude of vibration along the reference axis, and Q_{EFF} is the effective quality factor of the system.

The ENEΩ of the microgyroscope depends on the minimum detectable capacitance (ΔC_{MIN}) of the sense channel interface electronics and the *mechanical* scale factor (F/$°/h$).

$$ENE\Omega = \frac{\Delta C_{MIN}}{\text{Capacitive sensitivity}} \text{ (Units: } °/h) \qquad (30.3)$$

30.2.1.1 Scale Factor

Coriolis-induced sense-mode deflections of the proof masses are detected through capacitive, piezoresistive, piezoelectric, or optical means. The scale factor of a microgyroscope is the ratio of a change in output to a change in the input rotation rate. Scale factor is generally evaluated as the slope of the straight line that can be fit by the method of least squares to input–output data [3] and is commonly expressed in units of V/$°/s$ (volts per degree per second).

In a vibratory gyroscope, the sense-mode deflections are proportional to the effective quality factor (Q_{EFF}) and the drive amplitude, q_{Drive}. The Q_{EFF} depends significantly on the separation between the drive and sense-resonant mode frequencies. Increasing sense-mode deflections while maintaining a high aspect ratio for the capacitive gaps for sensing allows for large changes in the sense capacitance (ΔC), thereby generating a larger electrical pick-off signal for a given input rotation rate. Therefore, high effective quality factors, large drive amplitudes, low parasitics, and a high sense capacitance aspect ratio, all contribute to a high gyro scale factor, as shownthe following equation:

$$\text{Scale factor} \propto \frac{2V_P C_{SO} Q_{EFF} q_{\text{drive}}}{\left(C_{SO} + C_{\text{parasitics}}\right) d_{so} \omega_o} \Omega_z \qquad (30.4)$$

30.2.1.2 Zero Rate Output and Bias Stability

Zero rate output (ZRO) is the output signal from the gyroscope in the absence of input rotation. The drift of this ZRO bias (referred to as bias drift and expressed in°/h) is an important metric that ultimately determines the long-term stability of a microgyroscope. Any long-term variations in the rate, of which the bias drift is an accurate indicator, add up and cause a large error in angle information when the gyro is used in an IMU. Modern gyroscope systems periodically calibrate themselves with GPS to ensure that the heading information is accurate [2]. However, a long interval between calibration sequences is crucial for applications like deep sea navigation and oil exploration where it is not possible to resurface often to calibrate with GPS. In such cases, the longer a system can function accurately without the need for calibration, the better and more accurate heading and orientation information it delivers.

The bias drift of a gyro is composed of systematic and random components. The systematic components arise due to temperature variations, linear accelerations, vibrations, and other environmental factors [3]. The random component has been found to have a $1/f$ characteristic and it depends significantly on the noise floor of the gyroscope. The Allan variance technique is used to specify the drift of a microgyroscope [3]. Empirical studies [4] have formulated an expression that can predict drift in gyroscopes:

$$N_B \propto \frac{\omega_o^2}{Q_{EFF} \text{Area}_{\text{Electrodes}}} \qquad (30.5)$$

30.2.1.3 Bandwidth and Dynamic Range

The bandwidth of the microgyroscope determines the response time of the system. This is the time required for the output to settle within a certain range of the expected value for an input step function. The bandwidth requirements for a gyroscope depend on its application. Applications that demand very low rate resolutions typically require small bandwidths ($\sim$1 Hz); e.g. for use in gyrocompass navigation, where settling times of 1 min are tolerable. However, other applications like automotive rollover detection require larger bandwidth to detect the high yaw rates associated with vehicular skidding. Dynamic range refers to the range of input values over which the output is detectable. Typically, it is computed as the ratio between the maximum input rotation rate (full-scale range) that the sensor can tolerate and the system noise floor.

30.3 REVIEW OF MICROMACHINED GYROSCOPES

In the late 1980s, after successful demonstration of batch-fabricated silicon accelerometers, extensive efforts were initiated to replace quartz with silicon in micromachined vibratory gyroscopes [1]. Over the last two decades, a number of vibratory gyroscope structures were demonstrated which include (i) tuning fork structures, (ii) vibrating beams or proof masses, (iii) vibrating shells, and (iv) frame gyroscopes.

The first batch-fabricated silicon-micromachined rate gyroscope was demonstrated by the Charles Stark Draper Laboratory in 1991. This silicon bulk-mechanical device was a double gimbal vibratory gyroscope supported by torsional flexures [5] and a rotation rate resolution of 4°/s was realized using this structure.

Subsequently, in 1993, Draper Labs reported a silicon-on-glass tuning fork gyroscope (TFG) [6], as shown in Figure 30.2. This gyroscope was electrostatically vibrated in its plane using a set of

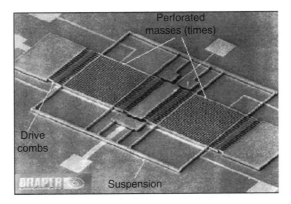

FIGURE 30.2 Draper Labs silicon-on-glass TFG. (From Bernstein, J., Cho, S., King, T., Kourepenis, A., Maciel, P., and Weinberg, M., *Proc. IEEE Micro Electro Mechanical Systems Workshop (MEMS '93)*, Fort Lauderdale, FL, pp. 143–148, 1993. With permission.)

interdigitated combs. A rotation signal normal to the drive mode would excite the out-of-plane rocking mode of the structure, which is capacitively monitored. The noise equivalent rate observed for this structure was $0.1°/s$ in a 60 Hz bandwidth.

Berkeley's z-axis vibratory rate gyroscope [7] resembled a vibrating beam design consisting of an oscillating mass, which was electrostatically driven into resonance using comb-drives. Any deflections resulting from Coriolis acceleration were detected differentially along the sense mode using interdigitated combs, as shown in Figure 30.3. This device, 1 mm across, was integrated with a transresistance amplifier on a single die using the Analog Devices BiMEMS process and demonstrated a resolution of $1°/s/\sqrt{Hz}$.

Currently, a popular commercial micromachined silicon gyroscope is the Analog Devices ADXRS iMEMS gyroscope series [8]. The sensor is implemented using a surface micromachining process, which integrates both the mechanical elements and electronics on a single die, as shown

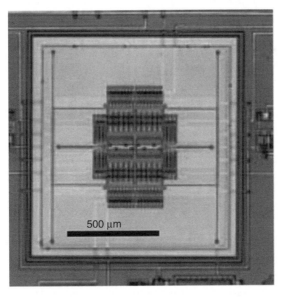

FIGURE 30.3 Berkeley's z-axis single chip gyroscope. (From Clark, W.A. and Howe, R.T., *Solid State Sensors and Actuator Workshop*, Hilton Head Island, SC, pp. 283–287, 1996. With permission.)

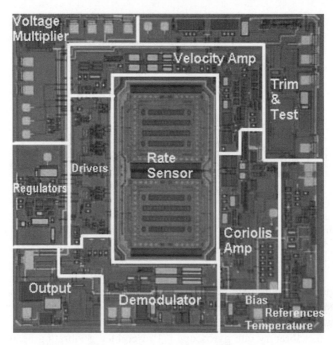

FIGURE 30.4   Die picture of an Analog Devices ADXRS series microgyroscope. (From Geen, J. Shermen, S.J., Chang, J.F., and Lewis, S.R., *IEEE J. Solid State Circuits*, 37(12), 1860, 2002. With permission.)

in Figure 30.4. The ADXRS gyroscope displays a rate sensitivity of $12.5\,\text{mV}/^\circ/\text{s}$ and an estimated Allan deviation (bias drift) of $50^\circ/\text{h}$.

Surface-micromachined gyroscopes suffer from thin-film residual stress, squeeze-film damping, and other problems associated with low mass. Therefore, significant research is focused toward developing bulk-micromachined technologies that yield greater mass-per-unit-area.

In 1999, Murata reported a DRIE gyroscope with decoupled sense and drive nodes with a resolution of $0.07^\circ/\text{s}$ in a $10\,\text{Hz}$ bandwidth [9]. Also reported by Samsung was a $40\,\mu\text{m}$ thick single-crystal silicon gyroscope fabricated using silicon-on-insulator (SOI) technology [10]. This device demonstrated a resolution of $0.015^\circ/\text{s}$ in $25\,\text{Hz}$ bandwidth, a bias drift of $500^\circ/\text{h}$ and a sensitivity of $145\,\text{mV}/^\circ/\text{s}$.

In 2000, researchers at the University of Michigan developed a high aspect ratio poly and single-crystalline silicon (HARPSS) process [11], capable of producing thick MEMS structures while implementing submicron high aspect ratio sensing and tuning capacitive gaps. A polysilicon HARPSS vibrating ring gyroscope [12] demonstrated a rate sensitivity of $0.2\,\text{mV}/^\circ/\text{s}$, full-scale range of $\pm 250^\circ/\text{s}$ and a measured noise floor of $0.1^\circ/\text{s}/\sqrt{\text{Hz}}$ (Figure 30.5).

In 2002, the University of Michigan reported a high-resolution vibrating ring gyroscope, implemented using (111) single-crystalline silicon structural material [13]. The $150\,\mu\text{m}$ structural device was defined by deep reactive ion etching and displayed a noise floor of $10^\circ/\text{h}/\sqrt{\text{Hz}}$ in a $2\,\text{Hz}$ bandwidth with a reported rate sensitivity of approximately $130\,\text{mV}/^\circ/\text{s}$.

30.4   ELECTRONIC CONTROL SYSTEMS IN GYROSCOPES

As silicon vibratory gyroscopes attain navigation-grade performance, the interface electronics that actuate, sense, and control these micromechanical structures are key elements in determining the overall performance of the microgyro system. Automotive and consumer product applications require rate noise floors in the order of 100°–$1000^\circ/\text{h}$, and must be able to sense rotation rates as large as $\pm 500^\circ/\text{s}$. Navigation-grade gyroscopes have similar full-scale ranges, but the noise floor specifications are on

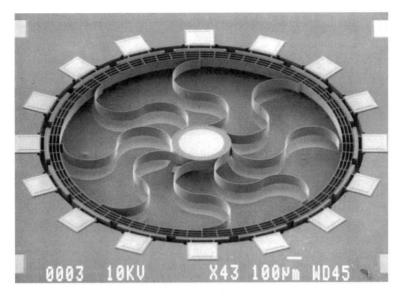

FIGURE 30.5 Implementation of the polysilicon vibrating ring gyroscope using HARPSS. (From Ayazi, F. and Najafi, K., *IEEE/ASME J. Microelectromechanical Syst.*, 10, 169, 2001. With permission.)

the order of $0.1°/h$. This is approximately three orders of magnitude lower than the requirements for the commercial counterparts. Since vibratory microgyroscopes, like micromachined accelerometers, are capacitive sensors, this translates to the need for ultralow-noise front-ends, able to detect subatto Farad capacitance changes [14,15]. In addition, while mechanical structures can typically attain dynamic ranges in excess of 120 dB, designing front-end electronics with such large dynamic range is challenging. High-performance electronics for the actuation, sensing, and control of the Coriolis vibratory gyroscope (CVG) are therefore essential in order to realize navigation-grade performance. Figure 30.6 below shows a schematic overview of the constituent system blocks in a typical microgyroscope system.

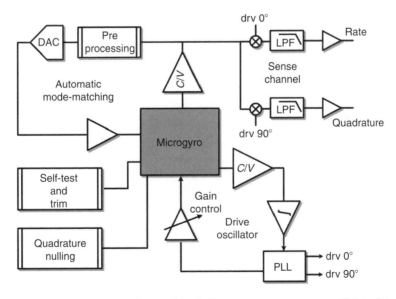

FIGURE 30.6 Overview of a typical micromachined vibratory gyroscope system. (Original drawing by Ajit Sharma.)

Based on functionality, the electronics for any vibratory microgyroscope can be divided into the following subsystems.

30.4.1 DRIVE LOOP

The drive loop electronics are responsible for starting and sustaining vibrations along the reference axis at constant amplitude. It is essential that constant drive amplitude is maintained, as any variations in the drive amplitude manifests as a change in velocity of the mechanical structure, resulting in false rate output. The drive loop uses an automatic level control (ALC) circuit to achieve and maintain constant drive amplitude. There are two approaches to implement the drive loop:

- An electromechanical oscillator: Here the drive mode oscillations are started and sustained by using a positive feedback loop that satisfies the Barkhausen's criteria (loop gain $= 1$, loop phase shift $= 0°$). The gyroscope forms the frequency-determining element of the electromechanical oscillator. A high mechanical quality factor for the drive resonant mode (Q_{DRV}) can significantly ease the design of the drive oscillator, enabling the drive oscillations to be built up, and sustained, using much smaller AC voltage levels.
- A phase-locked loop (PLL) oscillator: Here the PLL center frequency and capture range are set close to the drive resonant frequency of the gyroscope. On power up, the PLL locks on to the output of the front-end I–V converter. The PLL output is amplified or attenuated to achieve the desired voltage amplitude and is used to drive the microgyroscope. The PLL-oscillator relies on the precise $90°$ phase shift occurring at drive resonance. A variable gain amplifier is used to implement ALC.

30.4.2 QUADRATURE NULLING

Fabrication imperfections of the mechanical structure results in off-axis movement of the proof mass, causing a residual displacement along the sense axis even in the absence of rotation [7]. This is referred to as ZRO or quadrature error. A number of techniques have been used to null the quadrature error. Some of the earliest work involved trimming and bucking to control quadrature error [16,17], but suffered from the fact that this did not track over temperature and sensor lifetime. Other techniques involved servo-mechanisms, where a force was applied to the mass so as to null displacement that is in phase with position [7]. The Analog Devices gyroscope uses a set of optimally designed levers [8] to reduce quadrature error to less than 1 ppm by improving the selectivity of the suspension flexures. Yet another technique that has been implemented is the use of torque cancellation electrodes [18,19] to correct for any misalignment between the drive and sense axes.

30.4.3 MODE-MATCHING

CVG relies on the energy transfer between two resonant modes to sense rotation. Gyroscope performance is enhanced when the mechanical sense frequency is matched to the drive resonant frequency due to the mechanical amplification provided by the effective quality factor (Q_{EFF}). Again, due to the limits imposed by fabrication tolerances, the drive and sense frequencies are seldom equal. Therefore, it is necessary to use electronic or mechanical means to decrease, and eventually null, the separation between the two resonant frequencies [18]. This mode-matching is usually achieved by varying the mechanical bias voltages on the MEMS structure until the frequencies are equalized [20].

30.4.4 SENSE CHANNEL

The amplitude of sense-mode deflections is modulated by the applied input rotation rate. The primary function of the sense channel is to extract the input rotation information from the gyroscope output. Phase-sensitive demodulation allows for rejection of the interfering mechanically generated quadrature error due to the inherent 90° phase difference between the quadrature and Coriolis signals. The AM Coriolis output is synchronously demodulated using the drive oscillator signal. The lowpass-filtered signal is proportional to input rotation rate, and may be amplified if necessary at a later stage.

30.4.5 SELF-TEST AND TRIM

To be viable in a large-scale manufacturing environment, microgyroscope systems are equipped with a self-test capability to ensure quality and reliability. Apart from significantly reducing test times at the production facility, this allows for calibration of the sensor in the field. Currently most gyroscope systems use a certain amount of postfabrication trimming to account for microfabrication imperfections. With the development of more advanced lithographic tools and etching systems, these trimming procedures can be minimized.

30.5 CASE STUDY—MODE-MATCHED TUNING FORK GYROSCOPE

Figure 30.7 shows the scanning electron micrograph (SEM) view of a 1.5 mm × 1 mm, in-plane, mode-matched tuning fork gyroscope (M^2-TFG) [20] fabricated on 40 μm thick SOI substrate, using a simple two-mask process similar to one used for microgravity accelerometers reported in Ref. [15]. The gyroscope is comprised of two proof masses, supported by flexural springs and anchored at a central post. Actuation, sensing, quadrature nulling, and tuning electrodes are distributed around the proof masses and flexures.

The sensor structure is maintained at a DC polarization voltage (V_P) to provide the bias for capacitive transduction to prevent frequency doubling of the drive force [21], and for electrostatic frequency control. The proof masses are driven at resonance along the x-axis using interdigitated comb-drives. When the sensor undergoes a rotation about the z-axis, the resultant Coriolis acceleration causes the proof masses to vibrate along the y-axis [1]. This rotation-induced proof mass motion causes the gap between the sense electrode and the proof mass to change, proportional to the applied rate, and is detected electronically. The gyroscope is a resonant sensor, and the drive and sense modes have been designed to yield mechanical quality factors in excess of 40,000 [20]. The resonant frequencies are in the range of 10–20 kHz. Figure 30.8 shows the complete implemented interface electronics for the M^2-TFG presented in this case study.

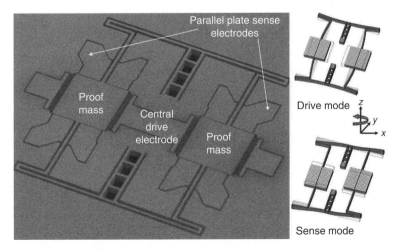

FIGURE 30.7 SEM of the M2-TFG and illustration of the mode-shapes. (From Sharma, A., Zaman, M.F., and Ayazi, F., *IEEE J. Solid-State Circuits*, 42, 1790, 2007. With permission.)

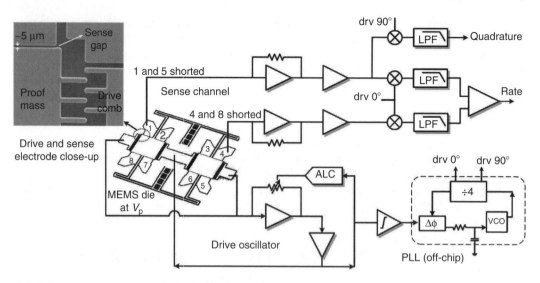

FIGURE 30.8 Implemented M2-TFG interface electronics showing drive oscillator and sense channel. (Inset) Close-up SEM showing the 5 μm sense gap and drive comb fingers. (From Sharma, A., Zaman, M.F., and Ayazi, F., *IEEE J. Solid-State Circuits*, 42, 1790, 2007. With permission.)

30.5.1 CHALLENGES AND TRADE-OFFS IN MICROGYRO INTERFACING

The minimum detectable rotation rate depends on the Brownian noise floor of the sensor and the electronic noise floor of the front-end interface. Since a fixed DC potential (V_P) is maintained across the sense gap (d_{so}), Coriolis-induced y-axis displacement of the proof mass in response to input rotation Ω_Z changes the sense rest capacitance (C_{so}), generating a motional current I_{SENSOR}, given by the following equation:

$$I_{SENSOR} = \frac{2V_P C_{so} Q_{EFF} q_{drive}}{d_{so}} \Omega_Z \text{ (Units: A)} \tag{30.6}$$

For a parallel plate capacitive transducer, the minimum detectable capacitance change (ΔC_{MIN}) is proportional to the input-referred current noise of the interface electronics integrated over the bandwidth of interest, as given by the following equation:

$$\Delta C_{MIN} = \frac{I_{noise}\sqrt{BW}}{\omega_o V_P} \text{ (Units: F/}\sqrt{Hz)} \tag{30.7}$$

While use of high aspect ratio micromachining techniques like HARPSS [22] can lower the ENEΩ by increasing mechanical sensitivity, the focus of this case study will be to present circuit techniques to reduce the total input-referred current noise of the sensor electronics. The theoretical MNEΩ of the sensor used here is 0.5°/h/$\sqrt{Hz}$, which means the electronic front-end must be able to detect a proof mass displacement as small as 0.1 Å, or resolve a capacitance change of 0.02 aF/$\sqrt{Hz}$ at the sensor operating frequency (~15 kHz).

The drive resonant mode of the TFG can be modeled as a two-port series RLC circuit [23,24]. The reactive elements determine the mechanical frequency of resonance and the *motional* resistance ($R_{MOT-DRV}$) represents the transmission loss element. The value of R_{MOT} is obtained by equating the mechanical energy dissipated per cycle to the electrical energy supplied by the sustaining sources. In order to avoid lateral snap-down and maximize the drive displacement, the gap, g, between adjacent comb electrodes must be increased [25]. To achieve drive amplitudes of about 4–5 μm, the gap between adjacent combs must be at least 7 μm, which results in this microgyroscope having a drive motional impedance of about 16 MΩ in vacuum.

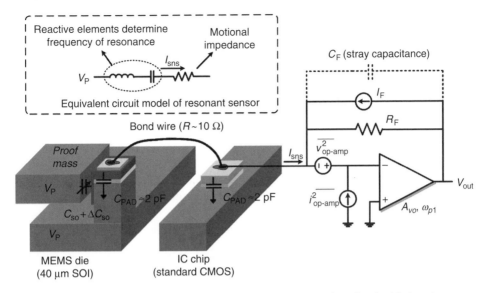

FIGURE 30.9 Schematic representation of a TIA (with noise sources) interfaced with the microgyroscope. (Inset) Series RLC model of a resonant microstructure. (From Sharma, A., Zaman, M.F., and Ayazi, F., *IEEE J. Solid-State Circuits*, 42, 1790, 2007. With permission.)

Large motional impedances require a large gain to be provided by the sustaining circuitry in the drive oscillator loop. These also require a higher AC drive voltage to be applied to the comb-drive electrodes, thereby dissipating more power. Circuits that achieve large on-chip gains for capacitive detection, with low power and area overheads, are therefore necessary.

The M^2-TFG is fabricated using a bulk-micromachining technology, which allows for the fabrication of MEMS structures with narrow capacitive gaps and large inertial mass [20]. The sensor is fabricated on a different substrate and is connected directly to the integrated circuit (IC) via wire bonds as shown in Figure 30.9. A two-chip implementation allows decoupling of the MEMS design and fabrication from the design of the interface electronics. Sensor performance can be improved considerably, unlike in Ref. [8], by leveraging the benefits of high aspect ratio mixed-mode processes [11]. Secondly, standard CMOS processes can be used, which significantly lowers cost and allows the electronics to be optimized for low-power dissipation, speed, and reliability. However, the front-end analog interface must be strategically chosen to ensure that the sub-picoampere level motional currents can be detected even in the presence of the increased parasitics.

30.5.2 Review of Microgyro Front-Ends

Several techniques have been used in electronic front-ends to sense the small capacitive displacements in MEMS gyroscopes. Charge integration using switched capacitor front-ends with correlated double sampling was initially used for static MEMS accelerometers [14,15], but has recently been used for microgyroscopes [26,27]. These schemes are best suited for microgyroscopes with low operating frequency (<5 kHz) because of the power budget associated with the switching and clock generation [28]. In addition, the effects of the capacitive loading of these front-ends on the microgyroscope quality factor have not been studied. Further, the use of such front-ends necessitates a switching voltage to be applied to the mechanical structure, which results in significant feed-through and parasitic electrical coupling.

In Ref. [12] a unity gain, CMOS source follower amplifier was used as the front-end to detect capacitance changes in a vibrating polysilicon ring gyroscope. The DC bias at the pick-off electrode was set using a minimum geometry diode. The noise injected by the diode at the input significantly degrades performance [19]. Further, special techniques such as internal bootstrapping and feedback need to be applied to minimize the capacitance of the input transistor. Finally, the use of a unity gain buffer does not allow independent control of the signal-to-noise ratio (SNR) of the electronic front-end.

Continuous time (CT) charge integrator front-ends are attractive for sensing capacitive displacements in microgyroscopes [7,8], because at typical operating frequencies, much larger AC impedances can be generated in a standard CMOS process using capacitors rather than resistors. Additionally, since these capacitors are not switched, there is no kT/C noise associated with them. However, the CT charge integrator requires the use of a large resistor to bias the input node. Various techniques like the use of controlled impedance field-effect transistors (FETs) [8] and subthreshold mixed oxide semiconductor field-effect transistors (MOSFETS) [7] have been proposed in the literature to implement these feedback resistors. The thermal noise of this resistor forms the dominant noise contributor of the front-end and determines overall performance.

30.5.3   TRANSIMPEDANCE FRONT-ENDS FOR MOTIONAL CURRENT DETECTION

Figure 30.9 shows a schematic of a CT Transimpedance Amplifier (TIA) interfaced with a microgyroscope. This work differs significantly from Refs. [7,8] in that the TIA has been optimized for noise is used as the front-end in *both* the drive loop, as well as for subatto Farad capacitive detection in the sense channel. Further, the gain of the TIA is variable and the proof mass is maintained at a constant DC potential unlike Refs. [7,26,27]. In Figure 30.9, R_F is the feedback resistance and C_F is the associated stray capacitance. The lumped parasitic capacitance at the inverting terminal of the op-amp, referred to as C_{TOT}, is composed of the electrode to substrate capacitance on the MEMS die ($C_{PAD\text{-}MEMS} \sim 1.5\,pF$), the interface IC pad capacitance ($C_{PAD\text{-}ASIC} \sim 1.5\,pF$), and the gate capacitance of the input differential pair transistors ($C_{GS\text{-}IN} \sim 0.5\,pF$) of the op-amp.

The high open loop DC gain of the op-amp ensures that the inverting terminal is a good virtual ground, and the shunt–shunt feedback presents low input impedance to the high-impedance sensor pick-off node. This makes the signal path relatively insensitive to the total parasitic capacitance ($C_{TOT} = C_{PAD\text{-}MEMS} + C_{PAD\text{-}ASIC} + C_{GS\text{-}IN}$), preventing significant signal loss. The low input impedance provided by the shunt–shunt feedback further helps reduce the loading that the sustaining electronics will have on the mechanical quality factor of the gyroscope drive mode.

The TIA interface allows the proof masses to be maintained at a constant DC potential unlike Refs. [25–27], where an AC capacitance bridge configuration is used. Applying a switching signal to the proof masses as in Ref. [27] increases the amount of electronic coupling into the ZRO of the gyroscope. By maintaining the proof mass at a constant DC potential, any spurious signal coupling into the sensing electrodes is eliminated, and the number of demodulation and filtering stages required are minimized, thereby lowering power consumption.

Figure 30.9 shows the main noise contributors in the transimpedance front-end, where $\overline{v_{op\text{-}amp}^2}$ and $\overline{i_{op\text{-}amp}^2}$ are the input-referred voltage and current noise of the core op-amp respectively, and $I_F = 4k_B T/R_F$ represents the thermal noise power of the feedback resistor (R_F). Since the sensor output is a current proportional to proof mass displacement, the total input-referred current noise of the TIA front-end determines the minimum detectable capacitance (Equation 30.5), and hence resolution of the microgyroscope. The equivalent input noise current ($I_{N\text{-}TOT}$) [29] for a TIA front-end is given by Equation 30.8, which includes effects of both the total parasitic capacitance seen at the input node (C_{TOT}), and the input resistance of the core amplifier, $R_{IN\text{-}op\text{-}amp}$. The noise contributions of $\overline{i_{op\text{-}amp}^2}$ and of $R_{IN\text{-}op\text{-}amp}$ are ignored for succinctness:

$$\overline{i_{N\text{-}TOT}^2} \approx \frac{4k_B T}{R_F} + \overline{v_{op\text{-}amp}^2}\left(\frac{1}{R_F^2} + \omega^2 C_{TOT}^2\right) \qquad (30.8)$$

In a bandwidth of 10 Hz about the sensor operating frequency, the equivalent input noise spectrum is assumed white and thermal noise of the feedback resistor forms the dominant noise contributor.

The advantage of using a TIA front-end is evident when we consider the SNR of the front-end interfaced with a microgyroscope. A TIA with transimpedance gain R_F yields an output signal voltage of $I_{SNS} \times R_F$ (for input motional current I_{SNS}) and output noise voltage of $\sqrt{4k_B T R_F}$. The amount of displacement current $i_{Brownian}$ due to the random Brownian motion of the proof mass along the sense axis can be derived by applying the equipartition theorem [22,30] to the M^2-TFG *at resonance* and computing the noise displacement x_n [22,24]. The sense-resonant mode of the microgyroscope can be modeled as a second-order system with an equivalent series RLC representation, very similar to that presented for the drive mode. The Brownian noise displacement is related to the mechanical motional resistance of the sense mode ($R_{MOT\text{-}SNS}$) and the equivalent Brownian noise current is derived:

$$\overline{i_{Brownian}^2} = \omega_o^2 V_P^2 \left(\frac{\partial C_{so}}{\partial x} \right)^2 \overline{x_n^2} = \frac{4k_B T}{R_{MOT\text{-}SNS}} \tag{30.9}$$

By using $I_{SNS} = i_{Brownian}$, the overall SNR is derived to be

$$\mathrm{SNR} = \sqrt{\frac{R_F}{R_{MOT\text{-}SNS}}} \tag{30.10}$$

Therefore increasing R_F improves the total SNR of an angular rate sensor. From Equations 30.8 and 30.10, it is evident that a large R_F for capacitive detection is beneficial not only in terms of increased transimpedance gain, but also for better SNR and lower input current noise. Therefore, the basis of this case study is to focus on strategies that yield large on-chip transimpedance.

30.5.4 LOW-NOISE WIDE DYNAMIC RANGE T-NETWORK TIA

Large transimpedance gains can be implemented on-chip in a number of ways [7,8,31,32]. In Ref. [8], the transresistance was implemented using a controlled impedance FET. In Ref. [31], long MOSFETs biased in the linear regime using a constant voltage were used. MOS-bipolar pseudoresistors are used in Ref. [32] for generating large resistances, but the maximum bandwidth obtained for the neural amplifier was 7.2 kHz. The main disadvantage of these approaches is that real-time control of the transresistance gain is not possible. Variation of the transresistance is possible to some extent using the approach proposed in Ref. [8], but it involves variation of the duty cycle used to switch the controlled impedance FET. A variable duty cycle–controlled resistance contributes to the bias drift of the sensor, and was therefore not adopted for this work. The strategy adopted in this work is to implement the feedback resistor in a TIA using a T-network of resistors. The implemented T-network TIA provides both high gain and low-noise for subatto Farad capacitive detection in an area- and power-efficient manner. Further, it allows for a simple analog control of the transimpedance without excessive phase shift, and any DC offset can be nulled by varying the bias voltage at the base of the T.

30.5.4.1 Design Considerations

Figure 30.10 shows the complete schematic of the implemented T-network TIA front-end, interfaced for capacitive detection. The op-amp used is a noise optimized two-stage miller-compensated operational transconductance amplifier (OTA). The equivalent transimpedance of the T-network TIA ($R_{F\text{-}EQ}$) is given by (30.11), where the voltage divider formed by R_2 and R_3 in the feedback path provides an amplification of the equivalent transimpedance.

$$\frac{V_{out}}{I_S} \equiv R_{F\text{-}EQ} = R_1 \left(1 + \frac{R_2}{R_3} \right) + R_2 \tag{30.11}$$

The primary advantage of using the T-network is that it reduces the resistance levels to be placed on-chip, making on-chip integration tractable. In this work, R_1 is implemented as a long MOS

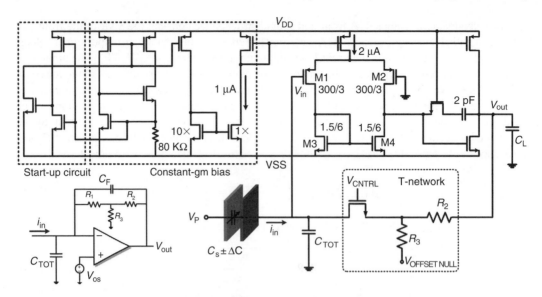

FIGURE 30.10 Circuit schematic T-network TIA interfaced for capacitive detection. (From Sharma, A., Zaman, M.F., and Ayazi, F., *IEEE J. Solid-State Circuits*, 42, 1790, 2007. With permission.)

transistor operating in the triode or deep-triode regions, and R_2 and R_3 were on-chip polyresistors. The MOSFET adds a degree of gain control to the transimpedance, which can be used for temperature compensation or for ALC applications. The resistances are designed such that $R_1 \gg R_2$ and R_3. From Equation 30.11, it might seem that arbitrarily high transimpedance can be obtained by increasing the R_2/R_3 ratio. However, in practice, bandwidth, noise, offset, and stability trade-offs limit the choice of this ratio [33].

For the case of the T-network TIA interfaced to a capacitive sensor, the SNR of the front-end degrades by a factor of $\sqrt{(1+R_2/R_3)}$ as given by Equation 30.12. This places a limit on the maximum transimpedance that can be used in the front-end:

$$\text{SNR} = \sqrt{\frac{R_{\text{F-EQ}}}{R_{\text{MOT-SNS}}\left(1+\frac{R_2}{R_3}\right)}} \tag{30.12}$$

The relationship to prevent excessive noise increase due to the T-network's amplification of the op-amp's voltage noise is therefore given by

$$\frac{R_2}{R_3} \leq \frac{C_{\text{TOT}}}{C_{\text{F}}} \tag{30.13}$$

where C_{TOT} for the two-chip solution varies between 2 and 5 pF. The stray feed-thorough capacitance (C_{F}) between the input and output is typically around 100–500 fF. An R_2/R_3 ratio of 2 was therefore chosen as it allowed for sufficient amplification of the transimpedance gain, without excessively increasing the noise gain.

DC offset restricts the maximum output signal swing, thereby determining the upper limit of the dynamic range. The expression for the output DC voltage due to finite offset (V_{OS}) for a T-network TIA interfaced directly with a capacitive sensor is given by Equation 30.14, which interestingly is the value of the noise gain at DC. Optimizing the R_2/R_3 ratio for SNR automatically minimizes effects of DC offset:

$$V_{\text{O}} = \left(1+\frac{R_2}{R_3}\right)V_{\text{OS}} \tag{30.14}$$

30.5.4.2 Characterization Results

The transimpedance gain of the T-network TIA was characterized for different values of gate control voltage (V_{CNTRL}) and is plotted in Figure 30.11a. At 10 kHz, the transimpedance gain can be varied between 0.2 and 22 MΩ by varying the gate control voltage of the MOS resistor in the feedback T. A transimpedance as large as 25 MΩ has been implemented on-chip, in a fraction of the area consumed otherwise. Optimizing the R_2/R_3 ratio has ensured that there is no gain peaking at the frequencies of interest, as evident from Figure 30.11a.

In reality, it is never possible to directly measure the input-referred noise of a circuit. The total output noise of the T-network TIA was measured using a spectrum analyzer for different values of V_{CNTRL}. The measured output voltage noise is divided by the measured transimpedance to yield the total input-referred current noise of the TIA. Figure 30.11b plots the measured total input-referred current noise of the T-network TIA for different values of transimpedance. Figure 30.11b clearly shows that with increasing R_F, the current noise floor decreases, as predicted by Equation 30.8. Therefore, a larger R_F lends to a lower noise floor and hence smaller minimum detectable capacitance. In the region between 1 and 10 kHz, flicker noise is still significant and accounts for the $1/f$ characteristic. The noise gain peaking due to the effect of the capacitance at the input node (C_{TOT}) is clearly visible from the plot. However, this noise gain peaking occurs beyond the sensor operating range (10–20 kHz) and there is a minima in the noise floor for the case of $R_F = 1.6$ MΩ within the sensor-operating range [34].

When interfacing CMOS front-ends with high-Q narrow-band resonant MEMS sensors, the spot noise of the interface at the sensor resonant frequency determines the minimum detectable capacitance. At 15 kHz, the T-network TIA with a V_{CNTRL} of 0.96 V has a transimpedance gain of 1.6 MΩ and an input-referred current noise of 88–100 fA/$\sqrt{\text{Hz}}$. This corresponds to a capacitive resolution of 0.02–0.04 aF/$\sqrt{\text{Hz}}$ at 15 kHz ($V_P = 40$ V). This is an order of magnitude better than that reported for the CT integrator of Ref. [31] and of the same order as the transcapacitance amplifier of Ref. [8]. Further this is comparable to the capacitive resolution of the chopper stabilized front-end interface of Ref. [28] has been attained without any power-consuming switching and further does not require any clock generation.

The maximum dynamic range provided by the front-end T-network TIA for sensing is therefore of interest. The maximum dynamic range is defined as [29]

$$DR_{MAX} = \frac{\text{Maxoutput signal}}{\text{Noise floor} \times \text{Bandwidth}} \tag{30.15}$$

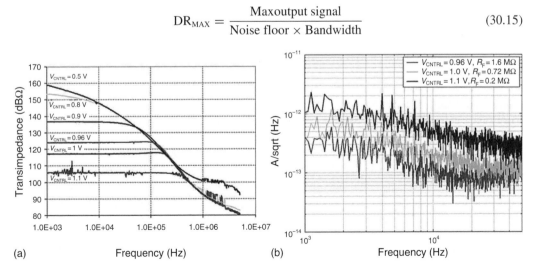

(a)   Frequency (Hz)   (b)   Frequency (Hz)

FIGURE 30.11 (a) Transimpedance gain characterization of the T-network TIA. (b) Measured input-referred current noise for the front-end T-network TIA as a function of R_F. (From Sharma, A., Zaman, M.F., and Ayazi, F., *IEEE J. Solid-State Circuits*, 42, 1790, 2007. With permission.)

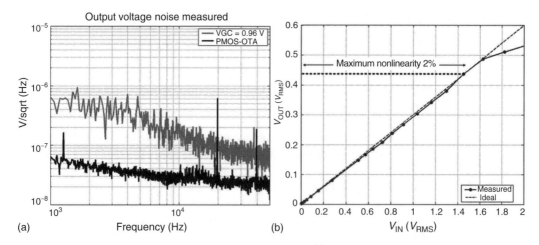

(a)   (b)

FIGURE 30.12 (a) Measured output voltage noise of the core amplifier and the T-network TIA for R_F of 1.6 MΩ. (b) Measured SNR plot of the T-network TIA for an R_F of 1.6 MΩ at 10 kHz. (From Sharma, A., Zaman, M.F., and Ayazi, F., *IEEE J. Solid-State Circuits*, 42, 1790, 2007. With permission.)

From Figure 30.12a, the measured output spot voltage noise of the T-network TIA at 10 kHz is about 250–300 nV/$\sqrt{\text{Hz}}$ for the case of $R_F = 1.6$ MΩ ($V_{CNTRL} = 0.96$ V). This is slightly higher than the thermal noise from an ideal 1.6 MΩ resistor, which must be expected because of the noise gain of the T-network. This noise is integrated over a bandwidth of 10 Hz and is used to determine the lower bound of the dynamic range.

DC offset limits the output swing, thereby determining the maximum linear range of the front-end interface. In order to find the maximum (nondistorted) SNR, i.e. signal-to-(noise + distortion) ratio (SNDR) for the circuit, the input voltage level was swept upwards until the output signal was found distorted. The maximum linear output swing of the TIA with TZ gain of 1.6 MΩ at a frequency of 10 kHz is limited to about 0.4 V_{RMS}, as shown in Figure 30.12b. Beyond this level the nonlinearity in the output exceeds 2%, which is unacceptable. The maximum dynamic range computed for a 10 Hz bandwidth and is found to be at least 104 dB at the sensor-resonant frequency.

30.5.5 DRIVE AND SENSE CHANNELS

Reference vibrations of the proof masses along the drive (x)-axis are excited using a PLL oscillator circuit that locks on to the mechanical drive resonance frequency of the TFG. The T-network TIA is used as the front-end to convert the motional current from the drive combs into a voltage [34]. An off-chip PLL [35] locks on to this voltage signal and provides the required phase shifted signals to sustain electromechanical drive oscillations as well as to perform the various signal processing operations in the sensor system. The output square wave signal from the PLL is attenuated and applied to the central comb-drive electrode. The level of the PLL output is set so as to achieve the desired drive amplitude (q_{DRIVE}). The voltage controlled oscillator (VCO) center frequency and the track/capture range is set such that drive oscillations are maintained over a wide range of temperature.

The rotation-induced Coriolis acceleration deflects the proof masses along the y-axis, exciting the sense-resonant mode of the M^2-TFG. The sense-resonant output (carrier signal) is amplitude-modulated by the magnitude of the input rotation signal. The rotation information is extracted from the sense output by performing a synchronous demodulation using the PLL output that is proportional to the proof mass velocity. A CMOS Gilbert multiplier with 200 kΩ on-chip poly load resistors is used for the multiplication. Figure 30.13 shows the mixer schematic and input and output test waveforms.

The output of the Gilbert multiplier is lowpass-filtered to yield an analog signal proportional to the rotation rate. The integrated, active first-order lowpass filter uses a 1.5 nF off-chip capacitor to

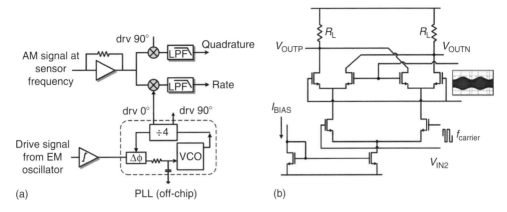

(a)   PLL (off-chip)   (b)

FIGURE 30.13 Schematic of the sense channel showing (a) $I - Q$ demodulation, and (b) Gilbert cells used for multiplication. (From Sharma, A., Zaman, M.F., and Ayazi, F., *IEEE J. Solid-State Circuits*, 42, 1790, 2007. With permission.)

set the cutoff frequency to 100 Hz and has a low pass gain of 2. The rate signal from the two channels is converted to a single-ended signal using an off-chip instrumentation amplifier.

The M^2-TFG was placed on a rate table and its scale factor was characterized as shown in Figure 30.14. The measured scale factor from one of the channels is $2 \, \text{mV}/^\circ/\text{s}$, with a maximum nonlinearity of 3% over the measured range. The sources of nonlinearity in the sense channel signal processing chain are the slight difference in the capacitive gaps on the MEMS structure, the nonlinearity of the front-end TIA, and more significantly, the incomplete cancellation of the higher order harmonic terms as the Gilbert cell was operated in a single-ended configuration.

Figure 30.14 also shows the sensor response to a 1.5 Hz sinusoidal input rotation as well as the response of the microgyroscope to both positive (CCW) and negative (CW) input step rotations.

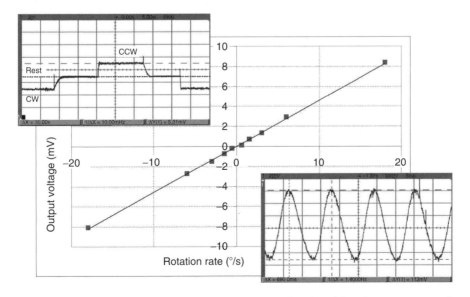

FIGURE 30.14 Scale factor of the microgyroscope and measured output of the microgyroscope to sinusoidal and step input rotations. (From Sharma, A., Zaman, M.F., and Ayazi, F., *IEEE J. Solid-State Circuits*, 42, 1790, 2007. With permission.)

30.5.5.1 System Integration

The noise floor and long-term stability of the M^2-TFG interfaced with electronics was characterized by performing an Allan variance analysis [3] on the ZRO. For microgyroscopes, the root Allan variance is the preferred means of specifying the noise floor rather than the power spectral density and is the method adopted in this work. The ZRO from one of the channels was buffered with an off-chip amplifier with a gain of 10 and sampled every 100 ms for a period of 12 h using a digital multimeter. The root Allan variance plot obtained without applying any prewhitening or filtering is shown in Figure 30.15 and the inset shows a time slice of the sampled ZRO. The slope at short cluster times (τ) yields the angle random walk (ARW), which is a measure of the white noise in the system. The ARW is $0.045°/\sqrt{h}$, which corresponds a measured noise floor of $15\,\mu V/\sqrt{Hz}$ ($-96\,dBV/\sqrt{Hz}$) over the signal bandwidth (1–10 Hz) for the entire microsystem. The output-referred total equivalent noise density (MEMS plus electronics) is therefore $2.7°/h/\sqrt{Hz}$. This is about an order of magnitude better than commercially available gyroscopes [8] and is one of the lowest recorded for a silicon vibratory gyroscope.

The second significant performance metric is the bias drift, which is a measure of the long-term stability of the microgyro system. The minimum of the Allan variance plot gives the value of the bias drift of the system [3], which for this case is $1°/h$. This about $50\times$ better than Ref. [8] and is one of the lowest recorded for MEMS gyroscopes till date.

The increase in the root Allan variance at large cluster times indicates the presence of a rate random walk component [3,36]. In this case, it is attributed to the incomplete nulling of the quadrature error in the MEM structure. Despite a low-noise TIA front-end, the measured spot rate noise floor

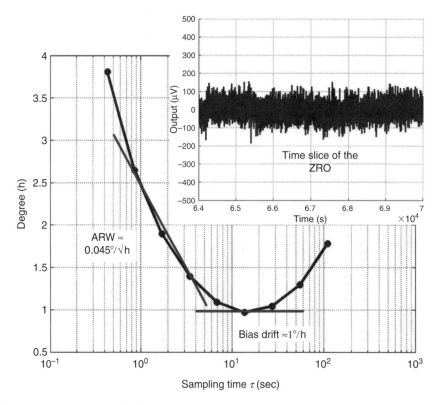

FIGURE 30.15 Root-Allan variance of the M2-TFG interfaced with electronics showing a bias drift of $1°/h$ (Inset) Time slice of the recorded ZRO. (From Sharma, A., Zaman, M.F., and Ayazi, F., *IEEE J. Solid-State Circuits*, 42, 1790, 2007. With permission.)

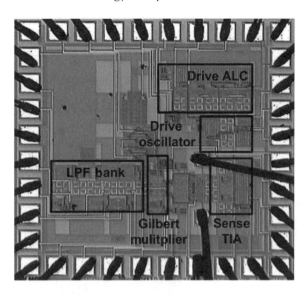

FIGURE 30.16 0.6 μm CMOS ASIC for gyroscope drive and sense channels. (From Sharma, A., Zaman, M.F., and Ayazi, F., *IEEE J. Solid-State Circuits*, 42, 1790, 2007. With permission.)

at the output of the microgyroscope system is slightly higher than the theoretical noise floor of the MEM sensor itself $0.5°/h/\sqrt{Hz}$. This is primarily due to the amplification of the TIA noise by the gain of the remaining portion of the sense signal chain. Additionally, the noise contribution of the subsequent signal processing stages, which include the multiplier, lowpass filters and external buffers adds to the overall output-referred noise floor. This can be significantly decreased by the use of bipolar stages [8] or by the use of low-noise chopper stabilization techniques [37] in the final output stage.

Figure 30.16 shows the micrograph of the 3 V 0.6 μm CMOS IC that is interfaced to the M^2-TFG using wire bonds on a custom printed circuit board [34]. The key microsystem parameters are summarized in Table 30.1.

30.6 FUTURE DIRECTIONS AND CONCLUSIONS

Most of the research and development of silicon-micromachined gyroscopes over the last decade has been driven by the automotive industry. The performance of micromachined gyroscopes has

TABLE 30.1
Summary of Key Sensor and IC Parameters

Parameter	Measured Value
Sensor capacitive sensitivity	802 aF/°/s
Amplitude of drive voltage applied	130 mV
Range of minimum detectable ΔC (at 15 kHz)	0.02–2 aF/rt(Hz)
Linear dynamic range of front-end T-TIA	104 dB
Rate sensitivity of gyro + IC	2 mV/°/s
Die area (0.6 μm CMOS)	2.25 mm²
Total power consumption	15 mW (±1.5 V)
Output voltage noise level at 10 Hz	15 μV/rt(Hz)
Measured rate noise floor	2.7°/h/rt(Hz)
Sensor + electronics bias stability	1°/h

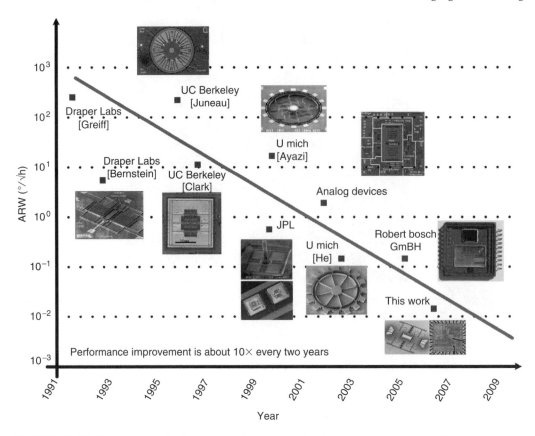

FIGURE 30.17 Performance of micromachined gyroscopes has improved by a factor of 10 every two years since 1991. (Original drawing by Ajit Sharma.) (Gomez, U., Kuhlman, B., Classen, W., and Bauer, W., *Tech. Digest Int. Conf. on Solid-State Sensors and Actuators (Transducers '05)*, Seoul, South Korea, pp. 184–187, 2005. Juneau, T., and Pisano, A., Tech. Dig. Solid-State Sensors and Actuator Workshop, Hilton Head Island, SC, pp. 299–302, 1996. Bae, S., Hayworth, K., Shcheglov, K., Yee, K., and Wiberg, D., *Tech. Dig. Solid-State Sensors and Actuators Workshop*, Hilton Head Island, SC, 2002. With permission.)

significantly improved over a rather short period, as illustrated in Figure 30.17, which is based on a sample of devices reported in the literature. Since 1991, the resolution of micromachined gyroscopes, indicated by the random angle walk, has improved by a factor of 10× every two years as shown below. It is anticipated that a continuing improvement in the performance of micromachined gyroscopes (resolution, bias stability, dynamic range, etc.) will be seen over the next decade. This will open up new avenues and market segments for micromachined gyroscopes, specifically inertial navigation for military and civilian applications.

Performance improvement is but one aspect of microgyroscope development. Considerable effort is underway for mass production of these micromachined sensors. Packaging and reliability are the two most important factors that must be considered before micromachined gyroscopes can be viably commercialized. To facilitate low-cost production of microgyroscopes, there is an increased emphasis on the design-for-manufacture and design-for-test [40]. The Analog Devices iMEMS gyroscopes are hermetically sealed, but operate at atmospheric pressure. Although this might suffice for rate-grade sensors, robust vacuum packaging that can withstand temperature variations will be required for navigation-grade gyroscopes [41]. Researchers at the University of Michigan and Georgia Tech are developing hermetically packaged, environmentally protected, near-inertial-grade micromachined silicon gyroscopes. This work aims at developing wafer-scale micromachining technologies with high reliability, long lifetime, low power consumption, and high immunity against shock and environmental variations such as temperature [42]. While high aspect ratio micromachining provides significant improvement in sensor performance, it also enables the development of MEMS gyroscopes that can operate at lower voltages. The MEMS devices can then leverage the benefits offered by state-of-the-art CMOS processes and operate off a low-voltage battery.

REFERENCES

1. Yazdi N, Ayazi F, and Najafi K, Micromachined inertial sensors, Invited paper, *Proc. of the IEEE*, pp. 1640–1659, August 1998.
2. Lawrence A, *Modern Inertial Technology: Navigation, Guidance, and Control*. Springer-Verlag, New York, 1993.
3. IEEE Recommended Practice for Inertial Sensor Test Equipment, Instrumentation, Data Acquisition, and Analysis, IEEE Std 1554–2005, pp. 1–103, 2005.
4. Watson WS and Henke TJ, Coriolis gyro configuration effects on noise and drift performance. *Symposium of Gyro Technology*, Stuttgart, Germany, 2002.
5. Greiff P, Boxenhorn B, King T, and Niles L, Silicon monolithic micromechanical gyroscope. *Tech. Dig. 6th Int. Conf. on Solid-State Sensors and Actuators (Transducers '91)*, San Francisco, CA, pp. 966–968, 1991.
6. Bernstein J, Cho S, King T, Kourepenis A, Maciel P, and Weinberg M, A micromachined comb-drive tuning fork rate gyroscope, *Proc. IEEE Micro Electro Mechanical Systems Workshop (MEMS '93)*, Fort Lauderdale, FL, pp. 143–148, 1993.
7. Clark WA and Howe RT, Surface micromachined Z-axis vibratory rate gyroscope, *Solid State Sensors and Actuator Workshop*, Hilton Head Island, SC, pp. 283–287, 1996.
8. Geen J, Sherman SJ, Chang JF, and Lewis SR, Single-chip surface micromachined integrated gyroscope with 50°/h Allan deviation, *IEEE Journal of Solid State Circuits*, 37(12), 1860–1866, 2002.
9. Park KY, Lateral gyroscope suspended by two gimbals through high aspect ratio ICP etching. *Tech. Dig. 10th Int. Conf. on Solid-State Sensors and Actuators (Transducers '99)*, Sendai, Japan, pp. 972–975, 1999.
10. Tanaka K, Mochida Y, and Sugimoto M, A micromachined vibratory gyroscope, *Sensors and Actuators A (Physical)*, 50, 111–115, 1995.
11. Ayazi F and Najafi K, High aspect-ratio combined poly and single-crystal silicon (HARPSS) MEMS technology, *IEEE/ASME Journal of Microelectromechanical Systems*, 9(5), 288–294, 2000.
12. Ayazi F and Najafi K, A HARPSS polysilicon vibrating ring gyroscope, *IEEE/ASME Journal of Microelectromechanical Systems*, 10, 169–179, 2001.
13. He G and Najafi K, A single-crystal silicon vibrating ring gyroscope. *Proc. of IEEE MEMS 2002*, Las Vegas, NV, pp. 718–721, 2002.
14. Lemkin MA, Ortiz M, Wongkomet N, Boser B, and Smith J, A 3-axis surface micromachined $\Sigma\Delta$ accelerometer. *Proc. IEEE Int. Solid-State Circuits Conf.*, pp. 202–203, 1997.
15. Amini BV and Ayazi F, A 2.5 V 14-bit sigma-delta CMOS-SOI capacitive accelerometer, *IEEE Journal of Solid State Circuits*, 2467–2476, 2004.
16. Barnaby R and Reinhardt A, U.S. Patent 2 544 646, 1951.
17. Barnaby R and Morrow C, U.S. Patent 2 753 173, 1956.
18. Sharma A, Zaman M, and Ayazi F, A 0.2°/h microgyro with automatic CMOS mode-matching. *Technical Digest of ISSCC 2007*, San Francisco, USA, 2007.
19. Putty M and Najafi K, A micromachined vibrating ring gyroscope, *Digest, Solid-State Sensors and Actuators Workshop*, pp. 213–220, 1994.
20. Zaman MF, Sharma A, and Ayazi F, High-performance matched-mode tuning fork gyroscope, *Proc. IEEE MEMS*, pp. 66–69, 2006.
21. Senturia, SD, *Microsystem Design*, 4th ed., Kluwer Academic, Dordrecht, 2002.
22. Ayazi F, A high aspect-ratio high-performance polysilicon vibrating ring gyroscope, PhD dissertation, The University of Michigan, 2000.
23. Pourkamali S, Hashimura A, Abdolvand R, Ho G, Erbil A, and Ayazi F, High-Q single crystal silicon HARPSS capacitive beam resonators with sub-micron transduction gaps, *IEEE Journal of Microelectromechanical Systems*, 12(4), 487–496, 2003.
24. Nguyen CT-C and Howe RT, An integrated high-Q CMOS micromechanical resonator-oscillator, *IEEE Journal of Solid-State Circuits*, 34(4), 440–455, 1999.

25. Tang W, Nguyen TH, Judy MW, and Howe RT, Electrostatic comb-drive of lateral polysilicon resonators, *Journal Sensors and Actuators A*, 21, 328–331, 1990.
26. Jiang X, Seeger JI, Kraft M, and Boser BE, A monolithic surface micromachined Z-axis gyroscope with digital output. *Symp. on VLSI Circuits Dig. of Tech. Papers*, Honalulu, Hawaii, June 2000.
27. Petkov V and Boser B, A fourth-order $\Sigma\Delta$ interface for micromachined inertial sensors, *IEEE Journal of Solid State Circuits*, 39(5), 722–730, 2005.
28. Wu J, Fedder GK, and Carley R, A low-noise low-offset capacitive sensing amplifier for a 50-μg$\sqrt{\text{Hz}}$ monolithic CMOS MEMS accelerometer, *IEEE Journal of Solid State Circuits*, 39(5), 722–730, 2004.
29. Gray PR, Meyer RG, Hurst PJ, and Lewis SH, *Analysis and Design of Analog Integrated Circuits*, John Wiley, Hoboken, NJ, 2001.
30. Gabrielson T, Mechanical-thermal noise in micromachined acoustic and vibration sensors, *IEEE Transactions on Electronic Devices*, 40(5), 903–909, 1993.
31. Saukoski M, Aaltonen L, Halonen K, and Solo T, Fully integrated charge sensitive amplifier for readout of micromechanical capacitive sensors, *Proc. ISCAS 2005*, pp. 5377–5380, 2005.
32. Harrison R and Charles C, A low-power low-noise CMOS amplifier for neural recording applications, *IEEE Journal of Solid-State Circuits*, 38(6), 958–965, 2003.
33. Graeme J, *Photodiode Amplifiers: Op Amp Solutions*, McGraw Hill, New York, 1996.
34. Sharma A, Zaman MF, and Ayazi F, A 104 dB dynamic range transimpedance-based CMOS ASIC for tuning fork microgyroscopes, *IEEE Journal of Solid-State Circuits*, 42(8), 1790–1802, 2007.
35. PLL HC4046 data sheet from Phillips Semiconductor.
36. Gomez U, Kuhlman B, Classen W, and Bauer W, New surface micromachined angular rate sensor for vehicle stabilizing systems in automotive applications. *Tech. Dig. Int. Conf. on Solid-State Sensors and Actuators (Transducers '05)*, Seoul, South Korea, pp. 184–187, 2005.
37. Enz CC, and Temes GC, Circuit techniques for reducing the effects of op-amp imperfections: autozeroing, correlated double sampling, and chopper stabilization. *Proc. of the IEEE*, pp. 1584–1614, 1996.
38. Juneau T, and Pisano A, Micromachined dual input axis angular rate sensor. *Tech. Dig. Solid-State Sensors and Actuator Workshop*, Hilton Head Island, SC, pp. 299–302, 1996.
39. Bae S, Hayworth K, Shcheglov K, Yee K, and Wiberg D, JPL's MEMS gyroscope—fabrication, 8-electrode tuning and performance results. *Tech. Dig. Solid-State Sensors and Actuators Workshop*, Hilton Head Island, SC, 2002.
40. Geen JA, Progress in Integrated Gyroscopes, IEEE PLANS 2004.
41. HERMIT site, DARPA, http://www.darpa.mil/into/programs/hermit/index.html.
42. Chae J, Giachino JM, and Najafi K, Wafer-level vacuum package with vertical feed throughs. *Proc. of IEEE Int. Micro Electro Mechanical Systems (MEMS '05)*, Miami, FL, pp. 548–551, 2005.

31 Analog Front End for a Micromachined Probe Storage Device

Christoph Hagleitner, Tony Bonaccio,
Hugo Rothuizen, Jan Lienemann,
Dorothea Wiesmann, Giovanni Cherubini,
Jan G. Korvink, and Evangelos Eleftheriou

CONTENTS

31.1 INTRODUCTION

The exponential growth of the capacity of mobile storage devices over the past few years was one of the key enabling factors for a large number of digital applications, including MP3 players, digital cameras, and smart phones. Today, a single NAND-flash chip has a capacity of up to 2 Gb [1], and many applications including, e.g. video cameras, require mobile storage devices with capacities in the range of 50–100 Gb. Probe-storage technology has the potential to reach densities of several Tb/in^2 and, therefore, enable such devices at reasonable cost. In probe-storage devices, a tip (probe) with dimensions in the nanometer range is used to read/write information on a storage medium (e.g. polymer films, ferroelectric films). Probe storage offers the advantage that the minimum feature

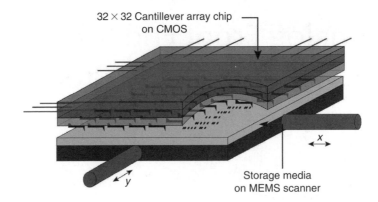

FIGURE 31.1   The concept of probe-based storage. (From Hagleitner, C., Bonaccio, T., Rothuizen, H., Lienemann, J., Wiesmann, D., Cherubini, G., Korvink, J., and Eleftheriou, E., *IEEE J. Solid-State Circuits*, 42, 1779,
2007. With permission.)

size is defined by the tip radius or the electrical field generated by the tip and not by the limits of
expensive state-of-the-art lithography.

This chapter describes the modeling, design, and verification of an experimental analog front
end (AFE) chip for a thermomechanical scanning-probe-based data storage concept that uses thin
polymer layers to store information. The concept, which is internally dubbed "millipede" [2,3],
combines ultrahigh density, small form factor, and high data rates. Ultrahigh storage densities beyond
1 Tb/in^2 have been demonstrated by using local probe techniques to write, read back, and erase data
in very thin polymer films [4]. High data rates are achieved by parallel operation of large 2D arrays
with thousands of micro/nanomechanical cantilevers/tips that can be batch-fabricated by silicon
surface-micromachining techniques. Figure 31.1 illustrates the "millipede" concept.

The high precision required to navigate the storage medium over the probe tips is achieved
by microelectromechanical system (MEMS)-based x/y actuators that position the medium relative
to the large array of probe tips for parallel write/read/erase operations [5–7] (see Figure 31.2). A
precision of a fraction of 1 nm is required to store and retrieve information reliably at an areal
density of greater than 1 Tb/in^2. The navigation control is based on two different types of position
information [8]. The absolute position is determined using a thermal position sensor that measures
the resistance change of a fixed heater that overlaps the moving scanner [5]. Position information
relative to the center of a servo track is obtained from dedicated servo channels [6].

Thermomechanical writing is achieved by applying a local force through the cantilever/tip to the
polymer layer and simultaneously softening the polymer layer by local heating to create indentations
that represent the written bits. To read the written information, the same cantilever is used to scan the
surface. The change in the thermal flow between a second heating element and the medium surface as
the tip moves into an indentation leads to a change in the electrical resistance of the heating element
(see Figure 31.3) [9].

The parallel operation of hundreds of cantilevers makes the CMOS integration of the AFE imperative. To this end, a 3D assembling technology has been developed to transfer the cantilevers onto
the CMOS read-out circuitry [10]. The small relative resistance change ($\Delta R/R < 10^{-3}$ for a typical
indentation) renders the design of the AFE's read channel a challenging task. Section 31.2 describes
the overall architecture of the chip and some building blocks in detail. In addition, the design methodology that has been used to simulate and verify the AFE chip throughout the design is presented.
Section 31.3 focuses on the multidomain model of the cantilever, and Section 31.4 presents its experimental verification. In Section 31.5, the design of the critical input stage of the read channel based
on cosimulation of the AFE circuitry and the cantilever model is presented. Experimental results
obtained with the complete 32-channel chip that was wirebonded to a cantilever array are given in
Section 31.6, and Section 31.7 contains a summary.

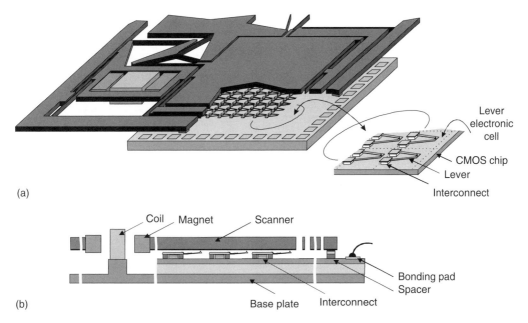

(a)

(b)

FIGURE 31.2 (a) Schematic of miniaturized scanner and cantilever array chip with integrated electronics and (b) side-view of scanner, cantilever array, and CMOS electronics. (From Eleftheriou, E., Bächtold, P., Cherubini, G., Dholakia, A., Hagleitner, C., Loeliger, T., Pantazi, A., Pozidis, H., Albrecht, T., Binnig, G., Despont, M., Drechsler, U., Duerig, U., Gotsmann, B., Jubin, D., Häberle, W., Lantz, M. A., Rothuizen, H., Stutz, R., Vettiger, P., and Wiesmann, D., *Proceedings of International Conference on Very Large Data Bases*, Berlin, Germany, pp. 3–7, 2003. With permission.)

31.2 ANALOG FRONT-END CHIP ARCHITECTURE

Figure 31.4 shows the block diagram of the AFE chip [11]. The chip includes 32 data channels, six servo channels, a programmable parameter memory, and test interfaces to make important analog and digital signals available for off-chip characterization. At the input of each data channel, there is a high-voltage switch-matrix that selects the actual operation of the corresponding cantilever

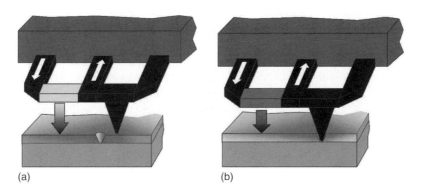

(a)   (b)

FIGURE 31.3 Principle of thermal sensing. The read-out circuitry measures the current through a low-doped heater that is placed close to the cantilever tip. When the cantilever tip drops into an indentation, the thermal conductivity between read resistor and medium surface increases. This leads to a temperature decrease of the silicon and as a consequence a decrease of the read resistance. (From Hagleitner, C., Bonaccio, T., Rothuizen, H., Lienemann, J., Wiesmann, D., Cherubini, G., Korvink, J., and Eleftheriou, E., *IEEE J. Solid-State Circuits*, 42, 1779, 2007. With permission.)

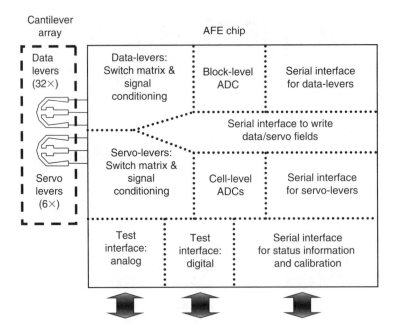

FIGURE 31.4 Block-diagram of AFE chip. (Reprinted from Hagleitner, C., Bonaccio, T., Pantazi, A., Sebastian, A., and Eleftheriou, E., *Proceedings of Symposium on VLSI Circuits*, Honolulu, HI, pp. 70–71, 2006. With permission.)

(read/write/erase/inactive). The switches are designed to operate at voltages of up to $2 \times V_{DS_{max}}$ of the individual transistors available in the IBM 0.25 μm RF CMOS technology that was used for the prototype design.

The circuitry to control the write operation is shared between the data and the servo channel because the writing of data information and servo information never occurs at the same time. The write circuitry enables independent control of the three most important parameters of the write pulse: (i) pulse duration, (ii) pulse height, and (iii) electrostatic force voltage. Figure 31.5 shows the basic circuit that is used to control the write pulse as well as a simple timing diagram. The pulse duration can be programmed through the parameter interface between 20 ns and 1.3 ms, with an accuracy of 20 ns. The chip also allows individual finetuning (± 5.12 μs) of the pulse duration relative to the programmed value. For each channel, one out of four programmable timing and voltage offsets can be selected.

The signal-conditioning circuitry amplifies and filters the input signals before all data channels are multiplexed to a single, 3.2 MS/s Flash-ADC (100 kS/s per channel). A serial interface transmits the samples to an off-chip data processor (FPGA, Field Programmable Gate Array). The design of the individual read channels is detailed in Section 31.2.1.

Dedicated servo channels are required for the retrieval of position and timing information to enhance navigation and to control the sampling of the data channels [1]. The six servo channels are similar to the data channels, but have individual 10 bit analog-to-digital converters (ADCs), which are sampled at an up to $8\times$ higher rate than the data channels. This enables the extraction of accurate position and timing information.

31.2.1 READ CHANNEL

Figure 31.6 shows the complete analog part of the read channel. The current from the input stage is converted to a differential voltage signal. More details about the design and optimization of the critical input stage can be found in Section 31.5. After the input stage, a high-pass filter is used to remove the remaining bias current and low-frequency noise. The high-pass filter has a bypass

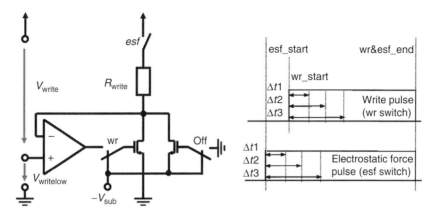

FIGURE 31.5 Schematic and timing diagram of the circuit used to control the write process. The timing of the two switches can be programmed individually. When the upper *esf* switch is closed, the cantilever is pulled into contact by the electrostatic force due to the voltage difference between the negative substrate. After closing the *wr* switch, the tip is heated up to the write temperature by the write resistance R_{write} of the cantilever. When the cantilever is neither in read nor in write mode, the off-switch is activated to set the potential of the cantilever to $-V_{sub}$, which corresponds to zero electrostatic force. In this configuration, heating without electrostatic force is not possible.

option to enable calibration of the input offset for optimum power supply rejection (PSR). Two instrumentation amplifiers with programmable gain and bandwidth ensure the flexibility needed for adapting the transfer function of the front end to a wide range of operating conditions (different cantilever designs, scanning speeds, bias voltages). An additional offset-compensation stage between the second and the third amplifier is used to cancel the input offset of the amplifiers in high-gain configurations. A sample-and-hold stage is used for synchronous sampling of all data channels before the outputs are sequentially multiplexed to the ADC.

31.2.2 THE MODELING APPROACH

A three-stage approach was used to design and verify the AFE. First, the basic operation of the device was simulated using a detailed cantilever model (see Section 31.3) and idealized circuitry building blocks described in Verilog-A (opAmps, sources). These simulations were used to establish

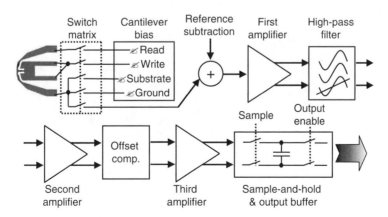

FIGURE 31.6 Block diagram of the read channel. (Reprinted from Hagleitner, C., Bonaccio, T., Pantazi, A., Sebastian, A., and Eleftheriou, E., *Proceedings of Symposium on VLSI Circuits*, Honolulu, HI, pp. 70–71, 2006. With permission.)

the theoretical performance of the device and provide input for the optimization of the cantilever design [12]. Next, transistor-level simulations of the input stage together with the cantilever were used to optimize the design of the input stage for minimum power consumption as described in Section 31.5. Finally, simulations of the complete parallel channels together with the ADCs were performed for functional verification of the chip. A mixed-signal simulation (Spectre-Verilog) that employs a simplified behavioral model of the cantilever was used to reduce the simulation time for the functional verification to reasonable values, i.e. approximately 24 h for 1 ms of simulation time for the data channels. This was sufficient to simulate start-up and settling of the cantilever and all analog circuits as well as all digital functions, including transmission of a few samples from each channel through the serial interface.

31.3 CANTILEVER MODEL

Information is stored in the mechanical topography of a thin polymer film as described above. The micromechanical cantilever acts as a transducer to convert topographical variations into electrical energy during the read process and vice versa during the write process. For the design, simulation and optimization of the electrical transducer interface circuit, i.e. the AFE, a transducer model that accurately captures the feedback mechanisms, and interactions between the electrical, thermal, and mechanical domain is required [12].

The Verilog-A analog hardware description language was chosen to implement a low-level, parameterized model of the micromechanical cantilever. This language is widely used by the circuit design community to create behavioral modules that can easily be instantiated into netlists for most circuit simulation programs. In addition, Verilog-A provides direct and consistent access to the electrical, thermal, and mechanical energy domains. This allows straightforward implementation of the associated nonlinear differential equations, and avoids non-intuitive detours through equivalent circuit descriptions.

Figure 31.7 shows a fabricated cantilever with the read resistor and the write tip. The cantilever model is based on the physical parameters of the design (dimensions and doping levels), a few technology-specific fit parameters that do not vary between the different designs, and one noise parameter that varies for different cantilever designs (see Section 31.3.4). Thus, the model is easily applied

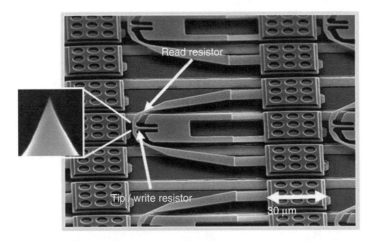

FIGURE 31.7 Scanning electron microscopy image of a section of a fabricated cantilever array and a close-up of cantilever tip (diameter ≈ 20 nm). (From Hagleitner, C., Bonaccio, T., Rothuizen, H., Lienemann, J., Wiesmann, D., Cherubini, G., Korvink, J., and Eleftheriou, E., *IEEE J. Solid-State Circuits*, 42, 1779, 2007. With permission.)

to co-design AFE circuitry together with any new generation of cantilever arrays. The main goal of the model is, on the one hand, to provide high accuracy for the small-signal analysis of the read channel in which the tip reads information from the medium surface and, on the other hand, to provide sufficient accuracy for large-signal operation, e.g. for pulling the cantilever tip electrostatically into contact with the medium surface from a distance in order to write indentations or to start a read operation.

The overall cantilever model is split into two main parts (see Figure 31.8). The electrical and mechanical models are connected using a schematic entry. To calculate the electrical and thermal resistances, the distance outputs of the reduced behavioral model are fed back into the lumped-element model used to describe the electrical and thermal domains. The electrical and thermal flows are mostly one-dimensional (horizontally along the cantilever and vertically through the air into the polymer medium) and, therefore, a small number of identical, parameterized cantilever segments are sufficient. The mechanical model describes the movement of the cantilever due to the electrostatic

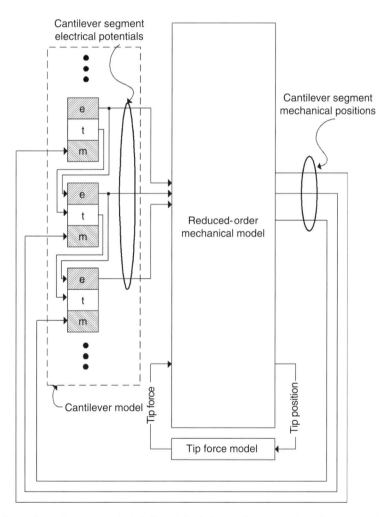

FIGURE 31.8 Full cantilever model. A full model of the cantilever consists of an array of interconnected segments, a reduced-order mechanical model, and a tip force model. The reduced-order electrostatic/mechanical model calculates the vertical position of each cantilever segment as a function of its electrical potential. The tip model calculates the force at the tip, which nominally is the only point at which the cantilever contacts the substrate. (From Hagleitner, C., Bonaccio, T., Rothuizen, H., Lienemann, J., Wiesmann, D., Cherubini, G., Korvink, J., and Eleftheriou, E., *IEEE J. Solid-State Circuits*, 42, 1779, 2007. With permission.)

force between the cantilever and the silicon substrate underneath the polymer medium. This is a reduced-order model which is automatically extracted from a finite element (FE) model describing interactions between the mechanical and the electrostatic domains, and is represented in the form of code (see Section 31.2.2). The voltages at the output of the electrical/thermal cantilever segments are the inputs to the mechanical/electrostatic model. The positions of the cantilever segments at the outputs of the mechanical/electrostatic model are fed back to the electrical/thermal model, where they are used to update the electrical and thermal conductivities and capacitances self-consistently. The contact between the surface and the tip is modeled by another Verilog-A module that calculates the force on the tip as a function of the tip–sample distance (see Section 31.3.3).

The mechanical behavior of the cantilever is also captured using Verilog-A as described in Section 31.2.2. The mechanical model is a reduced-order model (ROM) of the electrostatic force on the cantilever, which is extracted from a full FE mesh analysis of the cantilever.

31.3.1 ELECTRICAL/THERMAL MODEL

The electrical/thermal model captures the electrical resistances and capacitance of the cantilever elements and the thermal conduction through the silicon as well as through the air toward the recording-medium surface. Both the electrical and thermal conductivities of silicon exhibit a strong, nonlinear dependence on the temperature, geometric parameters, and doping. The theory of electrical and thermal conduction in silicon and air is well-developed and allows an accurate calculation of the conductivity over a wide range of geometries, temperature, and doping levels [13–16]. Therefore, the model can be parameterized to enable parameter sweeps and Monte Carlo simulations of process variations.

A lumped-element approach was chosen in which the cantilever is segmented into small equivalent elements that are parameterized with the local size, doping, and noise constants of the corresponding element (see Figure 31.9). Each of the segments has inputs/outputs toward the neighboring segments and toward the electrostatic/mechanical model in all three domains as shown in Figure 31.10.

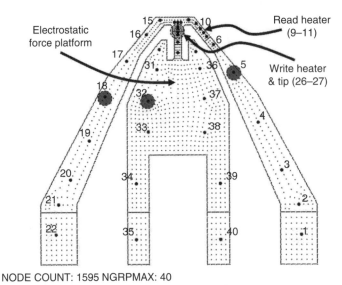

NODE COUNT: 1595 NGRPMAX: 40

FIGURE 31.9 FEM cantilever model showing the 40 distance outputs that are passed from the mechanical to the electrical/thermal model. A subset of 16 nodes is used as voltage inputs to the mechanical model. The large dots indicate the nodes that were chosen to compare the results of the FEM simulations with the simulations performed with the ROM. The 40 distance outputs indicate the centers of the lumped-elements used for the electrical/thermal model of the cantilever.

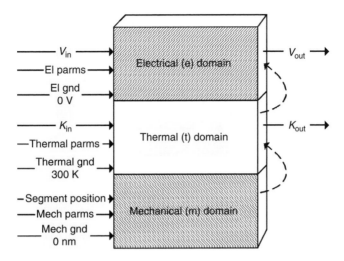

FIGURE 31.10 Cantilever segment model. Each segment contains electrical, thermal, and mechanical modeling domains. Physical characteristics such as the geometry and the doping of the segment are parameterized. Each segment has input and output terminals for electrical, thermal, and mechanical signals. These signals are used to build a full model of the cantilever from individual segments. (From Hagleitner, C., Bonaccio, T., Rothuizen, H., Lienemann, J., Wiesmann, D., Cherubini, G., Korvink, J., and Eleftheriou, E., *IEEE J. Solid-State Circuits*, 42, 1779, 2007. With permission.)

Figure 31.11 shows the equivalent circuit for the electrical and thermal domains of a single element. The heat source J_{pwr} in Figure 31.11 models the power dissipated in the electrical resistance. The electrical conductivity of silicon is a function of its temperature T, doping concentrations N_D and P_D, and is given by

$$\sigma_{Si} = 1.6e^{-19} \left(\mu_N [T, N_D, P_D] \, n [T, N_D, P_D] + \mu_P [T, N_D, P_D] \, p [T, N_D, P_D] \right) \qquad (31.1)$$

where it is important to model the mobilities μ_N, μ_P [13] accurately over wide temperature and doping concentration ranges to prevent wrong conclusions from the parametric simulations and convergence problems in the simulator.

The thermal conductivity of silicon is dependent not only on the temperature, choice of dopant, and doping concentration, but also on the layer thickness [16]:

$$k = \frac{1}{3} \sum_{j=L,T,TU} v_j^2 \int\limits_{0}^{\theta_j/T} C_{V,j}(x,T) \left[\tau_j(x,T) \times F(\delta) \right] dx \qquad (31.2)$$

where the subscripts $j = L, T, TU$ refer to the longitudinal and two transverse phonon modes, respectively, v_j is the phonon group velocity, θ_j is the Debye temperature, and x is the nondimensional phonon frequency. The specific heats $C_{V,j}$ and phonon scattering rates τ_j account for temperature dependence, while layer thickness effects appear through the boundary scattering reduction function $F(\delta)$, wherein $\delta = d_s/\Lambda_b$ is the ratio of the layer thickness and of the appropriate phonon mean free path. In this model for the thermal conductivity, both the effects of doping and of the reduced thickness are accounted for semi-empirically by decreasing the mean free path Λ_b compared with its value in bulk silicon.

The thermal resistance of air also depends on temperature:

$$\sigma_{air} = -3.933 \times 10^{-4} + T \times 1.02 \times 10^{-4} - T^2 \times 4.86 \times 10^{-8} + T^3 \times 1.52 \times 10^{-11} \qquad (31.3)$$

where T is the absolute temperature. To capture gradients more accurately, the air between the cantilever and the medium surface is split into three sequential thermal resistance elements. Finally, at

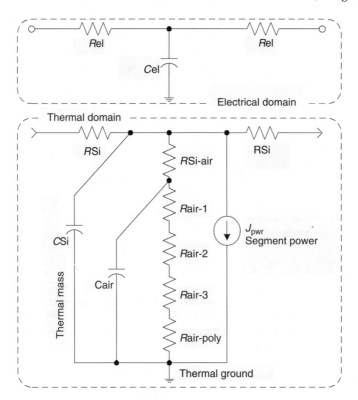

FIGURE 31.11   Electrical/thermal model of cantilever section. (From Hagleitner, C., Bonaccio, T., Rothuizen, H., Lienemann, J., Wiesmann, D., Cherubini, G., Korvink, J., and Eleftheriou, E., *IEEE J. Solid-State Circuits*, 42, 1779, 2007. With permission.)

dimensions approaching the molecular mean free path in air ($\Lambda_{air} \approx 60$ nm), the interface resistances at the silicon–air interface and the air–medium interface must be taken into account [9].

31.3.2   MECHANICAL/ELECTROSTATIC MODEL

To support the cantilever design, a commercial FEM tool (ANSYS) is used to enable accurate predictions of its coupled electrostatic and structural behavior and to ensure an adequate stability margin under operating conditions, e.g. to preclude premature electrostatic collapse of the cantilever legs. The structural aspects of the cantilever were modeled by meshing the quasiplanar thin lever regions with structural shell elements [17], and the pyramid-shaped tip was modeled with volume elements such that the physical tip apex is represented by a single FE node. The electrostatic actuation was implemented by connecting each node of the meshed lever to a fictitious counter-electrode using two-node transducer elements (see Figure 31.12) that operate in the direction orthogonal to the electrode plane and couple a voltage difference to the mechanical degree of freedom, i.e. displacement,

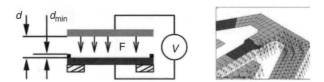

FIGURE 31.12   Schematic drawing of a transducer element (left). The upper movable capacitor electrode is connected to a single node of the cantilever structure (right). When applying a voltage, a force emerges that is proportional to V^2/d^2.

in this direction [17]. Whereas some amount of damping must be included in the model to moderate its dynamic response, squeeze-film nonlinearities are not expected to be strong because of the moderate surface-to-gap aspect ratio of the cantilever and because the cantilever's motion is stopped by the presence of the tip well before the gap closes (for most practical cases at approximately half the initial distance to the sample plane). A conventional proportional Rayleigh damping model is used [17]. The Rayleigh damping constants α and β are set to 0 and 1.0×10^{-7}, respectively, yielding a quality factor Q of ~20 at the typical first natural frequency of the levers (~68 kHz).

The interface of the electrostatic/mechanical model to the electrical/thermal model has already been described at the beginning of Section 31.3. There are, however, substantially more transducer nodes with voltage inputs in the FE model than there are segments in the electrical/thermal model. Therefore, the nodes in the vicinity of the centers of the electrical/thermal segments are combined into "leverage groups," which are held at a common potential (see Figure 31.9). The leverage groups match the sections of the lumped element model for the electrical/thermal domain. The potential at the output of each lumped element is an external input to the mechanical/electrostatic model as shown in Figure 31.8.

The complexity of the resulting FE model, which entails thousands of nonlinear differential equations, precludes its use in a circuit design environment, especially for dynamic simulations. Instead of developing an independent model, we applied symbol isolation and model-order reduction (MOR) to transform the full FE model automatically into a reduced version with a complexity that can be handled by a circuit simulator [18,19]. MOR takes the system of second-order ordinary differential equations (ODEs) obtained after spatial FEM discretization (meshing) of the cantilever layout and returns a much smaller system of ODEs, as described in Figure 31.13. The smaller system is chosen such that its transfer function is close to that of the original system in the neighborhood of a specific frequency.

For linear systems, a large number of MOR results have been published, and also automatic MOR has made significant progress [20]. Nonlinear MOR, however, still requires further research to find a general, automatic procedure [21,22]. Our approach to automate the nonlinear MOR is described in Refs. [12,19]. In this approach, the Taylor expansion of the nonlinear $1/x^2$ function that determines the force as a function of the node distance leads to a large number of coefficients in the load matrix. This can be avoided by performing symbol isolation [23] and by splitting the model into a linear and a nonlinear part, which are coupled by a greater number of terminals. The MOR is performed on the linear model only. Figure 31.14 schematically shows this approach, in which a separate Verilog-A model is used to calculate a normalized force as a function of the position output and the input voltage. Using this method, the approximation is very accurate, even for load values close to the collapse regime. The splitting reduces the size of the reduced system to O ([no. reduced degrees of freedom]^2), while the additional Verilog-A code describing the $1/x^2$ function is straightforward.

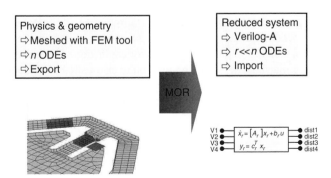

FIGURE 31.13 Principle of MOR. (From Hagleitner, C., Bonaccio, T., Rothuizen, H., Lienemann, J., Wiesmann, D., Cherubini, G., Korvink, J., and Eleftheriou, E., *IEEE J. Solid-State Circuits*, 42, 1779, 2007. With permission.)

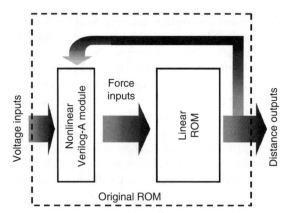

FIGURE 31.14 The electrostatic/mechanical model is split into a linear ROM and a Verilog-A block implementing the nonlinear force–distance relationship of the electrostatic force between the cantilever and the substrate underneath the medium. (From Hagleitner, C., Bonaccio, T., Rothuizen, H., Lienemann, J., Wiesmann, D., Cherubini, G., Korvink, J., and Eleftheriou, E., *IEEE J. Solid-State Circuits*, 42, 1779, 2007. With permission.)

This approach allows the application of linear MOR approaches with either guaranteed error bounds or high computational efficiency [20]. A very high accuracy can then be achieved for the linear part of the model.

The original system can be described by

$$M\ddot{x}(t) + E\dot{x}(t) + Kx(t) = Bu(t) \tag{31.4}$$

where the linear load vector $u(t)$ is calculated by the external model. M, E, and K are the mass, damping, and stiffness matrices, respectively, and B is the input matrix used to distribute a force applied to a leverage group on FEM nodes. The goal is to find a smaller ODE, i.e. one with $n_r \ll n$ equations,

$$M_r\ddot{x}_r(t) + E_r\dot{x}_r(t) + K_rx_r(t) = B_ru(t) \tag{31.5}$$

which exhibits an almost identical transient behavior.

The first simplification concerns the number of contact elements. In the full model, each node of the cantilever model is attached to a contact element as the transducer elements also have a "minimal gap" feature. During normal operation, no collapse of the capacitive platform occurs. Therefore, we only consider the contact element at the tip. The collapse of the cantilever can be detected by observing the out-of-plane deflection of the cantilever nodes. The contact element at the tip is handled externally (symbol isolation), and coupling is provided through an additional input $u_i(x_{\text{tip}})$.

Then the Arnoldi procedure [24] is employed to generate an orthonormal reduced basis V for the linear part of the system, where V is based on the Taylor expansion of the transfer function and can be written as rectangular matrix. Therefore, the method is also called a moment-matching method, as the first q coefficients (moments) of the transfer function's Taylor expansions match. The ODE (including the polynomial parts) is then projected onto this subspace:

$$M_{ij}V_{im}V_{jn}\ddot{x}_{r,n} + E_{ij}V_{im}V_{jn}\dot{x}_{r,n} + K_{ij}V_{im}V_{jn}x_{r,n} - B_{ij}V_{im}u_j = 0 \tag{31.6}$$

The advantage of this method is that the size of the original ODE has no influence on the size of the reduced model.

MOR allows the designer to make a trade-off between simulation complexity and accuracy by simply choosing the dimension of the reduced matrix V. The resulting system of second-order differential equations is automatically mapped into a Verilog-A description by using the symbolic computer algebra package in Mathematica and exporting an ASCII file. A comparison of the results obtained from the reduced model with those obtained from the original FEM simulations is made in Section 31.4.2.

31.3.3 MODEL OF TIP–SURFACE CONTACT

To capture the change in strains within the lever associated with tip touchdown, the contact between the tip and a sample is emulated by allowing a small amount of penetration of the node representing the tip into a fictitious plane representing the sample surface and exercising a reaction force onto the tip node that is proportional to this penetration. The value of the contact stiffness thus defined has little effect on the model results; it should, however, be much larger than the bending stiffness of the lever, but small enough to avoid convergence issues associated with contact chattering on a timescale commensurate with the transient step size. For the geometries currently being investigated, a stiffness value of 100 N/m is used.

31.3.4 NOISE MODEL

While it is straightforward to model the thermal noise of the silicon resistor, the $1/f$-noise contribution is more difficult to predict. In piezoresistive cantilevers, this conductance fluctuation noise is well described by a model developed by Hooge [27] that relates the $1/f$ noise to the number of carriers in the bulk of the resistor and assumes that it is proportional to the dissipated power [25–28]:

$$S_R/R^2 = \alpha/(fN) \qquad (31.7)$$

where

S_R is the power spectral density of the noise
α is the technology-dependent Hooge factor
N is the number of carriers.

The experimentally determined Hooge factors in piezoresistive cantilevers have been found to depend strongly on the fabrication process, especially on the annealing conditions [26], with values as low as 3×10^{-6} [28]. For the cantilever described in this paper, we obtained an α-factor of 8×10^{-6}.

31.4 VERIFICATION OF CANTILEVER MODEL

31.4.1 ELECTRICAL/THERMAL MODEL

The electrical/thermal lumped-element model is verified by measuring the I–V characteristic of the cantilever. The conductivity of silicon exhibits a doping-dependent minimum that can be used to correlate the power to the temperature. Figure 31.15 shows the good agreement between simulation results and measured values. The sensitivity of the cantilever is defined as the relative resistance change per nanometer of indentation depth for a given power. The model also matches the measured sensitivity values ($4.5 \times 10^{-5} \, \mathrm{nm}^{-1}$ @ 1.7 mW) to within 10% over a wide power range.

31.4.2 ELECTROSTATIC/MECHANICAL MODEL

Figure 31.16 compares the behavior of the full FE model with that of the linear ROM with symbol isolation using a reduced system with 19 ODEs (simulated with Spectre in the Cadence design environment). The diagram shows the movement of the tip and the two nodes on the supply leads subject to a voltage step of 7.45 V. The voltage step in the Spectre-simulation was scaled to 7.25 V. The nodes on the supply leads move about half the distance from their equilibrium positions.

Figure 31.17 shows the results from a simulation of the collapse of the cantilever legs onto the polymer medium. The voltage on the substrate underneath the medium is slowly ramped up from

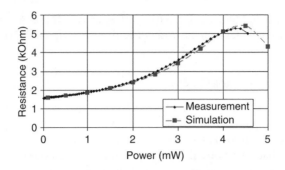

FIGURE 31.15 Simulation/measurement of resistance versus power. (From Hagleitner, C., Bonaccio, T., Rothuizen, H., Lienemann, J., Wiesmann, D., Cherubini, G., Korvink, J., and Eleftheriou, E., *IEEE J. Solid-State Circuits*, 42, 1779, 2007. With permission.)

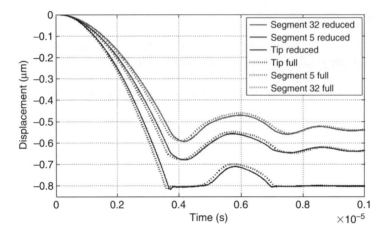

FIGURE 31.16 Comparison of the full FE model with the reduced model for a voltage step of 7.45 V on the medium. The initial tip-medium distance is 0.8 μm.

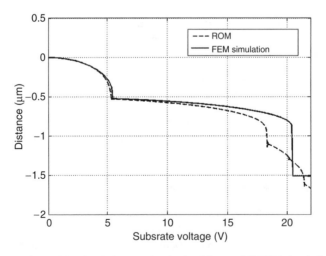

FIGURE 31.17 Comparison of the simulation results obtained from a full FEM simulation with the ROM for a collapse of the cantilever. (From Hagleitner, C., Bonaccio, T., Rothuizen, H., Lienemann, J., Wiesmann, D., Cherubini, G., Korvink, J., and Eleftheriou, E., *IEEE J. Solid-State Circuits*, 42, 1779, 2007. With permission.)

zero until one or all parts of the cantilever collapse onto the surface. For the full FEM simulation, the cantilever is pulled into contact at around 5.4 V and collapses at around 20.5 V. Whereas the pull-in voltage matches to within 50 mV, the collapse of the cantilever occurs at 18.5 V, which corresponds to an error of approximately 10%. This is sufficiently accurate for a system-level simulation in which it is only important to verify that the cantilever never collapses under any operating condition.

31.5 ANALOG FRONT END DESIGN

A block diagram of the complete read/write channel was already introduced in Figure 31.6 and a brief description of the signal-conditioning circuitry after the first amplification stage can be found in Section 31.2.1. In the following, the mode selection as well as the input stage will be described in more detail. A switch matrix is used at the input to select one out of three operating modes of the cantilever and the corresponding interface circuitry: (i) read mode, (ii) write mode, and (iii) inactive mode, during which the circuitry has to keep the cantilever at the same potential as the moving medium table to avoid accidental pull-in of the cantilever to the surface. The cantilever model is essential for the design of all parts of the cantilever interface circuitry. This includes verification of the different states of the switch matrix, timing optimization of the write-channel circuitry, and optimization of the read-channel functions. The design of the input stage of the read-channel will be described in detail in the following section.

31.5.1 INPUT STAGE

Traditionally, resistive sensors are read out by including them either in a Wheatstone bridge configuration [29,30], by constant voltage bias with an inverting amplifier configuration and some form of reference subtraction [31], by differential constant-current bias [32], or by employing direct readout of the current through the sensing resistor [33]. For our probe-storage device, we need to operate 300 channels in parallel to achieve video data rates. Therefore, power consumption is critical, and the power budget dictates the use of only a single sensing element per channel. The complete interface circuitry must consume less than 200 μW per channel while operating continuously (no power supply cycling). The power consumption of a Wheatstone bridge with its four sensing elements or an inverting amplifier configuration exceeds the power budget by far. Direct conversion of the cantilever current (no reference) cannot be used because of insufficient PSR.

Figure 31.18 shows a block diagram of the input stage designed. A constant, differential voltage is supplied to the cantilever using two OpAmps. While OpAmp $OP1$ at the top could be shared to supply several cantilevers, the output transistor $T1$ of OpAmp $OP2$ carries the signal current of an individual cantilever. The small sensitivity of the thermal sensing scheme ($\Delta R/R < 10^{-3}$ for a typical indentation) leads to a large offset current, which is subtracted by means of an identically biased reference element directly in the input stage to minimize power. Therefore, only a small fraction of the bias current is mirrored to the first amplification stage. In this configuration, the reference element is shared between 16 cantilevers. Using this approach, the contribution of the reference element to the overall power consumption is minimized.

Figure 31.19 shows the first amplification stage, which converts the input current to a differential output signal. Fully differential signal processing is crucial to achieve the required PSR in this mixed-signal design. A small bias current, approximately 5% of the current I_{ref} in the reference element, is added to the signal current in the input stage to guarantee stable operation of OpAmp $OP2$. I_{ref} is also derived from the current through the reference element to maximize PSR. The matching of the cantilever resistance to the resistance of the reference element is assumed to be better than 5%. A 7 bit current-steering digital-to-analog converter is used to adjust I_{ref}, therefore improving the matching to below 0.1%. This current is then mirrored to two current sources in a bridge configuration ($T2b$ and

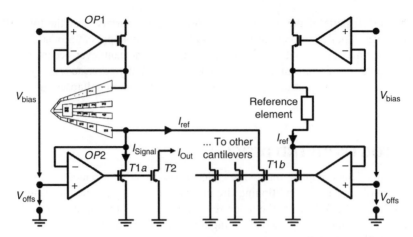

FIGURE 31.18 Schematic of AFE input stage. (From Hagleitner, C., Bonaccio, T., Rothuizen, H., Lienemann, J., Wiesmann, D., Cherubini, G., Korvink, J., and Eleftheriou, E., *IEEE J. Solid-State Circuits*, 42, 1779, 2007. With permission.)

$T5$), where the other two current sources in the bridge ($T4$ and $T7a$) are biased at a constant current of $0.05 \times I_{ref}$. Therefore, the signal current passes through the programmable resistor $R1$ and generates a differential output signal. Transistors $T9$–$T12$ are used for the common-mode feedback [34]. Figure 31.20 shows a transient simulation of the tip movement and the output of the input amplifier in the initial phase of a read operation. At time $t = 1\,\mu s$, the read voltage is switched on, and at $t = 50\,\mu s$ the voltage of the medium table is decreased to $-4\,V$ to pull the cantilever into contact. Next, the cantilever tip follows the shape of a "101" pattern applied to the topography input of the cantilever model.

31.5.2 Model-based Optimization of Power/Signal-to-Noise Ratio

The power consumption of the read channel is dominated by the power consumed in the current path through the cantilever. The sensitivity and noise of the cantilever both strongly depend on the power

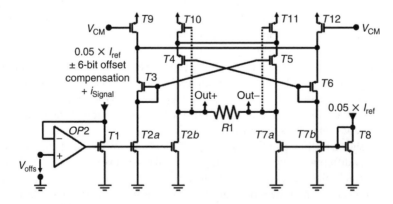

FIGURE 31.19 First amplifier of AFE, which converts the input current to a fully differential output voltage. Load resistor $R1$ is digitally programmable in four steps between 4 and $250\,k\Omega$. (From Hagleitner, C., Bonaccio, T., Rothuizen, H., Lienemann, J., Wiesmann, D., Cherubini, G., Korvink, J., and Eleftheriou, E., *IEEE J. Solid-State Circuits*, 42, 1779, 2007. With permission.)

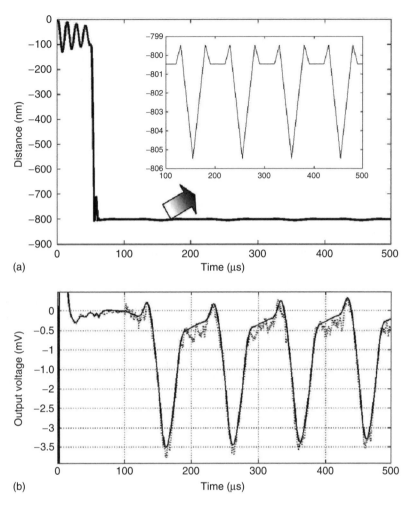

FIGURE 31.20 (a) Transient simulation of the tip movement (with close-up of bit tracking in the inset). (b) The output of the input amplifier (with and without noise sources) during the initial phase of a read operation. The read voltage is switched on at $t = 1\,\mu$s, and the medium voltage is switched on at time $t = 50\,\mu$s. (From Hagleitner, C., Bonaccio, T., Rothuizen, H., Lienemann, J., Wiesmann, D., Cherubini, G., Korvink, J., and Eleftheriou, E., *IEEE J. Solid-State Circuits*, 42, 1779, 2007. With permission.)

dissipated in the heating element. The $1/f$ noise of transistors $T1a$, $T1b$ and the differential pair at the input of the $OP2$ are the predominant noise sources of the read-channel circuitry, and its drain-source voltage VDS_{1b} adds considerably to the power consumption. An analytic evaluation of the resulting trade-offs is difficult because of the numerous feedback mechanisms and nonlinear effects. Extensive parameterized simulations were used to optimize the overall signal-to-noise ratio (SNR) for a given power budget. Assuming a power supply of 2.5 V and a Hooge factor of 1.0×10^{-5}, the best trade-off is to use around 75% of the power for the biasing of the cantilever because the increase in signal due to the higher temperature of the heating element outweighs the increased noise in transistor $T1b$ as long as the $1/f$ noise of the cantilever is not dominant. The output noise is determined by the noise contributions from the read channel in this case. The simulations predict a resolution of $<0.5\,$nm in the 1 to $100\,$kHz signal bandwidth for a heating power of $1.6\,$mW (resolution is derived from equivalent input-referred noise).

31.6   EXPERIMENTAL RESULTS

The AFE chip was characterized using a flexible prototyping environment that implements all functions required for a complete probe-storage device. A prototype system that includes the cantilever array mounted on top of the x/y scanner, the AFE chip, an FPGA interface board, discrete component test circuitry, and a digital signal processing (DSP) board was used to characterize the AFE chip. A discrete-component AFE is used to interface the thermal position sensors and drive the voice-coil actuators of the assembly, which consists of the cantilever array mounted on the microscanner. Both the assembly and the integrated AFE chip were mounted on a printed circuit board (PCB) board and connected using wirebonds. An FPGA was used to implement the low-level functions, e.g. the serial interfaces to the AFE chip monitoring functions for debugging. A standard DSP board was used to implement the high-level functions, including the servo controller and the data controller. In this way, the setup could be easily adapted for testing of the AFE chip.

Figure 31.21 compares the simulated output noise spectrum of the complete channel with the results obtained from measurements. To establish a valid reference, the output of a passive high-pass filter after the first amplification stage was shorted, and the thermal noise level in the simulations was adjusted to account for gain variations in the subsequent amplification stages [11]. Then the complete channel was measured, and the results show that the overall noise is dominated by the input stage. The simulation results match the measurements within the fabrication tolerances.

Figure 31.22 shows the simultaneous read-back signal obtained from four channels. The periodic patterns have been written earlier using the same AFE chip. Note that the noise is dominated by medium noise, which is in agreement with the simulations that predict a larger SNR if only electrical noise is taken into account.

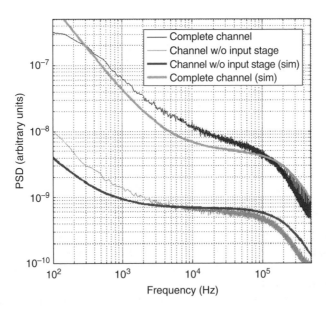

FIGURE 31.21   Comparison of the simulated output-noise spectrum of the channel with the results obtained from measurements. (From Hagleitner, C., Bonaccio, T., Rothuizen, H., Lienemann, J., Wiesmann, D., Cherubini, G., Korvink, J., and Eleftheriou, E., *IEEE J. Solid-State Circuits*, 42, 1779, 2007. With permission.)

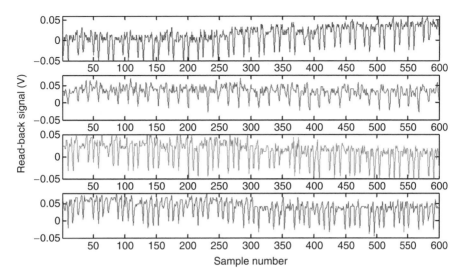

FIGURE 31.22 Read-back signal simultaneously recorded from four cantilevers. The AFE chip implements a high-pass and a second amplification stage after the input amplifier. (From Hagleitner, C., Bonaccio, T., Pantazi, A., Sebastian, A., and Eleftheriou, E., *Proceedings of Symposium on VLSI Circuits*, Honolulu, HI, pp. 70–71, 2006. With permission.)

31.7 SUMMARY

We have developed a 32-channel AFE chip for a parallel probe-storage device. The design of the read/write channel is based on a behavioral model of a micromachined scanning probe. The model accurately predicts both the small-signal and the large-signal behavior in the electrical, mechanical, and thermal domains. The use of automatic MOR enables fast simulation and verification of new cantilever-array designs obtained from FEM simulation. Because the model uses the physical dimensions of the cantilevers as parameters, it is straightforward to obtain them by automatic layout extraction. This enables design and verification of the complete system in a standard mixed-signal design flow.

ACKNOWLEDGMENTS

The authors wish to thank the entire "millipede" project team at the IBM Zurich Research Laboratory for their support. Special thanks go to M. Despont and U. Drechsler for the design and fabrication of the cantilever arrays, to A. Pantazi and A. Sebastian for their help with the prototype measurements, and P. Bächtold for the design of the PCB boards. A. Greiner and E. Rudnyi at the University of Freiburg (Germany) are acknowledged for their help with the MOR, and the foundry team at IBM Burlington for the fabrication of the AFE chip.

REFERENCES

1. Y. Choi, 16-Gbit MLC NAND flash weighs in, *EEtimes*, 2007.
2. P. Vettiger, G. Cross, M. Despont, U. Drechsler, U. Duerig, B. Gotsmann, W. Häberle, M. A. Lantz, H. E. Rothuizen, R. Stutz, and G. K. Binnig, The "millipede"—nanotechnology entering data storage, *IEEE Transactions on Nanotechnology*, 1, 39–55, 2002.
3. E. Eleftheriou, T. Antonakopoulos, G. K. Binnig, G. Cherubini, M. Despont, A. Dholakia, U. Duerig, M. A. Lantz, H. Pozidis, H. E. Rothuizen, and P. Vettiger, Millipede—a MEMS-based scanning-probe data-storage system, *IEEE Transactions on Magnetics*, 39, 938–945, 2003.

4. D. Wiesmann, U. Duerig, B. Gotsmann, A. Knoll, H. Pozidis, F. Porro, and R. Vecchione, Ultra-high storage densities with thermo-mechanical probes and polymer media, in *Proc. Innovative Mass Storage Technologies*, Enschede, the Netherlands, 2007.

5. M. A. Lantz, G. K. Binnig, M. Despont, and U. Drechsler, A micromechanical thermal displacement sensor with nanometre resolution, *Nanotechnology*, 16, 1089–1094, 2005.

6. A. Pantazi, A. Sebastian, G. Cherubini, M. A. Lantz, H. Pozidis, H. Rothuizen, and E. Eleftheriou, Control of MEMS-based scanning-probe data-storage devices, *IEEE Transactions on Control Systems Technology*, 15, 824–841, 2007.

7. E. Eleftheriou, P. Bächtold, G. Cherubini, A. Dholakia, C. Hagleitner, T. Loeliger, A. Pantazi, H. Pozidis, T. Albrecht, G. Binnig, M. Despont, U. Drechsler, U. Duerig, B. Gotsmann, D. Jubin, W. Häberle, M. A. Lantz, H. Rothuizen, R. Stutz, P. Vettiger, and D. Wiesmann, A nanotechnology-based approach to data storage, in *Proc. International Conference on Very Large Data Bases, Berlin, Germany*, pp. 3–7, 2003.

8. A. Pantazi, A. Sebastian, H. Pozidis, and E. Eleftheriou, Two-sensor based H_∞ control for nanopositioning in probe storage, in *Proc. IEEE Control and Decision Conference*, Seville, Spain, pp. 1174–1179, 2005.

9. U. Dürig, Fundamentals of micromechanical thermoelectric sensors, *Journal of Applied Physics*, 98, 449.6.1–449.6.14, 2005.

10. M. Despont, U. Drechsler, R. Yu, H. B. Pogge, and P. Vettiger, Wafer-scale microdevice transfer/interconnect: From a new integration method to its application in an AFM-based data-storage system, in *Proc. Transducers'03*, Boston, MA, pp. 1907–1910, 2003.

11. C. Hagleitner, T. Bonaccio, A. Pantazi, A. Sebastian, and E. Eleftheriou, An analog frontend chip for a MEMS-based parallel scanning-probe data-storage system, in *Proc. Symposium on VLSI Circuits*, Honolulu, HI, pp. 70–71, 2006.

12. C. Hagleitner, T. Bonaccio, H. Rothuizen, J. Lienemann, D. Wiesmann, G. Cherubini, J. Korvink, and E. Eleftheriou, Modeling, design, and verification for the analog front-end of a MEMS-based parallel scanning-probe storage device, *IEEE Journal of Solid-State Circuits*, 42, 1779–1789, 2007.

13. S. M. Sze, *Physics of Semiconductor Devices*. New York: Wiley, 1981.

14. M. Asheghi, B. Behkam, K. Yazdani, R. Joshi, and K. E. Goodson, Thermal conductivity model for thin silicon-on-insulator layers at high temperatures, in *Proc. IEEE Int. SOI Conf.*, Williamsburg, VA, pp. 51–52, 2002.

15. M. Asheghi and K. E. Goodson, Thermal conductivity model for nearly pure and doped thin Silicon layers at high temperatures, in *Proc. ASME International Mechanical Engineering Congress*, Washington, DC, 2003.

16. W. Liu and M. Asheghi, Thermal conductivity measurements of ultra-thin single crystal silicon layers, *ASME J. Heat Transfer*, 128, pp. 75–83, 2006.

17. Documentation of ANSYS Version 9.0, The elements used are SHELL181 and SOLID95 (structural), TRANS126 (transducer), and CONTA175 (tip contact). Canonsburg, PA: Ansys Inc.

18. J. Lienemann, Complexity reduction techniques for advanced MEMS actuators simulation, in IMTEK Freiburg, Germany, University of Freiburg, Ph.D thesis, 2006.

19. J. Lienemann, J. G. Korvink, C. Hagleitner, and H. Rothuizen, Nonlinear model order reduction of electrostatically actuated MEMS cantilever, *Proc. of the International Conference on Structural Engineering, Mechanics and Computation*, Cape Town, South Africa, pp. 449–454, 2007.

20. E. B. Rudnyi and J. G. Korvink, Review: Automatic model reduction for transient simulation of MEMS-based devices, *Sensors Update*, 11, 3–33, 2002.

21. J. R. Phillips, Projection-based approaches for model reduction of weakly nonlinear, time-varying systems, *IEEE Transactions on Computer-Aided Design of Integrated Circuits and Systems*, 22, 171–187, 2003.

22. M. Rewienski and J. White, A trajectory piecewise-linear approach to model order reduction and fast simulation of nonlinear circuits and micromachined devices, *IEEE Transactions on Computer-Aided Design of Integrated Circuits and Systems*, 22, 155–170, 2003.

23. G. Shi, B. Hu, and C.-J. R. Shi, On symbolic model order reduction, *IEEE Transactions on Computer Aided Design of Integrated Circuits and Systems*, 25, 1259–1272, 2006.

24. R. W. Freund, Krylov-subspace methods for reduced order modeling in circuit simulation, *Journal of Computational and Applied Mathematics*, 123, 395–421, 2000.

25. J. A. Harley and T. W. Kenny, High-sensitivity piezoresistive cantilevers under 1000 Å thick, *Applied Physics Letters*, 75, 289–291, 1999.

26. J. A. Harley and T. W. Kenny, $1/f$ noise considerations for the design and process optimization of piezoresistive cantilevers, *Journal of Microelectromechanical Systems*, 9, 226–235, 2000.

27. F. N. Hooge, $1/f$ noise sources, *IEEE Transactions on Electronic Devices*, 41, 1926–1935, 1994.
28. X. Yu, J. Thaysen, O. Hansen, and A. Boisen, Optimization of sensitivity and noise in piezoresistive cantilevers, *Journal of Applied Physics*, 92, 6296–6301, 2002.
29. D. McCartney, A. Sherry, T. Meany, T. Cummins, D. Brannick, and L. MacManus, A fully integrated sensor interface chip, in *Proc. European Solid-State Circuits Conference*, Duisburg, Germany, pp. 222–225, 1999.
30. U. Schoeneberg, B. J. Hosticka, and F. V. Schnatz, A CMOS readout amplifier for instrumentation applications, *IEEE Journal of Solid-State Circuits*, 26, 1077–1080, 1991.
31. M. Grassi, P. Malcovati, and A. Baschirotto, A 0.1% accuracy 100 Ω–20 MΩ dynamic range integrated gas sensor interface circuit with 13 + 4 bit digital output, in *Proc. European Solid-State Circuits Conference*, Grenoble, France, 2005, pp. 351–354.
32. W. Claes, M. De Cooman, W. Sansen, and R. Puers, A 136 µW/channel autonomous strain-gauge datalogger, *IEEE Journal of Solid-State Circuits*, 38, 2280–2287, 2003.
33. M. Malfatti, D. Stoppa, A. Simoni, L. Lorenzelli, A. Adami, and A. Baschirotto, A CMOS interface for a gas-sensor array with a 0.5%-linearity over 500 kΩ-to-1 GΩ range and 2.5 °C temperature control accuracy, in *Proc. IEEE International Solid-State Circuits Conference*, San Francisco, CA, pp. 1131–1140, 2006.
34. F. Krummenacher and N. Joehl, A 4-MHz CMOS continuous-time filter with on-chip automatic tuning, *IEEE Journal of Solid-State Circuits*, 23, 750–758, 1988.

Index

A

Active inductors
  Bode plots of input impedance, 137–138
  CMOS, 138
  configuration and characteristics, 136–137
  frequency range, 137–138
  quality factor of, 138
  Wu current reuse, 139
Actively body-bias controlled (ABC)
  sense amplifier and memory array, 64
  transistors, 57
Active 3T and 4T pixel, 450–453
Active transformers
  characterization of, 139–142
  CMOS, 144–145
  configuration of, 138–139
  current ratio of, 142
  nonideal, 142–144
  primary winding of, 140
  quality factor, 144
Adaptive bias control, 171–172
Adjacent channel leakage power ratio (ACLR), 319
Adjacent channel power ratio (ACPR), 319
Agilent, 378
Alexander PD, full-rate CDR circuit, 404–405
All-digital PLL (ADPLL), 365, 367
Analog front end (AFE) chip
  cantilever model, 592–594
  components of, 589–592
  input stage, 601–602
  modeling approach, 591–592
  parameters of, 590
  power dissipation and, 431–432
  prototype system, 604
  read channel, 590–591
  smart image sensor architecture and, 425–426
  switch matrix, 601
  thermomechanical scanning-probe-based data
    storage concept, 588
Analog/mixed-signal ICs, 46–47
Analog-to-digital converter (ADC), 187, 213, 225,
    468–469
  algorithmic and oversampling, 497–498
  architecture
    power efficiency of, 226–227
    successive approximation converter (SAR), 226
    time-based, 220–221
    time-inter leaved, 222
    topologies for, 226, 240–242
  asynchronous, design example
    architecture of, 230–231
    block diagrams of, 230
    critical digital blocks, 236
    digital calibration scheme, 234
    dynamic comparator and ready signal, 231–232

high speed digital logic of, 236
    nonbinary successive approximation, 232
    Nyquist frequency and, 236–238
    scaling trend of, 238–240
    series nonbinary capacitive ladder, 232–234
    variable duty-cycled clock, 234–236
  asynchronous processing
    concept of, 228
    conventional implementation for, 227
    T_{async}/T_{sync} ratio, 229–230
    total resolving time, 228
    V_{res} profile cases, 229
  circuits, 12
  $\Delta\Sigma$-modulated
    multiplying, 493–494
    weighted averaging, 494–496
  digital calibration techniques of, 218–219
  high-speed subsampling and oversampling,
    216–217
  multichannel IC, solutions of, 550–552
  PDD ASIC, 557–558
  performance scaling
    characterization and design, 214
    dynamic range, 215–216
  power dissipation, 422, 432
  series nonbinary capacitive ladder, 233
Antilogarithmic converter, 513, 515–516
Application programming interfaces (APIs),
    504–506
Application specific integrated circuit (ASIC)
  detector readout, 542, 552, 555
  PDD, 557–558
  RX64 layout, 553–554
Asymmetric T-coil peaking, 123–125
ATC-peaked amplifier, 126
Automatic level control (ALC) circuit, 572
Auto-zeroing and CDS techniques, 109

B

Backend-of-the-line (BEOL) platforms, 40–44
Backplane system, 384–385
Bandwidth extension ratio (BWER), 120
Bandwidths (BWs), filter topologies, 246
Berkeley's z-axis single chip gyroscope, 569
Bias switching, 103–105
Binary complementary metal oxide semiconductor
    (BiCMOS), 269
  technology, 79–80
Bipolar CMOS DMOS (BCD) technology, 7
Bipolar field-effect transistor technologies (BiFET), 334
Bipolar junction transistors (BJT), 177–178
Bitline circuit technique, 150
  leakage robustness, 153–154
  PMOS, 153